T0211705

# Lecture Notes in Computer Science     11945

More information about this series at http://www.springer.com/series/7407

Sheng Wen · Albert Zomaya ·
Laurence T. Yang (Eds.)

# Algorithms and Architectures for Parallel Processing

19th International Conference, ICA3PP 2019
Melbourne, VIC, Australia, December 9–11, 2019
Proceedings, Part II

 Springer

*Editors*
Sheng Wen
Department of Computer Science
and Software Engineering
Swinburne University of Technology
Hawthorn, Melbourne, VIC, Australia

Albert Zomaya
School of Computer Science
The University of Sydney
Camperdown, NSW, Australia

Laurence T. Yang
Department of Computer Science
St. Francis Xavier University
Antigonish, NS, Canada

ISSN 0302-9743 ISSN 1611-3349 (electronic)
Lecture Notes in Computer Science
ISBN 978-3-030-38960-4 ISBN 978-3-030-38961-1 (eBook)
https://doi.org/10.1007/978-3-030-38961-1

LNCS Sublibrary: SL1 – Theoretical Computer Science and General Issues

This Springer imprint is published by the registered company Springer Nature Switzerland AG
The registered company address is: Gewerbestrasse 11, 6330 Cham, Switzerland

# Preface

Welcome to the proceedings of the 19th International Conference on Algorithms and Architectures for Parallel Processing (ICA3PP 2019). ICA3PP is with the series of conferences started in 1995 that are devoted to algorithms and architectures for parallel processing.

The conference of ICA3PP 2019 will be organized by Swinburne University of Technology, Australia, and was held in Melbourne, Australia. The objective of ICA3PP 2019 was to bring together researchers and practitioners from academia, industry, and governments to advance the theories and technologies in parallel and distributed computing. ICA3PP 2019 follows the traditions of the previous successful ICA3PP conferences held in Hangzhou, Brisbane, Singapore, Melbourne, Hong Kong, Beijing, Cyprus, Taipei, Busan, Melbourne, Fukuoka, Vietri sul Mare, Dalian, Japan, Zhangjiajie, Granada, Helsinki, and Guangzhou.

ICA3PP focuses on two broad areas of parallel and distributed computing: architectures, algorithms, and networks, and systems and applications. This conference is now recognized as the main regular event of the world that is covering the many dimensions of parallel algorithms and architectures, encompassing fundamental theoretical approaches, practical experimental projects, and commercial components and systems. As applications of computing systems have permeated in every aspect of daily life, the power of computing system has become increasingly critical. This conference provides a forum for academics and practitioners from countries around the world to exchange ideas for improving the efficiency, performance, reliability, security, and interoperability of computing systems and applications.

ICA3PP 2019 attracted 251 high-quality research papers highlighting the foundational work that strives to push beyond the limits of existing technologies, including experimental efforts, innovative systems, and investigations that identify weaknesses in existing parallel processing technology. Each submission was reviewed by at least two experts in the relevant areas, based on their significance, novelty, technical quality, presentation, and practical impact. According to the review results, 73 full papers were selected to be presented at the conference, giving an acceptance rate of 29%. We also accepted 29 short papers. In addition to the paper presentations, the program of the conference included three keynote speeches and two invited talks from esteemed scholars in the area, namely: (1) Y. Thomas Hou from Virginia Tech (USA), talking about "GPU-Based Parallel Computing for Real-Time Optimization," (2) Ying-Dar Lin from National Chiao Tung University (Taiwan), giving us a speech "5G Mobile Edge Computing: Research Roadmap of the H2020 5G-Coral Project," (3) Wanlei Zhou from University of Technology Sydney (Australia), giving us a talk "AI Security: A Case in Dealing with Malicious Agents," and (4) Hai Jin from Huazhong University of Science and Technology (China), giving us a talk "Evening Out the Stumbling Blocks for Today's Blockchain Systems." We were extremely honored to have had them as the conference keynote speakers and invited speakers.

ICA3PP 2019 was made possible by the behind-the-scene effort of selfless individuals and organizations who volunteered their time and energy to ensure the success of this conference. We thank all participants of the ICA3PP conference for their contribution. We hope that you will find the proceedings interesting and stimulating. It was a pleasure to organize and host the ICA3PP 2019 in Melbourne, Australia.

December 2019

Sheng Wen
Albert Zomaya
Laurence T. Yang

# Organization

## Honorary Chair

Yong Xiang       Deakin University, Australia

## General Chairs

David Abramson       The University of Queensland, Australia
Yi Pan       Georgia State University, USA
Yang Xiang       Swinburne University of Technology, Australia

## Program Chairs

Albert Zomaya       The University of Sydney, Australia
Laurence T. Yang       St. Francis Xavier University, Canada
Sheng Wen       Swinburne University of Technology, Australia

## Publication Chair

Yu Wang       Guangzhou University, China

## Publicity Chair

Jing He       Swinburne University of Technology, Australia

## Steering Committee

Yang Xiang (Chair)       Swinburne University of Technology, Australia
Weijia Jia       Shanghai Jiaotong University, China
Yi Pan       Georgia State University, USA
Laurence T. Yang       St. Francis Xavier University, Canada
Wanlei Zhou       University of Technology Sydney, Australia

## Program Committee

Marco Aldinucci       University of Turin, Italy
Pedro Alonso-Jordá       Universitat Politècnica de València, Spain
Daniel Andresen       Kansas State University, USA
Danilo Ardagna       Politecnico di Milano, Italy
Man Ho Au       The Hong Kong Polytechnic University, Hong Kong, China
Guillaume Aupy       Inria, France

| | |
|---|---|
| Joonsang Baek | University of Wollongong, Australia |
| Ladjel Bellatreche | LIAS/ENSMA, France |
| Siegfried Benkner | University of Vienna, Austria |
| Jorge Bernal Bernabe | University of Murcia, Spain |
| Thomas Boenisch | High performance Computing Center Stuttgart, Germany |
| George Bosilca | University of Tennessee, USA |
| Suren Byna | Lawrence Berkeley National Laboratory, USA |
| Massimo Cafaro | University of Salento, Italy |
| Philip Carns | Argonne National Laboratory, USA |
| Arcangelo Castiglione | University of Salerno, Italy |
| Tania Cerquitelli | Politecnico di Torino, Italy |
| Tzung-Shi Chen | National University of Tainan, Taiwan |
| Kim-Kwang Raymond Choo | The University of Texas at San Antonio, USA |
| Jose Alfredo Ferreira Costa | Federal University of Rio Grande do Norte, Brazil |
| Raphaël Couturier | University of Burgundy - Franche-Comté, France |
| Masoud Daneshtalab | Mälardalen University, KTH Royal Institute of Technology, Sweden |
| Gregoire Danoy | University of Luxembourg, Luxembourg |
| Saptarshi Debroy | City University of New York, USA |
| Casimer Decusatis | Marist College, USA |
| Eugen Dedu | FEMTO-ST Institute, University of Burgundy - Franche-Comté, CNRS, France |
| Frederic Desprez | Inria, France |
| Juan-Carlos Díaz-Martín | University of Extremadura, Spain |
| Christian Esposito | University of Napoli Federico II, Italy |
| Ugo Fiore | University of Napoli Federico II, Italy |
| Franco Frattolillo | University of Sannio, Italy |
| Marc Frincu | West University of Timisoara, Romania |
| Jorge G. Barbosa | University of Porto, Portugal |
| Jose Daniel Garcia | University Carlos III of Madrid, Spain |
| Luis Javier García Villalba | Universidad Complutense de Madrid, Spain |
| Harald Gjermundrod | University of Nicosia, Cyprus |
| Jing Gong | KTH Royal Institute of Technology, Sweden |
| Daniel Grosu | Wayne State University, USA |
| Houcine Hassan | Universitat Politècnica de València, Spain |
| Sun-Yuan Hsieh | National Cheng Kung University, Taiwan |
| Xinyi Huang | Fujian Normal University, China |
| Mauro Iacono | Università degli Studi della Campania Luigi Vanvitelli, Italy |
| Shadi Ibrahim | Inria Rennes Bretagne Atlantique Research Center, France |
| Yasuaki Ito | Hiroshima University, Japan |
| Edward Jung | Kennesaw State University, USA |
| Georgios Kambourakis | University of the Aegean, Greece |

| | |
|---|---|
| Helen Karatza | Aristotle University of Thessaloniki, Greece |
| Gabor Kecskemeti | Liverpool John Moores University, UK |
| Muhammad Khurram Khan | King Saud University, Saudi Arabia |
| Sokol Kosta | Aalborg University, Denmark |
| Dieter Kranzlmüller | Ludwig Maximilian University of Munich, Germany |
| Peter Kropf | University of Neuchâtel, Switzerland |
| Michael Kuhn | University of Hamburg, Germany |
| Julian Martin Kunkel | University of Reading, UK |
| Algirdas Lančinskas | Vilnius University, Italy |
| Che-Rung Lee | National Tsing Hua University, Taiwan |
| Laurent Lefevre | Inria, France |
| Kenli Li | Hunan University, China |
| Xiao Liu | Deakin University, Australia |
| Jay Lofstead | Sandia National Laboratories, USA |
| Paul Lu | University of Alberta, Canada |
| Tomas Margalef | Universitat Autònoma de Barcelona, Spain |
| Stefano Markidis | KTH Royal Institute of Technology, Sweden |
| Barbara Masucci | University of Salerno, Italy |
| Susumu Matsumae | Saga University, Japan |
| Raffaele Montella | University of Naples Parthenope, Italy |
| Francesco Moscato | University of Campania Luigi Vanvitelli, Italy |
| Bogdan Nicolae | Argonne National Laboratory, USA |
| Anne-Cécile Orgerie | CNRS, France |
| Francesco Palmieri | University of Salerno, Italy |
| Dana Petcu | West University of Timisoara, Romania |
| Salvador Petit | Universitat Politècnica de València, Spain |
| Riccardo Petrolo | Konica Minolta Laboratory Europe |
| Florin Pop | University Politehnica of Bucharest, National Institute for Research and Development in Informatics (ICI), Romania |
| Radu Prodan | University of Klagenfurt, Austria |
| Suzanne Rivoire | Sonoma State University, USA |
| Ivan Rodero | Rutgers University, USA |
| Romain Rouvoy | University of Lille, Inria, IUF, France |
| Antonio Ruiz-Martínez | University of Murcia, Spain |
| Francoise Sailhan | CNAM, France |
| Sherif Sakr | The University of New South Wales, Australia |
| Ali Shoker | HASLab, INESC TEC, University of Minho, Portugal |
| Giandomenico Spezzano | CNR, Italy |
| Patricia Stolf | IRIT, France |
| Peter Strazdins | The Australian National University, Australia |
| Hari Subramoni | The Ohio State University, USA |
| Frederic Suter | CC IN2P3, CNRS, France |
| Andrei Tchernykh | CICESE Research Center, Mexico |
| Massimo Torquati | University of Pisa, Italy |

# Contents – Part II

**Big Data and Its Applications**

## Distributed and Parallel Algorithms

## Applications of Distributed and Parallel Computing

## Service Dependability and Security

# Contents – Part I

## Distributed and Parallel and Network-Based Computing

## Big Data and Its Applications

## Distributed and Parallel Algorithms

## Applications of Distributed and Parallel Computing

## IoT and CPS Computing

## Performance Modelling and Evaluation

# Parallel and Distributed Architectures

# SPM: Modeling Spark Task Execution Time from the Sub-stage Perspective

Wei Li, Shengjie Hu, Di Wang, Tianba Chen, and Yunchun Li[✉]

Beijing Key Lab of Network Technology, School of Computer Science
and Engineering, Beihang University, Beijing, China
{liw,hushengjie,fdjwd,chentb,lych}@buaa.edu.cn

**Abstract.** Tasks are the basic unit of Spark application scheduling, and its execution is affected by various configurations of Spark cluster. Therefore, the prediction of task execution time is a challenging job. In this paper, we analyze the features of task execution procedure on different stages, and propose the method of prediction of each sub-stage execution time. Moreover, the correlative time overheads of GC and shuffle spill are analyzed in detail. As a result, we propose SPM, a task-level execution time prediction model. SPM can be used to predict the task execution time of each stage according to the input data size and configuration of parallelism. We further apply SPM to the Spark network emulation tool SNemu, which can determine the start time of each shuffle procedure for emulation effectively. Experimental results show that the prediction method can achieve high accuracy in a variety of Spark benchmarks on Hibench.

**Keywords:** Spark · Task-level execution time prediction · Regression model · Network emulation

## 1 Introduction

In recent years Spark [1] has been widely used in many fields to process massive data, which leads to the researches on Spark performance. In-memory cluster computing help Spark avoid unnecessary transmission compared to Hadoop [2]. However, when the wide dependency transformation of RDD [3] happens, the shuffle transmission is unavoidable. The network emulation tool can help researchers to implement large-scale Spark network transmission behavior research at a lower cost.

For the emulation of the procedures of network transmission in Spark cluster, obtaining the time of shuffle transmission is a key technical problem. The transmission time of shuffle procedures are related to different computing operators. Therefore, it's important to find the connection between computing and shuffle transmission to get the time of shuffle. According to the scheduling mechanism of Spark, tasks pull shuffle data at the beginning of execution. Meanwhile, after the completion of a task, the executor node running this task will have an idle core. In order to achieve work conservation, the driver node will immediately dispatch the locality optimal task in the scheduling queue to this core. Therefore, we can predict the task execution time of each stage in the application and emulate the task scheduling process of Spark. Then, the

S. Wen et al. (Eds.): ICA3PP 2019, LNCS 11945, pp. 3–10, 2020.
https://doi.org/10.1007/978-3-030-38961-1_1

start time of each task is obtained and it can be treated as the time when the task pulls shuffle data.

In this paper, we explore the relation between the execution time of each sub-stage (The subphase of the spark task execution process) and the payload of the task. As a result, a task-level execution time prediction model is proposed. We make the following contributions:

- We propose SPM, a task execution time prediction model for Spark, and prove its effectiveness in multiple benchmarks.
- We quantize the relation of extra time overheads associated with GC and shuffle spill during the computation stage.
- We apply SPM to the network emulation tool of Spark to provide shuffle start time prediction, and verify the effectiveness of SPM.

The rest of this paper is organized as follows: Related works are shown in Sect. 2. We introduce the prediction model, SPM, in Sect. 3. Experimental results are illustrated in Sect. 4. Finally, Sect. 5 concludes this paper.

## 2   Related Work

To implement a high reliability Spark network emulation tool, it's important to model the computing process and network load for trace input. One part of the job is to extrapolate shuffle start time through predicting task execution time. Task execution time prediction is usually presented as a part of performance analysis and optimization. Gu et al. [4] use prediction job duration from a random forest performance model to drop the cost of configuration optimization for Spark Streaming. PREDIcT [5] uses sample runs for capturing the algorithm's convergence trend and uses an experimental methodology to predict the execution time of algorithms with different iterative features. Nguyen et al. [6] qualitatively analyze the impact of Spark configuration changes on multiple performance metrics through an execution process-driven model. FiM [7] implements a performance approximation method to simulate the performance of multi-stage iterative applications through stochastic markov model. Ernest [8] uses samples in small scale to fit linear regression model for application duration and help predict the duration in large scale. Wang et al. [9] use the results of partial input in a small-scale environment to predict the performance of complete input in a large-scale environment. Both of the above two works [8, 9] model execution time coarsely (application duration and total task execution time respectively), which may ignore the randomness factor in the actual task execution procedure.

In our work, we analyze Spark's task scheduling and execution mechanism and then divide the execution process into different sub-stages. Then, the relationship between the execution time of each sub-stage and the specified configurations is fitted by quantitative analysis of the performance. Finally, the task execution time can be obtained by adding the predicted time of all sub-stages. This fine-grained prediction can fully consider the characteristics of each sub-stage, so as to predict the execution time more accurately.

# 3 Modeling Task-Level Execution Time

## 3.1 Parameters for Prediction

Spark tasks are performed in a sequence of sub-stages, which are described as follow: (a) Scheduler process. In this process, task is distributed from the scheduler to the executor. (b) Task Deserialization. Executor deserializes the task closure. (c) Shuffle Read. Executor pulls the intermediate result from previous stages. (d) Computing. Executor processes computing work. (e) Shuffle Write. Executor then writes the result of the task to the local disk. (f) Result Serialization. Finally, executor serializes the metadata of the task result, which will be sent to the driver node. (g) In addition, there are also some potential processes during task execution, like GC and Shuffle Spill. On this basis, we propose Spark task execution time prediction model SPM.

In SPM, the variables primarily considered in the prediction of task execution time are the amount of input data of applications and the configuration of parallelism. We predict the time overhead of each sub-stage during the task execution rather than the total duration. Moreover, we predict the potential extra overheads of GC and shuffle spill. All these parameters are shown in Table 1. The target of our prediction is the average execution time of tasks. It can be modeled by adding all the sub-stage overheads up, like formula (1).

$$T_{Task} = T_{SD} + T_{TD} + T_{SR} + T_{CP} + T_{SW} + T_{RS} + T_{GC} + T_{SS} \tag{1}$$

**Table 1.** Parameters in the prediction model

| Parameter type | Parameter name | Parameter explanation | Parameter name | Parameter explanation |
|---|---|---|---|---|
| Input parameters | *InputSize* | The amount of input data | *Parallelism* | Configuration of parallelism |
| | $pl_{init}$ | $InputSize \times \frac{1}{Parallelism}$ | | |
| General parameters | $T_{SD}$ | Task Scheduler Delay | $T_{CP}$ | Task Computing Time |
| | $T_{TD}$ | Task Deserialization Time | $T_{SW}$ | Task Shuffle Write Time |
| | $T_{SR}$ | Task Shuffle Read Time | $T_{RS}$ | Result Serialization Time |
| Extended parameters | $T_{GC}$ | Task GC time | $T_{SS}$ | Task Shuffle Spill Time |
| Prediction target | $T_{Task}$ | Task execution time | | |

## 3.2 Prediction of Sub-stage Execution Time

**Payload Sensitive Sub-stages.** For Computing Time and Shuffle Write Time, we build a linear regression model between execution time and payload of each task via non-negative least squares (NNLS). NNLS can avoid over-fitting and ensure the result which can be used for analysis is meaningful. Both of Computing Time and Shuffle Write Time are positively correlated to the payload of the task. Thus, the independent variable in the model should be $pl_{init}$. Taking potential system overhead into consideration, we add bias to correct the value. The prediction model of Computing Time and Shuffle Write Time are shown in formula (2) and formula (3).

$$T_{CP} = \theta_0^{CP} + \theta_1^{CP} * pl_{init} \tag{2}$$

$$T_{sw} = \theta_0^{SW} + \theta_1^{SW} * pl_{init} \tag{3}$$

$\theta_0^{CP}$ and $\theta_0^{SW}$ represent the potential fixed overheads of Computing Time and Shuffle Write Time respectively. $\theta_1^{CP}$ and $\theta_1^{SW}$ represent the correlation coefficient between execution time and payload.

**Payload Insensitive Sub-stages.** For the payload insensitive fixed overheads (Scheduling Delay, Deserialization Time, Shuffle Read Time and Result Serialization Time), the weight of these times in the whole execution time is very small. We use the average of samples as the prediction results, which is shown in formula (4).

$$T_P = \frac{1}{n} * \sum_{i=1}^n T_P^i, \ T_P \in \{T_{SD}, T_{TD}, T_{SR}, T_{RS}\} \tag{4}$$

$T_P$ is the predictive value of specific sub-stage execution time. $T_P^i$ is the execution time of specific sub-stage for the $i$h sample. $n$ is the number of samples.

**Analysis of Time Compensation.** In SPM, we consider three special factors that affect the execution time, namely GC Time, Shuffle Spill and first batch tasks. This section presents solutions for these three cases.

*GC Time.* In SPM, we only predict the overhead of minor GC. We treat the tasks which contain Full GC as outliers (In fact, the proportion of these tasks is quite small). We still use NNLS to fit the prediction model of GC Time. The main parameters we consider are (a) $pl_{init}$, caused by the characteristics of the Spark in-memory cluster computing. (b) $pl_{init}^2$, to model the situations where intermediate data needs to be cached. (c) bias, which is the potential system overhead. The formula (5) describes the predict of GC Time.

$$T_{GC} = \theta_0^{GC} + \theta_1^{GC} * pl_{init} + \theta_2^{GC} * pl_{init}^2 \tag{5}$$

*Shuffle Spill Time.* Shuffle spill often occurs when the payload of a task is large. Therefore, we make the predict of shuffle spill an extension module. If shuffle spill occurs on most tasks, we analyze the correlation coefficient between the Computing Time $T_{CP}$ of the sample and the payload $pl_{init}$ to find the point where the correlation coefficient suddenly increases. Then we use it as the threshold, $p_{ss}$, to enable shuffle spill module. Later we use samples which are either below the threshold or above the threshold to construct model respectively via formula (2). Finally, the shuffle spill time, $T_{ss}$, can be calculated by the difference between the correlation coefficient and the fixed overhead. The process is shown in formula (6).

$$T_{ss} = \begin{cases} 0, pl_{init} \leq p_{ss} \\ (\theta_0^s - \theta_0^{cp}) + (\theta_1^s - \theta_1^{cp}) * pl_{init}, pl_{init} > p_{ss} \end{cases} \tag{6}$$

$T_{ss}$ is the time overhead of shuffle spill. $\theta_0^s$, $\theta_1^s$ are the fixed overhead and correlation coefficient of the payload during shuffle spill. $\theta_0^{CP}$, $\theta_1^{CP}$ are the fixed overhead and correlation coefficient of the payload under the same configuration without shuffle spill. $p_{ss}$ is the threshold of payload for shuffle spill.

*First Batch Tasks.* First batch tasks (The first task on each core) of each stage perform information deserialization and prepare the worker node for computing. Their execution time is usually greater than the execution time of subsequent tasks. Therefore, we perform a separate prediction model for first batch tasks. It is similar to the model of subsequent tasks. The only difference is that the training set is the first batch tasks.

## 4 Experiments

### 4.1 Experimental Setup and Data Acquisition

Our experimental cluster consists of four nodes. One of them is the Spark master node and HDFS name node. Others are used as Spark slave nodes and HDFS data nodes. Each node is configured with 4 CPU cores at 3.6 GHz, 8G RAM, and 1T HDD. The four nodes are connected with a gigabit switch. We deploy ubuntu 16.04, Hadoop 2.9.0 and Spark 2.3.2 in each node.

We use Hibench [10] as the benchmark tool and select WordCount, PageRank and TeraSort as benchmarks. The changes of benchmark configuration in the experiments are shown in Table 2. In each set of experiments, 80% of samples are used as the training set for the prediction model, and the remaining 20% as the test set to test the accuracy of the prediction.

**Table 2.** Configuration changes of benchmark in the experiments

| Benchmark | Input data amount | Parallelism | Total samples |
|---|---|---|---|
| WordCount | $3.2 \times 10^8$–$3.2 \times 10^9$ records (Step $1.6 \times 10^8$) | 24–84 (Step 12) | 120 |
| PageRank | $2.5 \times 10^5$–$3.75 \times 10^6$ records (Step $2.5 \times 10^5$) | 12–120 (Step 12) | 150 |
| TeraSort | $3.2 \times 10^6$–$3.2 \times 10^7$ records (Step $3.2 \times 10^6$) | 24–84 (Step 12) | 120 |

## 4.2   Experimental Results

**Accuracy.** We first evaluate the accuracy of the prediction. We use *Accuracy* as a measure, and it is used to indicate how close the test value of all test set data is to the actual value, as defined by formula (7).

$$Accuracy = 1 - \frac{1}{n} \sum\nolimits_{i=1}^{n} \left| \frac{T_{pred}^{i} - T_{test}^{i}}{T_{test}^{i}} \right| \tag{7}$$

$n$ is the number of test set. $i$ is the sequence number of test data. $T_{test}^{i}$ is the actual value of the $i$th test data, and $T_{pred}^{i}$ is the corresponding predicted value.

Figure 1 shows the prediction accuracy of task execution time in each stage of PageRank, WordCount and TeraSort. The red bars in the figure represent the prediction accuracy of the first batch tasks, and the blue ones represents the following batches tasks. As can be seen from the results, the prediction accuracy of each stage is above 90%. Given the fact that the execution time may fluctuate during the actual task execution, it can be considered that the execution time prediction of the task is accurate.

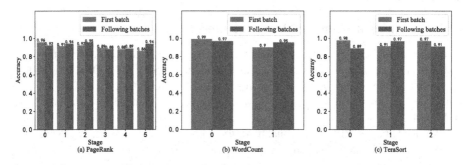

**Fig. 1.** The accuracy of the prediction (Color figure online)

**Efficiency of SPM.** Subsequently, we verify that SPM is more efficient than the coarse-grained model for task execution time. As a comparison, we implement a coarse-grained prediction model according to the method in Ernest [8], which directly establishes the relation between task payload and task execution time. The accuracy of the two methods is compared using PageRank and TeraSort data. The result is shown in Fig. 2. It can be seen that SPM effect in most stages is significantly better than the coarse-grained model. This is mainly because the coarse-grained prediction model cannot describe the overheads of GC and shuffle spill well.

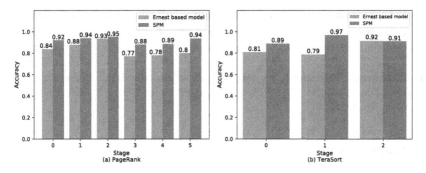

**Fig. 2.** Comparing SPM with coarse-grained prediction model

**Application in The Emulation Tool.** Finally, we evaluate the performance of the prediction model in SNemu [11], a data-driven spark cluster network emulation tool based on Mininet. Taking PageRank as an example, we set the parallelism to 12 and the iteration to 3, which means there are 6 stages (5 shuffle transmission) and one bitch tasks in each stage. The result is shown in Fig. 3 and the 5 larger data transmission represent the 5 shuffle transmission. It can be seen that the emulation result is very close to the real trace in the start time distribution of 5 shuffle transmission. The traffic that occurs in the real trace at around 10000 ms is the data transmission in the preparation stage of application execution (getting the jar file, reading the input data from the HDFS node, etc.). Since SNemu focuses on the transmission of shuffle, there is no emulation of the traffic in the preparation stage.

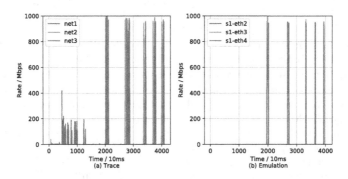

**Fig. 3.** Emulation network transmission compared with the real trace

## 5 Conclusion

In this paper, we propose an execution time prediction model, SPM, for Spark cluster. It is able to predict the sub-stage execution time and potential time overheads (GC, shuffle spill) of Spark task based on the payload. As a result, SPM can efficiently predict the average task execution time of Spark applications. Experimental results show that our prediction model can well predict the task execution time of various benchmarks. Further, we combine SPM with a Spark network emulation tool. Emulation result shows that the prediction of task execution time can effectively determine the time of shuffle transmission.

**Acknowledgement.** This work is supported by the National Key Research and Development Program of China (Grant No. 2016YFB1000304) and National Natural Science Foundation of China (Grant No. 1636208).

# References

1. Zaharia, M., Chowdhury, M., Franklin, M.J., Shenker, S., Stoica, I.: Spark: cluster computing with working sets. HotCloud **10**(10–10), 95 (2010)
2. Hadoop Homepage. http://Hadoop.apache.org/. Accessed 4 Sept 2019
3. Zaharia, M., et al.: Resilient distributed datasets: a fault-tolerant abstraction for in-memory cluster computing. In: Proceedings of the 9th USENIX Conference on Networked Systems Design and Implementation, p. 2. USENIX Association (2012)
4. Gu, J., Li, Y., Tang, H., Wu, Z.: Auto-tuning spark configurations based on neural network. In: 2018 IEEE International Conference on Communications (ICC), pp. 1–6. IEEE (2018)
5. Popescu, A.D., Balmin, A., Ercegovac, V., Ailamaki, A.: Predict: towards predicting the runtime of large scale iterative analytics. Proc. VLDB Endow. **6**(14), 1678–1689 (2013)
6. Nguyen, N., Khan, M.M.H., Albayram, Y., Wang, K.: Understanding the influence of configuration settings: an execution model-driven framework for apache spark platform. In: 2017 IEEE 10th International Conference on Cloud Computing (CLOUD), pp. 802–807. IEEE (2017)
7. Bhimani, J., Mi, N., Leeser, M., Yang, Z.: FIM: performance prediction for parallel computation in iterative data processing applications. In: 2017 IEEE 10th International Conference on Cloud Computing (CLOUD), pp. 359–366. IEEE (2017)
8. Venkataraman, S., Yang, Z., Franklin, M., Recht, B., Stoica, I.: Ernest: efficient performance prediction for large-scale advanced analytics. In: 13th USENIX Symposium on Networked Systems Design and Implementation (NSDI 2016), pp. 363–378 (2016)
9. Wang, K., Khan, M.M.H.: Performance prediction for apache spark platform. In: 2015 IEEE 17th International Conference on High Performance Computing and Communications, 2015 IEEE 7th International Symposium on Cyberspace Safety and Security, and 2015 IEEE 12th International Conference on Embedded Software and Systems, pp. 166–173. IEEE (2015)
10. Huang, S., Huang, J., Dai, J., Xie, T., Huang, B.: The hibench benchmark suite: characterization of the mapreduce-based data analysis. In: 2010 IEEE 26th International Conference on Data Engineering Workshops (ICDEW 2010), pp. 41–51. IEEE (2010)
11. SNemu. https://github.com/lab821/SNemu. Accessed 4 Sept 2019

# Improving the Parallelism of CESM on GPU

Zehui Jin[1], Ming Dun[1], Xin You[1], Hailong Yang[1(✉)], Yunchun Li[1], Yingchun Lin[2], Zhongzhi Luan[1], and Depei Qian[1]

[1] School of Computer Science and Engineering, Beihang University,
Beijing 100191, China
`hailong.yang@buaa.edu.cn`
[2] Fourth Research Institute of Telecommunications Technology Corporation,
Xi'an 710061, Shaanxi, China

**Abstract.** Community Earth System Model (CESM) is one of the most popular climatology research models. However, the computation of CESM is quite expensive and usually lasts for weeks even on high-performance clusters. In this paper, we propose several optimization strategies to improve the parallelism of three hotspots in CESM on GPU. Specifically, we analyze the performance bottleneck of CESM and propose corresponding GPU accelerations. The experiment results show that after applying our GPU optimizations, the kernels of the physical model achieve significant performance speedup respectively.

**Keywords:** CESM · GPU · Performance optimization

## 1 Introduction

Scientists have been striving to understand and even predict the earth's climate system since the early days. The effort led by the National Center for Atmosphere Research (NCAR) gave birth to the open-source simulation model named Community Climate Model (CCM) [4]. As the successor of CCM, NCAR released The Community Earth System Model (CESM), which becomes one of the cutting-edge climate simulation models nowadays.

The CESM simulation model is constituted of five geophysical components, including atmosphere, land, ocean, sea-ice and land-ice component. There is also a coupler in CESM to incorporate the above geophysical components. However, both the computation and memory cost when running CESM is overwhelmingly high on CPU due to the complex mathematical equations adopted in the simulation.

There have already been several works optimizing CESM by accelerating certain kernels in CESM on GPU. Korwar et al. [5] implement the long wave and short wave radiation kernels on GPU. Carpenter et al. [3] implement the spectral element based dynamical core (HOMME) on GPU. Sun et al. [10] optimize a solver in the chemistry procedure of the atmospheric component and implement

© Springer Nature Switzerland AG 2020
S. Wen et al. (Eds.): ICA3PP 2019, LNCS 11945, pp. 11–18, 2020.
https://doi.org/10.1007/978-3-030-38961-1_2

it on GPU for higher performance. However, none of the existing works simultaneously accelerates three hotspot kernels on GPU including aerosol masses conversion, longwave radiation, and solar radiation.

This work focuses on optimizing the performance of the atmospheric component, which is one of the most time-consuming components of CESM. First, we comprehensively analyze the performance bottlenecks of CESM. Then we optimize the three hotspot kernels on GPU. We demonstrate the effectiveness of our optimizations with experiment results.

In short, this paper makes the following contributions:

- We conduct a comprehensive performance analysis of CESM and identify three hotspot kernels in the atmospheric component.
- We implement the identified hotspot kernels on GPU and optimize the kernels for better performance.

The rest of the paper is organized as follows. The background of CESM and existing parallel methods are presented in Sect. 2. We perform the bottleneck analysis of CESM and present the performance optimization strategies in Sect. 3. In Sect. 4, the detailed experiment results are given. We present the related work in Sect. 5 and conclude this paper in Sect. 6.

## 2  Background

### 2.1  CESM Overview

CESM is mainly developed by the NCAR and its predecessor is known as the CCSM. CESM is comprised of several coupled components: atmosphere, ocean, land, land-ice, sea-ice and a central coupler component. Once the simulation began, the coupler will swap the two-dimensional data. The cutting-edge earth system simulation software provides support for plenty of options in different combinations of component configurations and resolutions [11].

### 2.2  The Atmosphere Component

The atmosphere component of CESM is called the Community Atmosphere Model (CAM). CAM 4.0 mainly constitutes four kinds of dynamical cores and a physical model which is an aggregation of the simulation of the physical process. The total parameter package of physical model can be indicated as Eq. 1, where M means (Moist) precipitation processes, R for radiation and clouds, S for the abbreviation of the Surface model, and T for Turbulent mixing [8].

$$P = \{M, R, S, T\} \tag{1}$$

## 2.3   Parallelization of CESM

**Computation Parallelization.** Within almost every geophysical part of CESM, the system is divided into hierarchical grids and the grids are allocated to different processes to compute. To the perspective of a standalone component, the atmosphere, ice, ocean, and land component support two levels of parallelization: the distributed memory parallelism(MPI) and the shared memory parallelism model (OpenMP) [13]. For the whole CESM, some geophysical components can run concurrently but there remain constraints. For instance, the atmosphere component cannot run with ice or land component.

**I/O Parallelization.** CESM usually use netCDF [9] for reading and writing data. There are two ways to parallel the I/O procedure. One way is by enabling the parallelism version of netCDF and the other way is to replace the default netCDF with pnetCDF [6]. Paralleled netCDF can read input data through multiple processors which can contribute to I/O acceleration.

# 3   Performance Optimization Strategies

## 3.1   Bottleneck Analysis of CESM

We probe the hotspots of CESM by using HPCToolkit on the platform whose software and hardware details are shown in Sect. 4. The hotspot analysis is shown in Fig. 1. From the analysis result, the conclusion can be drawn that the atmosphere component costs most of the execution time, as the dynamical part spends approximately 48.8% of total running time while the physical part takes up about 33% of the overall execution time.

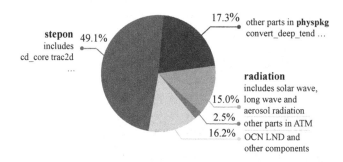

**Fig. 1.** The performance bottleneck analysis of CESM.

It can be figured out that the MPI communication and *cd_core* are two hotspots in the dynamic part. However, since the optimization of MPI can be done by alternating MPI library which is much easier than algorithm optimization and *cd_core* consists of numerous loops that only cost about 0.1% of execution time each, we focus on optimizing the hotspots in physical part in this paper.

In the physical model, it can be revealed from the result that the radiation routine is a hotspot which takes up 15% of execution time approximately. The radiation routine includes solar radiation, longwave radiation and aerosol masses conversion, which contains computation-intense loops.

## 3.2   Solar Radiation Optimization

In the original solar radiation routine in CESM, there is a computation-intense triple cycle which contains $nswband \times (pver + 1) \times Nday$ iterations. To enhance the performance in its GPU implementation, we assign every iteration to a thread in GPU to improve parallelization, the stream number and the size of the grid are shown in Algorithm 1. Once all the threads finish liquid and ice valid computation, they get the boundaries of solar radiation and then compute radiative properties of layers. After all the computation is done, the results will be sent to the host.

---

**Algorithm 1.** The optimized solar radiation kernel on GPU

---
1: *malloc memory for temporary vars on GPU*
2: *malloc memory for input vars on GPU and copy values*
3: */ * stream number = 1 * /*
4: */ * grid size = 4 × 4 × 1 * /*
5: */ * block size = 64 × 4 × 1 * /*
6: */ * total threads amount = Nday × (pver + 1) × nswbands * /*
7: **for** *each thread* **do**
8:     *initialize global constant parameters*
9:     **if** $(blockIdx.y \times blockDim.y + threadIdx.y)\%(pver + 1) \neq 0$ **then**
10:         *calculate liquid and ice valid*
11:     **end if**
12:     *synchronize*
13:     *get boundaries of solar wave spectral*
14:     *compute layer radiative properties*
15: **end for**
16: *copy out results and free memory on GPU*

---

## 3.3   Longwave Radiation Optimization

We focus on optimizing the absorptivities computation kernel when improving the performance of the longwave radiation routine. In the original kernel, there is a triple cycle which iterates $ncol \times (pverp - ntoplw + 1) \times (pverp - ntoplw + 1)$ times. To implement this kernel to GPU, we assign the threads to each iteration as shown in Algorithm 2. For each thread, it computes the band-dependent indices for non-window and window. Next, the threads calculate the absorptivity for non-nearest layers and finally calculate total absorptivity based on previous steps.

**Algorithm 2.** The optimized longwave radiation kernel on GPU

1:  *Initialize constants as local parameters*
2:  / * *stream number* : 1 * /
3:  / * *grid size* : 5 × 4 × 1 * /
4:  / * *block size* : 4 × 64 × 1 * /
5:  / * *total threads amount* : *ncol* × (*pverp* − *ntoplw* + 1) × (*pverp* − *ntoplw* + 1) * /
6:  **for** *each thread* **do**
7:      / * $H_2O$ *Continuum path for* 0 − 800 *and* 1200 − 2200 $cm^{-1}$ * /
8:      *Band* − *dependent indices for non* − *window*
9:      / * *Line transmission in* 800 − 1000 *and* 1000 − 1200 $cm^{-1}$ *intervals* * /
10:     *Get* $O_3$ 9.6 *micrometer band*
11:     *Get* $CO_2$ 15 *micrometer band system*
12:     *Calculate absorptivity for non nearest layers*
13:     *Calculate total absorptivity based on previous five steps*
14: **end for**

## 3.4 Aerosol Masses Conversion Optimization

In the GPU implementation of aerosol masses conversion routine whose processing logic is shown in Algorithm 3. After the stream and thread are set up by GPU, the device will get *opticstype* which indicates the type of aerosol to be computed. There are four types of aerosols: hygroscopic aerosols, non-hygroscopic aerosols, volcanic aerosols, and volcanic radius aerosols. Once the *opticstype* is received, the threads will calculate the corresponding optical properties for those aerosols.

**Algorithm 3.** The optimized aerosol masses conversion kernel on GPU

1:  *Malloc memory for temporary variables on GPU*
2:  *Create CUDA stream*
3:  / * *total threads* = *pver* × *pcols* × *nswbands* * /
4:  / * *stream number* = 1 * /
5:  / * *grid size* = 8 × 4 × 1 * /
6:  / * *block size* = 64 × 4 × 1 * /
7:  **for** *iaerosol* = 1 → *numaersols* **do**
8:      *Get opticstype* : *optics* ← *StringToInt*(*optistype*)
9:      **if** *optics* == 1 **then**
10:         *for each thread, calculate optical properties for hygroscopic aerosols*
11:     **else if** *optics* == 2 **then**
12:         *for each thread, calculate optical properties for non* − *hygroscopic aerosols*
13:     **else if** *opitcs* == 3 **then**
14:         *for each thread, calculate optical properties for volcanic aerosols*
15:     **else if** *optics* == 4 **then**
16:         *for each thread, calculate optical properties for volcanic radius aerosols*
17:     **else**
18:         *return error message* / * *default* * /
19:     **end if**
20:     *copy results to host's memory*
21:     *free memory on GPU*
22: **end for**

## 4    Evaluation

### 4.1    Experiment Setup

The experiments are conducted on a single server. The server is equipped with
2× Intel Xeon E5-2680v4 2.40 GHz 14 cores and two NVIDIA Volta 32 GB
V100. The operating system installed is CentOS v7.6 with Linux kernel v3.10.0-
957.el7.x86_64. We use CESM v1.2.2 compiled with Intel compiler v2018.5.274
as our baseline. CUDA version is 10.1, and netCDF is v4.6.2 enabled parallelization. For the simulation input, we choose two representative datasets, *E1850CN*
and *F*, with corresponding simulation setups. For each dataset, we run CESM
on two commonly used resolutions, 0.47 × 0.63 and 0.23 × 0.31.

### 4.2    Performance Analysis

In Fig. 2(a), the orange bars represent the average running time of the aerosol
masses conversion kernel in the original CESM, whereas the purple ones represent the average running time of our GPU optimization. It is obvious that our
GPU optimization of the kernel reduces the execution time significantly. The
average performance speedup on dataset *E1850CN* and *F* is 6.91× and 9.14×
respectively.

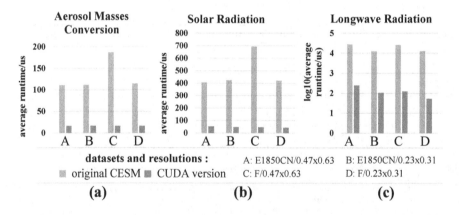

**Fig. 2.** The performance speedup of the three kernels. (a) for aerosol masses conversion
kernel, (b) for solar radiation and (c) for longwave radiation kernel. (Color figure online)

We can see from Fig. 2(b) that the experiment results on solar radiation
kernel show the similar trend. The average performance speedup on dataset
*E1850CN* and *F* is 8.36× and 12.38× respectively. The higher performance
speedup achieved by solar radiation kernel is due to its high computation intensity that benefits from the massive parallelism of GPU.

The performance speedup on longwave radiation kernel is much higher than
the previous two kernels. As shown in Fig. 2(c), due to the computation-intensive

nature of the longwave radiation kernel, the original implementation takes more than $10^4$ microseconds to complete, so we show the logarithmic results. The average performance speedup on dataset *E1850CN* and *F* is $114.95\times$ and $222.4\times$ respectively.

## 5   Related Work

Most of the relevant works port some of the kernels in CESM to GPU. Korwaret et al. [5] transfer the long-wave and solar radiation routines to GPU by OpenACC and develop a CPU-GPU asynchronous scheme. In the meantime, the spectral element based dynamical core(HOMME) is implemented to GPU by Carpenter et al. [3]. Later, Werkhoven et al. [12] modify the block-partitioning strategies in the ocean component of CESM named POP into a hierarchical scheme to optimize load balance and reduce MPI communication. Furthermore, Sun et al. [10] accelerate the second-order Rosenbrock solver by implementing it to GPU, and it turns out that a higher performance will be reached when the block size is equal to the warp size and memory is continuous. Moreover, as a part of the OpenACC based project Energy Exascale Earth System Model(E3SM) whose atmospheric component is a derivative of CAM5 in CESM1, Bertagna et al. [2] rewrite the HOMME with C++ and Kokkos which makes the core portable to GPU, which results in better performance. Since the inappropriate parameters can cause severe performance deterioration, an optimal execution configuration is desirable. Nan et al. [7] develop a framework named CESMTuner which can detect the best configuration automatically. Balaprakash et al. [1] develop a static parameter-probing model based on machine learning and apply it to the ice component of CESM.

## 6   Conclusion

In this paper, we optimize three hotspot kernels in CESM on GPU and achieve significant performance speedup. We use the HPC toolkit to analyze the performance of CESM and identify three hotspot kernels. Then we implement these hotspot kernels on GPU and optimize them correspondingly. The experiment results show that our GPU optimizations achieve $11.25\times$, $15.02\times$ and $237\times$ performance speedup for aerosol masses conversion, solar radiation and longwave radiation respectively.

**Acknowledgement.** This work is supported by National Key Research and Development Program of China (Grant No. 2016YFB1000304) and National Natural Science Foundation of China (Grant No. 61502019).

# References

1. Balaprakash, P., Alexeev, Y., Mickelson, S.A., Leyffer, S., Jacob, R., Craig, A.: Machine-learning-based load balancing for community ice code component in CESM. In: Daydé, M., Marques, O., Nakajima, K. (eds.) VECPAR 2014. LNCS, vol. 8969, pp. 79–91. Springer, Cham (2015). https://doi.org/10.1007/978-3-319-17353-5_7
2. Bertagna, L., et al.: HOMMEXX 1.0: a performance portable atmospheric dynamical core for the energy exascale earth system model. Technical report, Sandia National Lab. (SNL-NM), Albuquerque, NM (United States); Sandia ... (2018)
3. Carpenter, I., et al.: Progress towards accelerating homme on hybrid multi-core systems. Int. J. High Perform. Comput. Appl. **27**(3), 335–347 (2013)
4. Kiehl, T., Hack, J., Bonan, B., Boville, A., Briegleb, P., Williamson, L., Rasch, J.: Description of the NCAR community climate model (CCM3) (1996)
5. Korwar, S.K., Vadhiyar, S., Nanjundiah, R.S.: GPU-enabled efficient executions of radiation calculations in climate modeling. In: 20th Annual International Conference on High Performance Computing, pp. 353–361. IEEE (2013)
6. Li, J., et al.: Parallel netCDF: a high-performance scientific I/O interface. In: SC 2003: Proceedings of the 2003 ACM/IEEE Conference on Supercomputing, pp. 39–39. IEEE (2003)
7. Nan, D., Wei, X., Xu, J., Haoyu, X., Zhenya, S.: CESMTuner: an auto-tuning framework for the community earth system model. In: 2014 IEEE International Conference on High Performance Computing and Communications, 2014 IEEE 6th International Symposium on Cyberspace Safety and Security, 2014 IEEE 11th International Conference on Embedded Software and Systems (HPCC, CSS, ICESS), pp. 282–289. IEEE (2014)
8. Neale, R.B., et al.: Description of the NCAR community atmosphere model (CAM 4.0) (2010)
9. Rew, R., Davis, G.: NetCDF: an interface for scientific data access. IEEE Comput. Graph. Appl. **10**(4), 76–82 (1990)
10. Sun, J., et al.: Computational benefit of gpu optimization for the atmospheric chemistry modeling. J. Adv. Model. Earth Syst. **10**(8), 1952–1969 (2018)
11. Vertenstein, M., et al.: CESM user's guide (CESM1.2 release series user's guide). NCAR technical note (2013)
12. van Werkhoven, B., et al.: A distributed computing approach to improve the performance of the parallel ocean program (v2.1). Geosci. Model Dev. **7**(1), 267–281 (2014)
13. Worley, P.H., Mirin, A.A., Craig, A.P., Taylor, M.A., Dennis, J.M., Vertenstein, M.: Performance of the community earth system model. In: Proceedings of 2011 International Conference for High Performance Computing, Networking, Storage and Analysis, p. 54. ACM (2011)

# Parallel Approach to Sliding Window Sums

Roman Snytsar[1]([✉]) and Yatish Turakhia[2]

[1] Microsoft Corporation, One Microsoft Way, Redmond, WA 98052, USA
Roman.Snytsar@microsoft.com
[2] Stanford University, Stanford, CA 94305, USA

**Abstract.** Sliding window sums are widely used for string indexing, hashing, time series analysis and machine learning. New vector algorithms which utilize the advanced vector extension (AVX) instructions available on modern processors, or the parallel compute units on GPUs and FPGAs, would provide a significant performance boost.

We develop a generic vectorized sliding sum algorithm with speedup for window size $w$ and number of processors $P$ is $O(P/w)$ for a generic sliding sum. For a sum with commutative operator the speedup is improved to $O(P/log(w))$. Implementing the algorithm for the bioinformatics application of minimizer based k-mer table generation using AVX instructions, we obtain a speedup of over $5\times$.

## 1 Introduction

### 1.1 Prefix Sum

Parallel algorithms are often constructed from a set of universal building blocks. One of the hardest to identify, but extremely useful is the concept of a *prefix sum*, and the accompanying *scan* algorithm. A prefix sum is a transformation that takes an operator $\oplus$, and a sequence of elements

$$x_0, x_1, \ldots, x_k, \ldots$$

and returns the sequence

$$y_i = \sum_{j=0}^{i} x_j = x_0 \oplus x_1 \oplus \ldots \oplus x_i \tag{1}$$

or in recurrent form

$$y_{i+1} = y_i \oplus x_{i+1} \tag{2}$$

Despite the data carry dependency, the first $N$ elements of the prefix sum with an associative operator could be computed in $O(log(N))$ parallel steps using scan algorithm, as shown by [3].

© Springer Nature Switzerland AG 2020
S. Wen et al. (Eds.): ICA3PP 2019, LNCS 11945, pp. 19–26, 2020.
https://doi.org/10.1007/978-3-030-38961-1_3

## 1.2   Sliding Window Sum

Sliding window sum (sliding sum) takes a window size $w$ in addition to an operator $\oplus$, and a sequence of elements, and returns the sequence

$$y_i = \sum_{j=i}^{i+w-1} x_j = x_i \oplus x_{i+1} \oplus \ldots \oplus x_{i+w-1} \tag{3}$$

where each sum is defined in terms of the operator $\oplus$ and contains exactly $w$ addends. The asymptotic complexity of a naive sliding sum algorithm is $O(wN)$ where $N$ is the length of the source sequence.

Every sum defined by Eq. 3 is a prefix sum with operator $\oplus$ and input sequence $x_i \ldots \oplus x_{i+w-1}$. Many useful operators are associative, so the prefix scan algorithm is applicable here, reducing complexity of every sum in Eq. 3 to $O(log(w))$ and, trivially, the overall sliding sum complexity to $O(Nlog(w))$ parallel steps.

Sliding window operators are widely used in the high frequency data mining [2]. Using elaborate queue-based data structures, these streaming applications achieve $O(N)$ complexity, which is the current state of the art [11]. In machine learning, the sliding window approach is used for validating the time series models [5]. While working on the bioinformatics applications, we have used sliding window sums to represent the minimizer seeds.

## 1.3   The Seed-filter-extend Paradigm

Most heuristics to local sequence alignment are based on the *seed-filter-extend* paradigm, which was first popularized by the BLAST algorithm [1]. In aligning a reference sequence $R$ with a query sequence $Q$, the *seeding* stage finds small local matches, called *seed hits*, of size $k$ (also called k-mer, typically 10–19 basepairs in size) between $R$ and $Q$. The *filtering* stage itself may consist of several smaller sub-stages, which further reduces the search space by a combination of techniques, such as ungapped extension [1,4] or chaining multiple seed hits in a diagonal band [7,12]. The *extension* stage typically performs the compute-intensive dynamic programming step, usually employing the Smith-Waterman equations [9].

## 1.4   Seed Tables and Minimizers

Heuristics based on the seed-filter-extend paradigm often maintain a *seed table*— a data structure that enables fast lookup of seed hits in reference, $R$. Figure 1 shows an example reference sequence and seed table for seed size $k = 2$. Seed table maintains two tables: (i) a seed pointer table and (ii) a seed position table. For each of the $4^k$ possible seeds (16 seeds in Fig. 1), lexicographically sorted, the seed pointer table points to the beginning of a list of hits in the seed position table. In Fig. 1, lookups to 'CG' and 'CT' in the seed pointer table give the start and end addresses in the seed position table for hits of 'CT' in the reference.

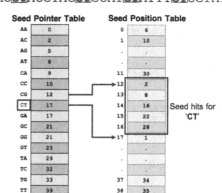

REFERENCE: AGCTTTACCTACGTAGCTGCATCTATTTCTCGTATTTAGC

**Fig. 1.** An example reference sequence and seed table used in D-SOFT.

Starting with $R = r_0, r_1, ...r_n$, we can define k-mers of $R$ as a sliding sum over window size k, string concatenation operator, and $R$. If we introduce operation $substr(K, i)$, a substring of string K starting from position $i$, a reverse operation to the string concatenation, it is possible then to generate k-mers recursively in $O(N)$ time:

$$K_{i+1} = substr(K_i, 1) \oplus r_{i+k} \tag{4}$$

*Minimizer seeds* (or *minimizers* for short), an idea originally proposed for compressing large seed tables in 2004 by Roberts et al. [8], have seen a recent revival in bioinformatics with the advent long read alignment [7] and metagenomics [13]. Minimizers can greatly reduce the storage requirements for the seed position table by storing only a subset of the seeds with only a small drop in sensitivity of the aligner.

Figure 2 illustrates how minimizers can be used to build a seed position table with an example. In addition to the seed size $k$, minimizers require a parameter $w$, the *minimizer window size*. In Fig. 2, $k = 3$ and $w = 3$. In each position $p$ of the reference $R$, a window $w$ consecutive seeds of size $k$ (k-mers) starting from position $p$ in $R$ are used to find the lexicographically minimum seed $s$ and its position $p'$, which is recorded in the seed position table. Adjacent windows can share the same minimizer (i.e. the $(s, p')$ pair), which reduces the storage requirement for the seed position table. Figure 2a shows the minimizers for four consecutive positions 0–3 in $R$ and the corresponding entries in the seed position table in Fig. 2b. Windows at positions $p = 1$ and $p = 2$ share the same minimizer ('CTT', 2), which is stored only once in Fig. 2b. Moreover, as seen in Fig. 2b, seeds at position 0, 2 and 5 are stored in the seed position table but those at positions 1, 3 and 4 are dropped. Roberts et al. [8] have shown that with a minimizer window of size $w$, a new minimizer occurs every $w/2$ bases on average.

Minimizers are a key innovation in Minimap [6] and its successor Minimap2 [7], both of which achieve an order of magnitude speedup over prior

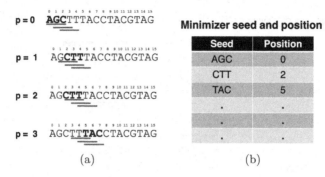

(a)                                      (b)

**Fig. 2.** Illustration of minimizer seeds using ($k = 3$, $w = 3$). (a) An example reference sequence with a minimizer window sliding over 4 positions. The three seeds within the window are underlined in red and the minimizer seed within the window is highlighted in bold. (b) Minimizer seed-position pairs as constructed from (a). (Color figure online)

techniques, most speedup resulting from fewer seed hits per read due to minimizers. We have found that turning off minimizers (using $w = 1$ instead of the default $w = 10$) slows down the seeding and filtering stage of Minimap2 by nearly 7× with only 0.5% higher sensitivity for sequencing reads from Pacific Biosciences. Constructing seed tables can take several hours for the *de novo* assembly of a human genome [12]. In this paper, we take a closer look at the connection between sliding sums and prefix sums, and attempt to supersede the linear complexity achieved by previous approaches.

## 2    Methods

### 2.1    Vector Algorithms

Our first algorithm is a vector-friendly way of calculating sliding sum assuming the input sequence elements become available one by one and are processed using the vector instructions of width $P > w$:

Vector Y is initialized to the suffix sums with the number of elements decreasing from $w - 1$ to 0. Then in a loop every incoming element $x_k$ is broadcast to the first $w$ elements of vector X. After vector addition the zeroth element of Y contains the next sliding sum. Next, the vector Y is shifted left by one element, as denoted by operator ⋘, and the state is ready for the next iteration.

Asymptotic complexity of the scalar input algorithm is $O(N)$ with no additional requirements on the operator ⊕.

This result could be improved if we assume that the input sequence arrives packed in vectors of width $P > w$.

At every iteration $P$ input elements are placed into vector X. X1 is filled with the prefix sums of up to $w$ addends, and Y1 is filled with the suffix sums constructed from the elements of $X$, as shown on the Fig. 3. Then the vector sum of $Y$ and $X1$ yields the next P output elements. Finally, the suffix sums

**Algorithm 1.** Scalar Input

**procedure** SCALARINPUT($x_0 \ldots x_{n-1}$)

$$Y \leftarrow \Big( \underbrace{\sum_{j=0}^{w-2} x_j, \sum_{j=1}^{w-2} x_j, \ldots, x_{w-3} \oplus x_{w-2}, x_{w-2}, 0, \ldots, 0}_{w-1} \Big)$$

**for** $i = w - 1$ to N **do**

$\quad X \leftarrow \Big( \underbrace{x_i, x_i, \ldots, x_i}_{w}, 0, \ldots, 0 \Big)$

$\quad Y \leftarrow Y \oplus X$

$\quad y_{i-w+1} \leftarrow Y[0]$

$\quad Y \leftarrow Y \lll 1$

**end for**

**end procedure**

**Algorithm 2.** Vector Input

**procedure** VECTORINPUT($x_0 \ldots x_{n-1}$)

$$Y \leftarrow \Big( \underbrace{\sum_{j=0}^{w-2} x_j, \sum_{j=1}^{w-2} x_j, \ldots, x_{w-3} \oplus x_{w-2}, x_{w-2}, 0, \ldots, 0}_{w-1} \Big)$$

**for** $i = w - 1$ to N step P **do**

$\quad X \leftarrow \Big( x_k, x_{k+1}, \ldots, x_{k+p-1} \Big)$

$\quad X1 \leftarrow \Big( \underbrace{X_0, X_0 \oplus X_1, \ldots, \sum_{j=0}^{w-2} X_j}_{w-1}, \sum_{j=0}^{w-1} X_j, \ldots, \sum_{j=p-w}^{p-1} X_j \Big)$

$\quad Y1 \leftarrow \Big( 0, \ldots, 0, \underbrace{\sum_{j=p-w}^{p-1} X_j, \sum_{j=p-w}^{p-2} X_j, \ldots, X_{p-w}}_{w-1} \Big)$

$\quad Y \leftarrow Y \oplus X1$

$\quad y_{k-w+1} \ldots y_{k-w+p} \leftarrow Y[0] \ldots Y[p-1]$

$\quad Y \leftarrow Y1 \lll (P - w)$

**end for**

**end procedure**

from $Y1$ are shifted into proper positions in vector $Y$, and it is ready for the next iteration.

The asymptotic complexity thus is $O(N \cdot w/P)$ with the parallel speedup $O(P/w)$ for any operator $\oplus$. If $\oplus$ is associative, the prefix/suffix sums could be computed in parallel using the algorithm in [3], and the complexity is reduced to $O(N \cdot log(w)/P)$ with the speedup improving to $O(P/log(w))$.

For example, since $min$ is an associative operator, the sliding window minimum can be computed using the faster version of the vector input algorithm.

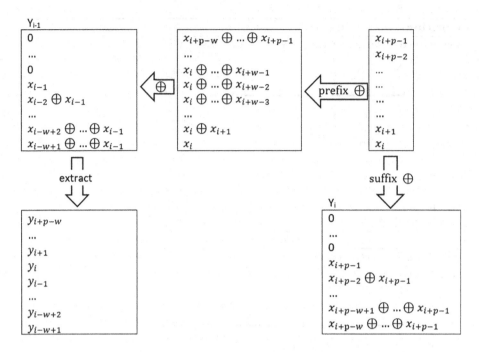

**Fig. 3.** Data flow of the vector input sliding sum algorithm.

## 3   Results

We tested the performance of various sliding minimum algorithms using the hashed 15-mers of the reference human genome assembly (GRCh38) from the Genome Reference Consortium. The test imitates a minimizer based seed table construction by a long-read aligner, such as Minimap2 [7], GraphMap [10] or Darwin [12]. Figure 4 compares the performance of the naïve array-based algorithm, linear dequeue-based algorithm, and our proposed vector algorithm.

The computer platform is an Intel Xeon Platinum 8168 system with 16 cores running at 2.7 GHz and 32 GB of RAM. The test for every data point has been run 3 times, and presented numbers are the averages of the observed run times.

Deque-based algorithm performance is indeed independent of the window size. It comes, however, at the cost of a significant overhead of managing the deque data structure and unpredictable branching.

Array-based algorithm, despite the worst asymptotic complexity, is simple to implement, and benefits from the automatic compiler vectorization. It is clear how the times drop when the window size is aligned with the SIMD vector width ($P = 4$, 8, and 16). For small window sizes the array algorithm is competitive with the deque approach.

Our vector sliding sum algorithm beats both previous implementations by a factor of 5×. With the SSE/AVX instruction set, any window size requires the same number of instructions as the closest (larger) power of 2. So the perfor-

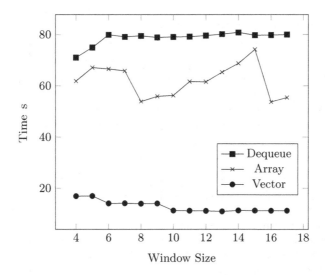

**Fig. 4.** Performance of the sliding minimum algorithms.

mance of our vector implementation does not change linearly with $w$ but drops when we switch to the different SIMD vector width $P$ at $w = 5, 9, 17$. Also, prefix sum computation across wider vectors incurs additional latencies for cross-lane data exchanges, resulting in the speedup less than theoretical $2\times$.

## 4 Conclusion

We introduced a family of algorithms for parallel evaluation of sliding window sums. The parallel speedup for window size $w$ and number of processors $P$ is $O(P/w)$ for a generic sliding sum. For a sum with a commutative operator the speedup is improved to $O(P/log(w))$. This gives the developer a choice of fast branchless algorithms suitable for implementation on any modern parallel architecture including modern CPUs with instruction-level parallelism, pipelined GPUs, or FPGA reconfigurable hardware.

## References

1. Altschul, S.F., Gish, W., Miller, W., Myers, E.W., Lipman, D.J.: Basic local alignment search tool. J. Mol. Biol. **215**(3), 403–410 (1990)
2. Basak, A., Venkataraman, K., Murphy, R., Singh, M.: Stream Analytics with Microsoft Azure: Real-Time Data Processing for Quick Insights Using Azure Stream Analytics. Packt Publishing Ltd., Birmingham (2017)
3. Blelloch, G.E.: Prefix sums and their applications. In: Synthesis of Parallel Algorithms. Morgan Kaufmann (1993)
4. Harris, R.S.: Improved pairwise alignment of genomic DNA. ProQuest (2007)

5. Kotu, V., Deshpande, B.: Data Science: Concepts and Practice. Morgan Kaufmann, Burlington (2018)
6. Li, H.: Minimap and miniasm: fast mapping and de novo assembly for noisy long sequences. Bioinformatics **32**(14), 2103–2110 (2016)
7. Li, H.: Minimap2: pairwise alignment for nucleotide sequences. Bioinformatics **1**, 7 (2018)
8. Roberts, M., Hayes, W., Hunt, B.R., Mount, S.M., Yorke, J.A.: Reducing storage requirements for biological sequence comparison. Bioinformatics **20**(18), 3363–3369 (2004)
9. Smith, T.F., Waterman, M.S.: Identification of common molecular subsequences. J. Mol. Biol. **147**(1), 195–197 (1981)
10. Sović, I., Šikić, M., Wilm, A., Fenlon, S.N., Chen, S., Nagarajan, N.: Fast and sensitive mapping of nanopore sequencing reads with graphmap. Nat. Commun. **7**, 11307 (2016)
11. Tangwongsan, K., Hirzel, M., Schneider, S.: Constant-time sliding window aggregation. IBM, IBM Research Report RC25574 (WAT1511-030) (2015)
12. Turakhia, Y., Bejerano, G., Dally, W.J.: Darwin: a genomics co-processor provides up to 15,000 x acceleration on long read assembly. In: Proceedings of the Twenty-Third International Conference on Architectural Support for Programming Languages and Operating Systems, pp. 199–213. ACM (2018)
13. Wood, D.E., Salzberg, S.L.: Kraken: ultrafast metagenomic sequence classification using exact alignments. Genome Biol. **15**(3), R46 (2014)

# Rise the Momentum: A Method for Reducing the Training Error on Multiple GPUs

Yu Tang, Lujia Yin, Zhaoning Zhang$^{(\boxtimes)}$, and Dongsheng Li

Science and Technology on Parallel and Distributed Laboratory,
National University of Defense Technology, Changsha, China
`zzningxp@gmail.com`

**Abstract.** Deep neural network training is a common issue that is receiving increasing attention in recent years and basically performed on Stochastic Gradient Descent or its variants. Distributed training increases training speed significantly but causes precision loss at the mean time. Increasing batchsize can improve training parallelism in distributed training. However, if the batchsize is too large, it will bring difficulty to training process and introduce more training error. In this paper, we consider controlling the total batchsize and lowering batchsize on each GPU by increasing the number of GPUs in distributed training. We train Resnet50 [4] on CIFAR-10 dataset by different optimizers, such as SGD, Adam and NAG. The experimental results show that large batchsize speeds up convergence to some degree. However, if the batchsize of per GPU is too small, training process fails to converge. Large number of GPUs, which means a small batchsize on each GPU declines the training performance in distributed training. We tried several ways to reduce the training error on multiple GPUs. According to our results, increasing momentum is a well-behaved method in distributed training to improve training performance on condition of multiple GPUs of constant large batchsize.

**Keywords:** Multiple GPUs · Batchsize · Distributed training · Momentum

## 1 Introduction

Over the past few years, deep learning has made great progress in plenty of fields, such as object detection [3,5,6,8], semantic segmentation [17–19], and image classification [2,7,9]. As the number of layers of neural networks continue to increase and the size of data continues to expend, training deep neural networks places higher demands on the computing power of computers. The development and applying of GPU satisfies this requirement which training deep learning model needs, making it possible to training deep neural networks within an acceptable time. However, factors such as GPU memory also limits large-scale neural network training.

S. Wen et al. (Eds.): ICA3PP 2019, LNCS 11945, pp. 27–41, 2020.
https://doi.org/10.1007/978-3-030-38961-1_4

On the other hand, distributed training architecture provides a new idea of training deep neural networks, including some new ideas about distributed training [1]. Distributed training declines the training time through parallelism. Parameter Sever [10] is a new kind of distributed training architecture, through which we could get an ideal reduction of time without doing much harm to the accuracy.

Stochastic Gradient Decent(SGD) [16] is a common used algorithm in deep learning. It utilizes a random subset of training dataset to update weights of the loss function. The size of the subset which is often viewed as batchsize is an important factor in deep learning training, as has shown in [16]. It can not only affect the training speed but also have non-negligible influence on the convergence of neural networks. Nowadays, large-scale training has become a research hotspot [25]. Large batchsize can be used to improve parallelism of SGD and reduce training time [15,22,23,26] while smaller batchsize may get a better training performance at the cost of time.

To fully understand these issues, we trained Resnet50 [4] on CIFAR-10 dataset and conducted several experiments. In this paper, We define a new variable **momentum-like factor**: factors controlling the former gradient influence on parameter updates in each iteration. We showed how batchsize influences training performance and convergence in Parameter Server [10]. Also, we analyzed the influence of GPU numbers in the Parameter Server for the number of GPUs plays a key role in distributed training. We kept the total batchsize constant in Parameter Server and according to the results, large batchsize and large number of GPUs do harm to the training performance in distributed training. Finally, based on our analysis we changed the momentum-like factor in different optimizers, such as SGD, Adam and NAG, and explored how they affected the training performance. Besides, we tried two different ways to increase the validation accuracy and found out one well-behaved way of declining the accuracy loss caused by multiple GPUs training. A key contribution of our work is introducing momentum-like factors and exploring how they improve training performance in distributed training.

The rest of this paper is organized as follows. Section 2 describes some previous work related to our work including Parameter server [10] and Stochastic Gradient Descent [16] and its variants. In Sect. 3, we conduct mathematical derivation analysis of linear learning rate strategy [14] and batch normalization [20]. In multi-GPUs training, we give an understanding of increasing momentum. We conduct our experiments on CIFAR-10 in the way of controlling batchsize and the number of GPUs respectively and present the accuracy results in Sect. 4. Finally in Sect. 5 conclusions are drawn.

## 2   Related Work

### 2.1   Parameter Server

The development of deep neural network causes the expansion of parameters' scale of some common models, which increases the difficulty of getting an expected training performance within some limited time. Parameter Server (PS) [10] aims to increase training efficiency of large models while maintaining the accuracy or getting an acceptable accuracy loss.

In Parameter Server, there are two kinds of nodes, namely *workers* and *servers*. They play different roles in this architecture. Workers are responsible for calculating the data allocated to itself and updating the corresponding parameters. Servers utilize distributed storage and each one stores a part of the global parameters. Workers and servers communicate with each other through *Push* and *Pull* operation. After computation, workers send updated parameters to servers through *Push* operation and servers receive the parameter query and parameter update request of workers.

Parameter Server has the following features in distributed training. First, communicate efficiently. It mainly adopts the asynchronous communication mode, which reduces the delay in the training process, network traffic and communication overhead. Second, flexible consistency. Parameter Server allows the users to design the experiment according to their own requirements. Third, scalability. Parameter Server has good scalability and it does not require restarting the system when added a server into this system. Fourth, good fault tolerance. If there happens to be a fault in the system, we don't need to interrupt the computation which prevents a lot of trouble. Fifth, simplicity. Parameters in the parameter server architecture can be expressed in various forms to facilitate the development and application of machine learning algorithms.

## 2.2   Stochastic Gradient Descent and Its Variants

Stochastic Gradient Descent(SGD) [16] is one of the simplest first-order algorithms for full-batch training, which introduces noise into the gradient and block optimization in training progress [32]. It is an expansion of Gradient Descent. Gradient descent uses all samples to update parameters, which is very slow when there are billions of samples. SGD randomly selects one sample or a random sample set at a time for parameter updating, so that updating the parameters does not create redundancy. When the data size is large, it can effectively accelerate the training.

The key insight of SGD is that we can view the gradient as the expectation and expect to use small-scale sample approximations. SGD can be parameterized in a training set or a subset of training sets, i.e. mini-batch. The size of mini-batch, namely batchsize, is an important superparameter in deep learning. How to choose a suitable batch decides the direction of the gradient of the loss function. If the dataset is small enough, we can set full dataset as a batch. By this mean, we can get a better representation of dataset and reduce training time. But for larger dataset, if batchsize is equal to the full dataset, it may exceed GPU's memory, which results in training difficulty or bad training performance. Generally, training is performed by selecting a subset of the training data set. It has been proved that if the dataset is sufficient, training using a subset of the training dataset can theoretically achieve the same effect as training with the entire dataset.

[12] stated that SGD should be interpreted as integrating stochastic equation. They also presented the scale of random fluctuation in the SGD dynamics,

$$g = \xi(\frac{N}{M} - 1) \tag{1}$$

where $\xi$ is the learning rate, $N$ is the size of training set and $M$ is the batchsize. This fluctuation scale $g$ could also be considered as *noise factor*. If we decay the learning rate $\xi$, the noise scale $g$ falls, enabling to get a better training performance. On the other hand, when we keep learning rate $\xi$ constant, we could also increase batchsize $M$ to lower the negative impact of noise scale. In contrast, small batchsize raises the noise scale and does harm to training performance. According to [21], calculating the mean and variance values over a batch makes the loss calculated for a particular example dependent on other examples in the same batch. Therefore, if the batchsize is large, high dependency between samples in the batch leads to lower training performance. When $M \ll N$, applying linear scaling rule [14] maintains the mean SGD weight update constant per training sample. More specific description about linear scaling rule [14] is shown in Sect. 3.1.

There have been many optimization methods for SGD in recent years, such as [24, 27]. A common one is SGD with momentum [12]. [12] extended traditional SGD to include momentum, and found the noise factor $g$ changed into

$$\begin{aligned} g &= \frac{\xi}{1 - m}(\frac{N}{M} - 1) \\ &\approx \frac{\xi N}{M(1 - m)} \end{aligned} \tag{2}$$

where $m$ is the momentum. This degenerates into the vanilla SGD when $m = 0$. If linear lscaling rule is adopted, then $\xi/M$ is constant. Then we get $g \propto 1/(1-m)$. Increasing $m$ results in the increasing of $g$, which may cause the drop of generalization performance. However this analysis is different from the results of our experiments in multi-GPUs training. While SGD algorithm improves training speed effectively compared with GD, it is difficult to choose a well-behaved learning rate. Besides, for some optimization problem or some convex problem, it cannot get the global optimal solution, only the local optimal solution can be obtained. Also, it is not easy to be achieved in parallel.

Adam [33] combines the advantages of two optimization algorithms, AdaGrad [35] and RMSProp [34]. It evaluates the First Moment Estimation of the gradient and the Second Moment Estimation and then calculates the update step. Adam is a second-order optimization method, which adjusts the learning rate for each parameter by performing smaller updates for frequent and larger updates for infrequent parameters. In Adam algorithm, $\beta_1$ and $\beta_2$ are set to control the influence of gradients and the square of gradients on the parameter update. They play the same role as momentum in SGD with momentum method. So they could be viewed as those momentum-like factors.

NAG (Nesterov accelerated gradient) [36] is an improved way of Momentum [37]. It updates with the gradient by "looking ahead" instead of the current gradient. In this method, the momentum-like factor controls the former gradient weight in each iteration [38]. Besides, it calculates variety of gradients with respect to last one. Utilizing these values, it updates the parameters in the training process. Therefore, NAG is a second-order optimizer.

# 3 Multi-GPUs Training

## 3.1 Linear Scaling Rule [14]

Assuming a deep neural network model with parameters $\theta$, and its corresponding loss function $L(\theta)$. So $L(\theta)$ is defined as the average of the total loss over the training dataset. The formula is as follows.

$$L(\theta) = \frac{1}{M} \sum_{i=1}^{M} L_i(\theta) \tag{3}$$

$L_i(\theta)$ is the loss of the *ith* training example. $M$ is the size of training dataset. As stated in Sect. 2.2, SGD uses one stochastic sample or one stochastic sample set to get the approximation of the gradient of the loss funstion $L(\theta)$. For batch $\mathcal{B}$ including $m$ training examples, its batchsize is $m$. Its corresponding weight update rule is

$$\theta_{k+1} = \theta_k + \xi \Delta \theta_k \tag{4}$$

$$\xi \Delta \theta_k = -\frac{\xi}{m} \sum_{i=1}^{m} \nabla_\theta L_i(\theta_k) \tag{5}$$

where $\xi$ denotes the learning rate and $k$ is the $kth$ epoch.

From formula (4), we can get $\mathcal{E}\{\xi \Delta \theta\} = -\xi \mathcal{E}\{\nabla_\theta L(\theta)\}$, which means the average of SGD weight update. So for batch $\mathcal{B}$ whose batchsize is $m$, the mean value of SGD weight update is proportional to $\xi/m$. Adopting *linear scaling rule*[14] is to keeps the mean SGD weight update per training example constant.

So if we want to keep the the average of SGD algorithm weight update between two adjacent training samples constant, we are supposed to set learning rate $\xi$ and batchsize $m$ follow *linear scaling rule*[14], which plays an important role in the subsequent experiments.

This linear scaling rule is adopted widely in [12,14,29–31].

## 3.2 Batch Normalization

Batch Normalization (BN) [20] is commonly deployed in modern deep neural networks. It has shown excellent achievements in improving training performance. However it also causes the performance decline in multi-GPU distributed training.

Considering a batch $\mathcal{B} = \{x_0, x_1, ..., x_{M-1}\}$ of batchsize $M$, its mini-batch mean value is

$$\mu_\mathcal{B} = \frac{1}{M} \sum_{i=0}^{M-1} x_i \tag{6}$$

where $x_i$ is one sample of set $\mathcal{B}$. Its variance is

$$\delta_{\mathcal{B}}^2 = \frac{1}{M} \sum_{i=0}^{M-1} (x_i - \mu_{\mathcal{B}})^2 \tag{7}$$

Then the samples are normalized by

$$\hat{x}_i = \frac{x_i - \mu_{\mathcal{B}}}{\sqrt{\delta_{\mathcal{B}}^2 + \epsilon}} \tag{8}$$

where $i$ changes from 1 to $M$ and $\epsilon$ is a small enough to avoid the denominator of formula 8 is zero. The mathematical expectation is

$$\mathcal{E}_1 = E(\delta_{\mathcal{B}}^2) = (M - 1)\delta^2 \tag{9}$$

After Batch Normalization, the samples are following the normal distribution whose variance is $\delta^2$. These values are fed into some layers in the deep neural networks to get better training results.

Now, we are considering training on multiple GPUs. Assuming that there are $P$ GPUs in the distributed training system and batch $\mathcal{B} = \{x_0, x_1, ..., x_{M-1}\}$ whose batchsize is $M$. So, the batch on per GPU is a subset of batch $\mathcal{B}$ and its batchsize is $K = M/P$. On GPU $j$($j$ changes from 0 to $P - 1$), the mean value of its training samples is

$$\mu_j = \frac{1}{K} \sum_{i=0}^{K-1} x_{ji} = \frac{P}{M} \sum_{i=0}^{M/P-1} x_{ji} \tag{10}$$

and its corresponding variance is

$$\delta_j^2 = \frac{1}{K} \sum_{i=0}^{K-1} (x_{ji} - \mu_j)^2 = \frac{P}{M} \sum_{i=0}^{M/P-1} (x_{ji} - \mu_j)^2 \tag{11}$$

where $x_{ji}$ is the $ith$ training sample on GPU $j$. The mean value of all samples of batch $\mathcal{B}$ is

$$\hat{\mu}_{\mathcal{B}} = \frac{1}{P} \sum_{j=0}^{P-1} \mu_j \tag{12}$$

The sum of every variance of all GPUs in the system is

$$S_e = \sum_{j=0}^{P-1} \frac{M}{P} \delta_j^2 = \frac{M}{P} \sum_{j=0}^{P-1} \delta_j^2 \tag{13}$$

The variance of all training batches among GPUs is

$$S_a = \sum_{j=0}^{P-1} \frac{M}{P} (\mu_j - \hat{\mu}_{\mathcal{B}})^2 = \frac{M}{P} \sum_{j=0}^{P-1} (\mu_j - \hat{\mu}_{\mathcal{B}})^2 \tag{14}$$

So, the total variance $S_t$ is

$$S_t = S_e + S_a = \frac{M}{P} \sum_{j=0}^{P-1} \delta_j^2 + \frac{M}{P} \sum_{j=0}^{P-1} (\mu_j - \hat{\mu}_\mathcal{B})^2 \tag{15}$$

When $P = 1$, formula 15 degenerate into formula 7.

Based on statistical knowledge, the expectation of formula 15 is

$$\begin{aligned}
\mathcal{E}_2 &= E(S_t) \\
&= (M-1)\delta^2 + \frac{M(P-1)}{P}\delta^2
\end{aligned} \tag{16}$$

where $\delta^2$ has the same meaning as that in formula 9.

Comparing formula 9 with formula 16, we get $\mathcal{E}_2 \geq P\mathcal{E}_1$ ignoring dependencies between samples when $P \geq 1$, which declines input features after normalization. Therefore, the variance of training on multiple GPUs raises which results in performance degradation.

As the variance of Batch Normalization increases, the gradient calculated on each worker in Parameter Server tends to be unstable. The gradient in multiple-GPU training update more than that of training on a single GPU. If we use SGD optimizer, considering a good method namely training on multiple GPUs with momentum value $m$, which means the proportion of former gradient in the training process, according to [13], the momentum update rule is

$$A = -(1-m)A + \frac{dL(\theta)}{dw} \tag{17}$$

$$\Delta w = -\Delta A \xi = [(1-m)A + \frac{dL(\theta)}{dw}]\xi \tag{18}$$

where $A$ denotes the "accumulation" and $\frac{dL(\theta)}{dw}$ is the average gradient of per training example. From formulas 17 and 18, $\Delta w$ declines if increasing momentum, which means slowing down the attenuation of weights. This correspondingly reduces training error on multiple GPUs in the same iteration.

## 4   Experiments

### 4.1   Experiment Setup

In order to explore the impact of the batchsize and the number of GPUs on the experimental results in the parameter server architecture, we mainly set up the experiments using MXNet [11], training Resnet50 on the CIFAR-10 dataset to evaluate the acceleration and effectiveness. To explore the influence of different batchsizes on the experimental results, we set the batchsize to 32, 64, 128, 256, 512, 1024, and 2048 respectively. At the same time, we explored the impact of the number of GPUs on the experimental results and use 1, 2, 4, 8, 16, 32 GPUs and test the validation accuracy of CIFAR-10 dataset in the case of different batchsizes and GPU numbers. Finally we presented our experiments in the following subsections. For the sake of convenience, in Fig. 1 we didn't show the validation accuracy when batchsize is 32 and the number of GPUs is 32, for the validation accuracy is 0.199.

## 4.2  SGD

We show our experiment results in Fig. 1. These values are also showed in Table 2 in Appendix A.

**Batchsize.** We set momentum 0.9 and adopt *linear scaling rule* [14]: when the mini-batch size is multiplied by $k$, multiply the learning rate by $k$. The learning rates corresponding to different batchsize are displayed in Table 1.

**Table 1.** The learning rates corresponding to different batchsizes

| Batchsize | Learning rate |
|---|---|
| 32 | 0.0125 |
| 64 | 0.025 |
| 128 | 0.05 |
| 256 | 0.1 |
| 512 | 0.2 |
| 1024 | 0.4 |
| 2048 | 0.8 |

In distributed training, the choice of batchsize is critical. According to our experiment, when batchsize is 32 and 32 GPUs are used, the validation accuracy is 0.199, which means that in distributed training such a small batchsize can't make the training process converge. There are 8 hosts in this case. The batchsize on each host is 4, which is relatively small and prevents it from convergence. From Fig. 1, we get a downtrend when the batchsize is rising. When the batchsize exceeds 512, the effect of increasing the batchsize on the experimental results is small, and the maximum drop is 0.008 (16 GPUs).

**GPU.** Multi-GPU training is a common way in distributed training. As the number of GPUs increases, the time for distributed training decreases correspondingly. Theoretically, the linear acceleration ratio can be obtained.

When the number is 1, 2, 4, we only use one single host. According to Fig. 1, within a single host, the GPU number from 1 to 4, the accuracy difference is relatively small, among which the maximum is 0.018 (batchsize 512: 0.925 and 0.907 respectively; batchsize 1024: 0.927 and 0.909 respectively). When using multiple GPUs, especially 32 GPUs, the accuracy can drop by 0.069 (batchsize 64) on the condition that only considering the training process is convergent. Because it is not convergent when the batchsize is 32 using 32 GPUs.

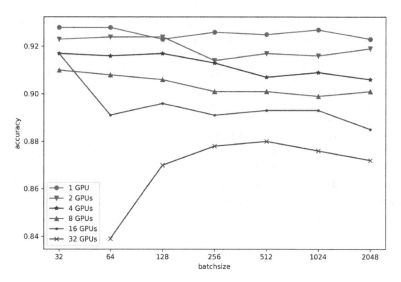

**Fig. 1.** Utilizing SGD, validation accuracy of Resnet50 on CIFAR-10 of different batch-sizes on multiple GPUs in Parameter Server. In these experiments, all hyperparameters except batchsize and the number of GPUs are set default.

**Momentum.** We discussed the increasing of GPU number in Parameter Server caused the decline of training accuracy when the batchsize remains constant before. In order to improve the accuracy of this situation, we increased the momentum [15], and compare the improvement of different momentums given a large batchsize of 1024. The results are shown in Fig. 2.

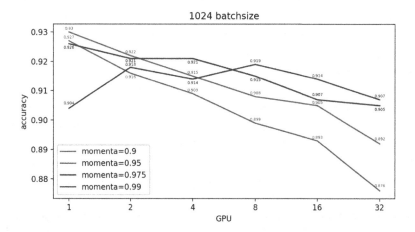

**Fig. 2.** Validation accuracy of different momentums on multiple GPUs. In these experiments, we set the momentum values 0.9, 0.95, 0.975 and 0.99 respectively. (Color figure online)

From Fig. 2, when the batchsize is 1024 and the momentum is 0.9, the final validation accuracy is 0.927 on 1 GPU while it drops to 0.876 on 32 GPUs (green line). When the momentum is 0.95, the validation accuracy is 0.93 and 0.892 on 1 GPU and 32 GPU, respectively(red line). It increases by 0.003 and 0.016 compared with that when the momentum is 0.9.

When the momentum increases to 0.99, the accuracy is 0.904 on 1 GPU and 0.907 on 32 GPUs (purple line). Compared with the green line (momentum is 0.9), accuracy drops by 0.026 on 1 GPU but increases 0.031 on 32 GPUs. From the curve trend in Fig. 2, it can be analyzed that increasing the momentum factor can effectively alleviate the accuracy degradation caused by the increase in the number of GPUs in the parameter server architecture, but the lifting effect is obvious for multi-GPU scenarios. However when the number of GPUs is small (less than 4), the effect is not obvious, and even leads to performance degradation.

### 4.3   Adam

To investigate the influence of increasing momentum, we also conduct experiments by increasing one of the momentum-like factors, $\beta_1$, in Adam algorithm, which could be regarded as increasing momentum in SGD. Regularly, $\beta_1$ and $\beta_2$ are set 0.9 and 0.999 respectively. We display our results of Resnet50 on CIFAR-10 in Fig. 3. These result values are also showed in Table 3 in Appendix B.

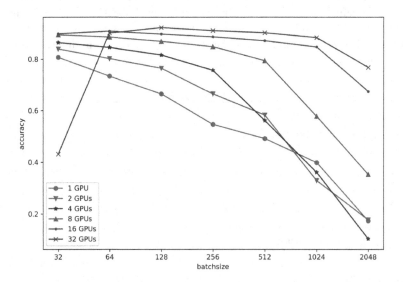

**Fig. 3.** Utilizing Adam, validation accuracy of Resnet50 on CIFAR-10 of different batchsizes on multiple GPUs in Parameter Server. The hyperparameters are set the same as those in the former SGD experiments.

Besides, we increase $\beta_1$ from 0.9 to 0.95, 0.975, 0.99 to explore its influence in distributed training given a large batchsize. We compare different $\beta_1$ values when

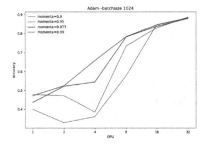

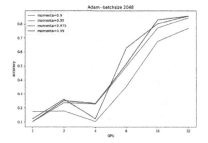

**Fig. 4.** Validation accuracy of different $\beta_1$ values in Adam method on multiple GPUs given batchsize $= 1024$

**Fig. 5.** Validation accuracy of different $\beta_1$ values in Adam method on multiple GPUs given batchsize $= 2048$

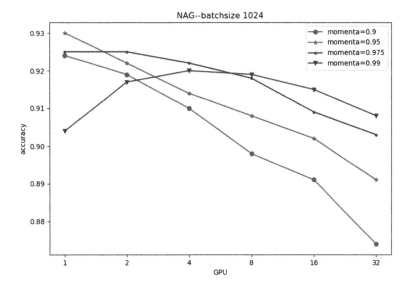

**Fig. 6.** Utilizing NAG, validation accuracy of Resnet50 on CIFAR-10 of different batchsizes on multiple GPUs in Parameter Server. The hyperparameters are set the same as those in the former SGD and Adam experiments.

batchsize is 1024 and 2048. From Figs. 4 and 5, we can identity that increasing $\beta_1$ can improve performance in distributed training. However, these improvements are not obvious in Adam method.

## 4.4 NAG

In Fig. 6, we show the validation accuracy of batchsize 1024 with different momentums, 0.9, 0.95, 0.975, 0.99. From Fig. 6, we find that in distributed training using more GPUs, the validation accuracy drops when momentum is 0.9, 0.95 and 0.975. When momentum is 0.99, the validation accuracy improves. Larger momentum alleviate the negative influence better.

### 4.5    Warm-Up Strategy

In addition to increasing momentum to improve training performance, we also adopt the warm-up strategy. The learning rate follows the linear strategy and the number of warm-up epoch is 5. We display the experimental results of batchsize 1024 and different momentum values, as shown in Fig. 7. From Fig. 7, in the case of 32 GPUs, the validation accuracy raises from 0.695 to 0.816 when the momentum changes from 0.9 to 0.99, increasing by 0.125.

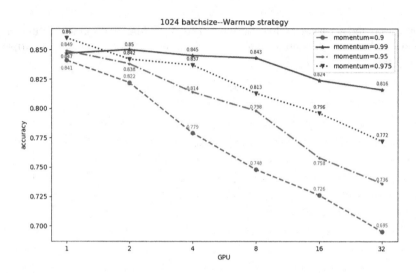

**Fig. 7.** Validation accuracy of different momentums on multiple GPUs with Warm-up

## 5    Conclusion

In this paper, we explored the impact of batchsize and the number of GPU on performance in distributed training. Using the parameter server as the basic architecture, we trained Resnet50 on the CIFAR-10 dataset. According to the experimental results, we found, to some degree, that we could increase the efficiency of distributed training and reduce the training time by increasing the batch size and increasing the number of GPUs. When increasing the number of GPUs to achieve distributed training, on one hand, increase the batchsize to ensure training convergence. On the other hand, we conclude that, for SGD-like optimizers, increase the momentum-like factors to alleviate the performance loss caused by the increase of the number of GPUs in distributed training or even improve the accuracy. Increasing momentum-like factors not only benefit for first-optimizers such as SGD, but also good for second-order optimizers, such as Adam and NAG. Besides, warm-up strategy is not a good way to improve distributed training. We tried other ways, such as Layer Normalization [28], and found Layer Normalization is unable to train convergence.

**Acknowledgement.** This work is sponsored in part by the National Key R&D Program of China under Grant No. 2018YFB2101100 and the National Natural Science Foundation of China under Grant No. 61932001 and 61872376.

# A   Appendix A

**Table 2.** SGD's results of Resnet50 on CIFAR-10 of different batchsizes on multiple GPUs in Parameter Server

| Batchsize | 1 GPU | 2 GPUs | 4 GPUs | 8 GPUs | 16 GPUs | 32 GPUs |
|-----------|-------|--------|--------|--------|---------|---------|
| 32 | 0.928 | 0.923 | 0.917 | 0.91 | 0.917 | 0.199 |
| 64 | 0.928 | 0.924 | 0.916 | 0.908 | 0.891 | 0.839 |
| 128 | 0.923 | 0.924 | 0.917 | 0.906 | 0.896 | 0.87 |
| 256 | 0.926 | 0.914 | 0.913 | 0.901 | 0.891 | 0.878 |
| 512 | 0.925 | 0.917 | 0.907 | 0.901 | 0.893 | 0.88 |
| 1024 | 0.927 | 0.916 | 0.909 | 0.899 | 0.893 | 0.876 |
| 2048 | 0.923 | 0.919 | 0.906 | 0.901 | 0.885 | 0.872 |

# B   Appendix B

**Table 3.** Adam's Results of Resnet50 on CIFAR-10 of different batchsizes on multiple GPUs in Parameter Server

| Batchsize | 1 GPU | 2 GPUs | 4 GPUs | 8 GPUs | 16 GPUs | 32 GPUs |
|-----------|-------|--------|--------|--------|---------|---------|
| 32 | 0.807 | 0.839 | 0.864 | 0.894 | 0.898 | 0.432 |
| 64 | 0.734 | 0.802 | 0.845 | 0.885 | 0.908 | 0.901 |
| 128 | 0.665 | 0.765 | 0.816 | 0.869 | 0.897 | 0.922 |
| 256 | 0.547 | 0.665 | 0.757 | 0.848 | 0.886 | 0.91 |
| 512 | 0.492 | 0.583 | 0.562 | 0.794 | 0.871 | 0.902 |
| 1024 | 0.399 | 0.329 | 0.361 | 0.578 | 0.847 | 0.883 |
| 2048 | 0.173 | 0.177 | 0.103 | 0.353 | 0.675 | 0.768 |

# References

1. Li, D., et al.:HPDL: towards a general framework for high-performance distributed deep learning. In: Proceedings of 39th IEEE International Conference on Distributed Computing Systems (IEEE ICDCS) (2019)
2. Szegedy, C., Ioffe, S., Vanhoucke, V., et al.: Inception-v4, Inception-ResNet and the impact of residual connections on learning. In: AAAI, vol. 4, p. 12 (2017)
3. Chollet, F.: Xception: deep learning with depthwise separable convolutions. arXiv preprint (2016)

4. He, K., Zhang, X., Ren, S., et al.: Deep residual learning for image recognition. In: Proceedings of the IEEE Conference on Computer Vision and Pattern Recognition, pp. 770–778 (2016)
5. Huang, G., Liu, Z., Weinberger, K.Q., et al.: Densely connected convolutional networks. In: Proceedings of the IEEE Conference on Computer Vision and Pattern Recognition, vol. 1, no. 2, p. 3 (2017)
6. Liu, W., Anguelov, D., Erhan, D., Szegedy, C., Reed, S., Fu, C.: SSD: single shot multibox detector. arXiv:1512.02325v2 (2015)
7. Russakovsky, O., et al.: ImageNet large scale visual recognition challenge. Int. J. Comput. Vis. **115**(3), 211–252 (2015)
8. Dai, J., Li, Y., He, K., Sun, J.: R-FCN: object detection via region-based fully convolutional networks. In: NIPS, pp. 379–387 (2016)
9. Qin, Z., Zhang, Z., Chen, X., et al.: FD-MobileNet: improved MobileNet with a fast downsampling strategy. arXiv preprint arXiv:1802.03750 (2018)
10. Li, M., et al.: Scaling distributed machine learning with the parameter server. In: Proceedings of OSDI, pp. 583–598 (2014)
11. Chen, T., et al.: MXNet: a flexible and efficient machine learning library for heterogeneous distributed systems. arXiv preprint arXiv:1512.01274 (2015)
12. Smith, S.L., Le, Q.V.: A Bayesian perspective on generalization and stochastic gradient descent. arXiv preprint arXiv:1710.06451 (2017)
13. Smith, S.L., Kindermans, P.-J., Le, Q.V.: Don't decay the learning rate, increase the batch size. arXiv preprint arXiv:1711.00489 (2017)
14. Krizhevsky, A.: One weird trick for parallelizing convolutional neural networks. arXiv preprint arXiv:1404.5997 [cs.NE] (2014)
15. Nitish, S.K., Mudigere, D., Nocedal, J., Smelyanskiy, M., Tang, P.T.P.: On large-batch training for deep learning: generalization gap and sharp minima. arXiv preprint arXiv:1609.04836 (2016)
16. Goyal, P.,: Accurate, large minibatch SGD: training imagenet in 1 hour. arXiv preprint arXiv:1706.02677 (2017)
17. Girshick, R., Donahue, J., Darrell, T., Malik, J.: Rich feature hierarchies for accurate object detection and semantic segmentation. In: Proceedings of the IEEE Conference on Computer Vision and Pattern Recognition, pp. 580–587 (2014)
18. Long, J., Shelhamer, E., Darrell, T.: Fully convolutional networks for semantic segmentation. In: Proceedings of the IEEE Conference on Computer Vision and Pattern Recognition, pp. 3431–3440 (2015)
19. Dai, J., He, K., Sun, J.: Instance-aware semantic segmentation via multi-task network cascades. arXiv:1512.04412 (2015)
20. Ioffe, S., Szegedy, C.: Batch normalization: accelerating deep network training by reducing internal covariate shift. In: Proceedings of 32nd International Conference on Machine Learning, ICML15, pp. 448–456 (2015)
21. Masters, D., Luschi, C.: Revising small batch training for deep neural networks. arXiv preprint arXiv:1804.07612 (2018)
22. You, Y., Gitman, I., Ginsburg, B.: Scaling SGD batch size to 32k for ImageNet training. arXiv preprint arXiv:1708.03888 (2017a)
23. Akiba, T., Suzuki, S., Fukuda, K.: Extremely large minibatch SGD: training ResNet-50 on ImageNet in 15 minutes. arXiv preprint arXiv:1711.04325 (2017)
24. Chaudhari, P., Choromanska, A., Soatto, S., LeCun, Y.: Entropy-SGD: biasing gradient descent into wide valleys. arXiv preprint arXiv:1611.01838 (2016)
25. You, Y., Zhang, Z., Hsieh, C.-J., Demmel, J., Keutzer, K.: ImageNet training in minutes. CoRR, abs/1709.05011 (2017)

26. Balles, L., Romero, J., Hennig, P.: Coupling adaptive batch sizes with learning rates. arXiv preprint arXiv:1612.05086 (2016)
27. Li, Q., Tai, C., Weinan, E.: Stochastic modified equations and adaptive stochastic gradient algorithms. arXiv preprint arXiv:1511.06251 (2017)
28. Ba, J.L., Kiros, J.R., Hinton, G.E.: Layer normalization. arXiv preprint arXiv:1607.06450 [stat.ML] (2016)
29. Chen, J., Pan, X., Monga, R., Bengio, S., Jozefowicz, R.: Revisiting distributed synchronous SGD. arXiv preprint arXiv:1604.00981 [cs.LG] (2016)
30. Bottou, L., Curtis, F.E., Nocedal, J.: Optimization methods for large-scale machine learning. arXiv preprint arXiv:1606.04838 [stat.ML] (2016)
31. Jastrzębski, S., et al.: Three factors influencing minima in SGD. arXiv preprint arXiv:1711.04623 [cs.LG] (2017)
32. Ghadimi, S., Lan, G., Zhang, H.: Mini-batch stochastic approximation methods for nonconvex stochastic composite optimization. Math. Program. **155**(1–2), 267–305 (2014)
33. Kingma, D., Ba, J.: Adam: a method for stochastic optimization. In: ICLR (2015)
34. Tieleman, T., Hinton, G.: Lecture 6.5-rmsprop, coursera: neural networks for machine learning. University of Toronto, Technical report (2012)
35. Duchi, J., Hazan, E., Singer, Y.: Adaptive subgradient methods for online learning and stochastic optimization. J. Mach. Learn. Res. **12**(Jul), 2121–2159 (2011)
36. Nesterov, Y.: A method for unconstrained convex minimization problem with the rate of convergence o$(1/k^2)$. Doklady ANSSSR (Transl. Soviet. Math. Docl.), **269**, 543–547 (1983)
37. Qian, N.: On the momentum term in gradient descent learning algorithms. Neural Netw.: Off. J. Int. Neural Netw. Soc. **12**(1), 145–151 (1999)
38. Ruder, S.: An overview of gradient descent optimization algorithms. arXiv preprint arXiv:1609.04747 [cs.LG] (2017)

# Pimiento: A Vertex-Centric Graph-Processing Framework on a Single Machine

Jianqiang Huang[1,2], Wei Qin[1], Xiaoying Wang[2], and Wenguang Chen[1,2(✉)]

[1] Department of Computer Science and Technology, Tsinghua University,
Beijing 100084, China
{hjq16,tanw16}@mails.tsinghua.edu.cn, cwg@tsinghua.edu.cn
[2] Department of Computer Technology and Applications, Qinghai University,
Xining 810016, China
Wangxiaofu163@163.com

**Abstract.** Here, we describe a method for handling large graphs with data sizes exceeding memory capacity using minimal hardware resources. This method (called Pimiento) is a vertex-centric graph-processing framework on a single machine and represents a semi-external graph-computing system, where all vertices are stored in memory, and all edges are stored externally in compressed sparse row data-storage format. Pimiento uses a multi-core CPU, memory, and multi-threaded data preprocessing to optimize disk I/O in order to reduce random-access overhead in the graph-algorithm implementation process. An on-the-fly update-accumulated mechanism was designed to reduce the time that the graph algorithm accesses disks during execution. Our experiments compared external this method with other graph-processing systems, including GraphChi, X-Stream, and FlashGraph, revealing that Pimiento achieved $7.5\times, 4\times, 1.6\times$ better performance on large real-world graphs and synthetic graphs in the same experimental environment.

**Keywords:** Vertex-centric · Graph processing · Semi-external · Passing message · Asynchronous update accumulation

## 1 Introduction

With the rapid development of the internet and the big-data era, there is a need to analyze large volumes of data. As an abstract data structure, graphs are used by many applications to represent large-scale data in real scenarios, and graph data structures are used to describe the relationships among data, such as mining relationships in social networks, goods recommendations in e-commerce systems, and analysis of the impact of traffic accidents on road networks. Additionally, many types of unstructured data are often transformed into graphs for post-processing and analysis. Research into large-scale graph-processing has increased in both academia and industry, and recently, numerous systems and state-of-the

© Springer Nature Switzerland AG 2020
S. Wen et al. (Eds.): ICA3PP 2019, LNCS 11945, pp. 42–56, 2020.
https://doi.org/10.1007/978-3-030-38961-1_5

art techniques for graph processing have emerged, including distributed systems and heterogeneous systems. Such systems present new computing models or highlight the design of high-performance runtime systems used to adapt to the features of graph data, such as its large scale, ability to dynamically change, and its high efficiency when processing big graph data.

Examples of these systems include distributed graph-computing systems, such as pregel [1], GraphLab [2], PowerGraph [3], and Gemini [4], which can theoretically deal with any large-scale graph data by deploying clusters with good extensibility and computational efficiency; however, there remain problems, including maintenance of load balance between nodes and communication latency.

Other systems include single graph computing system, such as GraphChi [5], X-Stream [6], FlashGraph [12], GridGraph [8], and other external graph-processing systems [7,9–11,13,14,22], which can reduce random disk-read and disk-write operations, avoid high communication overhead, and use parallelization technology to fully exploit multi-core computing resources to address large-scale graph data. Compared with distributed systems, these exhibit lower hardware cost and power consumption.

GraphChi is a single graph-computing system using a vertex-centric calculation model and multi-threaded parallel computing to improve computing performance. It utilizes parallel sliding-window (PSW) [5] technology to reduce random access to the disk and supports asynchronous computations. GraphChi processes graphs in three stages: (1) loading graph data from the disk to memory, (2) updating the values of vertices and edges, and (3) writing updates to disk.

GraphChi exhibits good platform usability and computing performance; however, its preprocessing requires sorting of the source vertex of the edges, which is costly. Moreover, computing processes and disk I/O access are executed in serial, and the parallelism between disk I/O and the CPU is not fully utilized to overlap computing and I/O in order to further improve computing performance. By contrast, X-Stream uses an edge-centric computing model, where all states are stored in the vertex.

To address these issues, we propose Pimiento, a vertex-centric graph-processing framework that combines asynchronization with efficiency. The outline of our paper is as follows: Sect. 2 introduces disk-based graph-computation challenges, describes system design and implementation in Sect. 3, Sect. 4 describes evaluation of Pimiento on large problems (graphs with billions of edges) using a set of algorithms, such as single source shortest path (SSSP), PageRank, and breadth-first search (BFS).

The main contributions of this paper are as follows:

– We describe the use of a vertex-centric computing model with effective graph-storage structure that adopts an innovative asynchronous update-accumulation mechanism. This enables update and repeat visits to any vertex to occur in memory in order to avoid a large number of random I/O and repeat I/O operations generated by frequent updates and reads of disk data.

- Pimiento implements a semi-external asynchronous graph-processing framework to maximize on-the-fly updates via thread optimization of computing and I/O, thereby reduced access to I/O data.
- Our evaluation showed that Pimiento outperformed current state-of-the-art techniques.

## 2    Disk-Based Graph Computation

A graph is a data structure that describes the complex relationship between data and comprises vertices and edges usually expressed as $G = (V, E)$, where the vertex set, $V$, represents an object or entity, and the edge set, $E$, represents the relationship between objects or entities. Each vertex $v \in V$ will have a vertex value. Given a directed edge from vertex $u$ to vertex $v$, $e = (u, v)$, $e$ is the in-edges of $v$ and the out-edges of $u$, where $u$ represents the in-vertex of $v$, and $v$ represents the out-vertex of $u$.

In a vertex-centric calculation model for iterative calculations, the value of each update vertex usually involves only the input vertex value. Once a vertex value is updated, a new message is sent to the output side, and the value of the output side is updated. This dynamic update of the iterative process is terminated when a convergence condition is satisfied. As framework [5] shows, a vertex-centric calculation model can address a broad range of problems. The method proposed in this paper is based on asynchronous calculations using a vertex-centric value-calculation model. Combined with the on-the-fly accumulation of the update mechanism, it promotes an effective graph-storage and calculation models. Based on the effective management of graph data, it can minimize disk data traffic and make full use of the parallel update of memory and CPU resources in order to improve computational efficiency.

### 2.1    Maintaining Specification Integrity

We divided vertex set $V$ of graph into $P$ intersecting intervals (see Fig. 1(b)). Each interval correlates with a shard that contains information needed to update the vertex calculation. As a result of the asynchrony of cumulative iterative computations, the graph partition has little effect on performance. This method only supports hash or range partitions based on a graph vertex number.

The system described by Pearce et al. [19] uses a CSR storage format to store the graph on a disk and is equivalent to storing the graph as an adjacency table, where the edges in each edge shard are sorted according to a source vertex. We call these edge data, which are stored continuously in contiguous blocks on the disk.

Suppose that the vertex set in Fig. 1(a) is divided into three intervals (interval1 = [1, 2], interval2 = [3, 4], and interval3 = [5, 6]), each of which is associated with a shard, including the edge-and vertex shards. All Vertex shards will cascade into a vertex table in order to initialize the vertex information, and all edge

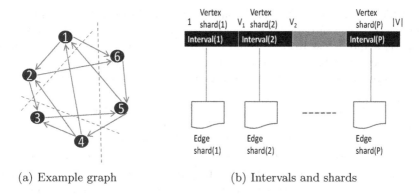

(a) Example graph                    (b) Intervals and shards

**Fig. 1.** Intervals and shards in the graph.

shards will cascade into an edge data stream in order to flow updates to the vertex information.

This graph-storage structure addresses the following three problems:

- To improve the parallelism of single-machine graph calculation, the graph-storage structure of the shard is used to render each executing thread responsible for one or more shards for parallel calculation;
- Because random access is more than an order of magnitude slower than sequential access to a disk, and given that the number of vertices in real-world graph data is smaller than the number of edges, we used memory for constant iterative updates of vertex data and secondary storage for edge lists in order to make full use of the random read-write capability of memory and the large capacity of secondary storage;
- To avoid secondary storage of random I/O, we organized edge data to ensure that access to graph data involves sequential I/O.

## 2.2 Computational Model

In incremental iterative calculations, graph data include read-only data by constantly updating vertex value $V$, as the vertex value of the cumulative value $\Delta V$. We found that $\Delta V$ is involved in the update of adjacent vertices and will usually be accessed many times. I/O represents a bottleneck to disk-based methods, and in order to avoid frequent updates and reads of disk $V$ and $\Delta V$, thereby causing repeated random I/O and I/O, read-only edge data are detached from the variable-peak value of $V$ and $\Delta V$, and the read-only edge data are continuously stored on the edge shard disk.

We combined the cumulative iterative computations and the cache of all of the vertices values for $V$ and $\Delta V$ into the memory. Because the space occupied by vertices values $V$ and $\Delta V$ are less than the space occupied by the edge data, the memory capacity of the modern computer can meet the requirements. Pimiento uses flow calculation, and the space occupied by the edge list in memory is dynamically balanced and controllable, which also proves the desirability of caching vertex data into memory. Due to the cumulative nature of the algorithm, the updates and access to peak value $V$ and $\Delta V$ can be performed in memory. At this point, updating each interval requires only one sequential scan of the corresponding read-only edge list to minimize the I/O overhead of graph data access.

This paper is based on the traditional incremental iteration theory [15] and presents a graph-computing model in a parallel environment for application for stand-alone large graph data processing. The parallel-computing model is adopted in the framework of general graph computing, where each execution thread is responsible for one or more shards, as well as each subdivision, including the vertex shard and a corresponding edge shard. Additionally, smaller vertex shards are loaded into memory to support frequent updates, and larger edge shards are placed on the disk to save memory.

The computing framework of the diagram is shown in Fig. 2. During the implementation process of the iterative calculation, each execution thread reads the edge information sequentially from disk and updates the neighbor vertex state based on the state of vertices $V$ and $\Delta V$ in the local Vertex shard. The communication between threads involves passing $\Delta V$. There are two main overheads in this model: I/O overhead for reading graph data from the disk and the overhead of interthread communication. This computing model uses cumulative iterative computation to greatly reduce these two overhead issues.

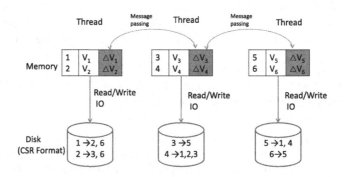

**Fig. 2.** Memory and secondary storage in the graph

## 2.3   Update Scheme

Algorithm 1 describes the implementation of the cumulative iterative-computing model in a single-machine parallel-computing environment. First, edge data is sequentially read for any vertex i, from edge shard data from the disk, and the information record of this vertex, i ($V_i$ and $\Delta V_i$), in the memory vertex shard is positioned according to the source vertex number of the edge data. When the vertex, i, edge data is loaded into memory, the algorithm determines whether the vertex information is a valid change (i.e., whether $\Delta V_i$ indicates 0) for the effective information ($\Delta V_i \neq 0$). First Algorithm 1: pseudo-code of the vertex update function for weighted PageRank.

---

**Algorithm 1** :Pseudo-code of the vertex update function for weighted PageRank.

---

**Input:** All intervals vertex-shards and edge-shards of graph G, optional initialization data.
**Output:** Desired output results.

 1: **function** UPDATE(vertex)
 2:     Initialize(vertex-shards);
 3:     **repeat**
 4:         $v[i] \leftarrow$ read values of out-edges of vertex i ;
 5:         $vertex.value \leftarrow f(v[i])$ ;
 6:         **if** $\Delta f(v[i]) \neq 0$ **then**   $f(v[i]) \leftarrow \Delta f(v[i]) + f(v[i])$ ;
 7:             **for** each edge of vertex **do**
 8:                 $edge.value \leftarrow f(vertex.value,\ edge.value))$;
 9:                 $\Delta f(v[i]) \leftarrow 0$ ;
10:             **end for**
11:         **end if**
12:     **until**
13:     $PassingMessage(vertex)$ ;
14:     remove outgoing edges of i
15: **end function**

---

Accumulate $\Delta V_i$ to vertex i and perform an update operation to use the update of $\Delta V_j$ of the neighbor vertex, j, followed by resetting the change of information in vertex i. When the operation on vertex i is completed, the edge data of vertex i is deleted from memory to free memory space for other uncomputed vertex edge data. This activity is repeated until the algorithm converges.

**Table 1.** Notations of a graph

| Notation | Meaning |
|---|---|
| G | A graph G = (V, E) |
| V | Vertices in G |
| E | Edges in G |
| n | Number of vertices in G, n = [V] |
| m | Number of edges in G, m = [E] |
| P | Number of intervals |
| $B_a$ | Size of a vertex attribute in bytes |
| $B_v$ | Size of a vertex id in bytes |
| $B_e$ | Size of an edge in bytes |
| $B_M$ | Size of available memory budget in bytes |
| B | Size of a disk block accessed by an I/O unit |

## 2.4   Analysis of the I/O Costs

During an iteration, GraphChi [5] processes each shard in three steps: (1) load the sub-graph from the disk; (2) update the vertex and edge values; and (3) write the updated values to the disk. In steps 1 and 3, each vertex is loaded and written back to the disk once, and the $nB_v$ data volume is read and written. For each edge data, in the worst case, each edge is accessed twice (once in each direction). The amount of data $2m(B_v + B_e)$ will be read in step 1, the updated edge value will be calculated in step 2, and the amount of data $2m(B_v + B_e)$ will also be written in step 3. During the entire calculation, the total amount of data in GraphChi read and written is $2m(B_v + B_e) + nB_v$. During each iteration, PSW [5] generates $P^2$ random reads and writes, whereas in during the entire calculation process, the number of I/O read and write events for the PSW is $(2m(B_v + B_e) + nB_v)/B + P^2$, Table 1 shows the Notations of a graph.

In X-Stream [6], an iteration is divided into: (1) a mixed scatter/shuffle phase and (2) a gather phase. In phase 1, the X-Stream loads all vertex and edge data, updates each edge, and writes the updated edge data back to disk. Because the edge data after update are used to pass values between adjacent vertices, we assume that the size of an updated piece of edge data is Be; therefore, for phase 1, the amount of data read is $nB_v + mB_e$, and the amount of data written is $mB_e$. In phase 2, the X-Stream loads all updated edge data and updates each vertex; therefore, for phase 2, the amount of data read is $nB_v$ and the amount of data written is $nB_v$. Therefore, for an iterative-calculation process, the total amount of data read by X-Stream is $(B_v + B_e)m + nB_v$, the total data amount written is $nB_v + mB_e$, the number of I/O reads is $(m(B_v + B_e) + nB_v)/B$, and the number of I/O writes is $nB_v/B + mB_v log_{Bm/B}^{P/B}$.

In FlashGraph [12], during the entire computation process, the number of I/O reads by Pimiento is $(mB_e + nPB_v)/B$, and the number of I/O writes is $nB_v/B$.

In Pimiento, the entire computation process loads all of the vertex shares once. During each iteration, all edge shares are loaded from disk in turn, and the entire computation process requires reading the amount of data $(mB_e + nB_a)$. After the computation, the vertex data value will be written back to disk, and the amount of data in $nB_v$ needs to be written. Note that the edge shard is read-only. To analyze the I/O cost, we use $B$ to represent the size of the disk block accessed by an I/O unit. According to a previous report, $B$ is 1MB on the SSD. During the entire computation process, the number of I/O reads by Pimiento is $(mB_e + nB_a)/B$, and the number of I/O writes is $nB_a/B$.

## 3   System Design and Implementation

Based on the asynchronous incremental-update model, we implemented the Pimiento system with C++. Pimiento divides each graph-processing task into three steps:

- Graph data shard and vertex information in memory are initialized;
- Stream-load edge data into memory, update vertex information, and clear edge data in order to free memory;
- Write the final result in memory back to disk.

Optimization techniques implemented in this paper include: I/O thread optimization, memory resource monitoring, and automatic switching of memory-external memory computing.

### 3.1  *I/O* Thread Optimization

Pimiento initiates parallel processing by executing threads that need to read edge data on the edge shard before they can perform subsequent vertex updates, which results in a lot of I/O. Because there is no synchronization between execution threads, computation and update speeds are very fast. However, it is often necessary to wait for the end of the I/O operation; therefore, I/O represents the Pimiento performance bottleneck.

A thread execution includes an I/O operation and an update operation. The I/O operation loads edge data into memory, and the update operation updates the vertex using edge data. However, this binds the I/O operation to the update operation in a thread of execution. In this case, I/O operations and update operations are synchronized more frequently, resulting in lower I/O throughput and CPU-resource utilization.

To address these problems, Pimiento separates the I/O operation from the update operation, creating multiple update threads responsible for each vertex-update operation while creating multiple I/O threads responsible for loading edge data into memory, thereby more reasonably allocating I/O and computing resources. However, if there are too many I/O threads relative to update threads, there will be too much cache data, and the update thread will not be able to execute, which will cause the cache to rapidly expand and fill memory. If the I/O thread is too small relative to the update thread, the update thread will execute too quickly while the I/O thread will be too small to keep up with the influx of data, resulting in an idle update thread while it waits for I/O.

To avoid these situations, Pimiento allows users to set the I/O- and update thread allocations according to resource and application features in order to use a memory monitoring strategy to ensure balance between the update and I/O threads to maintain saturation of I/O and CPU resources and maximize system performance.

### 3.2  Memory Resource Monitoring

In Pimiento, the I/O thread reads edge data and caches it in memory while and the update thread digests the edge data to update the graph vertex state, after which memory is freed when graph edge data is used. Because the I/O thread executes in parallel with the update thread, the I/O operation is not controlled by the update thread, which could result in a mismatch between the throughput of the graph edge data in during update thread processing and throughput of the graph edge data during I/O thread reading. If I/O throughput is too fast, this will result in increased caching of edge data loaded from disk into memory, which will eventually lead to memory overflow. If I/O throughput is too slow, this will result in the update thread remaining in a waiting state, leading to CPU-resource waste.

To address this problem, Pimiento uses a memory resource-monitoring thread to monitor memory usage. When memory for cached data is running low, the monitoring thread signals individual I/O threads to block I/O threads to prevent edge data loading in order to wait for the update thread to process the edge data and release memory. When the monitoring thread detects that memory overflow is no longer a possibility, it signals the individual I/O threads to continue loading edge data. The memory resource-monitoring strategy increases Pimiento memory efficiency, maximizes memory utilization to improve computing speed, and avoids memory overflow. The memory monitoring thread perfectly coordinates the update thread with the I/O thread, making the system more robust and coordinated while performing parallel computations and disk I/O operations.

## 4    Experimental Evaluation

We implemented and evaluated a wide range of applications in order to demonstrate the applicability of Pimiento to multi-domain problems. Despite the restrictive external memory setting, Pimiento retains the expressivity of other external graph-processing frameworks.

### 4.1    Test Setup

All experiments used a commercial server equipped with an e5-2670@v3 processor, which has two sockets running at 2.3 GHz, 32 MB L3 cache, with 12 cores per socket, and a disabled CPU hyper-threading feature. The commercial server was equipped with 32 Gbyte of memory and 1 Tbyte of disk (SSD), and the operating system was 64-bit Ubuntu 14.04 LTS. We evaluated Pimiento using the applications described in Section and analyzed its performance on a selection of large graphs (Table 2).

**Table 2.** Real-world and synthetic graphs data used in the experiments

| Dataset | Twitter  [16] | UK-2007  [17] | Rmat27  [18] |
|---|---|---|---|
| Vertex num | 41.6M | 134M | 128M |
| Edge num | 1.5B | 5.5B | 2B |
| Avg deg | 35.3 | 41.2 | 16 |
| Max outdeg | 770K | 22.4K | 123K |
| Size | 25 GB | 93 GB | 32 GB |

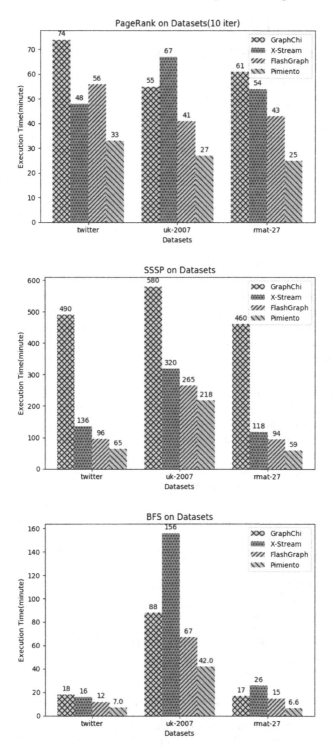

**Fig. 3.** Comparison of execution time when performing PageRank, SSSP and BFS over different data sets.

## 4.2 Comparison with Other Systems

### 4.2.1 Propagation-Based Algorithms

First, we evaluated graph-propagation-based traversal, such as that using BFS and SSSP. Figure 3 shows that Pimiento performed better on SSD than GraphChi and X-Stream. Compared with GraphChi, X-Stream, and FlashGraph on Twitter, Uk-2007, and Rmat27, respectively, Pimiento was 1.6 times to 7.5 times faster. There are mainly two reasons for the acceleration:

- Pimiento reads edge data sequentially from disk, thereby reducing random access to the disk
- Pimiento can reduce the amount of data written back to disk, effectively avoiding a data race.

### 4.2.2 Iteration-Based Algorithms

We then evaluated graph iteration-based algorithms, such as PageRank, and confirmed that PageRank is representative of a cumulative algorithm. When computing a PageRank value, each vertex should first collect all values from its source vertices in order to compute a sum. Pimiento uses a vertex-centric on-the-fly update model.

We compared four systems: Pimiento, GraphChi, X-Stream, and FlashGraph. In each iteration, the graph-processing system computed the new PageRank value for each vertex and selects the largest one. The iteration stops when the maximum PageRank value reaches a stable state (i.e., when the maximum change in PageRank value between iterations is less than the threshold value, computing is assumed to have converged and ends).

As shown in Fig. 3, Pimiento performed better on different data sets than GraphChi, X-Stream, and FlashGraph. Because Pimiento uses sequential disk access, it is multi-fold faster than GraphChi and X-Stream. Specifically, Pimiento is 2.3 times faster than GraphChi and 1.5 times faster than X-Stream on a Twitter dataset. The primary reason for this is that values of all vertices are sent to destination vertices along outer edges for cumulative updates, and there is no need to write the values of destination vertices back to disk. To evaluate the improved performance of Pimiento, we analyzed the total amount of I/O performed by the BFS, SSSP, and PageRank algorithms on different graphs (see Fig. 4). Specifically, compared with GraphChi, S-Stream, and FlashGraph, the I/O-data volume of Twitter, Uk-2007, and Rmat27 was reduced by a range of 30% to 98%, because the status values of all vertices were updated instantly, precluding the need to write the vertex state back to disk.

### 4.3 Optimization of the Update- and I/O Thread Proportions

When using SSD, we open multiple I/O threads in order to increase the storage capacity of data reading and computational efficiency. To explore the effect of

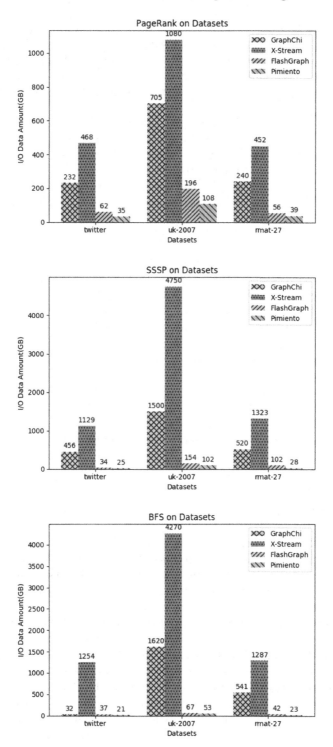

**Fig. 4.** Comparison of overall I/O data amount when performing PageRank, SSSP and BFS over different data sets.

the update- and I/O thread number selection on the performance of Pimiento, we compared the convergence speed of Pimiento in executing the iteration algorithm under different proportions of update and I/O threads. Figure 5 shows the average time for PageRank to converge relative to Pimiento, revealing that the convergence speed first increased and then decreased after peaking at a proportion of 4:1.

Our analyses showed that when the I/O thread was busier that the update thread, too much cache-structure data would require processing, precluding execution of the update thread. However, if the amount of data going to the I/O thread was less than that to the update thread, the update thread would execute too rapidly while the I/O would need to starve in order to maintain pace with the data input. These two situations would result in the output described in Fig. 5, which should be avoided.

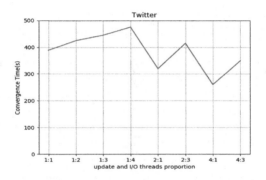

**Fig. 5.** Update and I/O threads proportion

# 5   Related Work

Here, we proposed improvements in single-computer-processing power and storage capacity using a graph-processing model. Such systems demonstrate adequate graph-processing performance, and compared with distributed systems, their obvious advantages include low hardware cost and low power consumption.

TurboGraph [9] makes full use of multi-core concurrency and the I/O performance of Flash SSD [20] to parallelize CPU processing and I/O processing in order to support rapid graph data storage. VENUS [14] is a point-centric streamlining graph-processing model that introduces a more efficient model for storing and accessing disk graph data using a cache strategy. FlashGraph [12] is a single-machine graph-processing system that can handle trillions of nodes on a solid-state hard-disk array while providing a dynamic load balancer to solve CPU-idle results from uneven computing tasks. GridGraph [21] supports selective scheduling, which can greatly reduce I/O and improve computing performance in algorithms, such as BFS and weakly connected components. NXgraph [13] provides three update strategies: (1) sort by the target vertex of each sub-shard edge; (2) based on the size of the graph and the available memory resources, the fastest execution strategy for different graph problems is adaptively selected to take full advantage of memory space and reduce data transmission; and (3) to solve the problem of large graphs fully loaded into memory, a previous study described the design of a disk-based single graph-processing platform using MMap [10] in Linux memory management. MMap maps a file or other pair to memory, where a process can access the file just as it accesses a normal memory without using operations, such as $read()$ and $write()$.

## 6   Conclusions

There currently numerous studies focused on addressing large graph-processing problems using high-performance single-server systems. The existing single-server graph-processing system has limitations, including poor locality, heavy synchronization cost, and frequent I/O access. Our study compared out-of-core graph-computing systems, including GraphChi, X-Stream, and FlashGraph, with Pimiento, revealing that Pimiento achieved $7.5\times, 4\times, 1.6\times$ better performance on large real-world graphs and synthetic graphs in the same experimental environment.

**Acknowlegements.** This paper is partially supported by "QingHai Province High-end Innovative Thousand Talents Program-Leading Talents", The National Natural Science Foundation of China (No. 61762074, No.61962051), The Open Project of State Key Laboratory of Plateau Ecology and Agriculture, Qinghai University (No. 2020-ZZ-03), and National Natural Science Foundation of Qinghai Province (No. 2019-ZJ-7034).

## References

1. Malewicz, G., et al.: Pregel: a system for large scale graph processing. In: Proceedings of the 2010 International Conference on Management of Data, SIGMOD 2010, pp. 135–146 (2010)
2. Low, Y., Bickson, D., Gonzalez, J., Kyrola, A., Hellerstein, J.M.: Distributed GraphLab: a framework for machine learning and data mining in the cloud. In: Proceedings of the VLDB Endowment, pp. 716–727 (2012)
3. Gonzalez, J.E., Low, Y., Gu, H., Bickson, D.: PowerGraph: distributed graph-parallel computation on natural graphs. In: Proceedings of the 10th USENIX Conference on Operating Systems Design and Implementation, OSDI 2012, pp. 17–30 (2012)
4. Zhu, X., Chen, W., Zheng, W., Ma, X.: Gemini: a computation-centric distributed graph processing system. In: Proceedings of the 12th USENIX Symposium on Operating Systems Design and Implementation, OSDI 2016, pp. 301–316 (2016)

5. Kyrola, A., Blelloch, G., Guestrin, C.: GraphChi: large-scale graph computation on just a PC. In: Proceedings of the 10th USENIX Conference on Operating Systems Design and Implementation, OSDI 2012, pp. 31–46 (2012)
6. Roy, A., Mihailovic, I., Zwaenepoel, W.: X-stream: edge-centric graph processing using streaming partitions. In: Proceedings of the Twenty-Fourth ACM Symposium on Operating Systems Principles, pp. 472–488 (2013)
7. Shao, Z., He, J., Lv, H., Jin, H.: FOG: a fast out-of-core graph processing framework. Int. J. Parallel Prog. 45(6), 1259–1272 (2017)
8. Zhu, X.W., Han, W.T., Chen, W.G.: Grid graph: large-scale graph processing on a single machine using 2-level hierarchical partitioning. In: Proceedings of the 2015 USENIX Conference on USENIX Annual Technical Conference, pp. 375–386 (2015)
9. Han, W.-S., et al.: TurboGraph: a fast parallel graph engine handling billion-scale graphs in a single PC. In: Proceedings of the 19th ACM SIGKDD International Conference on Knowledge Discovery and Data Mining, pp. 77–85 (2013)
10. Lin, Z., Kahng, M., Sabrin, K.M., Chau, D.H.P., Lee, H., Kang, U.: Mmap: fast billion-scale graph computation on a PC via memory mapping. In: IEEE International Conference on Big Data, IEEE, pp. 159–164 (2014)
11. Yuan, P., Zhang, W., Xie, C., Jin, H., Liu, L., Lee, K.: Fast iterative graph computation: a path centric approach. In: International Conference for High Performance Computing, Networking, Storage and Analysis, pp. 401–412. IEEE Computer Society (2014)
12. Zheng, D., Mhembere, D., Burns, R., Vogelstein, J., Priebe, C.E., Szalay, A.S.: FlashGraph: processing billion-node graphs on an array of commodity SSDs. In:13th USENIX Conference on File and Storage Technologies (FAST 2015) USENIX Association, pp. 45–58 (2015)
13. Chi, Y., Dai, G., Wang, Y., Sun, G., Li, G., Yang, H.: NXgraph: an efficient graph processing system on a single machine. In: Proceedings of the 32nd International Conference on Data Engineering, ICDE 2016, pp. 409–420 (2016)
14. Cheng, J., Liu, Q., Li, Z., Fan, W., Lui, J.C.S., He, C.: VENUS: vertex-centric streamlined graph computation on a single PC. In: Proceedings of the 31nd International Conference on Data Engineering, ICDE 2015, pp. 1131–1142 (2015)
15. Zhang, Y., Gao, Q., Gao, L., Wang, C.: Maiter: an asynchronous graph processing framework for delta-based accumulative iterative computation. IEEE Trans. Parallel Distrib. Syst. 25(8), 2091–2100 (2014)
16. Kwak, H., Lee, C., Park, H., Moon, S.: What is Twitter, a social network or a news media? In: Proceedings of the 19th International Conference on World Wide Web, pp. 591–600 (2010)
17. Boldi, P., Santini, M., Vigna, S.: A large time-aware web graph. SIGIR Forum 42(1), 78–83 (2008)
18. The graph 500 list (2014). http://www.graph500.org/
19. Pearce, R., Gokhale, M., Amato, N.: Multithreaded asynchronous graph traversal for in-memory and semi-external memory. In: SuperComputing (2010)
20. Badam, A., Pai, V.S.: SSDAlloc: hybrid SSD/RAM memory management made easy. In: Proceedings of the 8th USENIX conference on Networked Systems Design and Implementation. USENIX Association, p. 16 (2011)
21. Zhu, X., Han, W., Chen, W.: GridGraph: largescale graph processing on a single machine using 2-level hierarchical partitioning. Proceedings of the 2015 USENIX Annual Technical Conference, pp. 375–386 (2015)
22. Vora, K., Xu, G., Gupta, R.: Load the edges you need: a generic I/O optimization for disk-based graph processing. In: Proceedings of the 2016 USENIX Annual Technical Conference, pp. 507–522 (2016)

# Software Systems and Programming Models

# Parallel Software Testing Sequence Generation Method Target at Full Covering Tested Behaviors

Tao Sun[(⊠)], Xiaoyun Wan, Wenjie Zhong, Xin Guo, and Ting Zhang

College of Computer Science of Inner Mongolia University,
Hohhot 100021, Inner Mongolia, China
cssunt@imu.edu.cn, 15690582836@163.com,
zhongwenjie@mail.imu.edu.cn, guoxin_2017@163.com,
ml5561836616@163.com

**Abstract.** Parallel software system testing is very difficult because the number of states expands sharply caused by parallel behaviors. In this paper, a testing sequence generation method is proposed, target at full covering tested behaviors and related behaviors over all execution paths. Because of state spaces of parallel software systems are often large, this paper focuses on the state space subgraph, contained all firing of tested and related behaviors, of system model. Firstly, the software system is modeled with Colored Petri Net (CPN), called system model (SM), and every tested behavior or related behavior is modeled also with CPN, called Behavior Model Unit (BMU). Then, the method proposes mapping operation, intersection operation, and so on to finally realize the generation of test sequence. Practices show that this method is efficient, which could achieve full coverage.

**Keywords:** Parallel software · Testing sequence generation · Full covering · Tested behaviors · Colored Petri Net (CPN)

## 1 Introduction

Many traditional Model-based testing technologies [1, 2] cannot work effectively for parallel systems with masses of states. Some formal languages, like Finite State Machine, are not suitable for parallel software modeling. Some languages are suitable for parallel software modeling, but the state space of the model is not clear. So it is difficult to achieve full covering testing for tested behaviors all over execution paths.

This paper argues model-based testing for parallel software based on Colored Petri Net (CPN), because CPN is very suitable for parallel behaviors modeling, and the state space graph of the model could be calculated automatically. Methods in literature [3–5] are based on simply searching or traversal for the state space of CPN models, which will generate many redundant test sequences and the testing efficiency is low. Literature [6] proposed a test sequence generation method target at full covering sequence of linear tested behaviors, however, the method has limitation because tested behaviors must be linear.

In this paper, a testing sequence generation method is proposed, target at full covering tested behaviors and related behaviors over all execution paths. Firstly, the

S. Wen et al. (Eds.): ICA3PP 2019, LNCS 11945, pp. 59–67, 2020.
https://doi.org/10.1007/978-3-030-38961-1_6

software control flow is modeled with CPN. Then the testing purpose is described as tested behaviors and related behaviors. Tested behaviors we mean behaviors in the testing purpose, and related behaviors we mean behaviors related to tested behaviors in data flow. Testing sequences should cover all execution paths including these behaviors, so that all execution possible sequences in the software are covered, and we can know whether tested behaviors are correctly implemented.

Due to the state spaces of parallel software systems are often large, this paper focuses on the state space sub-graph of SM, which containing all firing of tested behaviors and related behaviors. The software system modeled with CPN is called system model (SM), and every tested behavior or related behavior modeled also with CPN is called Behavior Model Unit (BMU). State spaces of SM and all BMUs are calculated, the mapping operation between SM and a BMU is proposed, by which the state space sub-graph of SM containing all firing of the BMU is obtained. The intersection operation between these sub-graphs is used, BMUs with intersection are grouped into the same group, and union sub-graph of overlap sub-graphs in a group are obtained. Then testing sequences are generated in union sub-graphs, after that, repetitive removing operation and testing sequences connecting operation will be used. Sequences got by this method are fully covering tested behaviors and related behaviors over all execution paths in the parallel software system. Before this method, initial marking of SM has been given aiming at tested behaviors and related behaviors, which is described in our other papers [7].

The rest of this paper is organized as follows. Section 2 gives key definitions of the method. Section 3 describes sequence generation algorithm. Section 4 describes some practical applications of the method, and we conclude the paper in the last section.

## 2 Key Operations

This section presents some key operations in the sequence generation method. The definition about CPN and marking and state space are as same as common definitions. CPN model is supported by CPN Tools. There are seven key operations in the method.

**Definition 1.** Mapping(SM, BMU) operation.

This operation consists of two parts, Projection operation and Get Graph operation respectively and the result is the state space sub-graph of SM containing all firing of the BMU, which is written as BMU.mp. This operation should be used on all BMUs.

Projection operation P(SM, BMU). Projection operation is used between SM and a state $M_1$ in BMU: $P(SM, M_1) = \{m \mid m \in M_{SM}, \forall p \in P_{BMU}, M_1(p) \subseteq m(p)\}$. The set of states (i.e. markings) in SM is called $M_{SM}$. The set of places in BMU is called $P_{BMU}$. The result of $P(SM, M_1)$ is a state set of SM, whose states are all containing the same tokens in the same places of state $M_1$ in BMU.

Projection operation is used between SM and BMU: $P(SM, BMU) = (SM_{BMUI}, SM_{BMUE})$, $SM_{BMUI} = P(SM, BMUI)$, $SM_{BMUE} = P(SM, BMUE)$. $P(SM, BMU)$ returns $(SM_{BMUI}, SM_{BMUE})$. The initial state and the end state of the BMU model are denoted by BMUI and BMUE. $SM_{BMUI}$ is a state set of SM, whose states are all containing the same tokens in the same places of state BMUI in BMU, $SM_{BMUE}$ is a state set of SM, whose states are all containing the same tokens in the same places of state BMUE in BMU. Sub-graph between $SM_{BMUI}$ and $SM_{BMUE}$ contains all the firing of the BMU.

Get Graph operation GetGraph($SM_{BMUI}$, $SM_{BMUE}$). Depth-First algorithm is used in this operation. SMBMUI and SMBMUE are two sets of states, so operations Max(M) and Min(M) are used. Sub-graph between Max($SM_{BMUI}$) and Min($SM_{BMUE}$) is obtained in Get Graph operation, which is same as sub-graph between $SM_{BMUI}$ and $SM_{BMUE}$.

Any state-set $M \subseteq M_{SM}$, the number of states in M is n, and $M_1, M_2, ..., M_n \in M$. If $i \in 1 ... n$, and $M_1, ... , M_n$ are not precursor of $M_i$, then $M_i$ is the max-state of M. The set of all max-states in M is called max-states set, denoted by Max(M). Similarly, if $i \in 1 ... n$, and $M_1, ..., M_n$ are not successor of $M_i$, then $M_i$ is the min-state of M. The set of all min-states in M is called min-states set, denoted by Min(M).

**Definition 2.** Intersection(S1, S2) operation, written as S1 $\cap$ S2.

Every BMU.mp is covering only one of tested behaviors or related behaviors so the intersection operation is used between two BMU.mps. The set of nodes in sub-graph s is written as s.NS, and the set of arcs in sub-graph s is written as s.AS. S1 and S2 are two sub-graphs of SM state space. The result of Intersection(S1, S2) is $\varnothing$ or a sub-graph called ISG, which is obtained by set intersection operation of NS and AS.

Intersection(S1, S2) = $\varnothing$ iff S1.AS $\cap$ S2.AS = $\varnothing$, Intersection(S1, S2) = ISG iff S1. AS $\cap$ S2.AS $\neq \varnothing$, and ISG.NS = S1.NS $\cap$ S2.NS, and ISG.AS = S1.AS $\cap$ S2.AS.

For two BMUs called BMU1 and BMU2,

(1) If BMU1.mp $\cap$ BMU2.mp = $\varnothing$, then BMU1 and BMU2 are not parallel behaviors.
(2) If BMU1.mp $\cap$ BMU2.mp $\neq \varnothing$, then BMU1 and BMU2 are parallel behaviors.

**Definition 3.** Grouping(BMU[]) operation.

The array BMU[] contains all BMUs and every element of the array is a BMU. The array BMUGSet[] contains all sets of BMUs and every element of the array is a set of BMUs. The result of this operation is putting intercross execution BMUs into the same set. If two BMUs have intercross execution sub-graph, they must be parallel behaviors, so they should be in the same set. Particularly, a set may contain more than two BMUs, direct or indirect parallel behaviors are all in the same group.

**Definition 4.** Union(S1, S2) operation, written as S1 $\cup$ S2.

The result of Union(S1, S2) is the union sub-graph of S1 and S2 called USG, which is obtained by set Union operation of NS and AS. Union(S1, S2) = USG, and USG. NS = S1.NS $\cup$ S2.NS, and USG.AS = S1.AS $\cup$ S2.AS. S1 and S2 are two sub-graphs of SM state space. BMUs in a set of BMUGSet[] are parallel behaviors. The intersection operation is used between two BMU.mps. When the number of BMUs in the set is larger than two, there will be more than one ISGs for BMUs in the set. Then the union sub-graph for all the ISGs in a set should be obtained, called USGForSet.

**Definition 5.** Generating(S) operation.

S is a sub-graph of SM state space. The result of this operation is a path set which contains all paths in S. For a group, the union sub-graph contains all the intercross execution of BMUs in the group. This operation is used on union sub-graph, which generates sequences full covering tested behaviors and related behaviors in the group. Other BMUs are not parallel with BMUs in the set, so sequences are also full covering BMUs in the set over all the system.

**Definition 6.** Repetitive-removing(PathSet, BMUGSet).

PathSet is a set of paths. BMUGSet is an element of BMUGSet[], which is a set of BMUs. The result of this operation is another set of paths, which been removed repetitive paths. There are two additional definitions in the operation: (1) Mapping testing sequences operation MPTS(p, BMUGSet), which returns the projection sequence of path p on BMUs in BMUGSet. The projection sequence is p retained BMUs in BMUGSet but removed other behaviors, and the sequence element kept the same order as p. (2) MPTS[] is an array used to store mapping testing sequences.

**Definition 7.** Connecting(PathSet1, PathSet2), PathSet1 × PathSet2.

PathSet1 and PathSet2 are two sets of paths. The result of this operation is another set of paths which connects all paths of PathSet1 and all paths of PathSet2. Paths in different PathSets may not be consecutive, then this operation will connect them with one of the paths between them in the state space of SM. For every BMUGSet, the PathSet without repetition has been obtained. Then PathSets of different BMUGSet should be connected, so that generating full-paths of SM, which begins with initial marking and ends with end markings. The set of full-paths is called FullPathSet, the set of SM initial marking is called {MI}, and the set of SM end markings is called {ME}. FullPathSet is obtained by: {MI} × BMUGSet [1].path × …×BMUGSet[i].path × …×{ME}.

## 3    Testing Sequence Generation Algorithm

The main algorithm of testing sequence generation method is shown in Fig. 1, based on the operations shown in Sect. 2.

```
{
  Begin
    For each BMU in BMU[]
      BMU.mp = Mapping(SM, BMU);
    Next
    BMUGSet[] = Grouping (BMU[]);
    For each BMUGSet in BMUGSet[]
     USGForSet = Ø;
     For each (BMUᵢ, BMUⱼ) in BMUGSet
       If (BMUᵢ.mp∩BMUⱼ.mp≠ Ø)
         USGForSet = USGForSet ∪ (BMUᵢ.mp∩BMUⱼ.mp);
       End If
     Next
     PathSet = Generating(USGForSet);
     BMUGSet.pathset=Repetitive-removing(PathSet,BMUGSet);
    Next
    FullPathSet=MI}×BMUGSet[1].pathset×…×BMUGSet[i].pathset×…×{ME};
  End
}
```

**Fig. 1.** Testing sequence generation algorithm

There are three key steps in the algorithm.

(1) According to projection operation definition between SM and BMU makes that all firing of the BMU must be in the sub-graph, otherwise, it will be inconsistent with the definition of projection operation. To focus on the state space sub-graph of SM, mapping operation should be used on all BMUs. The result sub-graph of the operation achieves full coverage of the BMU and more efficiency for testing sequence generation.

(2) For two BMUs, $BMU_i$ and $BMU_j$, the result of intersection operation, $BMU_i$.mp $\cap$ $BMU_j$.mp, is the intercross execution sub-graph of the two BMUs, which is guaranteed by the definition of projection operation and intersection operation. If two BMUs have intercross execution sub-graph, they must be parallel behaviors, so they should be in the same group. When the number of BMUs in a group is larger than two, there will be more than one intersection sub-graphs for BMUs in the group. Then the union sub-graph for all the intersection sub-graphs in a group should be obtained, which is full covering BMUs in the group. Other BMUs are not parallel with BMUs in the group, so sequences generating for the union sub-graph are also full covering BMUs in the group over all the system.

(3) Generating operation, repetitive-removing operation and connecting operation are traversing all the sequences of union sub-graphs in all groups, so sequences generated from this algorithm are full covering all tested and related behaviors. The search scope of the sequence generating is cut by mapping operation, the repetitive sequences are removed by repetitive-removing operation, finally sequences generated from this algorithm are full covering all tested and related behaviors.

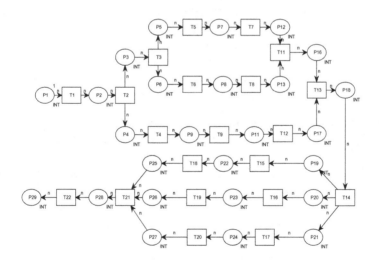

**Fig. 2.** CPN model of a simple project contract approval system

## 4  Testing Example and Result Analysis

The Fig. 2 is the CPN model of a simple project contract approval system. The Table 1 is the transition description of the model in Fig. 2. In the table head of Table 1, T means Transition and E means Explanation. The Fig. 3 is the state space of model in Fig. 2. We chose three test purposes for the experiment. Among them, the testing process interpretation and the result analysis were carried out for test purpose1, and the test purpose2 and test purpose3 only carried out the result analysis. In test purpose1, tested behaviors are T5 T6 T18 and T19. T20 is data related with T19, so related behavior is T20.

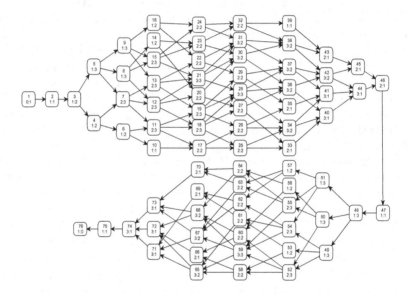

**Fig. 3.** State space of model in Fig. 2

**Step1:** The result of Mapping(SM, BMU) is shown in Table 2. In the table head of Table 2, T means Transition, P means initial and ending place.
**Step2:** T5.mp ∩ T6.mp ≠ ∅, T18.mp ∩ T19.mp ≠ ∅, T18.mp ∩ T20.mp ≠ ∅, T19.mp ∩ T20.mp ≠ ∅, so BMUGSet[1] = {T5, T6}, BMUGSet[2] = {T18, T19, T20}.

**Table 1.** Transition description of model in Fig. 2

| T | E | T | E | T | E |
|---|---|---|---|---|---|
| T1 | Application submitting | T8 | Review 2 | T16 | Review of the budget |
| T2 | Contract classification | T9 | Review 3 | T17 | Procurement approval |
| T3 | Resource classification | T11 | Comments 1 | T18 | Feedback comment1 |
| T4 | Submission to the security department | T12 | Comments 2 | T19 | Feedback comment2 |
| T5 | Submission to the funding department | T13 | Submission to the competent department | T20 | Feedback comment3 |
| T6 | Submission to the technical department | T14 | Submission to sub-departments | T21 | Comments summary |
| T7 | Review 1 | T15 | Review of legal materials | T22 | Supervisor approval |

**Step3:** The result of union operation. BMUGSet[1]: USGForSet = (M5, M34), BMUGSet[2]: USGForSet = (M52/M54/M55, M74)

**Step4:** The result of Generating(USGForSet). For BMUGSet[1], there are 4 paths, M5-M9-M15-M21-M28-M34, M5-M7-M13-M19-M28-M34, M5-M8-M15-M21-M28-M34, M5-M7-M12-M21-M28-M34. For BMUGSet[2], There are 36 paths. And the PathSet of BMUGSet[2] is not shown because of the limit of the length.

**Table 2.** Mapping(SM, BMU) operation on T5, T6, T18, T19, T20

| T | P | Initial and ending sets of projection states | Sub-graph |
|---|---|---|---|
| T5 | P5 | {M5, M7, M8, M11, M12, M14, M17, M18, M20, M25, M27, M33} | (M5, M40) |
| | P7 | {M9, M13, M15, M19, M21, M23, M26, M28, M30, M34, M36, M40} | |
| T6 | P6 | {M5, M7, M9, M11, M13, M16, M17, M19, M22, M26, M29, M35} | (M5, M41) |
| | P8 | {M8, M12, M15, M18, M21, M24, M25, M28, M31, M34, M37, M41} | |
| T18 | P22 | {M50, M52, M55, M58, M59, M63, M65, M67, M71} | (M50, M74) |
| | P25 | {M56, M60, M64, M66, M68, M70, M72, M73, M74} | |
| T19 | P23 | {M51, M54, M55, M59, M61, M64, M65, M68, M72} | (M51, M74) |
| | P26 | {M57, M62, M63, M67, M69, M70, M71, M73, M74} | |
| T20 | P24 | {M49, M52, M54, M59, M60, M62, M67, M68, M73} | (M52, M74) |
| | P27 | {M53, M58, M61, M65, M66, M69, M71, M72, M74} | |

**Table 3.** Final result of FullPathSet

| Paths |
|---|
| M1-M2-M5-M9-M15-M21-M28-M34-M40-M44-M46-M47-M48-M50-M55-M59-M65-M71-M74-M75-M76 |
| M1-M2-M5-M9-M15-M21-M28-M34-M40-M44-M46-M47-M48-M50- M52-M59-M67-M73-M74-M75-M76 |
| M1-M2-M5-M9-M15-M21-M28-M34-M40-M44-M46-M47-M48-M50- M52-M60-M66-M72-M74-M75-M76 |
| M1-M2-M5-M9-M15-M21-M28-M34-M40-M44-M46-M47-M48-M50- M52-M59-M68-M73-M74-M75-M76 |
| M1-M2-M5-M9-M15-M21-M28-M34-M40-M44-M46-M47-M48-M50- M55-M59-M67-M71-M74-M75-M76 |
| M1-M2-M5-M9-M15-M21-M28-M34-M40-M44-M46-M47-M48-M50- M52-M59-M65-M72-M74-M75-M76 |
| M1-M2-M5-M8-M15-M21-M28-M34-M40-M44-M46-M47-M48-M50-M55-M59-M65-M71-M74-M75-M76 |
| M1-M2-M5-M8-M15-M21-M28-M34-M40-M44-M46-M47-M48-M50- M52-M59-M67-M73-M74-M75-M76 |
| M1-M2-M5-M8-M15-M21-M28-M34-M40-M44-M46-M47-M48-M50- M52-M60-M66-M72-M74-M75-M76 |
| M1-M2-M5-M8-M15-M21-M28-M34-M40-M44-M46-M47-M48-M50- M52-M59-M68-M73-M74-M75-M76 |
| M1-M2-M5-M8-M15-M21-M28-M34-M40-M44-M46-M47-M48-M50- M55-M59-M67-M71-M74-M75-M76 |
| M1-M2-M5-M8-M15-M21-M28-M34-M40-M44-M46-M47-M48-M50- M52-M59-M65-M72-M74-M75-M76 |

**Table 4.** Testing sequences of three testing purposes

| Test purpose | Description | Number of sequences |
|---|---|---|
| 1 | T5, T6, T18, T19, T20 | 12 |
| 2 | T5, T6, T9 | 6 |
| 3 | T18, T19, T20 | 6 |

**Step5:** The result of Repetitive-removing(PathSet, BMUGSet).
There are 2 paths, M5-M9-M15-M21-M28-M34, M5-M8-M15-M21-M28-M34, in BMUGSet[1].pathset and 6 paths, M55-M59-M65-M71-M74, M52-M59-M67-M73-M74, M52-M60-M66-M72-M74, M52-M59-M68-M73-M74, M55-M59-M67-M71-M74, M52-M59-M65-M72-M74, in BMUGSet[2].pathset.

**Step6:** FullPathSet after connecting operation. FullPathSet =

$$\{M1\} \times \left\{ \begin{array}{l} M5 - M9 - M15 - M21 - M28 - M34 \\ M5 - M8 - M15 - M21 - M28 - M34 \end{array} \right\} \times \left\{ \begin{array}{l} M55 - M59 - M65 - M71 - M74 \\ M52 - M59 - M67 - M73 - M74 \\ M52 - M60 - M66 - M72 - M74 \\ M52 - M59 - M68 - M73 - M74 \\ M55 - M59 - M67 - M71 - M74 \\ M52 - M59 - M65 - M72 - M74 \end{array} \right\} \times \{M76\}.$$

The final result of FullPathSet contains 12 testing sequences shown in Table 3. Testing Sequences of three testing purposes is shown in Table 4.

Experimental results show that the sequence generated by this method completely covers all the tests and related behaviors and has efficiency.

# 5    Conclusion

In this paper, a testing sequence generation method is proposed, target at full covering tested behaviors and related behaviors over all execution paths. Due to state spaces of parallel software systems are large, this paper focuses on the state space sub-graph of SM, which containing all firing of tested behaviors and related behaviors. Generating operation, repetitive-removing operation and connecting operation are traversing all the sequences of USG in all groups, and the search scope of the sequence generating is cut by projection operation in mapping, the repetitive sequences are removed by repetitive-removing operation. Finally, experiments show that the sequence generated by this algorithm completely covers all tested and related behaviors.

**Acknowledgment.** This work was supported by the National Natural Science Foundation of China under Grant No. 61562064 and No. 61661041.

# References

1. Dalal, S.R., Jain, A., Karunanithi, N., et al.: Model-based testing in practice. In Proceedings of the 21st International Conference on Software Engineering, pp. 285–294 (1999)
2. Yan, J., Wang, J., Chen, H.: Survey of model-based software testing. Comput. Sci. **31**(2), 184–187 (2004)
3. Watanabe, H., Kudoh, T.: Test suite generation methods for concurrent systems based on coloured petri nets. In: Proceedings of the 2nd Asia-Pacific Software Engineering Conference, pp. 242–251 (1995)
4. Desel, J., Oberweis, A., Zimmer, T., et al.: Validation of information system models: petri nets and test case generation. In: Proceedings of the 10th IEEE International Conference on Systems, Man, and Cybernetics, pp. 3401–3406 (1997)
5. Farooq, U., Lam, C.P., Li, H.: Towards automated test sequence generation. In Proceedings of the 19th Australian Conference on Software Engineering, pp. 441–450 (2008d)
6. Sun, T., Ye, X., Liu, J.: A test generation method based on model reduction for parallel software. In: Proceedings of the International Conference on Parallel and Distributed Computing, Applications and Technologies, pp. 777–782 (2012)
7. Sun, T., Zhang, L., Ma, H.: An automatic generation method for condition expressions of CPN model focus on tested behaviors. In: processing of the 10th International Conference on Security, Privacy and Anonymity in Computation, Communication and Storage Workshops, pp. 271–285 (2017)

# Accurate Network Flow Measurement with Deterministic Admission Policy

Hongchao Du[1], Rui Wang[2], Zhaoyan Shen[1], and Zhiping Jia[1(⊠)]

[1] School of Computer Science and Technology, Shandong University,
Qindao 266237, China
mrdu@mail.sdu.edu.cn, {shenzhaoyan,jzp}@sdu.edu.cn
[2] State Grid Shandong Electric Power Research Institute, Jinan 250002, China
wangruiwell@foxmail.com

**Abstract.** Network management tasks require real-time visibility of current network status to perform the appropriate operations. However, the resource limitation of network devices and the real-time requirements make it difficult to provide accurate network measurement feedbacks. To reduce the error and inefficiencies caused by random operations in existing algorithms, we propose an efficient measurement architecture with the *Deterministic Admission Policy (DAP)*. DAP provides accurate large-flow detection and high network measurement precision by making full use of the information belong to large flows and small flows, and dynamically filtrating small flows as the network status evolves. To make the algorithm easy to implement on hardware, we propose *d-Length DAP* by replacing the global optimality with local optimality. Experimental results show that our algorithm can reduce the measurement error by 3 to 25 times compared to other algorithms.

**Keywords:** Network measurement · Top-K flow detection · Sketch

## 1 Introduction

Network measurement is an indispensable part of network management. Network managers support tasks such as anomaly detection, traffic engineering, and load balancing by collecting information of the network at different levels [1–10]. Thus, how to accurately measure the network situation is of vital importance. Currently, the analysis of network flows is the most commonly used network measurement method [1,3,7,10–12]. Network packets can be divided into flows based on the specific characteristics of the header. By counting the number of packets per flow, we can grasp the basic state of the current network.

The network measurement schemes consume DRAM, SRAM, and TCAM resources for counting network flows. However, today's network equipment has limited storage resources which cannot deal with a large number of network flows [1,7–10]. Meanwhile, network measurements also require real-time processing [2,4]. Both these constraints make the accurate network flow measurement

© Springer Nature Switzerland AG 2020
S. Wen et al. (Eds.): ICA3PP 2019, LNCS 11945, pp. 68–81, 2020.
https://doi.org/10.1007/978-3-030-38961-1_7

very challenging. To concur these issues, the state-of-the-art networks usually adopt two types of policies: counters or sketches.

The *counter based* methods allocate a counter and an ID to each flow and update it when a packet belonging to this flow arrives [1,7]. However, it is impractical to assign a counter and an ID to all network flsows, since recording a large number of flows can be very space consuming [2]. Therefore, some methods rely on the heavy-tail characteristic of network traffic propose only to count the large flows and ignore those small flows [1,13–15]. These methods trade off space overhead and data integrity. Unlike the counter based methods, the *sketch based* methods no longer retain the ID information of the flow, and the counter is shared by several flows, which significantly reduces the space overhead. Sketch mainly uses hash functions to maintain the mapping relationship between flow ID and counter. The sharing of counters occurs between the flows that have hash collisions [16]. The sketch methods use a fixed size space to count the dynamically changing traffic while providing provable tradeoffs of memory and accuracy [6].

The counter and sketch methods both have their shortcomings. The counter methods only count the large flow, resulting in complete loss of small flow information [2]. And the measurement accuracy is limited by the performance of the large flow detection algorithm [1]. To save space, the sketch methods don't support storage ID, so it cannot actively feedback the flow information, and can only passively accept query or offline analysis [3,16]. This also led to the lack of large-flow definition capabilities required for many tasks. Furthermore, existing large flow monitoring algorithms have involved random operations, which reduces measurement accuracy [1,13–15]. We intend to propose a new measurement architecture. Our goal is to provide accurate measurement of all flows with limited memory resource while providing large flow identification.

To achieve this, we deal with the large flow and the small flow separately. For large flows, we assign separate counters. For smaller flows, we count them with shared counters. Only the IDs of the large flows are recorded. In this way, we can provide large-flow monitoring capability in a limited space without losing small flow information. We propose the *Deterministic Admission Policy* (DAP) to dynamically distinguish between the large and small flows. Specifically, only the flow that becomes large enough in the small flow part is likely to enter the large flow part. Compared with other algorithms to randomly select a flow, our replacement algorithm provides the deterministic characteristics, which significantly improves the accuracy and reduces error. To make the algorithm easy to implement on hardware devices, we also proposed *d-Length DAP* (dL-DAP) by replacing the global optimality with local optimality. For better accuracy, in dL-DAP, we adopt a one-dimensional loop array that uses linear probing to reduce hash collisions. Experimental results show that DAP and dL-DAP efficiently improve the network measurement accuracy compared with the state-of-the-art algorithms.

Our main contributions are concluded as follows:

- We propose a novel accurate network measurement scheme to deal with large flows and small flows separately.

- We propose a DAP algorithm to distinguish between the large flows and small flows, and dynamically detect the large flows with the network traffic evolutes.
- To make the algorithm to be easily adopted to network hardware devices, we propose a dL-DAP with low hash collisions.
- We implement a prototype of our algorithm, and the evaluation results with real network data sets prove the validity and efficiency of our algorithm.

## 2   Background and Motivation

With the rapid development of the Internet, the emergence of new requirements has challenged the management of the network. Most network management tasks require the measurements of network status to perform the further operations. Due to the limited storage space and computing power of network devices, it is a crucial issue to provide accurate measurement results effectively with low overhead [1–10]. Among various network measurement methods, flow-based analysis is a typical measurement method [1,3,7,10–12]. A flow is a set of network packets with specific characteristics, such as IP 5-tuple. We can get the basic state of the current network by counting the number of packets per flow. By whether or not the counters are shared, we divide the measurement methods for flows into counter-based and sketch-based.

Due to the limited number of counters, many counter-based methods focus on calculating top-K flows, which account for the vast majority of the entire traffic. The ability to identify large flows is required by many management tasks. How to determine which flow is a top-K flow is key to such a method. In the existing methods, *Lossy Counting* [14] and *Frequent* [13] enable the newcomer to enter the top-K flow by periodically decrementing the counter value and filtering out the flow whose counter value is zero; *Space Saving* [15] replaces the minimum flow in the current counter with each new incoming flow. The disadvantage of these methods is the accuracy of the measured flow is affected by the unmeasured flow. Therefore, Ran et al. proposed *RAP* [1], a randomized admission policy. Each unmeasured flow packet has a probability of $\frac{1}{C_m+1}$ replacing the minimum

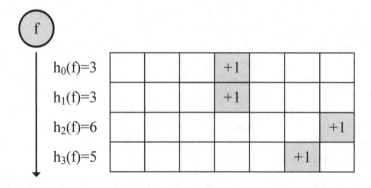

**Fig. 1.** The Count Min sketch with $d = 4$

flow with a counter value of $C_m$. RAP effectively reduces the measurement error and improving the accuracy of the large flow detection. It is state of the art in this type of methods.

Sketch uses shared counters to count all flow information with limited space. For example, *Count Min* sketch [16] performs $d$ hash operations on each flow ID, and updates the corresponding counter, where the minimum value is used as an estimate of the size of the flow (Fig. 1). Most sketches are only available for one type of task. *UnivMon* [5] proposes a single universal sketch to provide general support for measurement tasks while ensuring comparable accuracy. *SketchVisor* [4] uses a fast path technique to handle high traffic loads and improves robustness. The disadvantage of these methods is that they are computationally intensive and can not effectively detect the large flow. *Elastic sketch* [2] designs a flexible heavy part and light part to enable efficient measurements in different network environments. But its large flow detection algorithm is not efficient enough, resulting in reduced accuracy.

By carefully analyzing the above counter and sketch methods, we believe that a proper measurement architecture should meet the following points: (1) large flow detection capability (2) maintain information integrity (3) lower computation and resource overhead. To achieve this, we try to design a differentiated measurement architecture through separate processing of large and small flows. The key to this architecture is to distinguish between large flows and small flows accurately. The work most similar to us is the Elastic sketch. But the large-flow detection algorithm it uses does not converge, resulting in huge errors. The most effective way to detect large flows is RAP. However, in RAP algorithm and other large-flow monitoring algorithms, any flow may be regarded as a large flow, which leads to many invalid replacements. To this end, we propose DAP, and only the flow that has a hash conflict with the real large flow can be regarded as a large flow so that the number of candidate large flows can be reduced to improve accuracy.

## 3 Architectural Overview

### 3.1 Design of DAP

The design idea of DAP is to divide the measurement space into two parts: large flow part and small flow part, and only the flow that is large enough in the small flow part can enter the large flow part. As shown in Fig. 2, The measurement architecture we designed consists of two parts, separate counters for measuring large flows ($L$) and shared counters for measuring small flows ($S$). Each item in $L$ includes $ID$ and $Value$, which are used to record the flow ID and the number of packets, respectively. $S$ is a shared counter array; each counter is shared by flows that have hash collisions at this location. Based on this architecture, the DAP algorithm includes update, replace, and query operations.

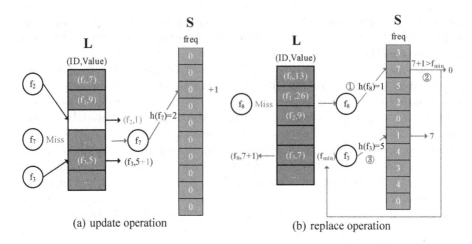

**Fig. 2.** The operations of DAP.

**Update Operation.** In the initial stage of DAP, both $L$ and $S$ are empty, and all new incoming flows are treated as a large flow. The ID used to define the flow is extracted from the packet, and $(ID, 1)$ is inserted into $L$. Then all packets belonging to this flow will increment the counter. If $L$ has no extra items, the new incoming flow will update $S$, which is an improvement of CountMin sketch at $d = 1$. Although setting a more significant $d$ will make the measurement more accurate, $d = 1$ is already precise enough. The update operation is shown in Fig. 2(a): the flow $f_2$ is not in the $L$ part but there is still a empty item, so $(f_2, 1)$ is inserted into $L$. With no empty item for $f_7$, the counter with index of $h(f_7) = 2$ in $S$ is updated. The flow $f_3$ take a position in $L$, so it just increases the counter.

**Replace Operation.** The first incoming flow is not necessarily a large flow, so when the real large flow appears and the L part has no empty item, we need to perform the replacement operation. The idea behind replace operation is as follows: if the true large flow is not in the L part, the value of the counter corresponding to it in the S part must be greater than the minimum value in the L part. So when updating the S part, we judge whether the value of the counter exceeds the minimum value of the L part, and if it is satisfied, replace the minimum flow of L with the current flow. The specific process is shown in Fig. 2(b): the minimum value of the L part is $(f_3, 7)$, the current packet is $f_8$, and the S part counters corresponding to $f_3$ and $f_8$ are respectively $S_1 = 7$, $S_5 = 1$. Because $S_1 + 1 = 7 + 1 > f_{min} = 7$, the $(f_8, 7 + 1)$ replaces $(f_3, 7)$ in L part with $S_1 = 0, S_5 = 7$.

**Query Operation.** The query operation is similar to the update. First, search the $L$ part according to the ID, and return the Value if hit. Otherwise, calculate the hash value of ID and return the frequency of the corresponding counter in the $S$ part.

## 3.2  Accuracy Analysis

Given two parameters $\epsilon$ and $\delta$, let $w = \lceil \frac{e}{\epsilon} \rceil, d = \lceil \ln(\frac{1}{\delta}) \rceil.d$ and $w$ are the number of rows and the number of counters per row in $L$. The error of the DAP algorithm for flow size estimation satisfies the following theory:

**Theorem 1.** *For any flow, set $f$ to the true size of the flow, Then the size $\hat{f}$ estimated by DAP satisfies the following formula with a probability of at least $1 - \delta$:*

$$\hat{f} \leq f + \epsilon N_S < f + \epsilon N$$

*Where $N_S$ is the sum of counters in $S$, and $N$ is the number of all the packets.*

*Proof.* First, we prove that the worst case satisfies the above bound, that is, $f$ is in the S part and has never performed a replacement operation. In fact, $\hat{f}$ is the CM sketch's estimate of $f$. And the error of CM sketch satisfies $f < \hat{f} \leq f + \epsilon N_S$ [16], so the theory is established. The counters in the S where the replacement has occurred can be divided into two cases, one is to filter into the L, and the other is to be replaced by a large flow. For the former, the counter is cleared to 0, which satisfies the theory. For the latter, assume that the original counter value is $c_1$, and the value of the large flow that is replaced is $v_1$. The estimate of CM sketch for this case is $c_1 + v_1$, and we let $\hat{f} = v_1 \leq c_1 + v_1$, so it satisfies the theory. For the flow of L part, because each flow has a counter exclusively, we think that its error is much smaller than the error of the CM sketch, and the longer the time a flow stays in the L, the smaller the error.

Next, we prove that our replacement operation can minimize the error caused by hash conflicts. Specifically, it is the replacement operations for S part counters that appear in our algorithm. We use the value of L to replace the counter in S directly. We emphasize that for several flows where a hash collision occurs in a counter of L, the error caused by returning the maximum value is much smaller than the error caused by the return sum. We have the following theory:

**Theorem 2.** *For CM sketch, if the value of each counter is no longer the sum of all the flows mapped to this counter, but the maximum, then we estimate the error for flow $f$ will satisfy the following formula with a probability of at least $1 - \delta$:*

$$f < \hat{f} < \epsilon N$$

*Proof.* This theorem can be demonstrated by a method similar to that shown by CM sketch [16], as long as the summation operation is replaced by the maximum operation. The expectation of sum is less than $f + \epsilon N$, and the expectation of the maximum is less than $\epsilon N$, from which it can be proved.

According to Theory 2, we can prove that directly replacing the value of the $S$ part counter with the amount of the large flow can reduce the error of the CM sketch. Because the value of the large flow is larger than the sum of the other flows of the $L$ counter, that is, the value of the large flow is equivalent to the maximum value of all flows.

# 4    dL-DAP

The $L$ part of the DAP needs to maintain a minimum while supporting fast finds and updates. These operations can be implemented efficiently using some data structures [17,18]. However, these data structures are complex and not suitable for implementation on hardware [1]. In this section, we introduce *d-Length Deterministic Admission Policy*, which is an variant of DAP that implements on hardware. The critical idea of dL-DAP is to use a hash table to maintain the mapping between flows and counters for fast lookups and updates. The key issue is that we need a strategy to resolve hash collisions, which is also required to maintain the minimum. We first briefly introduce common strategies for resolving hash conflicts in Sect. 4.1.

## 4.1    Hash Collision Resolution

The hash table uses the hash function to map keys to a bucket in the table. All the keys assigned to the same bucket have hash collisions. There are several strategies for resolving hash conflicts [19].

**Separate Chaining.** In the method known as separate chaining, each bucket is independent and has some sort of list of entries with the same index. The time for hash table operations is the time to find the bucket (which is constant) plus the time for the list operation.

**Open Addressing.** In another strategy, called open addressing, all entry records are stored in the bucket array itself. When a new entry has to be inserted, the buckets are examined, starting with the hashed-to slot and proceeding in some probe sequence, until an empty slot is found. When searching for an entry, the buckets are scanned in the same sequence, until either the target record is found, or an unused array slot is found, which indicates that there is no such key in the table.

## 4.2    Design of dL-DAP

dL-DAP uses a structure similar to a hash table to implement the L part of the DAP, enabling fast lookups and updates. To resolve hash conflicts, we can use separate chaining or open addressing strategy. However, based on actual conditions, we must impose certain restrictions on these two methods. If the length of the chaining list or buckets array is too large, the efficiency of the algorithm is significantly reduced. Therefore, we use $d$ to limit the range of the list or array of the two methods.

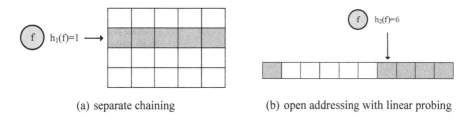

(a) separate chaining                    (b) open addressing with linear probing

**Fig. 3.** Different ways to resolve hash conflicts where $d = 5$.

dL-DAP implemented using separate chaining as shown in Fig. 3(a): all counters are organized into a two-dimensional array with the number of counters per row being $d$. Each packet is mapped to a row by a hash function. Both lookups and updates are made in this line. Similarly, the global minimum is replaced with the local minimum in this row, so that all operations can be performed linearly, and the execution speed is positively correlated with $d$. The disadvantage of using separate chaining is that the number of large flows mapped to each row is not uniform, resulting in a high large flow collision rate, which causes incorrect replacements. Using an open addressing strategy can alleviate this problem.

An example of using an open addressing strategy is shown in Fig. 3(b): all counters form a one-dimensional loop array, and each packet is mapped to one of its locations. In this figure, the remaining candidate positions are determined by a linear probing method, that is, starting from the hashed-to slot, linearly looking backward for d counters, all operations are performed on the d counters. Since the range of hash is expanded from $\frac{n}{d}$ to $n$, the large-flow hashes collision rate is reduced. We can use a more random distribution method to reduce further the collision rate of large flows, such as using quadratic probing or double hashing. However, the latter techniques destroy the principle of locality, so we use the open addressing method with linear probing. The comparison of these methods is shown in Fig. 4. We use counters from $2^7$ to $2^{12}$ to count the same number of flows, and the recall value is the ratio of the number of flows finally obtained to the total.

# 5   Evaluation

## 5.1   Experimental Setup

**Trace.** We use the CAIDA 2015 data set [20] as experimental data. This dataset contains anonymized passive traffic traces from CAIDA's Equinix-Chicago monitor on high-speed Internet backbone links. We divide the data set into 1M packets per part and extract the IP 5-tuple as the ID of the flow.

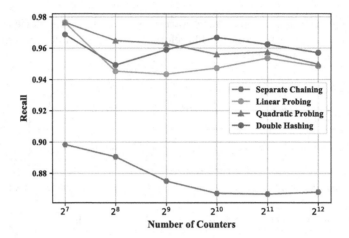

**Fig. 4.** Comparison of different strategies for solving hash conflicts.

**Metrics.** A flow that occupies a certain percentage of the overall packets is a large flow, such as a flow that accounts for more than 1%, 0.5%, or 0.01% of the total flow. Define the number of flows that satisfy such conditions as $K$. Let the number of returning large flows be $N$, where the number of real large streams is $k$. We compared the following metrics:

- Recall Rate (RR): $RR = \frac{k}{K}$
- Precision Rate (PR): $PR = \frac{k}{N}$
- $F_1$ score: $F_1 = \frac{2 \times RR \times PR}{RR + PR}$
- Mean Error Square (MSE): $MSE = \frac{1}{N} \sum_{i=1}^{n} (\hat{f}_i - f_i)^2$
- Replacement Number (RN): the number of times the replace operation occurs

### 5.2  Experimental Result

**Large Flow Detection.** First, we verify the validity of DAP by comparing with RAP and Elastic sketch. We set the size of the L part counter to 16 bits and the S part to 8 bits. We test the RR, PR, and $F_1$ scores for large-flow detection when the number of counters in the large flow part is from $2^7$ to $2^{12}$. The small flow part of DAP and Elastic Sketch is sized to be the same as the large flow part. The results of setting the large flow threshold to 0.1%, 0.05% and 0.01% are shown in Fig. 5.

When the threshold is 0.1%, we can find that even if only a small number of counters DAP are used, the RR value can still reach 100%, which is higher than the other two algorithms. This is due to the deterministic characteristics of DAP. As long as the flow is large and located in the S, it's counter value will exceed the minimum of the L, and it is very likely to be detected. However, when the number of counters is small, the hash collision rate of the S portion is also high. This leads to a false positive phenomenon. Many small flows add

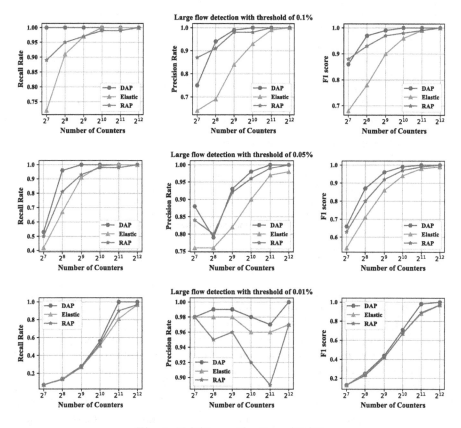

**Fig. 5.** Validity verification of DAP

up to the value of the large flow. This results in a lower PR value for the DAP, but in most cases, the PR value of the DAP is still better than the other two algorithms. The results of the $F_1$ score also prove this: DAP is optimal in most cases, only slightly worse than RAP when the number of counters is $2^7$ because the number of counters is insufficient and the hash collision rate is high.

In the case of the threshold is 0.5% and 0.01%, the RR values of three algorithms are not high when the number of counters is minimal, such as the threshold is 0.5% with the number of counters is $2^7$, the threshold is 0.01% with the number of counters less than $2^{11}$. This is because the number of large flows caused by the small threshold is more than the number of counters. We use this situation to test which algorithm performs best when space is relatively limited. The results show that the DAP algorithm outperforms the other two algorithms with all parameters, which proves the validity of our algorithm.

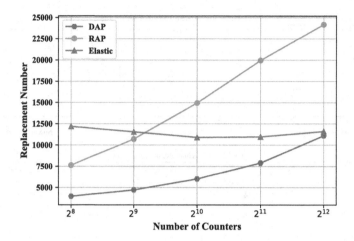

**Fig. 6.** Replacement number of different algorithms.

**Replacement Strategy Effectiveness.** As shown in Fig. 6, we also compared the number of replacement operations during the execution of different algorithms. Since the replacement operation is often the most complex operation in each case, this can reflect the time overhead of the algorithm to some extent. It also shows the effectiveness of the replacement: if the replacement operation is accurate, then there is no need to replace it again to correct the error, and also ensure the measurement accuracy. The results show that our algorithm performs replacement operations much less often than other algorithms, further demonstrating the effectiveness of deterministic over random. At the same time, we noticed that although the Elastic sketch is replaced more times than DAP, its performance seems to be more stable. This is because the replacement operation of the Elastic sketch is only related to the number of packets and parameter $\lambda$, which leads to large measurement errors [2]. We will confirm this in the next experiment.

**Large Flow Accuracy of dL-DAP.** To further test our measurement algorithm, we compared *dL-DAP* and *dW-RAP* and *Elastic* sketch from the perspective of easy implementation and practical use. We set the $d$ of the three algorithms to 8. Since dW-RAP has no small flow part, we assign it twice the counter, and the final result considers the most significant half. The result is shown in Fig. 7. The experimental results show that even if the L part has only half of the counter, the RR value of our algorithm is still the highest, although the PR value is worse than dW-RAP. However, our algorithm is still optimal in most of the cases according to the more comprehensive evaluation criteria F1 score.

Finally, we compare the MSE of the top-K flow for the three algorithms, where K is the number of L parts. The result is shown in Fig. 8. It shows that our algorithm obtains more accurate measurements. We can see that when the number of counters is $2^8$, the MSE of 8L-DAP is still about 2.5 times lower than the one of 8L-RAP. In all situations, the MSE of 8L-DAP is over ten times lower than the one of Elastic sketch.

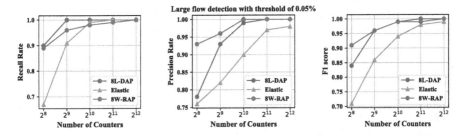

**Fig. 7.** Comparison of three algorithms in the case of $d = 8$.

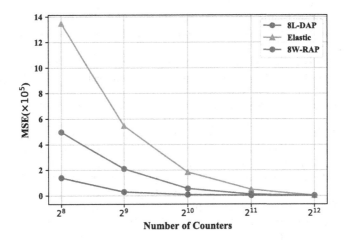

**Fig. 8.** Error comparison of three algorithms.

## 6   Conclusion

Accurate measurement of network traffic is an essential part of current network management tasks. Providing versatility support for a variety of network tasks is challenging under the hardware limitations of network devices. Traditional counter-based and sketch-based methods are subject to random operations and high hash collision rates, resulting in low accuracy. To solve this problem, we propose an efficient measurement architecture and offer a deterministic algorithm DAP for detecting large flows, thus achieving accurate measurements with less overhead. Also, we have introduced a hardware friendly algorithm dL-DAP replacing the global optionality with the local optionality. Experiments show that our algorithm significantly reduces the error compared to other algorithms and improves the accuracy of the measurement.

**Acknowledgments.** This research is sponsored by National Key R&D Program of China (2017YFB0902600); State Grid Corporation of China Project (SGJS0000DKJS1700840) Research and Application of Key Technology for Intelligent Dispatching and Security Early-warning of Large Power Grid.

# References

1. Basat, R.B., Einziger, G., Friedman, R., Kassner, Y.: Randomized admission policy for efficient top-k and frequency estimation. In: IEEE INFOCOM Conference on Computer Communications, pp. 1–9. IEEE (2017)
2. Yang, T., et al.: Elastic sketch: adaptive and fast network-wide measurements. In: Proceedings of the Conference of the ACM Special Interest Group on Data Communication, pp. 561–575. ACM (2018)
3. Huang, Q., Lee, P.P.C., Bao, Y.: Sketchlearn: relieving user burdens in approximate measurement with automated statistical inference. In: Proceedings of the Conference of the ACM Special Interest Group on Data Communication, pp. 576–590. ACM (2018)
4. Huang, Q.: Sketchvisor: robust network measurement for software packet processing. In: Proceedings of the Conference of the ACM Special Interest Group on Data Communication, pp. 113–126. ACM (2017)
5. Liu, Z., Manousis, A., Vorsanger, G., Sekar, V., Braverman, V.: One sketch to rule them all: rethinking network flow monitoring with UnivMon. In: Proceedings of the ACM SIGCOMM Conference, pp. 101–114. ACM (2016)
6. Yu, M., Jose, L.,Miao, R.: Software defined traffic measurement with OpenSketch. In: Presented as part of the 10th USENIX Symposium on Networked Systems Design and Implementation (NSDI 2013), pp. 29–42 (2013)
7. Zhou, Y., Zhou, Y., Chen, S., Zhang, Y.: Highly compact virtual active counters for per-flow traffic measurement. In: IEEE INFOCOM Conference on Computer Communications, pp. 1–9. IEEE (2018)
8. Assaf, E., Basat, R.B., Einziger, G., Friedman, R.: Pay for a sliding bloom filter and get counting, distinct elements, and entropy for free. In: IEEE INFOCOM Conference on Computer Communications, pp. 2204–2212. IEEE (2018)
9. Xiwen, Y., Hongli, X., Yao, D., Wang, H., Huang, L.: CountMax: a lightweight and cooperative sketch measurement for software-defined networks. IEEE/ACM Trans. Netw. (TON) **26**(6), 2774–2786 (2018)
10. Basat, R.B., Einziger, G., Friedman, R., Kassner, Y.: Optimal elephant flow detection. In: IEEE INFOCOM Conference on Computer Communications, pp. 1–9. IEEE (2017)
11. Basat, R.B., Einziger, G., Friedman, R., Kassner, Y.: Heavy hitters in streams and sliding windows. In: IEEE INFOCOM - The 35th Annual IEEE International Conference on Computer Communications, pp. 1–9. IEEE (2016)
12. Nyang, D.H., Shin, D.O.: Recyclable counter with confinement for real-time per-flow measurement. IEEE/ACM Trans. Netw. (TON) **24**(5), 3191–3203 (2016)

13. Demaine, E.D., López-Ortiz, A., Munro, J.I.: Frequency estimation of internet packet streams with limited space. In: Möhring, R., Raman, R. (eds.) ESA 2002. LNCS, vol. 2461, pp. 348–360. Springer, Heidelberg (2002). https://doi.org/10.1007/3-540-45749-6_33
14. Manku, G.S., Motwani, R.: Approximate frequency counts over data streams. In: VLDB 2002: Proceedings of the 28th International Conference on Very Large Databases, pp. 346–357. Elsevier (2002)
15. Metwally, A., Agrawal, D., El Abbadi, A.: Efficient computation of frequent and top-k elements in data streams. In: Eiter, T., Libkin, L. (eds.) ICDT 2005. LNCS, vol. 3363, pp. 398–412. Springer, Heidelberg (2004). https://doi.org/10.1007/978-3-540-30570-5_27
16. Cormode, G., Muthukrishnan, S.: An improved data stream summary: the count-min sketch and its applications. J. Algorithms **55**(1), 58–75 (2005)
17. Einziger, G., Friedman, R.: Tinyset–an access efficient self adjusting bloom filter construction. IEEE/ACM Trans. Netw. **25**(4), 2295–2307 (2017)
18. Einziger, G., Friedman, R.: Counting with tinytable: every bit countscounting with tinytable: every bit counts! IEEE Access (2019)
19. Hash table. https://en.wikipedia.org/wiki/Hash_table
20. The CAIDA UCSD anonymized internet traces 2015 - February 19th. http://www.caida.org/data/passive/passive_dataset.xml

# A Comparison Study of VAE and GAN for Software Fault Prediction

Yuanyuan Sun[1,2,3](✉), Lele Xu[3](✉), Lili Guo[3], Ye Li[3], and Yongming Wang[2](✉)

[1] School of Cyber Security, University of Chinese Academy of Sciences, Beijing, China
sunyuanyuan@csu.ac.cn
[2] Institute of Information Engineering, Chinese Academy of Sciences, Beijing, China
wangyongming@iie.ac.cn
[3] Key Laboratory of Space Utilization, Technology and Engineering Center for Space Utilization, Chinese Academy of Sciences, Beijing, China
xulele@csu.ac.cn

**Abstract.** Software fault is an unavoidable problem in software project. How to predict software fault to enhance safety and reliability of system is worth studying. In recent years, deep learning has been widely used in the fields of image, text and voice. However it is seldom applied in the field of software fault prediction. Considering the ability of deep learning, we select the deep learning techniques of VAE and GAN for software fault prediction and compare the performance of them. There is one salient feature of software fault data. The proportion of non-fault data is well above the proportion of fault data. Because of the imbalanced data, it is difficult to get high accuracy to predict software fault. As we known, VAE and GAN are able to generate synthetic samples that obey the distribution of real data. We try to take advantage of their power to generate new fault samples in order to improve the accuracy of software fault prediction. The architectures of VAE and GAN are designed to fit for the high dimensional software fault data. New software fault samples are generated to balance the software fault datasets in order to get better performance for software fault prediction. The models of VAE and GAN are trained on GPU TITAN X. SMOTE is also adopted in order to compare the performance with VAE and GAN. The results in the experiment show that VAE and GAN are useful techniques for software fault prediction and VAE has better performance than GAN on this issue.

**Keywords:** Deep learning · VAE · GAN · Software fault prediction

## 1 Introduction

In software project, software fault is an inescapable problem. Software fault may be incurred by internal defects of software or external attacks. Many cases show that software fault can cause huge loss and catastrophic consequences. For example, in 1962, the famous software fault resulted in the failure of Mariner rocket to Venus.

© Springer Nature Switzerland AG 2020
S. Wen et al. (Eds.): ICA3PP 2019, LNCS 11945, pp. 82–96, 2020.
https://doi.org/10.1007/978-3-030-38961-1_8

In 2003, the blackouts of the Northeastern United States were also because of software fault. In 2009, attackers launched offensive to the video software. This caused extensive software fault that people of 6 provinces in China could not access internet.

Software fault is closely related to security, reliability, maintainability of system [1]. Especially for high-risk system, software fault can lead to serious consequences. In software project, it is quite difficult for testers to find all the software faults. Researchers focus on software fault prediction, which can help tester estimate the number and distribution of fault reasonably. Researchers have studied on the metrics which are used to represent attributes of software. These attributes are quite helpful for software fault prediction, which can be used as features to predict software fault. The classical metrics include LOC count, McCabe [2, 3] and Halstead [2, 4].

Machine learning is always used in software fault prediction. 22 classifiers based on machine learning were used for software fault prediction in [5]. L Kumar set up the model of Least Square Support Vector Machine (LSSVM) for software fault prediction [6]. DR Ibrahim used random forest based on improved feature for software fault prediction [7]. An approach of decision tree for software fault prediction was proposed by Rathore [8]. Logistic Regression was compared with decision tree to enhance the result of software fault prediction [9].

Though these machine learning techniques are applied for software fault prediction, an important problem of software fault is ignored. That is imbalanced data [10]. Taking software fault for example, the amount of non-fault data (majority) is always well above the amount of fault data (minority) in software project. Especially, the fault data (minority) will always be predicted to the non-fault data (majority). How to resolve the difficulty of imbalanced data? There are usually two kinds of ways to deal with this problem. They are under-sampling [11] and over-sampling [12]. Under-sampling random reduces the amount of majority to balance the class of majority and minority. But it will bring out useful information loss. Usually, over-sampling is adopted. SMOTE (synthetic minority over-sampling technique) [13] is a famous over-sampling technique, which is very useful to resolve the problem of imbalanced data.

Can we utilize deep learning techniques for software fault prediction? In recent years, deep learning is widely used in many fields, such as image recognition, natural language processing [14] and voice recognition. It has achieved a resounding success. While up to present, it is seldom applied in the domain of software fault prediction.

It can be found that most applications of Variational Autoencoder(VAE) are used for image processing [15, 16]. Some of them are used for text generating [17]. Similarly, GAN (Generative Adversarial Networks) is always used for images [18, 19]. The framework of VAE is a generative model [20]. The framework of GAN is combined by a generative model and a discriminative model [21]. Both VAE and GAN have the ability to generate new synthetic samples which obey the distribution of real data.

Few researches involve deep learning techniques for software fault prediction. Here, we have the inspiration of utilizing the ability of generating synthetic samples of VAE and GAN to generate new fault samples. The new samples can be used to balance the class. In our previous work [22], we adopted VAE for software fault prediction and compared its performance with no-sampling method. In this paper, furthermore, both VAE and GAN are used and compared for software fault prediction. As we known, GAN has better ability to generate new image samples compared to VAE [23].

Intuitively, we get the idea of adopting both VAE and GAN to deal with the issue of imbalanced data and finding out which one is better in software fault prediction. SMOTE is also used in order to compare the performance with VAE and GAN.

In this paper, we find that deep learning techniques of VAE and GAN are useful in the field of software fault prediction. VAE has better performance than GAN and SMOTE. GAN outperforms SMOTE on some datasets.

The main contributions in this paper are as follows:

– Software fault data are multivariable data which are different with image data. The models of VAE and GAN are designed to fit for this type of data. The deep architectures of VAE and GAN are realized on GPU TITAN X by the framework of Keras.
– As far as we know, it is the first time we do research on both VAE and GAN for software fault prediction and the performance of VAE, GAN and SMOTE are compared. The results of experiment not only demonstrate that VAE and GAN are useful for software fault prediction, but also show that VAE outperforms GAN and SMOTE. It can be inferred that VAE has better ability than GAN on generating multivariable data of software fault, though GAN has better performance than VAE for generating image.

The rest part of the paper is structured as follows: in Sect. 2, the background knowledge is described; in Sect. 3, the methods of experiment are demonstrated; in Sect. 4, the results of experiment are given out; a conclusion is drawn in Sect. 5.

## 2 Background

### 2.1 Variational Autoencoder (VAE)

In 2014, Kingma proposed the theory of VAE. VAE is the theory that combines statistics learning and deep learning techniques [20]. VAE can generate new samples which obey the probability distribution of $Z$. Assuming $Z$ is subject to Gaussian distribution $p(Z)$. Random sampling $Z$ from $p(Z)$, new samples can be created on the basis of $p(Z/X)$. Within VAE model, assuming $p(Z/X)$ is subject to normal distribution. Supposing the input is $Xk$. $Xk$ obeys distribution of $p(Z|Xk)$. A generator, $G = g(Z)$, is trained. $Gk$ can be generated by sampling $Z$ from $p(Z|Xk)$.

The mathematic theory of VAE is complicated, while its realization is not hard to understand in engineering [24]. The implementation of VAE is shown in Fig. 1. VAE is combined with an encoder and a decoder (generator). The input data enter the encoder, and then the encoder outputs the latent variable's mean and logarithmic variance. After that, the outputs of encoder are transformed to obey standard normal distribution. It is implemented by formula (1) and (2). By sampling $\varepsilon$ from the distribution of $N(0, 1)$, $Z$ is acquired. The model of VAE is trained to minimize the loss of KL divergence. The VAE network can be trained by Stochastic Gradient Descent (SGD). $Gk$ is the generated data by decoder (generator).

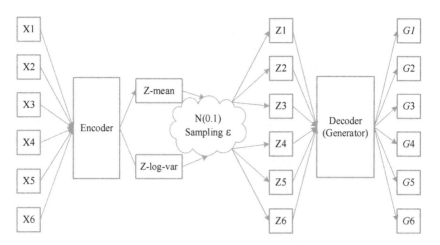

**Fig. 1.** The realization diagram of VAE

$$\varepsilon = (z - u)/\sigma \tag{1}$$

$$Z = \mu + \varepsilon \times \sigma \tag{2}$$

## 2.2 Generative Adversarial Networks (GAN)

Generative Adversarial Networks (GAN) was proposed in 2014 by Goodfellow [21]. GAN is a hot topic in recent years. It is widely used in the fields of image translation, Super-Resolution and semantic segmentation etc. The basic structure of GAN is illustrated in Fig. 2. GAN contains two parts. One is the generator G, the other is the discriminator D. The generator learns the distribution of real samples. Random noise is the input of the generator. The generator can utilize both random noise and the real sample's distribution to produce fake samples in order to simulate real samples. Both real samples and fake samples go into the discriminator. The discriminator tries to determine the input is real or fake. In short, the generator can be seen as a team of counterfeiter who tries to make fake currency, and use it freely without being found. The discriminator can be seen as police who tries to find the fake currency made by counterfeiters. The generator tries to cheat the discriminator and the discriminator tries to see through the fraud.

In the paper of Goodfellow, the generator and the discriminator are composed of multilayer perceptrons. The objective function can be seen in Eq. (3).

$$\min_{G} \max_{D} V(D, G) := E_{x \sim px}[\log D(x)] + E_{x \sim pg}[\log(1 - D(G(z)))] \tag{3}$$

The output of D is a single scalar, $D(x)$ is the probability of denoting $x$ from real samples. D is trained to maximize the probability of giving correct label to real sample and fake sample. G is also trained simultaneously to minimize $\log(1 - D(G(z)))$. The

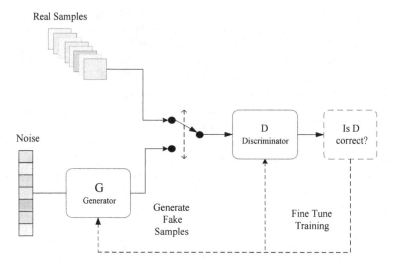

**Fig. 2.** Basic structure of GAN: the noise in latent space is the input of G. G generates fake samples. The real samples and fake samples go into D respectively. D will determine the sample is real or fake. The determination of D will be compared with the ground truth. The result of comparison will be sent back to G and D. G and D begin to adjust the parameters of networks by fine tune training.

generator and the discriminator compete with each other. This is a problem of min-max game. At last the generator and the discriminator reach Nash equilibrium.

### 2.3    Synthetic Minority Over-Sampling Technique (SMOTE)

SMOTE (Synthetic Minority Over-sampling Technique) was presented in 2002 by NV Chawla. It is the improvement of random over-sample. Random over-sample just increases samples by copying original samples. This always brings out the problem of poor generalization. SMOTE can improve the generalization ability. It can analyze the minority and generate synthetic samples for minority. In fact, the core of the technique is based on the idea of interpolation. The realization of SMOTE is as follows:

- Given a sample $x_i$ in minority, $i \in \{1, \ldots, T\}$, $T$ is the amount of samples in minority; Computing the Euclidean distance to each sample in the set of minority, $k$ neighbors are achieved. $x_{i(near)}$, $near \in (1, \ldots, k)$
- A sample $x_{i(nn)}$ from $k$ neighbors is chosen randomly. New sample is synthetized by the following formula.

$$x_{i1} = x_i + \zeta \cdot (x_{i(nn)} - x_i), \quad \zeta \in (0, 1) \tag{4}$$

- Repeating N times, N new samples are generated from sample $x_i$. $x_{inew}$, $new \in 1, \ldots N$.

# 3 Experimental Methodology

The experimental methodology is demonstrated in this section. As can be seen from the flow chart of Fig. 3, the main idea of the experiment is to balance the software fault data by the methods of VAE, GAN and SMOTE, which are used to generate synthetic fault samples to increase the amount of samples for minority. New fault samples generated by different methods will be added into original software fault data respectively. This can make the amount of fault samples approach the amount of non-fault samples. The flow chart of experiment will be explained in further details.

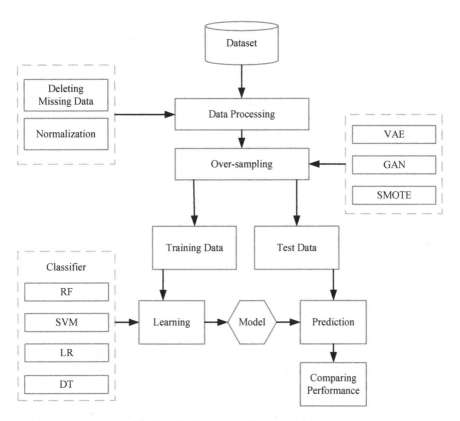

**Fig. 3.** The flow chart of the experiment

At beginning, data are processed. Data processing includes deleting missing data and normalization. And then the models of VAE and GAN are trained to generate new fault samples. SMOTE is also adopted to generate synthetic fault samples. The samples generated by VAE, GAN and SMOTE are added into original fault data (minority) respectively. These methods are called "VAE", "GAN" and "SMOTE" in the flow chart of Fig. 3. Four classifiers, such as RF (Random Forrest), SVM (Support Vector Machine), LR (Logistic Regression) and DT (Decision Tree) are adopted to get the results of software fault prediction. The measures of AUC, MCC, recall and F1-measure are selected to evaluate the results of classifiers. The performance is compared between the methods of VAE, GAN and SMOTE. The results of VAE and GAN in this experiment are also compared with the results of paper [5]. The experiment is implemented on the GPU of TITAN X. The runtime environment is as follows:

- Python 3.6
- Keras 2.1.6
- Tensorflow 1.4.1.

## 3.1   Data Processing

Data processing is the premise of the experiment. After deleting missing data, normalization is carried out in the process of data processing. Normalization has the ability to make values of different dimension to the same scope. Min-Max scaling and Z-score are classical techniques for normalization. Min-Max scaling is always adopted in neural network. In this paper, VAE and GAN are designed by the frameworks of MLP. We choose Min-Max scaling for normalization. The transformation of Min-Max scaling can be realized by the following formula (5).

$$Z = \frac{x_i - Min(x_i)}{Max(x_i) - Min(x_i)} \tag{5}$$

As for the missing data, there are several ways to deal with them. For example, let missing data be 0, 1 or the means of feature. Here, in order to reduce uncertainty, the missing instances are deleted. This is done by Pandas.

## 3.2   The Design of VAE

It can be seen from Table 1, the architecture of VAE is designed by MLP (multilayer perceptron).The dimension of input data for encoder is the number of code attributes of software fault data. In this experiment, the dimension of input is 21. The number of neurons of hidden layer is set to 100. In fact, in the process of training, it can be found that when we set the number of neuron of hidden layer to 100, the value of loss reduced rapidly. The dimension for output of encoder is 2. The output is mean and logarithmic variance of latent variable. As for the decoder (generator), the number of neurons of the first dense is also 100, and the output is fault instance of simulation. The dimension of the generator's output is also 21. RMSProp is selected as the optimizer for the model of VAE. The loss function of the model is KL divergence.

**Table 1.** The architecture of VAE

| Structure | Units | Non linearity | Dropout |
|-----------|-------|---------------|---------|
| Encoder |  |  |  |
| Dense | 100 | Tanh | 0 |
| Dense | 2 | Linear | 0 |
| Decoder (Generator) |  |  |  |
| Dense | 100 | Tanh | 0 |
| Dense | 21 | sigmoid | 0 |

## 3.3   The Design of GAN

The most difficult problem in the experiment is training the module of GAN. MLP is selected to set up generator and discriminator. After several times of failed attempt, a successful model is achieved for software fault data. The architecture of GAN can be seen in Table 2. The loss curves of generator and discriminator are as expected, which are shown in Fig. 4. Adam is selected as the optimizer for the model of GAN. The Leaky ReLU slope is set to 0.2.

**Table 2.** The architecture of GAN

| Structure | Units | Non linearity | Dropout |
|-----------|-------|---------------|---------|
| Generator G(z) |  |  |  |
| Dense | 42 | Leaky Relu | 0 |
| Dense | 42 | Leaky Relu | 0 |
| Dense | 42 | Leaky Relu | 0 |
| Dense | 21 | Tanh | 0 |
| Discriminator D(x) |  |  |  |
| Dense | 42 | Leaky Relu | 0 |
| Dense | 21 | Leaky Relu | 0 |
| Dense | 1 | sigmoid | 0 |

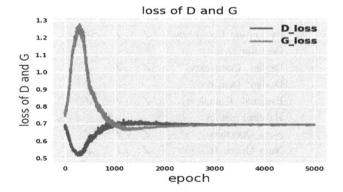

**Fig. 4.** The loss curves of generator and discriminator

**3.4   Evaluation**

In the experiment, AUC, MCC, recall and F1-measure are selected to evaluate the results of the experiment. AUC is the area under ROC (ROC represents receiver operating characteristic curve).

$$recall = \frac{TP}{TP + FN} \tag{6}$$

$$F1 = \frac{2 \times (recall \times precision)}{recall + precision} \tag{7}$$

$$MCC = \frac{(TP \times TN) - (FP \times FN)}{\sqrt{(TP \times FP) + (TP \times FN) + (TN \times FP) + (TN \times FN)}} \tag{8}$$

MCC: Matthews correlation coefficient;
F-measure: the harmonic mean of recall and precision;
TP: True Positives; FP: False Positives;
TN: True Negatives; FN: False negatives.

## 4   The Results of Experiment

In the experiment, three datasets of JM1, KC1 and KC2 are selected from Promise Repository. The datasets are public and can be downloaded from the Internet [25]. The three datasets are about software fault of spaceflight from NASA. In the three datasets, the number of code attributes is 21. The code attributes include the metrics of McCabe, Halstead, LOC and Miscellaneous which can be found in Table 3.

**Table 3.**  The metrics in JM1, KC1, and KC2

|  | JM1 | KC1 | KC2 |
|---|---|---|---|
| McCabe metrics |  |  |  |
| Cyclomatic_Complexty | ✓ | ✓ | ✓ |
| Decision_Density |  |  |  |
| Design_ Complexty | ✓ | ✓ | ✓ |
| Design_ Density |  |  |  |
| Essential_ Complexty | ✓ | ✓ | ✓ |
| Halstead metrics |  |  |  |
| Num_Operators | ✓ | ✓ | ✓ |
| Num_Operands | ✓ | ✓ | ✓ |
| Num_Uniq_ Operands | ✓ | ✓ | ✓ |
| Num_Uniq_ Operators | ✓ | ✓ | ✓ |

*(continued)*

**Table 3.** (*continued*)

|  | JM1 | KC1 | KC2 |
|---|---|---|---|
| Length | ✓ | ✓ | ✓ |
| Volume | ✓ | ✓ | ✓ |
| Level | ✓ | ✓ | ✓ |
| Difficulty | ✓ | ✓ | ✓ |
| Content | ✓ | ✓ | ✓ |
| Error_Estimate | ✓ | ✓ | ✓ |
| Programming_time | ✓ | ✓ | ✓ |
| Programming Effort | ✓ | ✓ | ✓ |
| LOC based metrics |  |  |  |
| LOC_Total | ✓ | ✓ | ✓ |
| LOC_comments | ✓ | ✓ | ✓ |
| LOC_executalble | ✓ | ✓ | ✓ |
| LOC_Blank | ✓ | ✓ | ✓ |
| LOC_Code_and_comment | ✓ | ✓ | ✓ |
| Number_of_lines | ✓ | ✓ | ✓ |
| Miscellaneous |  |  |  |
| Branch_count | ✓ | ✓ | ✓ |
| Number of code attributes | 21 | 21 | 21 |

**Table 4.** Data in the experiment

| NASA software fault datasets | Original data | | VAE/GAN/SMOTE | |
|---|---|---|---|---|
|  | *Normal* | *Anomaly* | *Normal* | *Anomaly* |
| JM1 | 8777 | 2103 | 8777 | 8424 |
| KC1 | 1783 | 326 | 1783 | 1630 |
| KC2 | 415 | 107 | 415 | 400 |

The amount of normal and anomaly instances used by the methods of VAE, GAN, and SMOTE can be seen in in Table 4. Anomaly instances represent fault data which belong to the minority. We focus on the performance of different methods for fault prediction in the experiment. The measures of AUC, MCC, recall and F1-measure are compared. In the three datasets, 90% of data are used for training set and 10% of data are used for test set. Three methods including VAE, GAN and SMOTE are adopted in the experiment. The results of AUC and MCC are shown in Table 5.

For JM1, in the three methods, VAE has the best AUC and MCC by classifier of Random Forest. The values of AUC and MCC are 0.92 and 0.78 respectively. The best AUC and MCC of GAN are 0.92 and 0.77 by classifier of Random Forest, which are higher than the best AUC and MCC of SMOTE. The best AUC and MCC of SMOTE are 0.89 and 0.73 respectively.

For KC1, in the three methods, VAE has the best AUC and MCC by classifier of Random Forest. The values of AUC and MCC are 0.94 and 0.81 respectively. The best AUC and MCC of GAN are 0.87 and 0.65 by classifier of SVM, which are lower than the best AUC and MCC of SMOTE. The best AUC and MCC of SMOTE are 0.88 and 0.69 respectively.

For KC2, in the three methods, VAE has the best AUC and MCC by classifier of Logistic Regression. The values of AUC and MCC are 0.94 and 0.83 respectively. The best AUC and MCC of GAN are 0.93 and 0.78 by classifier of Random Forest, which are higher than the best AUC and MCC of SMOTE. The best AUC and MCC of SMOTE are 0.92 and 0.76 respectively.

**Table 5.** AUC and MCC of VAE, GAN and SMOTE

| | JM1 AUC | JM1 MCC | KC1 AUC | KC1 MCC | KC2 AUC | KC2 MCC |
|---|---|---|---|---|---|---|
| VAE | | | | | | |
| RF | **0.92** | **0.78** | **0.94** | **0.81** | 0.92 | 0.74 |
| SVM | 0.87 | 0.68 | 0.87 | 0.67 | 0.94 | 0.80 |
| LR | 0.87 | 0.67 | 0.86 | 0.62 | **0.94** | **0.83** |
| DT | 0.88 | 0.71 | 0.92 | 0.76 | 0.93 | 0.79 |
| GAN | | | | | | |
| RF | **0.92** | 0.77 | 0.84 | 0.59 | 0.93 | 0.78 |
| SVM | 0.89 | 0.72 | 0.87 | 0.65 | 0.88 | 0.635 |
| LR | 0.88 | 0.69 | 0.85 | 0.59 | 0.88 | 0.614 |
| DT | 0.88 | 0.69 | 0.80 | 0.47 | 0.90 | 0.680 |
| SMOTE | | | | | | |
| RF | 0.89 | 0.73 | 0.88 | 0.69 | 0.92 | 0.76 |
| SVM | 0.71 | 0.27 | 0.72 | 0.27 | 0.92 | 0.757 |
| LR | 0.71 | 0.27 | 0.71 | 0.24 | 0.91 | 0.707 |
| DT | 0.86 | 0.65 | 0.85 | 0.60 | 0.89 | 0.681 |

The comparison of the best average recall in the three datasets by the methods of VAE, GAN and SMOTE can be seen in Fig. 5. For JM1, the recall of VAE is 0.89, which is the highest in the three methods. The recall of GAN is 0.88 and it is higher than that of SMOTE. The recall of SMOTE is 0.86. For KC1, the recall of VAE is 0.90, which is the highest in the three methods. The recall of GAN is 0.83 and it is lower than that of SMOTE. The recall of SMOTE is 0.85. For KC2, the recall of VAE is 0.92, which is the highest in the three methods. The recall of GAN is 0.89 and it is higher than that of SMOTE. The recall of SMOTE is 0.88.

The comparison of the best average F1-measure in the three datasets by the methods of VAE, GAN and SMOTE can be seen in Fig. 6. It is the same with the comparison of recall. VAE has the highest F1-measure in the three methods. GAN has better F1-measure than that of SMOTE on the datasets of JM1 and KC2.

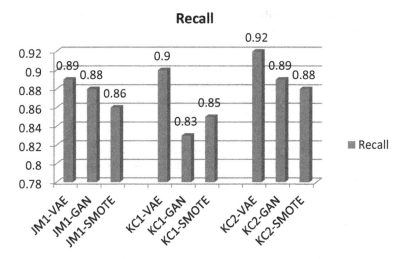

**Fig. 5.** The best average recall on three datasets by VAE, GAN and SMOTE

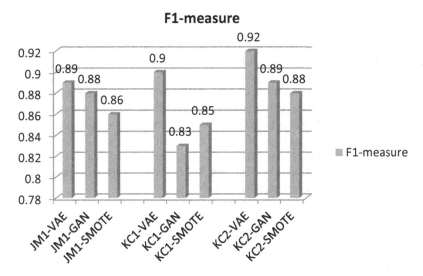

**Fig. 6.** The best average Fl-Measure on three datasets by VAE, GAN and SMOTE

The results of AUC of two datasets including JM1 and KC1 can be found in paper [5]. We compare the best AUC in paper [5] with the best AUC of VAE and GAN in this experiment. Comparison is shown in Table 6. It can be found that the methods of VAE and GAN have higher AUC than that of the best results in paper [5].

From comparison of the experiment, we find that VAE has better performance than GAN and SMOTE on the three datasets of JM1, KC1 and KC2. GAN has better performance than SMOTE on the datasets of JM1 and KC2. The best AUC of VAE and GAN outperform the best AUC acquired in paper [5]. It can be said that VAE and GAN

**Table 6.** Comparison of AUC

|          | JM1 AUC | KC1 AUC |
|----------|---------|---------|
| VAE      | **0.92** | **0.94** |
| GAN      | **0.92** | 0.87 |
| Paper [5] | 0.76   | 0.78 |

are useful methods for software fault prediction. Compared with VAE, GAN usually has better ability to generate image samples. While in this experiment, VAE outperforms GAN for generating software fault samples.

## 5  Conclusion

In this paper, we utilize deep learning techniques of VAE and GAN for software fault prediction and compare the performance of them. The architectures of VAE and GAN are designed to fit for the multivariable data of software fault. The ability of VAE and GAN to generate new fault samples is used to balance the normal and anomaly class. An experiment is implemented to verify the scheme we proposed. Typical datasets of JM1, KC1 and KC2 are selected, which are from NASA's software projects of spaceflight. Four classifiers are used for the experiment. We find that the scheme of VAE has better performance than the schemes of GAN and SMOTE for software fault prediction. Though GAN usually has better ability to generate image compared with VAE, it does not have better performance than VAE for generating software fault data in this experiment. GAN has better performance than SMOTE on the datasets of JM1 and KC2. Comparing the results of VAE and GAN with the results in paper [5], it can be found that VAE and GAN have better AUC. It can be inferred that it is practicable to apply deep learning techniques of VAE and GAN for software fault prediction, and VAE has better performance compared to GAN.

**Acknowledgement.** This work is supported by the National Natural Science Foundation of China (No. 61901454), and the Foundation of key Laboratory of Space Utilization, Technology and Engineering Center for Space utilization Chinese Academy of Sciences (No. CSU-QZKT-2018-08).

## References

1. Sharma, D., Chandra, P.: Software fault prediction using machine-learning techniques. Smart Comput. Inform. **78**, 541–549 (2018)
2. Curtis, B.: Measuring the psychological complexity of software maintenance tasks with the halstead and McCabe metrics. IEEE Trans. Softw. Eng. SE **5**(2), 96–104 (1979)
3. Yahya, N., Bakar, N.S.A.A.: McCabe's complexity and CK metrics on the internal quality of test first implementation in Malaysian education settings. Adv. Sci. Lett. **24**(2), 1201–1205 (2018)

4. Bailey, C.T., Dingee, W.L.: A software study using Halstead metrics. In: ACM Workshop/symposium on Measurement and Evaluation of Software Quality, pp. 189–197 (1981)
5. Lessmann, S.: Benchmarking classification models for software defect prediction a proposed framework and novel findings. IEEE Trans. Softw. Eng. **34**(4), 485–496 (2008)
6. Zhang, P., Chang, Y.T.: Software fault prediction based on grey neural network (2012)
7. Kanmani, S.: Object-oriented software fault prediction using neural networks. Inf. Softw. Technol. **49**(5), 483–492 (2007)
8. Shanthini, A., Vinodhini, G., Chandrasekaran, R.M.: Bagged SVM classifier for software fault prediction. Int. J. Comput. Appl. **62**(15), 21–24 (2013)
9. Ibrahim, D.R., Ghnemat, R., Hudaib, A.: Software defect prediction using feature selection and random forest algorithm. In: International Conference on New Trends in Computing Sciences (2017)
10. He, H., Garcia, E.A.: Learning from imbalanced data. IEEE Trans. Knowl. Data Eng. **21**(9), 1263–1284 (2009)
11. Donoho, D.L., Tanner, J.: Precise undersampling theorems. Proc. IEEE **98**(6), 913–924 (2010)
12. Last, F., Douzas, G., Bacao, F.: Oversampling for imbalanced learning based on K-Means and SMOTE (2017)
13. Chawla, N.V., et al.: SMOTE: synthetic minority over-sampling technique. J. Artif. Intell. Res. **16**(1), 321–357 (2002)
14. Lecun, Y., Bengio, Y., Hinton, G.: Deep learning. Nature **521**(7553), 436 (2015)
15. Walker, J., Doersch, C., Gupta, A., Hebert, M.: An uncertain future: forecasting from static images using variational autoencoders. In: Leibe, B., Matas, J., Sebe, N., Welling, M. (eds.) ECCV 2016. LNCS, vol. 9911, pp. 835–851. Springer, Cham (2016). https://doi.org/10. 1007/978-3-319-46478-7_51
16. Simonovsky, M., Komodakis, N.: GraphVAE: towards generation of small graphs using variational autoencoders. In: Kůrková, V., Manolopoulos, Y., Hammer, B., Iliadis, L., Maglogiannis, I. (eds.) ICANN 2018. LNCS, vol. 11139, pp. 412–422. Springer, Cham (2018). https://doi.org/10.1007/978-3-030-01418-6_41
17. Semeniuta, S., Severyn, A., Barth, E.: A hybrid convolutional variational autoencoder for text generation (2017)
18. Ding, Z., et al.: TGAN: deep tensor generative adversarial nets for large image generation (2019)
19. Gurumurthy, S., Sarvadevabhatla, R.K., Babu, R.V.: DeLiGAN: generative adversarial networks for diverse and limited data. In: IEEE Conference on Computer Vision and Pattern Recognition. IEEE Computer Society (2017)
20. Kingma, D.P., Welling, M.: Auto-encoding variational bayes. In: Conference Proceedings: Papers Accepted to the International Conference on Learning Representations. arXiv.org (2014)
21. Goodfellow, I.J., et al.: Generative adversarial nets. In: International Conference on Neural Information Processing Systems (2014)
22. Sun, Y., Xu, L., Li, Y., et al.: Utilizing deep architecture networks of VAE in software fault prediction. In: 2018 IEEE International Conference on Parallel and Distributed Processing with Applications, Ubiquitous Computing and Communications, Big Data and Cloud Computing, Social Computing and Networking, Sustainable Computing and Communications (ISPA/IUCC/BDCloud/SocialCom/SustainCom), pp. 870–877. IEEE (2018)

23. Lesort, T., Stoian, A., Goudou, J.-F., Filliat, D.: Training discriminative models to evaluate generative ones. In: Tetko, I.V., Kůrková, V., Karpov, P., Theis, F. (eds.) ICANN 2019. LNCS, vol. 11729, pp. 604–619. Springer, Cham (2019). https://doi.org/10.1007/978-3-030-30508-6_48
24. SOHU. https://www.sohu.com/a/226209674_500659. Accessed 25 June 2019
25. Promise homepage. http://promise.site.uottawa.ca/SERepository/datasets-page.html. Accessed 25 June 2019

# A Framework for Designing Autonomous Parallel Data Warehouses

Soumia Benkrid[1]([⊠]) and Ladjel Bellatreche[2]([⊠])

[1] Ecole nationale Supérieure d'Informatique (ESI), Algiers, Algeria
s_benkrid@esi.dz
[2] LIAS/ISAE-ENSMA, Poitiers, France
bellatreche@ensma.fr

**Abstract.** Parallel data platforms are recognized as a key solution for processing analytical queries running on extremely large data warehouses (*DW*s). Deploying a *DW* on such platforms requires efficient data partitioning and allocation techniques. Most of these techniques assume a priori knowledge of workload. To deal with their evolution, *reactive strategies* are mainly used. The BI 2.0 requirements have put *large batch and ad-hoc user queries at the center*. Consequently, reactive-based solutions for deploying a *DW* in parallel platforms are not sufficient. Autonomous computing has emerged as a paradigm that allows digital objects managing themselves in accordance with high-level guidance by the means of proactive approaches. Being inspired by this paradigm, we propose in this paper, a proactive approach based on a query clustering model to deploying a *DW* over a parallel platform. The query clustering triggers partitioning and allocation processes by considering only evolved query groups. Intensive experiments were conducted to show the efficiency of our proposal.

**Keywords:** Partitioning · Allocation · Common sub-expressions · Utility maximization · Autonomous system · Workload clustering

## 1 Introduction

With the invention of the Web 2.0, the traditional BI got impacted and automatically moved to BI 2.0, where the BI does not rely only on enterprise internal sources anymore but also makes heavy use of external data sources. One of the fundamental recommendations of BI 2.0 is to put the decision-makers at the center of the BI applications. This implies the development of supports and solutions covering software, hardware and platforms to deal with a large set of ad-hoc queries issued by these users in a batch fashion. This situation strongly influences the different design phases of a *DW*. The main particularity of these phases is that their algorithms are driven by two main entries representing (1) *the functional and non-functional requirements* and (2) *the data sources.*

© Springer Nature Switzerland AG 2020
S. Wen et al. (Eds.): ICA3PP 2019, LNCS 11945, pp. 97–104, 2020.
https://doi.org/10.1007/978-3-030-38961-1_9

The analysis of traditional *DW* design state-of-art shows that the large majority of the proposed solutions dedicated to all design phases are based on the KDK principle: *Knowing that Designers Know* the two entries. This principle is an instantiation of the *Knowing that One Knows thesis* [7]. The result of the design following this principle is static and cannot be reproduced in the context of BI 2.0 that needs adaptive design solutions. To illustrate this point, we consider in this paper the deployment phase which aims to find the adequate platform to deploy a *DW*. Once the platform is chosen, complex processes including data partitioning and fragment allocation have to be performed accompanied by query processing policy. Traditional parallel *DW* design problem is formalized as a Constraint Optimization Problem [1], this formalization follows the KDK principle. When the parallel *DW* does not meet the fixed non-functional requirements, reactive adaptation strategies are triggered. They are based on *adjustment techniques* using *Machine Learning* [5] or *utility-driven* techniques [9]. An extreme alternative is to forecast queries of the workload. The query forecasting has recently been studied in the context of self-driving DBMS [10], data partitioning [4,5] and sub-common expression selection [9]. These studies have three main limitations: (i) they do not focus on parallel *DW* design, (ii) they do not deal with batch queries and (iii) they forecast a small set of queries.

To overcome the limitations of KDK and forecasting principles, we propose a new framework to design parallel *DW*, inspired by humans. It includes an off-line learning phase that plays the role of knowledge base and an on-line phase responsible to enable proactive adaptation of parallel *DW* design. This principle is largely used in designing autonomous systems [8]. Initial partitioning and allocation schemes are obtained by the training set of OLAP queries used by the off-line phase. By making the parallel to human daily where activities are clustered into several classes, the queries of the training set are clustered based on their similarity measured by the degree of their shared common expressions. Each cluster is used to partition the *DW*. This augments the chance that all queries participate in the partitioning process and most probably future queries can be classified into the clusters generated by the off-line phase. The updated clusters will trigger partitioning and allocation processes.

The paper is structured as follows: fundamental notions and related work are presented in Sect. 2. In Sect. 3, we describe our framework in designing autonomous parallel *DW*. Section 4 presents the algorithm for fragment allocation. Section 5 provides our experimental results. Finally, Sect. 6 summarizes the main findings of our research.

## 2  Background and Related Work

Given the complexity of the subject that covers several problems, we are obliged to mix the related works and the background.

**Common Expression Sharing between Queries.** Generally speaking, OLAP workloads are a windfall of query sharing. This is because a typical OLAP and reporting workloads are overlapping. This overlap can occur for several reasons [13]: (i) queries might overlap in the kind of analysis they perform since different queries could compute different aggregates over a join (with selection(s)) of the same set of tables. The problem of identifying common expressions is known as Multi-Query Optimization (MQO). It has been largely studied in 80's [14] and then in all database generations without any exception. This is because it aims at optimizing the global performance of queries collectively instead of individually. The identified common query subexpressions may be candidates for caching, materializing [9], reusing, indexing, partitioning [2,4], ... etc.

**Workload Clustering.** The workload clustering consists in splitting large workload to a number of groups such that queries in the same group are related to each other. To ensure this clustering, feature selection is needed to represent the queries. Three types of features are used to represent queries: (1) lexical features represented by the logical structures of SQL string (query type, tables, columns, join clauses, selection predicates, aggregations) [4], (2) physical features characterized mainly by runtime metrics [6] and (3) arrival rate history represented by the past arrival rates of queries [10].

**Autonomous System.** It is a system able to automatically manage its behaviors in accordance with its internal state and its environment. IBM [8] defines an autonomous system as a system with 4 basic properties: self-configuration, self-optimization, self-healing, and self-protection. To achieve this type of systems, reference models have been proposed [3], we quote mainly MAPE-K (Monitor-Analyze-Plan-Execute-Knowledge), PLA (Proactive Latency-aware Adaptation) and Multi-Systems-Agents. We notice that the database community focused on self-optimization propriety [11].

# 3 Our Framework for Autonomous Parallel DW Design

Our challenge is to design an autonomous PDW in two steps phases: off-line and on-line. The first phase consists in determining the best data placement scheme, whereas the second one plays the role of the adaptation manager of our target deployed $DW$. The functioning of the off-line phase is quite similar to the traditional parallel $DW$ design [1]. The sole difference between the existing state-of-art solutions is that the training set of queries $TQ$ is clustered into a set of disjoint homogeneous queries groups $\mathcal{TQ} = \{Q_1, \ldots, Q_h\}$ that will be used to partition the $DW$. Once all the partitioning schemes are built, they will be merged into a single partitioning schema. This merging may reduce the number of final fragments. Finally, the so-generate fragments are allocated over the database cluster using an allocation algorithm. Each step aims to maximize the reuse of common sub-expression and minimize the cost of the workload $Q$. We emphasize that the processing of ad-hoc queries, in the on-line phase, is considered by the workload clustering model. Specifically, when a new query occurred, we assign it to the best group by calculating the similarity of the new query with existing groups.

## 3.1  Workload Clustering

To ensure our clustering, we use a bag-of-words approach. To do that, we use a lexical-semantic workload clustering algorithm following three main steps:

- **Selection of Query Features.** The goal of this step is to find the relevant clustering model for the predefined workload (past) and future queries (present and future). The only shared elements between the past, present, and the future represent the common sub-expressions. In this work, we consider the set of common sub-expressions as relevant features using Apriori algorithm because selecting the optimal set of common sub-expressions patterns is equivalent to finding the *frequent itemsets mining*.
- **Query representation.** Each query is represented by a binary vector in the m-dimensional space, where $m$ represents the number of selected item-sets $\{FI_1, \ldots, FI_m\}$. The vector representing a query takes the form of a binary vector, where for each selected common sub-expression $FI_j$, the corresponding weight value $w_{ij}$ equals 1 if $FI_i$ occurred in the query $Q_j$ and 0 otherwise.
- **Queries clustering.** Unsupervised learning algorithms family includes a large set of examples (K-means, Mean-Shift, DBSCAN, ...). We use DBSCAN clustering, in which it is not necessary to specify the number of clusters to be generated.

## 3.2  Our Data Partitioning Algorithm

We focus on range partitioning that splits each partitioning attribute into several sub-domains [1]. Any partitioning algorithm used for this purpose has to exploit as much as possible the common shared expression as in [2]. Merging the local partitioning schemes may generate numerous fragments of the fact table and consequently violates the maintenance constraint. To satisfy this constraint, we formalize the problem of scheme merging as a *Set-union Knapsack Problem (SUKP)* [12]: Given:

- Let $\mathcal{A} = \{A_1, \ldots \ldots, A_Z\}$ be the set of fragmentation attributes. Due to the range partitioning, the domain of each attribute $A_i$ is decomposed into $n_i$ sub-domains $\{x_{11}, x_{12}, \ldots, x_{1n_i}\}$. Each $x_{ij}$ has a weight $w_{ij}$ (number of the fragments). Let $SD$ be the union of all sub-domains of fragmentation attributes $SD = \{x_{11}, x_{12}, \ldots, x_{1n_1}, \ldots, x_{L1}, x_{L2}, \ldots, x_{Zn_Z}\}$.
- A fragmentation maintenance constraint $W$,
- A target profit $d$ which represents the minimum desired utility.
- A collection of generated partitioning schemes $SF = \{S_1, S_2, ..., S_m\}$, where each $S_i$ ($S_i \subseteq SD$) is associated to a profit $p_i$.

The *SUKP* involves finding a sub-set $SF^*$ of $SF$ such as:

$$P(SF^*) = max(\sum_{j=1}^{t} p_j \geq d) \qquad (1)$$

where $t$ is the cardinal of $SF^*$ and under the following constraint:

$$W(SF^*) = \prod_{j=1}^{t} w_j \leq W \tag{2}$$

This problem is known to be NP-hard [12]. We propose a greedy algorithm with two steps to solve it as follows:

- *Initialization phase.* For each generated partitioning attribute $A_i$, we first extract from the so-generated partitioning schemes ($S$), a set of sub-domain attributes. Then, we keep only those that occur frequently together using Close algorithm in order to reduce the size of the search space.
- *Exploration phase.* The main purpose is to find sub-domains that have a high "utility" for workload optimization. For this end, we propose an algorithm to incremental selecting partitioning attributes to deal with the on-line phase. This algorithm ranks partitioning attributes according to their occurrence frequency in the generated fragmentation schemes and selects the best schema that maximizes the profit (number of inputs-outputs). The idea is to select, first, sub-domains according to the partitioning attribute selected by the majority of query groups (the first one is denoted $p_1$). Then, the sub-domains that scores the maximum value of utility are chosen. The second attribute must maximize the utility of the attribute set $p_1, p_2$, and so on until the $NF$ fragments are chosen. Some queries clusters are so-called victims because they will not benefit from the new scheme, we treat them in an individual way where necessary.
- *Evaluation.* Primarily, for each query cluster $Q_i$ ($1 \leq i \leq h$), we estimate the number of inputs/outputs (IOs) required for executing $Q_i$ (denoted $InitialIO_i$). Next, we evaluate the number of IOs resulting of the execution of $Q_i$ under the current partitioning schema (denoted $CurrentIO_i$) and we calculate the utility as follows:

$$p_i = InitialIO_i - CurrentIO_i \;\; \forall i \;\; 1 \leq i \leq h \tag{3}$$

The utility of the global partitioning schema is the sum of utilities $p_i$ and it must exceed a threshold $d$ fixed by the designer.

$$Utility = \sum_{i \leq h} p_i \tag{4}$$

## 4    Generating Allocation Schema

The fragment allocation problem on a database cluster can be formalized as an *clustering problem* [1] Our allocation procedure is defined as follows:

- **Fragments representation in $\Re^d$.** First of all, we describe a set of $NF$ fragments $\mathcal{F} = \{F_1, F_2, \ldots, F_{NF}\}$ by a set of $m$ features $\mathcal{A} = \{A_1, A_2, \ldots, A_L\}$ of queries. In this work, our fragment set $\mathcal{F}$ is represented as a matrix (denoted $\mathcal{FM}$). A value $FM(F_i, C_j)$ is defined as follows:

$$FM(F_i, C_j) = \begin{cases} 1 & Q_i \text{ involves only the fact fragment } F_j \\ \text{join} & Q_i \text{ involves } F_j \text{ and dimension tables} \\ 0 & \text{otherwise} \end{cases}$$

– **Construction of Cluster Membership Matrix.** $\mathcal{CCM}$ indicates the degree to which fragments belong to each cluster. Our fragments allocation policy is based on the following idea: *"strongly and positively correlated fragments belong to the same class"*. For this, we combine a dimension reduction method such as the ACP and a classification method like K-means++. The value $CCM[i][j]$, such that $(1 \leq i \leq NF)$ and $(1 \leq j \leq M)$, belongs to the interval $[0, 1]$. This value corresponds to the distance between a fragment $F_i$ and the centroid of the class $C_j$ (each class represents a node).
– **Construction of Fragment Placement Matrix.** Once $CCM$ is created, we tag fragments into clusters by allocating each fragments class on one node of the database cluster. The outcomes of this step is a binary matrix, called $MP$ where $MP[i][m] = 1$ if the fragment $F_i$ is allocated on the node $N_m$, otherwise $MP[i][m] = 0$.

## 5  Experimental Results

In this section, we present the set of experiments that we conduct to evaluate the effectiveness of our proposal. The experiments were conducted on a database cluster of 10 processing nodes, where each node has a 3, 33 GHz Intel Core i7 processor, a 16 GB main memory, and a Microsoft SQL Server 2016 like DBMS. Our algorithms are implemented in Java. We use the datasets of the Star Schema Benchmark (SSB) with a scale factor of 1 (SF = 1). For the Workload, 1000 queries are randomly generated from the 13 original SSB queries.

In the first experiment, we study the impact of workload clustering on the quality of the obtained deployment. To do so, we compare the efficiency of our lexico-semantic features against logical features. Clustering using logical features gives 6 queries classes, whereas the lexical-semantic features selection provides 5, 8 and 13 queries classes for a minsup of $\geq 30\%$, 20%, 10% respectively. The results in Fig. 1 show that increasing the number of queries clusters improves the workload performance significantly. This experiment shows that our features selection method has successfully contributed to improving the efficiency of the deployed $DW$. This result confirms that the batch execution improves the performance of small workloads.

In the second experiment, we study the quality of our proposed approach in terms of its speed-up. For a partitioning threshold of 100 and a workload categorized into 5 query clusters, we vary the number of nodes from 1 to 10. For each value, we calculate the speed up. As shown in Figure 2, our approach scale linearly, but it is not ideal. That is due to the fact that we use a multi-level partitioning based on the splitting of the attribute's domain. To solve this

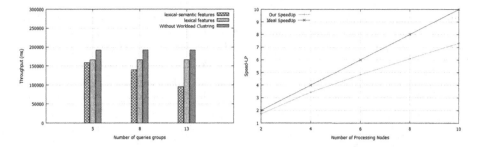

**Fig. 1.** Workload clustering performance            **Fig. 2.** SpeedUp

issue, we can improve the query processing by migrating fragments from highly loaded nodes to the lowly one. Another potential solution is to select relevant materialized views and/or replicas.

Finally, in the third experiment, we study the performance of our utility-based partitioning approach. we vary the fragmentation threshold in the interval [100–300] and we calculate the throughput workload execution. For that, we compare our approach based mainly on the workload clustering and a partitioning approach without workload clustering (we use the same approach used for partitioning each queries class). As depicted in Fig. 3, increasing the fragmentation threshold improves query performance. In addition, when the number of groups of queries is relatively large, the global fragmentation scheme favors more queries. This implies better utility for each queries group.

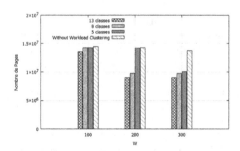

**Fig. 3.** Performance of our partitioning approach based on utility

# 6   Conclusion

In this paper, we attempt to integrate the fundamental characteristics of autonomous systems designing parallel data warehouses. We spent a lot of efforts in analyzing the existing studies and showed that they are based on the principle of Knowing that Designers Know the two entries representing workload and data sources. We rapidly identified its limitations in designing advanced data warehouses in the B2.0 era known by its batch and ad-hoc queries. The integration of this dimension in the design is ensured by proposing an adaptive technique. To do so, we proposed a comprehensive framework with a proactive off-line approach to designing a parallel *DW* over a database cluster. The off-line step is the core of design since it represents the basic knowledge base composed of a training set of OLAP queries. This set is clustered based on new feature selection method guided by lexical and semantic of the queries captured by their shared common nodes. Our framework is instantiated by proposing dynamic algorithms for data partitioning and fragment allocation. Intensive experiments were conducted and show the effectiveness and efficiency of our proposal.

# References

1. Benkrid, S., Bellatreche, L., Cuzzocrea, A.: A global paradigm for designing parallel relational data warehouses in distributed environments. Trans. Large-Scale Data-Knowl.-Cent. Syst. **15**, 64–101 (2014)
2. Boukorca, A., Bellatreche, L., Benkrid, S.: HYPAD: hyper-graph-driven approach for parallel data warehouse design. In: Wang, G., Zomaya, A., Perez, G.M., Li, K. (eds.) ICA3PP 2015, Part IV. LNCS, vol. 9531, pp. 770–783. Springer, Cham (2015). https://doi.org/10.1007/978-3-319-27140-8_53
3. Cámara, J., et al.: Self-aware computing systems: related concepts and research areas. In: Kounev, S., Kephart, J., Milenkoski, A., Zhu, X. (eds.) Self-Aware Computing Systems, pp. 17–49. Springer, Cham (2017). https://doi.org/10.1007/978-3-319-47474-8_2
4. Du, J., Miller, R.J., Glavic, B., Tan, W.: DeepSea: progressive workload-aware partitioning of materialized views in scalable data analytics. In: EDBT, pp. 198–209 (2017)
5. Durand, G.C., et al.: GridFormation: towards self-driven online data partitioning using reinforcement learning. In: First International Workshop on Exploiting Artificial Intelligence Techniques for Data Management, pp. 1–7 (2018)
6. Ghosh, A., Parikh, J., Sengar, V.S., Haritsa, J.R.: Plan selection based on query clustering. In: VLDB, pp. 179–190 (2002)
7. Hintikka, J.: 'Knowing that one knows' reviewed. Synthese **21**(2), 141–162 (1970)
8. Horn, P.: Autonomic computing: IBM\'s perspective on the state of information technology. IBM (2001)
9. Jindal, A., Karanasos, K., Rao, S., Patel, H.: Selecting subexpressions to materialize at datacenter scale. Proc. VLDB Endow. **11**(7), 800–812 (2018)
10. Ma, L., Van Aken, D., Hefny, A., Mezerhane, G., Pavlo, A., Gordon, G.J.: Query-based workload forecasting for self-driving database management systems. In: ACM SIGMOD, pp. 631–645 (2018)
11. Nehme, R. Bruno, N.: Automated partitioning design in parallel database systems. In: Proceedings of the 2011 ACM SIGMOD International Conference on Management of Data, SIGMOD 2011, pp. 1137–1148. ACM, New York (2011)
12. Goldschmidt, O., Nehme, D., Yu, G.: Note: on the set-union knapsack problem. Nav. Res. Logist. (NRL) **41**(6), 833–842 (1994)
13. Roy, P., Sudarshan, S.: Multi-query optimization. In: Encyclopedia of Database Systems, 2nd edn (2018)
14. Sellis, T.K.: Multiple-query optimization. ACM Trans. Database Syst. **13**(1), 23–52 (1988)

# Distributed and Parallel and Network-based Computing

# Stable Clustering Algorithm for Routing Establishment in Vehicular Ad-Hoc Networks

Jieying Zhou, Pengfei He[(✉)], Yinglin Liu, and Weigang Wu

School of Data and Computer Science, Sun Yat-sen University,
Guangzhou, China
{isszjy,wuweig}@mail.sysu.edu.cn,
{hepf3,liuylin6}@mail2.sysu.edu.cn

**Abstract.** With regard to the complex and varied urban scenes in Vehicular Ad-Hoc Network, such as the vehicle nodes with fast speed, unstable links and frequent changes in network topology, this paper proposed a stable clustering algorithm to establish routing for VANET. In this algorithm, clustering was first formed according to the Similar Neighbor Node Table. Then on the basis of the Highest Connectivity Algorithm, parameters such as position and speed of the node were introduced to calculate the Selection Priority which were used to produce the preferred and the alternative cluster head node. The introduction of alternative cluster head nodes improved the stability of clustering to a certain extent. Finally, the clustering was established and maintained for six special scenarios. Compared with the traditional clustering algorithm in VANET, it had the characteristics of lower end-to-end delay, higher packet delivery rate and lower change rate of preferred cluster head nodes, which greatly improved the communication quality of VANET.

**Keywords:** Vehicular Ad-Hoc Network · Routing protocol · Clustering algorithm

## 1 Introduction

People increasingly place their hope of alleviating or even completely solving traffic problems on intelligent transportation system and vehicle self-organizing network technology [1]. The Vehicular Ad-hoc Networks (VANET) builds a centerless, multi-hop, self-organizing mobile communication network with many nodes [2, 3]. And clustering algorithm is the key content that determines the performance of the VANET [4].

However, at present, there is no clear standard for the routing protocol or clustering algorithm in VANET industry. The clustering algorithm in VANET usually draws lessons from the clustering algorithm in the Mobile Ad-hoc Network (MANET). The representative algorithms are: Minimum ID Algorithm (ID-Lowest) [5], Highest Connectivity Algorithm (HC) [6], WCA [7], AOW [8], ALM [9] and so on. The advantages of ID-Lowest are as follows: the implementation process is simple and convenient; the change rate of cluster head nodes is low. However, its shortcomings are also obvious: because only the minimum ID is considered as the cluster head, it will be

© Springer Nature Switzerland AG 2020
S. Wen et al. (Eds.): ICA3PP 2019, LNCS 11945, pp. 107–115, 2020.
https://doi.org/10.1007/978-3-030-38961-1_10

unfair and unstable, and there is no good maintenance mechanism to manage clustering. The HC [6] is similar to ID-Lowest. The disadvantage of HC is that if there is no limit on the number of members in the cluster, it can be a lot. If the number of clusters becomes smaller, then the channel utilization will become very low, the throughput will decrease rapidly.

Although VANET and MANET have some similarities, there are still some differences between VANET in traffic roads. In this paper, according to the characteristics of urban traffic conditions, and its research and analysis, a stable clustering algorithm is designed, which is more suitable for VANET. Compared with HC, ID-Lowest and other clustering algorithms, it has smaller preferred cluster head node change rate and more lasting preferred cluster head node maintenance time, which improves the packet delivery rate, reduces the topology change rate, and can reduce the network transmission delay. The packet delivery rate is improved, which greatly improves the communication quality in the VANET.

## 2  Scenario of the Stable Clustering Algorithm

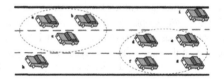

**Fig. 1.** The cluster of vehicles in local road sections

As shown in Fig. 1, in a road section with three lanes, a cluster was formed because vehicles $a$, $b$ and $c$ were close, and their average speed was similar. Because the speed of vehicle $h$ was fast, even if it was added to the cluster, it would get out of the cluster quickly, so the vehicle $h$ was excluded from the cluster. Similarly, vehicles $d$, $e$, $f$ and $g$ formed a cluster. Because the average speed of $i$ was slow, it was also excluded from cluster. Vehicle $i$ needed to find other vehicles to form a cluster. Based on this clustering idea, it designs a stable clustering algorithm, which can effectively reduce the network size in the urban scene, reduce the change rate of network topology, and improve the communication quality of the network.

## 3  Mechanism of the Stable Clustering Algorithm

In this stable clustering algorithm, it is assumed that all vehicle nodes in the VANET can get the following information:

- The direction, speed, the longitude and latitude of the location and other information of the vehicle by related devices, such as GPS, on-board sensor devices, etc.
- The HELLO messages used to maintain the topology of the network from the vehicle nodes in their communication range.

- The Beacon message within limited number of hops which includes vehicle ID, position, direction, speed and speed standard deviation, etc.

### 3.1 Clustering for the Similar Neighbor Nodes

In the range of communication between vehicular nodes, the neighbor vehicle nodes obtain each other's basic information through Beacon message. Only the vehicle nodes whose driving speed is within a certain confidence interval are selected to form a similar cluster of neighbor vehicle nodes. The traffic roads in the city will be quite straight and regular. In order to facilitate analysis, only the vehicles driving in the same direction are selected to form a cluster. It is defined that vehicle node $m$ has the following related parameters:

- $j$: the number of neighbor vehicle nodes of the vehicle node $m$;
- $\Delta T$: the sampling period of obtaining the speed of the vehicle node;
- $v_m(t)$, $\sigma_m^v$: the speed at a certain $t$ time and the standard deviation of speed;
- $\bar{v}_m$: the average speed over the periodic sampling interval;
- $\sigma_{max}^v$: the maximum standard deviation of speed in neighbor nodes;
- $V_{min}^{mj}$, $V_{max}^{mj}$: the minimum and maximum driving speed of vehicle nodes;
- $\mu$: the adjustable coefficient of the size of the speed range.

The vehicle node $m$ records its speed in the time interval $\Delta T$. Calculate the average speed and the standard deviation of the vehicle node $m$:

$$\bar{v}_m = \frac{1}{\Delta T} \sum\nolimits_{t=0}^{\Delta T} v_m(t) \tag{1}$$

$$\sigma_m^v = \sqrt{\frac{1}{\Delta T} \sum\nolimits_{t=0}^{\Delta T} (v_m(t) - \bar{v}_m)^2} \tag{2}$$

Since it is assumed that the number of neighbor vehicle nodes of the vehicle node $m$ is $j$, the vehicle node $m$ needs to find a maximum speed standard deviation among the $(j + 1)$ vehicle nodes. Calculate the maximum speed standard deviation:

$$\sigma_{max}^v = max\{\sigma_m^v, \sigma_k^v | k = 1, 2, \ldots, j\} \tag{3}$$

The confidence interval is used to exclude a few vehicle nodes whose speed is very different from that of most vehicle nodes. On the basis of (3), an adjustable coefficient $\mu$ is introduced, and the minimum and the maximum speed of vehicle nodes in the confidence interval of Gaussian Mixture Model can be obtained:

$$V_{min}^{mj} = \bar{v}_m - \mu\sigma_{max}^v \tag{4}$$

$$V_{max}^{mj} = \bar{v}_m + \mu\sigma_{max}^v \tag{5}$$

The value of $\mu$ can be adjusted according to the average speed of the vehicle and the actual traffic environment. From the properties of the Model, the range of the confidence interval and the value of the adjustable coefficient $\mu$ can be obtained (Fig. 2):

$$\begin{cases} P\{|X - \bar{v}_m| < c\} = 2\Phi\left(\frac{c}{\sigma}\right) - 1 \\ P\{|X - \bar{v}_m| < 2.58\mu\} = 2\Phi\left(\frac{2.58\mu}{\sigma}\right) - 1 \approx 99\% \\ P\{|X - \bar{v}_m| < 1.96\mu\} = 2\Phi\left(\frac{1.96\mu}{\sigma}\right) - 1 \approx 95\% \\ P\{|X - \bar{v}_m| < 1.64\mu\} = 2\Phi\left(\frac{1.64\mu}{\sigma}\right) - 1 \approx 90\% \end{cases} \quad (6)$$

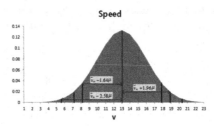

**Fig. 2.** The confidence probability of Gaussian Mixture Model

It can be obtained from (6) that some special nodes can be excluded by selecting the appropriate confidence interval. After the confidence interval is set, all the vehicle nodes that satisfy the driving speed in the interval $V_{min}^{mj} \leq V_j^{mj} \leq V_{max}^{mj}$ are formed into a new set in the neighbor table. Then a similar neighbor node table (SNNT) is created to store and record the information of all vehicle nodes in this collection.

### 3.2   Selection of Preferred Cluster Head Node

The parameters such as the number of neighbor vehicle nodes, speed and position spacing are modeled as a Gaussian Mixture Model. When the number of neighbor vehicle nodes is larger and the driving speed and vehicle position spacing are closer to the average value of the overall vehicle nodes in the table, the greater the selection priority (SP), the greater the probability that the vehicle node will become the preferred cluster head. It is defined that vehicle node $m$ has the following related parameters:

- $h$, $v_m$: the number of vehicle nodes in *SNNT* and the speed of the vehicle node $m$;
- $v_n$: the speed of each vehicle node in the SNNT, n = 1, 2, ..., h;
- $\bar{v}_{mean}$: the average speed of vehicle nodes in SNNT;
- $x_m$, $y_m$: the position parameter of the vehicle node $m$;
- $x_n$, $y_n$: the location parameters of each vehicle node in SNNT.
- $P_n$, $\bar{P}_{mean}$: the Euclidean distance and the average Euclidean distance between $m$ and the vehicle nodes in the SNNT;
- $\sigma_v$, $\sigma_p$: the standard deviation of the speed and the Euclidean distance $P$ in SNNT;
- $N_v$, $N_p$ : the normalization of the speed and the Euclidean distance $P$;
- $SP$: the selection priority of the preferred cluster head node.

Calculate the average speed of vehicle nodes, the Euclidean distance and the average of Euclidean distance between $m$ and the vehicle nodes in its SNNT:

$$\bar{v}_{mean} = \frac{1}{h}\sum_{n=1}^{h} v_n \qquad (7)$$

$$P_n = \arccos((\sin(Lat_m)\sin(Lat_n)) + (\cos(Lat_m)\cos(Lat_n)\cos(Lng_m - Lng_n))) * R \qquad (8)$$

$$\bar{P}_{mean} = \frac{1}{h}\sum_{n=1}^{h} P_n \qquad (9)$$

In (8), $Lng$ and $Lat$ represent longitude and latitude, respectively. By using the Z-score standardization method, it is necessary to calculate the standard deviation of speed and Euclidean distance of vehicle nodes in SNNT:

$$\sigma_v = \sqrt{\frac{1}{h}\sum_{n=1}^{h}(v_n - \bar{v}_{mean})^2} \qquad (10)$$

$$\sigma_p = \sqrt{\frac{1}{h}\sum_{n=1}^{h}(P_n - \bar{P}_{mean})^2} \qquad (11)$$

Then, the normalization of speed and Euclidean distance can be calculated:

$$N_v = (v_m - \bar{v}_{mean})/\sigma_v \qquad (12)$$

$$N_p = (P_n - \bar{P}_{mean})/\sigma_p \qquad (13)$$

Finally, the selection priority of $m$ is calculated:

$$SP = \frac{h}{e^{(|N_v| + |N_p|)}} = \frac{h}{e^{\left(\left|\frac{v_m - \bar{v}_{mean}}{\sigma_v}\right| + \left|\frac{P_n - \bar{P}_{mean}}{\sigma_p}\right|\right)}} \qquad (14)$$

### 3.3    Selection of Alternative Cluster Head Node

After the cluster is formed, the most suitable cluster member should be selected as the alternative cluster head node, which makes the cluster structure more stable during the period of maintenance. It is defined that vehicle node $m$ has the following related parameters:

- $H_n$, $N$: the Nth cluster and the number of member nodes in the cluster;
- $\phi_p$: a set of neighbor vehicle nodes of a member node in a cluster;
- $\mu_i$, $\theta_i$: the ID of member node in cluster, $i$ is a sort index in the range of $1 - N$;
- $C(\mu_i)$: the coverage cardinality of member nodes in a cluster;
- $\eta(\theta_i)$, $p$: the $SP$ of member nodes in a cluster and a member node in the cluster.

The coverage cardinality (CC) represents the number of same nodes in the cluster as the neighbor vehicle nodes of a member node $p$:

$$CC_p = \{|H_n \cap \phi_p|, \ \forall p \in H_n\} \tag{15}$$

Define a set $\Phi_{I_C}$, which consists of the $C(\mu_i)$ of all the member nodes in the cluster $H_n$, with a total of $N$ member nodes:

$$\Phi_{I_C} = \{C(\mu_1), C(\mu_2), C(\mu_3), \ldots, C(\mu_N)\} \tag{16}$$

The ID $\mu_i$ of all the member nodes in the cluster is sorted within the set $\Phi_{I_C}$. As $i$ gets bigger and bigger, the $CC$ $C(\mu_i)$ are sorted from large to small, corresponding to the ID $\mu_i$ of the member node. The sort set of the member node ID can be represented as:

$$I_C = \{\mu_1, \mu_2, \mu_3, \ldots, \mu_N | C(\mu_k) \geq C(\mu_j), \forall k < j\} \tag{17}$$

The ID $\theta_i$ of all the member nodes in the cluster are sorted in the set $\eta(\theta_i)$. As $i$ gets bigger and bigger, the $SP$ of all the member nodes in the cluster $H_n$ are sorted from large to small. The sort set $I_P$ of the member node ID can be represented as:

$$I_P = \{\theta_1, \theta_2, \theta_3, \ldots, \theta_N | \eta(\theta_k) \geq \eta(\theta_j), \forall k < j\} \tag{18}$$

A new alternative cluster head node has to meet the following two conditions: the vehicle node with a larger $CC$ $C(\mu_i)$ in (17), that is, the member node in the front of the set; the $SP$ $\eta(\theta_i)$ of the selected vehicle node is greater than a set threshold $\eta_T$. This threshold is selected in the set $I_P$ of (18) and satisfies the following formula:

$$\eta_T = \eta(\theta_i) \quad i = 1, 2, \ldots, N \tag{19}$$

Therefore, a new alternative cluster head node is represented as follows:

$$CH_B = \arg max\{\mu_i | \eta(\mu_i) \geq \eta_T\} \tag{20}$$

### 3.4   Maintenance of Neighbor Cluster

The maintenance of neighbor cluster is summarized into the following six scenarios, and targeted treatment is made to make the cluster stable.

1. The member nodes in the cluster are separated from the cluster: the ordinary member node in the cluster fails to receive *HELLO* message of the separated node, all the information of the node is deleted from its *SNNT* and the topology structure is updated. The separated node will clear the *SNNT* and change the state to the undetermined state, and regenerate the *Beacon* message to request to join other clusters.

2. The alternative cluster head node is separated from cluster: the member node in the cluster initiates the selection process of the alternative cluster head node and immediately selects a new alternative cluster head node.
3. The preferred cluster head node is separated from cluster: the alternative cluster head node immediately becomes the preferred cluster head node and manages all the members of the cluster. And then re-select an alternative cluster head node.
4. The merging of neighbor clusters: the two preferred cluster head nodes exchange information through *Beacon* messages, and then calculate the degree $D$ of dissimilarity between the two nodes. When $D$ is smaller than a certain threshold, the cluster merge is carried out.

$$D = \theta_d \frac{|d_{cb}|}{d_{com}} + \theta_v \frac{|v_{cb}|}{v_c} \tag{21}$$

5. Adding the undetermined vehicle nodes to the cluster: when an undetermined vehicle node $u$ is in the communication range of a cluster, $u$ will send its *Beacon* message to the preferred cluster head node of the cluster. After receiving the message, the preferred cluster head node $H$ calculates $D$, according to (21), between itself and u, and then decides whether to accept $u$ or not.
6. Re-establishment of cluster: the preferred cluster head node needs to obtain the vehicle node information from the *SNNT*, and then calculates its *SP* according to the (14). Finally, according to the (22), it is decided whether the vehicle node continues to be the preferred cluster head node:

$$SP = \frac{h}{e^{\left(\left|\frac{v_m - \bar{v}_{mean}}{\sigma_v}\right| + \left|\frac{P_n - \bar{P}_{mean}}{\sigma_p}\right|\right)}} < \frac{\pi h}{4e^{\left(\left|\frac{1}{\sigma_v}\right| + \left|\frac{1}{\sigma_p}\right|\right)}} \tag{22}$$

## 4  Simulation and Result Analysis

A routing protocol, the Stable Clustering-Based Routing Protocol in Vehicular Ad-Hoc Networks (SCBR), is proposed based on the stable clustering algorithm which is simulated and tested with the NS2 and the VanetMobiSim.

The driving speeds of vehicle nodes are 5, 10, 15, 20, 25 and 30 m/s, respectively. The number of vehicle nodes is set to 200. According to the simulation results, the influence of different vehicle node speed on the performance index is evaluated.

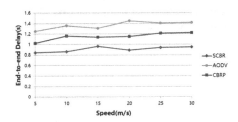

**Fig. 3.** The effect of speed on end-to-end delay

As shown in Fig. 3, when the number of vehicle nodes is the same, the speed of vehicle nodes has little effect on the end-to-end delay. The main factors affecting the end-to-end delay are the distribution and density of vehicle nodes, the average number of packets forwarding hops and the routing algorithm mechanism of the protocol. When a routing error occurs, the CBRP protocol can be repaired locally, so the end-to-end delay is lower than that of the AODV protocol. The stable clustering algorithm of SCBR protocol can improve the stability of communication link and clustering structure, and ensure that the data packet can reach the destination node successfully. Therefore, using the SCBR protocol, the driving speed of the vehicle node has the least effect on the end-to-end delay.

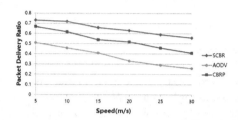

**Fig. 4.** The effect of speed on packet delivery rate

As shown in Fig. 4, when the vehicle node speed increases, the packet delivery rate of AODV decreases the fastest, while the SCBR decreases more slowly, which is less affected by the vehicle node speed. The main reason is that even when the vehicle nodes are driving at high speed, as long as they can maintain relative stillness with the neighbor vehicle nodes, the cluster structure can be relatively stable and the vehicle nodes can communicate with each other normally. Compared with AODV and CBRP, SCBR has higher packet delivery rate, and its performance index is the best.

## 5    Conclusion

SCBR protocol based on stable clustering algorithm proposed has good performance compared with other relative protocols in urban traffic roads. Clustering algorithm for suburban or high-speed environment could be further studied in the future so that the protocol could be applied to multiple environments simultaneously.

**Acknowledgment.** This work is supported by the National Key R&D Program of China (2018YFB0203803), the National Natural Science Foundation of China (U1801266), and the Program of Science and Technology of Guangdong (2015A010103007).

# References

1. Jin, M.: A brief talk on the development of China's intelligent transportation. Transp. Sci. Tech. **20**(2), 140–142 (2013). (in Chinese)
2. Zhang, Y., Wang, H., Peng, L.: Single-layer and cross-layer routing protocols in vehicular ad-hoc networks. Commun. Technol. **50**, 2279–2284 (2017)
3. Cheng, J., Ni, W., Wu, W.: A survey of the application research of vehicle ad hoc networks in intelligent transportation. Comput. Sci. **36**(1), 1–10 (2014). (in Chinese)
4. Menouar, H., Filali, F., Lenardi, M.: A survey and qualitative analysis of MAC protocols for vehicular ad-hoc networks. IEEE Wirel. Commun. **13**(5), 86–94 (2011)
5. Lin, C.R., Gerla, M.: Adaptive clustering for mobile wireless networks. IEEE J. Sel. Areas Commun. **15**(7), 1265–1275 (1997)
6. Li, S., Xiao, X.: Vehicle ad-hoc network routing protocol based on road segmentation. Comput. Eng. **45**(2), 32–37 (2019)
7. Chatterjee, M., Das, S.K., Turgut, D.: WCA: a weighted clustering algorithm for mobile ad hoc networks. IEEE J. Cluster. Comput. **5**(2), 193–204 (2002)
8. Wang, Z.: A vehicle ad-hoc network opportunity routing protocol in 3D urban scene. Electron Technol. **47**(7), 60–63 (2018)
9. Togou, M.A., Hafid, A.: Stablecds-based routing protocol for urban vehicular ad hoc networks. IEEE Trans. Intell. Transp. Syst. **17**(5), 1298–1307 (2017)

# Utility-Based Location Distribution Reverse Auction Incentive Mechanism for Mobile Crowd Sensing Network

Chunxiao Liu[1($\boxtimes$)], Huilin Wang[1], Yanfeng Wang[2], and Dawei Sun[3]

[1] College of Information Science and Technology,
Bohai University, Jinzhou, People's Republic of China
xiaoxiao198525@163.com
[2] School of Engineering, Huzhou University,
Huzhou, Zhejiang, People's Republic of China
[3] School of Information Engineering, China University of Geosciences,
Beijing, People's Republic of China

**Abstract.** In the mobile crowd sensing network, the existing research does not consider the completion quality factor of the task and the individualized difference of the participant's ability. The location distribution of the participant will affect the quality of the task and the timeliness of obtaining the sensing task information. Participants in good positions can improve the completion rate of tasks, while participants with good reputation values can ensure the quality of the tasks. In this paper, the distance between the sensing point and the worker is used as one of the criteria for selecting the sensing task object. A utility-based location-distribution reverse auction incentive mechanism (ULDM) is proposed, which comprehensively considers budget constraints, worker's reputation, and location characteristics in the sensing model, define the distance correlation and time correlation to evaluate the utility of the data collected by the winner. Finally the experimental results show that the successful package delivery rate, average delay and energy consumption are used as evaluation parameters, which improves the quality of task completion and suppresses the selfish behavior of selfish workers, which proves that ULDM has better incentive effect than reputation incentive mechanism.

**Keywords:** Sensing task · Distance · Winner · Utility

## 1 Introduction

In mobile crowd sensing, for the sensing platform, it is hoped to recruit more participants with the least cost or controllable cost, and is the provider of high quality credibility data [1]. In addition, for participants who sensing tasks, most participants

This work is supported by Social Science Foundation of Liaoning Province (L18AXW001), Huzhou Public Welfare Application Research Project (2019GZ02).

S. Wen et al. (Eds.): ICA3PP 2019, LNCS 11945, pp. 116–127, 2020.
https://doi.org/10.1007/978-3-030-38961-1_11

definitely want to get a certain return through their own sensing behavior, but some participants may upload false data in order to get more rewards.

Wu [2] pointed out that it is difficult to ensure the high-quality completion of the sensing tasks because the user's participation rate is improved. Therefore, the incentive mechanism should also stimulate the user's sensing behavior according to the task requirements and improve the quality of the sensing data. Ning [3] proposed a user participation incentive mechanism based on credibility model. The initial value of the user's credibility set in the task allocation process to select high credibility of the user to participate in task processing, and set the factor to reduce the cost of the user's cost, so that improve the efficiency of task processing on the basis of effective control of the budget. Therefore, the incentive mechanism also needs to consider many incentives for participants to sensing behavior and improve the credibility of participants [4, 5]. It can be seen that for the sensing platform, a reasonable incentive mechanism must not only ensure that the sensing task has a higher level of participation, but also stimulate the quality of the task completion. In this paper, the distance between the sensing point and the worker is used as one of the criteria for selecting the sensing task object. A utility-based location-distribution reverse auction incentive mechanism (ULDM) is proposed, which comprehensively considers budget constraints, worker's reputation, and location characteristics in the sensing model, select the winner to complete the sensing task, define the distance correlation and time correlation to evaluate the utility of the data collected by the winner. The sensing model in this paper improves the quality of the task completion, strengthens the incentive effect, and makes the winner receive more reasonable compensation, and eliminates the problem that the winner's own bid is unreasonable.

## 2   Sensing Model

Completing the sensing task is the main goal of the existence and development of mobile crowd sensing. This paper divides the complex sensing task with long time span and multiple sensing points from time and space to form a single task set of multiple single sensing points with short time spans. In this way, a large-scale complex sensing task can be transformed into a simple task set. The task set is represented as $\tau = \{\tau_1, \tau_2, \ldots, \tau_n\}$, and $\tau_i \in \tau$ represents one of the tasks. One of the tasks is represented in the form of a multi-group, as shown in (1):

$$\tau_i = <x, p, r, d, s, Q, v, m> \tag{1}$$

Where $x$ is the number of potential participants who issued the task, and $p$ is the sensing data center point specified by the sensing task $\tau_i$, but the worker cannot completely reach the collection location for data collection.

Therefore, the sensing platform setting distance $r$ represents the coverage of the collection location $p$, that is, in the range where $p$ is the center and $r$ is the radius, the sensing platform will consider the data collected by the worker to be valid. The sensing task $\tau_i$ will specify the optimal acquisition time $d$ of the task, but it is difficult to ensure that each worker collects data at the optimal time. Therefore, the sensing platform setting $s$ indicates the invalid time of the task, that is, within the $d \pm s$ time range, the data collected by the worker will be considered as the effective time of the task. For workers, its historical reputation value is also very important. It needs to have a good reputation when accepting a certain sensing task. The reputation value is represented by $Q$ in the sensing task. $v$ is the budget cost estimated by the sensing platform before the sensing task begins, and $m$ is the total remuneration paid by the sensing platform after the task is completed.

## 2.1  Release Task Stage

The publisher generates the sensing data needs, submits the task to the sensing platform, and after the sensing platform evaluates the task, feeds back the budget of the task to the publisher, and the publisher determines the budget and sends it to the sensing platform. The sensing platform publishes tasks, using pre-assessed task budgets to attract potential participants to participate in sensing tasks and become task workers.

*Definition 1.* Potential Participant Set
During the valid period of the sensing task, all online mobile users are represented by the set as $X = \{x_1, x_2, \ldots, x_n\}$, $x_i \in X$ represents one of the potential participants. Each potential participant is represented by a two-dimensional array, that is, $x_i = (w_i, c_i)$.

$$w_i = \begin{cases} 1 \\ 0 \end{cases} \tag{2}$$

Where $w_i$ indicates whether to accept the task, $w_i$ and $c_i$ indicates the bid of the potential participant $x_i$. $w_i = 1$, means accepting the task, and willing to report the payment budget. $w_i = 0$, means not accepting the task, not reporting the payment budget.

*Definition 2.* Task Budget Assessment
The sensing platform analyzes the data of the online sign-in data of the LBSN [6], and then evaluates the task with the time and place attribute according to the analysis result to obtain the budget for executing the task. There are two influencing factors in evaluating the task budget: regional heat and time heat. The regional heat is a measure of the frequency of user access in the area where the task is located. The time heat is a measure of the frequency of user access during the effective time of the task.
(1)  Regional Heat
Calculate the regional heat of the task by analyzing the participants in the LBSN online data and their corresponding sign-in locations. The regional heat increases with the number of participants in the task coverage area and the number of sign-in. The regional heat of the task area pi specified by the publisher is represented by $H(p_i)$, and

the change of the regional heat is between 0 and 1. The regional heat defined here is the embodiment of the number of times of sign-in in the unit area and the diversity of the sign-inusers. Calculate the regional heat $H(p_i)$ as (3) in the literature [13].

$$En(U_{T_i,p_i}) = -\sum_{y \in Y} (p_{T_i,p_i}(y) \times \log_2 p_{T_i,p_i}(y)) \tag{3}$$

$$H(p_i) = \frac{\sum\limits_{y \in Y, p_i \in P} U_{T_i,p_i(y)}}{\sum\limits_{y \in Y} U_{T_i,P(y)}} \times En(U_{T_i,p_i}) \tag{4}$$

$U_{T_i,p_i}(y)$ indicates the number of times the $T_i$ participant $y$ has signed in at the task point $p_i$ during the same time period, $P$ indicates the set of all the sign-in locations, and $Y$ indicates the set of all participants. represents the total number of sign-in of participants $y$ at all sign-in locations $U_{T_i,P}(y)$ during the same time period $T_i$, represents the time period $T_i$, the information entropy of the participants in the area where the task point $p_i$ is located. indicates $En(U_{T_i,p_i})$ the probkability that the participant $y$ will sign in the task point $p_i$, that is, the proportion of the number of times the participant $y$ has signed in the task point $p_i$ to the number of times of sign-in $p_{T_i,p_i}(y)$ all areas. The information entropy calculation formula is as (4) in the literature [13].

(2)  Time Heat

The number of sign-in in the LBSN data in a certain period of time is proportional to the total number of participants in the period, so the ratio of the number of sign-in of the period to the total number of sign-in in all time periods can fully represent the ratio of the total number of participants in the period to the total number of participants in all time periods, that is time heat. The time heat is expressed by $H(T_i)$ and is calculated as (5) in the literature [8].

$$H(T_i) = \frac{\sum\limits_{T_i \in T, y \in Y} R_{T_i}(y)}{\sum\limits_{y \in Y} R_T(y)} \tag{5}$$

(3)  Task Budget Assessment

The LBSN online sign-in data analysis is used to obtain the regional heat and time heat of the task, and then the task budget is evaluated according to these two attributes. denotes the average of the regional heat that a participant has signed in all areas $\overline{H(p_i)}$ by (3) represents the average of the time heat that all participants has signed by (5). When the regional heat is and the time heat is $\overline{H(T_i)}$, the task cost is 1, which is the unit cost of the task. For any task, the task budget $v_i$ is calculated as (6) in the literature [13], where the formula indicates $\overline{H(p_i)}$ that the task budget is inversely proportional to the regional heat and time heat of the region.

$$v_i = \frac{\overline{H(T_i)}}{H(T_i)} \times \frac{\overline{H(p_i)}}{H(p_i)} \tag{6}$$

## 2.2    Select Worker Stage

In the selection worker stage, the primary goal of the sensing platform is to screen out the potential participants from the sensing task to identify workers who agree to participate in the sensing task. Each worker can accept and complete multiple tasks, the premise is that the time to complete each task does not conflict. After the release task stage is over, the worker set who accept the sensing task is represented by $W$ as = $\{worker_1, worker_2, ..., worker_n\}$, where $worker_i \in W$ represents a worker.

*Definition 3.* Task Generation
If the number of workers is $worker_i \geq 1$, it means that the task is generated. If the number of workers is $worker_i = 0$, it means that the task is unmanned execution, and the sensing platform only needs to count such tasks and feed back to the publisher.

*Definition 4.* Task Assignment
Suppose that this model selects only one worker as the winner, all tasks $\tau = \{\tau_1, \tau_2, ..., \tau_n\}$, There are two situations in which the task $\tau_i$ is completed by one person:
1) The task has only one worker, that is, $worker_i = 1$ and $worker_i$ meets the task budget condition, it is the minimum condition for the task to be completed, and the winner $W_s = \{worker_i\}$.
2) There are multiple workers in the task, that is, $worker_i \geq 2$, and it is necessary to select a $worker_i$ from the set $W = \{worker_1, worker_2, ..., worker_i\}$ and satisfy the winner condition, which is an ideal condition for the task to be completed, and the winner $W_s = \{worker_i\}$.

## 2.3    Selecting the Winner Stage

The worker updates the location information to the sensing platform, and uses the distance from the sensing point as the criterion for screening the winner. The winner selects the effective worker first, thereby, the sensing task cannot be completed in time, the waste of invalid workers is avoided, and more energy and bandwidth overhead are avoided.

*Definition 5.* The Winner Condition
In the number of workers is $worker_i \geq 2$, the $worker_i$ becomes the winner of the task to meet the following conditions:

$$D = \begin{cases} 0, k(p_i, wor\ ker_i) > r \\ 1, k(p_i, wor\ ker_i) \leq r \end{cases} \tag{7}$$

$r$ is the acceptable threshold for the distance between the *worker*$_i$ and the sensing point $p_i$, that is, the coverage of the sensing point. $k(p_i, worker_i)$ is the linear distance between the sensing point $p_i$ and the *worker*$_i$. When the value is not greater than $r$, the value of $D$ is 1, which is a valid worker, and has the opportunity to complete the sensing task as the winner. When the value is not greater than $r$, the value of $D$ is 0, which is an invalid worker, who cannot participate in the sensing task.

The reputation value reflects the quality of the submitted data during the worker's historical execution of the task. This paper assumes that workers with high reputation values have relatively high quality of sensing data. In order to ensure the quality of the submitted data, the reputation of the worker is taken into account when selecting the winner of the task, and the reputation value is reordered among the remaining effective workers. When the task $\tau_i$ is processed, the sensing platform selects the workers with the highest reputation value (that is, the largest $Q_i$) to become the winner to complete the task $\tau_i$, and obtains the final winner $W_s$.

Set a separate threshold [9, 11] (that is, $Thres_i$) for each active worker, which is confidential to other workers and platforms, and set a reputation value $Q_i$ based on the ratio of the threshold of each active worker and the threshold of all workers. The formula is as (7) in the literature [3].

$$Q_i = \frac{\sum Thres}{Thres_i} \tag{8}$$

The task bid refers to the reference of the worker's task budget calculated by (6), first calculating the cost of completing the task with own condition by (10), and then submitting the psychological price to the sensing platform. Under the assumption of a rational worker, the worker provides a bid that is lower than the task budget to increase the probability of becoming a winner. Therefore, after the worker submits the data bid, the article selects the winner based on the reverse auction, that is, selects the worker with the lowest bid and pays for it.

***Definition 6.*** Task Budget Condition
This model uses reverse auction, sensing platform is willing to choose the low budget of the worker as the winner. When the number of workers is $worker_i = 1$, the completion of all tasks $\tau$ requires agreement with the task budget condition, which is expressed as (9) and (10):

$$a_i < c_i < v_i \tag{9}$$

$$m \le \sum_{k=1}^{L} v \tag{10}$$

The cost $a_i$ must be less than the final bid of the reverse auction $c_i$ is less than the budget cost $v_i$ estimated by the sensing platform before the sensing task begins. $m$ is the total reward for the last payment of the sensing platform after the task is completed, $L$ represents the number of tasks in the task $\tau$.

**Definition 7.** The Cost Of The Worker

Considering that each task $\tau_i$ may have multiple workers and only one winner, the winner $W_s$ will incur a cost when processing the task $\tau_i$. In this paper, for a task $\tau_i$, the cost function is expressed by $a_i = f\{g_i, k_i, Q_i\}$, that is, the cost of the worker is affected by the assigned task size $g_i$, the distance between the sensing point and the worker $k_i$, and the worker's reputation value $Q_i$. It is proportional to the reputation value and the task size, and inversely proportional to the distance between the sensing point and the worker. Assuming that the two workers have the same reputation value, the farther away the worker is from the sensing point, the higher the cost. The formula is as shown in (11).

$$a_i = \alpha g_i^{Q_i} \times \beta 0.5^{k_i} \tag{11}$$

Where $\alpha$ and $\beta$ are two factors and $\alpha + \beta = 10$.

## 2.4    Submit Data and Payment Stage

In order to increase the user's participation and maintain a certain number of partici-pants, the publisher pays the reward and task compensation to the winner and the loser according to (12). For the winner, the publisher pays the bid in the reverse auction, and for the loser, the remaining budget after paying the winner is distributed to each user on average.

$$m_i = \begin{cases} c_i, & W_s \text{ is the winner} \\ \frac{v_i - c_i}{n-1}, & W - W_s \text{ is the loser} \end{cases} \tag{12}$$

Where $c_i$ refers to the task bid submitted by the winner $W_s$ during the reverse auction process, that is, the reward that the winner wants to submit data to the sensing platform. $v_i$ refers to the publisher's task budget. $n$ means the number of workers.

## 2.5    Evaluate the Data Stage

In this model, the sensing data acquired by the sensing platform for the winner is to be evaluated, because the value of sensing data in the sense of mobile crowd sensing cannot be measured by the quality of pixels, sharpness, etc., the utility value [7] of the sensing data is measured in multiple dimensions. The data of the inefficient value cannot satisfy the sensing platform's demand for the sensing data. If the sensing platform collects a large amount of inefficient sensing data, it will not only affect the sensing platform's analysis of the sensing task, at the same time, it will also cause waste of resources such as communication, calculation and payment of the sensing platform.

This paper calculates the relevance of sensing data from the two dimensions of time and distance [8], express the sensing data as $data = \{VD(k), VT(t)\}$, where $t$ represents the acquisition time and $k$ represents the distance between the collection location and the task point.

**(1) Time correlation**

To calculate the time correlation in the sensing data utility multi-group, first set the optimal acquisition time d that contains the sensing data in the sensing task. The time correlation of the sensing data is determined by the relationship between the time $t$ at which the winner collects the data and the optimal acquisition time $d$ of the sensing task. For example, if there are two sensing data, that is, $data_1$ and $data_2$, except for $t_1 \neq t_2$ in $data_1$ and $data_2$, all the other elements are the same, and $|t_1 - d| < |t_2 - d|$, and $t_1, t_2 \in [d - s, d + s]$, then the time correlation of $data_1$ is higher than the time correlation of $data_2$. Therefore, it can be seen that the time correlation of the sensing data is centered on the optimal acquisition time $d$ of the sensing task, and is attenuated on both sides. When the winner collects data at time $t \in [d - s, d + s]$ of the sensing data, the time correlation of the sensing data is 1; When the winner collects data at time $t \notin [d - s, d + s]$ of the data is sensed, the time correlation of the sensing data is 0. The time correlation $VT(t)$ is calculated as (13) in the literature [8].

$$VT(t) = 2 \times \text{sgn}(|t - d|) \times \frac{1}{1 + e^{(|d - t|)}} + \text{sgn}(-|t - d|)$$

$$\text{sgn}(x) = \begin{cases} 1, & x > 0 \\ \frac{1}{2}, & x = 0 \\ 0, & x < 0 \end{cases} \tag{13}$$

**(2) Distance correlation**

The distance correlation of the sensing data is determined by the distance between the location where the winner collects the sensing data and the sensing task point, that is, the larger the value of $k$ is, the smaller the distance correlation of the sensing data is, for example, if there are two sensing data, that is, $data_1$ and $data_2$, except for $k_1 \neq k_2$ in $data_1$ and $data_2$, all the other elements are the same. If $k_1 < k_2$, and $k_1, k_2 \leq r$ ($r$ is the coverage of the sensing point), then the distance correlation of $data_1$ is higher than the distance correlation of $data_2$. Therefore, it can be seen that the change in the distance correlation of the sensing data is in the range of $[0, r]$, and decreases with the increase of $k$. When the winner collects the distance $k \in [0, r]$ between the location of the sensing data and the sensing point, the distance correlation is 1, and when the winner collects the distance $k \notin [0, r]$ between the location of the sensing data from the sensing point, the distance correlation is 0. The calculation of the distance correlation $VD(k)$ is as shown in (14).

$$VD(k) = \begin{cases} 1, & k \in [0, r] \\ 2 \times \text{sgn}(-(k - r)) \times \frac{1}{1 + e^{-(k - r)}} \times \text{sgn}(k - r), & k \notin [0, r] \end{cases} \tag{14}$$

(3)  sensing data utility value

In a sensing task, the utility value of the sensing data is represented by *Utility*, which reflects the value of the sensing data provided by the winner in the sensing task. The calculation of the utility value is as (15).

$$Utility = score \times (A \times VT(t) + B \times VD(k))  \tag{15}$$

Use the scoring system to reflect the publisher's satisfaction with the winner's submission of data, that is, the degree of correlation between the submitted data and the task requirements. The value interval for defining the evaluation score is [0, 1]. When the *score* is 0, the *Utility* is always 0, $A$ and $B$ are the weight values of the time correlation and the distance correlation, respectively, and $A + B = 1$. The sensing platform can adjust the weight value according to the task requirements to change the sensing data utility. For example, for time-sensitive task, the platform can increase the weight of the time correlation. For distance-sensitive task, the platform can increase the weight of the distance correlation.

### 2.6  ULDM Incentive Algorithm

Input: The task set is represented as $\tau = \{\tau_1, \tau_2, ..., \tau_n\}$, potential participant set $X = \{x_1, x_2, ..., x_n\}$

Output: the reward value $m$

Step 1, After the publisher issues a sensing task, the sensing platform will release the sensing task to potential participants in the network, Count the number of potential participants $x_i$ of each task $\tau_i$.

Step 2, through whether the potential participants $x_i$ accept the task $\tau_i$, determine the worker set $W$.

Step 3, the sensing platform prioritizes the screening of workers who are too far away, and then combines their own reputation value with the proposed the final bid of the reverse auction to determine the final winner $W_s$,

Step 4, the publisher pays the winner, and the sensing platform gives a little reward $m_i$ to the remaining workers who actively participate in the sensing task.

Step 5, The sensing platform then defines the distance correlation and time correlation, finally evaluates the *utility* of the winner's data collection.

## 3  Experimental Analysis

In order to measure the validity of the incentive mechanism proposed in this paper, we choose the package delivery rate, average latency and energy consumption as the evaluation parameters.

### 3.1  Simulation Environment and the Corresponding Parameter Settings

We use the D2DCrowd simulation platform to create a $500 \times 500$ m$^2$ simulation scenario. The simulation time of each group of experiments is 1000 s, assigning 10 sensing points and 10 workers collecting data, and the worker receiving the task moves to the target position at a speed of 7 m/s and a lower limit of 3 m/s. The worker will stay at a certain sensing point for a random time from zero to the maximum dwell time. The worker's minimum dwell time is 0 s and the maximum dwell time is 60 s. After completing a sensing task, the worker moves to a new randomly generated destination. The initial energy of the mobile phone device is set to 100, 200, 300 ... 1000 J, respectively. The distance threshold r is 25.

### 3.2   Simulation Results Analysis

In the context of AODV routing algorithm, there are selfish workers in the network. Analyze the simulation results from the following two incentive mechanisms:

① A+Reputation: indicates that the Reputation incentive mechanism is running on the AODV routing protocol.

② A+ULDM: Utility-based location distribution reverse auction incentive mechanism proposed in this paper runs on the AODV routing protocol.

(1)  The impact of two kinds of incentive mechanisms in the network on package delivery rate.

As can be seen from Fig. 1, as the number of selfish workers in the network increases, the successful delivery rate of A+Reputation decreases. When the proportion of selfish workers exceeds 60%, the incentive effect of the A+Reputation decreases rapidly, and the change of the delivery rate of A+ULDM is relatively flat, indicating that A+ULDM is still ideal in the case of more selfish workers.

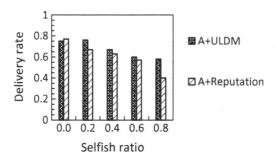

**Fig. 1.** Delivery rate under different selfish ratios

(2)  The impact of two kinds of incentive mechanisms in the network on energy consumption.

According to Fig. 1, the A+Reputation mechanism has poor incentive effect, and the reason why the energy consumption is less is that when the selfish worker gradually increases, Its selfish behavior leads to unwillingness to forward data, which saves power consumption, and the delivery rate is more. The smaller the delivery rate, the less energy consumption, both of which are inversely proportional to the proportion of selfish workers, and with the increase of selfish proportion, the energy consumption of A+ULDM is still much, indicating that the proposed incentive mechanism has a good incentive effect, as shown in Fig. 2.

**Fig. 2.** Energy consumption under different selfish ratios

(3) The impact of two kinds of incentive mechanisms in the network on average latency.

As can be seen from Fig. 3, the average delay of A+Reputation will gradually increase with the increase of the number of selfish workers, mainly because the incentive effect is weak, and the worker is still selfish and the message is not delivered to the destination in time.

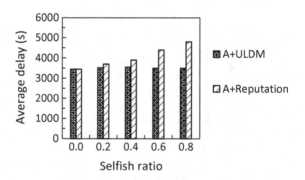

**Fig. 3.** Average delay under different selfish ratios

## 4   Conclusion

In order to ensure the network scale and user participation, the incentive mechanism is an indispensable technology in the mobile crowd sensing network. Since the user participates in the sensing task and needs a certain time and energy cost, the incentive mechanism needs to provide a certain economic reward to encourage the user [10], thereby improving the participation of the entire mobile crowd sensing network. At the same time, the quality of the completion of the sensing task is also very important, which will affect the actual working conditions of the entire sensing network. This paper conducts targeted research on this issue, the sensing platform is only interested in maximizing its own utility. Auction theory [12] is the perfect theoretical tool for designing incentive mechanisms for user-centered models. We proposed a utility-based location-distribution reverse auction incentive mechanism that takes the bids submitted by workers, the reputation of workers, and the location distribution of sensing points as inputs, and gives the winners reasonable and effective rewards, giving some compensation to other workers. Finally, the utility data is evaluated for its utility, and the sensing utility value is used to measure the value of the sensing task data, thereby updating the reputation value of the winner. The experimental results show that ULDM improves the quality of the task completion, strengthens the incentive effect, and makes the winner receive more reasonable compensation, and eliminates the problem that the winner's own bid is unreasonable.

# References

1. Luo, T., Das, S.K., Tan, H.P., et al.: Incentive mechanism design for crowdsourcing: an all-pay auction approach. ACM Trans. Intell. Syst. Technol. J. **7**(3), 1–26 (2016)
2. Wu, Y., Zeng, J.R., Peng, H., et al.: A review of research on incentive mechanism of crowd sensing. J. Softw. J. **27**(8), 2025–2047 (2016)
3. Ning, Z., Hua, S., Xue, M.S., et al.: User participation incentive mechanism of crowd sensing network based on reputation model. Comput. Appl. Softw. J. **2017**, 119–122 (2017)
4. Wang, H.: Resarch on incentive technology in participatory sensing network. Harbin Engineering University, Harbin (2014)
5. Zhao, L.M.: The research and implement of incentive mechanism based on participatory sensing. Beijing University of Posts and Telecommunication, Beijing (2015
6. Zheng, Y.: Tutorial on location-based social networks. In: Proceedings of the 21st International Conference on World Wide Web, WWW 2012, vol. 12 (2012)
7. Zhao, D., Li, X.Y., Ma, H.: How to crowdsource tasks truthfully without sacrificing utility: online incentive mechanisms with budget constraint. In: 2014 Proceedings of IEEE INFOCOM, pp. 1213–1221. IEEE (2014)
8. Tao, D., Zhong, S., Luo, H.: Staged incentive and punishment mechanism for mobile crowd sensing. Sens. J. **18**(7), 2391–2452 (2018)
9. Zhang, Y., Van der Schaar, M.: Reputation-based incentive protocols in crowdsourcing applications. In: 2012 Proceedings of IEEE INFOCOM, pp. 2140–2148. IEEE (2012)
10. Gao, L., Hou, F., Huang, J.: Providing long-term participation incentive in participatory sensing. In: Computer Communications, pp. 2803–2811. IEEE (2016)
11. Liu, J., Issarny, V.: An incentive compatible reputation mechanism for ubiquitous computing environments. In: International Conference on Privacy, pp. 297–311 (2016)
12. Zhu, G., Wei, X.J.: Research progress of multi-attribute reverse auction mechanism. J. Beijing Univ. Inform. Sci. Technol. (Nat. Sci.) J. **31**(2), 40–49 (2016)
13. Nan, W.Q., Guo, B., Chen, H.H., et al.: Dynamic incentive model of crowd sensing based on multi-interaction of cross-space. Chin. J. Comput. J. **38**(12), 2412–2425 (2015)

# Safeguarding Against Active Routing Attack via Online Learning

Meng Meng, Ruijuan Zheng$^{(\boxtimes)}$, Mingchuan Zhang, Junlong Zhu,
and Qingtao Wu

College of Information Engineering, Henan University of Science and Technology,
Luoyang 471023, China
zhengruijuan@haust.edu.cn

**Abstract.** The Border Gateway Protocol (BGP) is a vital protocol on
the Internet. However, the BGP is susceptible against the prefix interception attack, how to seek a secure route against prefix interception
attacks is an important problem. For this reason, we propose a novel
and effective router selection method on the Internet. First, we evaluate
resilience of autonomous systems against prefix interception attacks. Second, we evaluate security risk of next-hop routers and we also obtain the
historical performance of next-hop routers via online learning. Finally,
we compromise the two performance metrics of routers' resilience and
historical performance to choose a secure route. Our method is verified
by network simulations. The results show that the proposed method has
more resilience against prefix interception attacks.

**Keywords:** Online learning · Prefix interception · Routing attacks

## 1 Introduction

The Border Gateway Protocol (BGP) is a standard for changing network routes
between autonomous systems (ASes). The BGP does not validate route updates
via security mechanisms although it is vital to forward traffic on the Internet.
The BGP is vulnerable to the previously unknown active BGP routing attacks.
The BGP routing attacks including prefix hijacks and interception attacks [1].

The first prefix attack is succeed on January 22, 2006 [2]. AS-27506 mistakenly declared the IP prefix which is a part of AS-19758. Because the routers
don't want to verify exactly the rightful origin of every prefix. So they received
the announcement from the false/true origin which is based on route policies
and other standards. Finally, some ASes sent the traffic to the false origin takes
the place of the true origin. This is a prefix hijack attack. The prefix interception
attack is more advance than prefix hijack, the false origin AS will send the traffic
to the true origin AS after receive the traffic. The prefix interception attack does
not cause a great influence on the Internet. So prefix interception attacks are
hard to detect on the Internet.

© Springer Nature Switzerland AG 2020
S. Wen et al. (Eds.): ICA3PP 2019, LNCS 11945, pp. 128–134, 2020.
https://doi.org/10.1007/978-3-030-38961-1_12

Several methods have been proposed to improve the safety of BGP, which broadly fall into three categories: encipherment, anomaly mitigation, and anomaly detection. Cryptographic methods [3] generally use the Public Key Infrastructure to insure the routing announcements and guard against BGP attacks. Anomaly mitigation [4] propose to demote dubious route once it is detected. Anomaly detection approaches [5–8] discover anomalous information or behaviors in the BGP announcement and give alarm. Those methods aim to solve BGP routing attacks but they are weakness for defend prefix interception attacks.

To overcome this weakness, we represent a novel and effective route selection method against prefix interception attacks in this paper. First, we introduce the concept of resilience to evaluate the resilience of ASes against prefix interception attacks. The resilience of routers ensure the routes' reachability. Second, we evaluate the historical performance of routers via online learning and it ensures routes' security. And the risk value of prefix interception attacks is defined by mathematics first time. Finally, we compromise the two performance metrics of routers' resilience and historical performance to ensure the security and reachability of routes. We propose the method can significantly improve the security of routes by verified network simulations.

The paper is organized as follows. Section 2 describes the resilience evaluation of prefix interception attacks. We present a novel router selection method base on resilience and historical performance in Sect. 3. Section 4 uses network simulations on the Internet topology to evaluate the performance of our method and Sect. 5 concludes this paper.

## 2   Observation Resilience for the Prefix Interception Attacks

In this section, the concept of resilience is introduced for evaluating the ability of ASes against the prefix interception attack in AS-level.

Resilience: We introduce the resilience [9] to observe the ability of routers to resist prefix interception attacks. If the source AS $j$ is not hoaxed by a false origin source AS $f$ and still sends it's traffic to the true origin source AS $t$. We thought the source AS $j$ is resilience to the prefix interception attack initiated by the false origin AS $f$ on the true origin AS $t$.

Each node can be attacked success or failure, so we set $\alpha(t, j, f)$ to measure the resilience of nodes be attacked. Then, we have

$$\alpha(t, j, f) = \begin{cases} 1, & \text{hoaxed success;} \\ 0, & \text{otherwise.} \end{cases} \tag{1}$$

Each node has multiple paths leading to the false AS $f$ and the true AS $t$. Therefore, the resilience of a node is calculated as follows:

$$\bar{\alpha}(t, j, f) = \frac{p(j, t)}{p(j, t) + p(j, f)}. \tag{2}$$

In this formula, $p(j,t)$ is the number of paths from the AS $j$ to the true origin AS $t$. $p(j,f)$ is the number of paths from the AS $j$ to the false origin AS $f$. Further, we can calculate the resilience of the AS $j$ when the true origin ASes $t$ is decided and the false origin AS $f$ is not decided. Then, the resilience of AS $j$ is calculated as follows

$$R(j,t) := \sum_{f \in \mathcal{H}} \frac{\bar{\alpha}(t,j,f)}{(H-2)}. \tag{3}$$

$\mathcal{H}$ is the set of all ASes and the $H$ is the total quantity of ASes.

Route prediction: The route is decided via two conditions, which has shown in [10]: (1) Local Priority: The priority order is customer route, peer route and provider route; (2) Shortest Path: The path with the shortest hops will be preferred. And there are two cases to take into account if the invalid path be declared. In the first case, the existing routes from the false origin AS $f$ to the true origin AS $t$ are via peer or customer route, then its existing route to the true origin AS $t$ will not be effected. In the second case, the existing routes from the false origin AS $f$ to the true origin AS $t$ are via provider route, then it only can make the wrong announcement to its peers and customers.

Based on the above conditions, we use Algorithm 1 to measure the resilience of routers against the prefix interception attack.

---

**Algorithm 1.** Resilience to prefix interception attacks for ASes.

**Function** CalcInterceptResilience(graph $G$, node $j$, node $t$).
CalcPathsFromNode($G, j, t$).
  $R(j,t) = 0 \,\forall\, AS\ j, t$.
**for** each reachable node $v$ from node $j$ **do**
    $h \leftarrow$ num. of less preferred nodes than node $t$,
    $\mathcal{H} \leftarrow$ set of more preferred node than node $t$.
    **if** existing route $j$ to $v$ is provider route **then**
        $\mathcal{H} \leftarrow \mathcal{H} \bigcap \mathcal{M}$ where $\mathcal{M}$ contains all nodes for which $j$ to $m$ is provider route.
    **end if**
    $\mathcal{F} \leftarrow$ set of equally preferred node as node $t$.
    **if** existing route $j$ to $v$ is provider route **then**
        $\mathcal{P} \leftarrow \mathcal{F} \bigcap \mathcal{M}$ where $\mathcal{M}$ contains all nodes $m$ for which $j$ to $m$ is provider route,
        $\mathcal{F} \leftarrow \mathcal{F} - \mathcal{P}$.
    **end if**
    $R(j,t) \leftarrow h + len(\mathcal{H}) + len(\mathcal{P}) + \sum_{v \in \mathcal{V}} \bar{\alpha}(v,j,t)$
**end for**
$H \leftarrow$ number of nodes in $G$.
**return** $[R(j,t)/(H-2)$ for each node $j, t$ in $R]$.
**end Function**

---

## 3  Safeguarding Against the Prefix Interception Attack via Online Learning

In this section, we evaluate the historical performance of routers via online learning [11,12]. And we assume each AS has only one BGP router.

Consider a network with a set of $\mathcal{N} = \{1, 2, \ldots, N\}$ routers, the next-hop router is denoted by a set of $\mathcal{J} = \{1, 2, \ldots, J\}$ and the round is denoted by a set of $\mathcal{T} = \{1, 2, \ldots, T\}$. If an AS announces a prefix that it doesn't own, what is an AS launched prefix interception attacks. The false origin AS will send packets back to the true origin AS in a prefix interception attack. Because the false AS announcement a prefix which is same to packets' prefix of destination address. The false origin AS send the traffic to the true origin AS will be detected via neighbour nodes. Let $r_t(j)$ denotes the security risk of next-hop router $j$, we have

$$r_t^s(j) = \frac{\sum_{t=1}^{T} \mathbb{1}(\text{packets prefix} \neq \text{true BGP prefix})}{T}. \tag{4}$$

The prefix interception attack is time-sensitive, $r_t^s$ should change with time and the formula is valid during time interval $\mathcal{D}$.

In this paper, we aim to chosen a security route. Consistent with this criterion, we assume router $s$ is allowed to randomly select next-hop router from a given distribution $a_t^s \sim p_t^s \in \mathbb{R}^J$. To summarize, we have

$$\min_{\{p_t^s \in \Delta \mathcal{J}_t^s, \forall t, s\}} \sum_{t=1}^{T} \sum_{s=1}^{S} (p_t^s)^\top r_t^s \tag{5}$$

The $a_t^s$ denotes the set of selected next-hop router from the router $s$ in past round. $J_t^s$ represents the set of next-hop router that $s$ can select in round $t$, and $\mathbb{1}$ denotes the indicator function. $\Delta \mathcal{J}_t^s$ is defined as

$$\Delta \mathcal{J}_t^s := \left\{ p \in \mathbb{R}_+^J \,\middle|\, \sum_{j \in \mathcal{J}_t} p(j) = 1; p(j) = 0, j \notin \mathcal{J}_t^s \right\} \tag{6}$$

where $p(j)$ is denotes the $j$-th entry of $p$.

The historical performance of routers use information interaction assist the learning process. The information interaction between routers can be represented by a undirected graph. Consider a BGP router $s$ obtains the security risk of the next-hop router $r_t^s$, it also obtains the historical performance of other routers via information interaction. In this undirect graph $\mathcal{G}_t^s$, the node sets are the routers set $\mathcal{N}$. And we have $\mathcal{K}_t^s \subseteq \mathcal{N}$, $\mathcal{K}_t^s$ is not obligatorily a subset of $\mathcal{J}_t^s$ what is meaning that it's possible for router $s$ to acquire the historical performance of neighbor routers by information interaction. The historical performance of routers denote by

$$w_t^s(j) = \exp(-\eta_t^s \hat{R}_{t-1}^s(j)), \forall j \in \mathcal{J}. \tag{7}$$

The $\eta_t^s$ represents the stepsize, and $R_{t-1}^s(j)$ represents the accumulated security risk of router $j$, showed as

$$\hat{R}_{t-1}^s(j) = \sum_{\tau=1}^{t-1} \hat{r}_\tau^s(j), \forall j \in \mathcal{J}. \tag{8}$$

The $\hat{r}_\tau^s(j)$ is estimated security risk. The next-hop set $\mathcal{K}_t^s$ is showed, the possibility of the router $s$ choose the next-hop router $j$ is

$$p_t^s(j) = \frac{w_t^s(j)\mathbb{1}(j \in \mathcal{J}^s)}{\sum_{n \in \mathcal{J}_t^s} w_t^s(n)}. \tag{9}$$

Therefore, if $j \notin \mathcal{J}_t^s$, we have $p_t^s(j) = 0$. Following $p_t^s$, a next-hop router is chosen, and correspondent historical performance is showed after communication is over. We employ a biased and reduced-variance estimator in here, $\hat{r}_t^s(j)$ is defined as follow.

$$\hat{r}_t^s(j) = \frac{r_t^s(j)\mathbb{1}(j \in \{a_t^s \bigcup \mathcal{K}_t^s\})}{\mu_t^s + \Sigma_{(n,j) \in \mathcal{G}_t^s} p_t^s(n)}, \forall j \in \mathcal{J} \tag{10}$$

where $(n, j) \in \mathcal{G}_t^s$ denotes the whole edges between node $n$ and node $j$ in $\mathcal{G}_t^s$. The algorithm is conclusion in Algorithm 2.

---

**Algorithm 2.** The Security-Aware Router Selection under Stochastic client Attack

---

**Initialize:** weight $w_1^s = 1/J$, implicit explore factor $\mu_t^s$, learning rate $\eta_t^s$.
**for** $t = 1, 2, \cdots, T$ **do**
  Compute $w_t^s$ via (7).
  Reveal the available next-hop routers.
  Compute $p_t^s$ via (9) and choose next-hop $a_t^s \sim p_t^s$.
  Receive $r_t(a_t^s)$ and send $r_t(a_t^s)$ to other routers in $\{i \mid s \in \mathcal{K}_t^i\}$.
  Compute the security risk via (4).
  Estimate the security risk via (10)
**end for**

---

We describe two performance metrics of routers on above paper. The security of route can not ensure if we just care about the resilience of routers, and the reachability of route can not ensure if we just care about the historical performance. We compromise the two performance metrics of routers' resilience and historical performance to ensure the security and reachability of routes.

The steps are described as follows: First, we evaluate the resilience of routers use Algorithm 1. Second, we estimate the historical performance of routers via Algorithm 2. Finally, we offer a tunable parameter $\beta \in [0, 1]$ for compromise the two performance metrics,

$$W_j = \beta \times R(j, t) + (1 - \beta)w_t^s(j). \tag{11}$$

We obtain the best next-hop router by choosing the maximum value of $W_j$.

## 4    Simulation Tests

In this paper, we assess the router selection algorithm and compare it with counter-raptor (guard relay selection algorithm).

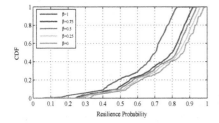
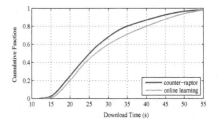

**Fig. 1.** Resilience probability.          **Fig. 2.** Download time for 5 MB data.

We compromise the two performance metrics of routers' resilience and historical performance with parameter $\beta$. We see the resilience probability has significantly improve when the historical performance participate in the next-hop router choose in Fig. 1. But with the increase of $\beta$, the increase of resilience probability is not significant. In the following work, we choose $\beta$ as 0.5 to evaluate the performance of network via measure the download time and the throughput of all nodes.

The network performance results shown this method has better performance from the network simulation. Figure 2 shows the download times of 5 MB data, and we can see that the latency has minor increase relative to guard relay selection algorithm for download 5 MB size data. Figure 3(a) and (b) show the 60s average receiver and sender throughput for all nodes, we can see that the online learning has almost the same throughput as counter-raptor during network simulation.

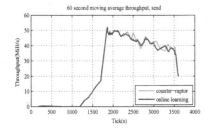
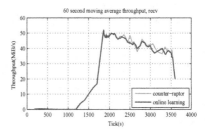

**Fig. 3.** Large-scale network performance evaluation.

# 5   Conclusion

In this work, we presented safeguard route against prefix interception attacks. First, we evaluate the resilience of routers for prefix interception attacks. Second, we measure the historical performance of routers for prefix interception attacks. Finally, we compromise the resilience and historical performance of routers to choose a best performance of next-hop router. The method successfully increases the probability of a AS being resilient to prefix interception attacks. Overall, our work is focus on proactively mitigating prefix interception attacks on the Internet.

**Acknowledgment.** This work was supported in part by the National Natural Science Foundation of China (NSFC) under Grants No. 61602155, No. U1604155, and No. 61871430, and in part by the basic research projects in the University of Henan Province under Grants No. 19zx010, and in part by the Science and Technology Development Programs of Henan Province under Grant No. 192102210284.

# References

1. Ballani, H., Francis, P., Zhang, X.: A study of prefix hijacking and interception in the internet. In: Proceedings of the ACM SIGCOMM 2007 Conference on Applications, Technologies, Architectures, and Protocols for Computer Communications, pp. 265–276 (2007)
2. Lad, M., Oliveira, R., Zhang, B., Zhang, L.: Understanding resiliency of Internet topology against prefix hijack attacks. In: The 37th Annual IEEE/IFIP International Conference on Dependable Systems and Networks, pp. 368–377 (2007)
3. Huston, G., Michaelson, G.: Validation of Route Origination Using the Resource Certificate Public Key Infrastructure and Route Origin Authorizations (2012). https://www.tools.ietf.org/html/rfc6483
4. Bellovin, S.M., Bush, R., Ward, D.: Security requirements for BGP path validation (2011). http://tools.ietf.org/html/draft-ymbk-bgpsec-reqs-02
5. Johann, S., Ralph, H., Quentin, J., Georg, C., Ernst, B.W.: HEAP: reliable assessment of BGP hijacking attacks. IEEE J. Sel. Areas Commun. **34**, 1849–1861 (2016)
6. Rahul, H., Niklas, C., Nahid, S.: Collaborative framework for protection against attacks targeting BGP and edge networks. Comput. Netw. **122**, 120–137 (2017)
7. Shi, X., Xiang, Y., Wang, Z., Yin, X., Wu, J.: Detecting prefix hijackings in the Internet with argus. In: Proceedings of the 12th ACM SIGCOMM Internet Measurement Conference (2012)
8. Zhang, Z., Zhang, Y., Mao, H.Y.C., Morley, Z., Randy, B.: ISPY: detecting IP prefix hijacking on my own. IEEE/ACM Trans. Netw. **6**, 1815–1828 (2010)
9. Sun, Y., Anne, E., Nick, F., Mung, C., Prateek, M.: Counter-RAPTOR: safeguarding Tor against active routing attacks. In: 2017 IEEE Symposium on Security and Privacy (2017)
10. Gao, L., Rexford, J.: Stable Internet routing without global coordination. IEEE/ACM Trans. Netw. **9**, 681–692 (2001)
11. Elad, H.: Introduction to online convex optimization. The Computing Research Repository (2019)
12. Li, B., Chen, T., Giannakis, G.B.: Secure mobile edge computing in IoT via collaborative online learning. The Computing Research Repository (2018)

# Reliability Aware Cost Optimization for Memory Constrained Cloud Workflows

E Cao[1], Saira Musa[1], Jianning Zhang[1], Mingsong Chen[1(✉)], Tongquan Wei[1],
Xin Fu[2], and Meikang Qiu[3]

[1] Shanghai Key Lab of Trustworthy Computing, East China Normal University,
Shanghai, China
mschen@sei.ecnu.edu.cn
[2] Department of Electrical and Computer Engineering, University of Houston,
Houston, USA
[3] Department of Computer Science, Texas A&M University, Commerce, TX, USA

**Abstract.** Due to the increasing number of constituting jobs and input
data size, the execution of modern complex workflow-based applications
on cloud requires a large number of virtual machines (VMs), which makes
the cost a great concern. Under the constraints of VM processing and
storage capabilities and communication bandwidths between VMs, how
to quickly figure out a cost-optimal resource provisioning and scheduling
solution for a given cloud workflow is becoming a challenge. The things
become even worse when taking the infrastructure-related failures with
transient characteristics into account. To address this problem, this paper
proposes a soft error aware VM selection and task scheduling approach
that can achieve near-optimal the lowest possible cost. Under the reliability and completion time constraints by tenants, our approach can
figure out a set of VMs with specific CPU and memory configurations
and generate a cost-optimal schedule by allocating tasks to appropriate
VMs. Comprehensive experimental results on well-known scientific workflow benchmarks show that compared with state-of-the-art methods, our
approach can achieve up to 66% cost reduction while satisfying both
reliability and completion time constraints.

**Keywords:** Workflow scheduling · Cost optimization · Reliability
constraint · Soft error · Evolutionary algorithm

## 1 Introduction

Along with the increasing popularity of cloud services in a pay-as-you-go manner,
more and more enterprises and communities adopt cloud platforms to deploy

Supported by the grants from National Key Research and Development Program of
China (No. 2018YFB2101300), Natural Science Foundation of China (No. 61872147)
and National Science Foundation (No. CCF-1900904, No. CCF-1619243, No. CCF-
1537085 (CAREER)).

S. Wen et al. (Eds.): ICA3PP 2019, LNCS 11945, pp. 135–150, 2020.
https://doi.org/10.1007/978-3-030-38961-1_13

their commercial or scientific workflows to facilitate the distribution of data and computation-intensive applications [1]. However, as modern workflows grow rapidly in terms of the number of constituting jobs and input data size, their task allocation and scheduling complexity is skyrocketing. The scheduling of workflows requires large number of virtual machines (VMs), which makes the workflow execution cost a great concern to cloud service providers.

Since the resource allocation problem for cloud workflows is NP-complete [1], various heuristics have been proposed to find near-optimal schedules quickly. However, as more and more data center servers adopt CMOS-based processors, few of existing methods take transient faults (i.e., soft errors) [2,3] into account. Typically, a modern CMOS processor consists of billions of transistors where one or more transistors form one logic bit to hold the logic value 0 or 1. Unfortunately, various phenomena (e.g., high energy cosmic particles, cosmic rays) can result into the notorious soft error where the binary values held by transistors are changed by mistake, and the probability of incorrect results or system crashes during cloud workflow execution becomes increasingly higher.

As a reliable fault-tolerance mechanism, the checkpointing with rollback-recovery [4] has been widely adopted to improve the reliability of cloud workflow execution. By periodically saving VM execution states in some stable storage at specified checkpoints, the rollback-recovery can restore the system with the latest correct state to enable re-execution when an execution error is detected. However, the unpredictable overhead of checkpointing with rollback-recovery operations prolonged the execution time of workflow jobs due to re-execution which not only cause severe temporal violations [1], but also increase the overall execution cost.

To achieve increasing profit in the fierce cloud computing market, cloud service providers need to explore efficient cloud workflow schedules involving both resource provisioning (i.e., a set of VMs with specific processing and storage configurations) and allocation (i.e., assignment of workflow tasks to the VMs without violating VM memory constraints) to minimize the execution cost. In this paper, we propose a novel approach that can generate cost-optimal and soft error resilient schedules for workflow applications considering the overhead of both checkpointing with rollback-recovery and inter-VM communications. This paper makes following three major contributions:

- Under the constraints of VM memory size and overall workflow makespan, we formalize the cost-optimization problem of task scheduling for cloud workflows considering the overhead of both checkpointing with rollback-recovery and inter-VM communication.
- Based on two-segment group genetic algorithm (TSG-GA), we propose a soft error aware cost-optimized workflow scheduling approach that can quickly figure out a schedule with cost-optimal resource provisioning and task-to-VM allocation for a given workflow application.
- We evaluate our approach on well-known complex scientific benchmarks and show the effectiveness of the proposed approach.

The rest of this paper is organized as follows. Section 2 presents Section the related work. Section 3 formalizes the cost optimization problem for cloud workflow scheduling considering both resource and reliability constraints. Section 4 details our proposed approach, and Sect. 5 presents the corresponding experimental results on well-known benchmarks. Finally, Sect. 6 concludes the paper.

## 2    Related Work

Despite all the advantages of cloud computing, task scheduling in cloud workflows with minimum completion time and reduced cost while maintaining high reliability have become a major challenge, which have attracted great attention from researchers and industry. For instance, Topcuoglu et al. [5] proposed a Heterogeneous Earliest Finish Time (HEFT) algorithm which assigns the task with the highest priority to the VM, in order to achieve the earliest finish time. Panday et al. [6] presented a scheduling heuristic based on Particle Swarm Optimization (PSO) to minimize the total execution cost of application workflows on cloud computing environments while balancing the task load on the available resources. Since, the faster cloud services are normally more expensive, therefore, users face a time-cost trade-off in selecting services. As any delay in completion time can produce negative impacts on cost optimization of workflow scheduling. A general way to address this trade-off is to minimize monetary cost under a deadline constraint. Nonetheless, only a few approaches have been presented to address this issue in the literature [7–10], which solve the workflow scheduling problem on the Infrastructure as a Service (IaaS) platform. Aforementioned literatures can effectively minimize the makespan or cost but, none of them considered reliability during task scheduling.

In order to achieve the reliability, Wang et al. [11] proposed a LAGA (Look-Ahead Genetic Algorithm) to optimize the reliability and makespan of a workflow application simultaneously. An algorithm was designed and implemented in [12] by Wen et al. to solve the problem of deploying workflow applications over federated clouds while meeting the reliability, security and monetary requirements. Although the above work can guarantee the reliability but, they did not consider the soft error occurrences in cloud data centers. Wu et al. [3] proposed a soft error-aware energy-efficient task scheduling for workflow applications in DVFS-enabled cloud infrastructures under reliability and completion time constraints. However, the above work did not consider the cost optimization.

To our best knowledge, our work is the first attempt to minimize the execution cost of cloud workflows under makespan, reliability and memory constraints while considering soft errors in cloud data centers.

## 3    Scheduling Model and Problem Definition

In this section, we present VM model, workflow model and fault tolerance. Finally, the problem of cost optimization workflow scheduling in the cloud environment is defined.

## 3.1 Modeling of VM

IaaS cloud provider offers a set of VM configurations $C = \{C_0, C_1, ..., C_n\}$ to tenants by renting VMs on demand. The VM configuration $C_i$ is characterized by a four-tuple $(vn, bw, ram, price)$, where $vn(C_i)$, $bw(C_i)$, $ram(C_i)$ and $price(C_i)$ denote the number of vCPUs, the network bandwidth, the memory and the rental price per unit time of $C_i$, respectively. A running VM with certain configuration is treated as an instance and customers can purchase unlimited number of VM instances according to their requirements. The set of VM instances is denoted by $S = \{S_0, S_1, ..., S_i\}$, where $S_i$ is a VM instance with a certain configuration $\varphi(S_i)$ of $C$. We assume that all the tasks are parallelizable so that all vCPUs can be fully used and have same processing capacities. It is noteworthy that although cloud service providers have massive computing and memory resources, there are upper limits on the number of vCPU and the amount of memory for a single VM instance. In addition, the allocation of memory source of VM is usually discrete, i.e., $ram(S_i) = \alpha \cdot M$, where $M$ is the unit of memory which depends on cloud service providers and $\alpha$ is an integer.

## 3.2 Modeling of Workflow

A workflow $W = (T, E)$ as shown in Fig. 1 with dependent tasks is represented as the Directed Acyclic Graph (DAG), where $T = \{t_0, t_1, ..., t_n\}$ represents the task set and $E$ denotes the set of dependencies between tasks. For instance, $e_{uv} \in E$ indicates the dependency between task $t_u$ and $t_v$, where $t_u$ is the immediate predecessor of $t_v$, and $t_v$ is the immediate successor of $t_u$. We use a four-tuple $(referload, mem, pred, succ)$ to represent a task, where $referload(t_u)$, $mem(t_u)$, $pred(t_u)$ and $succ(t_u)$ denote the reference workload, the maximum memory required for task execution on a VM instance, the immediate predecessors and the immediate successors of task $t_u$, respectively. If $pred(t_u) = \emptyset$, then $t_u$ is an entry task and if $succ(t_u) = \emptyset$, then $t_u$ is an exit task. This article allows single entry and exit task, this can be assured by adding a pseudo entry task and a pseudo exit task. We assume the reference workload is task execution time on a VM instance whose vCPU number equals 1.

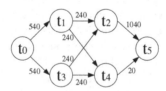

**Fig. 1.** A workflow example with 6 tasks

As shown in Fig. 1, each edge $e_{uv}$ have weight $wt_{u,v}$, which represents the amount of data that needs to be transmitted from $t_u$ to $t_v$. A task cannot start its execution until the input data has been received from all of its predecessors. If task $t_u$ and $t_v$ are assigned to VM $S_i$ and $S_j$, the communication cost $comm(t_u, t_v)$ can be calculated as follows:

$$comm(t_u, t_v) = \begin{cases} 0 & if\ S_i = S_j \\ \frac{wt_{uv}}{bw_{i,j}} & if\ S_i \neq S_j \end{cases} \tag{1}$$

We consider that the communication bandwidth $bw_{i,j}$ between $S_i$ and $S_j$ is the lower bandwidth, i.e., $bw_{i,j} = min(bw(\varphi(S_i)), bw(\varphi(S_j)))$.

### 3.3 Modeling of Tasks with Fault Tolerance

To ensure the reliability of workflow execution in a cloud environment, we use an equidistant checkpointing technique [4], where the lengths of checkpoint intervals are same. The execution state of task is stored in a secure device [3,13], guaranteeing that the task can read the latest correct state to re-execute when a soft error occurs.

Suppose that task $t_u$ is assigned to VM $S_i$, so the best case execution time of task $t_u$ without any soft error can be formulated as

$$ET_{best}(t_u, S_i) = \frac{referload(t_u)}{vn(\varphi(S_i))} + N(t_u, S_i) \cdot O_i, \tag{2}$$

where $N(t_u, S_i)$ is the number of checkpoint of task $t_u$ on VM $S_i$ and $O_i$ is the time overhead of checkpointing. Checkpoint interval length of task $t_u$ assigned to VM $S_i$ is formulated as

$$Seg(t_u, S_i) = \frac{referload(t_u)}{vn(\varphi(S_i))} \cdot \frac{1}{N(t_u, S_i) + 1}. \tag{3}$$

Let $Fmax$ denotes the maximum number of fault occurrences during task execution. Therefore, with $F_{max}$ soft error occurrences, the worst case execution time of task $t_u$ on VM $S_i$ can be expressed as

$$ET_w(t_u, S_i) = \frac{referload(t_u)}{vn(\varphi(S_i))} + 2 \cdot N(t_u, S_i) \cdot O_i + Seg(t_u, S_i) \cdot F_{max}, \tag{4}$$

where $2 \cdot N(t_u, S_i)$ indicates the accumulative overhead of $Fmax$ checkpoint saving and retrieval operations, and $Seg(t_u, S_i) \cdot F_{max}$ represents the fault tolerance overhead.

In order to minimize the worst case execution time $ET_w(t_u, S_i)$, we use the optimal number of checkpoint $N_{opt}(t_u, S_i)$ [4], which can be calculated as

$$N_{opt}(t_u, S_i) = \sqrt{\frac{Fmax}{O_i} \cdot \frac{referload(t_u)}{vn(\varphi(S_i))}} - 1. \tag{5}$$

We assume that the average arrival rate of soft error $\lambda_i$ of the VM instance $S_i$ is consistent with Poisson distribution [3]. Therefore, the probability of $F$ soft error occurrences on VM $S_i$ can be formulated as

$$Pr(t_u, S_i, F) = \frac{e^{-\lambda_i \cdot ET_w(t_u, S_i)} \cdot (\lambda_i \cdot ET_w(t_u, S_i))^F}{F!}. \tag{6}$$

Task reliability is defined as the probability that a task can be successfully executed in the presence of soft errors. The probability of successful recovery of $F$ faults can be calculated as

$$Pr_{succeed}(F, S_i) = e^{-\lambda_i \cdot F \cdot Seg(t_u, S_i)}. \tag{7}$$

Hence, the reliability of task $t_u$ on VM $S_i$ can be calculated as

$$R(t_u, S_i) = \sum_{F=0}^{F_{max}} Pr(t_u, S_i, F) \cdot Pr_{succeed}(F, S_i). \tag{8}$$

### 3.4    Problem Definition

A binary tuple ($Task\_VM$, $VM\_VMC$) is used to represent a workflow scheduling scheme $P$, where $Task\_VM$ represents the mapping of tasks to VM instances, and $VM\_VMC$ represents the VM instances to VM configurations mapping. $VM\_VMC_i$ is used to represent the configurations of VM instance $S_i$, i.e., $\varphi(S_i) = VM\_VMC_i$. $Task\_VM_u$ indicates the VM instance to which task $t_u$ is assigned. Let $ST_{i,u}$ and $FT_{i,u}$ denote the start time and finish time of task $t_u$ on VM $S_i$, respectively. We get the start time of the task $t_u$ on VM $S_i$ as follows:

$$ST_{i,u} = \begin{cases} 0 & \text{if } pred(t_u) = \emptyset \\ \max\limits_{t_w \in previous(t_u)} \{ \max\limits_{t_v \in pred(t_u)} \{ FT_{i,u} + \\ comm(t_u, t_v) \}, FT_{i,w} \} & \text{if } pred(t_u) \neq \emptyset \end{cases} \tag{9}$$

$FT_{i,w}$ is the finish time of the task $t_w$ executed before task $t_u$ on the same VM instance $S_i$ and $previous(t_u)$ represents the tasks executed before $t_u$ on $S_i$. Note that if there are multiple tasks that can be executed at the same time on the same VM instance, we select a task according to the order in which the tasks are scheduled to the VM instance. Therefore, $FT_{i,u}$ is formulated as

$$FT_{i,u} = ST_{i,u} + ET_w(t_u, S_i). \tag{10}$$

The makespan of workflow $W$ can be obtained as,

$$makespan(W, P) = \max_{t_u \in T(W) \& S_i \in S(P)} \{ FT_{i,u} \}, \tag{11}$$

where $T(W)$ is the task set of workflow $W$, $S(P)$ is the set of VM instances obtained by scheduling scheme $P$. In Sect. 3.3 we have obtained the reliability of a task, to calculate the workflow reliability we find the cumulative product of the reliability $R(W, P)$ of all the workflow tasks [3,14], such as

$$R(W, P) = \prod_{t_u \in T(W) \& S_i \in S(P)} R(t_u, S_i). \tag{12}$$

Let $t_i$ denote the tasks assigned to the VM $S_i$, i.e., $t_i = \{t_u | Task\_VM_u = S_i\}$. The start time $VM_{ST_i}$ and end time $VM_{FT_i}$ of VM $S_i$ can be formulated as

$$VM_{ST_i} = \min_{t_u \in t_i} \{ST_{i,u}\}, \tag{13}$$

$$VM_{FT_i} = \max_{t_u \in t_i} \{FT_{i,u}\}. \tag{14}$$

Finally, the cost of workflow $W$ scheduling can be formulated as

$$Cost(P, W) = \sum_{S_i \in S(P)} price(VM\_VMC_i(P)) \cdot (VM_{ST_i} - VM_{FT_i}). \tag{15}$$

Considering a workflow $W$ and a set of VM configurations $C$, we need to find a suitable scheduling scheme $P$ to minimize the cost of workflow scheduling while satisfying the makespan constraint $D_{goal}$, memory constraint, and reliability constraint $R_{goal}$. Therefore, the problem to be solved in this paper can be formally defined as the minimization problem:

$$Minimize : Cost(W, P) \tag{16}$$
$$Subject\ to : R(W, P) \geq R_{goal}, \tag{17}$$
$$makespan(W, P) \leq D_{goal}, \tag{18}$$
$$mem(t_u) < ram(\varphi(S_i)), \quad t_u \in T(W) \ \& \ S_i \in S(P). \tag{19}$$

Equation (19) describes the memory constraints of workflow scheduling. The memory required for a task should not exceed the RAM of the VM instance on which the task it assigned. The problem presented in this paper is a typical combinatorial optimization problem. It is worth noting that although assigning tasks to the powerful VM instances can reduce the makespan of a workflow, but the idle time caused by data dependencies on the powerful VM instances will increase the cost and excessive memory resources can also impose costly penalties. Meanwhile, the reliability of task can also be influenced by the processing capability of VM instances, which makes the scheduling problem more complex.

## 4   Our Evolutionary Approach

Genetic algorithm (GA) has the characteristics of powerful global search ability, excellent concurrency and strong robustness, which is easy to combine with other methods and has become a universal optimization method [15]. It is widely used in workflow scheduling and cloud computing [8,16,17]. Since the existing GA is difficult to apply directly to the workflow scheduling problem, we explore the two-segment group genetic algorithm (TSG-GA) for cost optimized workflow scheduling with makespan, reliability and memory constraints. The algorithm consists of encoding, initial population generation, selection, crossover, mutation, elitism, fitness function and chromosome modification.

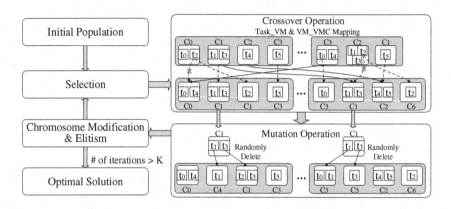

**Fig. 2.** The execution process of TSG-GA

The overall execution process of TSG-GA is shown in Fig. 2. We first randomly generate the initial population according to the target encoding, and select individuals with better fitness from initial population for crossover, mutation, and modification operations. Here the modification operation is used to satisfy the memory constraint for each chromosome. Then, we use the elitism strategy to preserve the best individual generated during the process of evolution. After a certain number of iterations ($K$), the final best individual (global optimal solution) is returned as the workflow scheduling solution.

### 4.1 Encoding

As discussed in Sect. 3.4, the encoding of our approach is a two segment integer encoding based on task grouping. For example, we use *ind* to represent a chromosome, i.e., an individual in the population, and *ind* consists of two segment: *ind. Task_VM* and *ind. VM_VMC*.

The encoding example is shown in Fig. 3. *Task_VM* is a group-based integer encoding, grouping tasks according to their corresponding VM instances. The encoding length of *Task_VM* is equal to the number of tasks. Gene index in *Task_VM* encoding represents the task, and the value represents the corresponding VM instance, For example, $Task\_VM(1) = 1$ indicates that task $t_1$ is assigned to VM instance $S_1$. *VM_VMC* is encoded as a variable length integer encoding, the length of which is the maximum index value of the VM instances in *Task_VM*.

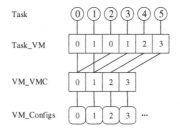

**Fig. 3.** An example of encoding

Similar to *Task_VM* encoding, $VM\_VMC(1) = 2$ indicates that the configuration of VM Instance $S_1$ is $C_2$. The two-segment integer encoding method designed in this paper is simple and intuitive.

### 4.2  Fitness Function

Fitness function is used to evaluate the fitness of solution in the evolution process. In this paper, we use fitness function to minimize the cost $Cost(W, P)$ as described in Eq. (15). In order to satisfy makespan and reliability constraints, penalty parameters $\gamma$ and $\delta$ are introduced. How to satisfy memory constraints will be described in Sect. 4.4. The fitness of a chromosome deteriorates if it does not meet the makespan or reliability constraint. Let $\xi$ be the set of constraints $\{D_{goal}, R_{goal}\}$, the *Fitness* function can be formulated as

$$Fitness(W, P) = \begin{cases} Cost(W, P), & \xi \ is \ satisfied \\ \gamma \cdot \delta \cdot Cost(W, P), & otherwise \end{cases} \quad (20)$$

$\gamma$ and $\delta$ are the real numbers greater than 1 if $D_{goal}$ and $R_{goal}$ are not satisfied, respectively. The goal of TSG-GA is to minimize the $Fitness(W, P)$.

### 4.3  Crossover and Mutation

The designed crossover operator can ensure that the original task grouping and VM configurations information of chromosomes will not be lost, which uses the tasks in the same VM instance as the crossover unit to avoid the problem that the direct crossover for *Task_VM* may destroy the task grouping information.

Figure 4 shows the execution process of the crossover operator. Suppose *ind1* and *ind2* are parents. Firstly, an empty chromosome *L1* is created as the offspring, and then a segment of genes of *ind1.Task_VM* are randomly selected. As shown in Fig. 4, *ind1.Task_VM(3)* to *ind1.Task_VM(4)* are selected. Genes in the same groups as the selected genes are copied into the offspring *L1* with the associated VM configurations, i.e., *ind1.Task_VM(1)*, *ind1.Task_VM(3)*, *ind1.Task_VM(4)*, *ind1.VM_VMC(1)*, and *ind1.VM_VMC(2)*. The length of the selected genes is limited to length of *ind1.VM_VMC* to avoid duplicating genes

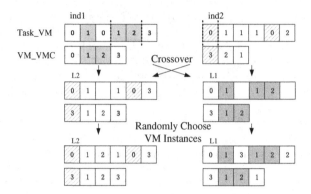

**Fig. 4.** An example of *Task_VM* and *VM_VMC* crossover

from *ind1* too much which destroy the grouping information of *ind2*. Then copy the groups and corresponding VM configurations of *ind2* that do not overlap with the previously copied genes of *ind1* into *L1*. Here *ind2.Task_VM*(2) and *ind2.Task_VM*(5) cannot be copied because the corresponding VM instances overlap with the previously copied VM instances. At this time, there are some fragments of *L1.Task_VM* are not filled. We simply assign the tasks in these fragments to the existing or new VM instances and the crossover operation is completed. Same operation is performed by swapping the roles of *ind1* and *ind2*, and we get the offspring individual *L2*.

For mutation operator, as shown in Fig. 2, it marks a chromosome as a mutated individual according to mutate rate, where we randomly delete one of the VM instance and assign the tasks of the VM instance to the existing or new VM instances. This paper argues that splitting and reorganizing the tasks in the VM instance with the most tasks is beneficial to jump out of the local optimum.

### 4.4   Chromosome Modification

The cloud workflow scheduling problem in this paper includes memory constraint that genetic algorithm does not have ability to deal with. In the process of evolution, if memory constraint is not satisfied, chromosome modification algorithm is called to satisfy the memory constraint.

As shown in Algorithm 1, for the chromosome that does not satisfy memory constraint, lines 3–15 search for an alternative VM configuration for each VM instance (traversing from $k = 0$) that does not satisfy memory constraint. Line 8 uses *GetAvailVmConfig* function to get an available VM configuration for the VM instance which need to increment the RAM from VM configurations $C$. The available configuration should satisfy memory constraint and have the same or similar processing capability with the original VM instance (the number of vCPUs is close to the original VM instance). Then line 9 uses the available VM configuration to replace the configuration of original VM instance, and line 12

---

**Algorithm 1:** Chromosome Modification

---

**Input**: i) $ind$, the chromosome to be modified;
ii) $vmConfigs$, available VM configuration set;
**Output**: $new\_ind$, modified chromosome

1 **ReformIndividual**($ind, vmConfigs$) **begin**
2     $new\_ind = ind$;
3     **for** $k = 0$ *to* $ind.VM\_VMC.size()$ **do**
4        $v = ind.VM\_VMC[k]$;
5        $candidate = []$;
6        $tasks = GetTasks(ind, k)$;
7        **if** $FindMaxMem(tasks) > vmConfigs[v].ram$ **then**
8           **for** $vmConfig \in GetAvailVmConfig(vmConfigs[v], tasks)$ **do**
9              $new\_ind.VM\_VMC[k] = vmConfig$;
10              $candidate\_ind = new\_ind$;
11              $candidate.add(candidate\_ind)$;
12           **end**
13           $new\_ind = FindElitis(candidate)$;
14        **end**
15     **end**
16     **return** $new\_ind$;
17 **end**

---

adds the modified chromosome to the candidate individual set. Line 13 selects the best individual from the candidate individual set. In the end, the entire chromosome is modified and the memory constraint is satisfied.

**Table 1.** Price of custom machine types provided by Google Cloud

| Charge items | Cost |
|---|---|
| vCPU | $0.033174/vCPU hour |
| Memory | $0.004446/GB hour |

## 5 Performance Evaluation

The effectiveness of the proposed method was evaluated through thorough experiments based on WorkflowSim [18] using well-known workflows [19] such as CyberShake, Inspiral and Epigenomics. Three workflows with two sets of tasks for each workflow were used in the experiment, which were generated by the toolkit Workflow-Generator [20] based on its default configurations. To reflect memory constraints, we randomly add memory attributes (1–8 GB) to the tasks of the generated workflows. Furthermore, HEFT and PSO algorithms with makespan and VM idle time minimization were compared with the proposed method. The HEFT algorithm assigns a task to the VM instance to achieve the earliest finish time according to task priority, yielding shortest workflow makespan. The PSO is an evolutionary computational algorithm which is widely used in the research of task scheduling for workflow application in the cloud. For comparison with our approach, the HEFT and PSO algorithm were modified to make sure an unlimited number of VM instances can be created with the same configuration. We also added memory constraints in both HEFT and PSO algorithm, and minimized the size of the memory of VM instance to reduce its execution cost. In addition, we calculated the reliability and cost of workflow scheduling according to Eqs. (12) and (15), respectively. All the experiments were performed on a desktop with 3.10 GHz Intel Core i5 CPU and 8 GB RAM.

## 5.1 Experimental Setting

The VM configuration and price were set by referring to custom machine types of Google Cloud. According to the characteristics of workflows and the VM configurations, a total of 40 VM configurations are selected with vCPU 1–4 and memory of 1–10 GB (1 GB, 2 GB, ..., 10 GB). Moreover, the network bandwidth between VM instances is 10 Mbps. The price of custom machine types provided by Google Cloud is shown in Table 1:

Note that although the price shown in Table 1 is in hour, Google Cloud can charge each VM instance in seconds. We assume that the soft error occurs independently in each VM instance and it is in accordance with the Poisson distribution. Supposing soft error occurrence rate $\lambda_i$ of each VM instance is the same for a workflow, we have taken different soft error rates for different workflows, and the value of $\lambda_i$ ranges from $10^{-6}$ to $10^{-3}$. We set the maximum soft error tolerance number $F_{max}$ to 1, and the overhead of each checkpoint $O_i$ to 0.1 s. The size of the population of TSG-GA is 100, the number of generation, the crossover rate, and the mutation rate are 100, 0.8 and 0.1, respectively. For PSO, the size of the population is 100 and the number of generation is 100. While the learning factors c1 = 2, c2 = 2, and inertia weight is 0.9.

To evaluate the effectiveness of the proposed TSG-GA under makespan, reliability and memory constraints, we set makespan constraint to $D_{goal} = \theta \cdot MH$, where $MH$ is the workflow scheduling makespan obtained by HEFT algorithm and $\theta$ is a constant real number. In the experiment, we let $\theta$ take different values ($1 \leq \theta \leq 2$), which means that our approach should not make the makespan of a workflow scheduling $\theta$ times longer than the makespan obtained by HEFT algorithm, meanwhile, we set the workflow scheduling reliability constraint to $R_{goal} = RH - \beta$, where $RH$ is the workflow scheduling reliability obtained by HEFT algorithm and $\beta$ is the reliability margin. The number of soft error tolerances $F_{max}$ is fixed and assumed that the soft error rate of each VM instance is the same. Since the reliability of the task is directly related to the task execution time, it is more likely to encounter a soft error when the execution time becomes longer, which results in lower reliability. The HEFT algorithm can achieve approximate shortest makespan and highest task reliability as it finishes the task in the earliest time. Therefore, we used the reliability and makespan obtained by HEFT as references of $R_{goal}$ and $D_{goal}$ for each workflow.

## 5.2    Results and Analysis

In the experiment, the workflow scheduling generated by our approach always satisfies the memory constraint, because chromosome modification guarantees the memory constraints. We performed experiments five times on each workflow and finally took the average as the final result.

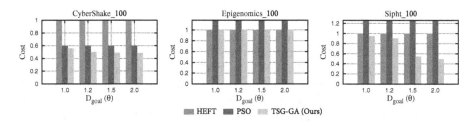

**Fig. 5.** Cost of large workflows with fixed reliability ($\beta = 0$) and varying makespan

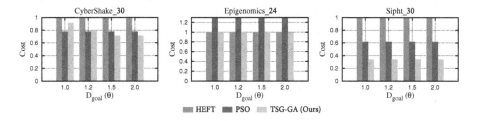

**Fig. 6.** Cost of small workflows with fixed reliability ($\beta = 0$) and varying makespan

**Results of Workflows with Fixed Reliability Constraint.** Firstly, we performed experiments using three workflows with two sets of task for each workflow under different makespan constraints, fixed reliability constraint, and memory constraint. Each set of tasks were defined as small and large workflows with 30 (or 24) and 100 tasks, respectively. We set $\theta$ to 1, 1.2, 1.5 and 2, and set $\beta$ to 0 (i.e., $RH$). Figures 5 and 6 show the cost results of the proposed approach in comparison with HEFT and PSO on large workflows (i.e., CyberShake_100, Sipht_100 and Epigenomics_100) and small workflows (i.e., CyberShake_30, Sipht_30 and Epigenomics_24). Note that we did not set any constraint for PSO method, and just get results of HEFT and PSO once for one workflow in the case of $\theta = 1$. Our approach spent around 11.20 s on average to generate one schedule on large workflows and 1.41 s on small workflows.

To facilitate performance comparison, we took HEFT method as baseline, and took scheduling costs divided by the cost of HEFT as the final costs for each workflow. Our approach always satisfied the constraints both on large and small workflows. From Figs. 5 and 6, it can be observed that our approach outperforms the HEFT and PSO algorithms. For example, compared to the HEFT algorithm, PSO can achieve 40.0% cost reduction on the CyberShake_100 while our approach can achieve 44.1% cost reduction when $\theta = 1$. When $\theta = 2$, PSO can achieve 39.1% cost reduction on Sipht_30 while our approach can achieve 66.0% cost reduction. If we compare the worse performance cases, the proposed approach TSG-GA only performs worse than PSO on CyberShake_30 in the case of $\theta = 1$. However, PSO performance is even worse than HEFT on half of the workflows. This is mainly because it tends to converge prematurely and falls into local optimum due to the lack of diversity of the population in the search space. The processing capability of the VM instance using custom machine type provided by Google Cloud is linear to the price in the experiment. Therefore, cost reduction lies in reducing the idle time of VM instances, while HEFT just finishes tasks as quickly as possible. Complex dependencies between tasks make tasks to wait for execution on the VM instances, making it impossible to guarantee a minimum idle time. Our approach can create an appropriate number of VM instances with appropriate configurations and schedule tasks reasonably according to the dependencies, while reducing idle time of instances to reduce costs. We can see that the cost optimization achieved on Epigenomics is only 2%. This is because Epigenomics workflow transmits less data and its data dependency is relatively simple, so its main cost comes from the vCPU usage time but not idle time. It can be seen that as $\theta$ increases, our approach can achieve better results due to the vast search space in genetic algorithm.

**Results of Workflows with Fixed Makespan Constraint.** We conducted experiments with the three workflows discussed in the previous section under fixed makespan constraint, different reliability constraints and memory constraint. We set $\theta$ to 1 and $\beta$ to 0.0000, 0.0001, 0.0002 and 0.0003. Figures 7 and 8 show the comparisons of workflow scheduling results obtained by our approach on those three workflows with HEFT and PSO algorithm.

Similarly, we used the HEFT as the benchmark. It is found that our approach always satisfies the constraints on these three workflows and outperforms the HEFT and PSO algorithms as depicted in Figs. 7 and 8. When $\beta = 0$, the reliability constraint of CyberShake_30 is 0.9987 and our approach can achieve 6.5% cost reduction compared to HEFT method. When $\beta = 0.0003$, the reliability constraint of CyberShake_30 is 0.9984, our approach can achieve 12% cost reduction compared to HEFT method. Reliability constraint make VM configurations to have strong processing capability to allow tasks to be completed as quickly as possible, which improves the reliability of the tasks. It can be seen from Figs. 7 and 8, when reliability constraint become loose, our approach can search for a better scheduling scheme.

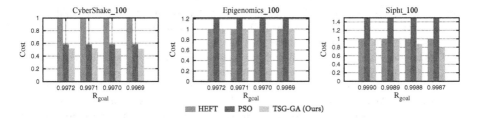

**Fig. 7.** Cost of large workflows with fixed makespan ($\theta = 1$) and varying reliability

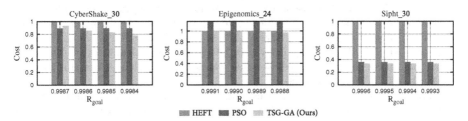

**Fig. 8.** Cost of small workflows with fixed makespan ($\theta = 1$) and varying reliability

## 6    Conclusions

Due to the increasing number of transistors on modern processors, the servers in data center is more susceptible to the notorious transient faults (i.e., soft errors). Although checkpointing with rollback-recovery mechanism is promising in tackling this problem to improve the reliability of cloud workflows, its overhead is too large to be neglected. The inevitable overhead will strongly affect the overall cost of workflow execution on cloud with a pay-as-you-go manner. To address this problem, this paper proposed a genetic algorithm based approach, known as TSG-GA, that can quickly figure out a cost-optimal schedule by considering both the overhead of checkpointing with rollback-recovery and the resource constraints (i.e., maximum number of vCPUs and available memory within a VM, network bandwidth) given by cloud workflow tenants. Comprehensive experimental results on well-known complex scientific benchmarks shows the effectiveness of our proposed approach.

## References

1. Liu, X., et al.: The Design of Cloud Workflow Systems. Springer, New York (2012). https://doi.org/10.1007/978-1-4614-1933-4
2. Vishwanath, K.V., Nagappan, N.: Characterizing cloud computing hardware reliability. In: Proceedings of ACM Symposium on Cloud Computing (SoCC), pp. 193–204 (2010)
3. Wu, T., Gu, H., Zhou, J., Wei, T., Liu, X., Chen, M.: Soft error-aware energy-efficient task scheduling for workflow applications in DVFS-enabled cloud. J. Syst. Archit. **84**, 12–27 (2018)

4. Wei, T., Chen, X., Hu, S.: Reliability-driven energy-efficient task scheduling for multiprocessor real-time systems. IEEE Trans. Comput.-Aided Des. Integr. Circuits Syst. (TCAD) **30**(10), 1569–1573 (2011)
5. Topcuoglu, H., Hariri, S., Wu, M.: Performance-effective and low-complexity task scheduling for heterogeneous computing. IEEE Trans. Parallel Distrib. Syst. (TPDS) **13**(3), 260–274 (2002)
6. Pandey, S., Wu, L., Guru, S.M., Buyya, R.: A particle swarm optimization-based heuristic for scheduling workflow applications in cloud computing environments. In: Proceedings of International Conference on Advanced Information Networking and Applications, pp. 400–407 (2010)
7. Qiu, M., Sha, E.H.M.: Cost minimization while satisfying hard/soft timing constraints for heterogeneous embedded systems. ACM Trans. Des. Autom. Electron. Syst. (TODAES) **14**(2), 1–30 (2009)
8. Zhang, M., Li, H., Liu, L., Buyya, R.: An adaptive multi-objective evolutionary algorithm for constrained workflow scheduling in Clouds. Distrib. Parallel Databases **36**(2), 339–368 (2018)
9. Sahni, J., Vidyarthi, D.P.: A cost-effective deadline-constrained dynamic scheduling algorithm for scientific workflows in a cloud environment. IEEE Trans. Cloud Comput. **6**(1), 2–18 (2015)
10. Chen, M., Huang, S., Fu, X., Liu, X., He, J.: Statistical model checking-based evaluation and optimization for cloud workflow resource allocation. IEEE Trans. Cloud Comput. 1 (2016)
11. Wang, X., Yeo, C.S., Buyya, R., Su, J.: Optimizing the makespan and reliability for workflow applications with reputation and a look-ahead genetic algorithm. Future Gener. Comput. Syst. **27**(8), 1–18 (2011)
12. Wen, Z., Cala, J., Watson, P., Romanovsky, A.: Cost effective, reliable, and secure workflow deployment over federated clouds. In: Proceedings of IEEE International Conference on Cloud Computing, pp. 604–612 (2015)
13. Han, L., Canon, L., Casanova, H., Robert, Y., Vivien, F.: Checkpointing workflows for fail-stop errors. IEEE Trans. Comput. **67**(8), 1105–1120 (2018)
14. Zhang, L., Li, K., Li, C., Li, K.: Bi-objective workflow scheduling of the energy consumption and reliability in heterogeneous computing systems. Inf. Sci. **379**, 241–256 (2016)
15. Whitley, D.: A genetic algorithm tutorial. Stat. Comput. **4**(2), 65–85 (1994)
16. Zhang, X., Wu, T., Chen, M., Wei, T., Zhou, J., Hu, S., Buyya, R.: Energy-aware virtual machine allocation for cloud with resource reservation. J. Syst. Softw. **147**, 147–161 (2019)
17. Gai, K., Qiu, M., Zhao, H.: Cost-aware multimedia data allocation for heterogeneous memory using genetic algorithm in cloud computing. IEEE Trans. Cloud Comput. 1 (2016)
18. Chen, W., Deelman, E.: WorkflowSim: a toolkit for simulating scientific workflows in distributed environments. In: Proceedings of International Conference on E-Science, pp. 1–8 (2012)
19. Bharathi, S., Chervenak, A., Deelman, E., Mehta, G., Su, M., Vahi, K.: Characterization of scientific workflows. In: Proceedings of International Workshop on Workflows in Support of Large-Scale Science, pp. 1–10 (2008)
20. Da Silva, R.F., Chen, W., Juve, G., Vahi, K., Deelman, E.: Community resources for enabling research in distributed scientific workflows. In: Proceedings of International Conference on e-Science, pp. 177–184 (2014)

# Null Model and Community Structure in Heterogeneous Networks

Xuemeng Zhai[1]([⊠])(iD), Wanlei Zhou[2], Gaolei Fei[1], Hangyu Hu[1], Youyang Qu[3], and Guangmin Hu[1]([⊠])

[1] University of Electronic Science and Technology of China, Chengdu 611731, China
zhaixuemeng@hotmail.com, hgm@uestc.edu.cn
[2] University of Technology Sydney, Ultimo, NSW 2007, Australia
[3] Deakin University, Burwood, VIC 3125, Australia

**Abstract.** Finding different types of communities has become a research hot spot in network science. Plenty of the real-world systems containing different types of objects and relationships can be perfectly described as the heterogeneous networks. However, most of the current research on community detection is applied for the homogeneous networks, while there is no effective function to quantify the quality of the community structure in heterogeneous networks. In this paper, we first propose the null model with the same heterogeneous node degree distribution of the original heterogeneous networks. The probability of there being an edge between two nodes is given to build the modularity function of the heterogeneous networks. Based on our modularity function, a fast algorithm of community detection is proposed for the large scale heterogeneous networks. We use the algorithm to detect the communities in the real-world twitter event networks. The experimental results show that our method perform better than other exciting algorithms and demonstrate that the modularity function of the heterogeneous networks is an effective parameter that can be used to quantify the quality of the community structure in heterogeneous networks.

**Keywords:** Heterogeneous network · Community detection · Modularity · Twitter network

## 1 Introduction

Network science is a fundamental tool to analyze the basic problems of the real-world complex systems, such as social networks, metabolic networks, computer networks and etc. [2]. Community detection has become a key research in network science during the past decades [4,8]. The community refers to the cluster of nodes that are connected densely and community detection focuses on finding the such clusters effectively in the networks. The modularity proposed by

This work was supported by National Natural Science Foundation of China No. 61571094 and Sichuan Science and Technology Program under Grant 2019YFG0456.

© Springer Nature Switzerland AG 2020
S. Wen et al. (Eds.): ICA3PP 2019, LNCS 11945, pp. 151–163, 2020.
https://doi.org/10.1007/978-3-030-38961-1_14

Newman based on the null model is the most famous parameter to quantify the quality of the community structure in the homogeneous single networks [9]. Based on the modularity function, effective algorithms of the community detection in homogeneous networks are proposed such as the famous BGLL algorithm [1]. The null model and the modularity function are also used in the research on the homogeneous multiplex networks [7,13]. However, most of the research just focus on the homogeneous networks and there is no effective function to quantify the quality of the community structure in heterogeneous networks.

Most of the real-world networks contain more than one type of the nodes and relationships. For example, in the DBLP networks, there are three types of nodes: authors, papers, and conferences [12] and in twitter event networks, there are two types of nodes: users and events [5]. Such networks are heterogeneous in nature. The community detection method is no longer available for those heterogeneous networks. Therefore, it is necessary to propose the suitable method to detect the communities in heterogeneous networks. The main problem of the heterogeneous community detection is how to deal with the heterogeneous relationships among the different types of the nodes. Researchers propose several method to detect the heterogeneous communities focused on the heterogeneous relationships. However, the basic community structure is ignored so that there is no effective function to quantify the quality of the community structure in heterogeneous networks like the homogeneous modularity function.

In this paper, we propose the null model and modularity function of the heterogeneous networks. The heterogeneous node degree is proposed to replace the node degree of homogeneous networks based on the heterogeneous relationships in the heterogeneous networks. Then we build the null mode of the heterogeneous networks with the same heterogeneous node degree distribution of the original network. The modularity function of the heterogeneous networks is built with the probability of there being an edge between two nodes in the null model. Based on our modularity function, a fast algorithm of heterogeneous community detection is proposed to demonstrate the effectiveness of the modularity. The experimental results show that the community structure of the heterogeneous networks can be exposed effectively through the modularity function. Our findings fill the gap in the field of null model of heterogeneous networks and provide a powerful tool for detecting communities in the complex systems with multiple objects and relationships in many general scientific fields.

The reminder of the paper is structured as follows: The Sect. 2 is the related work about our research. The heterogeneous node degree and null model of heterogeneous networks are introduced in Sect. 3. In the Sect. 4, we discuss the modularity function of the heterogeneous networks and the algorithm of the community detection is shown in Sect. 5. The experiments is presented on Sect. 6. The Sect. 7 is the conclusions.

## 2    Related Work

The null model in homogeneous networks has the same degree distribution with the original network. The modularity function proposed by Newman based on the null model is the most famous parameter that can be used to quantify the quality of the communities in homogeneous network [10]. The modularity refers to the number of edges within communities minus the expected number of such edges in the null model. Based on the modularity, Blondel et al. [1] propose a fast modularity optimization method called BGLL algorithm. They found the high modularity partitions of large networks in short time and unfolded a complete hierarchical community structure for the network. The method still focused on the homogeneous networks.

Compared with analysis for the homogeneous single networks and multiplex networks, the research on community detection in heterogeneous networks started relatively late. Cai et al. [3] propose a method to find the hidden community in heterogeneous social networks. They built the weighted matrix of different relationships according to the priori community detection results and used the optimized algorithm to calculate the optimal coefficient of each relationship matrix. The coefficient represented the influence of the different relationships on the result of the community detection. The method requires prior knowledge about community detection.

Zhao et al. [14] propose a framework of mining different types of communities from web based on the heterogeneity and evolution of web data. They gave the clearly definition of the heterogeneous networks and use a 8-tuple vector to represent them. The features of particular communities were extracted using the PopRank algorithm to build the SVM regression model for the prediction.

Comar et al. [6] use the multi-task learning to classify nodes and detect communities at the same time. They derived two homogeneous subnetworks form a heterogeneous network that contains two types of nodes, one subnetwork for classification and the other for community detection. The author classify the nodes and detect communities through the relevance of the two subnetworks. The methods requires the heterogeneous networks must be bipartite.

Qiu et al. [11] focus on the overlapping community detection of the heterogeneous social networks. They propose an algorithm called OcdRank (Overlapping Community Detection and Ranking) combining the overlapping community detection and community-member ranking together in directed heterogeneous social networks. The algorithm still works on bi-type heterogeneous social networks.

Our work differs from those found on the literature because the null model and modularity are the basic theory of the community detection in homogeneous networks. We propose the two basic conceptions of the heterogeneous networks and focus on the community structure itself with the considering of the heterogeneity in heterogeneous networks. The work is original and unprecedented.

# 3   Null Model of Heterogeneous Networks

## 3.1   Heterogeneous Networks and Heterogeneous Node Degree

We first introduce the basic conception of the heterogeneous networks. In this paper, we use the set of adjacency matrices to describe a heterogeneous networks $(HW)$ as $HW = \{A^S, ..., H^{SR}, ...\}, S, R \in T$, where $T$ refers to the type of nodes. $A^S = (a_{ij}^S)_{N_S \times N_S}$ donated as the adjacency matrix of the same-type nodes, where $N_S$ refers to the number of $S$-type nodes. $H^{SR} = (h_{ij}^{SR})_{N_S \times N_R}$ donated as the adjacency matrix of two different types of nodes. In the representation, we just separate the homogeneous nodes and heterogeneous nodes to ensure importance of the heterogeneous links in the community detection of the heterogeneous networks.

The existing null model of the homogeneous single network is proposed by Newman and has the same distribution of the node degree with the original network. To build the null model of the heterogeneous networks, we should first propose a new parameter to describe the basic connection among the different types of nodes in the heterogeneous networks like the node degree in the homogeneous networks. Therefore, we first define the heterogeneous node degree as follows:

**Definition 1.** Heterogeneous Node Degree: For each type of nodes, there are neighbors of the $S$-type node $i$. The heterogeneous node degree refers to the number of neighbors of different types from a node $i$. We give the $u_i^{SR}$ to represent the heterogeneous node degree of types $R$ for the $S$-type node $i$. When $S = R$, the heterogeneous node degree $u_i^{SS}$ becomes the homogeneous node degree $k_i$.

Therefore, the node degree in the heterogeneous networks is divided into two parts: the homogeneous node degree $k_i$ and the heterogeneous node degree of all types $\sum_R u_i^{SR} (S \neq R)$.

### 3.2   Null Model of the Heterogeneous Networks

With both homogeneous and heterogeneous node degree, we give definition of the null model of the heterogeneous networks:

**Definition 2.** Null Model of Heterogeneous Networks: The null model of the heterogeneous networks refers to those network models that has the same set of types of nodes $T$, number of homogeneous nodes $N$, number of heterogeneous $U$, distribution of homogeneous node degree $P(k)$ and distribution of heterogeneous node degree $P(u)$ with the original network, while otherwise is taken to be an instance of the random network.

For each two types of the nodes, there is a distribution of heterogeneous node degree. Therefore, there are $|T|^2 - |T|$ distribution of heterogeneous node degree in a heterogeneous network, where $T$ refers to the set of types of nodes.

### 3.3   Random Walk on Heterogeneous Networks

Here we use the random walk theory to build the null model of the heterogeneous networks. The process can be explained by the Laplacian Dynamics. Considering a homogeneous network, if there is an edge between node $i$ and node $j$, the two nodes are regarded as reachable. We suppose that there is a walker walking randomly among the nodes in the networks and each walk from one node to the other is completely independent and random. The process is actually a Markov process in which each walk has no relationship with the last time. Therefore, the probability of the walker walking from a arbitrary node $j$ to node $i$ and staying at node $i$, $\dot{p}_i$ is:

$$\dot{p}_i = \sum_j \frac{a_{ij}}{k_j} p_j - p_i \tag{1}$$

where $p_j$ refers to the probability of the walker staying at the node $j$. Differently, in a heterogeneous network, the edges among nodes is divided into homogeneous edges (edges between two same-type nodes) and heterogeneous edges (edges between two different-type nodes). Therefore, when the walker walks in the heterogeneous network, both homogeneous and heterogeneous edges should be considered. The probability of the walker walking from a arbitrary $R$-type node $j$ to $S$-type node $i$ and staying at node $i$, $\dot{p}_i^S$ is:

$$\dot{p}_i^S = \sum_{j,R} \frac{a_{ij}^S \delta_{SR} + h_{ij}^{SR} \bar{\delta}_{SR}}{\kappa_j^R} p_j^R - p_i^S \tag{2}$$

where $a_{ij}^S$ refers to the connection relationship between the two nodes $i$ and $j$ that belong to the same type $S$; $h_{ij}^{SR}$ refers to the connection relationship between $S$-type node $i$ and $R$-type node $j$; $\delta_{SR}$ is the reaction function; When $S = R$, $\delta_{SR} = 1$; When $S \neq R$, $\bar{\delta}_{SR} = 1$; $p_j^R$ refers to the probability of the walker staying at the $R$-type node $j$; $\kappa_j^R$ refers to total degree of the $R$-type node $j$, that is the sum of homogeneous degree and heterogeneous degree, donated as:

$$\kappa_j^R = k_j^R + \sum_S u_j^{RS} \tag{3}$$

Therefore, we give the conditional probability of the walker walking from $R$-type node $j$ to $S$-type node $i$ of in the null model of heterogeneous networks, donated as:

$$p(_i^S|j^R) = \frac{k_j^R}{\kappa_j^R} \frac{k_i^S}{2M_S} \delta_{SR} + \frac{u_j^{RS}}{\kappa_j^R} \frac{u_i^{SR}}{2M_{SR}} \bar{\delta}_{SR} \tag{4}$$

where $M_S$ refers to the edge number among the $S$-type nodes and $M_{SR}$ refers to the edge number between $S$-type nodes and $R$-type nodes. When the Markov process of random walk reaches steady state, the steady probability of the walks staying at the $R$-type node $j$ is donated as:

$$p_j^{R*} = \frac{\kappa_j^R}{2M} \tag{5}$$

where $M$ refers to the total number of the edges in the heterogeneous network. Therefore, the joint probability of the walker walking from $R$-type node $j$ to $S$-type node $i$ in the null model is:

$$p(Si, Rj) = p(^S_i|j^R) \times p^{R*}_j$$

$$= (\frac{k^R_j}{\kappa^R_j}\frac{k^S_i}{2M_S}\delta_{SR} + \frac{u^{RS}_j}{\kappa^R_j}\frac{u^{SR}_i}{2M_{SR}}\bar{\delta}_{SR})\frac{\kappa^R_j}{2M} \qquad (6)$$

$$= \frac{1}{2M}(\frac{k^R_j k^S_i}{2M_S}\delta_{SR} + \frac{u^{RS}_j u^{SR}_i}{2M_{SR}}\bar{\delta}_{SR})$$

The $p(Si, Rj)$ is the probability of there being an edge between $S$-type node $i$ and $R$-type node $j$ in the null model of the heterogeneous network. The equation is divided into two pasts: the homogeneous part and the heterogeneous part. The homogeneous part is the same with the probability in the homogeneous null model and the homogeneous part represents the heterogeneous relationships in the heterogeneous network. With this edge-building probability, we could build the modularity function of the heterogeneous network based on the null model.

## 4  Modularity Function of Heterogeneous Networks

The modularity function is first proposed by Newman in 2006. The modularity $Q = $ (the number of edges within communities-the expected number of such edges in the null model). The null model here is homogeneous and the modularity proposed by Newman is still built for the homogeneous networks. Similarly, when we replace the null model of homogeneous networks by the one of heterogeneous networks, we can build the modularity function of the heterogeneous networks. Here, we give the definition of the modularity function of heterogeneous networks:

**Definition 3.** Modularity Function of Heterogeneous Networks: The modularity function of heterogeneous networks $Q_h = $ (the number of edges within communities in heterogeneous networks-the expected number of such edges in the heterogeneous null model) and normalized by the total degree of the networks:

$$Q_h = \frac{1}{2M} \sum_{ijSR} [E(Si, Rj) - P(Si, Rj)]\delta(g_{Si}, g_{Rj}) \qquad (7)$$

where $E(Si, Rj)$ refers to the number of edges within communities in heterogeneous networks and $P(Si, Rj)$ refers to the expected number of such edges in the heterogeneous null model. $\delta(g_{Si}, g_{Rj}) = 1$ if the $S$-type node $i$ and $R$-type node $j$ belong to the same community, otherwise $\delta(g_{Si}, g_{Rj}) = 0$. In the Eq. 6, we obtain the probability of there being an edge between $S$-type node $i$ and $R$-type node $j$ in the null model of the heterogeneous network. Therefore, the $P(Si, Rj)$ is donated as:

$$P(Si, Rj) = p(Si, Rj) * 2M$$

$$= \frac{k^R_j k^S_i}{2M_S}\delta_{SR} + \frac{u^{RS}_j u^{SR}_i}{2M_{SR}}\bar{\delta}_{SR} \qquad (8)$$

The actual number of edges between two nodes in heterogeneous networks can be represented as:

$$E(Si, Rj) = A_{ij}^R \delta_{SR} + H_{ij}^{SR} \overline{\delta}_{SR} \tag{9}$$

Withe give the equation of the modularity function in details:

$$Q_h = \frac{1}{2M} \sum_{ijSR} [(A_{ij}^S - \frac{k_j^R k_i^S}{2M_S}) \delta_{SR} + \\ (H_{ij}^{SR} - \frac{u_j^{RS} u_i^{SR}}{2M_{SR}}) \overline{\delta}_{SR}] \delta(g_{Si}, g_{Rj}) \tag{10}$$

In the Eq. 10, we divide the modularity function of heterogeneous networks $Q_h$ into two parts, the homogeneous part $A_{ij}^S - \frac{k_j^R k_i^S}{2M_S}$ and the heterogeneous part $H_{ij}^{SR} - \frac{u_j^{RS} u_i^{SR}}{2M_{SR}}$. Therefore, the modularity can be understood as the sum of both homogeneous and heterogeneous part, which reflects the whole kinds of relationships in the heterogeneous networks.

## 5    Community Structure and the Fast Algorithm in Heterogeneous Networks

Similar with the homogeneous networks, there are also community structure in the heterogeneous networks, that is, the set of multi-type nodes that are connected closely. The modularity of heterogeneous networks can be used to quantify the quality of the heterogeneous community structure. When the modularity get max, the results of the community detection are the best.

We start from the basic structure of the networks to detect the heterogeneous communities. Therefore, we do not distinguish the type of nodes when detecting the communities. Which community a node belongs to is decided by the change of modularity function when it joins the community. The final results of each heterogeneous community will contain at least one type of nodes or more. It all depends on the maximum modularity function of the heterogeneous networks. After the community detection, we could extract the same-type nodes in each community to get the homogeneous node clusters.

Based on the modularity function of heterogeneous networks, we give a fast algorithm to detect the communities in the heterogeneous networks. The process of the algorithm in shown in Algorithm 1. The time complexity of Algorithm 1 (FAHCD) is $O(N \times max(\kappa_i))$. The algorithm is based on the famous fast algorithm BGLL of the homogeneous networks. In the large networks, the $max(\kappa_i)$ is far less than the the number of nodes $N$. Therefore, the time complexities of the Algorithm 1 is close to $O(N)$.

## 6    Experiments

We use the FAHCD to detect the heterogeneous communities of the twitter event networks we build through the real-world data. The twitter data is collected from

**Algorithm 1 .** Fast Algorithm of Heterogeneous Community Detection (FAHCD).

---

**Require:** The adjacency matrix set of heterogeneous network, $HW = \{A^S, ..., H^{SR}, ...\}, S, R \in T$

**Ensure:** The results of the heterogeneous communities, $C_h^K = \{C_{h_1}, C_{h_2}, ..., C_{h_n}\}$;

1: initial $HW^0 = HW$;

2: **repeat**

3:    Regarding each node in $HW^k$ as a community initially. $C_h^k = \{Node_1, Node_2, ..., Node_N\}$, where $N$ is the total number of nodes in $HW^k$;

4:    Computing the increment of modularity $\Delta Q_h^{ij}$ between each node $i$ and its each neighbor $j$ in the heterogeneous network;

5:    $\Delta Q_h^{ij} = \frac{1}{2M} \{ \sum_{z \in g_j} [(A_{iz}^S - \frac{k_i^S k_z^R}{2M_S}) \delta_{SR} + (H_{iz}^{SR} - \frac{u_i^{SR} u_z^{RS}}{2M_{SR}}) \overline{\delta}_{SR}] - \sum_{z \in g_i} [(A_{iz}^S - \frac{k_i^S k_z^R}{2M_S}) \delta_{SR} + (H_{iz}^{SR} - \frac{u_i^{SR} u_z^{RS}}{2M_{SR}}) \overline{\delta}_{SR}] \}$;

6:    For the node $j$ with the max $\Delta Q_h^{ij}$ with node $i$, adding the node $i$ into the community with node $j$;

7:    Updating the set of communities $C_h^k$ after the aggregation of $C_h^{k-1}$ in step 6;

8:    Regarding each type of nodes in the new community in $C_h^k$ as the new specific-type node; Regarding connections among the nodes as the self-loop of the new node with the weight of number of connections; Regarding edges between two different type of nodes as the new edge between two new nodes with the weight of number of edges; Generating a new heterogeneous network $HW^k$

9: **until** $\Delta Q_h < 0$ of all nodes;

10: **return** $C_h^K$;

---

the Twitter API. The MongoDB database is used to store the collected data. After pre-processing, including tweet language filtering, spam tweet filtering, useless field filtering and text content filtering, we obtained the valuable tweets and accounts. The Named Entity Recognition (NER) is used to extract the name of related people in each tweet and to extract the hashtag by the key symbol #. The twitter events are clustered based on the text similarity among the tweets. We cluster a large number of tweets with high text similarity to detect a twitter event that occur in the Twitter space. Then we build the twitter event networks with 5 type of nodes: Account, Tweet, Event, NameEntity and Hashtag. We capture 4 type of relationships among the 5 type of nodes. The networking rules are shown in Table 1.

We collected Twitter data about the UK elections from May 12nd, 2017 to June 10th, 2017. The Twitter event network in 30 days we built consisted of 70,536 account nodes, 32,593 tweet nodes, 2,618 event nodes, 1745 named entity nodes, and 1462 hashtag nodes. The twitter event network on May 12nd, 2017 is visualized in Fig. 1. There are 3,459 nodes and 4,329 edges including 2232 account nodes, 854 tweet nodes, 161 event nods, 134 named entity nods and 78 hashtag nodes. As shown in Fig. 1, the core-type of the nodes are the tweet nodes. They are connected with the rest other type of nodes to form the 4 types of the edges in the Twitter event networks. The rest 4 types of nodes are disconnected.

**Table 1.** The networking rules of the Twitter event network

| Type | Name | Description |
|------|------|-------------|
| Node | Account | The twitter accounts of the users |
| Node | Tweet | The short message written by the twitter users |
| Node | Event | The events detected in the Twitter space |
| Node | Named Entity | The name of related people detected in tweets |
| Node | Hashtag | The content tag for the tweets |
| Edge | Account and Tweet | Connected if the tweet is written by the account |
| Edge | Tweet and Event | Connected if the tweet belongs to the event |
| Edge | Tweet and NameEntity | Connected if the name appears in the tweet |
| Edge | Tweet and Hashtag | Connected if the tweet has the hashtag |

We detected the heterogeneous communities on the Twitter event networks we built in 30 days using FAHCD. The partial results of the community detection are visualized in Fig. 2. Nodes in the same color belong to the same communities. For visualization, we delete lots of nodes and edges of the network. The results show that different types of nodes could be divided into the same community because the dense connection among them such as the green nodes in the center in Fig. 2. Because of the different types of connection, different types of nodes could be divided into different communities, such as the orange nodes and the red nodes in the left top of Fig. 2. The orange nodes are account and the red nodes are hashtag. They are all connected with the tweet nodes in green but they are divided in to different communities just because the connections among them are heterogeneous. Therefore, our algorithm detects the communities using modularity based on the heterogeneous structure itself of the heterogeneous networks. It can not be replaced by transferring the heterogeneous networks into homogeneous networks and using the homogeneous community detection methods, which ignores the critical heterogeneous structure information.

The 504 communities are detected in the Twitter event network consists of 108,954 nodes. 84% communities contains the whole 5 types of nodes and only 4 communities contains just 2 types of the nodes. Such communities are small with less than 100 nodes and made up by the account nodes and tweet nodes. They are not connected with any other types of nodes. We manually labeled the election position of 1350 account nodes as the ground-truth to quantify the performance of our community detection. The results are shown in Table 2. Here we got three position about the UK election: proposition, neutral and opposition. The position of a community is determined by the position of most of its nodes. If the most of nodes are proposition in a community, all of nodes in the community are regarded as the proposition node. Therefore, we could calculate the accuracy as follows:

$$Accuracy = \frac{N_{correct}}{N_{total}}, \tag{11}$$

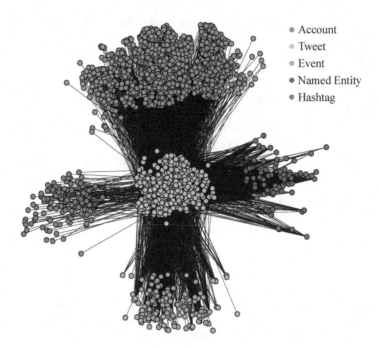

**Fig. 1.** The Twitter event network about UK elections.

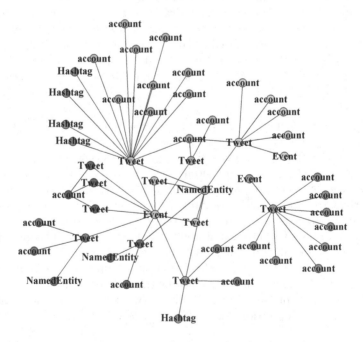

**Fig. 2.** The partial results of the heterogeneous community detection on Twitter event network. (Color figure online)

where $N_{correct}$ refers to the number of nodes with the correct position and $N_{total}$ refers to the total number of nodes in the communities of a same position. From the results, we could conclude that people may communicate with each other who has the same position in Twitter. Our algorithm detect the cluster of most people with the same position on the UK election from the Twitter event heterogeneous networks. The error less than 10% is caused by those active nodes and some junk accounts who may connect with people of any position.

**Table 2.** The performance of FAHCD based on ground-truth

| Election position | Number of communities | Accuracy |
|---|---|---|
| Proposition | 256 | 92.3% |
| Neutral | 127 | 94.7% |
| Opposition | 121 | 91.2% |

# 7 Conclusion

The results in the experiment section demonstrate the advantageous heterogeneous community detection performance on real-world Twitter event networks based on the null model of heterogeneous networks. Our method could deal with large-scale heterogeneous networks on a almost linear time complexity. Based on the FAHCD, we find the cluster of most people with the same position on the UK election from the Twitter event heterogeneous networks we built. The accuracy of all three position is over 90%, which show a great performance of our method on heterogeneous community detection. The community we detected contains more than one type of nodes based on the structure of heterogeneous networks and could be further divided into several homogeneous communities based on the type of each node.

In a general sense, the null model of heterogeneous networks is a general null model for any systems with multi-type of nodes including social networks. The rationality of the model can be explained by the traditional random-walk theory. The general significance of the model is that in addition to heterogeneous community structure, many other specific properties of heterogeneous networks can be revealed through the model. These properties, including motif identities, propagation-rate threshold, redundancy-distribution correlations and synchronization-state stability, have already been shown to be important in homogeneous network research. Additionally, the null model of heterogeneous networks can be used in directed networks based on in-and-out heterogeneous degree. Our future work is based on such extensions of our null model and its high-order representations, which may lead to some problems involving the applications of all systems with multi-type of nodes that can be described by heterogeneous networks.

Finally, the null model of heterogeneous networks provides a powerful tool for the structure analysis of complex systems with multi-type of nodes. Through comparisons, the specific nature of these systems can be exposed quantitatively by the model. We believe that the null model of heterogeneous networks can give rise to much stronger and more general applications in many areas, including social science, Internet topology, bioscience, engineering, economics, and education, where systems can be described by heterogeneous networks. To accomplish this, much more work needs to be done to gain a deeper understanding of the model and its high-order representations. We hope that many more attributes of the complex systems can be modelled and analysed through the null model of heterogeneous networks.

**Acknowledgment.** This work was supported by National Natural Science Foundation of China No. 61571094 and Sichuan Science and Technology Program under Grant 2019YFG0456. The data sets used to obtain the results in this manuscript are collected through Twitter API (https://dev.twitter.com/).

# References

1. Blondel, V.D., Guillaume, J.L., Lambiotte, R., Lefebvre, E.: Fast unfolding of community hierarchies in large networks. J. Stat. Mech. (2008). abs/0803.0476
2. Börner, K., Sanyal, S., Vespignani, A.: Network science. Ann. Rev. Inf. Sci. Technol. **41**(1), 537–607 (2007)
3. Cai, D., Shao, Z., He, X., Yan, X., Han, J.: Mining hidden community in heterogeneous social networks. In: Proceedings of the 3rd International Workshop on Link Discovery, pp. 58–65. ACM (2005)
4. Cao, Y., Zhang, G., Li, D., Wang, L.: Online energy management for smart communities with heterogeneous demands. In: 2018 IEEE Global Communications Conference (GLOBECOM), pp. 1–6. IEEE (2018)
5. Chen, F., Neill, D.B.: Non-parametric scan statistics for event detection and forecasting in heterogeneous social media graphs. In: Proceedings of the 20th ACM SIGKDD International Conference on Knowledge Discovery and Data Mining, pp. 1166–1175. ACM (2014)
6. Comar, P.M., Tan, P.N., Jain, A.K.: Simultaneous classification and community detection on heterogeneous network data. Data Min. Knowl. Disc. **25**(3), 420–449 (2012)
7. Mucha, P.J., Richardson, T., Macon, K., Porter, M.A., Onnela, J.P.: Community structure in time-dependent, multiscale, and multiplex networks. Science **328**(5980), 876–878 (2010)
8. Newman, M.E.: The structure and function of complex networks. SIAM Rev. **45**(2), 167–256 (2003)
9. Newman, M.E., Girvan, M.: Finding and evaluating community structure in networks. Phys. Rev. E **69**(2), 026113 (2004)
10. Newman, M.E., Strogatz, S.H., Watts, D.J.: Random graphs with arbitrary degree distributions and their applications. Phys. Rev. E **64**(2), 026118 (2001)
11. Qiu, C., Chen, W., Wang, T., Lei, K.: Overlapping community detection in directed heterogeneous social network. In: Dong, X.L., Yu, X., Li, J., Sun, Y. (eds.) WAIM 2015. LNCS, vol. 9098, pp. 490–493. Springer, Cham (2015). https://doi.org/10.1007/978-3-319-21042-1_47

12. Sun, Y., Yu, Y., Han, J.: Ranking-based clustering of heterogeneous information networks with star network schema. In: Proceedings of the 15th ACM SIGKDD International Conference on Knowledge Discovery and Data Mining, pp. 797–806. ACM (2009)
13. Zhai, X., et al.: Null model and community structure in multiplex networks. Sci. Rep. **8**(1), 3245 (2018)
14. Zhao, Q., Bhowmick, S.S., Zheng, X., Yi, K.: Characterizing and predicting community members from evolutionary and heterogeneous networks. In: Proceedings of the 17th ACM Conference on Information and Knowledge Management, pp. 309–318. ACM (2008)

# Big Data and Its Applications

# An Asynchronous Algorithm to Reduce the Number of Data Exchanges

Zhuo Tian$^{(\boxtimes)}$ ⓘ, Yifeng Chen, and Lei Zhang

HCST Key Lab, EECS, Peking University, Beijing 100871, China
{t.z,cyf,lei.z}@pku.edu.cn

**Abstract.** Communication or data movement cost is significantly higher than computation cost in existing large-scale clusters, for clusters having long network latency. For high-frequency parallel iterative applications, performance bottleneck is the long network latency caused by frequent data exchange. This paper presents an asynchronous algorithm capable of reducing the number of data exchanges among processes of parallel iterative applications. The proposed algorithm has been tested on a stencil-based parallel computation and compared with a BSP implementation of the same application. The asynchronous algorithm can effectively reduce the number of data exchanges at the expense of higher computation overhead and larger message size, performance can be improved up to 2.8x.

**Keywords:** Communication · Data exchange · Asynchronous · Stencil

## 1 Introduction

Existing supercomputer tends to have memory access delays and long network latency. We test the communication speed of existing large-scale infiniband clusters with manycore or GPU accelerators. It only executes peer-to-peer communication but no computation which means computation speed is very fast. Clusters can only deliver maximum 3000 32-way neighborhood send-receive short messages per second [1]. If the cost of data exchange is under 20% of total execution time, a cluster at most runs 600 time steps per second (20%*1 s = 0.2 s, $3000 msgs/s*0.2s = 600 msgs$), but it is too slow for scientists.

Therefore, the problem for existing clusters is the strong mismatch between computing performance and communication performance. Especially for high-frequency iterative applications, frequent data exchanges make communication be the major performance bottleneck.

Then, can we reduce the number of data exchanges for high-frequency parallel iterative applications? Actually: Yes.

---

Supported by National Key R&D Program of China (2017YFB0202001), and National Natural Science Foundation of China (61432018, 61672208).

S. Wen et al. (Eds.): ICA3PP 2019, LNCS 11945, pp. 167–174, 2020.
https://doi.org/10.1007/978-3-030-38961-1_15

We propose a new asynchronous algorithm, which is able to reduce the number of data exchanges at the expense of performing more floating-point arithmetics and exchanging larger messages, but doing fewer times of communications(and hence fewer messages).

Our experiment is based on stencil computation [2] which is a determinant component for the performance of seismic simulation [3], atmospheric modeling [4], gaseous wave propagation [5], etc. For stencil computation, the asynchronous algorithm can reduce the number of data exchanges to 1/4, and the actual performance is improved by 2.8 times.

## 2   Existing Parallel Models

Existing parallel models mainly include the Parallel in Time model (PiT) [6] and the Asynchronous Iterative Algorithm (AIA) [7,8] model. Both of them have certain limitations.

PiT algorithms start with a first coarse guess of the trajectory with long time steps. Then it runs fine propagator on each long time steps in parallel with shorter time steps [9]. PiT model is not effective in practice such as for protein folding. A coarse iteration likely deviates from its correct path after a few steps. Fine iterations starting from coarse states on incorrect slopes of surface will slide into wrong directions, rendering their entire computation useless or even counterproductive.

AIA model communicates asynchronously and computes speculatively with the outdated states of other processes. It does not guarantee convergence of trajectory. In molecular dynamics simulation [10], physical forces like van der Waals in close range are drastically non-linear in nature and sensitive to atomic displacement. In experiments inaccurate atomic positions can easily cause the iteration to diverge.

Our asynchronous algorithm in fact combines the characteristics of the two existing algorithmic models. It computes from coarsely speculated future states in parallel and communicates asynchronously. But the asynchronous algorithm can guarantee convergence of trajectory and does not deviate from its correct path.

## 3   Asynchronous Algorithm

### 3.1   Description of the Method

The asynchronous algorithm is based on multiple rounds of iteration. Specifically, in round 1, each process starts with the same initial states. In other steps of round 1, each process computes speculatively with the outdated states of other processes without data exchanging. Repeating the iteration over multiple rounds, all of the states will converge.

We compare the differences between BSP algorithm [11] and asynchronous algorithm(fewer data exchanges than BSP) as shown in Algorithm 1 and Algorithm 2. In BSP model, it computes one step at one time and exchanges data with others after one step. In asynchronous algorithm, during each round of iteration, every process computes its local states of all steps within some interval independently without performing any internode communications. After each round, all processes exchange states with each other so that remote states of the current round are actually the computed states of the last round; Such iteration and data exchange are repeated until no state changes.

---

**Algorithm 1.** BSP Algorithm
---
1: **for** $n = 1$ **to** $nsteps$ **do**
2:     $computation()$;
3:     $data\ exchange()$;
4: **end for**

---

**Algorithm 2.** Asynchronous Algorithm
---
1: **for** $n = 1$ **to** $nrnds$ **do**
2:     **for** $n = 1$ **to** $nsteps$ **do**
3:         $computation()$;
4:     **end for**
5:     $data\ exchange()$;
6: **end for**

---

Considering two processes in two rounds, as shown in Figs. 1 and 2. The left part is on process $p_0$ and the right part is on process $p_1$. Each process is responsible for updating local state $S_p(t, i)$ which occurs at time step $t$ during round $i$. As a basic rule, all processes share the same initial state, $S_{p_0}(0, i) = S_{p_1}(0, i)$.

In round 0, $S_{p_0}(0, 0) = S_{p_1}(0, 0)$. Thus, $S_{p_0}(1, 0)$ and $S_{p_1}(1, 0)$ are accurate in round 0. But $S_{p_0}(1, 0)$ will be sent asynchronously to process $p_1$ at the end of round 0. That means, process $p_0$ will get the accurate state $S_{p_1}(1, 0)$ in round 1, and can accurately compute the state $S_{p_0}(2, 1)$ in round 1.

## 4    The Optimization

We can shorten the message length by reducing the number of steps in each round. If the asynchronous algorithm converges to step $t$ ($t \geq k$) in round $k$, then in round $k+1$, it can be calculated from step $t+1$, rather than from step 0.

In addition, if the asynchronous algorithm converges to step $t$ in round $k$, then step $t+1$ will import small error. Based on this error, step $t+2$ will import larger error. An so on, the error will be magnified. We consider discarding some time steps with larger error for they have little effect on the precision.

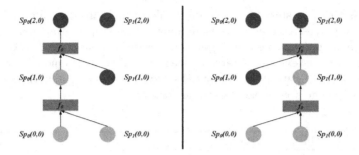

**Fig. 1.** The asynchronous algorithm in round 0.

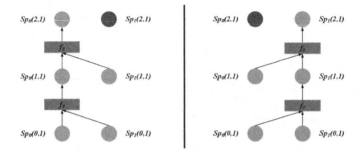

**Fig. 2.** The asynchronous algorithm in round 1.

The calculated interval can be called a window, and the size of window should not be set too small or too large. Smaller window means we have discarded more time steps. Thus, the asynchronous algorithm will not provide more speculative computations for next round and convergence speed will be very slow. But if it's too large, computing more time steps will slow down the computation speed. Thus, we should choose a suitable window size according to the computational complexity and the data length of different applications.

## 5  Implementation

In this section, we take Himeno benchmark [12] for example. The Himeno benchmark was developed by Dr. Ryutaro Himeno in 1996 at the RIKEN Institute in Japan. Since its introduction the benchmark has grown in popularity and is used throughout the HPC community, especially in Japan [13]. Our tests choose the following sets of $imax = 32$, $jmax = 32$, $kmax = 64$ and choose weights as standard values as defined in the benchmark. Algorithm 3 shows the asynchronous implementation of Himeno benchmark.

The asynchronous algorithm is composed of multiple rounds and each round is composed of multiple iterative time steps. All computed states $(p[n][i][j][k])$ should be communicated with other processes at the end of a round. During every round of iteration, each process iterates through multiple steps but is only responsible for updating local partition of global state, while states of other partitions come from asynchronous exchange of state updates with other processes. Such iteration and data exchange are repeated until no state changes.

---

**Algorithm 3.** Himeno Benchmark(Asynchronous Algorithm)

---

1: **for** $rnd = 0$ **to** NRNDS **do**
2:     **for** $n = 0$ **to** 4096 **do**
3:         **for** $i = 1$ **to** $32 - 1$ **do**
4:             **for** $j = 1$ **to** $32 - 1$ **do**
5:                 **for** $k = 1$ **to** $64 - 1$ **do**
6:                     $s_0 = p[n][i+1][j][k] + p[n][i][j+1][k]$
7:                         $+p[n][i][j][k+1] + p[n][i-1][j][k]$
8:                         $+p[n][i][j-1][k] + p[n][i][j][k-1]$
9:                     $ss = s_0 * a_3[i][j][k] - p[n][i][j][k];$
10:                     $sbuf[n][i][j][k] = p[n][i][j][k] + omega * ss;$
11:                 **end for**
12:             **end for**
13:         **end for**
14:     **end for**
15:     $send(sbuf);$
16:     $recv(rbuf);$
17:     $p[n][i][j][k] = rbuf[n][i][j][k];$
18: **end for**

---

## 6    Performance Results and Analysis

We define two concepts:

– samsara speed: the number of rounds to converge per second.
– iteration speed: convergence speed * samsara speed

By deferring the position of starting point, we can avoid repeating computation of steps that have converged. Deferring starting point does not affect the convergence speed of asynchronous algorithm. But the samsara speed will be influnced.

Advancing the position of ending point could change the size of window. In this section, we test the speedup, the samsara speed and the iteration speed at different window sizes. Figure 3 shows the convergence speed will be effected by the window size. Horizontal axis indicates the size of window and vertical axis denotes the convergence speed. The larger the window, the faster the convergence speed. Larger window can cover more time steps in each round, and the computed steps is more accurate.

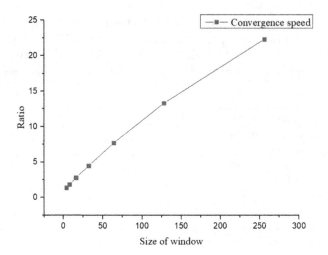

**Fig. 3.** The convergence speed at different window sizes.

If the size of windows is *nsteps* (the number of iteration steps in BSP), the covergence speed can reach 58x. But the computation strength and the message length is quite unreasonable, and the samsara speed will be very slow. Our goal of optimization is to reduce the window size from *nsteps* to a reasonable range. In the asynchronous algorithm, convergence speed is the theoretical maximum speed, but we should consider the influnce of samsara speed. Performance depends on the iteration speed which is the product of covergence speed and samsara speed.

Increasing the window size, although the convergence speed will increase as shown in Fig. 3, but the samsara speed will decrease as shown in Fig. 4. The reason is that the increased window size will lead to increase the computation strength and the message length. Thus, we should mainly concern the iteration speed which could directly reflect the speedup of the asynchronous algorithm.

As limited by cluster communication speed, iteration speed of BSP model does not exceed 6000 *times/s*. But, in asynchronous algorithm, the fastest iteration speed is 17528 steps/s when window size is 32, which is 2.8 times faster than BSP model as shown in Fig. 5.

Our work gives a solution for reducing the number of data exchanges on a stencil-based parallel computation. But, the asynchronous algorithm also can be applicable to other time-dependent problems such as solving sparse linear equations and graph algorithms where synchronization is often the bottleneck of performance.

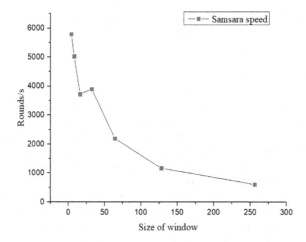

**Fig. 4.** The samsara speed at different window sizes.

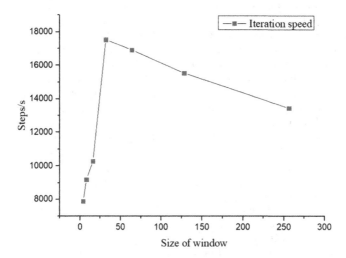

**Fig. 5.** The iteration speed at different window sizes.

## 7    Conclusion

In this paper, we propose an asynchronous algorithm which can effectively reduce the number of data exchanges in communication at the expense of higher computation length and larger message size. If the computation time does not exceed the time of data exchanges and the longer messages do not significantly increase the communication time, performance will be improved. The asynchronous algorithm implements 2.8x faster on stencil-based problem than the BSP model.

When problem size and network setup make communication latency the main performance bottleneck, asynchronous algorithm will demonstratehe performance advantage. The main challenge of implementing asynchronous algorithm lies in optimization of communications.

# References

1. Chen, Y., Huang, K., Wang, B., Li, G., Cui, X.: Samsara parallel: a non-BSP parallel-in-time model. In: Proceedings of the 21st ACM SIGPLAN Symposium on Principles and Practice of Parallel Programming, Barcelona (2016)
2. Ao, Y., et al.: 26 PFLOPS stencil computations for atmospheric modeling on sunway TaihuLight. In: IEEE International Parallel and Distributed Processing Symposium (IPDPS 2017). IEEE (2017)
3. Shield, C.K., French, C.W., Timm, J.: Development and implementation of the effective force testing method for seismic simulation of large-scale structures. Philos. Trans. Roy. Soc. London A: Math. Phys. Eng. Sci. 359(1786), 1911–1929 (2001)
4. Dennis, J.M., Edwards, J., Evans, K.J., et al.: CAM-SE: a scalable spectral element dynamical core for the community atmosphere model. Int. J. High Perform. Comput. Appl. 26(1), 74–89 (2012)
5. Dou, H.-S., Tsai, H.M., Khoo, B.C., Qiu, J.: Simulations of detonation wave propagation in rectangular ducts using a three-dimensional WENO scheme. Combust. Flame 154(4), 644–659 (2008)
6. Baffico, L., Bernard, S., Maday, Y., Turinici, G., Zerah, G.: Parallel-in-time molecular-dynamics simulations. Phys. Rev. E 66, 5 (2002)
7. Bahi, J.M., Contassot-Vivier, S., Couturier, R.: Evaluation of the asynchronous iterative algorithms in the context of distant heterogeneous clusters. Parallel Comput. 31(5), 439–461 (2005)
8. Blathras, K., Szyld, D.B., Shi, Y.: Timing models and local stopping criteria for asynchronous iterative algorithms. J. Parallel Distrib. Comput. 58(3), 446–465 (1999)
9. Lions, J.-L., Manday, Y., Turinici, G.: Resolution EDP par un schema en temps parareal. C. R. Acad. Sci. Numer. Anal. 332(7), 661–668 (2001)
10. Yu, Y.: Parallel implementation and performance optimization for refactoring GROMACS on the sunway many-core architecture. University of Science and Technology of China (2018)
11. Valiant, L.G.: A bridging model for parallel computation. SIAM J. Sci. Stat. Comput. 33, 103–111 (1990)
12. The Riken Himeno CFD Benchmark. http://accc.riken.jp/HPC/HimenoBMT/indexe.html
13. Phillips, E.H., Fatica, M.: Implementing the Himeno benchmark with CUDA on GPU clusters. In: IEEE International Symposium on Parallel and Distributed Processing IEEE (2010)

# Two-Stage Clustering Hot Event Detection Model for Micro-blog on Spark

Ying Xia$^{(\boxtimes)}$ and Hanyu Huang

School of Computer Science and Technology,
Chongqing University of Posts and Telecommunications, Chongqing 400065, China
xiaying@cqupt.edu.cn, hanyuhuang.hhy@gmail.com

**Abstract.** With the rapid development of micro-blog, it has become one of the main platforms to publish news and express opinions. Micro-blog analyzing for hot event detection is widely concerned by researchers. However, hot event detection is not easy because micro-blog blogs have the characteristics of large scale, short text and irregular grammar. In order to improve the performance of hot event detection, a two-stage clustering hot event detection model for micro-blog is proposed. The model is designed in spark environment and divided into two parts. First, K-Means method is improved by threshold setting and cosine similarity to cluster blogs. Then, the result of blogs clustering is clustered again to detect hot events by LDA (Latent Dirichlet Allocation) model. Sufficient experiments have been carried out in spark environment, it is shown that the proposed model gains higher accuracy and time efficiency for hot event detection.

**Keywords:** Micro-blog blogs · Hot event detection · Spark · Two-stage cluster · K-Means model · LDA model

## 1 Introduction

Hot event refers to event with high public discussion and widespread concern. Timely detection of hot event has great significance for society management and public safety maintenance. Micro-blog is an important online communication media, hot event can be considered when a large number of blogs discussing a same topic. To detect hot events, researchers have proposed different models. These models can be divided into two categories, keywords extraction model and topic model.

Keywords extraction model can analyze and extract keywords from blogs. Extracted keywords are used to cluster texts and then detect events. Early research [9] paid more attention to extract keywords. But only extract keywords may cause insufficient semantic information. To solve this problem, researchers combine related features and keywords to detect events. Stilo et al. [12] and Ozdikis et al. [11] used Hashtag to enhance accuracy of event detection. Furthermore, different features are added according to different research objectives.

© Springer Nature Switzerland AG 2020
S. Wen et al. (Eds.): ICA3PP 2019, LNCS 11945, pp. 175–183, 2020.
https://doi.org/10.1007/978-3-030-38961-1_16

Sun et al. [13] combined external knowledge base of related fields to detect events. Yilmaz et al. [17] and Zhong et al. [18] mixed geographical position and keywords to detect location-related events. In addition, in order to improve efficiency of event detection and ensure accuracy, keywords clustering process are focused. Ai et al. [1] proposed a TMHTD model in Spark, it detects events via calculating the similarity of keywords in two-layers structure.

Compared with keywords extraction model, topic model gains higher event detection accuracy but needs sufficient features. LDA model [2] is a representative topic model, it uses word bags to describe events and no need to consider words order in texts. Hao et al. [5] used LDA model to extract topics while identifying abnormal behavior sentences with each topic. Wang et al. [14] visualized topics after extracting topics from LDA model. In addition, some research try to improve accuracy of topic model by expanding feature space [3,4,6,7]. However, complex features are difficultly added due to the limitation of model structure. Some research extended semantic information to further improve accuracy. Yan et al. [16] and Kitajima et al. [8] used advanced semantic information like binary or triple sets instead of word bags for clustering. Xu et al. [15] proposed a TUS-LDA model which used pseudo-texts as the input of LDA model. The pseudo-texts were clustered by different topic types to expand semantic information.

Considered the advantages of the two categories of models, a two-stage clustering hot event detection model is proposed. These two stages are named text-cluster stage and semantic-cluster stage, respectively. In which, K-Means model and LDA model are involved in clustering process according to data characteristics of different stages, K-Means model is optimized by threshold setting and cosine similarity for keywords extraction, and a set of spark jobs is designed for large-scale data processing.

The rest of paper is organized as follows. Section 2 presents terminology definition. Section 3 proposes the two-stage clustering hot event detection model. Section 4 designs the optimized model in Spark environment. Section 5 evaluates accuracy and efficiency of proposed model. Section 6 draws a summary.

## 2   Terminology Definition

For easily understanding, related terminologies are presented here.

**Micro-blog Blogs:** Micro-blog blogs are stored by rows, each blog includes tags, content and related features. Related features mainly include timestamp, number of comments, number of forwards and number of likes.

**Word-bag:** Word-bag is a set of keywords with an id. The keywords are from the text corresponding to the id and can describe the text. This paper mainly uses word-bag to describe text.

**Heat:** Heat is used to evaluate the popularity level, which is calculated by the related features of blog posts. Heat of blogs can be abbreviated as $Heat\,(d_i)$ and

Heat of event can be abbreviated as $Heat(E)$. Specific definitions are as follows,

$$Heat(d_i) = sum(features_i) = comments + forwards + likes \qquad (1)$$

where $comments$ represents the number of comment, $forwards$ is the number of forwarding and $likes$ is the number of like in blogs. The sum of $Heat(d_i)$ represents the heat of the event $Heat(E)$.

## 3  Two-Stage Clustering Hot Event Detection Model

A two-stage clustering hot event detection model is proposed that contains both text-cluster and semantic-cluster. Because K-Means and LDA models will be used in each of two stages for improvement respectively, thus the proposed model is abbreviated as KMLDA.

### 3.1  Text-Cluster Stage

In text-cluster stage, micro-blog blogs are equally divided into slices to reduce the size of data. The blogs in each slice are divided into words by Jieba[1] and then converted to vectors using the Word2vec [10] method. Finally, K-Means is selected as a clustering method to cluster blogs of each slice. After text-cluster, many text clusters will be generated. Text clusters with a small number of blogs will be filtered because they represent insufficient discussion. Furthermore, K-Means is optimized to fit KMLDA model. Main optimizations are as follows.

(1) Cosine similarity is used as the measure distance of K-Means, and it is defined as Eq. (3),

$$
\begin{aligned}
dis &= 1 - \cos(\overrightarrow{w_i}, \overrightarrow{w_j}) \\
&= 1 - \frac{\overrightarrow{w_i} \cdot \overrightarrow{w_j}}{\|\overrightarrow{w_i}\| \cdot \|\overrightarrow{w_j}\|} \\
&= 1 - \frac{\sum\limits_{k=1}^{n} w_{i,k} \cdot w_{j,k}}{\sqrt{\sum\limits_{k=1}^{n} w_{i,k}^2} \sqrt{\sum\limits_{k=1}^{n} w_{j,k}^2}}
\end{aligned}
\qquad (2)
$$

where $\overrightarrow{w_i}$ represents weight vectors of blog $d_i$, $n$ is feature dimension, and $w_{i,k}$ is the kth weight of blog $d_i$. When $dis$ is smaller, the blog similarity will be higher.

(2) Set $AVG(dis)$ as a minimum similarity threshold. Because the accuracy of cluster-centers updating may be affected by large $dis$ in the K-Means training process. Specific definition as Eq. (3),

$$AVG(dis) = \frac{SUM(dis_{max})}{NUM(d)} = \frac{1}{n} \cdot \sum_{i=1}^{n} dis_{max,i} \qquad (3)$$

---

[1] "Jieba" (Chinese for "to stutter") Chinese text segmentation: built to be the best Python Chinese word segmentation module. GitHub: https://github.com/fxsjy/jieba/.

where $NUM(d)$ is the number of blogs, $SUM(dis_{max})$ is the sum of the cosine similarities $dis_{max,i}$ which are the maximum cosine similarity between text and cluster-centers.

$AVG(dis)$ as a threshold, $dis$ beyond this threshold will not participate in update of cluster-centers.

### 3.2  Semantic-Cluster Stage

LDA model is chosen for semantic-cluster because it can accurately detect hot events when semantic information is sufficient. LDA model is implemented through the spark machine learning package. Meanwhile, how to input the results of text-cluster into LDA model is designed in detail. The process of processing text clusters is divided into two steps: keyword extraction and vector transformation.

Keywords are extracted from text clusters and used as word-bags. In order to find words of widespread concern, blog heat which is introduced in Sect. 2 is used to extract keywords.

With word-bags, we need to convert them to vectors because the input of LDA model is a vectorized text. TF-IDF method is used to vectorize word-bags due to high effective and adapted the characteristics of word-bags. Specific definitions are shown in Eq. (4),

$$w_{i,k} = tf_{i,k} * idf_k = \frac{n_{i,k}}{n_i} * \log(\frac{N}{n_k+1}) \tag{4}$$

where $tf_{i,k}$ represents word frequency, $n_{i,k}$ is the number of the kth word in word-bags $wb_i$, and $n_i$ is the number of words in word-bags $wb_i$. $idf_k$ represents reverse text frequency, $N$ is the number of text clusters, $n_k$ is the number of text clusters which include the kth word. $w_{i,k}$ represents weight of the kth word of word-bags $wb_i$.

After transforming word-bags into vectors by TF-IDF method, the vectors are used as input to cluster hot events by LDA model. The hot events clustered are displayed in the form of word-bags and sorted by $Heat(E)$ which is the heat of the event.

## 4  Parallel Computing Design and Implementation

Spark is a popular memory-based large data processing framework, which processes and stores data based on the data structure RDD. In order to meet the requirement of large-scale blog processing, a set of spark jobs are designed for KMLDA model so that to ensure event detection efficiently. Meanwhile, a parallel processing framework on Spark is designed for text-clustering to reduce the size of data per RDD and improve computational efficiency.

## 4.1   Updating Cluster-Center

The KMLDA framework is divided into two parts. The first part corresponds to the text-cluster stage. Firstly, micro-blog blogs and related features are read into RDD. Then, the RDD is equally divided into multiple parts which represent as $\{RDD_1, RDD_2,..., RDD_n\}$. For each $RDD_i \in RDD$, the clustering operation in Sect. 3.2 is executed by parallel. The second part corresponds to the semantic-cluster stage. All RDD which have processed by test-cluster are merged into one RDD. The operation in Sect. 3.3 and Sect. 3.4 are used to process this RDD and detect events.

## 4.2   Spark Implementation

In the text-cluster stage, it is necessary to update cluster-center when training K-Means. Training data is in one RDD makes it difficult to update cluster centers, because data within the RDD is difficult to interoperate. To solve this problem, a flag is added to each blog after judging cluster-center of the blog. The flag is used to mark which cluster center the blog belongs to. Flags will be grouped to update cluster-centers.

In the semantic-cluster stage, word-bags are needed to be transformed into vector by TF-IDF method. However, the calculation of IDF value is limited by the size of data. In order to efficiently calculate IDF value, an inverted sorting method is designed. Words are used as keys to cluster blogs and calculate their number. A Hashmap containing words and the number of texts is made. The Hashmap makes it easy to calculate IDF values.

# 5   Experiment and Analysis

## 5.1   Experimental Preparation

In order to verify accuracy and efficiency of KMLDA model, experiments are performed on Sina Weibo. Totally 49.19 million micro-blog blogs are collected by Sina Weibo API. The data have no specific category and longer than three words. Among them, 17 million micro-blog blogs are marked with a single word, like cooking, football, Messi, Trump, etc. The other data has type labels, like weather, sports, life, etc. This part of data is used to train KMLDA model and marked data is used to verify the accuracy of KMLDA model.

Test environment is a Spark cluster which has two nodes, each node is CentOS7 and 256 GB memory. Spark-LDA [2] and TMHTD [1] are chosen as comparative models. Spark-LDA improves LDA model to run in the Spark environment. TMHTD is an event detection model with two-layer cosine clusters which running in the Spark environment.

## 5.2   Accuracy Evaluation

Recall rate, accuracy rate and event accuracy rate are used as evaluation indicators. Specific definitions are as follows,

$$recall = \frac{N_{reality}}{N_{all}} \tag{5}$$

$$accuracy = \frac{N_{right}}{N_{reality}} \tag{6}$$

$$eventAccuracy = \frac{E_{right}}{E_{all}} \tag{7}$$

where $N_{reality}$ represents the number of blogs after clustering, $N_{right}$ represents the number of blogs which are correctly clustered, $E_{all}$ represents the number of blogs before clustering, $E_{right}$ represents the number of events which are correctly detected, $E_{all}$ represents the number of events.

These indicators accuracy rate and recall rate are based on blog, and eventAccuracy rate is based on event. The marked blogs are extracted to different sizes for accuracy verification, results are shown in Table 1.

**Table 1.** Accuracy evaluation.

| Data size | Methods | Accuracy | Recall | EventAccuracy |
|-----------|---------|----------|--------|---------------|
| 64 MB | Spark-LDA | 0.7832 | 0.8912 | 0.94 |
|  | TMHTD | 0.8575 | 0.9131 | 0.94 |
|  | KMLDA | 0.8523 | **0.9254** | **0.96** |
| 128 MB | Spark-LDA | 0.7551 | 0.8543 | 0.90 |
|  | TMHTD | 0.8564 | 0.8856 | 0.93 |
|  | KMLDA | 0.8357 | **0.9133** | **0.96** |
| 512 MB | Spark-LDA | 0.7324 | 0.8102 | 0.85 |
|  | TMHTD | 0.8365 | 0.8772 | 0.89 |
|  | KMLDA | 0.8336 | **0.8935** | **0.93** |
| 2 GB | Spark-LDA | 0.6567 | 0.7154 | 0.75 |
|  | TMHTD | 0.7886 | 0.8225 | 0.85 |
|  | KMLDA | 0.7552 | **0.8543** | **0.91** |
| 4 GB | Spark-LDA | 0.5546 | 0.6625 | 0.63 |
|  | TMHTD | 0.7138 | 0.7856 | 0.81 |
|  | KMLDA | 0.6958 | **0.8127** | **0.86** |

The experiment uses data sets under different size, including 64 MB, 128 MB, 512 MB, 2 GB and 4 GB. As can be seen from Table 1, the indicators show a downtrend with the increase of data size. Compared with TMHTD and KMLDA, the downtrend of Spark-LDA is obvious. In addition, TMHTD is slightly better than KMLDA in accuracy rate, because TMHTD is supposed to calculate cosine similarity between any two blogs. For the recall rate, KMLDA is higher than the other models. The reason is that KMLDA is not strict in setting blog filtering conditions. Meanwhile, KMLDA considers keywords extraction according to the heat of blogs, and LDA model has better event detection ability, so that KMLDA performs better than Spark-LDA and TMHTD in eventAccuracy rate.

## 5.3   Running Time Comparison

Time efficiency verification is mainly divided into running time comparison and scalability verification. As shown in Fig. 1.

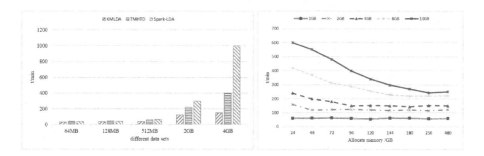

**Fig. 1.** Time efficiency evaluation.

The left figure represents the running time of each algorithm under different size of data. From figure, KMLDA has a significant improvement in running time which compared with Spark-LDA and TMHTD. It is proved that KMLDA which designs parallel calculation framework and linear algorithm complexity has high efficiency to detect events.

As shown in right figure, the scalability of KMLDA is tested by increasing memory and adjusting the size of data. From figure, the running time of KMLDA gradually decreases with the increase of memory size. However, by algorithm complexity, CPU resources and IO stream, the running time finally approaches to a stable value.

## 6   Summary

In this paper, a two-stage clustering hot event detection model KMLDA for micro-blog is proposed on Spark. This model considers the characteristics of blogs and time efficiency in big data environment. The process of KMLDA is divided into two stages. In text-cluster stage, data size can be reduced via slicing, and K-Means is adapted to improve the accuracy by threshold setting and cosine similarity. In semantic-cluster stage, word-bags are extracted from text clusters and then LDA model clusters word-bags to detect hot events. Experimental results show that KMLDA improves the accuracy and time efficiency of hot event detection in big data environment. In future work, how to integrate user characteristics and topic types to satisfy personalized event detection, and real-time data processing of micro-blog blogs will be considered.

**Acknowledgments.** This work was financially supported by the Natural Science Foundation of China (41571401).

# References

1. Ai, W., Li, K., Li, K.: An effective hot topic detection method for microblog on spark. Appl. Soft Comput. **70**, 1010–1023 (2018)
2. Blei, D.M., Ng, A.Y., Jordan, M.I.: Latent Dirichlet allocation. J. Mach. Learn. Res. **3**, 993–1022 (2003)
3. Cao, J.X., Xu, S., Chen, G.J., Zhao, L.Y., Zhou, T., Liu, B.: Discovering geographical topics in online social networks. Chin. J. Comput. **40**(7), 1530–1542 (2017)
4. Chen, X., Zhou, X., Sellis, T., Li, X.: Social event detection with retweeting behavior correlation. Expert Syst. Appl. **114**, 516–523 (2018)
5. Hao, Y., Zheng, Q., Chen, Y., Yan, C.: Recognition of abnormal behavior based on data of public opinion on the web. Comput. Res. Dev. **53**(3), 611–620 (2016)
6. Huang, F.L., Feng, S., Wang, D.L., Yu, G.: Mining topic sentiment in microblogging based on multi-feature fusion. Chin. J. Comput. **40**(4), 872–888 (2017)
7. Huang, F.L., Yu, G., Zhang, J.L., Li, C.X., Yuan, C.A., Lu, J.L.: Mining topic sentiment in micro-blogging based on micro-blogger social relation. J. Softw. **28**(3), 694–707 (2017)
8. Kitajima, R., Kobayashi, I.: A latent topic extracting method based on events in a document and its application. In: Proceedings of the ACL 2011 Student Session, pp. 30–35. Association for Computational Linguistics (2011)
9. Mathioudakis, M., Koudas, N.: TwitterMonitor: trend detection over the twitter stream. In: Proceedings of the 2010 ACM SIGMOD International Conference on Management of Data, pp. 1155–1158. ACM (2010)
10. Mikolov, T., Chen, K., Corrado, G., Dean, J.: Efficient estimation of word representations in vector space. Comput. Sci. (2013)
11. Ozdikis, O., Senkul, P., Oguztuzun, H.: Semantic expansion of hashtags for enhanced event detection in Twitter. In: Proceedings of VLDB 2012 Workshop on Online Social Systems, pp. 1–6 (08 2012)
12. Stilo, G., Velardi, P.: Temporal semantics: time-varying hashtag sense clustering. In: Janowicz, K., Schlobach, S., Lambrix, P., Hyvönen, E. (eds.) EKAW 2014. LNCS (LNAI), vol. 8876, pp. 563–578. Springer, Cham (2014). https://doi.org/10.1007/978-3-319-13704-9_42
13. Sun, R., Guo, S., Ji, D.H.: Topic representation integrated with event knowledge. Chin. J. Comput. **40**(4), 791–804 (2017)
14. Wang, Z.H., Chen, S.M., Yuan, X.R.: Visual analysis for microblog topic modeling. J. Softw. **29**(4), 1115–1130 (2018)
15. Xu, K., Qi, G., Huang, J., Wu, T., Fu, X.: Detecting bursts in sentiment-aware topics from social media. Knowl.-Based Syst. **141**, 44–54 (2018)
16. Yan, X., Guo, J., Lan, Y., Cheng, X.: A biterm topic model for short texts. In: Proceedings of the 22nd International Conference on World Wide Web, pp. 1445–1456. ACM (2013)

17. Yilmaz, Y., Hero, A.O.: Multimodal event detection in Twitter hashtag networks. J. Signal Process. Syst. **90**(2), 185–200 (2018)
18. Zhong, Z.M., Guan, Y., Li, C.H., Liu, Z.T.: Localized top-k bursty event detection in microblog. Chin. J. Comput. **41**(7), 1504–1516 (2018)

# Mobility-Aware Workflow Offloading and Scheduling Strategy for Mobile Edge Computing

Jia Xu[1], Xuejun Li[1], Xiao Liu[2(✉)], Chong Zhang[2], Lingmin Fan[1],
Lina Gong[1], and Juan Li[3]

[1] School of Computer Science and Technology, Anhui University, Hefei, China
[2] School of Information Technology, Deakin University, Geelong, Australia
xiao.liu@deakin.edu.au
[3] Computer Science and Technology School, Wuhan Institute of Technology,
Wuhan, China

**Abstract.** Currently, Mobile Edge Computing (MEC) is widely used in different smart application scenarios such as smart health, smart traffic and smart home. However, smart end devices are usually constrained in battery and computing power, and hence how to optimize the energy consumption of end devices with intelligent task offloading and scheduling strategies under constraints such as deadlines is a critical yet challenging topic. Meanwhile, most existing studies do not consider the mobility of end devices during task execution but in reality end devices may need to be constantly moving in a MEC environment. In this paper, motivated by a patient health monitoring scenario, we propose a Mobility-Aware Workflow Offloading and Scheduling Strategy (MAWOSS) for MEC which provides a holistic approach that covers the workflow task offloading strategy, the workflow task scheduling algorithm and the workflow task migration strategy. Comprehensive experimental results show that compared with others, MAWOSS is able to achieve the optimal fitness with lower energy consumption and smaller workflow makespan under the deadlines.

**Keywords:** Mobile Edge Computing · Mobility · Workflow · Task offloading · Task scheduling

## 1 Introduction

With the rapid enhancement of the computing power of mobile devices, many new smart applications such as smart health, smart traffic and smart home are being developed in the market. In a patient health monitoring scenario, it is necessary to analyze massive and heterogeneous medical business processes, and require timely collection of medical data [1, 2]. For example, if one medical data indicator contains 4 bytes and the data collection frequency is every 1 min, then 6 indicators need 24 bytes per minute, namely 12 MB per year. If the resident population is 5 million, then a massive 57.3 PB of data will be generated in a year [3]. For such a kind of smart health scenario, traditional cloud computing cannot meet its requirement of fast response time.

© Springer Nature Switzerland AG 2020
S. Wen et al. (Eds.): ICA3PP 2019, LNCS 11945, pp. 184–199, 2020.
https://doi.org/10.1007/978-3-030-38961-1_17

Meanwhile, Mobile Edge Computing (MEC) is gradually becoming the next generation information technology platform [4]. MEC comprehensively utilizes three different layers of computing resources including end devices, edge servers and cloud servers. Different from cloud computing, MEC does not need to transmit every task to the cloud. Instead, it makes fully use of the idle resources at the end device and edge servers by filtering and analyzing the characteristic of the tasks [5]. It can not only utilize the powerful computing capacity of cloud computing, but also meet the flexible requirements of distributed environment and real-time response of various smart applications [6]. Therefore, various smart health applications such as patient health monitoring can be deployed in the MEC environment so that the workflow tasks of smart health business process can meet the requirements of computing resources and real-time response constraints [7].

However, compared with standalone cloud computing, MEC has more heterogeneous computing resources and more complicated network structures. At present, many researchers focus on resource management problems in the edge computing environment, specifically the task offloading strategy and the task scheduling algorithm [8–11]. The task offloading strategy mainly tackles the limitation of battery and computing capacity of end devices by offloading tasks to the edge or cloud server [12]. The task offloading strategy aims to optimize the task execution time and the energy consumption of end devices in the MEC environment [13]. The task scheduling algorithm mainly tackles the task scheduling problem on the edge or cloud server with the aim to optimize the task execution time at different resource layers [14].

Currently, most existing research works on task offloading and task scheduling in MEC fail to consider the mobility of end devices [14, 15]. The location of the end device is assumed to be fixed during the period of task execution at the edge server. However, this is impractical in the real world as most mobile devices in the MEC environment are constantly moving. Therefore, task migration among edge servers needs to be considered when the location of mobile devices changes. Otherwise, the task execution result may fail to be delivered to the end device in time as the connection to the edge server which executes the task is seriously weakened or even lost completely. To address such a problem and using a patient health monitoring scenario as the motivating example, this paper proposes the novel Mobility-Aware Workflow Offloading and Scheduling Strategy (MAWOSS). We first define the location and moving path model of the end device in the MEC environment. Second, we construct the energy consumption and response time models of workflow tasks in the MEC environment. Finally, we present the detailed strategies and algorithms used in MAWOSS. Specifically, in the task offloading stage, MAWOSS can make the best offloading decisions for workflow tasks according to the characteristic of the task workload and data size. In the task execution stage, MAWOSS can find the best task scheduling plan with the minimum energy consumption of the moving end device under the deadlines. When the task execution stage finished, MAWOSS can select the best edge server for task migration according to the updated location of the end device. Comprehensive experimental results show that compared with other existing strategies, our proposed strategy is able to achieve the optimal fitness with lower energy consumption and smaller workflow makespan under the deadlines.

The major contributions of this paper include: (1) important models have been formulated for the mobility of the end device, the energy consumption of the end device, and the time of workflow tasks in the MEC environment; (2) the novel Mobility-Aware Workflow Offloading and Scheduling Strategy (MAWOSS) for MEC has been proposed as a holistic approach which covers the workflow task offloading strategy, the workflow task scheduling algorithm and the workflow task migration strategy; (3) comprehensive experiments have been conducted to evaluate our strategy and compare with five other representative strategies.

The rest of this paper is organized as follows. Section 2 presents the problem formulation with detailed models. Section 3 proposes our novel Mobility-Aware Workflow Offloading and Scheduling Strategy. Section 4 demonstrates the experimental results. Finally, Sect. 5 concludes this paper and points out some future work.

## 2    Problem Formulation

In this section, using a patient healthcare monitoring example scenario, we propose the major models used in this paper including the location and moving path model for the end device, the wireless signal model for the edge server, the energy consumption and task response time model for the end device, the edge server and the cloud server, and the fitness value model for measuring the quality of task offloading decisions.

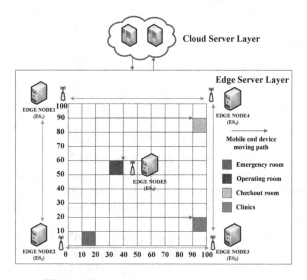

**Fig. 1.** The location and moving path model.

### A. The Location and Moving Path Model

For an example emergent treatment process, the starting location of the hospital bed (installed with many mobile monitoring devices) is the emergency department. The hospital bed then passes through the examination room, operating room and finally reaches to the ward. Based on these four key locations, the location and moving path model of the end device is shown in Fig. 1. The location model is a square with the side length of 100 meters which includes three different types of computing resources: the cloud server, the edge server and the mobile end device. The number of virtual machines in the cloud server is M, which are denoted as $CS_i = \{CS_1, CS_2, \ldots, CS_M\}$. The CPU frequency of the cloud server is $f_i^{cloud} = \{f_1^{cloud}, f_2^{cloud}, \ldots, f_M^{cloud}\}$. There are N edge servers in this area which are denoted as $ES_i = \{ES_1, ES_2, \ldots, ES_N\}$. Each edge server contains K virtual machines and the CPU frequency is $f_i^{edge} = \{f_1^{edge}, f_2^{edge}, \ldots, f_K^{edge}\}$. The CPU frequency of the end device is $f^{end}$. The end device moves through this area according to the moving path as depicted in Fig. 1 during the workflow task response time interval $[t_{start}, t_{end}]$. The location of $ES_2$ is considered as the original point and the line between $ES_2$ and $ES_3$ is considered as the X axis. The line between $ES_1$ and $ES_2$ is considered as the Y axis. Accordingly, the coordinate of the end device's position at time $t$ can be represented by $(x, y)$, where $0 \leq x \leq 100, 0 \leq y \leq 100$. The distance between the end device and edge servers is $distance_i^t = \{distance_1^t, distance_2^t, \ldots, distance_N^t\}$.

## B. The Wireless Signal Model

In the offloading process of workflow tasks, each task is transferred to the edge server or the cloud server for execution, or executed locally at the end device, according to the offloading decision. When the task is being offloaded to the edge server, the speed of data transfer is changing with the location of the end device. The larger the distance between the end device and the edge server, the lower the data transfer speed is, and vice versa. In this paper, without the loss of generality, the effective communication radius of each edge server is assumed to be 50 m, and the transmission speed is divided into three levels [14]. As illustrated in Fig. 2, with three circles centered at the edge server i $(ES_i)$, the radius of each circle are $r_1, r_2, r_3$ respectively. Furthermore, we partition the biggest circle into three parts which are represented by $l_1, l_2, l_3$ respectively. Specifically, when the end device is within the area of $l_1, l_2, or\, l_3$, the data transmission speeds are 2048 KB per second, 1024 KB per second, or 512 KB per second respectively.

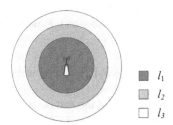

**Fig. 2.** The communication radius of each edge server.

## C. The Energy Consumption and Task Response Time Model

Here we will present the energy consumption and task response time models according to different task offloading decisions, viz. at the end device (namely no offloading), at the edge server, and at the cloud server. Please be noted that since the focus in MEC is to reduce the energy consumption of the end device so as to extend its battery life, the energy consumptions of the edge server and the cloud server are not considered in this paper. The task response time is the time between the task submitted by end device to the result data return to the end device, which includes transfer time before the task execution, task execution time and receiving time after task completion. Each workflow task $T_i$ can be represented by a quadruple, $T_i = \{load_i, transfer_i, receive_i, deadline_i\}$. Specifically, $load_i$ is the workload of $T_i$, $transfer_i$ is the size of data transferred before the task execution, $receive_i$ is the size of data the end device received after task completion, $deadline_i$ is the deadline constraint of $T_i$.

(1) *Energy Consumption and Task Response time at the End Device*

When the workflow task $T_i$ is not offloaded to the edge or cloud server, it will be executed locally at the end device. The energy consumption of the end device for the execution of task $T_i$ is calculated as follows:

$$E_i^{\text{end}} = P^{\text{end}} \times T_i^{\text{end}} \tag{1}$$

Where $E_i^{\text{end}}$ is the energy consumption for the end device to execute task $T_i$. The task execution power of end device is $P^{\text{end}}$. $T_i^{\text{end}}$ is the task response time at the end device which can be calculated as follows:

$$T_i^{\text{end}} = \frac{load_i}{f^{\text{end}}} \tag{2}$$

Where $load_i$ is workload of task $T_i$. $f^{\text{end}}$ is the CPU frequency of the end device.

(2) *Energy Consumption and Task Response time when offloaded to the Edge Server*

When task $T_i$ is offloaded to the edge server, the energy consumption is divided into three parts. The first part is the energy consumption for transferring the task data from the end device to the edge server. The second part is the energy consumption for receiving the task data back to the end device. The third part is the idle energy consumption of the end device when the task is being executed and migrated at the edge server. The total energy consumption for task $T_i$ when offloaded to the edge server is calculated as follows:

$$E_i^{\text{edge}} = P_{\text{tr}} \times T_i^{tran} + P_{\text{re}} \times T_i^{rece} + P_{\text{idle}} \times \left(T_i^{exec} + T_i^{migr}\right) \tag{3}$$

Where $P_{tr}$ is the data transfer power of the end device. $T_i^{tran}$ is the time of task data transfer. $T_i^{rece}$ is the time of task data receive. $T_i^{exec}$ is the task response time in edge server. $T_i^{migr}$ is the task migration time of the $T_i$. $P_{idle}$ is the idle power of the end device. $P_{re}$ is the task data receive power of the end device.

The total task response time includes four parts when task $T_i$ is offloaded to edge server. The first part is the time for task data transfer to the edge server. The second part is the actual response time for task execution at the edge server. The third part is for task data transfer back to the end device. The last part is the time for task migration when the task needs to be migrated to another edge server due to the movement of the end device. The total task response time in edge server is calculated as follows:

$$T_i^{edge} = T_i^{tran} + T_i^{exec} + T_i^{rece} + T_i^{migr} \tag{4}$$

$$T_i^{exec} = \frac{load_i}{f_j^{edge}} \tag{5}$$

$$T_i^{tran} = \frac{transfer_i}{R_i^{tran}} \tag{6}$$

$$T_i^{rece} = \frac{receive_i}{R_i^{rece}} \tag{7}$$

$$T_i^{migr} = \frac{receive_i}{R_i^{migr}} \tag{8}$$

Where $load_i$ is the workload of the task $T_i$. $f_j^{edge}$ is the CPU frequency of the virtual machine $j$ in edge server. $T_i^{migr}$ is the time of migration when task is migrated to another edge server. $R_i^{migr}$ is the task migration speed when task migrate to other edge server. $transfer_i$ is the data size of $T_i$ sent to the edge server. $R_i^{tran}$ is the data transfer speed when task offloading to edge server. $receive_i$ is data size of $T_i$ when task send back to end device. $R_i^{rece}$ is data receive speed when task return to end device. $R_i^{migr}$ is the data migration speed when task migrate to other edge server node.

(3) *Energy Consumption and Task Response time when offloaded to the Cloud Server*
Similarly, when task $T_i$ is offloaded to the cloud server, the energy consumption includes three parts which are data transfer to the cloud server, data transfer back to the end device and the idle energy consumption of the end device. The total energy consumption for task $T_i$ when offloaded to the cloud server is calculated as follows:

$$E_i^{cloud} = P_{tr} \times \frac{transfer_i}{R_{tran}^{cloud}} + P_{re} \times \frac{receive_i}{R_{rece}^{cloud}} + P_{idle} \times T_i^{cloud} \tag{9}$$

Where $R_{tran}^{cloud}$ is the data transfer speed to the cloud server. $R_{rece}^{cloud}$ is the data transfer speed back to the end device. $T_i^{cloud}$ is the task response time at the cloud server. $transfer_i$ is the size of data transferred before the task execution, $receive_i$ is the size of data the end device received after task completion. The total task response time at the cloud server is calculated as follows:

$$T_i^{\text{cloud}} = \frac{transfer_i}{R_{tran}^{\text{cloud}}} + \frac{load_i}{f_j^{\text{cloud}}} + \frac{receive_i}{R_{rece}^{\text{cloud}}} \tag{10}$$

Where $f_j^{cloud}$ is the CPU frequency of the virtual machine $j$ in the cloud server.

### D. The Fitness Value

The fitness value is designed to evaluate the quality of offloading decisions and task scheduling plan. The fitness function can evaluate the energy consumption of end device under workflow task deadlines [16]. The smaller fitness value, the lower energy consumption of the solution is. The fitness value can be calculated as follows:

$$fitness = (f_1 \times E_{\text{sum}}) + \left(f_2 \times 10 \times E_{\text{sum}} \times \frac{makespan}{deadline}\right) \tag{11}$$

Where $E_{sum}$ is the total energy consumption of the end device. *makespan* is the duration for the whole workflow. *deadline* is the workflow deadline constraint. The penalty coefficient for missing the deadline is set as 10, which is the same as in the previous work [16]. The total energy consumption of the end device for the whole workflow is calculated as follows:

$$E_{\text{sum}} = \sum_{i=1}^{\max} E_i^{\text{end}} + \sum_{j=1}^{\max} E_j^{\text{edge}} + \sum_{k=1}^{\max} E_k^{\text{cloud}} \tag{12}$$

$E_i^{end}$, $E_j^{edge}$, $E_k^{cloud}$ are the energy consumption of the end device when the task is executed at the end device, offloaded to the edge server or offloaded to the cloud server, respectively.

The fitness function consists of two parts: the first part is the total energy consumption of the end device when the workflow deadline is met $(f_1 = 1, f_2 = 0)$; the second part is the total energy consumption of the end device when the deadline is missed $(f_1 = 0, f_2 = 1)$. The idea is that when the workflow deadline is met, the energy consumption of the end device (the first part) is regarded as the fitness value. In this case, the strategy mainly focuses on optimizing the energy consumption of the end device. However, if the workflow deadline is missed, the penalty for missing the deadline is regarded as the fitness value (the second part). In this case, the strategy should focus on optimizing the energy consumption under the deadline constraint.

## 3    Mobility-Aware Workflow Offloading and Scheduling Strategy

Based on these models presented in Sect. 3, the novel Mobility-Aware Workflow Offloading and Scheduling Strategy (MAWOSS) is proposed in this section to solve the task offloading and scheduling problem for a moving end device in the MEC environment. As depicted in the pseudocode below, the process of MAWOSS has three main phases. Firstly, MAWOSS generates the offloading decisions for all workflow tasks (Lines 1–3). Secondly, after all offloading decisions are made, workflow scheduling is conducted for all types of resources allocated in the MEC environment (Lines 4–9). Finally, once the execution of a workflow task is completed, according to the updated location of the end device, the task migration strategy can select the best edge server for task migration to make sure the task execution result will be delivered successfully back to the end device (Lines 10–14). Detailed strategies and algorithms are presented in the subsequent sections.

---

**Overview of MAWOSS**

---

1  for i=1 to $task\_num$

2    generate the offloading decision of task $T_i$ according to Strategy 1;

3  end for

4  generate the task queues in all resources allocated in the MEC environment according to the task offloading decisions;

5  for i=1 to 3 //three different types of resources including the end device, the edge server and the cloud server

6    generate the scheduling plan in three different types of resource according to the Algorithm 1

7  end for

8  task offloading according to the generated task offloading decisions;

9  task scheduling and execution according to the generated task scheduling plan;

10  for i=1 to $n_{ready\_task}$

11    generate the best task migration plan for task $T_i$ according to Strategy 2;

12    migrate $T_i$ according to the generated task migration plan;

13    task execution result sent back to the end device;

14 End for

---

### 3.1  Mobility-Aware Workflow Task Offloading Strategy

Strategy 1 shows the Mobility-Aware Workflow Task Offloading Strategy for the end device. As described in the pseudocode below, the strategy consists of five major parts. The first part is to get the ready tasks according to the dependency of workflow tasks (Line 2). The second part is to calculate the task data transfer time and the execution time when the task is offloaded to the edge server (Lines 3–6). In the third part, the strategy makes the task migration decision and calculates the migration time and energy consumption of the end device (Lines 7–10). The fourth part is to calculate the energy consumption of the end device and total task response time with different task offloading decisions (Lines 11–12). The last part is to choose the best offloading decision which has the lowest energy consumption of the end device under the given deadline constraint and update the ready task queue according to the dependency of workflow tasks (Lines 13–22).

---

**Strategy 1: Mobility-Aware Workflow Task Offloading Strategy**

---

**Input:** Current time $t_{current}$, Tasks $T_i$, Cloud Server CS, Edge Server ES, End Device END, task deadline constraint *deadline*;

**Output:** the best task offloading decision $D_{energy-best}$;

1  while *task_num*! = 0

2    getting the ready task according to the dependency of workflow task and put into the ready task queue $Q_{ready\_task}$, calculate the number of ready task queue $n_{ready\_task}$;

3    for i=1 to $n_{ready\_task}$

4      calculate the distance between the end device to different edge servers $distance_j^t$ according to the current time $t_{current}$;

5    chose the edge server node $ES_i^{tran}$ which has the fastest transfer speed between the end device to edge server;

6    calculate the task $T_i$ execution time and transfer time when task offloading to edge server $ES_i$ by Formulas 5-6;

7    update the end device receive data time $t_{receive}$ according to Step 6 and calculate the distance between the end device to different edge servers $distance_j^t$;

8    chose the edge server node $ES_i^{rece}$ which has the fastest receiving speed between the edge server to end device;

9    compare the edge server $ES_i^{tran}$ which has the fastest transfer speed with the edge server $ES_i^{rece}$ which has the fastest receive speed and migrate the task to the best edge server ;

10   calculate the task migration time by Formula 8;

11   calculate the energy consumption of end device and task response time when task offloading to edge server by Formulas 3-4;

12   calculate the energy consumption of end device and task response time when task executed in end device or offloading to cloud server by Formulas 1-2, 9-10;

13   if $(T_i^{end} > deadline_i)\&\&(T_i^{edge} > deadline_i)\&\&(T_i^{cloud} > deadline_i)$

14     chose the lowest energy consumption offloading decision as the task $T_i$;

15   else

16     chose the lowest energy consumption under the deadline constraint offloading decision as the task $T_i$;

17   end if

18   update the current time $t_{current}$ according to the task offloading decision;

19   end for

20   $task\_num = task\_num - n_{ready\_task}$;

21   end while

22   return $D_{energy-best}$

---

## 3.2   PSO Based Task Scheduling Algorithm in the MEC Environment

Algorithm 1 describes the Particle Swarm Optimization based Task Scheduling Algorithm in the Mobile Edge Computing Environment. This algorithm mainly includes three parts: initialization of the task scheduling plan, iterative process of the algorithm, and return the best task scheduling plan for the three resource layers. The initialization of the task scheduling plan is to randomly generate the task scheduling plan and the search speed to find the initial global best task scheduling plan (Lines 1–8). The iterative process of the algorithm consists of two loops (Lines 9–18). The outer loop updates the task scheduling plan and other algorithm parameters (Lines 9–18). The inner loop calculates the fitness value of each task scheduling plan (Lines 11–14). In the inner loop, the algorithm first calculates the task energy consumption of end device and the task response time in three different types of computing resources (Line 12). Then, the algorithm calculates the total energy consumption of the end device, the total workflow makespan and the fitness value according to the scheduling plan (Line 13). Finally, we select the task scheduling plan with the lowest fitness value as the global best task scheduling plan (Line 15). When the iteration terminates, the algorithm returns the best task scheduling plan (Line 19). The details about the parameters and operators of the Particle Swarm Optimization (PSO) algorithm can be found in our previous work [17] and hence omitted here due to the space limits.

---

**Algorithm 1: Particle Swarm Optimization based Task Scheduling Algorithm in the Mobile Edge Computing Environment**

---

**Input:** maximum iterations: Iteration, tasks $T_i$, cloud server CS, edge server ES, end device END, task deadline constraint *deadline*;

**Output:** the best task scheduling plan $S_{energy-best}$;

1  for $i$=1 to $k$ do
2    initial the task scheduling plan $S_i$ and search speed $v_i$ randomly;
3  end for
4  for $i$=1 to $k$ do
5  calculate the energy consumption of the end device and the task response time in three different types of computing resources according to the task scheduling plan $S_i$.(Formulas 1-6);
6  calculate the total energy consumption of the end device, the total workflow makespan and the fitness value according to the scheduling plan $S_i$ (Formulas 11-12);
7  end for
8  select the global best task scheduling plan from scheduling plan $S_i$;
9  for $i$=1 to *Iteration* do
10  update the task scheduling plan according to the search speed $v_i$;
11    for $j$=1 to $k$ do
12  calculate the energy consumption of the end device and the task response time in three different types of computing resources according to the task scheduling plan $S_i$.(Formulas 1-6);
13    calculate the total energy consumption of the end device, the total workflow makespan and the fitness value according to the scheduling plan $S_i$ (Formulas 11-12);
14    end for
15    select the task scheduling plan $S_i$ with the lowest fitness value as the global best task scheduling plan;
16    update the inertia weight;
17    update the search speed $v_i$;
18  end for
19  return $S_{energy-best}$;

### 3.3 Mobility-Aware Task Migration Strategy

The Mobility-Aware Task Migration Strategy is shown in Strategy 2. This strategy mainly includes three parts: initialization of the finished task queue, the iterative process to find the best task migration decision, and the return of the task migration decision $M_{energy-best}$. Initialization of the finished task queue is to get the queue of the finished tasks at the current edge server from Algorithm 1 and then put it into the global finished task queue and calculate the number of finished tasks (Line 1). The iterative process is used to find the best task migration decision (Lines 2–8). In the iterative loop, the strategy first calculates the distances between the end device and different edge servers according to the task finish time and then chose the edge server having the fastest data receiving speed (Lines 3–4). Then, the strategy makes the task migration decision having the lowest migration energy cost $M_{energy-best}$ (Line 5). When the iteration terminates, the strategy return the best task migration plan (Line 9).

---

**Strategy 2: Mobility-Aware Task Migration Strategy**

**Input:** task finish time $t_{real}$, tasks $T_i$, edge server ES, task deadline constraint $deadline$;

**Output:** the best task migration plan $M_{energy-best}$;

1  get the finished task queue of the edge server from Algorithm 1 and put it into the global finished task queue $Q_{finish\_task}$, calculate the total number of finished tasks $n_{finish\_task}$;

2  for i=1 to $n_{finish\_task}$

3  calculate the distance between the end device and different edge servers $distance_j^t$ according to the task finish time $t_{real}$;

4  choose the edge server node $ES_i^{real}$ which has the fastest receiving speed between the edge server and the end device;

5  make the task migration decision $M_{energy-best}$ which migrates the task from its current edge server $ES_i^{tran}$ as in Strategy 1 to the selected edge server $ES_i^{real}$;

6  calculate the energy consumption of the end device and the task response time (Formulas 3-4);

7  update the current time $t_{real}$;

8  end for

9  return $M_{energy-best}$

---

## 4 Evaluation

In this section, simulation experiments are conducted to validate the effectiveness of our proposed MAWOSS strategy. We compare MAWOSS with five other task offloading strategies in terms of their fitness value, the energy consumption of the end device and the workflow makespan. The results on the time and energy consumption for task migration are also highlighted given the focus of this paper on the mobility of the end device. Specifically these strategies include: (1) the local-based partial reasonable task schedule construction algorithm (LOPRTC) [14]; (2) the energy efficient multi-resource computation offloading strategy (EMO) [17]; (3) the strategy which executes all tasks in the cloud server (ONLY-CLOUD); (4) the strategy which executes all tasks in the edge server (ONLY-EDGE); (5) the strategy which executes all tasks in the end device (ONLY-END). For the fairness of comparison, particle swarm optimization based task scheduling algorithm is used with all offloading strategies.

### 4.1 Experimental Settings

The simulation experiment runs on a PC with the following configurations: Intel Core i7 3.6 GHz CPU, 16 GB RAM, and Windows 10 OS. The simulations are developed in MatlabR2017b. The workflow structure is defined using the Montage workflow which is widely used for simulation experiments [18]. The task number of workflows ranges from 50 to 300. The workload of each task is generated between 30 and 30000 Mega Cycles randomly [14]. The upload and download data size of each task varies from 500 KB to 1000 KB [14]. The transmission, receiving, execution and idle power of end device are 100 mW, 50 mW, 700 mW, 3 mW respectively [19]. The network bandwidths of LAN and WAN are 100 Mbps and 10 Mbps respectively [14]. The number of virtual machines in the cloud server is set as 5. The number of virtual machines in the edge server is set as 3. The CPU frequency of the end device is 3 GHz. The CPU frequency of the virtual machines in the edge server varies from 3 GHz to 9 GHz. The CPU frequency of the virtual machines in the cloud server varies from 9 GHz to 15 GHz. The deadlines of workflow tasks are assigned as double the average response time of workflow tasks running a 9 GHz CPU [14]. The parameter settings of the particle swarm optimization algorithm are set according to the work in [17]. The location of the edge servers and the moving path of the end device are set the same as in the example patient healthcare monitoring scenario shown in Fig. 1. Under such a physical setting, the end device will always be covered by at least one effective edge server. Therefore, the failure of task migration (when the end device is outside the effective communication range of any edge servers) is not considered in our experiments, but we will investigate such abnormal situations in the future.

### 4.2 Evaluation Results

#### A. Fitness value
The fitness value represents the general effectiveness of these strategies. The results on the fitness are shown in Fig. 3. As the number of tasks grows, the fitness value of all

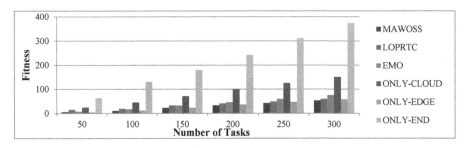

**Fig. 3.** Comparison of the fitness value.

methods increases. It can be seen that the fitness value of MAWOSS is much lower than the other five methods, which means that MAWOSS can always find the task offloading decision and scheduling plan with lower energy consumption of the end device under the given deadline constraint. For example, when the task number is 50, the fitness value of MAWOSS is 2% lower than LOPRTC. When the task number becomes 300, the fitness value of MAWOSS is 4% lower than LOPRTC. Therefore, in general, MAWOSS is the most effective task offloading and scheduling strategy for reducing the energy consumption under the given deadlines.

### B. Energy consumption of the end device

Figure 4 shows the results on the energy consumption of the end device. The energy consumption of the ONLY-EDGE strategy is always the lowest. However, the gap between ONLY-EDGE and our MAWOSS strategy is small. For example, when the task number is 150, the energy consumption of MAWOSS is 3% lower than LOPRTC. However, if comparing their workflow makespan as shown in Fig. 5, we can find that the workflow makespan of ONLY-EDGE is much higher than MAWOSS. For example, when the task number is 150, the workflow task makespan of ONLY-EDGE is 14% higher than MAWOSS. This means ONLY-EDGE's task offloading decision and scheduling plan may miss the given deadlines. In contrast, MAWOSS can always find the best task offloading decision and scheduling plan for reducing the energy consumption under the given deadlines.

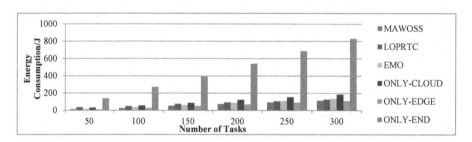

**Fig. 4.** Comparison of the end device's energy consumption.

### C. Workflow makespan

The results on the workflow makespan are shown in Fig. 5. It can be seen that the workflow makespan of ONLY-END is always the lowest, and MAWOSS is the

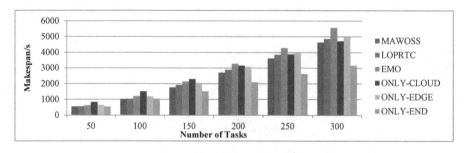

**Fig. 5.** Comparison of the workflow makespan.

second-lowest. However, if comparing their energy consumption of the end device as shown in Fig. 4, we can find that the workflow makespan of ONLY-END is much higher than MAWOSS. For example, when the task number is 100, the workflow makespan of ONLY-END is 9.8 times higher than MAWOSS. Therefore, it proves that MAWOSS can always find the best task offloading decision and scheduling plan for reducing workflow makespan and energy consumption under the given deadlines.

### D. Task migration time and energy consumption

Since task migration among edge servers directly deals with the mobility issue of the end device, here we present a detailed look at the results on time and energy consumption of the end device for task migration as shown in Figs. 6 and 7. As ONLY-CLOUD and ONLY-END do not have task migration, we compare MAWOSS with LOPRTC, EMO and ONLY-EDGE. It can be seen that the task migration time and energy consumption of MAWOSS are always lower than the other strategies. For example, when the task number is 50, the task migration time and energy consumption of MAWOSS is 6.3% lower than LOPRTC. When the task number becomes 300, the task migration time and energy consumption of the MAWOSS is 15.1% lower than LOPRTC. Therefore, MAWOSS is able to achieve the minimum task migration time and energy consumption.

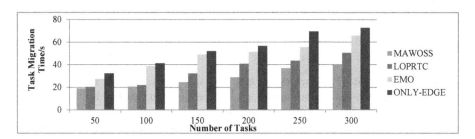

**Fig. 6.** Comparison of the task migration time.

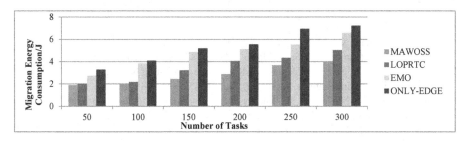

**Fig. 7.** Comparison of the task migration energy consumption.

## 5  Conclusion and Future Work

Most existing works on task offloading and task scheduling in the mobile edge computing environment overlooked the mobility issue of the end device, which could result in the failure of delivering the task execution results back to the end device in time and hence halt the workflow execution. In this paper, considering the mobility of the end device, we propose the Mobility-Aware Workflow Offloading and Scheduling Strategy (MAWOSS) as a holistic approach to minimize the energy consumption of the end device and the workflow makespan under the deadlines. Experimental results showed that the MAWOSS can always achieve the optimal fitness with lower energy consumption and smaller workflow makespan compared with other strategies.

This paper focused on the workflow offloading and scheduling problem for a single end device in the MEC environment. In the future, we can explore a more complicated scenario where multiple end devices are running multiple different workflows in MEC environment.

**Acknowledgement.** This work is the partially supported by the Humanities and Social Sciences of MOE Project No. 16YJCZH048, the National Natural Science Foundation of China Project No. 61972001, the Key Natural Science Foundation of Education Bureau of Anhui Province Project KJ2016A024, and the Nature Science Foundation of Hubei Province Project 2019CFB172.

## References

1. Azimi, I., Pahikkala, T., Rahmani, A., et al.: Missing data resilient decision-making for healthcare IoT through personalization: a case study on maternal health. Future Gener. Comput. Syst. **96**, 297–308 (2019)
2. Hamza, R., Yan, Z., Muhammad, K., et al.: A privacy-preserving cryptosystem for IoT E-healthcare. Inf. Sci. (2019, early access)
3. Forkan, A., Khalil, I., Atiquzzaman, M.: ViSiBiD: a learning model for early discovery and real-time prediction of severe clinical events using vital signs as big data. Comput. Netw. **113**, 244–257 (2017)
4. Roman, R., Lopez, J., Mambo, M.: Mobile edge computing, fog et al.: a survey and analysis of security threats and challenges. Future Gener. Comput. Syst. **78**, 680–698 (2018)
5. Shi, W., Dustdar, S.: The promise of edge computing. Computer **49**(5), 78–81 (2016)
6. Bouet, M., Conan, V.: Mobile edge computing resources optimization: a geo-clustering approach. IEEE Trans. Netw. Serv. Manag. **15**(2), 787–796 (2018)
7. Sodhro, A., Luo, Z., Sangaiah, A., et al.: Mobile edge computing based QoS optimization in medical healthcare applications. Int. J. Inf. Manag. **45**, 308–318 (2019)
8. Lyu, X., Tian, H., Ni, W., et al.: Energy-efficient admission of delay-sensitive tasks for mobile edge computing. IEEE Trans. Commun. **66**(6), 2603–2616 (2018)
9. Zhang, W., Zhang, Z., Zeadally, S., et al.: Efficient task scheduling with stochastic delay cost in mobile edge computing. IEEE Commun. Lett. **23**(1), 4–7 (2018)
10. Ning, Z., Dong, P., Kong, X., et al.: A cooperative partial computation offloading scheme for mobile edge computing enabled Internet of Things. IEEE Internet Things J. **6**(3), 4804–4814 (2018)

11. Mach, P., Becvar, Z.: Mobile edge computing: a survey on architecture and computation offloading. IEEE Commun. Surv. Tutor. **19**(3), 1628–1656 (2017)
12. Lyu, X., Tian, H., Jiang, L., et al.: Selective offloading in mobile edge computing for the green Internet of Things. IEEE Netw. **32**(1), 54–60 (2018)
13. Kuang, Z., Li, L., Gao, J., et al.: Partial offloading scheduling and power allocation for mobile edge computing systems. IEEE Internet Things J. (2019, early access)
14. Zhu, T., Shi, T., Li, J., et al.: Task scheduling in deadline-aware mobile edge computing systems. IEEE Internet Things J. **6**(3), 4854–4866 (2018)
15. Hu, M., Zhuang, L., Wu, D., et al.: Learning driven computation offloading for asymmetrically informed edge computing. IEEE Trans. Parallel Distrib. Syst. (2019, early access)
16. Hu, J., Jiang, M., Zhang, Q., et al.: Joint optimization of UAV position, time slot allocation, and computation task partition in multiuser aerial mobile-edge computing systems. IEEE Trans. Veh. Technol. (2019, early access)
17. Xu, J., Li, X., Ding, R., et al.: Energy efficient multi-resource computation offloading strategy in mobile edge computing. Comput. Integr. Manuf. Syst. **25**(4), 954–961 (2019)
18. WorkflowGenerator. https://confluence.pegasus.isi.edu/display/pegasus/WorkflowGenerator. Accessed 03 July 2019
19. Cao, S., Tao, X., Hou, Y., et al.: An energy-optimal offloading algorithm of mobile computing based on HetNets. In: 2015 International Conference on Connected Vehicles and Expo (ICCVE), pp. 254–258. IEEE, Shenzhen (2015)

# HSPP: Load-Balanced and Low-Latency File Partition and Placement Strategy on Distributed Heterogeneous Storage with Erasure Coding

Jiazhao Sun, Yunchun Li, and Hailong Yang[✉]

Sino-German Joint Software Institute, School of Computer Science and Engineering,
Beihang University, Beijing 100191, China
hailong.yang@buaa.edu.cn

**Abstract.** To speedup the accesses to massive amount of data, heterogeneous architecture has been widely adopted in the mainstream storage
system. In such systems, load imbalance and scheduler overhead are the
primary factors that slow down the I/O performance. In this paper, we
propose an effective file scheduling strategy HSPP that includes statistic
based file classification, partition with erasure coding and adaptive data
placement to optimize load balance and read latency on the distributed
heterogeneous storage system. The experiment results show that HSPP
is superior than existing strategies in terms of load balance, read latency,
and scheduling overhead.

**Keywords:** Data partition and placement · Erasure coding ·
Heterogeneous storage

## 1 Introduction

Large-scale storage and distributed file system have become the foundation of
the IT industry, such as Alibaba Network Attached Storage [1] and Google File
System [15]. In recent years, in-memory storage systems [2,5,6,10] have gradually
replaced traditional disk-based systems [3,11] for better I/O performance. The
need for large scale storage in production system conflicts with the high price
and volatility of memory, which leads to the design of heterogeneous storage
system. Optimizing the file partition and placement in heterogeneous storage
system becomes important to improve the I/O performance.

Many distributed storage systems [2,3,19,22,23] support integrating various
storage devices, where HDFS [3] and Alluxio [2] are wildly deployed systems
that can support storage devices such as memory, SSD and disk. Due to the difference of storage devices, there are many research works to optimize file placement on distributed heterogeneous storage system to improve overall I/O performance [18,30,36,37]. The erasure coding has been applied in distributed storage
systems to achieve the load balance and low latency [21,28,32]. Joshi et al. [21]

© Springer Nature Switzerland AG 2020
S. Wen et al. (Eds.): ICA3PP 2019, LNCS 11945, pp. 200–214, 2020.
https://doi.org/10.1007/978-3-030-38961-1_18

establish a model for the read process in the distributed storage system with erasure coding using the (n, k) fork-join queuing model, and demonstrate the mean latency of the proposed model is tightly bounded. Moreover, studies [32,34] develop this theory with the tradeoffs among the partition number, network communication overhead and the system straggler, where Yu et al. [34] observe the *elbow point* and apply an approximation method to determine the optimal partition number.

The file partition and placement strategies mentioned above can effectively balance the load in the homogeneous storage system. However, in the heterogeneous storage system, the load tends to aggregate on nodes with multiple storage layers, which enlarges the overall access latency. In addition, existing strategies incur significant scheduling overhead due to the coarse-grained file classification. Therefore, this paper proposes an effective file partitioning and placement strategy of heterogeneous storage system (named as HSPP), which optimizes load balance and read latency significantly. In addition, the Reed-Solomon (RS) erasure coding is used to resist system straggler under low redundancy overhead. The experimental results show that the load imbalance factor reduces by more than 9×, and the mean and tail latency reduces by more than 18% for read operation with HSPP compared with existing strategies.

Specifically, this paper makes the following contributions:

- We establish a model for the upper bound of mean read latency, and propose the exponential search algorithm to determine the optimal number of file partitions. In addition, the erasure coding is used to mitigate the mean and tail latency under system straggler.
- We propose a new file classification mechanism that place the files at the optimal storage layer, which significantly reduces the scheduling overhead and extends the lifespan of SSD.
- We implement the above strategies on Alluxio (named as HSPP) with the partition vectors. The vectors are generated on Master with the file popularity and size, the HSPP partition and place files adaptively to reduce both the scheduling and redundancy overhead.

The rest of this paper is organized as follows. Section 2 presents the background and research motivation. Section 3 describes the file partition and placement mechanism of HSPP. Section 4 presents the implementation details of HSPP. Section 5 compares HSPP with existing file partition and placement strategies on the distributed heterogeneous storage system. Section 6 presents the related works. We conclude this paper in Sect. 7.

## 2   Background and Motivation

In this section, we briefly introduce the storage heterogeneity and high skewed popularity and size distribution observed of the production load, which leads to the motivation of this paper.

## 2.1   Heterogeneity in Distributed Storage System

With the emergence of high-speed network equipment and the advances of network fabrics, the gap between network bandwidth and storage bandwidth is rapidly shrinking [14,16], and the performance bottleneck of distributed storage systems is shifting from network to storage I/O. To meet high-speed and massive files write and read, amounts of distributed storage systems apply heterogeneous architecture, consists of storage devices with different bandwidth and capacity.

Some distributed storage systems such as HDFS [3], OcotpusFS [22] and Alluxio [2] currently support the configuration and management of memory, SSD and disk. Table 1 summarizes the characteristics of the storage devices, where the write and read throughput is evaluation on Alluxio. It should be noted that the lifespan of SSD is much shorter than memory and disk, because the Program/Erase Cycle of NAND Flash is approximately 3,000 times [33].

**Table 1.** The characteristics of different storage devices.

| Tier | Write throughput (MB/s) | Read throughput (MB/s) | Price each GB (USD) | P/E Cycle |
|------|-------------------------|------------------------|---------------------|-----------|
| MEM  | 750                     | 1100                   | 4.500               | –         |
| SSD  | 600                     | 750                    | 0.310               | 3000      |
| HDD  | 450                     | 550                    | 0.003               | –         |

Existing studies [19,23,30,37] optimized the file placement with various storage devices based on cache eviction algorithms [20,25,26,31], by placing hot files on high throughput storage to improve overall I/O. For the limited P/E cycles of SSD, the LARC algorithm [18] is improved by ARC, which can reduce the SSD traffic. However, this kind of approaches that ignores load statistical characteristics can't describe the file popularity as a whole, resulting in frequent data migration between storage layers. The high scheduling overhead drags the overall throughput, which is verified in Sect. 5.

## 2.2   Load Imbalance

Load imbalance and system straggler are common in distributed storage systems. Main factors of the load imbalance is the high skewed popularity and the high imbalanced network traffic [9,28,34].

File popularity in production load generally subjects to the Zipf distribution [28,29]. The majority of file access is contributed by a handful of hot files. Figure 1 presents The distribution of file popularity and size in Yahoo! cluster [7]. Most files (77.64%) are cold with less than 10 access, while 2.29% of files are very hot with more than 100 access. We also found a positive correlation between file access popularity and size. The existing in-memory [28,34] and disk-based [9,17] solutions cannot adequately balance high skewed load. HSPP partitions the file with optimal number of (n, k) fork-join queuing model, and decide placement node with the weighted random algorithm, which effectively balancing the load of distributed heterogeneous storage system.

## 2.3  The (n,k) Fork-Join Queuing Model

Suppose that the file is divided into $k$ partitions, and $r$ redundant partitions are generated by erasure coding, then $n = k + r$ data partitions are placed on separate nodes. The file request can be modeled as the (n ,k) fork-join queuing system, and is completed when any $k$ out of $n$ tasks are served.

Assuming that the file read request as a Poisson process with rate $\lambda$, so the service delay of each node is exponentially distributed with mean $\mu$ [12,21,32]. Since each node holds $\frac{1}{k}$ of the file, so the service delay is exponentially distributed with mean $\frac{\mu}{k}$. The queue of each node can be modeled as an independent M/M/1 queue with the exponentially distributed service delay. But in fact, it is hard to describe the service delay in a single model, so the queue should be extend to the M/G/1 model. Studies [21,32] demonstrates that the mean latency upper bound of the (n, k) fork-join queuing model could tightly bound the practical latency.

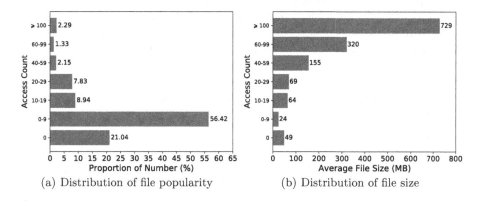

(a) Distribution of file popularity    (b) Distribution of file size

**Fig. 1.** The distribution of file popularity and size observed in Yahoo! cluster trace.

## 2.4  Motivation

Although there are research works on file placement strategies to improve I/O performance, the existing strategies fail to utilize the unique characteristics of heterogeneous storage system, and thus lead to severe load imbalance and high read latency when directly applied. The above shortcomings motivate our work to propose a new strategy to partition and place the files effectively in heterogeneous storage system. In addition, when the file popularity and size changes, the proposed strategy can adaptively adjust the placement horizontally and vertically to improve load balance and reduce read latency.

## 3   The Methodology of HSPP

The HSPP includes the methods for data partitioning, data classification and data placement, that together effectively address the issues of high read latency, load imbalance and poor utilization of top-layer storage in distributed heterogeneous storage.

### 3.1   Partition with Erasure Coding

HSPP partitions the file to $k$ evenly, and calls Intel ISA-L library [4] to generate $r$ redundant partitions by RS erasure encoding, and then place the $n = k + r$ data partitions on separate nodes with the weighted random algorithm. The file read request forks $n$ sub-requests, and any $k$ out of $n$ sub-requests are served, the file can be decoded to retrieve.

**Upper Bound of Mean Latency.** The service delay of file $i$ in server $s$ is represented by $D_{i,s}$, which consists of queuing delay and transfer delay. In the (n, k) fork-join model, if any $k$ data blocks arrive, the read request for file $i$ completes, so the service delay of file $i$ is determined by the delay of the $k^{th}$ arriving partition, so the mean service delay $T_{n,k}$ of file $i$ is described as Eq. 1. The (k, k) fork-join model is a particular state of the (n, k) system, since no redundant partitions, in which the read request for the file completes with all partitions arriving. Therefore, $T_{k,k}$ is the upper bound of $T_{n,k}$ [32,34].

$$T_{n,k} = E(D_{i,s}(k)) \leq T_{k,k} = E(\max D_{i,s}) \tag{1}$$

Xiang et al. [32] prove that on (k, k) fork-join queuing model, there is a tight upper bound of mean read latency by solving a convex optimization problem, as shown in Eq. 2.

$$\overline{T} \leq \widehat{T} = \min_{z \in R} \left\{ z + \sum_{s:C_s} \frac{1}{2}(E(D_{i,s}) - z) + \sum_{s:C_s} \frac{1}{2}\sqrt{(E(D_{i,s}) - z)^2 + Var(D_{i,s})} \right\} \tag{2}$$

Where $z$ is an auxiliary variable introduced to make upper bound tighter, $C_s$ denotes the set of servers on which the file $i$ is placed. It should be noted that the Eq. 2 is not a closed form and is solved as a convex optimization problem with the expectation and variance of $D_{i,s}$. With the Pollaczek-Khinchin transform, the expectations and variances can be calculated as Eqs. 3 and 4.

$$E(D_{i,s}) = \frac{S_i}{k_i B_t} + \frac{L_s \Gamma_s^2}{2(1 - \rho_s)^2} \tag{3}$$

$$Var(D_{i,s}) = \left(\frac{S_i}{k_i B_t}\right)^2 + \frac{L_s \Gamma_s^2}{3(1 - \rho_s)} + \frac{L_s (\Gamma_s^2)^2}{2(1 - \rho_s)^2} \tag{4}$$

Where $\Gamma_s^2$ and $\Gamma_s^3$ indicate the second and third moment of the service delay, which is exponentially distributed, so $\Gamma_s^2$ and $\Gamma_s^3$ are derived as Eqs. 5 and 6, where $t$ is the stored layer, $\rho_s$ indicates the request intensity, $L_s$ is the load of server $s$. $\mu_s$ is the mean transfer delay of server $s$ as shown in Eq. 7, $p_i$ is the popularity of file $i$ as described in Sect. 3.3.

$$\Gamma_s^2 = \sum_{t \in T_s} \sum_{i \in C_{s,t}} 2 \frac{p_i}{L_s} \left( \frac{S_i}{k_i B_t} \right)^2 \tag{5}$$

$$\Gamma_s^3 = \sum_{t \in T_s} \sum_{i \in C_{s,t}} 6 \frac{p_i}{L_s} \left( \frac{S_i}{k_i B_t} \right)^3 \tag{6}$$

$$\mu_s = \sum_{t \in T_s} \sum_{i \in C_{s,t}} \frac{p_i S_i}{k_i B_t} \tag{7}$$

**Determine the Optimal K and R.** To verify the mean read latency upper model, we deployed a distributed system with 20 heterogeneous storage nodes and wrote in 100k 10 MB files. The detailed setting of our experiment is given in Sect. 5. As shown in Fig. 2, we compare the upper bound and the measured mean read latency within three types of storage devices. When the partition number $k$ is small, the upper bound of memory and SSD could strictly limit the evaluated read latency. Since the addressing delay of HDD is not included in upper bound model due to the high unpredictability, the bounding effect is little weak.

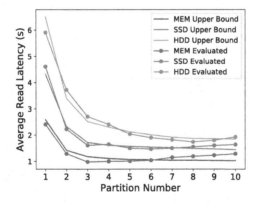

**Fig. 2.** Comparison of the theoretical upper bound and the measured mean read latency on heterogeneous storage system.

Although the upper bound decrease as the partition increases, in fact, the mean read latency will increase when partition number is large, because of excess network overhead and straggler effect, such as amount of TCP connections and the incast effect. HSPP resort to an *exponential search* algorithm to find the optimal $k$ on the model. Initially, the $k$ is setting to $\frac{N}{5}$, where $N$ is the number of storage nodes. And then inflates $k$ by 1.5× each iteration until the latency improvement is below 5%, that means the extra network overhead offsets the latency gain. The optimal partition number $k$ is negatively related to storage throughput, that is, the optimal partition number $k$ in memory is smaller than SSD and HDD. And it should be noted that the computational overhead is so limited, it takes only 5.54 s to determine optimal k for 100k files in our system.

To alleviate the severe impact of straggler on tail latency, HSPP applys redundant partitions and extra read threads. Since redundant partitions introduce storage overhead, the optimal number of redundant partitions requires a trade-off between redundant overhead of storage and read latency. We designed a set of experiments to observe the impact of redundant partition number $r$ on the mean and tail read latency with system straggler of various frequency. The straggler is generated with given frequency and in Microsoft Bing cluster pattern [8], and $k$ is set to 10.

As shown in Fig. 3, the evaluation was divided into three groups, and the straggler frequency was 0, 0.05 and 0.1 (very intensive). We found that redundant partitions can reduce the mean and tail latency. Even in the group without straggler, the tail latency decreases obviously. In the group with intensive straggler, the redundant partitions are most effective. These confirm that the redundant partitions could highly improve the system anti-straggler ability. However, redundant partitions introduce excess storage and network overhead, which will drag system performance. We found that the mean and tail latency is optimal when $r$ is 3, so HSPP sets the redundancy ratio to 30%. Also, we proposed a redundant migration mechanism to utilize the storage heterogeneity, which vertically migrates the redundant partitions to the lower layer to reduce the upper overhead, especially in improving the memory usage. The mechanism will be introduced in Sect. 4.

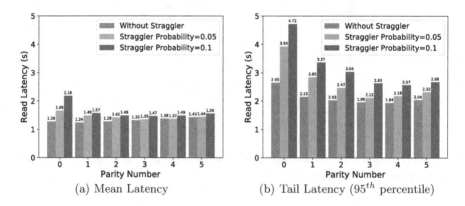

(a) Mean Latency                      (b) Tail Latency ($95^{th}$ percentile)

**Fig. 3.** The Mean (a) and tail (b) read latency under different frequency of straggler occurrence.

## 3.2  Statistic Based Data Classification

Analysis of read trace of Google [29] and Yahoo! [7], we found that the popularity of hot files subjects the heavy-tailed distribution over time. For example, the popularity of hot file gradually decrease over a long span, there is still a considerable amount of access in the span. So this type of files can be classified as the warm files. And we regard the files with power-low distributed popularity as the cold files.

HSPP describes the file popularity with *exponential decay* algorithm, which could effectively fit the re-access probability of files [13]. As shown in Eq. 8, $n$ is the count of file access, and $t_i$ represents the time of $i^{th}$ access. For long-running loads, HSPP needs to describe the popularity of the file on a large time scale, so the parameter $a$ is set to $10^{-6}$ in this paper.

$$P(t) = \sum_{i=1}^{n} e^{-a(t-t_i)} \tag{8}$$

The *exponential decay* algorithm is recursive that popularity can be updated only with the time interval. If the file is not accessed within the time interval $\Delta t$, the new popularity at time $t + \Delta t$ is calculated as Eq. 9. If it is accessed at time $t + \Delta t$, the new popularity is calculated as Eq. 10. The computational overhead for updating file popularity is so limited that takes only 90 ms to update the popularity of 100K files in our system.

$$P(t + \Delta t) = P(t) \times e^{-a\Delta t} \tag{9}$$

$$P(t + \Delta t) = P(t) \times e^{-a\Delta t} + 1 \tag{10}$$

HSPP periodically samples the file popularity and finds the 90 and 60 quantiles as the Hot and Warm thresholds, and then classify all the files with the thresholds as shown in Table 2. The files are divide to Hot, Warm and Cold, and writing in memory, SSD, and HDD preferentially. If the optimal layer of Alluxio worker does not have enough space, the data blocks are written to the lower layer. The classification method based on statistical characteristics reduce the scheduling overhead and traffic into SSD effectively.

**Table 2.** Data classification threshold and optimal storage layer.

| Type | Threshold | Optimal layer |
| --- | --- | --- |
| Hot | $p > p^{90}$ | MEM |
| Warm | $p^{90} > p > p^{60}$ | SSD |
| Cold | $p < p^{60}$ | HDD |

### 3.3 Adaptive Data Placement

This subsection presents the design detail of data placement in the distributed heterogeneous storage system.

HSPP utilize the partition vector <K, R, MEM, SSD, HDD> to maintain the file storage state, K and R represent the number of source partitions and redundant partitions respectively. MEM, SSD, and HDD represent number of partitions stored in memory, SSD, and Disk. The HSPP-Master periodically updates the partition vector of files, and the HSPP-Client responds to various update requests discriminatively. For example, the partition vector of file $i$ changes from <3, 1, 4, 0, 0> to <3, 1, 0,4,0>, only needs to migrate all the file partitions from memory to SSD. If a node has no SSD layer, the partition will be migrated to other node. If the number of file partitions has changed, such as vector changing from <3, 1, 4, 0, 0> to <4, 1, 5, 0, 0>, the HSPP-Client should rewrite the file.

The storage node adjusts the weights depend on its load. The load $L$ of server $s$ is as shown in Eq. 11, where $T_s$ is the layer set of server $s$, $C_{s,t}$ is the set of files which has place a partition at layer $t$ of server $s$, and $p_i$ represents the popularity of file $i$, and the file size is $S_i$. The weight of the node is as shown in Eq. 12. The lower the load, the higher the weight of that node, so the weight can be adjusted. Compared with the random placement algorithm adopted by EC-Cache [28] and Selective Partition [34], the load balance is significantly improved with HSPP. The experiment in Sect. 5 verifies that.

$$L_s = \sum_{t \in T_s} \sum_{i \in C_{s,t}} \frac{p_i S_i}{k_i} \tag{11}$$

$$weight = \frac{1}{L_s \sum \frac{1}{L_s}} \tag{12}$$

## 4    The Implementation of HSPP

We have implemented HSPP atop the Alluxio with HSPP-Master and HSPP-Client. The overview of system architecture is shown in Fig. 4. The HSPP-Master is responsible for metadata maintaining and management, including file popularity and storage status, and periodically updating file popularity and partition vectors. The HSPP-Client partitions and erasure encodes files according to the partition vector, and then place the partitions on separate nodes. The erasure coding is implemented by Intel ISA-L library [4]. The HSPP-Client periodically notifies the HSPP-Master of the file retrieve log. In particular, HSPP further reduce the read latency with the Dual-Way Service mechanism, and utilize the index K-Hit Ratio to adaptively migrate the redundant partitions vertically to spare upper layer storage.

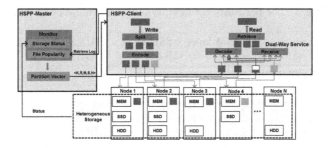

**Fig. 4.** The overview of HSPP implementation.

### 4.1 Dual-Way Service

Although the ISA-L library provides a highly optimized implementation of RS decoding, the decoding overhead is still not trivial. We found that the decoding overhead is positively correlated with file size as shown in Fig. 5. For example, retrieving a 300 MB file results in 800 ms overhead, and the missing source partitions may arrive during the decoding window. Therefore, we design a dual-way service mechanism to shorten the service delay further. As shown in Fig. 4, the erasure decoding is started with first $k$ partitions, and forking a new thread to receive the missing source partitions. If all the source partitions arrive, HSPP-Client will stop decoding and complete the file request. Compared with late binding, dual-way service mechanism could utilize the decoding window to reduce file read latency.

### 4.2 K-Hit Ratio

We consider the file retrieve with source partitions as a K-Hit event, and the K-Hit Ratio indicates the necessity of redundant partitions. In HSPP, if the K-Hit ratio reaches 80%, the redundant block will be migrated to the lower storage to release upper space. For instance, the partition vector <10, 3, 13, 0, 0> changes to <10, 3, 12, 1, 0>, it means that move a redundant partition from memory to SSD. This approach can reduce the redundancy overhead of the upper storage in the case of sparse straggler.

## 5  Evaluation

### 5.1 Experimental Setup

We evaluate on a distributed heterogeneous storage system with 20 nodes, which can be divided into four groups based on the hardware configurations in Table 3. Additional 5 nodes are used as the clients to continuously submit read requests following Poisson distribution. The load is synthesized based on the file popularity and size distribution of the Yahoo! cluster [27].

Table 3. The hardware configuration of the evaluation system.

| CPU | MEM layer (GB) | SSD layer (GB) | HDD layer (GB) | # nodes |
|---|---|---|---|---|
| Intel Xeon Phi 7210 | 20 | 50 | 100 | 4 |
| Intel Xeon E5-2650 | 5 | 20 | 100 | 7 |
| Intel Xeon E5620 | 10 | 0 | 200 | 5 |
| Intel Xeon E5620 | 5 | 0 | 100 | 4 |

We compare HSPP with three widely used file partition and placement strategies on distributed storage system:

**Selective Replication:** to keep up with the 30% redundancy overhead in HSPP, we generate three replicas for the top 10% popular files and use the consistent hashing algorithm to determine the file placement [24].

**EC-Cache:** to provide a fair comparison, we adopt the (13, 10) erasure coding scheme and determine the partition placement with the algorithm [28]. The file read is served with the late binding mechanism, which means, when any 10 partitions have arrived, it starts decoding and ignores the rest partitions.

**Selective Partition:** use the same latency upper bound model and the search algorithm for the optimal number of file partitions as HSPP, but without redundant partitions. In addition, the file partition placement is based on the algorithm [34].

## 5.2   Load Balance

We use the imbalance factor to indicate the level of load imbalance in a distributed storage system. As shown in Eq. 13, the smaller the $\xi$ is, the more balanced the system load is. As shown in Fig. 6, compared to the other three strategies, HSPP achieves the optimal load balance. Specifically, the $\xi$ of HSPP is 0.023, which is 35×, 11×, and 9× better than selective replication (0.822), EC-Cache (0.263) and selective partition (0.219) respectively. Due to copy and distribute the replications of hot files that easily generates hotspots in the system, the load under Selective Replication is most unbalanced.

EC-Cache and Selective Partition show good performance in the homogeneous storage system with random placement algorithm. However, in heterogeneous storage system, the load aggregates on nodes with multiple storage layers. Specifically, the load of EC-Cache is higher than Selective Partition, therefore the load imbalance is more severe. Based on the optimal file partition, HSPP applies the weighted random algorithm to determine placement, thus minimizing the load imbalance in heterogeneous storage system.

$$\xi = \frac{L_{max} - L_{min}}{L_{avg}} \tag{13}$$

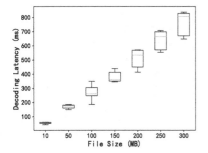

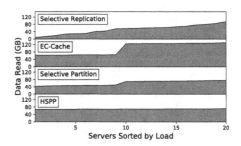

**Fig. 5.** The correlation between decoding window and file size.

**Fig. 6.** The load distribution under four different strategies.

### 5.3    Mean and Tail Latency

We evaluate the read mean and tail latency under different strategies by setting the frequency of straggler to 0.05, which represents the real situation on production cluster. As shown in Fig. 7, the mean and tail latency of HSPP is the lowest among all strategies. The mean latency of Selective Replication is higher than HSPP and EC-Cache. This is because the delay of file transfer is much higher than file partitioning. In addition, the load with highly skewed file popularity and size causes hotspots, which leads to long service queue and increases the read latency, especially the tail latency. Selective Partition achieves the highest mean latency due to the severe influence of straggler. Once a node becomes the straggler, the read latency of files with partition stored on that node increases significantly. Compared to EC-Cache, the mean and tail latency of HSPP reduces by 18% both. This is because the dual-way service mechanism can reduce the read latency by utilizing the decoding window. In addition, the adaptive redundant partitions can reduce the storage overhead so that the memory layer can store 1.3× more files than EC-Cache.

### 5.4    Scheduling Overhead

The scheduling overhead of the heterogeneous storage systems mainly comes from the migration of files between different storage layers. We define the amount of data written at all storage layers as scheduling overhead, and compare HSPP with other scheduling strategies, including LRU [31], LRFU (0.25), LRFU (0.025) [25] and ARC [26]. As shown in Fig. 8, the scheduling overhead of other strategies is 1.5× to 2.5× higher than HSPP. In particular, the SSD lifespan is strongly related to the amount of data written. Under the same load, the amount of data written at SSD under HSPP is much smaller than other strategies. This is because the exponential decay algorithm used in HSPP can accurately represent the file popularity and thus place the file at optimal storage layer correspondingly.

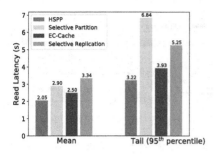
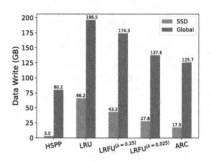

**Fig. 7.** The mean and tail latency of file read under four different strategies.

**Fig. 8.** The scheduling overhead under four different strategies.

## 6   Related Work

File replication and partition can achieve load balance and better utilization of the I/O bandwidth of distributed storage. Replication strategies applied on disk-based storage systems typically use consistent hashing [24] or self-tuning mechanism [9,17,27,35] to place replications on the system, such as Selective replication [9,17]. However, these methods introduce high storage overhead, and the load becomes unbalanced when the file popularity are highly skewed. Partition strategies are commonly used on homogeneous storage systems, such as Selective Partition [34] and EC-Cache [28], which partitions files in the non-redundant and low-redundant manner respectively, and then randomly distributes the partitions in the system for load balance. However, in the case of intensive straggler, the read latency of the above strategies increases significantly. Moreover, to provide high throughput I/O, distributed storage systems generally apply heterogeneous architecture such as HDFS [3], Alluxio [2] and OctopusFS [22], which drives the study of data schedule strategies [18,19,23,30,37]. Our work distinguishes from existing works by proposing an effective file partition and placement strategy for the heterogeneous storage system.

## 7   Conclusion

This paper proposes an effective file partition and placement strategy HSPP for heterogeneous storage system. In HSPP, we establish an upper bound model for the mean read latency and determine the optimal number of file partitions based on the exponential search algorithm. To mitigate the system straggler, the erasure coding is used to generate the redundant partitions. Moreover, a new file classification mechanism is proposed to enable the file placement at the optimal layer of the system. The experiment results show that HSPP is more effective to improve load balance and reduce read latency compared to existing strategies.

**Acknowledgement.** This work is supported by National Key Research and Development Program of China (Grant No. 2016YFB1000304) and National Natural Science Foundation of China (Grant No. 61502019).

# References

1. Alibaba network attached storage. https://www.alibabacloud.com/product/nas
2. Alluxio. http://alluxio.org/
3. Hdfs. https://hadoop.apache.org
4. Intel storage acceleration library (open source version). https://goo.gl/zkVl4N
5. Memcached. http://www.memcached.org
6. Redis. http://www.redis.io
7. Yahoo! webscope dataset. https://webscope.sandbox.yahoo.com
8. Ananthanarayanan, G., Kandula, S., Greenberg, A.G., Stoica, I., Harris, E.: Reining in the outliers in map-reduce clusters using mantri. In: Usenix Conference on Operating Systems Design & Implementation (2010)
9. Ananthanarayanan, G., et al.: Scarlett: coping with skewed content popularity in mapreduce clusters. In: Eurosys 2011, pp. 287–300 (2011)
10. Armbrust, M., et al.: Spark SQL: relational data processing in spark. In: ACM SIGMOD International Conference on Management of Data (2015)
11. Dean, J., Ghemawat, S.: MapReduce: simplified data processing on large clusters. Commun. ACM **51**(1), 107–113 (2008)
12. Fidler, M., Jiang, Y.: Non-asymptotic delay bounds for (k, l) fork-join systems and multi-stage fork-join networks (2015)
13. Floratou, A., Megiddo, N., Potti, N., Özcan, F., Kale, U., Schmitz-Hermes, J.: Tech. rep. IBM (2015)
14. Gao, P.X., et al.: Network requirements for resource disaggregation. In: Usenix Conference on Operating Systems Design & Implementation (2016)
15. Ghemawat, S., Gobioff, H., Leung, S.T.: The google file system. In: Proceedings of the Nineteenth ACM Symposium on Operating Systems Principles, pp. 29–43 (2003)
16. Han, S., Egi, N., Panda, A., Ratnasamy, S., Shi, G., Shenker, S.: Network support for resource disaggregation in next-generation datacenters. In: Twelfth ACM Workshop on Hot Topics in Networks (2013)
17. Hong, Y.J., Thottethodi, M.: Understanding and mitigating the impact of load imbalance in the memory caching tier. In: Symposium on Cloud Computing (2013)
18. Huang, S., Wei, Q., Chen, J., Chen, C., Feng, D.: Improving flash-based disk cache with lazy adaptive replacement. In: Mass Storage Systems & Technologies (2013)
19. Islam, N.S., Lu, X., Wasi-Ur-Rahman, M., Shankar, D., Panda, D.K.: Triple-H: a hybrid approach to accelerate HDFS on HPC clusters with heterogeneous storage architecture. In: IEEE/ACM International Symposium on Cluster, Cloud and Grid Computing, pp. 101–110 (2015)
20. Jiang, S., Zhang, X.: LIRS: an efficient low inter-reference recency set replacement policy to improve buffer cache performance. ACM SIGMETRICS Perform. Eval. Rev. **30**(1), 31–42 (2002)
21. Joshi, G., Liu, Y., Soljanin, E.: On the delay-storage trade-off in content download from coded distributed storage systems. IEEE J. Sel. Areas Commun. **32**(5), 989–997 (2014)

22. Kakoulli, E., Herodotou, H.: OctopusFS: a distributed file system with tiered storage management. In: Proceedings of the 2017 ACM International Conference on Management of Data, pp. 65–78. ACM (2017)
23. Krish, K.R., Anwar, A., Butt, A.R.: hatS: a heterogeneity-aware tiered storage for hadoop. In: IEEE/ACM International Symposium on Cluster, Cloud and Grid Computing, pp. 502–511 (2014)
24. Lakshman, A., Malik, P.: Cassandra: a decentralized structured storage system. SIGOPS Oper. Syst. Rev. **44**, 35–40 (2010)
25. Lee, D., et al.: LRFU: a spectrum of policies that subsumes the least recently used and least frequently used policies. IEEE Trans. Comput. **50**(12), 1352–1361 (2001)
26. Megiddo, N., Modha, D.S.: ARC: a self-tuning, low overhead replacement cache. In: Usenix Conference on File & Storage Technologies (2003)
27. Paiva, J., Ruivo, P., Romano, P., Rodrigues, L.: AUTOPLACER: scalable self-tuning data placement in distributed key-value stores. ACM Trans. Auton. Adapt. Syst. **9**(4), 19 (2014)
28. Rashmi, K.V., Chowdhury, M., Kosaian, J., Stoica, I., Ramchandran, K.: EC-Cache: load-balanced, low-latency cluster caching with online erasure coding. In: Usenix Conference on Operating Systems Design & Implementation (2016)
29. Reiss, C., Tumanov, A., Ganger, G.R., Katz, R.H., Kozuch, M.A.: Heterogeneity and dynamicity of clouds at scale: Google trace analysis. In: ACM Symposium on Cloud Computing (2012)
30. Shu, P., Gu, R., Dong, Q., Yuan, C., Huang, Y.: Accelerating big data applications on tiered storage system with various eviction policies. In: IEEE Trustcom/BigDataSE/ISPA (2016)
31. Weng, M., Shang, Y., Tian, Y.: The design and implementation of LRU-based web cache. In: International Conference on Communications and NETWORKING in China, pp. 400–404 (2013)
32. Xiang, Y., Lan, T., Aggarwal, V., Chen, Y.F.R.: Joint latency and cost optimization for erasurecoded data center storage. ACM SIGMETRICS Perform. Eval. Rev. **42**(2), 3–14 (2014)
33. Yu, C., Luo, Y., Haratsch, E.F., Mai, K., Mutlu, O.: Data retention in MLC NAND flash memory: characterization, optimization, and recovery. In: IEEE International Symposium on High Performance Computer Architecture (2015)
34. Yu, Y., Huang, R., Wang, W., Zhang, J., Letaief, K.B.: SP-cache: load-balanced, redundancy-free cluster caching with selective partition. In: SC18: International Conference for High Performance Computing, Networking, Storage and Analysis (2018)
35. Zaman, S., Grosu, D.: A distributed algorithm for the replica placement problem. IEEE Trans. Parallel Distrib. Syst. **22**, 1455–1468 (2011)
36. Zhou, J., Xie, W., Dai, D., Chen, Y.: Pattern-directed replication scheme for heterogeneous object-based storage. In: IEEE/ACM International Symposium on Cluster (2017)
37. Zhou, W., Feng, D., Tan, Z., Zheng, Y.: PAHDFS: preference-aware HDFS for hybrid storage. In: Wang, G., Zomaya, A., Perez, G.M., Li, K. (eds.) ICA3PP 2015. LNCS. Part II, vol. 9529, pp. 3–17. Springer, Cham (2015). https://doi.org/10.1007/978-3-319-27122-4_1

# Adaptive Clustering for Outlier Identification in High-Dimensional Data

Srikanth Thudumu[1]([⊠]), Philip Branch[1], Jiong Jin[1], and Jugdutt Jack Singh[2]

[1] Swinburne University of Technology, Hawthorn, VIC 3122, Australia
{sthudumu,pbranch,jiongjin}@swin.edu.au
[2] State Government of Sarawak, Kuching, Malaysia
jack.singh@sarawak.gov.my

**Abstract.** High-dimensional data brings new challenges and opportunities for domains such as clinical, scientific and industry data. However, the curse of dimensionality that comes with the increased dimensions causes outlier identification extremely difficult because of the scattering of data points. Furthermore, clustering in high-dimensional data is challenging due to the intervention of irrelevant dimensions where a dimension may be relevant for some clusters and irrelevant for others. To address the curse of dimensionality in outlier identification, this paper presents a novel technique that generates candidate subspaces from the high-dimensional space and refines the identification of potential outliers from each subspace using a novel iterative adaptive clustering approach. Our experimental results show that the technique is effective.

**Keywords:** Outlier detection · High-dimensionality problem · Adaptive clustering · Big data

## 1 Introduction

Large amounts of data and data sources have become ubiquitous in recent years and become available for analysis in many application domains. This availability is commonly referred to as "big data" comprising large-volume, heterogeneous, complex, unstructured data sets with multiple, autonomous sources growing beyond the ability of available tools. As Gartner [8] noted, big data demands cost-effective novel data analytics for decision-making that infer useful insights. In recent years, the core challenges of big data have been widely established. These are contained within the five Vs of big data volume, velocity, variety, veracity and value. However, such a definition ignores another important aspect: "dimensionality", that plays a crucial role in real-world data analysis. Research in the data analytics community has mostly been concerned with "volume", whereas "dimensionality" of big data has received lesser attention [19].

Dimensionality refers to the number of features, attributes or variables within the data. High-dimensionality refers to data sets that have a large number of

© Springer Nature Switzerland AG 2020
S. Wen et al. (Eds.): ICA3PP 2019, LNCS 11945, pp. 215–228, 2020.
https://doi.org/10.1007/978-3-030-38961-1_19

independent variables, components, features, or attributes within the data available for analysis. Data with high-dimensionality has become increasingly pervasive, and has created new analytical problems and opportunities simultaneously. The curse of dimensionality often challenges our intuition based on two and three dimensions [3]. Anomaly detection in high-dimensional data sets is computationally demanding and there is a need for more sophisticated approaches that are currently available. An important issue in big data is outlier or anomaly detection, outliers represent fraudulent activities or other anomalous events that are subject to our interest. The "curse of dimensionality", may negatively affect outlier detection techniques as the degree of data abnormality in fault-relevant dimensions can be concealed or masked by unrelated attributes. When dimensionality increases, the data set becomes sparse, and the conventional methods such as distance based, proximity based, density based and nearest neighbour becomes far less effective [6]. The average distance between a random sample of data points in a high-dimensional space is much larger than the typical distance between one point and the mean of the same sample in low-dimensional space.

While high-dimensionality is one measure of high volume big data, much recent work has focused on finding anomalies using methods that can only draw implicit assumptions from relatively low dimensional data [1]. Furthermore, when the available dimensions of the data are not relevant to the specific test point, the analysis quality may not be credible as the underlying measurements are affected by irrelevant dimensions. This result in a weak discriminating situation where all data points are situated in approximately evenly sparse regions of full dimensional space. However, computing the similarity of one data point to other data point is essential in the outlier detection process.

Clustering in high-dimensional data space is a difficult task due to the intervention of multiple dimensions. A dimension may be relevant for some specific clusters, but unrelated to others. However, clustering is an indispensable step for data mining and knowledge discovery; characterised by unsupervised learning that seeks to detect homogeneous groups of objects based on the values of their attributes or dimensions and grouping them based on similarity, to reveal the underlying structure of data. Conventional methods of clustering attempt to identify clusters constituted of similar samples based on some statistical significance such as distance measurement. The increase in dimensions facilitates similar distance points originated from sparsity triggered by irrelevant dimensions or other noise, aiding to difficulty in identifying accurate and reliable clusters with high quality. The existence of irrelevant attributes or noise in the subspaces critically impacts the formation of clusters. As a result, different subsets of features may be relevant for different clusters, in addition to which diverse correlations among attributes may tend to determine different clusters. Consequently, the curse of dimensionality has become the main challenge for data clustering in high dimensional data sets [7]. This challenge of the clustering process in high dimensional data makes a global dimensionality reduction process inappropriate to identify a subspace that encompasses all the clusters. Nevertheless, in high-dimensional space, meaningful clusters can be found by projecting data onto certain lower-dimensional feature subspaces and manifolds [9,10,12,17].

In this paper we propose a novel method of clustering that can identify possible outliers in the candidate subspaces of high-dimensional data. To effectively detect outliers in high-dimensional space, we integrate a technique based on our previous work [16] that explores locally relevant and low-dimensional subspaces using Pearson Correlation Coefficient (PCC) and Principal Component Analysis (PCA).

The structure of the paper is as follows: Sect. 2 presents the related work. Section 3 discusses the proposed algorithm. Section 4 discusses the proposed adaptive clustering framework for outlier identification in high-dimensional data. Section 5 presents the experimental results, followed by the conclusion and future work.

## 2    Related Work

The curse of dimensionality poses significant challenges for traditional clustering approaches, both in terms of efficiency and effectiveness. Tomašev et al. [18] have proved that hubness-based clustering algorithms perform well, whereas standard clustering methods fail due to the curse of dimensionality. Hubness is the tendency of data points to occur frequently to k-nearest-neighbor lists of other data points in a high-dimensional space. To address the challenges of clustering technique in high-dimensional data, Ertoz et al. [6] presented an algorithm that can handle multiple dimensions and varying densities, which automatically determines the number of clusters. The algorithm is more focused on identifying clusters in the presence of noises or outliers but not particularly on outlier detection. Deriving meaningful clusters from the data set is an important step because outliers are hidden due to the sparsity in high-dimensional space. Agrawal et al. [2] presented a clustering algorithm called CLIQUE that accurately finds clusters in large high-dimensional data sets. Schubert et al. [15] presented a framework for clustering by extracting meaningful clusters from uncertain data that visualizes and understand the impact of uncertainty by selecting clustering approaches with less variability.

Furthermore, subspace clustering is another technique that is proposed to address the limitations of traditional clustering, which aims to find clusters in all subspaces, but, it is not effective or scalable in case of increasing dimensionality. Liu et al. [13] proposed identifying subspace structures from corrupted data by an objective function that finds the lowest rank representation among all the candidates and can represent the data samples as linear combinations. Elhamifar and Vidal [4] proposed a method for clustering based on sparse representation from multiple low-dimensional subspaces. They have also proposed sparse subspace clustering algorithm [5] to cluster data points that fall in a union of low-dimensional subspaces. Zimek et al. [20] have discussed some important aspects of the 'curse of dimensionality' in detail by surveying specialized algorithms for outlier detection. Many researchers addressed important issues but the key issue of computationally feasible algorithms for anomaly detection in high dimensional space is still largely open. This paper attempts such an algorithm where outliers are derived from low-dimensional subspaces using a novel iterative clustering technique.

# 3   Proposed Algorithm

The objective of the proposed algorithm is the identification of outliers from the resulting candidate subspaces in the high-dimensional data. Details of elicitation of candidate subspaces are presented in [16]. However, we have included the approach in Algorithm 1 from steps for discovering candidate subspaces in high-dimensional data. The contribution of this paper is the technique based on adaptive clustering approach in the identification of fine-grained outliers from the candidate subspaces of high-dimensional data.

---

**Algorithm 1.** Fine-grained Outliers in High-dimensional data:

---

1: Apply Standardization or Normalization
2: **for** $i = 1$ to no. of dimensions **do**
3:     calculate correlation $r = \dfrac{n(\sum ab) - (\sum a)(\sum b)}{\sqrt{[n \sum a^2 - (\sum a)^2][n \sum b^2 - (\sum b)^2]}}$
4: **end for**
5: Calculate positive correlation to $CORR$
6: Calculate negative correlation to $UNCORR$
7: Apply PCA $X = W. \sum .W^T$ on CORR and generate $PC1_{corr}$ and $PC2_{corr}$
    by selecting two highest variances
8: **for** $i = 1$ to no. of dimensions in $UNCORR$ **do**
9:     Apply PCA on $PC1_{corr}$ , $PC2_{corr}$ and $i^{th}$ dimension of $UNCORR$
10:    Save to resultant subspaces $CS_i$
11: **end for**
12: **for** $i = 1$ to no. of candidate subspaces in $CS$ **do**
13:    Apply Clustering on each $CS_i$
14:    Generate optimal $j$ clusters using Elbow criterion
15:    **for** $j = 1$ to no. of clusters in each $CS$ **do**
16:        Calculate centroid of each cluster $(x_c, y_c)_i$
17:        **while** $k < threshold$ **do**
18:            Calculate the distance of centroid and each point in the cluster
                $D_i = \sqrt{(x_c - x_i)^2 + (y_c - y_i)^2}$
19:            Calculate the mean of all the distances
                $D_{mean} = \sum(D_i)/N$
20:            Use $D_{mean}$ as the equivalent radius to formulate a circle
21:            Exclude data points within the circle
22:        **end while**
23:    **end for**
24: **end for**
25: **for** $i = 1$ to no. of candidate subspaces in $CS$ **do**
26:    Append remainder data points
27: **end for**
28: Calculate the occurrences of data points in each CS

---

Algorithm 1 provides a step-by-step approach to the technique. Initially, a standardization technique is applied as a pre-processing step to rescale the range of features of input data set if the features of input data consist of large variances between their ranges. To check the correlation among the dimensions of the input data, a Pearson Correlation Coefficient (PCC) is applied to measure the strength of a linear association among the available dimensions. Highly correlated dimensions are combined to form a correlated subspace, and all the uncorrelated dimensions to an uncorrelated subspace, respectively. PCC calculates correlation coefficient of any two dimensions and generates a series of values between +1 to −1. Therefore the correlation coefficient of every dimension with all the other available dimensions in the dataset is calculated and summed up resulting in a final score. If the resultant final correlation score of any dimension is greater than zero, then that particular dimension belongs to CORR subspace or else it belongs to UNCORR subspace.

Principal component analysis (PCA) is applied on the correlated subspace to identifying two highest variances, called principal components, along which the variation in the data is maximal. The resultant principal components are iteratively combined with each dimension of uncorrelated subspace to populate Candidate Subspaces(CS). Every derived candidate subspace is applied with a K-means clustering technique. To find the optimal number of clusters, Elbow model is applied [11]. Based on the result, every CS generates the required number of clusters. In every cluster, a centroid is calculated along with the mean of the distances of available data points to the centroid which we call an "Equivalent Radius" (ER). A circle is formulated in the cluster, and the data points falling within the circle in each cluster are excluded, and the remainder of data points are carry forwarded to the next stage. A new centroid is calculated again in the next stage based on the remaining data points; mean of the distances among each available data points to the new centroid is calculated for a new ER. Then the data points falling within the circle established on the new ER are excluded again. The remainder of the data points is carried forward to the next stage. This process is repeated until the number of data points drops below a certain threshold. Once the threshold is reached, the data points in each CS are calculated for the number of occurrences.

# 4    Adaptive Clustering Framework

This section discusses the proposed framework based on the adaptive clustering approach. Figure 1 delineates the process of outlier identification from the candidate subspaces of the high-dimensional data.

## 4.1    Local Relevancy and Low-Dimensionality

The local relevant subspaces are defined by applying PCC to the data set that differentiates the correlated and uncorrelated dimensions as given in (1), for all the available dimensions $1...n$ in the data set where no two columns are equal ($a \neq b$). The resultant correlated dimensions are referred to a correlated subspace. Each dimension that is in the uncorrelated subspace is referred to a low-dimension.

$$r_{a \neq b} = \frac{n(\sum ab) - (\sum a)(\sum b)}{\sqrt{[n \sum a^2 - (\sum a)^2][n \sum b^2 - (\sum b)^2]}} \qquad (1)$$

PCA is applied on the subspace of correlated dimensions using eigen decomposition or singular value decomposition and we call this subspace as locally relevant subspace.

$$X = W. \sum .W^T \qquad (2)$$

### 4.2 Candidate Subspaces

The principal components resulted from the correlated subspace are combined with each of the low-dimension available from the uncorrelated subspace are the candidate subspaces of the original data. The intention behind combining every low-dimension of uncorrelated subspace with the principal components of the correlated subspace is to reveal the hidden outliers masked by the curse of dimensionality. Furthermore, data points appearing in more than one CS have the highest probability of being an anomaly or outlier.

### 4.3 Adaptive Clustering

A clustering on each CS is applied to exclude the data points falling within the definition. Section 5 discusses the importance of repetitive application of this technique and the reason we call as "Adaptive Clustering" on candidate subspaces of high-dimensional data.

**K-Means Clustering.** The proposed technique uses a k-means clustering algorithm that flows a simple and easy way to classify a given dataset through a certain number of clusters (K- clusters) fixed a priori [14].

$$\beta = \sum_{i=1}^{k} \sum_{j=1}^{m} (||a_i^{(j)} - c_j||)^2 \qquad (3)$$

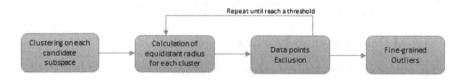

**Fig. 1.** Outlier identification from candidate subspaces

**Optimal Number of Clusters.** The number of clusters should match the data in the CS. An unfitting selection of the number of clusters may undermine the whole process. The best approach is to use Elbow criterion that interprets and validates the consistency within cluster analysis to find the optimal number of clusters [11]. The Elbow model is applied to each CS to deduce the optimal number of clusters in each CS.

**Equivalent Radius (ER).** The centroid for each cluster in each CS is computed. Then the centroid is used for estimating the mean of the distances between each data point within the cluster to its centroid. The resultant mean value is used to formulate a circle in the cluster. This process is repeated until the total number of data points in each CS are less than the given threshold.

$$D_i = \sqrt{(x_c - x_i)^2 + (y_c - y_i)^2} \tag{4}$$

Calculate the mean of all the distances

$$D_{mean} = \sum (D_i)/N \tag{5}$$

Use $D_{mean}$ as the equivalent radius to formulate a circle.

**Data Points Exclusion.** The data points inside the circle definition based on the calculation of ER are excluded, and the data points outside the circle are carried out to the next stage. A new centroid is calculated based on the new set of data points and latest ER is used to form another circle. This process of calculation of the new ER is carried out until a specific condition or threshold is reached. If the data points are less than the given threshold, the ER before the given limit is taken into consideration, and the resulting data points from each CS where the threshold is reached are analysed.

**Fine-Grained Outliers.** The calculation of the number of occurrences of each data point in all the CS are calculated based on the final iteration. The more number of times a particular data point appears, the more likely that data point is an outlier. This process is referred to as fine-graining of outliers. The next step is to trace back the fine-grained outliers to its original index.

## 5   Experimental Evaluation

We used a data set with 19 dimensions and 21000 rows, of which 17 are correlated, and 2 are uncorrelated when analysed with PCC. To verify the effectiveness of outlier identification, we have purposefully introduced synthetic anomalies into the data. The combination of correlated subspace with every dimension from uncorrelated subspace with the application of PCA results in two candidate subspaces, as seen in Fig. 2. We applied the proposed technique of adaptive clustering to both candidate subspaces to fine-grain the outliers in each CS.

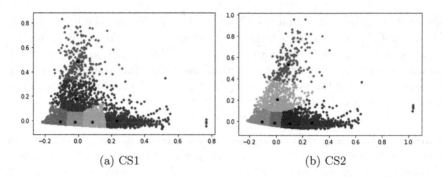

(a) CS1                                  (b) CS2

**Fig. 2.** Original candidate subspaces

In this section, we present the results of three experiments we have conducted to explain the effectiveness of the adaptive clustering approach in identifying outliers. Figure 2a represents the first candidate subspace and Fig. 2b represents the second candidate subspace.

**Table 1.** Equivalent radii and the associated data points

| Figure | Equivalent radius | Iterations | No. of data points |
|--------|-------------------|------------|--------------------|
| a | ER*1 | 0 | 15498 |
| b | ER*1.06 | 0 | 14951 |
| c | ER*3.0 | 0 | 2086 |
| d | ER*4.0 | 0 | 569 |
| e | ER*5.0 | 0 | 219 |

## 5.1  Data Points Exclusion Using a Large ER

Identifying anomalous data points from the candidate subspaces is difficult and may not reveal real anomalies as there are many data points in each CS as depicted in Fig. 2. Hence, an efficient technique is required to filter the possible outliers in each CS. In this experiment, we present a technique that finds outliers and evaluates the technique's effectiveness in outlier identification by taking one candidate subspace CS1 and a large ER, that excludes data points within the circle definition from every cluster. As mentioned in Sect. 4.3, an ER is computed from the mean of the distances of data points available within the cluster to its centroid. The computed ER is used to define a circle, and the data points within the circle definition are excluded from the CS. The motivation behind the proposed equivalent radius is to deselect the nearest points as to reveal hidden outliers.

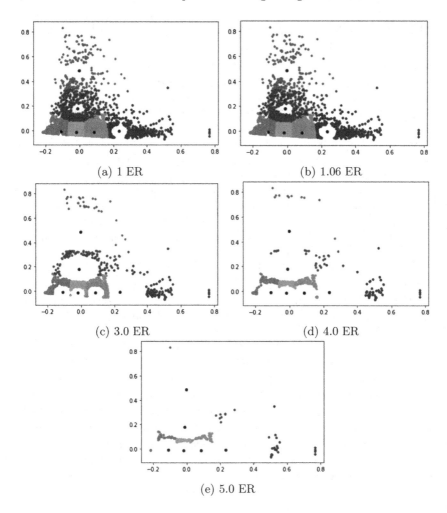

**Fig. 3.** Exclusion of data points using large radii

Table 1 presents the number of remaining data points after the exclusion of data points from the definition of a circle formed from the respective ER. The increase in ER leads to a decrease in the number of data points remaining. However, this approach is not effective when finding the outliers in each cluster of the CS. Furthermore, the increase in ER caused the circle to grow bigger, excluding even the possible outliers that may be hidden in the clusters. Figure 3 shows the exclusion of data points when the ER is increased progressively. Figure 3a represents the exclusion of data points when the computed mean is taken 1 ER, however, when we multiply 1 ER to 1.06 (ER*1.06) as in Fig. 3b, 3.0 as in Fig. 3c, 4.0 as in Fig. 3d and 5.0 as in Fig. 3e, the declination of data points or irrelevant grouping is observed. To address this issue and to fine-grain the outliers, we calculated the ER iteratively, defining a new circle after each exclusion and presented in the following section.

## 5.2   Data Points Exclusion with Iterative ER

In the second experiment, we evaluate the behaviour of outlier identification by iteratively calculating the ER depending on the new set of data points after the exclusion from the previous circle. The process of computing a new ER that forms a dependent circle is terminated when the total data points in the CS becomes lesser than the given threshold of 100 data points.

When the ER is computed to form a circle followed by the exclusion of data points within the respective circle definition, a centroid is calculated again based on the new set of data points upon which new ER is calculated, defining a respective circle area. The process was repeated until the total number of data points became less than 100. The process was terminated, and the most recent CS with before the threshold is benchmarked. If the ER is smaller with the fewer data points or no data points are excluded even with new iterations, an increase in ER value progressively is needed. Originally, second experiment has generated many graphs; however, we present the less results that exhibit the effectiveness of the technique. Table 2 presents the number of data points and iterations processed, along with the remaining data points within the circle definition formed by the respective ER. The results show that outliers can be effectively filtered when the ER is increased progressively and iterated until the threshold is met. Figure 4 shows the exclusion of data points when the ER is increased progressively. Figure 4a shows the remainder of 4413 data points when the computed ER is multiplied with 1.02 (ER*1.02) after 7 iterations, Fig. 4b shows the remainder of 4304 data points when the ER is increased to ER*1.03 after 7 iterations, we observe a slight decline of data points when there is an increase in ER. The data points decreased slightly with more iterations, Fig. 4c

**Table 2.** Iterative ER and the associated data points

| Figure | Equivalent radius | Iterations | No. of data points |
|--------|-------------------|------------|--------------------|
| a | ER*1.02 | 7 | 4413 |
| b | ER*1.03 | 7 | 4304 |
| c | ER*1.08 | 8 | 3895 |
| d | ER*1.06 | 9 | 3963 |
| e | ER*1.05 | 12 | 1869 |
| f | ER*1.11 | 18 | 502 |
| g | ER*1.1 | 21 | 359 |
| h | ER*1.04 | 22 | 608 |
| i | ER*1.12 | 23 | 116 |
| j | ER*1.09 | 36 | 123 |

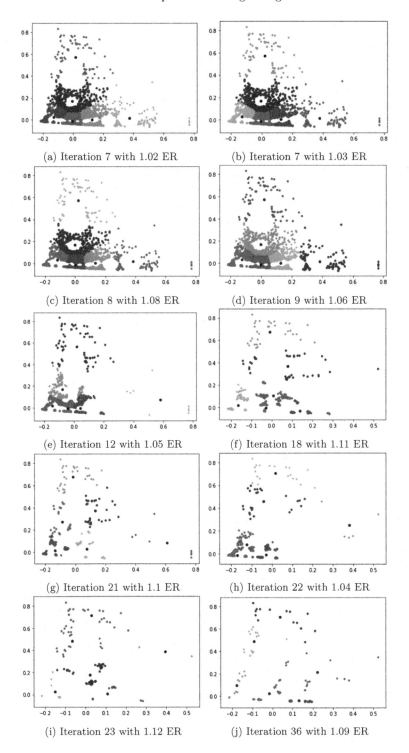

**Fig. 4.** Exclusion of data points using ER iteration approach

consists of 3895 data points with 8 iterations, however, Fig. 4d has 3963 data points with 1.06 ER even after 9 iterations. Hence, the increase in ER filters out more data points than more iterations. We continued the experimentation to observe the decline of the grouping of data points, at $12^{th}$ iteration with ER*1.05, we observed that there is a slight decline of irrelevant grouping as observed in Fig. 4e.

We continued observing the data points Fig. 4f with 18 iterations and ER*1.11, Fig. 4g with 21 iterations and ER*1.1, Fig. 4h with 22 iterations and ER*1.04, Fig. 4i with 23 iterations and ER*1.12, and Fig. 4j with 36 iterations and ER*1.09. Finally, when the threshold of less than 100 data points is reached, the iteration stops and resulting in Fig. 4i with the remainder of 116 data points and Fig. 4j with 123 data points. When compared the two final CS with the results generated when using one large ER presented in Sect. 5.1, the iterative approach of calculating ER outperforms the first experiment with one large ER.

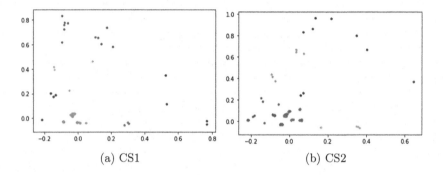

(a) CS1                         (b) CS2

**Fig. 5.** Final candidate subspaces

## 5.3   Calculation of Occurrences for Fine-Grain Outliers

To identify the synthetically introduced outliers, we integrated a technique for the occurrence calculation of each data point in the final candidate subspaces. The more times a particular data point appears in all of the candidate subspaces, the more likely the data point is an outlier. We call the most appeared data points fine-grained outliers. To verify the synthetically introduced outliers are in the final subspace, we traced back each data point to its original index location before evaluating the occurrence in each CS. As observed in Fig. 5, the final data points in CS1 (Fig. 5a) and CS2 (Fig. 5b), 90% of synthetically introduced outliers have appeared in both the candidate subspaces, and 10% of them appeared once. However, it is to be noted that all the introduced outliers are observed in the final candidate subspaces.

# 6    Conclusion and Future Work

This paper introduces Adaptive Clustering that identifies the outliers from the candidate subspaces of the high-dimensional data. To reduce the effect caused by the curse of dimensionality PCC and PCA are integrated to define locally relevant and low-dimensional subspaces. An equivalent radius in each cluster of the candidate subspace is calculated based on the mean of the distances between the centroid and the data points. An iterative application of equivalent radius is computed and used to exclude data points of no interest. To demonstrate that iterative calculation of equivalent radius is more effective, we evaluated the results from both large equivalent radii and iterative calculations of ER and showed that the iterative approach outperforms the other approach. Finally, the resultant data points in each candidate subspace are computed for the number of occurrences. The more times a data point appears, the more likely it is an outlier. In our future work, we will evaluate the performance and accuracy of the proposed technique by analysing the trade-off with respect to the volume and dimensionality to develop a big data framework.

# References

1. Aggarwal, C.C., Philip, S.Y.: An effective and efficient algorithm for high-dimensional outlier detection. VLDB J. **14**(2), 211–221 (2005)
2. Agrawal, R., Gehrke, J., Gunopulos, D., Raghavan, P.: Automatic subspace clustering of high dimensional data for data mining applications, vol. 27. ACM (1998)
3. Christiansen, B.: Ensemble averaging and the curse of dimensionality. J. Clim. **31**(4), 1587–1596 (2018)
4. Elhamifar, E., Vidal, R.: Sparse subspace clustering. In: 2009 IEEE Conference on Computer Vision and Pattern Recognition, pp. 2790–2797. IEEE (2009)
5. Elhamifar, E., Vidal, R.: Sparse subspace clustering: algorithm, theory, and applications. IEEE Trans. Pattern Anal. Mach. Intell. **35**(11), 2765–2781 (2013)
6. Ertöz, L., Steinbach, M., Kumar, V.: Finding clusters of different sizes, shapes, and densities in noisy, high dimensional data. In: Proceedings of the 2003 SIAM International Conference on Data Mining, pp. 47–58. SIAM (2003)
7. Gan, G., Ng, M.K.P.: Subspace clustering with automatic feature grouping. Pattern Recogn. **48**(11), 3703–3713 (2015)
8. Gartner, I.: Big data definition. https://www.gartner.com/it-glossary/big-data/. Accessed 6 Sept 2019
9. Jing, L., Ng, M.K., Huang, J.Z.: An entropy weighting k-means algorithm for subspace clustering of high-dimensional sparse data. IEEE Trans. Knowl. Data Eng. **8**, 1026–1041 (2007)
10. Jing, L., Ng, M.K., Xu, J., Huang, J.Z.: Subspace clustering of text documents with feature weighting $K$-means algorithm. In: Ho, T.B., Cheung, D., Liu, H. (eds.) PAKDD 2005. LNCS (LNAI), vol. 3518, pp. 802–812. Springer, Heidelberg (2005). https://doi.org/10.1007/11430919_94
11. Ketchen, D.J., Shook, C.L.: The application of cluster analysis in strategic management research: an analysis and critique. Strateg. Manag. J. **17**(6), 441–458 (1996)

12. Li, T., Ma, S., Ogihara, M.: Document clustering via adaptive subspace iteration. In: Proceedings of the 27th Annual International ACM SIGIR Conference on Research and Development in Information Retrieval, pp. 218–225. ACM (2004)
13. Liu, G., Lin, Z., Yan, S., Sun, J., Yu, Y., Ma, Y.: Robust recovery of subspace structures by low-rank representation. IEEE Trans. Pattern Anal. Mach. Intell. **35**(1), 171–184 (2012)
14. Mucha, H.J., Sofyan, H.: Nonhierarchical clustering (2011)
15. Schubert, E., Koos, A., Emrich, T., Züfle, A., Schmid, K.A., Zimek, A.: A framework for clustering uncertain data. Proc. VLDB Endow. **8**(12), 1976–1979 (2015)
16. Thudumu, S., Branch, P., Jin, J., Singh, J.J.: Elicitation of candidate subspaces in high-dimensional data. In: 2019 IEEE 21st International Conference on High Performance Computing and Communications. IEEE (2019, in press)
17. Tomasev, N., Radovanovic, M., Mladenic, D., Ivanovic, M.: The role of hubness in clustering high-dimensional data. IEEE Trans. Knowl. Data Eng. **26**(3), 739–751 (2014)
18. Tomašev, N., Radovanović, M., Mladenić, D., Ivanović, M.: Hubness-based clustering of high-dimensional data. In: Celebi, M.E. (ed.) Partitional Clustering Algorithms, pp. 353–386. Springer, Cham (2015). https://doi.org/10.1007/978-3-319-09259-1_11
19. Zhai, Y., Ong, Y.S., Tsang, I.W.: The emerging "big dimensionality" (2014)
20. Zimek, A., Schubert, E., Kriegel, H.P.: A survey on unsupervised outlier detection in high-dimensional numerical data. Stat. Anal. Data Min.: ASA Data Sci. J. **5**(5), 363–387 (2012)

# Penguin Search Aware Proactive Application Placement

Amira Rayane Benamer[1](✉), Hana Teyeb[2](✉),
and Nejib Ben Hadj-Alouane[2](✉)

[1] Faculty of Sciences of Tunis, University of Tunis El Manar, Tunis, Tunisia
amira.rayane.benamer@gmail.com
[2] National Engineering School of Tunis, OASIS Research Lab,
University of Tunis El Manar, Tunis, Tunisia
hana.teyeb@gmail.com, nejib_bha@yahoo.com

**Abstract.** With the huge proliferation of IoT devices, new challenges
have been raised. These IoT devices generate a huge amount of data,
instantly. In addition, they are time sensitive, geographically distributed,
require high bandwidth, and location awareness. In order to cope with
these challenges, recent studies have allowed exploring a new paradigm
so-called fog computing. This latter extends Cloud computing at the edge
of the network. Fog computing is an intermediate layer that facilitates
the deployment of IoT applications by leveraging new characteristics such
as support of mobility, location-awareness, and lower latency. However,
its limited resources arise the problem of resource provisioning which
has an impact on the application placement decisions. In this paper, we
focus on the mobile application placement problem in hybrid Cloud-Fog
environment. We have considered both delay-sensitive and delay-tolerant
applications. Hence, we propose an exact solution as well as a new approach based on penguin search metaheuristic named PsAAP to fulfill
the dynamic demands as well as the application's QoS requirements. To
evaluate the proposed approach, we introduce a mobile scenario including
three different types of applications. Moreover, we compare the suggested
policy with the exact solution, baseline algorithms, heuristic, and metaheuristic methods. Experiments have been conducted using CPLEX and
IfogSim-simulator. The final results show the effectiveness of the proposed approach.

**Keywords:** Placement · Latency · QoS · Fog · Cloud

## 1 Introduction

Due to the wide-spread popularity of Internet of Things (*IoT*) [17], and the
embedded applications, a huge amount of data will be generated. However, IoT
devices have a limited resource capacity and computational power, to process all
this data. This latter arises the need for external computing resources. Besides,

© Springer Nature Switzerland AG 2020
S. Wen et al. (Eds.): ICA3PP 2019, LNCS 11945, pp. 229–244, 2020.
https://doi.org/10.1007/978-3-030-38961-1_20

IoT applications may have different requirements in terms of Quality of Service ($QoS$) which includes latency, deadline, mobility awareness, etc. Hence, all these factors will impact the choice of the host nodes. In spite of its limitless capacity, cloud computing is not able, in some cases, to meet alone all these requirements [3], due to bandwidth constraints, and high number of competing demands. Thus, in order to bridge this gap, fog computing was introduced as an intermediate layer [5]. Its main concept consists of bringing back part of the back-end (i.e cloud) computation to the front-end devices. This paradigm is characterized by its location awareness, low latency, support of mobility and scalability [5]. Despite all these benefits, fog computing still rises new challenges due to its limited resources. As a result, a hybrid cloud-fog environment is crucial in order to satisfy all users demands. In this context, one of the main issues is the dynamic application resource provisioning. This latter consists of deciding where the application modules should run according to their requirements in terms of computing resources and $QoS$. Moreover, the problem becomes more challenging for heterogeneous mobile applications. In this case, we could distinguish several application classes: critical mission, sensitive, and delay-tolerant applications [10]. This classification has a huge impact, over time, on placement decisions. This is due to the variation of workloads, more applications may come and others may leave. To tackle this problem, throughout this paper, we investigate how to place applications' modules, and update the deployment of those already placed for optimized usage of fog stratum. We believe that application placement problem in a fog-cloud context is different from that already discussed only in a cloud context [7,13]. This is due to the wide distribution of fog nodes, resource capacity constraints and stringent applications' tolerable thresholds. Despite tremendous proposed solutions in a hybrid context for such a problem [1,4,14,18], placement application related issues still an open room for further optimization. As this problem is qualified to be NP-Hard [18] and large scale problem in terms of chosen nodes and applications to be deployed, metaheuristics may be the less expensive solution that can explore the space efficiently within a reasonable time frame [2].

Therefore, in this paper, first, we formally define the problem of dynamic resource provisioning and placement for different IoT application classes. We then solve it via the use of new metaheuristic recently proposed, so-called penguin search (PS) algorithm [9]. Our aim is to investigate a new metaheuristic, not yet used in literature, and explore its effectiveness to solve such a problem. Finally, we solve the exact problem formulation using CPLEX [8]. Then, we evaluate the adopted metaheuristic by calculating the gap with the exact solution. In addition, we evaluate our proposed approach using a well-known fog simulator IfogSim [11].

The remainder of this paper is organized as follows: Sect. 2 reviews relevant related works. In Sect. 3, we present the system model. As for Sect. 4, it describes the studied use case. Section 5 presents the exact solution as well as the proposed metaheuristic. Section 6 details and illustrates experiment results. Finally, we conclude in Sect. 7.

# 2   Related Work

In [18], a heuristic policy was proposed to allocate resources for application modules in a hybrid fog-cloud environment. In this work, the authors have considered only resource requirements while maximizing the utilization of fog stratum. [1] reduced the overall latency of an application by minimizing the latency between application components. However, they did not consider the maximization of fog resource utilization nor application sensitivity. In [4], the authors studied the application placement problem by considering dynamic heterogeneous application demands. In this work, they applied three baseline algorithms which are FCFS *(First Come First Served)*, Concurrent loading, Priority Delay Strategy. However, these three strategies were not efficient for a huge number of problem instances. The authors of [16] adopted the genetic algorithm to solve the problem. However, the experiments have been conducted only over small instance sizes and compared only with a baseline algorithm and cloud-based execution. By adopting the simulated annealing algorithm, [15] aimed at minimizing the energy, power consumption, and cost. However, the maximization of fog resource usage was not considered. In a Cloud environment, [12] solved the problem by formulating an integer linear programming (ILP) model, and an iterative sequential co-deployment algorithm. However, neither the application's priority nor resource optimization were considered while placing services. Other works [7,13] were proposed for the same objective, in Cloud context. However, these solutions cannot directly be adopted for fog computing. As known, fog nodes are widely distributed with a larger number and have a restrict resource capacity, in contrast to the fewer centralized cloud datacenters with limitless resources. Hence, in this paper, we aim to cope with such a problem differently from already studied in the literature. In the next section, we present the system model.

# 3   System Model

## 3.1   IoT-Fog-Cloud Architecture

IoT-Fog-Cloud computing architecture is composed of three-hierarchical layers as represented in Fig. 1. The front-end layer consists of a plethora of heterogeneous devices that act as user interfaces. The far-end layer represents the cloud with limitless capacity. The near-end layer is an intermediate layer, ensures an IoT-cloud service continuum, composed of a set of fog nodes. These nodes may collaborate among them to accomplish a task. Consequently, more efficient solutions could be offered in regards to user requirements. However, these collaborations may also impact the *QoS* if are not well managed. Therefore, we propose to integrate an orchestrator with a global view over the network, to receive user demands, ensures dynamically an effective collaboration among nodes, and placing applications regarding their requirements. This may result in improving the resource management of the fog nodes as well as the user's quality of experience *QoE*. In this context, the orchestrator has proactive behavior, as it replaces all the application demands periodically.

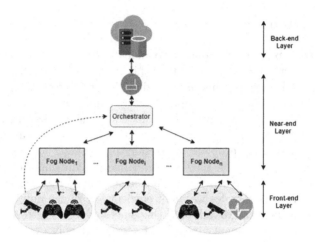

**Fig. 1.** IoT-Fog-Cloud architecture.

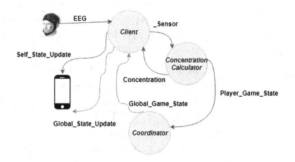

**Fig. 2.** EEG Tractor Beam Game.

## 3.2    IoT Application Model

In literature [11,14,18], IoT applications are modeled as a set of interdependent lightweight application modules, to facilitate their deployment on fog nodes.

Considering three genres of applications Figs. 2, 3, and 4, *EEG Tractor Beam Game* [11], the *Intelligent Surveillance Application VSOT* [11], and the *Health Care Application HC* [15], respectively.

The *EEG* application is a human-vs-human game that necessitates to be processed within a real-time frame. Its main module is *Concentration Calculator* that processes the incoming data sent from *Client* to calculate the concentration level and determining the user brain state. In its turn, this information will be displayed on Actuator. In addition, *Coordinator* performs in the global level to coordinate between all players and continuously sends the current sate of all players to the *Client* module.

The *Intelligent Surveillance Application VSOT* monitors a specified area by coordinating different cameras. The *Motion Detector* module should be placed

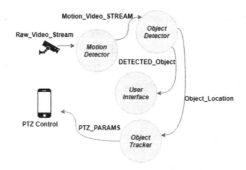

**Fig. 3.** Intelligent Surveillance Application VSOT.

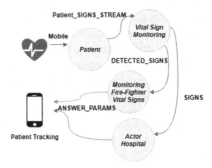

**Fig. 4.** Health Care Application.

on *Camera* for filtering the raw video streams and forwarding them, in case of detection of motion, to the *Object Detector* for further processing. The *Object Detector* recognizes the moving object, then it calculates its coordinates in order to sending them to the *Object Tracker*. This latter calculates the optimal PTZ to send it to the *PTZ Control* module.

The *Health Care Application HC* aims to monitor patient health. To this end, the *Patient* module which is placed on user phone collects data. Then, it forwards them to *Vital Sign Monitoring* for further processing. In case of detecting signs, the current module sends them to *Monitoring Fire-Fighter Vital Signs* which is responsible for sending instructions to the user. Otherwise, the *Actor Hospital* will be the last solution.

As shown, each application is composed of an aggregator module that collects data from sensors. This latter should be placed on the user device. In addition, a global coordinator module that maintains the global state, it should be placed on the cloud. To process the collected data, and track user demands, processing module should be placed according to a prior strategy. This latter is the critical module that should be placed carefully in regards to the application requirements i.e in terms of latency and resource requirements.

We assume that $HC$ applications are mission-critical because any module failure may cause a significant risk on patient safety. Hence, $HC$ requires less than 100 ms to work. $EEG$ requires a real time interactions, but without any high risk. Thus, it needs less than 200 ms; whereas the $VSOT$ application is tolerant to latency. In the next section, we illustrate and detail the studied use case.

## 4   Studied Use Case

In a smart nursing house, private and common rooms are covered by one, and a number of smart cameras, respectively. These smart cameras are running the $VSOT$ application. Besides, we distinguish other types of users. These users may be $EEG$ players, and/or $HC$ users. During daily peak times, several users move to a common room. Thus, further heterogeneous workloads need to be placed, subject to their latency and resource requirements. As a result, a prior strategy is required to manage the three types of applications and mapping them on fog-cloud in an effective manner without violating application requirements and by avoiding an over-provisioning or capacity wastage related issues. Therefore, in the next section, we present the exact and metaheuristic solutions to tackle such a problem.

## 5   Problem Formulation

### 5.1   Exact Solution

We denote by $A$ the set of applications where $a \in A$. We distinguish three categories of applications, in regards to their latency tolerable thresholds, critical mission, sensitive and tolerant applications. The threshold $C_a$ is defined by the user. To prioritize the applications per their $QoS$ requirements, we use $CP(a)$ as follows $CP(a) = \frac{1}{C_a}$. The $CP(a)$ reinforces, as much as possible, the placement of sensitive applications while maximizing the utilization of resources. If the node cannot host a sensitive application, the $CP(a)$ ensures the maximization of the resource utilization by accepting placement solutions of tolerant applications. Each application $a \in A$ is composed of a set of services so-called modules denoted $M_a$. We denote by $N$ a set of nodes that can host $M_a$. Let us define $P_{nr}$ the available capacity of the node $n \in N$, and $U_{ir}$ the requirements of the module $i \in M_a$, in terms of resources $r \in R$ (CPU, RAM, Storage).

In this formulation we use the following decision variable:

- $x_{ij}$ is equal to 1 if the module $i \in M_a$ is placed on the node $j \in N$, 0 otherwise.

Our purpose is to maximize the fog landscape usage by placing a maximum number of services while ensuring their $QoS$ requirements, as shown in expression (1). To do so, we place services on the most consumed fog nodes to the less ones via adoption of $U_{ir}/P_{nr}$.

The problem is defined as follows:

$$\max \sum_{a \in A} \sum_{i \in M_a} \sum_{j \in N} \sum_{k \in R} \frac{U_{ik}}{P_{jk}}.x_{ij}.CP(a) \tag{1}$$

$$\sum_{a \in A} \sum_{i \in M_a} U_{ir}.x_{in} \le P_{nr} \quad \forall n \in N, r \in R \tag{2}$$

$$\sum_{n \in N} x_{in} = 1 \quad \forall i \in M_a, \forall a \in A \tag{3}$$

$$x_{in} \in \{0, 1\} \quad \forall i \in M_a, \forall a \in A, \forall n \in N \tag{4}$$

The Eq. (2) aims to ensure node capacity constraints. Constraint (3) ensures that each module will be running on exactly one node.

## 5.2  Penguins Search Metaheuristic

*Penguins Search Optimization Algorithm* (PeSOA) [9] is a bio-inspired meta-heuristic based on the hunting behavior of penguins population. These penguins are divided into groups with varied sizes. Each group forages randomly in the sea looking for foods. In fact, this foraging process depends on penguins' oxygen reserves. Oxygen reserve of the penguin represents its health condition while foraging action. Throughout the search for food, the penguins change their locations according to the best local leader of each group (i.e group leader) as well as to their last position, as shown in Eq. 5.

$$X_{new} = X_{id} + rand()|X_{BestGlobal} - X_{id}| \tag{5}$$

$X_{new}$ is the new generated position for the penguin
$X_{id}$ is the last position for the penguin
rand() is a random number drawing from $(0, 1)$
$X_{BestGlobal}$ is the best position of the local leader

After a rough number of dives, penguins return to the ice for sharing their locations and the abundance of food sources with their groups via *intra-group* communications. In the case of poor groups, they can follow the best group $(X_{BestGlobal})$ via *inter-group* communications.

## 5.3  Penguin Search Aware Application Resource Provisioning

Typically, *Penguin Search Metaheuristic Aware Application Provisioning* (PsAAP) aims to find the combination of fog nodes that can maximize resource utilization. In our case, the sea corresponds to the space solution, which is the set of fog node candidates. So, the penguin searching purpose is to locate the best position showing the maximized fog resource utilization. Thus, the position of each penguin is a candidate solution for the application placement problem. In order to encode the individual penguin, a two-dimensional matrix is adopted.

As shown in Table 1, the size of each penguin is $m \times n$, where $m$, and $n$ refer to the number of all offloaded applications' modules, and all candidate resources, respectively. Furthermore, in the matrix, the intersection (i.e row with column) represents a placement decision of the module in row $i$ on the corresponding node in column $j$. This placement decision is represented by a check mark that corresponds to $x_{ij} = 1$, otherwise 0, in Sect. 5.1. The quantity of hunted food represents the fitness of our problem, expression (1). As input, PsAAP receives a set of different application's modules with their requirements and the list of node candidates (including cloud). As output, *PsAAP* gives a mapping scheme for all applications. Initially, we need to fix the oxygen reserve, the population size, and the number of generations. Afterward, we generate randomly a set of penguins so-called the initial population. In order to widely explore the search space, we assign each penguin to one independent group. So, search progress depends on the actual position of the penguin $X_{id}$ as well as on the position of the best group in the population $X_{BestGlobal}$ which has the maximum fitness value. This latter should be valid, which means it should fulfill the two problem constraints (i.e 2 and 3). By following the $X_{BestGlobal}$ in the population, penguins could converge rapidly to the global optimum within a reasonable number of iterations. So, the penguin updates its position using the adjusted Eq. 6.

$$X_{new} = X_{id} \oplus rand() \otimes |X_{BestGlobal} \ominus X_{id}| \qquad (6)$$

$X_{BestGlobal}$ is the best solution in the population
$X_{id}$ is the current solution
rand() is a binary random number

As the taken decision is binary (i.e 1 or 0), we have redefined the arithmetic operations of the Eq. 6. All present operations act on two operands bit by bit. The subtraction operation is defined as $\ominus$. This latter is to do OR exclusive between two bits. The multiplication is defined as $\otimes$. It acts as a Binary OR operator. The addition $\oplus$ is considered as Binary AND operator. The result of the equation is not necessarily valid in terms of the two aforementioned constraints 2, and 3. Furthermore, the chosen combination may not guarantee the satisfaction of *QoS* applications. So, to cope with this problem, we propose the *Allocation Repair Algorithm* (ARA). After repairing solutions using (ARA), the fitness of those solutions will be compared to the current ones. Whether the new solution maximizes the solution more than the current one, or the old solution does not satisfy problem constraints 2, 3, the penguin changes its place to the new one. Once the oxygen reserves are consumed, the penguins return to share the information via inter-group communications by choosing the best group whose chased a lot of fish (i.e fitness). The foraging process stills running until reaching the stopping criterion. Both concepts of PsAAP and ARA are summarized in Algorithm 1 and, Algorithm 2, respectively.

## 5.4   Allocation Repair Algorithm (ARA)

The allocation repair algorithm (ARA) is composed of three parts which are: capacity constraint, one placement constraint, and *QoS* constraint. First, we

---

**Algorithm 1.** Penguin Search Aware Application Provisioning

---

    **Input:** List<AppModule>M, List<Nodes>N
    **Output:** modulesMapping $[M][N]$
1: Initialize the population P, Iterations, RO2
2: Define $X_{BestGlobal}$                      ▷ individual valid with high fitness
3: **while** (iter<Iterations ) **do**
4:     **for** each individual i ∈ P **do**
5:         **while** (RO2>0 ) **do**
6:             Generate a new solution $X_{new}$ using Eq 6
7:             Repair $X_{new}$ using $ARA$ algorithm
8:             **if** (fitness($X_{new}$)>fitness($X_{id}$)) Or !checkConstraints($X_{id}$) **then**
9:                 Replace $X_{id}$ by $X_{new}$
10:            **end if**
11:         **end while**
12:     **end for**
13:     **if** (fitness($X_{id}$)>fitness($X_{BestGlobal}$)) **then**
14:         Replace $X_{BestGlobal}$ by $X_{id}$
15:     **end if**
16: **end while**
17: moduleMapping=$X_{BestGlobal}$
18: **return** $moduleMapping$

---

**Table 1.** Encoding scheme.

|   | 1 | 2 | 3 | 4 | .. | n |
|---|---|---|---|---|----|---|
| 1 | ✓ |   |   |   |    |   |
| 2 |   |   |   |   |    |   |
| 4 |   | ✓ |   |   |    |   |
| i |   |   |   |   |    |   |
| .. |  |   |   |   | ✓  |   |
| m |   | ✓ |   |   |    |   |

address the problem of excess of capacity, constraint 2, for each fog node candidate. In such a case, we choose randomly modules, assigned to the corresponding host, and forwarding them to the cloud. We choose the cloud as a rapid solution to the problem. Then, we test the obtained result, if it satisfies constraint 3. In such a case, there are two possibilities: several placements for one module or the module was not be placed. In the first case, we choose randomly one fog node between all nodes, and we remove the module from the others. In the other case, we place it on the cloud for further improvement. However, the cloud may not fulfill application $QoS$ requirements. For this reason, we sort the placed modules on the cloud by their sensitivity (i.e from sensitive modules to tolerant ones). In addition, we sort fog nodes per their available resources (i.e from the most consumed to the less one). Next, we choose the fog node that can satisfy module resource requirements as well as maximizing the objective function (i.e to

---

**Algorithm 2.** Allocation Repair Algorithm

---

**Variables:** violation: Boolean, count: Integer
**Input:** $X_{new}[M][N]$, List<AppModule>M, List<Nodes> N
**Output:** $X_{new}[M][N]$

1: **for** each node n $\in$ N **do**                                          ▷ constraint 2
2:     violation=true
3:     **while** (violation) **do**
4:         **if** $(n_{usedCapacity} > n_{availableCapacity})$ **then**
5:             Forward randomly one module to the cloud
6:             release space in n
7:         **else**
8:             violation=false
9:         **end if**
10:     **end while**
11: **end for**
12: **for** each module m $\in$ M **do**                                       ▷ constraint 3
13:     count=0
14:     **for** each node n $\in$ N **do**
15:         count+=$X_{new}[m][n]$
16:     **end for**
17:     **if** count > 1 **then**
18:         choose randomly one node and release other ones
19:     **end if**
20:     **if** count = 0 **then**
21:         place m on Cloud
22:     **end if**
23: **end for**
24: Sort modules placed on cloud per sensitivity
25: **for** each module placed on the cloud **do**                           ▷ QoS satisfaction
26:     Sort resources per their availability
27:     **for** each n in N **do**
28:         **if** (module $_{req}$ <=$n_{availableCapacity}$) **then**
29:             Place module on n
30:             Update $X_{new}$
31:             break
32:         **end if**
33:     **end for**
34: **end for**
35: **return** $X_{new}$

---

guarantee to meet $QoS$ requirements of applications). This process is repeated for all modules placed on Cloud. Eventually, we obtain a maximized usage of fog stratum and a maximum satisfaction percentage of all offloaded applications; whatever their classes[1].

---

[1] Oxygen reserves.

**Table 2.** Fog node Resources class.

| Class | Resources | | |
|---|---|---|---|
| | CPU (MIPS) | RAM (GB) | Storage (GB) |
| Small | 1000 | 1 | 2 |
| Medium | 2000 | 2 | 4 |
| Large | 4000 | 4 | 8 |

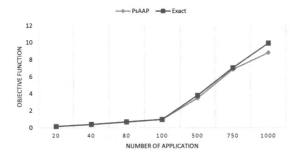

**Fig. 5.** Comparison of exact solution and PsAAP.

**Table 3.** Average optimality gap and execution time.

| Nbr of applications | Nbr of nodes | PsAAP execution time (s) | Exact solution execution time (s) | Gap % |
|---|---|---|---|---|
| 20 | 10 | 0.246 | 2.48 | 2.05 |
| 40 | 20 | 1.205 | 3.22 | 1.11 |
| 80 | 40 | 13.697 | 88 | 2.97 |
| 100 | 50 | 12.5 | 112 | 1.43 |
| 500 | 200 | 73.962 | 153.46 | 8.68 |
| 750 | 300 | 205.380 | 708.49 | 2.79 |
| 1000 | 500 | 874.206 | 2140 | 11.22 |

# 6    Simulation and Experiments

In this section, we present the result of experiments conducted on both proposed exact and heuristic solutions. A summary of different used parameters is shown in Tables 2, 4, 5, and 6. All parameter values are taken from the literature [4], [11], [18], and [16]. Metaheuristic parameter values (i.e $R02$, Population size, and iteration number) are obtained after completing some experiments on each variable.

**Table 4.** Application resource requirements.

| Application | Modules[a] | CPU (MIPS) | RAM (GB) | Storage (GB) |
|---|---|---|---|---|
| EGG | Concentration Calculator | 350 | 0.4 | 0.3 |
| | Coordinator | 100 | 0.1 | 0.2 |
| VSOT | Object Detector | 300 | 0.35 | 0.5 |
| | Object Tracker | 300 | 0.2 | 0.3 |
| HC | Vital Sign Monitoring | 350 | 0.2 | 0.5 |
| | Monitoring Fire Fighter | 350 | 0.35 | 0.3 |

[a]Processing module to be placed for each application [11,15]

**Table 5.** Metaheuristics parameters.

| Metaheuristic | Parameters | Value |
|---|---|---|
| PsAAP | Ro2 | 3 |
| | Generation | 50 |
| | Population Size | 20–200 |
| GA | Crossover rand | 0.5 |
| | Mutation rand | 0.7 |

**Table 6.** UpLink nodes latency

| Node type | Latency (ms) |
|---|---|
| End Device to Fog Node | 2 |
| Fog Node to Gateway | 4 |
| Gateway to Cloud | 100 |

## 6.1   Exact Solution Vs PsAAP

In order to evaluate the PsAAP's solution quality, we compare it with the exact solution benchmark using CPLEX [8].

Figure 5 shows the goal function results for metaheuristic and exact solution. As shown, PsAAP results are very closer to the optimal solution. In addition, we have calculated the average gap $G$ [6] among the exact $S^*$ and approximate $S$ solutions. This is achieved by varying the number of applications, the number of nodes and their sizes.

$$G = \frac{|S^* - S|}{max\{S^*; S\}} * 100$$

Table 3 shows the consumed time for generating solutions, as well as the gap for the two solutions. The results show that the overall gap for all configurations does not exceed 5%. According to [6], we can say that the solution provided by the PsAAP has a good quality.

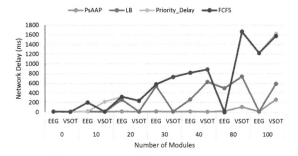

**Fig. 6.** Network Delay of EEG and VSOT.

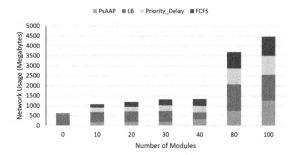

**Fig. 7.** Total network usage for EEG and VSOT.

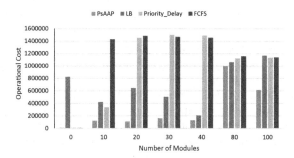

**Fig. 8.** Operational execution cost on cloud.

## 6.2 Simulations

Using IfogSim [11], the proposed PsAAP is compared with different approaches, baseline algorithms [4], heuristic [18], and metaheuristic [16], over the proposed scenario Sect. 4.

In this set of tests, a service execution delay, network usage and cloud execution cost, reported by IfogSim [11], are considered as performance metrics. By Analyzing the source code of Ifogsim Simulator[2], the operational cost is calcu-

---

[2] https://github.com/harshitgupta1337/fogsim.

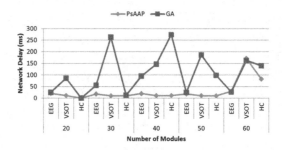

**Fig. 9.** Network Delay of EEG, VSOT, and HC.

lated in function of the total cost TC, the reported cloudsim clock CC, RPM refers to the rate per MIPS, the last utilization update time LUUT, and the total MIPS of the host THM, as represented in the Eq. 7:

$$Cost = TC + (CC + LUUT) * RPM * LU * THM \qquad (7)$$

Figure 6 shows the network delay of both applications *EEG* and *VSOT*. PsAAP keeps the latency reduced for all the configurations. This is because the orchestrator manages dynamically the placement of application modules regarding their tolerable thresholds while getting benefit from all available fog nodes. Hence, modules will stay closer to sensors, and response delay under tolerable thresholds. Meanwhile, [18] which is the closest to PsAAP manages modules regarding their resource requirements without considering their latency constraints.

In Fig. 7, PsAAP reduces network usage considerably compared to other approaches. PsAAP achieves a reduced network usage due to maximizing fog landscape usage while placing modules. Thus, a few numbers of modules will be sent to Cloud in case of no available capacity at the edge of the network. Figure 8 shows cloud operational cost results. Because PsAAP places modules at the rim of the network in contrast to all compared methods, and reduces network usage, consequently the amount of used cloud resources will be reduced. As a result, the operational cost on the cloud also will be reduced.

Figure 9 shows the latency of the three applications *EEG, VSOT, HC*, simultaneously. PsAAP minimizes the latency compared to the genetic algorithm (GA). This is due to ARA algorithm that helps Penguin search to refine its solutions regarding problem constraints. In the last configuration, the PsAAP prioritizes the HC and EEG applications than the VSOT which revealed the importance of priority management while mapping phase.

## 7    Conclusion and Future Work

Integrating fog computing as an intermediate layer has opened perspectives to support IoT applications' demands while reducing delays. However, due to its limited capacity, new issues have been risen in terms of optimal resource provisioning and application placement. Therefore, in this paper, we addressed this issue in a hybrid cloud-fog computing, for heterogeneous application genres. This was achieved by proposing both exact and metaheuristic solutions. Experiments were discussed in terms of optimality gap, network delay, operational cost and network usage.

In future work, we plan to investigate other metaheuristics, and compare them with the proposed policy. In addition, we plan to study a decentralized resource provisioning to deal with scalability issue. We also aim to focus on the scheduling problem of modules after placing them on devices.

# References

1. Benamer, A.R., Teyeb, H., Ben Hadj-Alouane, N.: Latency-aware placement heuristic in fog computing environment. In: Panetto, H., Debruyne, C., Proper, H., Ardagna, C., Roman, D., Meersman, R. (eds.) OTM 2018. LNCS, vol. 11230, pp. 241–257. Springer, Cham (2018). https://doi.org/10.1007/978-3-030-02671-4_14
2. Bianchi, L., Dorigo, M., Gambardella, L.M., Gutjahr, W.J.: A survey on metaheuristics for stochastic combinatorial optimization. Nat. Comput. **8**(2), 239–287 (2009)
3. Bittencourt, L., et al.: The internet of things, fog and cloud continuum: integration and challenges. Internet Things **3**, 134–155 (2018)
4. Bittencourt, L.F., Diaz-Montes, J., Buyya, R., Rana, O.F., Parashar, M.: Mobility-aware application scheduling in fog computing. IEEE Cloud Comput. **4**(2), 26–35 (2017)
5. Bonomi, F., Milito, R., Zhu, J., Addepalli, S.: Fog computing and its role in the internet of things. In: Proceedings of the First Edition of the MCC Workshop on Mobile Cloud Computing, pp. 13–16. ACM (2012)
6. Bovet, D.P., Crescenzi, P., Bovet, D.: Introduction to the Theory of Complexity. Citeseer (1994)
7. Cardellini, V., Grassi, V., Lo Presti, F., Nardelli, M.: Optimal operator placement for distributed stream processing applications. In: Proceedings of the 10th ACM International Conference on Distributed and Event-based Systems, pp. 69–80. ACM (2016)
8. Flatberg, T.: IBM Corporation ILOG CPLEX (2009). http://www.ilog.com/products/cplex/
9. Gheraibia, Y., Moussaoui, A.: Penguins search optimization algorithm (PeSOA). In: Ali, M., Bosse, T., Hindriks, K.V., Hoogendoorn, M., Jonker, C.M., Treur, J. (eds.) IEA/AIE 2013. LNCS (LNAI), vol. 7906, pp. 222–231. Springer, Heidelberg (2013). https://doi.org/10.1007/978-3-642-38577-3_23
10. Guevara, J.C., Bittencourt, L.F., da Fonseca, N.L.: Class of service in fog computing. In: 2017 IEEE 9th Latin-American Conference on Communications (LATIN-COM), pp. 1–6. IEEE (2017)
11. Gupta, H., Vahid Dastjerdi, A., Ghosh, S.K., Buyya, R.: iFogSim: a toolkit for modeling and simulation of resource management techniques in the internet of things, edge and fog computing environments. Softw.: Pract. Exp. **47**(9), 1275–1296 (2017)

12. Kang, Y., Zheng, Z., Lyu, M.R.: A latency-aware co-deployment mechanism for cloud-based services. In: 2012 IEEE 5th International Conference on Cloud Computing (CLOUD), pp. 630–637. IEEE (2012)

13. Leitner, P., Hummer, W., Satzger, B., Inzinger, C., Dustdar, S.: Cost-efficient and application SLA-aware client side request scheduling in an infrastructure-as-a-service cloud. In: 2012 IEEE Fifth International Conference on Cloud Computing, pp. 213–220. IEEE (2012)

14. Mahmud, R., Ramamohanarao, K., Buyya, R.: Latency-aware application module management for fog computing environments. ACM Trans. Internet Technol. (TOIT) (2018)

15. Rezazadeh, Z., Rahbari, D., Nickray, M.: Optimized module placement in IoT applications based on fog computing. In: Iranian Conference on Electrical Engineering (ICEE), pp. 1553–1558. IEEE (2018)

16. Skarlat, O., Nardelli, M., Schulte, S., Borkowski, M., Leitner, P.: Optimized iot service placement in the fog. Serv. Oriented Comput. Appl. **11**(4), 427–443 (2017)

17. Sun, X., Ansari, N.: EdgeIoT: mobile edge computing for the internet of things. IEEE Commun. Mag. **54**(12), 22–29 (2016)

18. Taneja, M., Davy, A.: Resource aware placement of IoT application modules in fog-cloud computing paradigm. In: 2017 IFIP/IEEE Symposium on Integrated Network and Service Management (IM), pp. 1222–1228. IEEE (2017)

# A Data Uploading Strategy in Vehicular Ad-hoc Networks Targeted on Dynamic Topology: Clustering and Cooperation

Zhipeng Gao[✉], Xinyue Zheng[✉], Kaile Xiao, Qian Wang, and Zijia Mo

State Key Laboratory of Networking and Switching Technology, Beijing University of Posts and Telecommunications, Beijing 100876, China
{gaozhipeng,zxy_feifei,wangqian1991, mozijia}@bupt.edu.cn, xiaokaile77@gmail.cn

**Abstract.** Vehicular Ad-hoc Network (VANET) is a special network composed of driving vehicles with dynamic topology. Data uploading from the VANET to the computation server is a challenging issue due to the high mobility of vehicles. By introducing the Mobile Edge Computing (MEC) server deployed on the roadside, this paper proposes a stable clustering strategy based on adjacency screening and designs an Intra-Cluster Data Uploading (ICDU) algorithm to improve the efficiency of data uploading in a dynamic environment. The connection lifetime between vehicles is taken as a key indicator for our proposed clustering strategy to form stable clusters. After the formation of clusters, the ICDU algorithm plans a stable path for vehicles in a cluster to upload data in a cooperative method. Extensive simulation results show that the proposed clustering strategy performs better in terms of the clustering stability compared with Vehicular Multi-hop algorithm for Stable Clustering (VMaSC) and the greedy clustering strategy. The results also prove that our proposed ICDU algorithm outperforms the self-uploading algorithm and can achieve a larger data uploading throughput in the dense scenario compared with the greedy-uploading algorithm.

**Keywords:** Vehicular Ad-hoc Network · Dynamic topology · Mobile Edge Computing · Clustering · Data uploading

## 1 Introduction

With the application of Internet of Things (IoT) in the automotive field, the concept of Intelligent Transportation System (ITS) has gradually arisen, and the Vehicular Ad-hoc network (VANET) has gradually entered our vision. Nowadays, under the circumstance that vehicles and roads are equipped with intelligent communication units, VANET makes it possible for us to have safer integrated services which are more efficient, comfortable and energy-saving. In the driving process of a vehicle, some real-time task requests will be generated constantly and need to be offloaded to the server for computation. In order to respond quickly to the real-time task requests, Mobile Edge Computing (MEC) with high bandwidth and low latency is widely used in this

© Springer Nature Switzerland AG 2020
S. Wen et al. (Eds.): ICA3PP 2019, LNCS 11945, pp. 245–260, 2020.
https://doi.org/10.1007/978-3-030-38961-1_21

scenario [1]. Since the amount of input data of a task in VANET is larger than that of output data [2], data uploading has become an important issue to study on. Massive data generated during the driving process need to be uploaded from the onboard unit (OBU) which is deployed on the vehicle to the roadside unit (RSU) which is deployed on the road. RSUs are equipped with MEC servers with limited resources to provide computation services for the driving vehicles. However, the mobility of vehicles in VANET makes it a complex and challenging problem to upload data, i.e., vehicles may pass by an RSU within an extremely short time while the considerable data to be uploaded may not be all uploaded to the same RSU completely. Specifically, if data from a moving vehicle is irregularly segmented and uploaded to several different RSUs along the driving trace, the data handoff between RSUs will occur frequently. The communication between adjacent RSUs is established by the wireless backhaul, which has an unpredictable delay and a low data rate due to the effect of the urban environment and the interference of wireless links [2]. As a result of the data handoff between RSUs, the delay of data uploading will be increased. Therefore, in this paper, we propose a stable clustering strategy targeted on the dynamic topology of VANET and an Intra-Cluster Data Uploading (ICDU) algorithm to avoid the data handoff between RSUs. Consequently, the efficiency of data uploading in the dynamic environment of VANET can be improved.

In VANET, the communications of data uploading process are called V2X communications, in which we mainly consider two methods: vehicle-to-infrastructure (V2I) communications and vehicle-to-vehicle (V2V) communications [3, 4]. V2I communications can provide connections between vehicles and roadside infrastructures (such as traffic lights or dedicated roadside units) via the fourth generation (4G) cellular network which can also be described as Long Term Evolution (LTE) network. V2V communications can establish connections between vehicles by using Dedicated Short-Range Communication (DSRC) technology [9]. DSRC uses the Wireless Access in Vehicular Environment (WAVE) which is also known as IEEE 802.11p protocol to build the communications. Vehicles in VANET are in a heterogeneous moving state, which results in the dynamic topology of the network. However, data dissemination in a dynamic topology in VANET still remains an unresolved issue worthy of study. One of the best strategies adopted for improving the scalability of data uploading in a dynamic topology is clustering [5]. A framework UFC is proposed to optimize the cluster performance by improving the efficiency of cluster formation, cluster changing rate and cluster stability [5]. But the identification of a cluster is ambiguous as a result of the overlapping between clusters in UFC. Vehicular Multi-hop algorithm for Stable Clustering (VMaSC) [6] is a clustering technique based on choosing the node with the least mobility as cluster head by controlling several different types of timers. Compared with VMaSC, VMaSC-LTE [7] combines the LTE and VMaSC to keep the usage of cellular network at a minimum level in order to achieve a high data packet delivery ratio. Another clustering strategy PMC ensures the coverage and stability of cluster based on the idea of multi-hop clustering [8]. There also exists a self-adaptive clustering method based on the iterative self-organizing data analysis technique to adjust the optimal number of clusters automatically [9]. In addition, some clustering strategies are based on the trajectory. A clustering algorithm using Affinity Propagation (AP) for VANET is proposed in [10] based on that all vehicles share their locations and trajectories with each other. A trajectory-aware edge node clustering (TENC) scheme is proposed in [11] for the edge nodes to form a cluster depending on the trajectory of a target vehicle. A two-level cooperative clustering scheme which uses K-means algorithm to determine the deployment of RSUs and uses the spectral clustering strategy to form a cluster is proposed in [12]. A density-based clustering algorithm is proposed in [13]. In urban city scenario, a new mobility-based and stability-based clustering algorithm (MSCA) is proposed in [14] to improve the stability of clusters. There are still many different clustering methods in [16–18].

However, most of the previous works only focused on the clustering strategies, lacking a clear description of the specific data uploading process after the formation of a cluster. To fill the gap, this paper proposes a stable clustering strategy based on adjacency screening and also designs an Intra-Cluster Data Uploading (ICDU) algorithm for vehicles in a cluster to upload data in a cooperative method. To ensure the stability and reliability of clusters, the proposed strategy forms clusters in terms of the V2V connection lifetime, which can lead to a lower frequency of the cluster reconstruction. The selection of cluster head (CH) is based on the average relative speed (ARS). The global ARS for each vehicle in a cluster will be taken into consideration in this paper, while in [6] the relative speed of each vehicle is calculated only based on its neighbors. In the previous work [9, 16, 18], only CH can exchange messages with RSUs, which results that CH is facing tremendous pressure of the network traffic flow. Different from previous works, in the proposed strategy, the responsibility of CH is to integrate the global structure of its cluster and then plan a cooperative path for the cluster member (CM) who requires to upload its data, which means that each vehicle in the path can exchange data with RSUs. Compared to the path planned by SDN server for the target vehicle to offload the whole task in [15], in the proposed algorithm, the path planned by CH for the target CM is used to divide its data into several segments so that each vehicle in this path can help the target CM vehicle to upload a part of the data segments collaboratively.

## 2  System Model

### 2.1  Scenario Architecture

In this paper, we consider the data uploading problem in VANET in which messages can be disseminated through V2I communications and multi-hop V2V communications. The communication range of V2I and V2V are denoted by $R_{v2i}$ and $R_{v2v}$, respectively. Since the transmitting power and antenna gain of V2I is usually higher than those of V2V, we assume that $R_{v2i}$ is wider than $R_{v2v}$. As shown in Fig. 1, vehicles run along the one-dimensional west-east road in both directions and each direction can have three lanes. Each vehicle will generate its task including a heterogenous amount of input data which needs to be uploaded to a proper RSU firstly. We assume that adjacent RSUs deployed on the road have an equal and fixed distance. Each RSU is equipped with a MEC server for calculation.

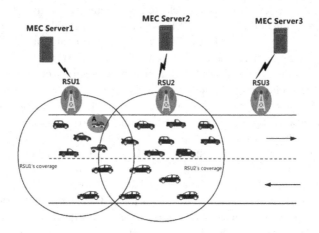

**Fig. 1.** Scenario architecture.

## 2.2   Cluster Formation

In the data uploading strategy proposed in this paper, when driving on the road, vehicles are organized into clusters based on adjacency screening in terms of the lifetime they can hold the V2V communication links among them. The adjacency screening will be introduced in detail in Sect. 3.1. In order to ensure that clusters don't need to be reconstructed frequently and the stability and reliability of clusters can be improved, vehicles in a cluster need to drive towards the same direction so that the cluster can have the longest lifetime. Each vehicle $n_i$ is equipped with the embedded GPS interfaces to get its location denoted by $l_i(t)$ at the time instant $t$. Then the relative distance between vehicle $n_i$ and vehicle $n_j$ can be denoted by $\Delta d_{ij}(t) = |l_i(t) - l_j(t)|$. At the same time, the velocity of vehicle $n_i$ is denoted as $v_i(t)$. In the scenario which we described in Sect. 2.1, since each direction on the road has three lanes, the direction of a driving vehicle may not simply be a straight west-east line. Therefore, the velocity of vehicle $n_i$ is regarded as a two-dimensional vector, which means it can be decomposed into two components in west-east direction and north-south direction expressed as

$$v_i(t) = \left( v_i^{WE}(t), v_i^{NS}(t) \right) \tag{1}$$

$$|v_i(t)| = \sqrt{\left[ v_i^{WE}(t) \right]^2 + \left[ v_i^{NS}(t) \right]^2} \tag{2}$$

where $v_i^{WE}(t) \in (-\infty, 0) \cup (0, +\infty)$ denotes the west-east velocity component, $v_i^{NS}(t) \in (-\infty, +\infty)$ denotes the north-south velocity component. When the vehicle $n_i$ is driving towards east, $v_i^{WE}(t) > 0$, otherwise $v_i^{WE}(t) < 0$. Similarly, when the vehicle $n_i$ is changing its lane towards south, $v_i^{NS}(t) > 0$ and $v_i^{NS}(t) < 0$ when towards north. It is noted that if the vehicle $n_i$ is driving on a straight line without changing lanes, $v_i^{NS}(t) = 0$. In comparison to velocity and for simply calculation, we can also decompose the scalar relative distance between vehicle $n_i$ and vehicle $n_j$ into the two corresponding components which can be denoted as

$$\Delta d_{ij}(t) = \sqrt{\left[\Delta d_{ij}^{WE}(t)\right]^2 + \left[\Delta d_{ij}^{NS}(t)\right]^2} \tag{3}$$

where $\Delta d_{ij}^{WE}(t)$ denotes the west-east relative distance component at the time instant $t$, while $\Delta d_{ij}^{NS}(t)$ denotes the north-south relative distance component. If the vehicle $n_i$ and vehicle $n_j$ are on the adjacent lanes, then $\Delta d_{ij}^{NS}(t) = d_l$ where $d_l$ is the width of a lane, and if there is a middle lane between vehicle $n_i$ and vehicle $n_j$, then we have $\Delta d_{ij}^{NS}(t) = 2d_l$. It is also noted that if the vehicle $n_i$ and vehicle $n_j$ are driving on the same lane, $\Delta d_{ij}^{NS}(t) = 0$.

In order to improve the stability of the cluster, we take the lifetime of each V2V link into consideration and as a key indicator to form a cluster. When vehicle $n_i$ drives out of the communication range of vehicle $n_j$ which equals to the V2V communication range $R_{v2v}$, the connection between them is broken. Therefore, there are three different situations which need to be considered to calculate the lifetime of the connection between two vehicles: (1) the vehicle in the rear has a faster velocity than the vehicle in the front, (2) the vehicle in the rear has a slower velocity than the vehicle in the front, (3) the vehicle in the rear has the same velocity with the vehicle in the front.

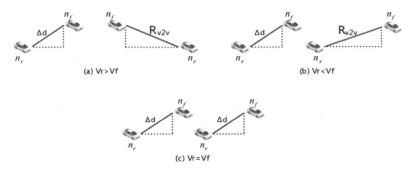

**Fig. 2.** Three different situations when calculating the V2V connection lifetime.

As shown in Fig. 2(a), vehicle $n_r$ is in the rear while $n_f$ is in the front, and the velocity of $n_r$ is faster than that of $n_f$. Then the connection life time between them denoted by $LT_{rf}(t)$ satisfies

$$LT_{rf}(t) = \Delta t_c + \Delta t_b \tag{4}$$

$$\Delta t_c = \frac{\Delta d_{rf}^{WE}(t))}{v_r^{WE}(t) - v_f^{WE}(t)} \tag{5}$$

$$\Delta t_b = \frac{\Delta d_{rf}^{WE}(t + \Delta t_c)}{v_r^{WE}(t + \Delta t_c) - v_f^{WE}(t + \Delta t_c)} \tag{6}$$

where $\Delta t_c$ is the chasing time for $n_r$ to catch vehicle $n_f$, $\Delta t_b$ is the subsequent connection-breaking time before $n_r$ drives out of vehicle $n_f$'s communication range $R_{v2v}$. Since the $\Delta d_{rf}^{NS}(t)$ can be easily achieved by their driving lanes, then $\Delta d_{rf}^{WE}(t) = \sqrt{\left[\Delta d_{rf}(t)\right]^2 - \left[\Delta d_{rf}^{NS}(t)\right]^2}$.

In Fig. 2(b), the velocity of vehicle $n_r$ is slower than that of vehicle $n_f$. In this scenario, there doesn't exist the chasing time $\Delta t_c$ on account that $n_r$ cannot catch up with vehicle $n_f$. Then the connection lifetime satisfies

$$LT_{rf}(t)v_f^{WE}(t) - \Delta d_{rf}^{WE}\left(t + LT_{rf}(t)\right) = LT_{rf}(t)v_r^{WE}(t) - \Delta d_{rf}^{WE}(t) \tag{7}$$

The connection lifetime in this situation can be obtained by a simplification of formula (7) as follows.

$$LT_{rf}(t) = \frac{\Delta d_{rf}^{WE}\left(t + LT_{rf}(t)\right) - \Delta d_{rf}^{WE}(t)}{v_f^{WE}(t) - v_r^{WE}(t)} \tag{8}$$

$$\Delta d_{rf}^{WE}\left(t + LT_{rf}(t)\right) = \sqrt{R_{v2v}^2 - \Delta d_{rf}^{NS}\left(t + LT_{rf}(t)\right)^2} \tag{9}$$

$$\Delta d_{rf}^{WE}(t) = \sqrt{\Delta d_{rf}(t)^2 - \Delta d_{rf}^{NS}(t)^2} \tag{10}$$

where $\Delta d_{rf}^{NS}\left(t + LT_{rf}(t)\right)$ and $\Delta d_{rf}^{NS}(t)$ can be directly obtained by the driving lanes of vehicles at the time instant respectively.

As shown in the Fig. 2(c), when vehicle $n_r$ and vehicle $n_f$ are driving at the same velocity, the connection between them can be maintained and its lifetime can be regarded as $\infty$ until they change their speed and then satisfy one of the two situations mentioned above. Thus, we can obtain that $LT_{fr}(t)$ equals to $LT_{rf}(t)$ and

$$LT_{rf}(t) = \begin{cases} \Delta t_c + \Delta t_b, & v_r(t) > v_f(t) \\ \frac{\Delta d_{rf}^{WE}\left(t + LT_{rf}(t)\right) - \Delta d_{rf}^{WE}(t)}{v_f^{WE}(t) - v_r^{WE}(t)}, & v_r(t) < v_f(t) \\ \infty, & v_r(t) = v_f(t) \end{cases} \tag{11}$$

## 2.3   Cluster Head Selection

Message can be disseminated between every two vehicles in a cluster by adopting multi-hop V2V communications. A cluster consists of two roles to construct its structure: a unique CH and several CMs. Each cluster must have a CH and each cluster is uniquely identified by its CH. Since the CH plays a critical role in a cluster, it is necessary to select a proper vehicle to be the CH after clustering vehicles by lifetime. CHs need to be stable so that clusters do not need to occupy extra bandwidth frequently to reconstruct their structures. To ensure the stability of CHs, the average relative speed with other vehicles in a cluster is calculated in our clustering strategy. The average relative speed of vehicle $n_i$ is denoted by

$$\text{ARS}_i(t) = \frac{\sum_{j=1}^{Re(n_i)} |v_i(t) - v_j(t)|}{Re(n_i)} \tag{12}$$

where $Re(n_i)$ is the number of remaining vehicles except for vehicle $n_i$ in the cluster, $v_i(t)$ is the velocity of $n_i$ while $v_j(t)$ is the velocity of vehicle $n_j$. Therefore, the vehicle $n_k$ who satisfies

$$\text{ARS}_k(t) = \min\{\text{ARS}_i(t)\} = \min\left\{ \frac{\sum_{j=1}^{Re(n_i)} |v_i(t) - v_j(t)|}{Re(n_i)} \right\} \tag{13}$$

will be selected as the CH in the current cluster.

## 2.4   Cluster Communication

In a cluster, CMs broadcast their hello packets including their locations and velocities periodically to its one-hop neighbors to have a calculation of the connection lifetime. It is noted that since velocity is a two-dimensional vector, there is no need to describe the driving directions which can be derived from the velocity. Then CMs broadcast beacons periodically to the CH which include their heterogeneous information, e.g., locations, velocities, relative distance and lifetime of connections with its neighbors. We assume that the period for vehicles to broadcast beacons to CHs is $\delta_b$ and vehicles drive at a constant velocity during $\delta_b$. The format of beacons from CMs to its CH is denoted as

$$b_i^M(t) = \left( l_i(t), v_i(t), NRD_i^M(t), NLT_i^M(t) \right) \tag{14}$$

where $l_i(t)$ is the location of CM $n_i$ at the time instant $t$, $v_i(t)$ is the velocity of CM $n_i$, $NRD_i^M(t)$ denotes the set that includes relative distance from CM $n_i$ to its one-hop neighbors in the cluster, $NLT_i^M(t)$ denotes the connection lifetime between CM $n_i$ and its one-hop neighbors. CH broadcast an overview beacon of its cluster to the CHs of neighboring clusters to identify its existence. We denote the beacons from CH $n_j$ to its neighbor CH $n_k$ as

$$b_{jk}^H(t) = \left( s_j(t), \bar{v}_j(t), CRD_{jk}^H(t) \right) \tag{15}$$

where $s_j(t)$ is the size of cluster $j$ which is uniquely identified by its CH $j$ and $s_j(t)$ is denoted by the number of vehicles in the cluster, $\bar{v}_j(t) = \frac{1}{s_j(t)} \sum_{i=1}^{s_j(t)} v_i(t)$ is the average velocity of cluster $j$ which is also a two-dimensional vector, $CRD_{jk}^H(t)$ denotes the relative distance between cluster $j$ and cluster $k$ which is denoted by the relative distance between their CHs. In addition, the CH will send a periodical message including the lifetime of the current cluster structure and the number of the CMs to the nearest RSU, then the RSU will send the message to the MEC server in order to let the server have a global overview of the current traffic structure in its coverage.

## 3    Clustering Strategy and Cooperative Uploading

### 3.1    Adjacency Screening

In this paper, a clustering strategy based on adjacency screening is proposed. Let the set $\mathcal{N} = \{n_i, i = 1, 2, \ldots N\}$ represents the vehicles on the road at the initial time. Each vehicle $n_i$ maintains an information table (VIT) including the connection lifetime with the adjacent vehicles. And the adjacent vehicles of $n_i$ are gathered in the set $AD_i$ after exchanging hello packets with its one-hop neighbors. Vehicles will be organized into clusters by achieving the maximum average adjacent connection lifetime of the current cluster denoted by $AALT_c$ and meanwhile constrained by the predetermined maximum numbers of hops denoted by $MAX\_HOP$ which means the number of hops between the two farthest vehicles in a cluster. When the first vehicle $n_i$ is chosen, it scans the connection lifetime in its VIT. If its adjacent vehicle $n_j$ which has the longest connection lifetime with $n_i$ does not join other clusters, then $n_j$ and $n_i$ will be organized into the same cluster. After the cluster has at least two vehicles in it, $n_i$ continuously scans its adjacent vehicles and calculates

$$AALT_c = \frac{\sum LT_{v2v}}{vehicle\_num} \tag{16}$$

where $LT_{v2v}$ is the connection lifetime between the vehicle and its selected adjacent vehicles, and $vehicle\_num$ is the number of vehicles whose connection lifetime are added in the molecule in the current cluster. If $AALT_c$ can be enlarged or maintained after introducing the adjacent vehicle $n_k$, then $n_k$ can be organized into the current cluster. After $n_i$ finished screening all the adjacent vehicles, the newly joined adjacent vehicles sequentially start to screen its adjacent vehicles and calculate $AALT_c$ to introduce vehicles into the cluster until the cluster satisfies the limitation of $MAX\_HOP$.

It is noted that when the vehicle $n_i$ is screening its adjacency, it prefers to select the vehicle who has the same velocity with it to join the cluster. Since the connection lifetime between vehicles with the same velocity is regarded as $\infty$ in formula (11), it will not be added in the molecule when calculating $AALT_c$.

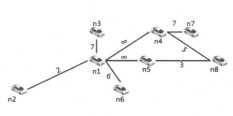

1. screen in AD1={n4,n5,n3,n6,n2},
choose n4,n5, then cluster:{n1,n4,n5}
2. choose n3,
AALT=7/2,
then cluster:{n1,n4,n5,n3}
3. choose n6, AALT'=13/3≥7/2,
then cluster:{n1,n4,n5,n3,n6}
4. choose n2, AALT'=15/4<13/3,
then cluster:{n1,n4,n5,n3,n6}
5. screen in AD4={n1,n7,n8}
6. choose n7, AALT'=20/4≥13/3,
then cluster:{n1,n4,n5,n3,n6,n7}
7. choose n8, AALT'=25/5≥20/4,
then cluster:{n1,n4,n5,n3,n6,n7,n8}
......

**Fig. 3.** An example of adjacency screening.

Taking an example in Fig. 3, the weight between two vehicles is their connection lifetime calculated by formula (11). The initial vehicle $n_1$ screens in its $AD_1$ and firstly selects $n_4$ and $n_5$ who have the same velocity with it and thus lifetime is $\infty$ to join its cluster. Then $n_1$ continuously screens in $AD_1$ and selects $n_3$ who has the longest countable connection lifetime to join its cluster and calculates $AALT_c$. Then $n_1$ scans $n_2$ and calculates $AALT_c$, it finds that if $n_2$ is introduced into the cluster, $AALT_c$ will be decreased. Thus, $n_2$ will not be accepted to join the current cluster. So far, $AD_1$ has been finished screening and the adjacency $AD_4$ of newly joined vehicle $n_4$ will be screened next in the same method.

### 3.2  Cooperative Uploading

In VANET, data are disseminated through two types of channels: control channel (CCH) and service channel (SCH). The CCH is utilized to transmit control data including the beacons from CMs to CH, while the SCH transmits the service data. In this section, we mainly consider the service data that need to be uploaded through SCH. When uploading data to the RSU, each vehicle will be allocated with a limited part of the bandwidth of the SCH. The massive data will cost a long time delay when uploaded through a small and limited bandwidth. In addition, the occurrences of data handoff between RSUs also have a significant impact on the uploading delay due to the low data rate between RSUs. Therefore, an Intra-cluster Cooperative Data Uploading (ICDU) algorithm which can reduce the delay and improve the efficiency of data uploading is proposed in this paper. In the proposed algorithm, after the cluster organization is completed, the CM with data to be uploaded will send a data uploading request to CH. CH received the request and responded CM with a stable intra-cluster data segmentation and collaboration path to divide data into segments for multiple vehicles in the path to upload segments collaboratively. It is noted that in our proposed algorithm, the responsibility of CH is to plan the cooperative vehicles and integrate the structure of the cluster. Unlike that only CH can communicate with RSU in the previous studies, each CM can upload data segments to RSU in the proposed algorithm.

The pseudo code of the proposed algorithm ICDU for CH is shown in Table 1. When CH received data uploading request from CM $n_i$, according to the location of the cluster, it firstly chooses a proper target RSU from the RSUs that can be access to. The standard of choosing RSUs is to allow more CMs to participate in collaborative data uploading process. Let the set $\{link_{ij}, n_j \in cluster\}$ denotes the set of connection lifetime of the path from vehicle $n_i$ to $n_j$. Let *Path* denotes the vehicle set which includes vehicles selected to participate in the data uploading process of $n_i$ in the coverage of the target RSU. In the proposed algorithm, the main idea of selecting vehicles to participate in the data uploading is that these vehicles need to have a long connection lifetime with each other. The reason is that the V2V connections among the cooperative vehicles will not break during the uploading process so that it can efficiently avoid the packet loss in data segmentation. After initializing the $\{link_{ij}\}$, CH scans the vehicle $n_j$ who has not been visited and has the maximum value in $\{link_{ij}\}$ to add into *Visited*. After $n_j$ is added in *Visited*, vehicles in adjacency $AD_j$ of $n_j$ and not in *Visited* will update the connection lifetime of the path from $n_i$ to themselves. It is noted that once one of the V2V connections is broken, the path is broken. Therefore, the connection lifetime of a path is not the sum of each V2V connection lifetime, it actually is the shortest V2V connection lifetime in this path.

**Table 1.** Intra-cluster cooperative data uploading algorithm for CH.

| Algorithm 1 | ICDU-CH |
|---|---|
| 1: | //CH received data uploading request from CM $n_i$ |
| 2: | choose the target RSU in terms of the location of the whole cluster |
| 3: | $PATH_i = false$ |
| 4: | **for** $n_j \in cluster$ and $n_j \neq n_i$ **do** |
| 5: | **if** $n_j \in AD_i$ **then** |
| 6: | $link_{ij} = LT_{ij}$ |
| 7: | **else** |
| 8: | $link_{ij} = -1$ |
| 9: | **end if** |
| 10: | **end for** |
| 11: | $Visited = \{n_i\}, Path = \{\}, pre_{plt} = plt = \{\}$ |
| 12: | **while** $PATH_i == false$ **do** |
| 13: | **if** $\exists n_j \notin Visited$ and $link_{ij} = \max\{link_{ij}, n_j \in cluster\}$ and $n_j$ is in the coverage of the target RSU **then** |
| 14: | $Visited = Visited \cup n_j$ |
| 15: | **for** $n_k \in cluster$ and $n_k \in AD_j$ **do** |
| 16: | **if** $n_k \in Visited$ and $\min\{link_{ij}, LT_{jk}\} > link_{ik}$ **then** |
| 17: | $link_{ik} = \min\{link_{ij}, LT_{jk}\}$ |
| 18: | **end if** |
| 19: | **end for** |
| 20: | **if** the $plt$ will not lead to a rollback to $pre\_plt$ **then** |
| 21: | $Path = Path \cup n_j$ |
| 22: | **else** |
| 23: | $PATH_i = true$ |
| 24: | **end if** |
| 25: | **end while** |

To illustrate the ICDU-CH algorithm, an example is depicted in Fig. 4. A cluster consists of the vehicle set $\mathcal{N} = \{n_i, i = 1, 2, \ldots 12\}$ and the CH is vehicle $n_5$. When vehicle $n_1$ sends a data uploading request to $n_5$, $n_5$ decides the target RSU which is $RSU_f$ in this example and uses ICDU-CH algorithm to obtain the Path $= \{n_1, n_4, n_5, n_7, n_9, n_{10}\}$. The lifetime of the data segmentation path equals to $LT_{57} = 7$. It is noted that when the CH scans the adjacency of $n_5$, it will finally choose $n_7$ rather than $n_8$ to participate in the path as a result that $LT_{8,10}$ is shorter than $LT_{79}$ and $LT_{9,10}$.

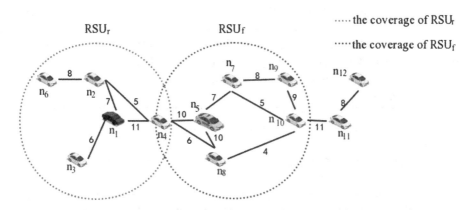

**Fig. 4.** An example of ICDU algorithm.

After receiving the respond including the target RSU and the data segmentation path from CH, CM $n_i$ will use the ICDU-CM algorithm (Table 2) to divide its data to the vehicles in the path. Specifically, when CM $n_i$ received the information of the target RSU, it firstly checks whether itself is in the coverage of the target RSU, which determines the number of segments it will divide its data into. Let *Path_num* denotes the number of vehicles in the *Path*. If CM $n_i$ is covered by the target RSU, $n_i$ will divide its data into *Path_num* $+ 1$ segments, otherwise data will be divided into *Path_num* segments for vehicles to upload cooperatively.

**Table 2.** Intra-cluster cooperative data uploading algorithm for CM.

| Algorithm 1 | ICDU-CM |
|---|---|
| 1: | //CM $n_i$ received the target RSU data segmentation path denoted by $Path$ from CH |
| 2: | **if** CM $n_i$ is in the coverage of target RSU **then** |
| 3: | divides its data into $Path\_num + 1$ segments |
| 4: | **else** |
| 5: | divides its data into $Path\_num$ segments |
| 6: | end if |
| 7: | disseminates the data segments to the corresponding vehicles |
| 8: | **if** any CM received data segment from other CMs **then** |
| 9: | upload the data segment to the target RSU |
| 10: | **end if** |

## 4   Simulations and Analysis

In this section, the proposed clustering strategy based on adjacency screening and the proposed ICDU algorithm are implemented. We compare the proposed clustering strategy with VMaSC designed in [7] and the clustering strategy which we regard as a greedy clustering strategy in [18] in which vehicles are organized into clusters as long as the distance between them is shorter than $R_{v2v}$. Then the proposed ICDU algorithm are compared with two benchmark algorithms. One is self-uploading without cooperation and another is greedy-uploading in which all the vehicles in the coverage of the target RSU will participate in the collaborative uploading process.

### 4.1   Basic Simulation Settings

In the simulation, a 3000 m $*$ 26 m west-east road segment which has 6 lanes with 3 lanes in each direction is modeled. The width of each lane is 4 m and there is a 2 m-wide isolation zone in the middle of the road. The locations of vehicles are generated randomly based on the Poisson distribution and the velocity of each vehicle varies from 10 m/s to 30 m/s. We assume that vehicles drive at a constant velocity during each time interval $\delta_b = 2$ s which means vehicles can only change their velocities at the start of each $\delta_b$. And the communication range of V2I and V2V are set as $R_{v2i} = 300$ m, $R_{v2v} = 100$ m respectively. In addition, the data rate of V2I communications is 3 Mbps and $MAX\_HOP = 5$.

## 4.2  Performance Analysis

**The Number of Clusters.** The number of clusters is reflected by the number of CHs among all the current vehicles. As shown in Fig. 5(a), when the density of vehicles is quite small at the beginning, the number of clusters is slightly larger than the minimum. The reason is that when the vehicle density is sparse, the isolated vehicle who has no neighbors in its V2V communication range will form itself as a cluster separately. In Fig. 5(a), the number of clusters formed by the proposed clustering strategy is more than that of the greedy clustering and less than that of VMaSC. The reason behind it is that the cluster formation in our proposed strategy is constrained by each V2V connection lifetime. However, the large number of clusters means that each cluster has fewer vehicles in it and the small number of clusters means that there are more vehicles in a cluster. When the number of vehicles in each cluster is small, the cluster may be less stable because the slight velocity change can impact the CH selection easily. Correspondingly, when the number of vehicles in a cluster is too large, it is a great burden for CH to exchange beacons with its CM which means that the CH may be overloaded.

**CH Change Rate.** The stability of a cluster can be represented by the CH change rate. We compared the CH change rate of the proposed clustering strategy with that of the greedy clustering and VMaSC. The result is shown in Fig. 5(b). It can be found that when the vehicles are sparse, the CH change rate is relatively high. The explanation is that the number of vehicles in each cluster is small so that the velocity change can change its CH easily. The performance proved that the CH change rate of proposed clustering strategy is much lower than that of the greedy clustering and VMaSC. The greedy clustering strategy in [18] chooses the vehicle who is closest with the target RSU as CH, which makes the CH change rate is large due to the mobility of vehicles. The CH selection in VMaSC is based on the average relative speed in a neighboring area while our proposed strategy takes the global relative speed into consideration. That is the reason why the CH change rate in our proposed strategy is lower than that of VMaSC.

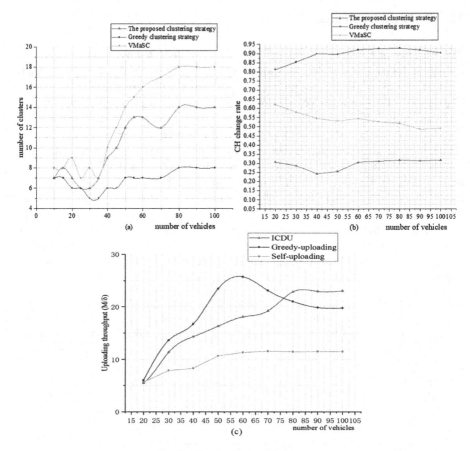

**Fig. 5.** The metrics performance.

**Data Uploading Throughput.** Throughput is a typical metric that can reflect the efficiency of data uploading. The data uploading throughput in each time interval $\delta_b$ is calculated in the simulation. The simulation results of the proposed ICDU algorithm are compared with that of self-uploading and greedy-uploading strategies. Figure 5(c) depicts that when the density of vehicles is in the medium level, the throughput of ICDU algorithm is larger than that of self-uploading but smaller than that of greedy-uploading. The explanation is that our proposed ICDU algorithm works in a constrained and cooperative method. With the help of cooperative vehicles to upload data at the same time, the data uploading bandwidth can be enlarged for the target vehicle. Thus, the data throughput of proposed algorithm performs better than self-uploading without cooperation. When the data to be uploaded do not achieve a saturated level in the network, the CH in greedy-uploading will choose all the vehicles in the coverage to help uploading data cooperatively, which can lead to a large throughput. However, when the density of vehicles is high, i.e., there exists a large amount of data to be uploaded, the greedy-uploading may lead to the failure of uploading. The reason is that it chooses cooperative vehicles without constraints, the connection lifetime may be extremely short between vehicles. It is more likely for greedy-uploading that the data segmentation path breaks frequently due to the short V2V connection lifetime, which may lead to the packet loss. And our proposed algorithm avoids the path from breaking frequently by selecting proper vehicles with long connection lifetime between each other to participate in the uploading process. Therefore, the proposed ICDU algorithm can provide a stably increasing data uploading throughput compared with two benchmark strategies and can achieve a larger throughput than greedy-uploading when vehicles are dense.

## 5 Conclusion

In this paper, a clustering strategy and a cooperative intra-cluster data uploading (ICDU) algorithm are proposed. We modeled the number of clusters, CH change rate, and the data uploading throughput in the simulation. The number of CHs and CH change rate of the proposed clustering strategy are compared with those of VMaSC and greedy clustering strategy. The outcome shows that the proposed clustering strategy outperforms the two previous works. The data uploading throughput of the proposed ICDU algorithm is also compared with self-uploading and greedy-uploading strategies. The result shows that ICDU performs better than self-uploading in all kinds of the vehicle densities and outperforms greedy-uploading when in the dense scenario.

As for future work, we aim to investigate the cooperation between different clusters to make the adjacent clusters work together in order to optimize the efficiency of global data uploading in VANET. The data uploading method based on trajectory prediction will be taken into consideration.

**Acknowledgement.** This work is supported by National Key Research and Development Program of China (2016YFE0204500), National Science and Technology Pillar Program (2015BAH03F02), and Industrial Internet Project of Ministry of Industry and Information Technology of PRC.

## References

1. Mach, P., Becvar, Z.: Mobile edge computing: a survey on architecture and computation offloading. IEEE Commun. Surv. Tutor. **19**(3), 1628–1656 (2017)
2. Zhang, K., Mao, Y.: Mobile-edge computing for vehicular networks: a promising network paradigm with predictive off-loading. IEEE Veh. Technol. Mag. **12**(2), 36–44 (2017)
3. Sun, F., Hou, F.: Cooperative task scheduling for computation offloading in vehicular cloud. IEEE Trans. Veh. Technol. **67**(11), 11049–11061 (2018)
4. Hou, X., Li, Y.: Vehicular fog computing: a viewpoint of vehicles as the infrastructures. IEEE Trans. Veh. Technol. **65**(6), 3860–3872 (2016)
5. Ren, M.: A unified framework of clustering approach in vehicular ad hoc networks. IEEE Trans. Intell. Transp. Syst. **19**(5), 1401–1414 (2018)

6. Ucar, S.: VMaSC: vehicular multi-hop algorithm for stable clustering in vehicular ad hoc networks. In: IEEE Wireless Communications and Networking Conference (WCNC): Network, pp. 2381–2386 (2013)
7. Ucar, S.: Multihop-cluster-based IEEE 802.11p and LTE hybrid architecture for VANET safety message dissemination. IEEE Trans. Veh. Technol. 65(4), 2621–2636 (2016)
8. Zhang, D.: New multi-hop clustering algorithm for vehicular ad hoc networks. IEEE Trans. Intell. Transp. Syst. 20(4), 1517–1530 (2019)
9. Wang, T.: Self-adaptive clustering and load-bandwidth management for uplink enhancement in heterogeneous vehicular networks. IEEE Internet Things J. 6(3) (2019)
10. Shahwani, H.: A stable clustering algorithm based on affinity propagation for VANETs. In: The 19th International Conference on Advanced Communications Technology (ICACT2017), February 2017
11. Lee, J.: Trajectory-aware edge node clustering in vehicular edge clouds. In: The 16th IEEE Annual Consumer Communications & Networking Conference (CCNC) (2019)
12. Calvo, J.A.L.: A two-level cooperative clustering scheme for vehicular communications. In: The 6th International Conference on Information Communication and Management (2016)
13. Kuklinski, S., Wolny, G.: Density based clustering algorithm for vanets. In: 2009 5th International Conference on Testbeds and Research Infrastructures for the Development of Networks Communities and Workshops. TridentCom 2009, pp. 1–6, April 2009
14. Ren, M.: A new mobility-based clustering algorithm for Vehicular Ad Hoc Networks (VANETs). In: 2016 IEEE/IFIP Network Operations and Management Symposium (NOMS 2016), April 2016
15. Huang, C.-M.: V2V data offloading for cellular network based on the software defined network (SDN) inside mobile edge computing (MEC) architecture. IEEE Access, 17741–17755 (2018)
16. Lakshmi Devi, R.: A cluster based authentic vehicular environment for simple highway communication. In: International Conference on Information and Network Technology (ICINT 2012), vol. 37 (2012)
17. Hande, R.S.: Comprehensive survey on clustering-based efficient data dissemination algorithms for VANET. In: International Conference on Signal Processing, Communication, Power and Embedded System (SCOPES) (2016)
18. Ni, Y.: Data uploading in hybrid V2V and V2I vehicular networks: modeling and cooperative. IEEE Trans. Veh. Technol. 67(5), (2018)

# Cloud Server Load Turning Point Prediction Based on Feature Enhanced Multi-task LSTM

Li Ruan[1,2]($\boxtimes$), Yu Bai[1,2], and Limin Xiao[1,2]

[1] State Key Laboratory of Software Development Environment, Beihang University, Beijing 100191, China
{ruanli,mr_by2017,xiaolm}@buaa.edu.cn
[2] School of Computer Science and Engineering, Beihang University, Beijing 100191, China

**Abstract.** Cloud workload turning point is either a local peak point which stands for workload pressure or a local valley point which stands for resource waste. The local trend on both sides of it will reverse. Predicting such kind of point is the premise to give warnings to the system managers to take precautious measures. Comparing with the value base workload predication approach, turning point prediction can provide information about the changing trend of future workload i.e. downtrend or uptrend. So more elaborate resource management schemes can be adopted for these rising and falling trends. This paper is the first study of deep learning based server workload turning point prediction in cloud environment. A well-designed deep learning based model named Feature Enhanced multi-task LSTM is introduced. Novel fluctuate features are proposed along with the multi-task and feature enhanced mechanisms. Experiments on the most famous Google cluster trace demonstrate the superiority of our model comparing with five state-of-the-art models.

**Keywords:** Cloud computing · Turning point prediction · Multi-task LSTM · Feature fusion · Time series

## 1 Introduction

The last decade has witnessed a surge of interest and commercial usage in Cloud computing. Better resource management strategy have been well studied [1]. With the development of artificial intelligence, machine learning based and deep learning based server load prediction models have been closely integrated with the resource management strategy. The more information we can get about the future workload the more efficient our resource management will be.

Although value based prediction methods have made great progress, due to the volatility of cloud host load, accurately predicting host load is still a challenge. Even if we can predict the load value at the next moment, it is difficult

S. Wen et al. (Eds.): ICA3PP 2019, LNCS 11945, pp. 261–266, 2020.
https://doi.org/10.1007/978-3-030-38961-1_22

to know the workload state and dynamics of the underlying trend, more specifically it's hard to know whether it is the potential peak point which stands for workload pressure or valley point which stands for resource waste. Because they could deliver very little information about future.

In this paper we focus on predicting the turning point of cloud server workload. A turning point is a local peak or valley point, the local trend on both sides of it will reverse. Figure 1(a) and (b) show a valley point and a peak point, The dotted lines on either side of the point represent local trends. Both of the two kinds are what we called turning point. Our main goal is to predict whether the current workload is a turning point.

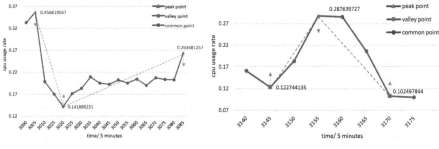

(a) Example 1 server workload with turning points

(b) Example 2 server workload with turning points

**Fig. 1.** Turning points in a real cloud server

The study most similar to us is [5], the authors use the state-of-the-art turning point prediction model in the stock market area called WSVM [4] to predict abrupt changes of the number of virtual machine requests. Comparing with the number of virtual machine requests, cloud server workloads have stronger nonlinearity and time variability, which are difficult for conventional machine learning model to capture.

So, in this paper we propose our cloud server workload turning point prediction model, which consists of three parts: workload repository, workload preprocessing, and Feature-Enhanced Multi-task LSTM.

The contributions of this paper are summarized as following:

- We introduce a Feature Enhanced Multi-task LSTM for cloud server workload turning point prediction. To the best of our knowledge, we are the first to propose a deep learning model to do turning point prediction in cloud environment.
- Four novel fluctuate features are proposed along with the multi-task and feature enhanced mechanisms.
- We conduct extensive experiments on the most famous Google production cluster trace, the result demonstrate the superiority of our model comparing with five state-of-the-art models.

## 2   Notation and Problem Definition for Workload Turning Point Prediction

### 2.1   Definition for Workload Turning Point

Given a workload time series $\mathbf{x} = (x_1, ...x_2, ...x_T)$, local trend is often captured by applying a piece wise linear time series segmentation (or called representation) algorithm (PLR) [2]. The boundaries of segments are the candidate set of turning points. After filtering the additional boundaries point we can get the labels. 1 means turning point and 0 means common point, and the other points which are not segment boundaries are also labeled as common point.

We can predict the turning point by learning a turning point indicative function $\mathbf{y_t} = f(\mathbf{x_t})$ where $\mathbf{y_t} \in (0, 1)$. 0 means $\mathbf{x_t}$ is a common point and 1 means $\mathbf{x_t}$ is a turning point.

## 3   Workload Turning Point Prediction for Cloud Server

Our model architecture is shown in Fig. 2, which consists of three parts: workload repository, workload preprocessing, and Feature-Enhanced Multi-task LSTM. The workload repository keeps the trace of cloud servers, including time series of CPU, memory and other workloads, which are the inputs our model. The workload preprocessing module is responsible for extracting features of the workload and generating labeled data for model training, both of them are sent to the subsequent prediction module. The prediction module uses the Feature Enhanced Multi-task LSTM deep model to predict turning point and output the corresponding predicted label.

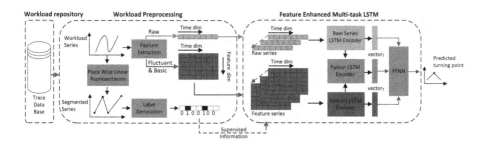

**Fig. 2.** Architecture of our cloud server workload turning point prediction model

In addition to the three basic statistical features (current workload, average of the workload, variance of the workload) reported in [5], we have proposed four novel fluctuant features, each of which can partially reflect recent load fluctuation:

– Relative strength of the workload:

$$f_{rsi} = x_{up}/(x_{up} + x_{down}) \tag{1}$$

$$x_{up} = \sum_{i=1}^{n-1} |x_{t-i+1} - x_{t-i}| I(x_{t-i+1} - x_{t-i} > 0) \tag{2}$$

$$x_{down} = \sum_{i=1}^{n-1} |x_{t-i+1} - x_{t-i}| I(x_{t-i+1} - x_{t-i} < 0) \tag{3}$$

– Absolute sum of the workload changes:

$$f_{asc} = \sum_{i=1}^{n-1} |x_{t-i+1} - x_{t-i}| \tag{4}$$

– Mean second derivative central of the workload:

$$f_{msdc} \frac{1}{n-2} \sum_{i=2}^{n-1} [(x_{t-i+2} - x_{t-i+1}) - (x_{t-i+1} - x_{t-i})]. \tag{5}$$

– Linear trend slope of the workload.

### 3.1    Feature Enhanced Multi-task LSTM

After the preprocessing steps, workload features along with the raw workload series are sent to our Feature Enhanced multi-task LSTM model to do the final prediction. The extracted feature sequences have different meanings from the original time series data. The feature sequence describes the local trend changes from a higher level. So a proper way is to model the raw workload series and its feature sequence separately, by using two LSTM. This is what we called multi-task. However, considering the combination of features and the raw workload series may generate new useful features (for example the intersection of the moving average sequence of the workload and the original workload reflects a new changing pattern of local workload trend), we use another LSTM layer to fusion the hidden representation of them. And this is what we called feature enhanced. We use two different LSTM as encoder to encoder the feature sequence and the raw workload series separately. Along with the generation of two kinds of representation at each time step, they are fed into the another LSTM which is played as the fusion encoder. Finally we will get three kinds of hidden representation at the last time step: the raw workload series representation $h_t^f$, the feature series representation $h_t^f$, the newly generated feature representation $h_t^u$. They are added together to get the final representation for workload turning point prediction.

## 4    Experiments

To evaluate the performance of the proposed model, we have conducted extensive experiments on a famous public trace *google cluster trace*[1]. We randomly select three machines with ID: 207776314 ($M1$), 908054 ($M2$), 1273805 ($M3$) as three of our experiment data sets. We divide each data set into a training set, a validation set, and a test set with ratio 0.8, 0.1 and 0.1. We refer our Feature Enhanced Multi-task LSTM model as FEMT-LSTM, and in order to verify the effectiveness of our method, we compared five state-of-the-art models. Logistic regression (LR), weighted SVM with basic features (WSVM-Basic) [3], WSVM-basic model with fluctuant features (WSVM-Fluctuant), LSTM with only workload series as input (P-LSTM), LSTM with workload series and the feature series as inputs but without a separate approach (S-LSTM). And We use the F1-score as the main metrics for model evelation.

**Table 1.** Comparing of 6 model performance in $M_1 - M_3$.

| Method | M1 | | | M2 | | | M3 | | |
|---|---|---|---|---|---|---|---|---|---|
| | F1 | Precision | Recall | F1 | Precision | Recall | F1 | Precision | Recall |
| $LR$ | 0.4428 | 0.3500 | 0.5900 | 0.3444 | 0.2716 | 0.4705 | 0.1597 | 0.2053 | 0.1306 |
| $SVM - Basic$ | 0.5 | 0.4201 | 0.6206 | 0.4487 | 0.3736 | 0.5614 | 0.3949 | **0.4405** | 0.3579 |
| $SVM - Fluctuant$ | 0.5237 | 0.4703 | 0.6012 | 0.4789 | **0.4715** | 0.4866 | 0.4201 | 0.4143 | 0.4261 |
| $P - LSTM$ | 0.4692 | 0.3097 | **0.9671** | 0.4964 | 0.4416 | 0.5668 | 0.4562 | 0.3651 | 0.6079 |
| $S - LSTM$ | 0.5144 | 0.3911 | 0.7511 | 0.5011 | 0.4395 | 0.5828 | 0.4212 | 0.3290 | 0.5852 |
| $FEMT - LSTM$ | **0.5660** | **0.4774** | 0.6948 | **0.5056** | 0.4346 | **0.6043** | **0.4706** | 0.3816 | **0.6136** |

The experiment results are shown in Table 1. As We can see, our FEMT-LSTM achieve the best F1-score in all the three data set, with a maximum improvement more than 4.2%. Note that the two LSTM baselines we propose can often beat the traditional WSVM based models, and achieve second place, so the 4.2% improvement is comparing with our baseline LSTM models. When comparing with the LR and WSVM based models, our model can improve 5% at most in F1-score.

The WSVM model with fluctuant features beats the WSVM without fluctuant features on three data set with a maximum improvement more than 5.3%. It means our features are effective for turning point prediction. The FEMT-LSTM model beats the P-LSTM with a maximum improvement more than 9.5%. After fusing the handcraft Fluctuant features we can get a better performance, It demonstrates the feature enhanced characteristic of our FEMT-LSTM model. The FEMT-LSTM model beats the S-LSTM with a maximum improvement more than 5.1%. Modeling the raw workload series and the features series separately will improve the performance. And this demonstrates the multi-task characteristic of our FEMT-LSTM model.

---

[1] https://github.com/google/cluster-data.

## 5   Conclusion

In this paper, we study for the first time the problem of workload turning point prediction in cloud environment. And we are the first to propose a deep learning model to do the prediction. Four novel fluctuate features and the specially designed Feature Enhanced Multi-task LSTM are introduced along with the multi-task and feature enhanced mechanisms for better prediction. Extensive experiments on the most famous real production cloud workload trace, Google cluster trace, demonstrate the superiority of our model comparing with five state-of-the-art models.

**Acknowledgements.** This work is by supported by the National Key R&D Program of China under Grant NO. 2017YFB0202004, the fund of the State Key Laboratory of Software Development Environment under Grant No. SKLSDE-2017ZX-10, and the National Science Foundation of China under Grant No. 61772053 and No. 61572377. Guangzhou Science and Technology Projects (Grant Nos. 201807010052 and 201610010092).

## References

1. Duggan, M., Shaw, R., Duggan, J., Howley, E., Barrett, E.: A multitime-steps-ahead prediction approach for scheduling live migration in cloud data centers. Softw. Pract. Exp. **49**(4), 617–639 (2019). https://doi.org/10.1002/spe.2635
2. Keogh, E.J., Chu, S., Hart, D.M., Pazzani, M.J.: An online algorithm for segmenting time series. In: Proceedings of the 2001 IEEE International Conference on Data Mining, 29 November–2 December 2001, San Jose, California, USA, pp. 289–296 (2001). https://doi.org/10.1109/ICDM.2001.989531
3. Luo, L., Chen, X.: Integrating piecewise linear representation and weighted support vector machine for stock trading signal prediction. Appl. Soft Comput. **13**(2), 806–816 (2013). https://doi.org/10.1016/j.asoc.2012.10.026
4. Luo, L., You, S., Xu, Y., Peng, H.: Improving the integration of piece wise linear representation and weighted support vector machine for stock trading signal prediction. Appl. Soft Comput. **56**, 199–216 (2017). https://doi.org/10.1016/j.asoc.2017.03.007
5. Xia, B., Li, T., Zhou, Q., Li, Q., Zhang, H.: An effective classification-based framework for predicting cloud capacity demand in cloud services, p. 1 (2018). https://doi.org/10.1109/TSC.2018.2804916

# Distributed and Parallel Algorithms

# Neuron Fault Tolerance Capability Based Computation Reuse in DNNs

Pengnian Qi[1], Jing Wang[1,2(✉)], Xiaoyan Zhu[1],
and Weigong Zhang[1,2]

[1] School of Information Engineering, Capital Normal University, Beijing, China
pengnianqi@gmail.com, {jwang,zwg771,5590}@cnu.edu.cn
[2] Beijing Advanced Innovation Center for Imaging Technology, Beijing, China

**Abstract.** For applications of speech and video, the consecutive inputs exhibit a high degree of similarity, hence, some results of previous execution can be reused. The technique of quantization can efficiently increase the similarity of consecutive inputs. However, when using quantization, the smaller the number of quantization bits the higher the similarity, as the inputs are constrained to a smaller set of values, but the larger the accuracy loss since input errors are increased. Therefore, we observe that existing reuse schema just applied unique the number of quantization bits in the entire network. If the number of quantization bits is too long, it will directly reduce the similarity between the inputs and thus reduce the reuse ratio. Hence, it is important that exploits the tradeoff among the number of quantization bits, reuse rate, and accuracy. There is an opportunity to significantly improve the performance and efficiency of DNN execution by use multiple quantization bits simultaneously according to the technique of neuron criticality analysis. To do so, we propose a novel reuse schema called *Mquans* based on neuron criticality analysis without accuracy loss. And evaluation results show that our proposed design achieves 2.7 speedups and 38% energy saving on average over the baseline.

**Keywords:** Deep learning · Computational reuse · Approximate computation

## 1 Introduction

An initial proposal [1] aims to reuse some results of previous executions, instead of computing the entire DNN. However, it's just applied unique the number of quantization bits in the entire network. If the number of quantization bits is too long it will directly reduce the similarity between the inputs and thus reduce the reuse ratio. On the contrary, if the number of quantization bits is too short, the loss of precision will be very large. Instead of using uniquely the number of quantization bits where one of the inputs can reuse the previous result. We opt to design a novel computational reuse schema called *Mquans* (applying multiple the number of quantization bits simultaneously) combination neurons criticality analysis [3]. The technique of neurons criticality analysis can simply explain that computing the partial derivative of each neuron from backward propagation in training phase, the larger the partial derivative, the fairly small fault can cause a large error, and the poor fault tolerance of the neuron is more

© Springer Nature Switzerland AG 2020
S. Wen et al. (Eds.): ICA3PP 2019, LNCS 11945, pp. 269–276, 2020.
https://doi.org/10.1007/978-3-030-38961-1_23

critical. In this paper, we calculate first the criticality factor of each neuron in the training phase based on ApproxANN [3]. Then, using computational results divide all neurons into different clusters by the k-medoids algorithm by computed value. We also define the criticality level of each cluster by computed value—finally, we applying a specific quantization bit for each cluster according to their criticality level. Evaluation results show that our proposed design achieves 2.7 speedups and 38% energy saving on average over the baseline reuse schema.

In summary, the main contributions of this work include:

- The quantization can further increase the similarity between consecutive inputs. However, the smaller the number of quantization bits the higher the similarity, as the inputs are constrained to a smaller set of values, but the larger the accuracy loss since input errors are increased. Therefore, it is important that exploits the tradeoff among the number of quantization bits, reuse rate, and accuracy.
- We propose a novel computation reuse scheme based on neuron critical analysis which employs multiple the number of quantization bits simultaneously in reuse case to increase reuse rate without accuracy loss.
- We implement a reuse-based inference technique with the software on top of state-of-the-art DNN accelerator. Evaluation results show that our proposed design achieves 2.7 speedups and 38% energy saving on average over the baseline reuse schema.

## 2  Background

### 2.1  Input Similarity Analysis

Figure 1 illustrates a simple case of speech recognition that classifies a sequence of audio frames in phonemes by DNN execution multiple times. We found that those consecutive frames have high similarity. On the other hand, some applications such as video processing also exhibit similar behavior; the consecutive images in a video tend to be very similar. It's mainly due to the following reasons. At first, the length of frames in the order of several milliseconds, and the speech signal almost stationary in each

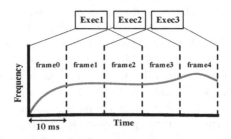

**Fig. 1.** [1] The audio signal is split in frames of 10 ms, in speech recognition. The DNN is executed multiple times to classify the frames in phonemes. In this example, each DNN execution takes as input windows of three frames.

short interval; therefore, adjacent frames exhibit a high degree of similarity. And then, the DNN uses neighbor frames to classify to each audio frame. Hence, successive execution operations overlapping windows of frames [1].

## 2.2 Reuse Principle

We apply a mechanism that calculates the outputs of each DNN layer by reusing the buffered results for the previous execution. Assuming that the output $z_1$ is the result of the first execution of DNN, it can compute as follow: $z_1 = i_{11}w_1 + i_{12}w_2 + i_{13}w_3 + b$, where $i$, $w$ and $b$ are the inputs, weights, and bias respectively. Same as $z_1$, $z_2$ is the output of second DNN execution of this neuron also computed as $z_2 = i_{21}w_1 + i_{22}w_2 + i_{23}w_3 + b$. However, if the first two inputs are the same in both we apply, the output can be computed more efficiently as $z_2 = z_1 + (i_{32} - i_{31})w_3$. It is obvious that we just need to subtract inputs that are different and multiplied by their respective weights, then, added the result of the previous execution. Note that in this case, only one weight has to be fetched instead of three, the bias is not required and only three floating-point computations are performed instead of six.

From the above, we can summarize the general conclusion as follow:

$$z'_o = z_o + \sum_{i=1}^{n} ((q'_i - q_i) \cdot W_{io}) \tag{1}$$

For each layer, we first compute the difference between the quantized value of previous and current ($d_i = q'_i - q_i$). For each output neuron $o$, we just need to multiply the weight and its corresponding difference $d_i$ and added the result of previous execution, only if di is not equal to zero. Otherwise, directly reuse previous execution result and skip all computations and data accesses.

## 2.3 Neurons Criticality Analysis

In [3], Zhang et al. proposed a theoretical neuron criticality analysis technique based on approximate computing, which can be efficiently applied to neural networks. We used that method to calculate the criticality factor of neurons in each output layer and hidden layers after the training phase while all the weights have been fixed. And then, we classified them into different clusters by k-medoids algorithm, according to the computed value. Note that the value of k decided how many criticality levels we classified for all neurons. In this work, we applied several k values (3, 6, 9, 12) to compare their different impact for reuse rate and accuracy loss.

## 3    Mquans Reuse Schema

### 3.1    Cluster-Wise Quantization

We define the quantization function $Q(i)$ as [2]:

$$Q(i) = \frac{1}{2^k - 1} round((2^k - 1) \bullet z_q)$$  (2)

Where $Z_i$ denotes the full-precision value of the input, $Q(i) \in [0, 1]$ denotes the quantized value. And same as [2], we first use a clip function $f(z_i) = clip(z_i, 0, 1)$ to bound the inputs to [0,1]. After that, we conduct quantize the inputs by applying the quantization function Q(i) on $f(z_i)$.

$$z_q = Q(f(z_i))$$  (3)

As shown in Table 1, we list different quantization strategies corresponding to the number of clusters. We used 32-bit quantization bits for the least fault tolerant clusters, mainly due to that is the longest quantization bits used in the baseline articles. The baseline also applied 16 bits, 12 bits, 8 bits, respectively. Besides, we employed other quantization bits in our experiment. And the experimental results showed that with the number of clusters increases computational reuse rate gradually drops and accuracy increase obviously. Note that 3 clusters have the highest reuse rate and the worst accuracy, the 12 clusters have the best accuracy and the worst reuse rate. However, reuse rates averaged 18% higher than baseline and accuracy loss less than 6% when we used 12 clusters. And it's the best tradeoff between accuracy and reuse rate which dividing all neurons into 6 clusters in our experiments.

**Table 1.** Quantize full-precision inputs into different bit-width based on the criticality of each cluster

| Cluster | The number of quantization bits for each cluster |
|---------|--------------------------------------------------|
| 3       | 32-16-12                                         |
| 6       | 32-26-20-16-12-8                                 |
| 9       | 32-28-24-20-16-12-8-6-4                          |
| 12      | 32-28-24-20-16-16-8-8-6-6-4-4                    |

### 3.2    Mquans Reuse Schema

As shown in Fig. 2, we design two buffers to support our reuse schema. The reuse buffer is used to store the data of applied in reuse schema. And we organized in three different areas for it. The first area used to store the quantization value that is required to verify whether an input approximates equal the previous execution one. The second area stores the output of all the layers where the computation reuse schema is exploited, and to be later reused in next DNN execution. The third area stores the intermediate data of layers (temporal activations) where our reuse schema is not applied. The Weights-Buffer has used to stores the weight of every neuron.

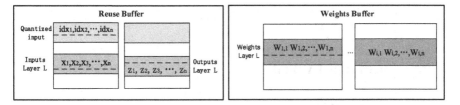

**Fig. 2.** The structure of reuse buffer and configurable buffer

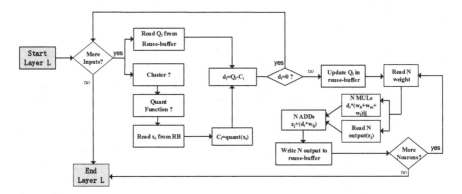

**Fig. 3.** The execution flow of Mquans reuse schema

Figure 3 shows the execution flow of our reuse schema. Note that first execution is different, as it must be calculated the DNN from scratch whereas subsequent executions can reuse previous results. The first execution can describe as follows: initially, the DNNs reads and quantizes the first input, and the quantization process is to select the corresponding the number of quantization bits according to the fault tolerance capability level of the cluster this neuron belonged. In parallel, DNN read N weights of different neurons from the buffer. The quantized value of the input is stored into Reuse-Buffer to preparation for reuse in the next execution. It performs N multiplications by input and weight, followed by N additions to accumulate the result of each output neuron. The next step is stored outputs into Reuse-Buffer, to be used by the next layer and to be reused by the same layer in the next DNN execution. Subsequent executions employ our **Mquans** reuse model. At first, the accelerator reads first input and the corresponding quantized value of previous execution input from Reuse-Buffer. The next step is quantized the current input and subtracted the quantized value of the previous execution input. If the outcome is zero, the DNN ignored current input and skipped all corresponding computations and memory accesses. In case the outcome is not zero, the quantized value of current, stored into Reuse-Buffer, as the target of next time compared. To this end, the DNN fetched N weights of different neurons from memory. Then, N multiplications performed with corresponding weight and the difference between the current input and the previous one. At the same time, DNN reads N outputs of previously executed from Reuse-Buffer. And then, computes the final result and updated in the Reuse-Buffer. DNN repeated this process until all neurons are corrected for changed input.

## 4   Experiment and Evaluation

### 4.1   Experimental Setup

We implement our design and baseline using a software simulator based on computational reuse schema shown in Sect. 3. We evaluate our reuse schema on three state-of-the-art DNNs from different application domains, including acoustic scoring, speech recognition and video classification. We first use MLP to implement acoustic scoring in the Kaldi toolkit that is a popular framework for speech recognition and trained by Librispeech dataset. Then, we use RNN to implement end-to-end speech recognition in Pytorch, and trained by TEDLIUM dataset. Finally, We employ C3D, an efficient video classification framework from Facebook implemented in Caffe. We use UCF10 dataset to evaluate C3D. We compare our reuse schema running on a high-performance CPU (Intel Core i7 4790), and a modern high-end GPU (NVIDIA GTX TiTan X). We use the RAPL library [4] to collect energy consumption in the CPU, and Nvidia-smi (NVIDIA System Management Interface) to measure GPU power dissipation.

### 4.2   Results Analysis

The reuse schema proposed by baseline achieved 1.9X and 2.1X speedup for Kaldi, EESEN, respectively. Hence, we choose these applications as our workloads. As shown in Fig. 4, our design can achieve consistent speedups for three DNNs that range from 2.2X (C3D) to 3.3 (EESEN). Furthermore, the overheads of performing the quantization and comparison two consecutive inputs are fairly small.

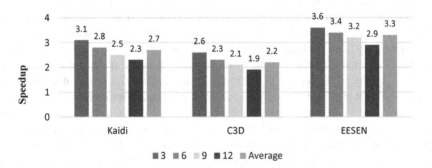

**Fig. 4.** Overall speedup among different applications and 3, 6, 9, 12 denotes that we classified all the neurons into 3, 6, 9, 12 clusters

Figure 5 illustrates the energy saving for each DNN that applied in our experiments. On average, our reuse schema reduces the energy consumption of the DNN by 38%. The energy saving mainly correlated with the diversity of quantization precision and optimization of inputs that cannot be reused in DNN execution.

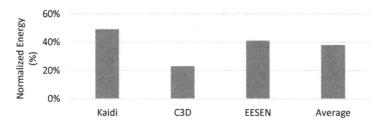

**Fig. 5.** Energy saving for each DNN

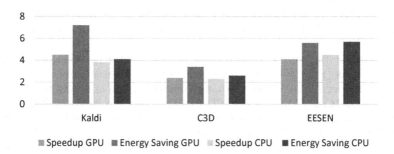

■ Speedup GPU  ■ Energy Saving GPU  ■ Speedup CPU  ■ Energy Saving CPU

**Fig. 6.** Speedup and energy saving for accelerator with software implementation running on GPU and CPU

As shown in Fig. 6, we compare speedup and energy saving for our reuse schema with software implementation running on CPU and GPU. Regarding the speedup, our accelerator outperforms both CPU and GPU in all workloads except C3D. This is because C3D achieves close to peak performance in the GPU. On average, the energy saving form CPU and GPU is 4.1X and 5.3X respectively.

## 5 Related Works

**CNN Accelerator.** DianNao [6] was the first work that includes its own on-chip SRAM buffers to reduce memory accesses energy consumptions, and DaDianNao [7] further improved this aspect by adding eDRAM to store the weights. Eyeriss [10] proposed a row stationary dataflow by exploiting local data reuse of filter weights and input neurons. Farabet et al. [5] propose a systolic architecture called NeuFlow by remaining weights in the register to reuse it as far as possible.

**Computational Reuse.** Several research [1, 8, 9], proposed various computational reuse technique to reduce the overhead of computations and memory accesses in DNN. The [1] designs an accelerator that exploits computational reuse by input similarity. Hegde et al. [8] proposed a CNN accelerator called Unique Weight CNN Accelerator (UCNN) which uses weight repetitions to reuse CNN sub-computations (e.g., dot productions) and to reduce model size. COREX [9] leverages datacenter redundancy by integrating a storage layer together with the accelerator processing layer and added layer stores the outcomes of previous computations. The preciously computed results are reused in the case of recurring computations, thus eliminating the need to re-compute them.

# 6   Conclusion

To improve the percentage of reuse without accuracy loss, we applied the technique of neurons criticality analysis into our reuse model and divided all neurons from each layer into different clusters by k-medoids algorithm, and each cluster uses a specific quantization precision. Evaluation results show that our proposed accelerator achieves 2.7 speedups and 38% energy saving on average over the baseline accelerator.

**Acknowledgment.** This work was supported by the National Natural Science Foundation of China (NSFC) under grants (61772350), Common Information System Equipment Pre-research Funds (Open Project, JZX2017-0988/Y300), the Construction Plan of Beijing High-level Tea-cher Team (CIT&TCD201704082, CIT&TCD20170322), the Open Project of State Key Lab-oratory of Computer Architecture (CARCH201607). The work is also supported by the Capacity Building for Sci-Tech Innovation Fundamental Scientific Research Funds (025185305000). Beijing Nova program (Z181100006218093), Research Fund from Beijing Innovation Center for Future Chips (KYJJ2018008).

# References

1. Riera, M., Arnau, J.-M., Gonzalez, A.: Computation reuse in DNNs by exploiting input similarity. In: ISCA (2018)
2. Zhou, S., Wu, Y., Ni, Z., et al.: DoReFa-net: training low bitwidth convolutional neural networks with low bitwidth gradients. arXiv preprint arXiv:1606.06160 (2016)
3. Zhang, Q., Wang, T., Tian, Y., Yuan, F., Xu, Q.: ApproxANN: an approximate computing framework for artificial neural network. In: DATE (2015)
4. Weaver, V.M., et al.: Measuring energy and power with PAPI. In: ICCP (2012)
5. Farabet, C., Martini, B., Corda, B., Akselrod, P., Culurciello, E., LeCun, Y.: NeuFlow: a runtime reconfigurable dataflow processor for vision. In: CVPRW (2011)
6. Chen, T., et al.: DianNao: a small-footprint High-throughput accelerator for ubiquitous machine-learning. In: ASPLOS (2014)
7. Chen, Y., et al.: Dadiannao: a machine-learning supercomputer. In: MICRO (2014)
8. Hegde, K., Yu, J., Agrawal, R., et al.: UCNN: exploiting computational reuse in deep neural networks via weight repetition. In: ISCA (2018)
9. Fuchs, A., Wentzlaff, D.: Scaling datacenter accelerators with compute-reuse architectures. In: ISCA (2018)
10. Chen, Y.-H., Emer, J., Sze, V.: Eyeriss: a spatial architecture for energy-efficient dataflow for convolutional neural networks. In ISCA (2016)

# Reliability Enhancement of Neural Networks via Neuron-Level Vulnerability Quantization

Keyao Li[1], Jing Wang[1,2(✉)], Xin Fu[3], Xiufeng Sui[4],
and Weigong Zhang[1,2]

[1] School of Information Engineering, Capital Normal University, Beijing, China
{2171002068, jwang, zwg771}@cnu.edu.cn
[2] Beijing Advanced Innovation Center for Imaging Technology, Beijing, China
[3] Department of Electrical and Computer Engineering, University of Houston,
Houston, USA
xfu8@central.uh.edu
[4] School of Information and Electronics, Beijing Institute of Technology,
Beijing, China
suixiufeng@bit.edu.cn

**Abstract.** Neural networks are increasingly used in recognition, mining and autonomous driving. However, for safety-critical applications, such as autonomous driving, the reliability of NN is an important area that remains largely unexplored. Fortunately, NN itself has fault-tolerance capability, especially, different neurons have different fault-tolerance capability. Thus applying uniform error protection mechanism while ignore this important feature will lead to unnecessary energy and performance overheads. In this paper, we propose a neuron vulnerability factor (NVF) quantifying the neural network vulnerability to soft error, which could provide a good guidance for error-tolerant techniques in NN. Based on NVF, we propose a computation scheduling scheme to reduce the lifetime of neurons with high NVF. The experiment results show that our proposed scheme can improve the accuracy of the neural network by 12% on average, and greatly reduce the fault-tolerant overhead.

**Keywords:** Neural network · Soft error · Reliability · Memory protection · Neuron Vulnerability Factor · Fault tolerance

## 1 Introduction

Neural networks (NN) is popular in many fields such as speech recognition, autonomous vehicles, and data centers as they can achieve unprecedented accuracy. Although NN has been increasingly used in many applications, the use of NN must comply with the application's requirements on precision, flexibility, reliability and safety. For example, neural networks greatly improves the performance of autonomouse driving during using to sense and learn the road conditions. But if the autonomous vehicle misclassifies a truck or a pedestrian as a flying object due to errors, the vehicle may not execute the brake operation to avoid the collision. Therefore, it is

---

The original version of this chapter was revised: the affiliation of the third author was corrected. The correction to this chapter is available at https://doi.org/10.1007/978-3-030-38961-1_58

necessary to protect the neural networks from errors that cause the malfunction of the self-driving system. As can be seen, the reliability investigation becomes imperative especially in some safety-critical fields, that is because any failure in output may lead to catastrophic consequences.

Unfortunately, the reliability of DNN is rarely investigated. The primary unreliable factor in modern systems mainly comes from the soft errors. Soft errors can change the data (e.g. bit flips [1]), which may further cause large deviation of standard system outputs [2]. Studies have shown that about 80% of application failures are caused by soft errors [3]. So in this study, we mainly focus on the soft error reliability for NN.

The conventional fault-tolerant methods, such as error correcting codes, can be applied to tolerate soft errors in DNNs. However, 100% accurate and totally fault-free may cause redundancy on design and significant energy overhead. Fortunately, NN has natural fault tolerance capability, and there have been some studies leverages NN's inherent fault-tolerant characteristics for error tolerance and correction, which compromises between accuracy and energy consumption to improve performance [4, 5]. However, different applications have different requirements on the accuracy (e.g., self-driving car has strict requirements on accuracy while data center [6] can tolerate considerable accuracy loss). It implies that different NN applications should be treated differently on soft error tolerance to achieve the optimal application-specific accuracy/performance/energy trade-offs [7, 8]. More interestingly, neurons in the NN usually exhibit different capability in fault tolerance [9, 10]. Applying uniform error protection mechanism while ignoring this important feature will lead to unnecessary energy and performance overheads.

In this study, we propose NVF, a quantitative method, to measure the effect of soft errors on network output, and propose a set of NVF-guided reliability optimization techniques for NN. The contributions are summarized as follows:

- We propose a neuronal sensitivity factor (NVF), which combines hardware structural and neuron's criticality and can provide guidance for error-tolerant techniques.
- We propose a computation scheduling scheme to reduce the lifetime of high NVF neurons, hence, improving the reliability of the network to tolerate soft errors.

## 2  Background

It is well known that not all soft errors will affect the correct execution of the program, and some soft errors can be masked, the processor architecture level can shield more than half of the soft errors. In order to predict the failure rate of the system caused by soft errors and quantify the fault shielding effect, an architectural vulnerability factor (AVF), also called soft error-sensitive factor, is proposed [11]. For example, if a single-bit fault in the committed program counter causes wrong instructions to be executed, which will certainly affect the program's result, then the AVF for the committed program counter is 100%. The key to computing AVFs is to determine which bits affect the final system output (i.e., which are the ACE bits) and which do not (i.e., the un-ACE bits). Then, the AVF of a hardware structure is equal to:

$$\frac{\Sigma \text{ residency (in cycles) of all ACE bits in a structure}}{\text{number of bits in hardware structure} \times \text{execution cycles}} \tag{1}$$

AVF can also measure the structural vulnerability factors of the neural networks. However, it will treat all neurons equally and uniformly in the network, which leading to in-accurate measurement of the NN's vulnerability to soft errors. This is because neural networks have some unique characteristics compared to general system structures. Studies have shown that errors that occur on different neurons have different effects [10].

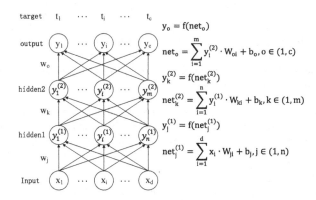

**Fig. 1.** Notations in neural network structure [10]

As the notations illustrated in Fig. 1, E represents the network's cost function can be described as:

$$E = \frac{1}{2} \sum_{k=1}^{c} \left( t_k - y_k^{(o)} \right)^2 \tag{2}$$

the *i-th* neuron's criticality, denoted by $\delta_i$, can be represented by the derivative of E with respect to $y_i$, as illustrated by Eq. (3):

$$\delta_i = \frac{\partial E}{\partial y_i} \tag{3}$$

As can be seen, naively applying AVF to measure the soft error vulnerability of the neural network fails to obtain the neuron level charactor. The importance of a neuron in the network, $\frac{\partial E}{\partial y}$, and combined with the characteristics of neural networks need to be considered as well.

## 3   NVF: Neuronal Soft-Error Vulnerability Factor

In order to accurately measure the soft error sensitivity factor of the neural network, we propose Neuronal soft-error Vulnerability Factor (NVF) which describes the probability that the output result is affected by soft errors in NN. We estimate NVF from hardware structural and neuron's criticality perspectives, and characterize them using two factors: AVF and $\frac{\partial E}{\partial y}$. And the following is the analysis of these factors.

**Hardware Structural Vulnerability:** AVF can calculate the component vulnerability of neural networks. As can be seen from Eq. (1), AVF considers the proportion of ACE bits in the structure and the execution time. Since operations of operators (CNN, Full Connection and Pooling, etc.) are fixed, and computation of each layer is always cyclical repetition, and the structure of neurons is usually similar in each layer, so the ratio of ACE is fixed. The major difference is the residency time or lifetime of each neuron in the network. AVF exactly reflects the impact of the neuron's residency time on vulnerability.

**Neuron's Criticality:** Since AVF does not consider uniqueness of the neural network and the difference among neurons, we take neuron's criticality into consideration as well, which could differentiate the criticality of the node. Theoretically critical analysis of the neurons is performed based on Eq. (3).

Based on the above hardware structural and neuron's criticality characteristics, we combine them together and define the neuronal vulnerability factor (NVF). The calculation formula of NVF of each neuron is:

$$NVF_{node} = AVF * \frac{\partial E}{\partial y} \qquad (4)$$

The NVF for one layer is:

$$NVF_{layer} = \frac{\Sigma AVF * \frac{\partial E}{\partial y}}{\text{number of nodes in the layer}} \qquad (5)$$

NVF could quantify the vulnerability of the neurons, and as well as the overall network. Soft error on neurons with lower NVF may induce no or little impact on NN accuracy due to the inherent fault-tolerant ability of neuron network. And it will seriously affect the output when a neuron with a high NVF value is attacked by a soft error. So reducing the NVF of the entire network can improve the NN reliability to soft errors.

# 4   NVF-Guided Reliability Enhancements for the NNs System

NVF describes the probability that a particle strike will result in erroneous outputs or accuracy loss. The factor AVF reflects the fraction of the neuron's lifetime during which the neuron contains ACE bits. To optimize the NN's reliability, we should reduce the lifetime of neurons with higher NVF that is more vulnerable to soft errors. Scheduling the computation sequence could change the lifetime of neurons, and improve the NN accuracy.

## 4.1   NVF-Guided Neuron Scheduling

Neurons with higher NVF value is more likely to be affected by soft errors, and the possibility of erroneous output induced by soft error will increase, with longer high NVF neurons lifetime. One efficient way to tackle this challenge is processing these vulnerable neurons as soon as possible to reduce the lifetime.

Our main idea is giving the neurons with high NVF higher priority to be calculated.

As shown in Eq. (5), if the neuron process sequecne is ranked according to their NVF values, the NVF of each layer will be reduced, hence, reducing the probability of the entire neural network being affected by soft errors.

Scheduling the neuron processing order can be viewed as transforming the NN during its training procedure, where the weight array is transformed under the guidance of NVF. As shown in Fig. 2, this transformation involves two key steps: (1) Calculation of NVF, the vulnerability is analyzed to identify how neurons impact output quality. (2) Reconstruct of weight array, in which the weigh array of each layer is re-arranged based on NVF, thus reducing the lifetime of high NVF neurons. In Fig. 3, (a) represents original NN, while (b) is the transformed NN, and the red number in green circle represents the NVF value, and the number in blue circle in (a) and red circle in (b) is the label number of neurons, in (b) a darker color implies a higher NVF. We take the first hidden layer as an example, and suppose the system perform calculation of one neuron each time, the grey neuron in figure represents nodes that finish operations. The average NVF of the first layer is 5.25 at beginning, then the average NVF of original process sequence is 6, 5.5 and 9, while the NVF after scheduling is 4, 2.5 and 2. As can be seen from the change of NVF value, the vulnerability is improved under our NVF-guided neuron scheduling.

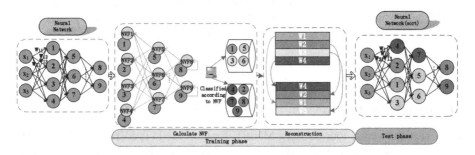

**Fig. 2.** NVF guided neuron scheduling flow chart

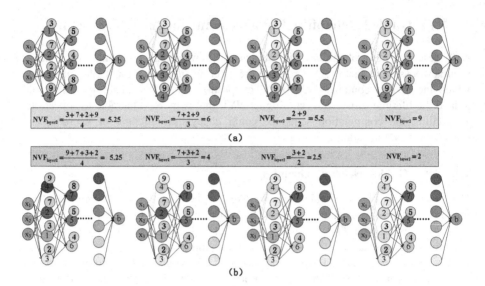

Fig. 3. Network NVF before (a) and after (b) scheduling

**NVF Calculation.** A neural network needs to be trained before it can be deployed for an inference or classification task. Training entails learning and updating the weights of each neuron of a neural network is by performing the backward propagation algorithms. We propose to utilize back propagation to characterize the criticality of each neuron. Back propagation apportions the error at the output of the NN to the outputs of individual neurons. Thereby, it provides a qualification of the error contributed by each neuron to the outputs of the network. Neurons that contribute the least to the global error are more resilient. Conversely, neurons contributing the highest error during back propagation are deemed sensitive.

For each instance in the training dataset, the error at the output of the neural network is computed using forward propagation. Then the errors are propagated back to the outputs of individual neurons and their average error contribution over all inputs in the training set is obtained. We use a statistical-based method to calculate NVF: First, for each input in the dataset, we calculate NVF of all neurons, and sort neurons by NVF. Then, we combine the ranking results of all input into an M × N matrix, where M denotes number of neurons, and N denotes number of inputs in the dataset. Finally, we analyze which neuron in each row of the matrix appears most frequently, and rearrangement the weight array based on the analysis.

## 5 Evaluation

### 5.1 Experimental Setup

In this section, we use four datasets: MNIST, Fashion-MNIST, SVHN, and CIFAR-10 to evaluate NVF guided scheduling and heterogeneous memory protection scheme. The network topology of Fashion-MNIST is $300 \times 200 \times 100$, and MNIST, SVHN, and CIFAR-10 is $256 \times 256 \times 256$. They are trained and tested on PyTorch which is a fast and flexible experimentation that has been widely used in both industry and academia.

## 5.2 Effectiveness of Neurons Scheduling Scheme

We evaluate the effectiveness of NVF guided neurons scheduling scheme on four data sets, and compare the improvment on accuracy over original process by using our NVF scheduling approach with (1) $\partial E/\partial y$: scheduling scheme only considers $\partial E/\partial y$, (2) NVF: scheduling scheme combines $\partial E/\partial y$ and AVF. The injected fault time and fault location are generated randomly, and the failure rate is set to 2%, 4%, 6%, and 8%, respectively. We conduct 25 runs of random fault injection experiments and present the averaged improvement on output accuracy.

As Fig. 4(a) shows, when fault rate is 4% and running SVHN dataset, compared with random processing, $\partial E/\partial y$ guided scheduling improves accuracy by 14%, and NVF guided scheduling improves accuracy by 21%. And compared with $\partial E/\partial y$ guided scheduling, NVF leading to accuracy improvements of 1%, 7%, 1% and 4% at different failure rates. This is because our NVF guided technique leverages the critical of neurons to effectively improve the reliability of neural network. In Fig. 4(b), while running Fashion MNIST dataset, the improvement on accuracy compared with $\partial E/\partial y$ is 1.3%, 0.7%, 2.8%, 4%. We can find the optimization effect is more obvious at higher fault rate. As Fig. 4(c) shown, while running MNIST dataset, the improvement on accuracy compared with $\partial E/\partial y$ is 3%, 7%, 1%, 5.4%. Finally, the improvement on Cifar-10 is lower than 2%, that is because the accuracy of Cifar-10 is lower compared with other datasets and the space for reliability improvement is small.

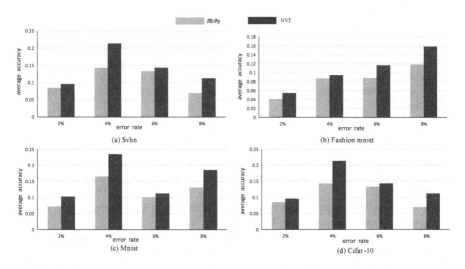

**Fig. 4.** Improvement on accuracy of the network using different scheduled schemes

The evaluation result shows, for all datasets, under different failure rates and different network configuration, $\partial E/\partial y$ scheduling performance better than random and original process sequence. Moreover, the performance of NVF scheme is better than $\partial E/\partial y$, and the best accuracy is obtained, which proves the effectiveness of our NVF guidance scheduling.

## 6  Conclusion

We leverage the observation that neurons in the NN usually exhibit different capability in fault tolerance, and present a neuron vulnerability factor (NVF) to quantify the neural network vulnerability to soft error. This paper also proposes an optimization scheme based on NVF: a NVF-guided scheduling scheme to reduce the lifetime of vulnerable neurons. The experiment result shows our scheme could effectively improve the overall accuracy and system reliability.

**Acknowledgment.** This work was supported by the National Natural Science Foundation of China (NSFC) under grants (61772350), Common Information System Equipment Pre-research Funds (Open Project, JZX2017-0988/Y300), the Construction Plan of Beijing High-level Teacher Team (CIT&TCD201704082, CIT&TCD20170322), the Open Project of State Key Laboratory of Computer Architecture (CARCH201607). The work is also supported by the Capacity Building for Sci-Tech Innovation Fundamental Scientific Research Funds (025185305000). Beijing Nova program (Z181100006218093), Research Fund from Beijing Innovation Center for Future Chips (KYJJ2018008).

## References

1. Sangchoolie, B., Pattabiraman, K., Karlsson, J.: One bit is (not) enough: an empirical study of the impact of single and multiple bit-flip errors. In: International Conference on Dependable Systems and Networks (2017)
2. Azizimazreah, A., et al.: Tolerating soft errors in deep learning accelerators with reliable on-chip memory designs. In: IEEE International Conference on Networking, Architecture and Storage (NAS) (2018)
3. Karlsson, J., Liden, P., Dahlgren, P., Johansson, R., Gunneflo, U.: Using heavy-ion radiation to validate fault-handling mechanisms. In: MICRO (1994)
4. Zhang, Q., Xu, Q.: Approxit: a quality management framework of approximate computing for iterative methods. IEEE Trans. Comput.-Aided Des. Integr. Circ. Syst. (2017)
5. Chen, X., Chen, D.Z., Hu, X.S.: moDNN: memory optimal DNN training on GPUs. In: DATE (2018)
6. Ma, Y., et al.: Optimizing loop operation and dataflow in FPGA acceleration of deep convolutional neural networks. In: FPGA (2017)
7. Savino, A., Vallero, A., Carlo, S.D.: ReDO: cross-layer multi-objective design-exploration framework for efficient soft error resilient systems. IEEE Trans. Comput. **67**, 1462–1477 (2018)
8. Reagen, B., et al.: Minerva: enabling low-power, highly-accurate deep neural network accelerators. In: ISCA (2016)

9. Venkataramani, S., Ranjan, A., Roy, K., Raghunathan, A.: AxNN: energy-efficient neuromorphic systems using approximate computing. In: ISLPED (2014)
10. Zhang, Q., Wang, T., Tian, Y., Yuan, F., Xu, Q.: ApproxANN: an approximate computing framework for artificial neural network. In: DATE (2015)
11. Mukherjee, S.S., Weaver, C., Emer, J., Reinhardt, S.K.: A systematic methodology to compute the are ehitectural vulnerability factors for a hish performance microprocessor. In: MICRO (2003)

# A Fault Detection Algorithm for Cloud Computing Using QPSO-Based Weighted One-Class Support Vector Machine

Xiahao Zhang and Yi Zhuang$^{(\boxtimes)}$

Department of Computer Science and Technology,
Nanjing University of Aeronautics and Astronautics,
Nanjing 211106, Jiangsu, China
{zxh, zyl6}@nuaa.edu.cn

**Abstract.** The complexity and diversity of cloud computing bring about cloud faults, which affect the quality of services. Existing fault detection methods suffer problems such as low efficiency and low accuracy. In order to improve the reliability of the cloud data center, a fault detection algorithm based on weighted one-class support vector machine (WOCSVM) is proposed to detect and identify the host faults in the cloud data center. Specifically, first, we conduct correlation analysis among monitoring metrics and select key ones for reducing the complexity. Second, for imbalanced monitoring dataset, one-class support vector machine is used to detect and identify host faults, and a weight allocation strategy is proposed to assign weights to the samples, which describes the importance of different sample points in order to improve detection accuracy on potential faults. Finally, for the purpose of increasing the accuracy further, the parameters are set via a parameter optimization algorithm based on quantum-behaved particle swarm optimization (QPSO). Furthermore, experiments by comprising with similar algorithms, demonstrate the superiority of our algorithm under different classification indicators.

**Keywords:** Cloud computing · Fault detection · Mutual information · One-class support vector machine · Quantum-behaved particle swarm optimization

## 1 Introduction

Cloud computing is a new computing model that provides computing, services, and applications as public facilities to users [1]. As more applications and services are deployed in clouds, the reliability of cloud computing becomes more important. Cloud computing has complex features such as resource virtualization, application hosting, rapid elastic architecture and multi-tenancy, which will bring about cloud faults from time to time. The faults not only have a huge bad impact on people's normal life and work, but also cause serious economic losses in business and society. For example, the Amazon website was down for about 45 min in 2013, which brought about 5.3 million losses as customers could not make purchases during that period [2]. Hence, the effective fault detection algorithm is one of the key techniques of guaranteeing the reliability of cloud computing systems.

© Springer Nature Switzerland AG 2020
S. Wen et al. (Eds.): ICA3PP 2019, LNCS 11945, pp. 286–304, 2020.
https://doi.org/10.1007/978-3-030-38961-1_25

Faults in cloud computing systems are often caused by hardware faults and software faults, which are usually randomly generated and are difficult to be reproduced. Although virtualization provides software faults isolation among different virtual machines, virtualization infrastructure including hypervisors and privileged virtual machines is still vulnerable to hardware errors [3]. Due to the complex structure and dynamic characteristics of the cloud computing systems, the detection efficiency and accuracy of existing fault detection methods need improvement. To track the running status of these hosts, the monitoring system needs to collect a large amount of monitoring data from different categories including processor, memory, disk and network of hosts. However, the fault detector analyzing such large-scale data will bring huge overhead. Therefore, it is urgent to propose a fault detection method that better adapts to the cloud computing environment to improve the reliability of cloud computing systems.

Recently, lots of methods have been developed to detect faults in the cloud data center. Threshold-based methods are most frequently used in practical applications. In terms of academia, more research is based on statistics and machine learning methods. The statistical methods [4] describe the system by a certain probability model in order to obtain the fault information. However, these methods are heavily dependent on probability distribution, which reduces the detection performance. Machine learning-based methods are mainly divided into supervised [5, 6] and unsupervised methods [7], these methods aims at identifying fault behavior by modeling historical data. However, supervised methods are not so suitable for fault detection in the cloud environment because the fault data is difficult to be obtained, and these methods usually suffer problems such as low accuracy and low efficiency.

Considering these problems, a fault detection algorithm based on weighted one-class support vector machine (WOCSVM) is proposed to detect the host faults in the cloud data center. Specifically, first, we select key metrics which reflect a system's running status in order to reduce the complexity of the model and improve detection efficiency. Second, we detect and identify the host faults by using one-class SVM (OCSVM) which only needs normal data. Then a weight allocation strategy is proposed to improve detection ability on potential faults, and use quantum-behaved particle swarm optimization (QPSO) to find the optimal parameters combination of the WOCSVM model. Finally, we simulate three fault types based on OpenStack in order to evaluate the proposed algorithm, and the proposed algorithm is compared with similar algorithms.

Our main contributions are summarized as follows.

- We design a key metric selection algorithm based on mutual information, for reducing the complexity of the model and improving detection efficiency.
- We propose a WOCSVM model to solve the problem of cloud computing fault detection, and we design a weight allocation strategy to improve detection ability on potential faults.
- We decide the optimal parameters combination of the WOCSVM model by using quantum-behaved particle swarm optimization (QPSO) for improving the accuracy of the detection model.
- We build a virtual platform based on OpenStack to evaluate the proposed algorithm, and we perform comparative experiments with similar algorithms for validating the effectiveness of our algorithm.

The remainder of this paper is organized as follows. Section 2 introduces the related work. Section 3 presents the QPSO-based weighted one-class support vector machine (QPSO-WOCSVM) model. The QPSO-based WOCSVM host fault detection algorithm is described in Sect. 4. Section 5 conducts performance evaluation. Eventually, the conclusion is summarized in Sect. 6.

## 2 Related Work

Fault detection methods can generally be divided into two categories including rule-based detection and anomaly detection [8]. Rule-based detection method defines the distinguishable features of the fault based on the phenomena represented by the historical fault, and then matches the observed phenomena with the defined fault features. When the match is successful, it is detected as faulty, otherwise, the system is considered to be operating normally. Qiang et al. [9] proposed a fault detection method based on Bayesian and decision tree. Firstly, the Bayesian model was used to detect the fault node. After the system administrator verified the abnormality, the data was marked and then the marked data was used to construct the decision tree, which is used to predict failures. Arefin et al. [5] proposed a framework called FlowDiff, which collects information from all entities including applications and infrastructure operated in the data center and continuously builds behavioral models of operations, and then analyzing the problem by matching the current information to a known fault model or rule. Rule-based fault detection methods require a large amount of information inside the system to describe the fault characteristics, while a large number of metrics need to be monitored and analyzed. It is difficult for system administrators to make rules based on experience, and rule-based detection methods cannot identify the faults that have not occurred.

Methods based on anomaly detection generally establish a model for the target system, and the system behavior is compared with the benchmark to determine whether the fault occurred or not. Liu et al. [10] proposed a fault detection model based on neural network, which uses genetic algorithm to optimize weights, and the proposed model had high precision and generalization ability. However, methods based on neural network require a large amount of training samples, and it is very difficult to collect sufficient fault samples. Modi et al. [11] proposed a fault detection algorithm which combining Snort and Bayesian for cloud computing systems. The algorithm uses Bayesian classifier to predict whether a given event is an attack by observing previously stored network events, while Snort is used to detect network faults in the cloud environment. Palm et al. [12] designed a Bayesian fault detection system for cloud computing systems, which uses Bayesian network to describe the complex relationships between fault factors to effectively diagnose and predict faults of cloud computing systems. However, Bayesian network is complicated to construct because it is necessary to consider both network structure and sample integrity, and are usually applied to the modeling of discrete attributes, so it might be not suitable for large-scale cloud computing systems. Dinh-Mao et al. [13] proposed a IaaS system fault detection algorithm based on fuzzy logic and prediction technology, which uses Gaussian Process Regression (GPR) to predict the host resource utilization and network metrics of next epoch, and then detect the faults by using fuzzy logical algorithm. However, GPR has huge overhead of calculation when the data set is large, which is not suitable for high real time cloud computing systems, and expert knowledge in the reasoning process is difficult to obtain for fault detection. Wang et al. [4] proposed a self-adaptive fault detection algorithm for cloud computing systems, which uses principal component analysis to describe the operating status of the system, and then evaluates the anomaly degree based on the cosine similarity. Finally, the reliability model was built based on the exponential distribution model to predict the time of fault to adjust the monitoring period. Adamu et al. [6] proposed a fault detection method for cloud data centers based on Support Vector Machine (SVM), which uses normal samples and fault samples to train the fault detection model, and then determined whether the hosts were abnormal or not by inputting the monitoring data into the model. The authors in [14] pointed out that SVM has a strong mathematical foundation and high reliability in many applications. However, SVM suffers problems such as difficulty in determining kernel function and hyperparameters, and SVM is not suitable for the cases where the training set is imbalanced.

To sum up, the existing fault detection methods for cloud computing have the following shortcomings.

- Some machine learning-based detection methods suffer problems such as difficulty in selecting hyperparameters and kernel function which will affect the detection accuracy.
- Statistical methods rely on certain probability distribution assumptions, so it is not suitable for complex and varying cloud environments.
- Supervised detection algorithms require labeled data to train models. However, in cloud environments, it is highly cost that using expert knowledge to distinguish and mark data [7]. Generally, the dataset is an imbalanced dataset containing a large amount of normal data and a small amount of abnormal data, which will reduce the detection accuracy while using two-category classification algorithms.

## 3   Methodology

Different from the two-category SVM, by just providing the normal monitoring data of hosts, OCSVM can create a fault detection model to detect the host faults. Therefore, OCSVM is suitable for cloud computing environment where normal monitoring data are easy to be obtained. The WOCSVM model combines weight allocation strategy and OCSVM algorithm, and a host fault identification method is proposed based on OCSVM, which improves the fault detection capability on different fault types. Furthermore, good accuracy can still be acquired without enough priori knowledge. Based on metrics gathered from hosts, we extract the key metrics, and the WOCSVM model is built for normal host operation, then the model is used to pinpoint anomalous events.

### 3.1   Dimensionality Reduction

In order to guarantee the reliability of the cloud services, the monitoring system needs to collect a large number of metrics to continuously track the running status of the hosts in the cloud data center. However, the metrics usually correlate strongly with each other. Therefore, it is necessary to select key metrics while satisfying the accuracy of detection model.

Let host monitoring metric vector set $MS = \{M_1, M_2, \cdots, M_j, \cdots, M_n\}$, where $M_j = \{m_1, m_2, \cdots m_Y\}$ contains $Y$ metrics which is collected regularly from the hosts. Especially, we mainly collect metrics related to CPU, memory, network and disk (e.g., resource utilization and packet loss rate). The similarity among system metrics can be assessed by the mutual information (MI) [15], which is an information metric for measuring the amount of shared information between the two random variables. Not limited to the linear relationship, MI is also applicable for evaluating the nonlinear correlation of two variables. The mutual information $I(m_i, m_j)$ for any two metrics $m_i$ and $m_j$ is defined as:

$$I(m_i, m_j) = \sum_{m_j \in M} \sum_{m_i \in M} p(m_i, m_j) \log \frac{p(m_i, m_j)}{p(m_i)p(m_j)}, \tag{1}$$

where, $I(m_i, m_j)$ represents the amount of information shared between $m_i$ and $m_j$, $p(m_i, m_j)$ is the joint probability function of $m_i$ and $m_j$, and $p(m_i)$ and $p(m_j)$ are their marginal probability distribution functions, respectively. $m_i$ and $m_j$ are independent with each other if there is no shared information between them. The stronger the correlation is, the higher the mutual information is.

In order to improve the detection efficiency, the correlation between every two metrics is obtained by Eq. (1), and then the key metrics is selected according to the correlation among the metrics. Let key metric vector set $KS = \{S_1, S_2, \cdots, S_j, \cdots, S_n\}$, where $S_j = \{s_1, s_2, \cdots, s_k\}$ is the key metric vector after extracting, and $k$ represents the number of key metrics. The key metric selection algorithm is proposed specifically in Algorithm 1, and we describe the algorithm as follows.

**Algorithm 1.** Key Metric Selection algorithm
Input: monitoring metric vector set $MS$.
Output: key metric vector set $KS$.
**Step 1:** Calculate the normalized mutual information by (1) between every two metrics in the same category (network, memory), for representing the correlation. Then, we use an undirected graph to describe the correlation among metrics. In the graph, each node represents a metric, and two metrics with strongly correlated should be connected. The weight of each node means the number of connected edges.
**Step 2:** Delete the edges and nodes connected to node which have the largest weight, and update the weight of each node. Repeat until the number of edges in the graph is zero.
**Step 3:** For all categories, perform the first two steps. Then, we obtain $KS$ according to remaining nodes in **Step 2**.

The process of the Algorithm 1 is shown in Fig. 1, and the key metrics $m_1$ and $m_6$ are finally obtained. In **Step 1**, the correlation between every two metrics is calculated to establish an undirected graph, and the time complexity is $O(x^2)$, where $x$ is the number of metrics. In **Step 2**, all nodes are sorted by the weight. After that the node

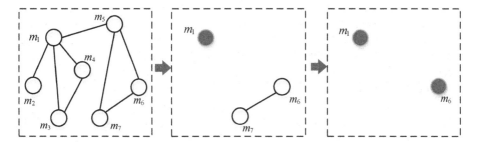

**Fig. 1.** The process of key metric selection.

with the largest weight is deleted, along with its adjacent edges and nodes. And then the remaining nodes are reordered by the weight again. Finally, the key metric is obtained, thus, the time complexity is $O(x^2 \log x)$.

## 3.2 Fault Detection and Identification Model

OCSVM is an unsupervised learning algorithm extended by Scholkopf et al. on traditional SVM algorithm, which has many advantages such as less calculation time and good generalization ability in processing small numbers of sample [16]. The fault detection model based on OCSVM generates a boundary in the kernel space that maximizes the interval between the normal data and the origin to determine the fault data, and we let as many cloud normal data points as possible fall inside the decision boundary.

In this paper, we choose *KS* as the training set. Let $y$ represents the detection function of the host fault, which is defined as:

$$y = \text{sgn}(f(S)) = \text{sgn}(\omega \cdot \phi(S) - \rho), \tag{2}$$

Where, $\omega$ and $\rho$ are hyperplane parameters, $\phi(S)$ maps the key metric vector from the low dimensional space to a high dimensional space where the samples can be linearly separable. The host is abnormal when $y = -1$, and $y = 1$ denotes the host is normal. According to the principle of structural risk minimization of OCSVM [14], the optimization problem of the cloud data center host fault detection function is as follows

$$\min_{\omega,b,\xi_i} \quad \frac{1}{2}\|\omega\|^2 + \frac{1}{vn}\sum_{i=1}^{n}\xi_i - \rho$$
$$s.t. \quad \omega^T\phi(S_i) \geq \rho - \xi_i$$
$$\xi_i \geq 0, \, i = 1, 2, \ldots, n \tag{3}$$

Here, $n$ denotes the number of training samples, $v \in (0, 1]$ is the regularization parameter which indicates the degree of the trade-off between maximum margin and minimum classification error. Since the training set may have fault records which should not fall inside the fault detection boundary, slack variable $\xi_i$ is introduced to let some fault points fall outside the fault detection boundary, which will improve the fault detection accuracy. The selection of the kernel function is critical to the performance of the fault detection model. Radial Basis Function (RBF) has good generalization ability and fast convergence [17], and it requires fewer parameters. Using RBF can reduce the complexity of the model and improve the detection accuracy. Therefore, we adopt RBF as the kernel function in this paper, then the decision function of the fault detection model can be solved as:

$$y = \text{sgn}(f(S)) = \text{sgn}(\sum_{i=1}^{n}\alpha_i\exp(-\frac{\|S_i - S\|^2}{2\sigma^2}) - \rho), \tag{4}$$

Where $\sigma > 0$ is the parameter of Gaussian kernel function, and the fault detection boundary is more compact when the $\sigma$ is smaller.

The above OCSVM-based fault detection model cannot identify the fault type. In this paper, a fault identification method is proposed for the monitoring data which falls outside the fault detection boundary. By calculating the contribution of each metric of the monitoring data to the deviation from the fault detection boundary, the fault type depends on the metric with the greatest contribution. For example, if the metric with the greatest contribution is a processor related metric, then the host fault type is determined to be a processor fault.

According to (4), the data falling outside the boundary has the greater anomaly when the Euclidean distance between the abnormal data and the support vector in the kernel space is larger. If a sample $S$ falling outside the boundary, then the contribution of metric $s_j$ to the deviation from the fault detection boundary is calculated as:

$$C(j) = \sum_{i=1}^{N}\frac{|sv_{ij} - s_j|^2}{\sum_{l=1}^{k}|sv_{il} - s_l|^2}, \tag{5}$$

Here, $N$ is the number of samples located on the detection boundary in the training set, and $sv_{ij}$ is the $j^{th}$ metric of the $i^{th}$ sample which lies on the boundary in $S$. According to (5), higher $C(j)$ indicates a greater contribution of metric $s_j$ to the deviation from the fault detection boundary, then the fault is identified according to the metric with the greatest $C(j)$.

### 3.3  Weight Allocation Strategy

The weight of the training sample has an important impact on detection performance. The existence of the outliers, which may be fault samples in the training set, will shift the fault detection boundary and reduce the generalization ability of the model. The detection model based on OCSVM is not robust to the outliers in the training set. The work in [18] pointed out that sample points far from the center have a small influence on the fault detection boundary and are given smaller weights, while bigger weights are given to sample points near the center so that the fault detection boundary is closer to the normal samples, thus reducing the influence of outliers on the hyperplane and more potential fault data can be found, and the sample weight is calculated as follows:

$$dis_i = \exp(-\frac{\hat{d}^2(S_i, C)}{2}),$$  (6)

Where, $C$ is the center of training set, and $\hat{d}(S_i, C)$ represents the normalized distance, which indicates the ratio of the distance between $S_i$ and $C$ to the maximum distance.

For the cases when the potential fault sample points are near to the normal sample points, these potential fault samples will be assigned higher weights, which reduces the sensitivity of the fault detection boundary to detect potential faults. In order to improve the detection ability on potential faults, we design a weight allocation strategy for training samples to determine the location of the fault detection boundary. The strategy considering both the distances between training samples to $C$ and their local density. The local density based weighting strategy considers that the sample points in the high-density region should be given a larger weight, and the local density weight $den_i$ of training sample $S_i$ is designated as:

$$den_i = \exp(-\frac{loc^2(S_i)}{2}),$$  (7)

Where, $loc(S_i)$ represents the local density of $S_i$, which is given by

$$loc(S_i) = \frac{\sum_{j=1}^{h} d(S_i, S_i(j)) - \min_{i=1,\dots n} \sum_{j=1}^{h} d(S_i, S_i(j))}{\max_{i=1,\dots n} \sum_{j=1}^{h} d(S_i, S_i(j)) - \min_{i=1,\dots n} \sum_{j=1}^{h} d(S_i, S_i(j))},$$  (8)

Where, $S_i(h)$ is the $h^{th}$ nearest neighbor of $S_i$. According to (7) and (8), when $S_i$ is in the high-density region, $loc(S_i)$ becomes smaller while $den_i$ becomes larger. The weight $w_i$ of sample $S_i$ we proposed is shown as:

$$w_i = u_1 den_i + u_2 dis_i,$$  (9)

Here, $u_1, u_2 \in [0, 1]$ are trade-off parameters of these two strategies, respectively. After setting weights for all samples, Eq. (3) was transformed into

$$
\begin{aligned}
&\min_{\omega, b, \xi_i} \quad \tfrac{1}{2} \|\omega\|^2 + \tfrac{1}{vn} \sum_{i=1}^{n} w_i \xi_i - \rho \\
&s.t. \quad \omega^T \phi(S_i) \geq \rho - \xi_i \\
&\qquad \xi_i \geq 0, \; i = 1, 2, \ldots, n
\end{aligned}
\tag{10}
$$

According to Eq. (9), the proposed weight allocation strategy combines the advantages of two strategies, which can assess the outlier degree of the training samples more accurately and improve the detection ability of the detection model on potential faults.

### 3.4  Parameter Optimization Algorithm Based on QPSO

The fault detection algorithm based on WOCSVM cannot handle the problem of parameter selection well, while inappropriate parameters will reduce the detection ability. In the cloud host fault detection problem in this paper, the parameters that need to be set in advance are the penalty factor $v$ in Eq. (3) and the Gaussian kernel function's parameter $\sigma$ in Eq. (4). The regularization parameter $v$ represents the degree of the trade-off between maximum margin and minimum classification error, and the Gaussian kernel function's parameter $\sigma$ denotes the bandwidth value which controls the radial extent of the function. Since the combination of these two parameters will limit the performance of the fault detection model, we consider the parameter selection problem as an optimization problem and adopt quantum-behaved particle swarm optimization algorithm to solve it.

Since the Particle Swarm Optimization (PSO) algorithm has fast convergence and is easy to fall into local optimum, Sun et al. proposed Quantum-behaved Particle Swarm Optimization (QPSO) [19]. In the QPSO algorithm, the particle system is assumed to be a quantum particle system where each particle has a quantum behavior, and the probability density function is used to represent the probability of a particle at a certain location. Compared with the PSO algorithm, the QPSO algorithm has a stronger power to search global optimal solutions and has been successfully applied in various practical fields [20, 21].

Let $r_i = (v_i, \sigma_i)$ indicates the position of the $i^{th}$ particle, and each position represents a parameter combination of the WOCSVM model. Firstly, the position of $q$ particles $\{r_1, r_2, \ldots r_q\}$ is initialized. Then, the fitness of each individual in the population is calculated as:

$$
f(r) = 1 - \frac{l_f}{l},
\tag{11}
$$

Where $l_f$ is the number of misclassified samples, and $l$ is the total number of samples. The fitness of the $i^{th}$ particle is greater when the detection accuracy of the model is higher. Therefore, we find the best parameter combination when the particle fitness is the global optimal. Next, each particle records its own historical optimal position $pbest_i$ and the global optimal position $gbest$ searched in the population, and the positions of all particles are updated by

$$
r_{ij}(t+1) = \left\{
\begin{array}{ll}
pbest_{ij}(t) + \beta(t) \cdot \left| Mbest_j(t) - r_{ij}(t) \right| \cdot \ln(\frac{1}{\mu}), & if\ \mu \geq 0.5 \\
pbest_{ij}(t) - \beta(t) \cdot \left| Mbest_j(t) - r_{ij}(t) \right| \cdot \ln(\frac{1}{\mu}), & if\ \mu < 0.5
\end{array}
\right\}, \quad (12)
$$

Where, $r_{ij}(t+1)$ is the $j^{th}$ position of the $i^{th}$ particle after updating, $Mbest_j(t)$ is the $j^{th}$ average optimal position of all particles at the $t^{th}$ iteration, $pbest_{ijt}(t)$ indicates the $j^{th}$ historical optimal position of the $i^{th}$ particle at the $t^{th}$ iteration, $\mu$ is random numbers in $[0, 1]$, and $\beta(t)$ is the control parameter, which is determined by [22]

$$
\beta(t) = \beta_{max} - \frac{t \cdot (\beta_{max} - \beta_{min})}{MaxITER}, \quad (13)
$$

Where $MaxITER$ is the maximum of iterations, $\beta_{max}$ and $\beta_{min}$ are the maximum and minimum of $\beta(t)$, respectively, and $t$ is the number of iterations currently. After updating the positions of all particles, we calculate the fitness of them by (11) and then update the $pbest$ of all particles and $gbest$. The $pbest_i$ is determined by [23]

$$
pbest_{id}(t+1) = c \cdot pbest_{id}(t) + (1 - c) \cdot gbest_d, \quad (14)
$$

Here, $c$ is a random number in $[0, 1]$. Finally, the iterations are performed until the fitness of one particle is within the expected error range or finishing the maximum of iterations, and the position of the global optimal particle we obtained is the best parameter combination of the WOCSVM model.

## 4    The QPSO-Based WOCSVM Fault Detection Algorithm

The QPSO-based WOCSVM (QPSO-WOCSVM) fault detection algorithm is divided into four parts, including key metrics selection, model training, fault detection and fault identification. In the first stage, we make correlation analysis to select key metrics, which reducing complexity and improving detection efficiency. In the model training phase, the samples are set weights according to the weight allocation strategy for improving the ability to detect the potential faults, and OCSVM is used to create a fault detection boundary for fault detection. At the same time, the QPSO-based parameter optimization algorithm is used to find the optimal parameter combination $(v, \sigma)$ of the WOCSVM model. In the fault detection stage, we input the monitoring data after extracting the key metrics to the decision function of the fault detection model, and we can determine whether there is abnormity in the host according to (2), if there is, identifying the fault type according to the result of (5). The architecture of the QPSO-WOCSVM fault detection algorithm is shown in Fig. 2.

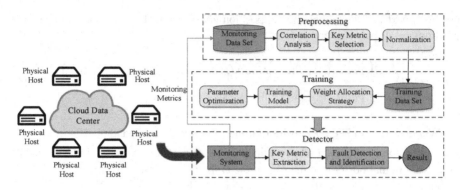

**Fig. 2.** The architecture of the QPSO-WOCSVM fault detection algorithm.

In the weight allocation phase, the distances between each node and its neighboring nodes need to be calculated, so the time complexity is $O(h \cdot n \cdot \log n)$, where $h$ is the number of neighboring nodes, and $n$ is the number of training samples. In the model training phase, the time complexity is between $O(sv^3 + n \cdot sv^2 + k \cdot n \cdot sv)$ and $O(k \cdot n^2)$ which is same as the OCSVM algorithm. Here, $sv$ is the number of the support vector and $k$ is the sample dimension. In the fault detection phase, the time complexity is $O(k \cdot n)$, and in the fault identification stage, the metric with the greatest contribution is calculated, thus, the time complexity is $O(k \cdot sv \cdot \log k)$. The pseudocode of QPSO-WOCSVM fault detection algorithm in the cloud data center is shown in Algorithm 2.

| **Algorithm 2** QPSO-WOCSVM fault detection algorithm |
|---|
| **Input** The host monitoring data set $MS$ |
| **Output** The detection result |
| 1.  **begin** |
| 2.      Get the training set $KS$ by Algorithm 1; |
| 3.      Normalize the $KS$ ; |
| 4.      Calculate the weights of training samples by (9); |
| 5.      Get parameter combination by QPSO; |
| 6.      **for** $i = 1$ to $n$ |
| 7.          WOCSVM.train( $KS.S_i$ ); |
| 8.      **end for** |
| 9.      **while** $i > n$ and $i < end$ **do** |
| 10.         $S_i =$ KeyMetric ( $M_i$ ); |
| 11.         Fault detection by (2) and (5); |
| 12.         $i++$ ; |
| 13.     **end while** |
| 14. **end** |

## 5    Experiments and Analysis

In order to verify the algorithm proposed in this paper, we build the simulation platform based on OpenStack and collect the monitoring data for comparison experiments. We get abnormal samples by injecting faults into compute node, and the experimental error is evaluated by using accuracy, precision, recall, and F-score. The experimental platform we built is shown in Fig. 3.

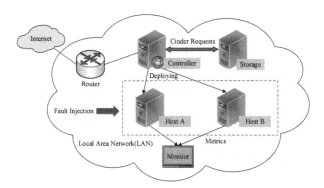

**Fig. 3.** Experimental platform setup.

### 5.1    Experimental Setting

According to Fig. 3, the virtual platform consists of four hosts including two compute nodes, one controller node and one storage node. The Experiment configurations is shown in Table 1.

**Table 1.** Experiment configurations.

| Name | Configuration |
| --- | --- |
| CPU | Intel Core i5-6500 |
| Memory | 8 GB |
| Disk | 512G |
| Operating system | Ubuntu 16.04 |
| OpenStack version | OpenStack Newton |

Reference to the paper [22, 24, 25], the parameter settings of the QPSO-WOCSVM fault detection algorithm are shown in Table 2.

**Table 2.** Parameters settings for the proposed algorithm.

| Parameter name | Value |
|---|---|
| Correlation threshold | 0.8 |
| Regularization parameter $v$ | 0.15–0.85 |
| Gaussian kernel function parameter $\sigma$ | 0.001–100 |
| Maximum of $\beta(t)$ | 1 |
| Minimum of $\beta(t)$ | 0.5 |
| Population scale | 20 |
| Number of iterations | 200 |

## 5.2  Analysis of Experimental Results

**Correlation Analysis.** There are six projects deploying on the compute node, and the controller node creates instances for running those. For collecting the monitoring data, we use the sysstat [26] to collect 36 system metrics in more than four categories including CPU, memory, disk, network and so on from the compute node every ten seconds. For selecting key metrics, we take network related metrics as an example. Table 3 shows the seven network related metrics we collect.

**Table 3.** Network related metrics.

| Metric's index | Description |
|---|---|
| rxpck/s | Number of received packets per second |
| txpck/s | Number of sent packets per second |
| rxmcst/s | Number of received multicast packets per second |
| rxerr/s | Number of errors in receiving packets per second |
| txerr/s | Number of errors in sending packets per second |
| rxdrop/s | Number of dropped in receiving packets per second |
| txdrop/s | Number of dropped in sending packets per second |

We calculate the correlation between each two metrics and connect two metrics with strong correlation, and an undirected graph is obtained as shown in Fig. 4.

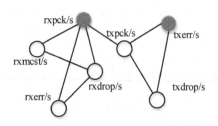

**Fig. 4.** Correlations among network-related metrics.

It can be seen from Fig. 4 that the most network related metrics correlate strongly with each other. Then, we obtain the key metrics (i.e., rxpck/s and txerr/s) by using Algorithm 1.

**Parameter Optimization.** By providing the parameter settings in Table 2, we find the optimal parameter combination of WOCSVM model according to Algorithm 2. The relation between fitness and the iteration number is shown in Fig. 5.

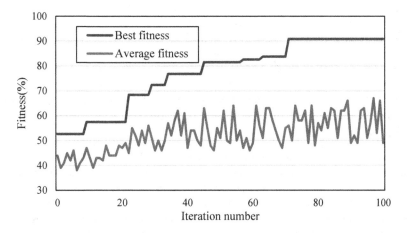

**Fig. 5.** The results of QPSO.

The iteration results show that QPSO reaches the optimal solution when the iteration number is 80. By using QPSO, the optimal parameters for WOCSVM are obtained as $v = 0.0016$, $\sigma = 6.25$. Then, the obtained optimal parameter combination are used to train the WOCSVM model.

**Algorithm Performance Analysis.** We compare the QPSO-WOCSVM fault detection algorithm with PSO-OCSVM [27] and IPSO-MSVM [28] algorithms. Since both OCSVM and SVM use radial basis kernel function and have good classification performance, they are often used in the fault detection field. We use sklearn package [29] for data processing and model training.

In our experiment, 5000 monitoring samples were selected from the compute node as the training data set, and we obtain abnormal samples by injecting faults into the compute node. Specifically, we choose three typical faults related with network, CPU and memory to inject with methods as follows.

1. Memory related faults: We use Stress [30] to simulate memory leak by constantly applying dynamic memory without releasing the space until exhausting memory resources.
2. CPU related faults: We use Stress to simulate CPU hog, which increasing computing resource load by adding CPU processes.

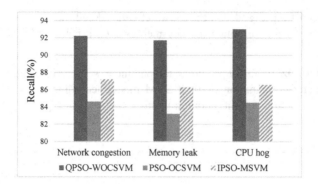

**Fig. 6.** Recall of three algorithms.

3. Network related faults: We use iPerf [31] to simulate network transmission congestion by sending data packets to other network nodes to occupy the network bandwidth of the compute node.

We use the same data set to train the QPSO-WOCSVM and PSO-OCSVM algorithms, and we set labels for the abnormal samples after fault injection, then 6000 abnormal samples consist of 2000 memory fault samples, 2000 CPU fault samples and 2000 network fault samples are selected into the data set to train the IPSO-MSVM algorithm. The three algorithms were tested with the same test sets. Figures 6, 7, 8 and 9 and Table 4 show the test results.

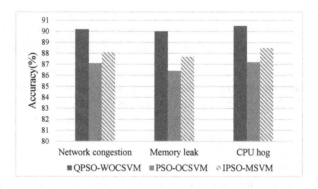

**Fig. 7.** Accuracy of three algorithms.

Figure 6 shows the recall of three algorithms under network, CPU, and memory related faults. Recall refers to the ratio of detected abnormal samples to total abnormal samples, which is the primary concern for fault detection. We can see that for network congestion, the recall of the proposed algorithm exceeds 92%, while the recall of other two algorithms are relatively lower, respectively 84.6% and 87.2%. For Memory leak and CPU hog, the recall of the proposed algorithm is also obviously better than the other two algorithms. It can be seen from Fig. 6 that the proposed algorithm has a stronger ability to detect the faults and a higher fault identification rate on these three fault types.

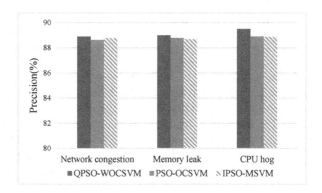

**Fig. 8.** Precision of three algorithms.

Figure 7 shows the accuracy of three algorithms, which is the ratio of correctly detected samples to all samples. We add normal samples into each fault test set for testing the detection accuracy of three algorithms, and it can be seen from Fig. 7 that compared with the other two algorithms, the proposed algorithm in this paper has better detection accuracy.

Precision is the ratio of the true abnormal samples to the samples which are predicted to be abnormal. As shown in Fig. 8, due to the similar detection accuracy on normal data by three algorithms, the precision of the three algorithms is similar, and the precision of the QPSO-WOCSVM fault detection algorithm is still better than the other two algorithms.

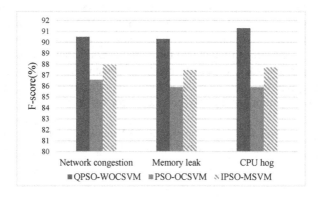

**Fig. 9.** F-score of three algorithms.

**Table 4.** The identification results.

| Fault type | Number of test samples | Number of correctly identified samples | | |
|---|---|---|---|---|
| | | QPSO-WOCSVM | PSO-OCSVM | IPSO-MSVM |
| Normal | 2000 | 1764 | 1782 | 1780 |
| Network congestion | 2000 | 1844 | 1692 | 1742 |
| Memory leak | 2000 | 1840 | 1664 | 1726 |
| CPU hog | 2000 | 1856 | 1690 | 1760 |
| Total | 8000 | 7304 | 6828 | 7008 |
| Mean accuracy | | 91.3% | 85.35% | 87.6% |

Figure 9 shows the F-score which indicates the harmonic mean of precision and recall of three algorithms, and the algorithm have a larger F-score only if both precision and recall are larger. The numerical identification results are shown in Table 4.

It can be seen from Table 4 that the identification results using QPSO-WOCSVM algorithm is better than the other two algorithms, and the mean accuracy reaches 91.3%, which confirms that the superiority of the proposed algorithm.

## 6  Conclusion

In this paper, a weighted one-class support vector machine (WOCSVM) model was proposed for the host fault detection and identification in the cloud data center. Aiming at the problem of higher overhead on analyzing a large number of monitoring metrics by the fault detector, we used mutual information to conduct correlation analysis and select key metrics for reducing the complexity of the proposed algorithm. One-class support vector machine algorithm was used to detect and identify the host faults in the case where the training set is imbalanced. The weight allocation strategy was used for improving the ability to detect the potential faults, and the quantum particle swarm optimization algorithm was used to optimize the parameters for improving detection performance. Finally, based on the WOCSVM model, a QPSO-WOCSVM fault detection algorithm was proposed to detect the host faults of the cloud data center. At the same time, we built a virtual platform based on OpenStack, and we compared the proposed algorithm with PSO-OCSVM and IPSO-MSVM algorithms. For training SVM and testing three algorithms, we obtained abnormal samples by injecting memory, CPU, and network related faults into the compute node, then verified the proposed algorithm by using accuracy, precision, recall and F-score. The experimental results showed that the proposed algorithm outperforms the other two similar algorithms. In the future, we will continue to modify our model for better efficiency.

**Acknowledgements.**  This work was supported by the National Natural Science Foundation of China (General Program) under Grant No.61572253 and the Aviation Science Fund under Grant No. 2016ZC52030.

# References

1. Bera, S., Misra, S., Rodrigues, J.J.P.C.: Cloud computing applications for smart grid: a survey. IEEE Trans. Parallel Distrib. Syst. **26**(5), 1477–1494 (2015)
2. Yousif, M.: Cloud computing reliability—failure is an option. IEEE Cloud Comput. **5**(3), 4–5 (2018)
3. Zhang, P.Y., Shu, S., Zhou, M.C.: An online fault detection model and strategies based on SVM-grid in clouds. IEEE/CAA J. Automatica Sinica **5**(2), 445–456 (2018)
4. Wang, T., Xu, J., Zhang, W., et al.: Self-adaptive cloud monitoring with online anomaly detection. Future Gener. Comput. Syst. Int. J. Escience **80**, 89–101 (2018)
5. Arefin, A., Singh, V.K., Jiang, G., et al.: Diagnosing data center behavior flow by flow. In: IEEE International Conference on Distributed Computing Systems 2013, pp. 08–11. IEEE, Philadelphia (2013)
6. Adamu, H., Mohammed, B., Maina, A., et al.: An approach to failure prediction in a cloud based environment. In: IEEE International Conference on Future Internet of Things & Cloud 2017, pp. 191–197. IEEE, Prague (2017)
7. Watson, M., Shirazi, N.: Malware Detection in Cloud Computing Infrastructures. IEEE Trans. Dependable Secure Comput. **13**(2), 192–205 (2016)
8. Tao, W., Zhang, W., Wei, J., et al.: Fault detection for cloud computing systems with correlation analysis. In: IFIP/IEEE International Symposium on Integrated Network Management 2015, pp. 652–658. IEEE, Ottawa (2015)
9. Qiang, G., Ziming, Z., Song, F.: Ensemble of bayesian predictors and decision trees for proactive failure management in cloud computing systems. J. Commun. **7**(1), 52–61 (2012)
10. Liu, Q., Feng, Z., Min, L., et al.: A fault prediction method based on modified Genetic Algorithm using BP neural network algorithm. In: IEEE International Conference on Systems 2017, pp. 4614–4619. IEEE, Budapest (2016)
11. Modi, C.N., Patel, D.R., Patel, A., et al.: Bayesian classifier and snort based network intrusion detection system in cloud computing. In: Third International Conference on Computing Communication & Networking Technologies 2012, Coimbatore, India (2012)
12. Palm, E., Mitra, K., Saguna, S.: A Bayesian system for cloud performance diagnosis and prediction. In: IEEE International Conference on Services Computing 2017, pp. 281–288. IEEE, Honolulu (2017)
13. Dinh-Mao, B., Thien, H.-T., Lee, S.: Early fault detection in IaaS cloud computing based on fuzzy logic and prediction technique. J. Supercomput. **74**(11), 5730–5745 (2018)
14. Rahulamathavan, Y., Phan, R., Veluru, S., et al.: Privacy-preserving multi-class support vector machine for outsourcing the data classification in cloud. IEEE Trans. Depend. Secur. Comput. **11**(5), 467–479 (2014)
15. Vinh, L.T., Lee, S., Park, Y.T., et al.: A novel feature selection method based on normalized mutual information. Appl. Intell. **37**(1), 100–120 (2012)
16. Scholkopf, B., Platt, J.C., Shawe-Taylor, J., et al.: Estimating the support of a high-dimensional distribution. Neural Comput. **13**(7), 1443–1471 (2001)
17. Shen, W., Guo, X., Wu, C., et al.: Forecasting stock indices using radial basis function neural networks optimized by artificial fish swarm algorithm. Knowl.-Based Syst. **24**(3), 378–385 (2011)
18. Yang, J., Deng, T., Sui, R.: An adaptive weighted one-class SVM for robust outlier detection. In: Jia, Y., Du, J., Li, H., Zhang, W. (eds.) Proceedings of the 2015 Chinese Intelligent Systems Conference. LNEE, pp. 475–484. Springer, Heidelberg (2016). https://doi.org/10.1007/978-3-662-48386-2_49

19. Sun, J., Xu, W.B., Feng, B., et al.: A global search strategy of quantum-behaved particle swarm optimization. In: IEEE Conference on Cybernetics and Intelligent Systems 2004, Singapore, vol. 1, pp. 111–116 (2004)
20. Liu, G.Q., Chen, W.Y., Chen, H.D.: Quantum particle swarm with teamwork evolutionary strategy for multi-objective optimization on electro-optical platform. IEEE Access **7**, 41205–41219 (2019)
21. Duca, A., Duca, L., Ciuprina, G.: QPSO with avoidance behaviour to solve electromagnetic optimization problems. Int. J. Appl. Electromagnet. Mech. **59**(1), 63–69 (2019)
22. Jia, Y., Gong, Q., Li, J., et al.: The power load combined forecasting based on CEEMDAN and QPSO-SVM. Electr. Meas. Instrum. **54**(1), 16–21 (2017)
23. Ghorbani, M.A., Kazempour, R., Chau, K.-W., et al.: Forecasting pan evaporation with an integrated artificial neural network quantum-behaved particle swarm optimization model. Eng. Appl. Comput. Fluid Mech. **12**(1), 724–737 (2018)
24. Thomaz, R., Carneiro, P., Bonin, J.E., et al.: Novel Mahalanobis-based feature selection improves one-class classification of early hepatocellular carcinoma. Med. Biol. Eng. Comput. **56**(5), 1–16 (2017)
25. Wuensch, K.L.: Straightforward statistics for the behavioral sciences. J. Am. Stat. Assoc. **91**(436), 1750 (1996)
26. SYSSTAT. http://sebastien.godard.pagesperso-orange.fr
27. Miandare, M.S., Jalili, S.: VoIP anomaly detection by combining OCSVM and PSO algorithm. In: International Symposium on Telecommunications (IST) with Emphasis on Information and Communication Technology 2012, ITRC, pp. 1038–1043. ICT Res Inst, Tehran, IRAN (2012)
28. Sun, Y.K., Xie, G., Cao, Y. et al.: A fault diagnosis method for train plug doors based on MNPE and IPSO-MSVM. In: International Conference on Control Automation and Information Sciences 2018, pp. 467–471. IEEE, Hangzhou (2018)
29. Scikit learn. https://scikit-learn.org/stable/index.html
30. Stresslinux. http://www.stresslinux.org/sl/
31. Iperf.fr. https://iperf.fr/

# ParaMoC: A Parallel Model Checker for Pushdown Systems

Hansheng Wei[1], Xin Ye[1,4], Jianqi Shi[1,2(✉)], and Yanhong Huang[1,3]

[1] National Trusted Embedded Software Engineering Technology Research Center,
East China Normal University, Shanghai, China
jqshi@sei.ecnu.edu.cn
[2] Hardware/Software Co-Design Technology and Application Engineering,
Research Center, Shanghai, China
[3] Shanghai Key Laboratory of Trustworthy Computing, Shanghai, China
[4] LIPN and Paris University 7, Paris, France

**Abstract.** Model checking on Pushdown Systems (PDSs) has been extensively used to deal with numerous practical problems. However, the existing model checkers for pushdown systems are executed on the central processing unit (CPU), the performance is hampered by the computing power of the CPU. Compared with the CPU, the graphics processing unit (GPU) has more processing cores, which are suitable and efficient for parallel computing. Therefore, it is very attractive to accelerate model checking of PDSs on the GPU. In this paper, we present a new parallel model checker, named ParaMoC, to speed up the performance of model checking problems for pushdown systems (PDSs). Moreover, we focus on how to use Graphics Processing Units (GPUs) to accelerate the reachability verification and the LTL model checking of PDSs. The ParaMoC running on a state-of-the-art GPU can be 100 times faster than the traditional PDS model checker.

## 1 Introduction

Model checking is an important formal verification technique using the state-space searching to explore all possible system states. In this way, this technique makes it possible to check whether a given system can satisfy some specific properties or not. A pushdown system (PDS) is a finite transition system equipped with a stack. Due to its stack, PDSs have been successfully used in malware detection [11] and data flow analysis [9]. However, there are also two major problems that limit the development of model checking of PDSs. Firstly, with the complexity of the program, one of the real challenges encountered in model checking is the well-known state explosion problem. This problem may make the execution time of model checking unbearable. Secondly, the efficiency of existing model checker for PDSs is always subject to the size of the state space. It is challenging to perform model checking more effectively with limited processors on the CPU.

S. Wen et al. (Eds.): ICA3PP 2019, LNCS 11945, pp. 305–312, 2020.
https://doi.org/10.1007/978-3-030-38961-1_26

By now, many techniques are proposed to solve the above problems, e.g., symbolic model checking approaches [4], symmetry reduction [3] and other technologies. Besides these typical techniques, parallel computing has recently shown unique advantages in large-scale computing tasks, which has attracted widespread attention from model checking researchers. Verification would be much more efficient if the procedure of the PDS model checking is parallel.

Parallel computing is a type of computation which can carry out calculations simultaneously [1]. The GPU has a tremendous potential advantage over alternatives. It runs thousands of threads in parallel and improves the performance with several orders of magnitude. In the past few years, accelerating computation on the GPU and other parallel frameworks has achieved a remarkable progress. The key to effective utilization of the GPU for model checking is the design and implementation of data transferring and synchronization.

To the best of our knowledge, there is none similar model checker addressing the model-checking problem of PDSs with the GPU. In this paper, we present a new parallel model checker called **ParaMoC** (A **Pa**rallel **Mo**del **C**hecker for Pushdown Systems). ParaMoC is based on CUDA [8] parallel architecture and employs the GPU to accelerate the reachability verification and the LTL model checking of PDSs. We also propose a parallel transition generation mechanism and a parallel transition transferring for better performance. According to the experimental results, it has been shown that our model checker can achieve up to around 100X speedup compared to Moped [10], one of the most widely used model checkers for PDSs.

The paper is structured as follows. Section 2 introduces a brief background of model checking on pushdown systems. The details of ParaMoC will be presented in Sect. 3. In Sect. 4, experimental results are shown. The conclusion is given in Sect. 5.

**Related Work.** The model checking problem for PDSs was considered in [5]. Our tool serves as an extension of those works in the parallel version. Moped [10] is a popular model checker to check reachability and LTL formulas for PDSs. However, this tool is only run on the CPU, which is not efficient when the model is too large. PDSolver [6] is a $\mu$-calculus model checker to check branching time properties for PDSs, but it is not suitable for reachability analysis and LTL formulas. Different from previous existing researches, our tool take advantage of the multithreading of the GPU to improve the performance of reachability analysis and LTL model checking for PDSs.

## 2   Model Checking on Pushdown Systems

In this section, we introduce a brief background of model checking on pushdown systems.

**Pushdown   Systems.**   A   pushdown   system   is   a   quadruple   $\mathcal{P}$   = $(P, \Gamma, \Delta, \langle p_0, w_0 \rangle)$, where $P$ contains the control locations and $\Gamma$ is the stack alphabet. A configuration $c$ is a pair $\langle p, w \rangle$, where $p \in P$ and $w \in \Gamma^*$. The transition rule set $\Delta$ is a finite subset of $(P \times \Gamma) \times (P \times \Gamma^*)$. If $\langle p, \gamma \rangle \hookrightarrow \langle q, w \rangle \in \Delta$, there is a successor relation $\langle p, \gamma w' \rangle \Rightarrow \langle q, ww' \rangle$ for every $w' \in \Gamma$. The head of a transition rule $\langle p, \gamma \rangle \hookrightarrow \langle p', w \rangle$ is the configuration $\langle p, \gamma \rangle$. A head $\langle p, \gamma \rangle$ is repeating if there exists $v \in \Gamma^*$ such that $\langle p, \gamma v \rangle$ can be reached from for some $v$ while visiting some accepting state along the way.

Reachability problem is the base of model-checking problems which can be solved in two ways: forward and backward. Given a pushdown system $\mathcal{P}$ and a set of configurations $C$, the reachability analysis consists of computing the predecessors of elements of $C$ (backward) and the successors of elements of $C$ (forward).

**Büchi Automaton.** A Büchi automaton $\mathcal{B}$ is a tuple $(Q, \Gamma, \delta, q_0, F)$ where $Q$ is a finite set of states, $\Gamma$ is a finite set of alphabet, $\delta \subseteq (Q \times \Gamma \times Q)$ is a set of transitions, $q_0 \in Q$ is the initial state and $F \subseteq Q$ is the set of final states. The language accepted by $\mathcal{B}$ is the set of all infinite sequences $w$ where $q_0 \xrightarrow{w} p$ for some $p \in F$.

**LTL Model Checking on Pushdown Systems.** Given a pushdown system and an LTL formula $\varphi$, the LTL model checking problem is to check whether the pushdown system satisfies the LTL formula. Let $\mathcal{B} = (Q, 2^{At(\varphi)}, \delta, q_0, F)$ be a Büchi automaton which accepts the negation of a LTL formula $\varphi$. A Büchi pushdown system $\mathcal{BP} = (P, \Gamma, \Delta, c_0, G)$ is the product of a Büchi automaton for $\neg\varphi$ and a pushdown system $\mathcal{P}$, where $P$ is the finite set of states, $\Gamma$ is a finite alphabet, $\Delta \in (P \times \Gamma) \times (P \times \Gamma^*)$ is a finite set of transition rules, $c_0$ is an initial configuration, and $G \subseteq P$ is a set of final states. The LTL model checking problem boils down to the emptiness problem of $\mathcal{BP}$ [10]. While the most important step to the emptiness problem of $\mathcal{BP}$ is computing the repeating heads with the reachability graph $\mathcal{G}$ of $\mathcal{BP}$. $\mathcal{G}$ is a directed labelled graph whose nodes are the head of $\mathcal{BP}$ and whose edges are labelled with 0 or 1.

# 3   Overview of ParaMoC

The ParaMoC model checker is an extension of Moped. In order to support the GPU, ParaMoC runs the model checking algorithms [10] on many threads in parallel. The challenge is how to effectively perform the model checking on the GPU. This section will introduce the details of ParaMoC.

## 3.1   Architecture and Implementation

ParaMoC is implemented in CUDA. Figure 1 presents the architecture of ParaMoC. Given a PDS and an LTL formula, ParaMoC outputs YES if the PDS satisfies the LTL formula. Otherwise, ParaMoC outputs NO and an counter example. ParaMoC consists of two components including BP Constructor and Model Checking Engines.

BP Constructor takes as input a PDS and an LTL formula and constructs them into a Büchi pushdown system. BP Constructor relies on Spin [7] and Moped [10]. Spin translates an LTL formula into a Büchi automaton. Moped

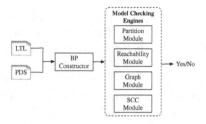

**Fig. 1.** The architecture of ParaMoC

synchronizes a PDS and a Büchi automaton to a Büchi pushdown system. Model Checking Engines are composed of four modules, respectively, Partition module, Reachability Module, Graph Module and SCC Module. Partition module partitions a Büchi pushdown system into sub Büchi pushdown systems. Reachability Module takes as input a sub Büchi pushdown system and computes predecessors. Graph Module takes as input the results of Reachability Module and builds the reachability graph. SCC Module computes the SCC (Strongly Connected Components) and the repeating heads with Tarjan Algorithm [13]. For reachability analysis of pushdown systems, ParaMoC only invokes the partition module and the reachability module to compute the predecessors and the successors.

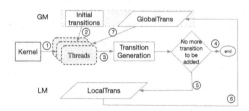

**Fig. 2.** Parallel transition generation mechanism

In the GPU, different types of memory have different access rates, memory costs and time consumptions. For data accessing, storage and synchronization, we propose a parallel transition generation mechanism. As shown in Fig. 2, GlobalTrans and LocalTrans are used to store transitions in different memories. GlobalTrans is Global Memory (GM), and LocalTrans is Local Memory (LM) in the GPU. Note that all threads share only one GlobalTrans, and each thread has a LocalTrans.

Our parallel transition generation works in the GPU as follows. The computation starts from ①. The kernel is launched from the CPU to start threads. Each thread independently obtains the initial transitions in ② and processes the transition generation in ③. If there are no more transition to be added, the transition generation ends and those threads terminate, as shown in ④. Otherwise, the transitions generated in each thread are stored in LocalTrans, as shown in

⑤. In ⑥, there is synchronization, which moves all of the LocalTrans to Global-Trans. Then, in ⑦, the threads will fetch the transitions from GlobalTrans again and start a new iteration.

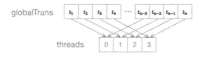

**Fig. 3.** Parallel transitions transferring

For transitions transferring, we should ensure that all of the threads can fetch transitions simultaneously. Figure 3 shows how threads get transitions from GlobalTrans. Assuming that there are n transitions in GlobalTrans, namely $t_1, t_2, ..., t_n$, and 4 threads in the GPU with thread numbers from 0 to 3. Threads with thread number from 0 to 3 share GlobalTrans. For example, when fetching the transitions, thread 0 takes the transitions $t_1, t_5, t_9, ..., t_{n-3}$, thread 1 takes the transitions $t_2, t_6, t_{10}, ..., t_{n-2}$, thread 2 takes the transitions $t_3, t_7, t_{11}, ..., t_{n-1}$, and thread 3 takes the transitions $t_4, t_8, t_{12}, ..., t_n$. They do not interact with each other and can get their respective transitions at the same time.

The transition generation involves many iterations, and each iteration requires a global synchronization. In this type of problem, one has to consider how the threads in different blocks communicate with each other. Instead of using locks, we take advantage of the new feature cooperative group of CUDA 9 to synchronize threads. There is a flag in each thread to mark whether the thread has completed its iteration. When all of the threads complete their respective iterations, there will be a global synchronization which moves all of the Local-Trans to GlobalTrans.

### 3.2 The Usage of ParaMoC

We provide ParaMoC as an executable file, so that it can be directly used without having to compile it. It requires at least CUDA9.0 and JDK1.8 for the operation of ParaMoC. The tool ParaMoC[1] is available for evaluation. It is possible to download the command-line version, together with the usage and some examples.

## 4 Experiment

This section gives the experimental evaluation of ParaMoC and its application in malware detection.

---

[1] ParaMoC is available at https://sites.google.com/view/ParaMoC.

## 4.1   Performance Evaluation

As a benchmark, we compared our tool to Moped version 1.0.16. The push-down systems are translated from Java programs. The tested data are from the DaCapo Benchmarks [2], which is a set of general purpose, realistic, freely available Java applications. We chose parts of the programs and then use Jimpleto PDSolver [6] to translate real Java programs into PDSs. Our experiments were conducted on a Linux platform (Ubuntu 16.04) with Intel i7-7820X CPU and NVIDIA GTX 1080TI GPU. To ensure the fairness of the experiment, the price of the CPU is similar to the price of the GPU. The compute capability of the GPU is 6.1 based on the Pascal architecture. The version of CUDA used is CUDA9.0. We mark the execution time of computation on both ParaMoC and Moped.

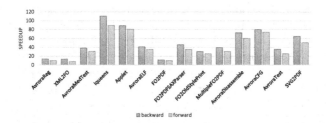

**Fig. 4.** Speedups of reachability analysis

The experimental results of reachability analysis are presented in Fig. 4. In Fig. 4, the y-axis is the speedup ratio, and the x-axis is the pushdown systems. We compare the performance of ParaMoC on the GPU with Moped on the CPU. The baseline for speedup is Moped on the CPU. Based on the performance comparison, our parallel tool is very efficient for the backward and forward reachability analysis of PDSs. In terms of these PDSs, our tool achieves up to 100X acceleration.

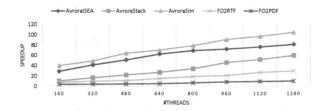

**Fig. 5.** Speedups of LTL model checking

Figure 5 shows the scalability of the multi-thread acceleration using the speedups of the ParaMoC to Moped. The y-axis is the speedup ratio of LTL model checking, and the x-axis is number of threads. Because there are 32 threads per warp in CUDA, the number of threads we used is a multiple of 32. In Fig. 5, the speedups of five different benchmarks using different numbers of threads are compared. Obviously, all of them demonstrate robust scalability, indicating that the speedup grows as the number of threads increases. The experimental results show that the LTL model checking of PDSs can be pleasingly parallelized with multi-thread.

## 4.2   Application

Many well-known malwares use stack operations for adding useless push and pop instructions, or hiding calls to the operating system. Using pushdown systems as program model allows to consider malwares' stack. LTL can express many malicious behaviors more precisely. For example, there is a statement that "A register r1 is assigned by 0, and then the content of this register is pushed onto the stack." This statement can be expressed in LTL as $mov(r1, 0) \land push(r1)$.

Our parallel model checker can be used to solve malware detection problems. We use the method proposed in [11] to model the binary program as PDS, where PDS's control locations correspond to the control points of the program, and the PDS's stack mimics the execution stack of the program. This method uses disassemble tools to get the control flow graph of the binary program and translate the control flow graph into a pushdown system. The LTL formulas we used to describe the malicious behaviors comes from [12].

**Table 1.** Detection of some real programs

| Dataset | #LOC | Time (s) | Result |
| --- | --- | --- | --- |
| Virus.Win32.Anar.b | 671 | 0.11 | Yes |
| Email-Worm.Win32.Mydoom.y | 26902 | 20.28 | Yes |
| Trojan-PSW.Win32.LdPinch.aar | 1245 | 18.59 | Yes |
| Trojan-PSW.Win32.LdPinch.ld | 6609 | 1.61 | Yes |
| Virus.Win32.Alcaul.b | 904 | 2.83 | Yes |
| Virus.Win32.Agent.ce | 8951 | 15.26 | Yes |
| shutdown.exe | 2524 | 13.87 | No |
| Regsvr32.exe | 1280 | 18.56 | No |
| Cmd.exe | 35887 | 45.95 | No |
| Java.exe | 21868 | 39.10 | No |

We carried out some experiments to detect some real binary programs from VX Heavens and Microsoft WindowsXP system, the results of which are presented in Table 1. Column Dataset shows the name of the program we checked. Column #LOC denotes the number of instructions in the program. The running times are given in seconds. The result Yes denotes that the program is detected as a malware. Otherwise, the result is No. As it can be seen, our tool can correctly perform malware detection.

## 5 Conclusions

In this paper, we presented a new parallel model checker called ParaMoC, which enhances the performance of model checking for PDSs with the GPU. Compared with the sequential model checker for the same purpose in the literature, our tool runs faster. For future study, one problem that we are aware of is the limited memory size of a single GPU device. We are going to solve this limitation by employing our approach on multi-GPUs. We also plan to integrate the unpacking tool into ParaMoC so that our tool can perform malware detection directly.

**Acknowledgements.** This work is partially supported by Shanghai Science and Technology Committee Rising-Star Program (No. 18QB1402000), National Natural Science Foundation of China (No. 61602178), China HGJ Project under Grant (No. 2017ZX01038102-002), and National Defense Basic Scientific Research Program of China (No. JCKY2016204B503).

## References

1. Almasi, G.S., Gottlieb, A.: Highly parallel computing (1988)
2. Blackburn, S.M., et al.: The DaCapo benchmarks: Java benchmarking development and analysis. In: ACM Sigplan Notices, vol. 41, pp. 169–190. ACM (2006)
3. Bošnački, D., Leue, S., Lafuente, A.L.: Partial-order reduction for general state exploring algorithms. Int. J. Softw. Tools Technol. Transf. **11**(1), 39–51 (2009)
4. Clarke, E.M., Mcmillan, K.L., Campos, S.V.A., Hartonasgarmhausen, V.: Symbolic model checking, pp. 419–427 (1993)
5. Esparza, J., Hansel, D., Rossmanith, P., Schwoon, S.: Efficient algorithms for model checking pushdown systems. In: Emerson, E.A., Sistla, A.P. (eds.) CAV 2000. LNCS, vol. 1855, pp. 232–247. Springer, Heidelberg (2000). https://doi.org/10.1007/10722167_20
6. Hague, M., Ong, C.-H.L.: Analysing mu-calculus properties of pushdown systems. In: van de Pol, J., Weber, M. (eds.) SPIN 2010. LNCS, vol. 6349, pp. 187–192. Springer, Heidelberg (2010). https://doi.org/10.1007/978-3-642-16164-3_14
7. Holzmann, G.J.: The model checker spin. TSE **23**(5), 279–295 (1997)
8. Nvidia: Nvidia CUDA compute unified device architecture (2010). https://developer.nvidia.com/cuda-toolkit/
9. Reps, T., Schwoon, S., Jha, S., Melski, D.: Weighted pushdown systems and their application to interprocedural dataflow analysis. SCP **58**, 206–263 (2005)
10. Schwoon, S.: Model-checking pushdown systems, pp. 73–84 (2002)
11. Song, F., Touili, T.: Efficient malware detection using model-checking. In: Giannakopoulou, D., Méry, D. (eds.) FM 2012. LNCS, vol. 7436, pp. 418–433. Springer, Heidelberg (2012). https://doi.org/10.1007/978-3-642-32759-9_34
12. Song, F., Touili, T.: LTL model-checking for malware detection. In: Piterman, N., Smolka, S.A. (eds.) TACAS 2013. LNCS, vol. 7795, pp. 416–431. Springer, Heidelberg (2013). https://doi.org/10.1007/978-3-642-36742-7_29
13. Tarjan, R.: Depth-first search and linear graph algorithms. SIAM J. Comput. **1**(2), 146–160 (1972)

# FastDRC: Fast and Scalable Genome Compression Based on Distributed and Parallel Processing

Yimu Ji[1,4,5,6], Houzhi Fang[1], Haichang Yao[1,2(✉)] [iD], Jing He[3],
Shuai Chen[1], Kui Li[1], and Shangdong Liu[1]

[1] School of Computer Science, Nanjing University of Posts
and Telecommunications, Nanjing 210023, China
2017040238@njupt.edu.cn
[2] School of Computer and Software, Nanjing Institute of Industry Technology,
Nanjing 210023, China
[3] School of Software and Electrical Engineering,
Swinburne University of Technology, Melbourne 3122, Australia
[4] Institute of High Performance Computing and Big Data,
Nanjing University of Posts and Telecommunications, Nanjing 210003, China
[5] Nanjing Center of HPC China, Nanjing 210003, China
[6] Jiangsu HPC and Intelligent Processing Engineer Research Center,
Nanjing 210003, China

**Abstract.** With the advent of next-generation sequencing technology, sequencing costs have fallen sharply compared to the previous sequencing technologies. Genomic big data has become the significant big data application. In the face of growing genomic data, its storage and migration face enormous challenges. Therefore, researchers have proposed a variety of genome compression algorithms, but these algorithms cannot meet the processing requirements for large amount of biological data and high processing speed. This manuscript proposes a parallel and distributed referential genome compression algorithm-Fast Distributed Referential Compression (FastDRC). This algorithm compresses a large number of genomic sequences in parallel under the Apache Hadoop distributed computing framework. Experiments show that the compression efficiency of the FastDRC is greatly improved when it compresses large quantities of genomic data. Moreover, FastDRC leads to the only distributed computing method known to us in the field of genome compression. The source code for FastDRC can be obtained from this link: https://github.com/GhostCCCatHenry/FastDRC.

**Keywords:** Genome compression · Distributed processing · Apache Hadoop · FASTA

## 1 Introduction

Nowadays, high-throughput genome sequencing technology has gradually matured, sequencing efficiency continues to increase while sequencing cost are gradually reduced. In the context of this development, genomic data accumulates at an

© Springer Nature Switzerland AG 2020
S. Wen et al. (Eds.): ICA3PP 2019, LNCS 11945, pp. 313–319, 2020.
https://doi.org/10.1007/978-3-030-38961-1_27

exponential rate. Faced with such a large amount of data, how to store them will become an important problem for researchers to solve [1].

Compression is the most direct and effective way to solve the problem. The genome compression algorithm based on the reference sequence effectively improves compression ratio. These algorithms select a sequence file as the reference file, and only the difference between the original file and the reference file is stored in the compressed file.

The FASTA format is a text-based format representing nucleotide or polypeptide sequences [2]. In this format, each sequence text will have a single-line description in the front, followed by multiple lines of sequence information, where the nucleic acid or amino acid information will be represented by a single letter. Compression for the FASTA format is the focus of genomic data compression.

## 2  Problem Statement and Methodology

At present, a large number of referential compression algorithms based on FASTA format sequences have been proposed [3–5]. In addition, until now FASTA format genome compression algorithms are all stand-alone algorithms. However, when dealing with terabytes of gene files, serial compression algorithms need to take dozens of hours to complete the compression of these files, which reduces the practicability of the compression algorithm. Therefore, the use of big data technology to achieve the compression of genomic big data is one of the research focuses in this field.

In this manuscript, a fast and scalable genome compression based on distributed and parallel processing (FastDRC) is proposed. The Hadoop's MapReduce programming paradigm [6] is used to improve FastRC. Use Yarn to isolate resources to adjust resource distribution parameters to achieve different degrees of optimization. Maximize the working potential of Hadoop to achieve optimal use. The detailed process and experimental verification are described in the following parts.

## 3  Method

### 3.1  Distributed Method

As a reference sequence compression algorithm, FastDRC is mainly divided into three steps: reading the sequence files, matching the compressed sequence, and serializing of the matching result in binary format. The entire calculation process is completed only in the Map tasks, and the Reduce task is deserted. Therefor the entire compression process is completed in the memory of the same node and the disk shuffle can be avoided.

In order to ensure high utilization of resources, the decisive factors of the degree of parallelism is further refined from the number of nodes. Yarn uses Linux's CGroups mechanism to divide each node resource into many containers. These computing containers do not affect each other. In summary, the number of calculations of FastDRC equals the total number of divided containers. The degree of parallelism

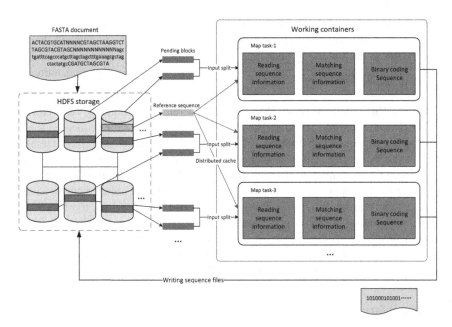

**Fig. 1.** FastDRC implementation process. Firstly, the gene files in FASTA format are stored in the HDFS, and these files are divided into blocks for storage [7]. When the MapReduce task is executed, these file blocks are divided into input splits and sent to the working container. All file scheduling follows the FIFO principle. After being compressed, the files output as binary files and they are saved to HDFS again.

determines the number of containers that execute the compression algorithm at the same time. The target files will be entered into their respective containers through the index queue for calculation. The overall process is shown in Fig. 1.

### 3.2 Implementation Steps

**Sequence Information Extraction.** Since these gene files are stored in HDFS, which is a distributed abstract storage system, it is necessary to take into account the size setting for input slice and the cache configuration of input files when processing input files.

*Map Input Slice.* In HDFS, large sequence files will be divided into several data blocks. If the capacity of the input splits is not adjusted before the Map task, the input slice will be processed by different Map tasks. Since the processing is independent of each other, it will eventually lead to compression errors. Before input, the range of the capacity of each input split $S_{split}$ shall be $S_{file} \leq S_{split} < S_{block} + S_{file}$, where $S_{block}$ represents the size of a data block in HDFS, and $S_{file}$ represents the size of the file to be compressed. If the size of the input single-sequence file exceeds the size of each HDFS block, the Map input fragmentation capacity should be adjusted to a multiple of the block size during the compression process. This ensures that all sequences of blocks can be completely read.

*MapReduce Distributed Cache.* Since the reference sequence is shared by all nodes, in order to reduce transmission costs, MapReduce distributed cache is adopted. Sending file on HDFS to each working node, and caching it in a specific directory as an index file. In this way, the number of network transmissions is only the total number of nodes.

*Information Extraction Process.* The extraction of reference sequence only takes the A, C, G, T characters into account. The lowercase characters of the sequences shall be converted into uppercase characters. The positions of lowercase characters that appear will be stored in a vector. This part is done in the Setup function of the Map class. Next, the sequence to be compressed is read as a line-by-line iteration process via the Map function from the input files. Replace the default TextInputFormat method with the KeyValueInputFormat. When <key, value> is divided for each line, the former does not need to calculate the offset position of each line as the latter, thereby reducing the operation time. The reading mode of the uppercase characters and the lowercase is similar to the reference sequence. The remaining character information (special characters such as N characters) is extracted using the same extraction method as lowercase characters.

**Sequence Information Matching.** At this stage, the procedure uses the static character array $G_{target}$ and $G_{reference}$ for greedy matching, not involving data iteration and is mainly done in the last custom function Cleanup in the Map class.

**Binary Encoding.** To output binary in MapReduce, it is necessary to encode the input in binary using the SequenceFile format. The serialized encapsulation format of the output is set to BytesWritble. Experimental Verification and Analysis.

## 4    Experimental Environment and Data

The experiment selects 6 servers for cluster construction and distributed computing. All working nodes are configured as 32 GB RAM with two six-core Intel(R) Xeon(R) CPU E5-2620 v2 2.10 GHz and a 3 TB hard disk and responsible for storing data blocks and performing calculations.

The experiment selects the current largest genome dataset 1000 Genome Project [8] full-quantity data, a total of 1092 human genome data. We also selected hg13, hg16, hg17, hg18, hg19, hg38, K131, K224, YH and HuRef as experimental subjects. The above 10 data are the standard test data of the gene compression algorithm.

### 4.1    Experimental Results and Analysis

FastDRC is an extensible algorithm that can increase the speed of operation by scale-out of cluster nodes and vertical expansion of computing resources of a single node. The Yarn [9] resource scheduling platform can be used to increase the number of working containers. This algorithm regulates the Yarn resource allocation in the Hadoop configuration file, and maximize cluster load. Degree of parallelism is set to

35. The parameters that can be used for improvement in terms of the compression algorithm are the K-mer read length $k$ and the hash length threshold of matched sequence $m$. In the phase of matching, $k$ is used to reduce the number of comparisons, and $m$ is used to ensure the length of the matched sequence. But too high $k$ and $m$ will have a negative impact on performance. After a lot of experiments, take $k$ as 12 and $m$ as 30. Then, overall comparison was done using the optimal parameters of FastDRC between FastDRC algorithm and FastRC algorithm. Finally, FastDRC is compared with other mainstream algorithms ERGC [4], HiRGC [10] for standard data sets.

**Overall Comparison.** This part is performed on the two methods in terms of compression time and compression ratio. The experimental compression test data set selects the chr1–8 chromosome of 1091 people in 1000 Project, and the reference sequence selects the chromosome of human with number HG0096. The experimental results are shown in Table 1. From this table, we can see that the compression time and the compression ratio of FastDRC has been greatly improved compared with FastRC (Table 1).

**Table 1.** Overall comparison of FastDRC and FastRC.

| Chromosome number | Compression time (s) | | Relative compression speed gain | Compression ratio | | Relative compression ratio gain |
|---|---|---|---|---|---|---|
| | FastDRC | FastRC | | FastDRC | FastRC | |
| Chr1 | 2538 | 61511 | 24.2 | 276 | 163 | 1.7 |
| Chr2 | 3194 | 55435 | 17.4 | 273 | 159 | 1.7 |
| Chr3 | 2086 | 42478 | 20.4 | 247 | 147 | 1.7 |
| Chr4 | 1959 | 28912 | 14.8 | 241 | 145 | 1.7 |
| Chr5 | 1920 | 29353 | 15.3 | 261 | 158 | 1.7 |
| Chr6 | 2157 | 26569 | 12.3 | 232 | 140 | 1.7 |
| Chr7 | 2300 | 25269 | 11.0 | 260 | 159 | 1.6 |
| Chr8 | 1789 | 20357 | 11.4 | 240 | 153 | 1.6 |

**Comparison with Popular Compression Algorithms.** To verify the competitiveness of FastDRC, compare it to the performance of today's mainstream algorithms. The experimental compression test data set selects 1000Project and standard performance test data sets. The reference sequence selects the chromosome sequence file with human number hg13. The experimental results are shown in Table 2. The experimental results show that FastDRC has far surpassed the two algorithms in compression time. However, in terms of compression ratio, it only surpasses ERGC, and there is still a gap with HiRGC.

**Table 2.** Comparison with HiRGC and ERGC.

| Chromosome number and original size (MB) | Compression time (s) | | | Compressed size (MB) | | |
|---|---|---|---|---|---|---|
| | FastDRC | HiRGC | ERGC | FastDRC | HiRGC | ERGC |
| Chr1(265231.7) | 4107 | 43023 | 92898 | 3129 | 2475 | 5760 |
| Chr2(258893.9) | 3685 | 37064 | 82602 | 2979 | 2365 | 4889 |
| Chr3(210796.8) | 3150 | 21441 | 75184 | 3508 | 430 | 3035 |
| Chr4(203284.7) | 1860 | 18395 | 52556 | 1885 | 1272 | 2625 |
| Chr5(192592.7) | 1801 | 19837 | 34552 | 1719 | 1187 | 2449 |
| Chr6(181202.1) | 1380 | 16263 | 21614 | 1104 | 619 | 1272 |
| Chr7(169370.1) | 1430 | 17457 | 24129 | 1327 | 892 | 1840 |
| Chr8(155750.5) | 1718 | 15850 | 21784 | 2127 | 1605 | 2638 |
| Chr9(150259.3) | 1330 | 12953 | 18387 | 1478 | 1001 | 1640 |
| Chr10(144293.6) | 1206 | 12524 | 15354 | 876 | 541 | 884 |
| Chr11(143680.6) | 1473 | 13455 | 19085 | 1470 | 1070 | 1755 |
| Chr12(142399.7) | 2422 | 24895 | 22362 | 3105 | 2651 | 4374 |
| Chr13(122600.1) | 780 | 9616 | 7628 | 600 | 333 | 626 |
| Chr14(114222.3) | 891 | 9409 | 10717 | 652 | 390 | 615 |
| Chr15(109143.7) | 945 | 11900 | 7558 | 1050 | 767 | 1104 |
| Chr16(96191.3) | 1748 | 15960 | 10459 | 1984 | 121 | 2294 |
| Chr17(86397.2) | 1838 | 19664 | 12285 | 2375 | 1784 | 3385 |
| Chr18(83120.9) | 860 | 9232 | 8026 | 660 | 402 | 758 |
| Chr19(62903.2) | 1175 | 10293 | 6505 | 278 | 222 | 415 |
| Chr20(67082.5) | 995 | 7901 | 4703 | 374 | 164 | 308 |
| Chr21(51209.9) | 479 | 4863 | 4984 | 698 | 420 | 795 |
| Chr22(54541.3) | 560 | 5540 | 8845 | 339 | 297 | 560 |
| ChrX(163753.4) | 1376 | 18625 | 20521 | 2107 | 1568 | 2967 |
| ChrY(29133.1) | 347 | 2357 | 2831 | 77 | 85 | 167 |

## 5   Conclusion

This manuscript proposes a genome compression algorithm improved by the big data parallel computing framework. This algorithm compensates for the gap in the field of parallel genome compression, and uses the computing resource cluster to shorten the original tens of hours of compression time to only tens of minutes, which greatly improves the efficiency of genome file compression. Through improving the algorithm, the compression ratio is improved. However, there is still much space for improvement. In the face of different data sets, the compression ratio of the algorithm will decrease, and in terms of compression performance, there is a gap between high-end algorithms such as HiRGC. In the next phase, it is necessary to improve these issues and think about the introduction of the Reduce method for secondary compression.

**Acknowledgements.** We would like to thank all reviewers for their valuable comments and suggestions to improve the quality of our manuscript.

**Funding.** This work was supported by the National Key R&D Program of China [2017YFB1401302, 2017YFB0202200], the National Natural Science Foundation of P. R. China [No. 61572260, 61872196], Outstanding Youth of Jiangsu Natural Science Foundation [BK20170100], Key R&D Program of Jiangsu [BE2017166], Postgraduate Research & Practice Innovation Program of Jiangsu Province [KYCX19_0906, KYCX19_0921], The Natural Science Foundation of the Jiangsu Higher Education Institutions of China [19KJD520006] and Modern Educational Technology Research Program of Jiangsu Province in 2019 [2019-R-67748].

# References

1. Kahn, S.D.: On the future of genomic data. Science **331**(6018), 728–729 (2011)
2. Pearson, W.R.: Rapid and sensitive sequence comparison with FASTP and FASTA. Methods Enzymol. **183**(1), 63–98 (1990)
3. Xie, X., Zhou, S., Guan, J.: CoGI: towards compressing genomes as an image. IEEE/ACM Trans. Comput. Biol. Bioinform. **12**(6), 1275–1285 (2015)
4. Deorowicz, S., Grabowski, S., Ochoa, I., et al.: ERGC: an efficient referential genome compression algorithm. Bioinformatics **31**(21), 3468–3475 (2015)
5. Wandelt, S., Leser, U.: FRESCO: referential compression of highly similar sequences. IEEE/ACM Trans. Comput. Biol. Bioinform. **10**(5), 1275–1288 (2014)
6. Wu, X.-D., Ji, S.-W.: Comparative study on MapReduce and spark for big data analytics. J. Softw. **29**(6), 1770–1791 (2018)
7. Shvachko, K., Kuang, H., Radia, S., Chansler, R.: The hadoop distributed file system. In: Proceedings of the 2010 IEEE 26th Symposium on Mass Storage Systems and Technologies (MSST), MSST 2010, pp. 1–10. IEEE Computer Society, Washington, DC (2010)
8. Abecasis, G.: The 1000 genomes project consortium. An integrated map of genetic variation from 1,092 human genomes. Nature **491**, 56–65 (2012)
9. Vavilapalli, V.K,, Murthy, A.C., Douglas, C., Agarwal, S., Konar, M., Evans, R., et al.: Apache hadoop YARN: yet another resource negotiator. In: Proceedings of the 4th Annual Symposium on Cloud Computing, p. 5. ACM, New York (2013)
10. Liu, Y.S., et al.: High-speed and high-ratio referential genome compression. Bioinformatics **33**(21), 3364–3372 (2017)

# A Parallel Approach to Advantage Actor Critic in Deep Reinforcement Learning

Xing Zhu$^{(\boxtimes)}$ and Yunfei Du

School of Data and Computer Science, Sun Yat-sen University, Guangzhou, China
zhux25@mail2.sysu.edu.cn

**Abstract.** Deep Reinforcement learning (DRL) algorithms recently still take a long time to train models in many applications. Parallelization has the potential to improve the efficiency of DRL algorithms. In this paper, we propose an parallel approach (ParaA2C) for the popular Actor-Critic (AC) algorithms in DRL, to accelerate the training process. Our work considers the parallelization of the basic advantage actor critic (Serial-A2C) in AC algorithms. Specifically, we use multiple actor-learners to mitigate the strong correlation of data and the instability of updating, and finally reduce the training time. Note that we assign each actor-learner MPI process to a CPU core, in order to prevent resource contention between MPI processes, and make our ParaA2C approach more scalable. We demonstrate the effectiveness of ParaA2C by performing on Arcade Learning Environment (ALE) platform. Notably, our ParaA2C approach takes less than 10 min to train in some commonly used Atari games when using 512 CPU cores.

**Keywords:** Deep reinforcement learning · Advantage actor critic · Parallelization · MPI · Scalable

## 1 Introduction

Reinforcement learning (RL) is an efficient approach able to solve sequential decision making problems. The problems are described as such a process that an agent continuously interacts with an environment by trial-and-error until reaching the target state. Deep reinforcement learning (DRL) combines deep neural networks (DNN) and RL algorithms, to become a very attractive direction. Recently for many applications and models, DRL algorithms still take a long time to train.

Specifically, the popular Actor-Critic (AC) algorithms in DRL combine value-based and policy-based methods that are widely mentioned. Among these AC algorithms, the basic advantage actor critic (Serial-A2C) [7] uses sequential observed data to learn, and it is strongly correlated as Serial-A2C is an online-learning-based DRL algorithm. In addition, due to the instability of generated data, the model updating is unstable in Serial-A2C. In order to mitigate the instability and the correlation, the previous work focused on the parallelization

© Springer Nature Switzerland AG 2020
S. Wen et al. (Eds.): ICA3PP 2019, LNCS 11945, pp. 320–327, 2020.
https://doi.org/10.1007/978-3-030-38961-1_28

and algorithmic optimization, such as asynchronous A2C (A3C) [9], and optimization approaches like PPO [11].

In order to further improve the algorithmic efficiency, we present ParaA2C, a parallel approach for serial-A2C. Our ParaA2C has the ability to reduce the training time significantly, as well as provide a good scalability for online-learning-based AC algorithms on a CPU cluster. In ParaA2C approach, we use multiple parallel actor-learners, where each actor interacts with an environment instance and each learner learns a model replication. Each actor-learner has an unique MPI process running on a CPU core. In this way, ParaA2C enables each actor-learner to fully utilize the computing capability of a CPU core, further exposing a high degree of parallelism. With 32 nodes and each node has 16 MPI processes, our ParaA2C approach can play some commonly used Atari games in less than 10 min.

## 2  Related Work

Asynchronous Serial-A2C (A3C) [9] performs the asynchronous parallel training for several days on a multi-core CPU. GPU-based A3C (GA3C) [3] accelerates the A3C using a GPU and performs training for over a dozen hours. Batched A3C (BA3C) [2] is a CPU implementation of GA3C and has similar performance. Distributed-BA3C [1] could play Atari games in 21 min, when using 64 distributed CPU nodes to train, with each node performs BA3C. A2C implemented in OpenAI baselines [6] is a synchronous Serial-A2C algorithm, which has similar performance with A3C. Parallel Advantage Actor Critic (PAAC) [5] is a GPU-based algorithm outperforming previous GA3C, but also takes more than ten hours to train. Further, Adam Stooke and Pieter Abbeel [12] evaluate the previous Serial-A2C-based algorithms. They perform the experiments on the state-of-the-art NVIDIA DGX-1 platform which has 40 CPU cores and 8 GPUs, and enable the training to coverage in about 10 min.

According to the above, the training of Serial-A2C-based algorithms can be accelerated on CPUs or GPUs. In this paper, we regard Serial-A2C algorithm as an entirety, and consider its parallelization. As the batch size of Serial-A2C is too small, our approach is not suitable for performing on a GPU. Thus the proposed ParaA2C scales Serial-A2C on a modern CPU cluster. At best, the algorithm is trained to converge in less than 10 min when adopting 512 CPU cores.

## 3  Our ParaA2C Approach

### 3.1  Coarse-Grained Parallelization

The implementation of Serial-A2C algorithm has two stages: sampling and training. In sampling stage, an actor interacts with an environment simulator and generates interactive data. In training stage, a learner trains the model network

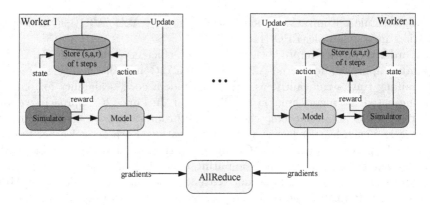

**Fig. 1.** The multiple actor-learners parallelization. Each worker executes Serial-A2C, and communicates gradients with each other.

using the interactive data. Both sampling and training stages can be accelerated by parallelization. Our work uses multiple actors to interact with multiple environment simulators, in order to generate a larger batched input data and accelerate the sampling stage. In addition, we use multiple learners to accelerate the training stage by learning multiple model replications with data parallelism.

In our proposed ParaA2C, we regard Serial-A2C algorithm as an entirety, and consider its coarse-grained parallelization. Specifically, we adopt multiple actor-learners and make each actor-learner as a worker to perform a Serial-A2C algorithm separately. Different from OpenAI baselines [6] that synchronizes the samples of all actors, ParaA2C synchronizes the gradients calculated by all actor-learner workers, and each worker has a model replication for inference and training.

**Multiple Actor-Learners.** ParaA2C assigns an actor and a learner to a worker, and uses multiple workers to perform training as shown in Fig. 1. In each worker, an actor uses a model replication to generates $t_{max}$-step interactive samples, and a learner in turn uses these samples to calculate its gradients. The gradients require to be communicated between all workers. Then new gradients are applied to the parameters of each model replication.

At each iteration, a worker $i$ generates a batch data $B_i$ consisting of $t_{max}$ samples to calculate its gradients $\bigtriangledown J_i(\theta)$. After completing an iteration, $N$ workers adopt All-Reduce patterns to communicate $\bigtriangledown J_i(\theta)$, and apply new gradients $\bigtriangledown J(\theta) = \frac{1}{N} \sum_{i=1}^{N} \frac{1}{t_{max}} \sum_{x \in B_i} \bigtriangledown l(x, \theta) = \frac{1}{Nt_{max}} \sum_{i=1}^{N} \sum_{x \in B_i} \bigtriangledown l(x, \theta)$ to update parameters. when $B = \{B_1, ..., B_N\}$, $\bigtriangledown J(\theta) = \frac{1}{Nt_{max}} \sum_{x \in B} \bigtriangledown l(x, \theta)$. It is intuitive that using $N$ workers and processing $t_{max}$ samples per worker is equivalent to one worker to copy with $Nt_{max}$ samples. Thus model replications on all workers are all the same at every iteration, further implementing the data parallelism.

**Algorithm 1.** ParaA2C algorithm

---

Set $P$ MPI processes.

Initialize model network parameters $\theta$.

Process 0 broadcasts its parameters to other processes.

Initialize total interaction steps $T_{total}$ and number of steps per iteration $t_{max}$.

Set number of updates $N_{update} = T_{total} / (t_{max} \times P)$.

**for** $j \in [1, N_{update}]$ **do**

    **for** $t \in [1, t_{max}]$ **do**

        Input $s_t$ into the model and get state value $v_t$

        Select an action $a_t$

        Execute the action $a_t$ and get reward $r_t$ and next observation $s_{t+1}$

        store $(s_t, a_t, r_t, v_t)$

    **end**

    Input $s_{t+1}$ into the model and store state value $v_{t+1}$

    **for** $t \in [t_{max}, 1]$ **do**

$$g_{gae} = \begin{cases} r_t - v_t, & s_{t+1} \text{ is terminal} \\ r_t + \gamma v_{t+1} - v_t + \gamma\lambda g_{gae}, & s_{t+1} \text{ is non-terminal} \end{cases}$$

$$R_t = g_{gae} + v_t$$

    **end**

    Compute loss function $l_i(s, \theta)$ using $R_t$

    Compute gradients $\triangledown J_i(w) = \frac{1}{t_{max}} \sum_{s \in B_i} \triangledown l_i(s, \theta)$

    Communicate gradients between $P$ processes using All-Reduce

    All $p$ processes get gradients $\triangledown J(\theta) = \frac{1}{Pt_{max}} \sum_{i=1}^{P} \sum_{s \in B_i} \triangledown l_i(s, \theta)$

    Update $\theta$ using $\triangledown J(\theta)$ and RMSprop optimizer

**end**

---

### 3.2 Scalability

To explore the scalability by using multiple CPU cores, we use $P$ actor-learner MPI processes in our ParaA2C approach. Each actor-learner MPI process performs Serial-A2C algorithm and communicates gradients after completing an iteration. The implementation details are shown in Algorithm 1.

Note that if we assign an actor-learner MPI process to a CPU core, it is able to handle $t_{max}$ samples quickly, and prevent resource contention between MPI processes. Especially, when $t_{max}$ is empirically set to 5, the utilization rate of a CPU core is close to 100%. In general, the value of $t_{max}$ will not be too large for most Serial-AC-based algorithms which use temporal-difference learning. If $t_{max}$ is too large, some instant rewards used for updating may be neutralized, and will cause a same problem with policy gradient methods. But if $t_{max}$ is too small, we cannot make full use of the computing capability of a CPU core in our experiments.

Our approach that assigns a MPI process to a CPU core has a good scalability. Once we increase the number of CPU cores, the batch size is enlarged, and will not affect the calculation of other cores. When 16 actor-learners work on a CPU node, we can scale our ParaA2C algorithm to 32 CPU nodes (512 cores), and at best, the algorithm can be trained to converge in less than 10 min during our experiments.

## 4   Evaluation

We perform the experiments on a CPU cluster to evaluate the performance of the ParaA2C approach. The ParaA2C approach is implemented using C++ programming language in the PyTorch framework. The CPU cluster has 32 nodes and each node has 2 Intel Xeon 12-core CPUs. We adopt environment simulators from the commonly used Arcade Learning Environment (ALE) platform [4] that includes over 60 Atari 2600 games. Especially, the preprocessing of data from ALE framework is described in [10]. We invoke many parallel workers to train a model synchronously, and each worker performs a actor-learner MPI process. For scalable comparisons, we use 16 cores per node, and assign a worker to a core in our experiments. In each ALE game, we carried out 6 experiments respectively, with the number of nodes $N_{node}$ doubling from 1 to 32, and the number of environment simulators $n_e$ ranging from 16 to 512 correspondingly. When $t_{max}$ set to 5, the batch size $bs$ increases from 80 to 2560.

### 4.1   Learning Rate Study

As the same to the previous Serial-A2C-based approaches, our ParaA2C approach is relatively sensitive to learning rate(LR) $\eta$. At first, we select two commonly used methods from distributed deep learning to adjust the learning rate of ParaA2C. Linear scaling rule increases LR with the batch size [8]. Sqrt scaling rule increases LR with the square root of the batch size [8]. Experimental results show that the sqrt scaling rule is better than the linear scaling rule in our ParaA2C approach. However, when the number of CPU cores scales to 256 or 512, the sqrt scaling rule cannot provide performance improvements in ParaA2C approach. Thus we conduct several experiments to adjust the learning rate $\eta$, and find that when we set $\eta = 0.001$ and keep it fixed, the performance is more stable and the acceleration is more significant. For comparison, we finally use $\eta = 0.001$ in all our experiments for scalable performance.

### 4.2   Performance Study

In order to get significant performance improvement and explore scalable parallelization, we compare the results of each ALE game with 16, 32, 64, 128, 256, and 512 CPU cores, respectively. Generally, for a fixed number of data for updating, when the batch size is larger, the number of iteration will be smaller. This is the advantage of data parallelism as the training time will be shorter. But this results in a scalability bottleneck as the optimal value may be neglected in the gradient descent. Thus we next focus on two performance metrics: one is number of iterations and the other is training time.

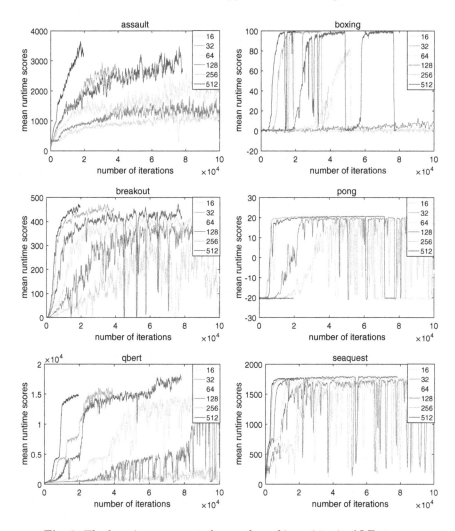

**Fig. 2.** The learning curves vs the number of iterations in ALE games.

Figure 2 shows the learning curves vs the number of iterations in ALE games, with the number of CPU cores ranging from 16 to 512. Experiment results show that when the number of CPU cores increases, our ParaA2C approach takes the less number of iterations $Iter_{max}$ to converge to a feasible solution. This is because of the increasement of batch size in the ParaA2C approach. With using 512 CPU cores, $Iter_{max}$ reduces more than four times comparing with that using 16 CPU cores. Thus under the current settings, our ParaA2C approach has a good performance speedup with the increasing number of CPU cores.

Figure 3 shows the learning curves vs training time in ALE games. Experiment results show that the more the number of CPU cores, the shorter the training time to reach the same optimal runtime reward, that shows a good

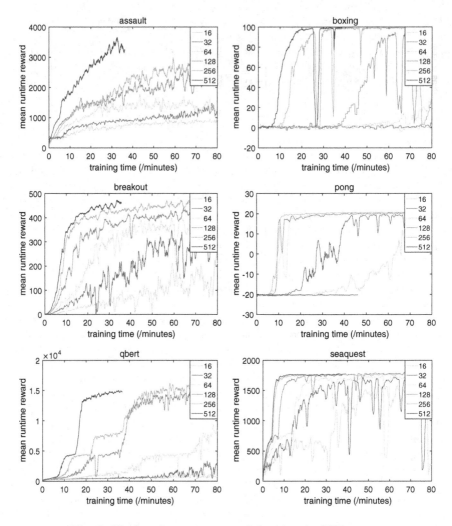

**Fig. 3.** The learning curves vs training time in ALE games.

scalability. In some better cases, the ParaA2C approach only takes about 10 min to complete the steepest learning phase and enter the convergence phase using 256 or 512 CPU cores. Especially with the *seaquest* game, it converges in less than 10 min. In addition, all test ALE games in our experiments retain a higher score in the process of scaling ParaA2C to more CPU cores, while suffering from a bottleneck or a sharp performance decline at a certain threshold.

## 5    Conclusions

In this paper we present ParaA2C, a parallel approach for Serial-A2C that provides the scalable performance speedup. Our ParaA2C approach is data parallelism by leveraging multiple actor-learners. Specifically, we assign each actor-learner MPI process to a CPU core, in order to prevent resource contention between MPI processes. At each iteration of training, these actor-learners communicate gradients using All-Reduce of MPI, and then apply new gradients to their own model parameters. When scaling to 32 nodes (512 CPU cores), our ParaA2C approach retains the best performance, that it takes less than 10 min to train in some ALE games to generate an optimized result.

**Acknowledgements.** This research was supported by the Natural Science Foundation of China under Grant NO. U1811464 and the Program for Guangdong Introducing Innovative and Enterpreneurial Teams under Grant NO. 2016ZT06D211.

# References

1. Adamski, I., Adamski, R., Grel, T., Jędrych, A., Kaczmarek, K., Michalewski, H.: Distributed deep reinforcement learning: learn how to play atari games in 21 minutes. In: Yokota, R., Weiland, M., Keyes, D., Trinitis, C. (eds.) ISC High Performance 2018. LNCS, vol. 10876, pp. 370–388. Springer, Cham (2018). https://doi.org/10.1007/978-3-319-92040-5_19
2. Adamski, R., Grel, T., Klimek, M., Michalewski, H.: Atari games and intel processors. In: Cazenave, T., Winands, M.H.M., Saffidine, A. (eds.) CGW 2017. CCIS, vol. 818, pp. 1–18. Springer, Cham (2018). https://doi.org/10.1007/978-3-319-75931-9_1
3. Babaeizadeh, M., Frosio, I., Tyree, S., Clemons, J., Kautz, J.: Reinforcement learning through asynchronous advantage actor-critic on a GPU. arXiv preprint arXiv:1611.06256 (2016)
4. Bellemare, M.G., Naddaf, Y., Veness, J., Bowling, M.: The arcade learning environment: an evaluation platform for general agents. J. Artif. Intell. Res. **47**, 253–279 (2013)
5. Clemente, A.V., Castejón, H.N., Chandra, A.: Efficient parallel methods for deep reinforcement learning. arXiv preprint arXiv:1705.04862 (2017)
6. Dhariwal, P., et al.: Openai baselines (2017). https://github.com/openai/baselines
7. Konda, V.R., Tsitsiklis, J.N.: Actor-critic algorithms. In: Advances in Neural Information Processing Systems, pp. 1008–1014 (2000)
8. Krizhevsky, A.: One weird trick for parallelizing convolutional neural networks. arXiv preprint arXiv:1404.5997 (2014)
9. Mnih, V., et al.: Asynchronous methods for deep reinforcement learning. In: International Conference on Machine Learning, pp. 1928–1937 (2016)
10. Mnih, V., et al.: Human-level control through deep reinforcement learning. Nature **518**(7540), 529 (2015)
11. Schulman, J., Wolski, F., Dhariwal, P., Radford, A., Klimov, O.: Proximal policy optimization algorithms. arXiv preprint arXiv:1707.06347 (2017)
12. Stooke, A., Abbeel, P.: Accelerated methods for deep reinforcement learning. arXiv preprint arXiv:1803.02811 (2018)

# Applications of Distributed and Parallel Computing

# Blockchain-PUF-Based Secure Authentication Protocol for Internet of Things

Akash Suresh Patil, Rafik Hamza, Hongyang Yan, Alzubair Hassan, and Jin Li[✉]

School of Computer Science and Cyber Engineering, Guangzhou University, Guangzhou 51006, People's Republic of China

**Abstract.** Devices constituting the Internet of Things (IoT) have become widely used, therein generating a large amount of sensitive data. The communication of these data across IoT devices and over the public Internet makes them susceptible to several cyber attacks. In this paper, we propose an efficient blockchain approach based on the secret computational model of a physically unclonable function (PUF). The proposed framework aims to guarantee authentication of the devices and the miner with a faster verification process compared to existing blockchain techniques. Furthermore, the combination of the blockchain and PUF allows us to propose an efficient framework that guarantees data provenance and data integrity in IoT networks. The proposed framework employs PUFs, which provide unique hardware fingerprints for establishing data provenance. Smart contracts based on the blockchain provide a decentralized digital ledger that is able to resist data tampering attacks.

**Keywords:** Blockchain · Physically Unclonable Function · Authentication · Internet of Things · Data integrity

## 1 Introduction

The rapid development and evolution of miniaturization and electronics as well as the massive deployment of communications and networking technologies have bestowed unprecedented advances onto the world [1]. This trend has emerged in a number of electronic devices in every field of work, reducing human effort and increasing the cost of productivity. This rapid development is facilitating a shift toward the digital world.

The Internet ensures fast and efficient communication, which facilitates societal advancement. In recent decades, digitalization has seen significant progress, with developments that can be achieved and implemented through the Internet and through the concept of the Internet of Things (IoT). IoT has emerged as an encapsulate of various technologies, from radio frequency identification to wireless sensors networks to physical sensors [2]. Indeed, IoT equipment with

© Springer Nature Switzerland AG 2020
S. Wen et al. (Eds.): ICA3PP 2019, LNCS 11945, pp. 331–338, 2020.
https://doi.org/10.1007/978-3-030-38961-1_29

micro-controllers, digital communication transreceivers and protocol stacks that allow communication has become an integral part of the Internet. IoT devices can be applied in many research areas as electronic devices, from wearable devices to physical hardware development platforms [3,4].

Security issues have become the most challenging problem facing IoT [5]. It is essential to provide security because IoT systems are directly involved in human safety. A large number of IoT devices are connected in a system and are not managed by a single controller [6]. The design of security protocols is complicated due to the aforementioned issues.

Most security protocols are reliable for the Internet; however, they are not satisfactory for IoT systems [7–9]. Modern security protocols must be resistant to physical and side channel attacks, in addition to preserving anonymity and privacy. Additionally, modern security protocols must be efficient when used in IoT devices because they have very low computational, power and memory resources [10–12]. Thus, new security protocols and frameworks are required for establishing a secure and reliable IoT system.

In this paper, we present a blockchain-based architecture for IoT security. Moreover, we propose an authentication framework between IoT devices and miners in a blockchain network. Our work contributes to achieving identity authentication, access control, replay attack resistance, DOS attack resistance and data integrity without incurring overhead or delays. The unique features of the physically unclonable function (PUF) offer hardware security for IoT devices because a PUF carries a unique identification number for every chip at the time of manufacture, thereby offering data provenance.

In the next section, we will present the background of the presented material and challenges. In Sect. 3, we present our system architecture and the proposed framework's information work-flow. In Sect. 4, Security evaluations are discussed. Finally, the conclusions are given in Sect. 5.

## 2   Background and Related Works

### 2.1   Physical Unclonable Function (PUF)

The PUF is defined as a digital fingerprint that offers unique identification for semiconductor devices such as microprocessors. PUFs are based on unique physical variations developed during manufacturing. In short, a PUF is a physical entity embodied in a physical structure [13]. A PUF is based on the concept that even though the mask and manufacturing process are the same for every integrated circuit (IC), each IC is quite different from other ICs due to normal manufacturing variability. PUFs hold this variability to derive secret information that is unique to the chip. PUFs are promising novel primitives that can be used for secret key storage and authentication without the need for expensive hardware [14,15]. PUFs derive their secrecy from the physical characteristics of the IC. Thus, there is no need to store secrets in digital memory.

## 2.2  Blockchain

The first record of blockchain technology came in 2008 by mysterious founders using the name Satoshi Nakamoto [16].

Basically, a blockchain is a time-stamped chain of blocks jointly maintained by every participating node. Each block is chained together cryptographically. Blocks are digitally signed and chained to a previous block using their hash values. Blockchain technology is completely distributed, restricted to the given contractual code, autonomous and fully traceable [17].

## 2.3  Challenges

The IoT environment encapsulates millions of devices, and each device should be authenticated to the network before establishing communication. Because there is no human intervention in an IoT system, every IoT device should be equipped with a way to identify and authenticate themselves. However, modern techniques require secret credentials to be stored in the device memory. Unfortunately, these modern techniques are not well suited for the physically unprotected devices that are part of IoT systems [18]. An adversary may use various physical attacks to manipulate the entire IoT system.

Another issue relates to physical and cloning attacks, where an adversary may attempt to imitate a genuine and authenticated IoT device by cloning other IoT devices by extracting secret information [15]. The main intention of an attacker is to manipulate and access the IoT data sent by the other IoT devices. An attacker can eavesdrop on the communication by introducing a new message, change or replay messages, or establish other identities.

# 3  Proposed Framework Based on PUF Model

## 3.1  System Architecture

In this section, we introduce our system architecture equipped with various entities, such as the IoT devices, blockchain networks and the data owner, as shown in Fig. 1.

In our system, we have three participants: the IoT devices, the blockchain network and the data user.

- Various physical objects are merged to become smart objects, along with sensors and actuators, allowing the collection and processing of data from the real world.
- A peer-to-peer network, where every node may be a high-resource device. This network imbues our system with a distributed nature. Every node in the blockchain acts as a server/miner and maintains the history of all transactions. The blockchain is the mechanism that allows transactions to verify and provide the distributed, immutable, transparent, auditable and secure features. Additionally, every node of the blockchain network is responsible for data storage and providing the required computations.

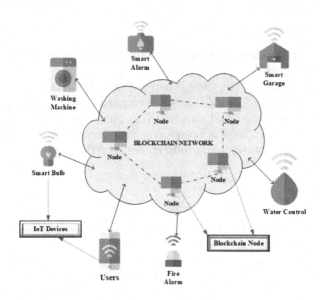

**Fig. 1.** The proposed system architecture

– A data user is an entity who has authority to provide their own personal data. They have full control over their data and set data access policies for intended purposes only. By adopting the blockchain and PUFs, data integrity, data provenance, and data tampering resistance can be achieved for the owner of the data.

## 3.2    Enrollment Phase

First, the IoT devices will request a node/miner in the blockchain network for registration. Then, this node/miner will store the registration details in local storage in a database. After successfully storing these details, the node/miner will be approved by the IoT devices (see Fig. 2). Simultaneously, the node/miner will broadcast the registration to the whole blockchain network, which will be confirmed to the user.

## 3.3    Verification Phase

When a user wants to interact with the node/miner, the user has to request permission; the node/miner will query this permission from local storage. Then, the node/miner will broadcast the request signed by the node/miner to the blockchain network. Once the blockchain network verifies the request, the user can interact with the IoT devices (see Fig. 3).

Our proposed work presents a unique approach to authentication using blockchain based smart contracts along with the PUF model, which offers a dense solution for secure authentication. Basically, our proposed solution will interact

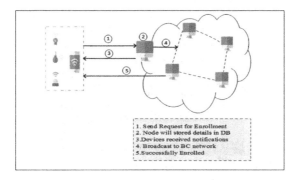

**Fig. 2.** Enrollment phase

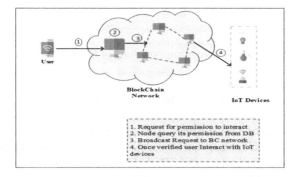

**Fig. 3.** Verification phase

with a blockchain based smart contract to ensure safe and secure communication. A smart contract is designed in such a way that the data coming from the IoT devices will interact with the distributed nodes in the blockchain network. Initially, all devices will need to enroll following their respective device ID and secret computational PUF model with the smart contract. Figure 4 presents the information flow of the entire IoT system.

## 4   Analysis

In this section, we illustrate some perspectives on evaluating the proposed PUF-based blockchain.

1. **Faster verification process**

   Existing PUF-based blockchain techniques require storing lists of challenge-response pairs (CRPs) in the database on the verifier side [19]. Therefore, collecting new CRPs from devices and storing them will become an exhaustive task and require unnecessarily storage and computations, especially under resource-constrained environments. We intend to overcome these problems by

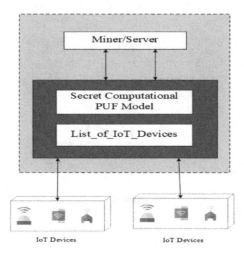

**Fig. 4.** The information flowchart of the IoT system.

proposing a new approach with minimally intensive processes on the database on the verifier side.

2. **Authentication of the devices and the miner**

   In the proposed PUF-blockchain model, IoT devices are authenticated with miner nodes after validating the *ID, MAC address, and secret computational model (PUF model)* parameters. The challenge-response protocol concatenation with hash ID transactions that will be verified with the stored hash *IDs* on the miner side at the time of enrollment. This guarantees a secure authentication of the devices in IoT networks.

   The server/miner will perform authentication using the device parameters that have been established in the previous transactions. The miner node will receive an *ID* from the users, and accordingly, it will send challenges or reject their requests. The parameters received from legitimate devices are checked and used to decide whether to continue or terminate the process.

3. **Data provenance and data integrity**

   The use of PUFs and the features of the blockchain ensure data provenance and data integrity for IoT environments [20]. The blockchain employs hash functions that confirm data integrity, while the unique *ID* in the PUF model guarantees data provenance for each IoT device. The blockchain holds a list of transactions between users, which are user data that are robust to tampering. This provides immunity from impersonation attacks. The blockchain offers an immutable chain of records; all the data transmitted previously are validated and then stored in the blockchain in such a way that these data cannot be altered by attackers.

4. **Resistance against replay attacks**

   Every step of the timestamp is used to check the data freshness, including all hash lists when transmitting the transactions. Thus, the proposed framework is well protected against replay attacks and can withstand such attacks.

5. **Resistance against man-in-the-middle attacks**

The proposed scheme guarantees resistance against man-in-the-middle attacks. The sender is validate and verified by the receiver before processing any transactions.

Additionally, the security evaluation of the proposed framework illustrates a high level of security. The smart contract stores the full list of registered device IDs, MAC addresses and PUF models during the enrollment phase. Thus, it is nearly impossible for attackers to apply well-known attacks such as denial of service attacks, distributed denial of service attacks, and impersonation attacks. The smart contract maintains the list of registered IoT devices and the respective PUF model, along with the MAC addresses of the IoT devices, thereby offering trustworthiness for user access policies and user data usage records.

## 5 Conclusion

In this paper, we present an emerging technology: the PUF-based blockchain. The proposed framework ensures authentication for users and data integrity in IoT systems. Our proposed method represents an efficient and secure method for interaction between IoT devices and a miner in a blockchain network. Distributed ledgers and smart contracts carried out on the blockchain guarantee data integrity and user privacy. Cryptographic operations are implemented to enhance the blockchain protocol and ensure secure, efficient and more reliable authentication protocols. The proposed PUF-based blockchain can be employed as an efficient solution to preserve data integrity and facilitate authentication as well as reduce the computation power for IoT devices as PUF equipped in it.

**Acknowledgment.** This work was supported by National Natural Science Foundation of China (No. 61702125, 61702126).

## References

1. Lu, Y., Xu, L.D.: Internet of Things (IoT) cybersecurity research: a review of current research topics. IEEE Internet Things J. **6**(2), 2103–2115 (2019)
2. Bedi, G., Venayagamoorthy, G.K., Singh, R., Brooks, R.R., Wang, K.: Review of Internet of Things (IoT) in electric power and energy systems. IEEE Internet Things J. **5**(2), 847–870 (2018)
3. Ikpehai, A., et al.: Low-power wide area network technologies for Internet-of-Things: a comparative review. IEEE Internet Things J. **6**(2), 2225–2240 (2019)
4. Udoh, I.S., Kotonya, G.: Developing IoT applications: challenges and frameworks. IET Cyber-Phys. Syst.: Theor. Appl. **3**(2), 65–72 (2018)
5. Frustaci, M., Pace, P., Aloi, G., Fortino, G.: Evaluating critical security issues of the iot world: present and future challenges. IEEE Internet Things J. **5**(4), 2483–2495 (2018)
6. Arif, M., Wang, G., Wang, T., Peng, T.: SDN-based secure VANETs communication with fog computing. In: Wang, G., Chen, J., Yang, L.T. (eds.) SpaCCS 2018. LNCS, vol. 11342, pp. 46–59. Springer, Cham (2018). https://doi.org/10.1007/978-3-030-05345-1_4

7. Patil, A.S., Tama, B.A., Park, Y., Rhee, K.-H.: A framework for blockchain based secure smart green house farming. In: Park, J.J., Loia, V., Yi, G., Sung, Y. (eds.) CUTE/CSA -2017. LNEE, vol. 474, pp. 1162–1167. Springer, Singapore (2018). https://doi.org/10.1007/978-981-10-7605-3_185

8. Granjal, J., Monteiro, E., Sá Silva, J.: Security for the Internet of Things: a survey of existing protocols and open research issues. IEEE Commun. Surv. Tutorials **17**(3), 1294–1312 (2015)

9. Muhammad, K., Hamza, R., Ahmad, J., Lloret, J., Wang, H., Baik, S.W.: Secure surveillance framework for IoT systems using probabilistic image encryption. IEEE Trans. Ind. Inf. **14**(8), 3679–3689 (2018)

10. Nguyen, V., Lin, P., Hwang, R.: Energy depletion attacks in low power wireless networks. IEEE Access **7**, 51915–51932 (2019)

11. Arif, M., Wang, G., Balas, V.E.: Secure vanets: trusted communication scheme between vehicles and infrastructure based on fog computing. Stud. Inform. Control **27**(2), 235–246 (2018)

12. Hamza, R., Yan, Z., Muhammad, K., Bellavista, P., Titouna, F.: A privacy-preserving cryptosystem for IoT e-healthcare. Inf. Sci. (2019)

13. Herder, C., Yu, M., Koushanfar, F., Devadas, S.: Physical unclonable functions and applications: a tutorial. Proc. IEEE **102**(8), 1126–1141 (2014)

14. Gao, Y., Ma, H., Abbott, D., Al-Sarawi, S.F.: Puf sensor: exploiting puf unreliability for secure wireless sensing. IEEE Trans. Circ. Syst. I: Regul. Pap. **64**(9), 2532–2543 (2017)

15. Mukhopadhyay, D.: Pufs as promising tools for security in Internet of Things. IEEE Des. Test **33**(3), 103–115 (2016)

16. Nakamoto, S.: Bitcoin: a peer-to-peer electronic cash system. http://bitcoin.org/bitcoin.pdf

17. Casino, F., Dasaklis, T.K., Patsakis, C.: A systematic literature review of blockchain-based applications: current status, classification and open issues. Telematics Inf. **36**, 55–81 (2019)

18. Mukherjee, A.: Physical-layer security in the Internet of Things: sensing and communication confidentiality under resource constraints. Proc. IEEE **103**(10), 1747–1761 (2015)

19. Javaid, U., Aman, M.N., Sikdar, B.: Blockpro: blockchain based data provenance and integrity for secure IoT environments. In: BlockSys@SenSys (2018)

20. Liang, X., Shetty, S., Tosh, D., Kamhoua, C., Kwiat, K., Njilla, L.: Provchain: a blockchain-based data provenance architecture in cloud environment with enhanced privacy and availability. In: 2017 17th IEEE/ACM International Symposium on Cluster, Cloud and Grid Computing (CCGRID), pp. 468–477, May 2017

# Selective Velocity Distributed Indexing for Continuously Moving Objects Model

Imene Bareche$^{(\boxtimes)}$ ⓘ and Ying Xia

School of Computer Science and Technology,
Chongqing University of Posts and Telecommunications, Chongqing 400065, China
imen.1993.bareche@gmail.com

**Abstract.** The widespread of GPS embedded devices has lead to a ubiquitous location dependent services, based on the generated real-time location data. This introduced the notion of continuous querying, and with the aid of advanced indexing techniques several complex query types could be supported. However the efficient querying and manipulation of such highly dynamic data is not trivial, processing factors of crucial importance should be carefully thought out such as accuracy and scalability. In this study we focus on Continuous KNN (CKNN) queries processing, one of the most well-know spatio-temporal queries over large scale of continuously moving objects. In this paper we provide an overview of CKNN queries and related challenges, as well as an outline of proposed works in the literature and their limitations, before getting to our contribution proposal. We propose a novel indexing approach model for CKNN querying, namely VS-TIMO. The proposed structure is based on a selective velocity partitioning method, since we have different objects with varying speeds. Our structure base unit is a comprised of a non overlapping R-tree and a two dimensions grid. In order to enhance performances, we design a compact multi-layer index structure on a distributed setting, and propose a CKNN search algorithm for accurate results using a candidate cells identification process. We provide a comprehensive vision of our indexing model and the adopted querying technique.

**Keywords:** Continuous KNN querying · Moving objects indexing · Distributed spatio-temporal indices · Velocity partitioning based index · Parallel processing

## 1 Introduction

Nowadays, we are experiencing a rapid growth in the scale of generated spatio-temporal data, which has inspired a series of new services called Location Aware Services (LAS); referring to location-related requests regardless to user' location, for example searching the list of available hotels in a city. Location dependent services LDS are a sub-class of LAS, intended to address people's social and societal needs based on their positioning data [1], for example searching the

© Springer Nature Switzerland AG 2020
S. Wen et al. (Eds.): ICA3PP 2019, LNCS 11945, pp. 339–348, 2020.
https://doi.org/10.1007/978-3-030-38961-1_30

nearest hospital to user' location. LDS has attracted substantial interests in both academia and industry. Business talking it has accumulated a huge collection of data of which we could make benefit in different real-world services to boost economy and improve people' life and safety. LDS are designed by Location Dependent Queries (LDQ), the result set will vary according to the user' location that might be moving. LDQs processing is more challenging in terms of accuracy since they include both the spatial and the temporal component over moving objects. Besides the response frequency feature, because objects might be static or moving leading to two main types of queries namely snapshot or continuous queries respectively [1]. Many studies in academia have addressed the problem of efficiently answering such a time-parametrized and position-related requests. However the research is not yet well established and still need to be investigated. Researchers have targeted two main points: indexing efficiency and querying efficiency. Various indexing structures and querying approaches have been proposed for different types of LDQ among which KNN queries. The indexing and querying of a snapshot KNN query has been extensively addressed in the last decades, while continuous KNN and LDQ indices in general still have more technical issues to deal with. The first technique dealing with the CKNNs was proposed in [3], it was designed for centralized setting such in most of existing works and can not be deployed in a distributed architecture. This is mainly due to the continuous updates process inspired from the snapshot KNN querying, which drastically reduces performances. Despite the rising number of spatial distributed platforms and frameworks, the continuous querying under distributed setting is not yet well addressed.

## 1.1   Contribution

Based on these observations we propose a distributed hybrid index for continuous queries namely CKNN and trajectory multi-attribute query. The proposed index called VS-TIMO (Velocity Saptio-Temporal Index for Moving Objects) aims to handle the problems of accuracy and querying complexity cost in a distributed setting. Our index is implemented in parallel distributed master-sleeve paradigm. Calculations over sleeve nodes are parallelized to reduce the complexity cost then summarized and sent to the master node for final results compilation. Our main contribution can be summarized as follows:

- We propose a velocity partitioning of objects based on selective similar velocities equation, leading to two distinct views of the sub-space.
- We design a region-based indexing for continuously moving objects in each view within a distributed setting.
- We implement a two layered compact index model, described in two base structures: tree-like and grid-based layout in a simple way to benefit from both at once. We integrate a multi-attribute trajectory querying structure on the top of the index.
- We propose an algorithm for dynamic CKNN query candidates selection.

The remainder of this paper is organized as follows: Sect. 2, consists of the research motivation, the corresponding main factors to consider, and an overview of some related works. In Sect. 3, we present the proposed index structure and the proper querying technique in details. Section 4 Concludes the paper with directions on the future works to be after implementation to enhance performances.

## 2  Background, Challenges and Related Works

In this section we first point the motivation and the background of MO processing, we discuss some basic concepts and major challenges; then we outline existing works. Ground intelligent transportation systems, Business Coverage Analysis, Location-based recommendation are examples of LDS applications over moving objects supported by complex LDQs, by means of advanced indexing techniques and novel appropriate spatial processing platforms. The complexity of location-related systems in general and continuously MO processing above all, is related to several factors, including: (a) Data nature and variability: besides of its significant amount the indexed data are highly volatile much more in continuous context. (b) Data' auto-correlation: since changes in the spatio-temporal component of objects occur smoothly over time. Besides the CKNN related works limitations including: (c) Velocity uncertainty: most of works on MO uncertain velocity focus exclusively on Euclidean spaces [11,13], but when considering objects within a road network the location accuracy [6] involves more repetitive query re-evaluation. (d) The undefined number of CKNN search iterations: most of search algorithms are based on unknown number of iterations to locate the region of the KNN, which results in an extra communication cost [11] in a distributed setting. (e) Data skewness: the skew distribution of data over nodes reduces the performances due to the nonuniform distribution over space. Proposed indexing paradigms to alleviate the complexity of MO querying can be classified into three main categories according to the index structure base unit: Tree-like indices, Grid-based indices, Hybridized indices; In regard to the target data model MO indexing techniques have been addressed under two primary categories: Euclidean space MO indexing and MO indexing within real-world road network. In this latter, we index the road network information, besides assuming movement patterns and calculating the distance between objects based on the road segments and not the euclidean distance equation. R-tree variants indices are very common in state-of-the-art and known for large amount of data support, and fast search but high maintenance cost. R-tree-like indices are based on the use of MBRs (Minimum Bounding Rectangle) for spatial information, then extended to fit the mobility nature of data by introducing velocity component like in TPR-tree [2] and its variants, which is a modified R*-tree used to index moving objects for KNN queries. The main drawbacks in this approach were the unconditional expansion of MBRs, and the maintenance cost. This latter was addressed later in TPR*-tree [3] while the former remained unsolved, until authors in [4] proposed $B^x$-tree, and have also investigated MO frequent updates. However the structure was sensitive to data skeweness and the querying approach reduced the accuracy. Unlike previous approaches authors in [11] consider MO within a road network constraints for CKNN queries, a two-layered index structure that indexes road information. Other approaches have investigated the problem of the repeated KNN search technique using an R-tree like in [6]which unfortunately lead to an expensive search process. Due to its simple and clear representation and low update cost, grid-based structures for MO continuous querying are numerous. Grid-based indices [1,7,9] divide the sub-space into units designed as vertical and/or horizontal splits that indexes the enclosed objects. Generally the KNN objects search processing begins from the query cell then accesses the next closest cells, in a rotating fashion drawing a circle with the query issuer as a centre. This process is performed repetitively for unknown number of iterations until K objects are found, which is one of the main drawbacks of grid indexing. In recent years we had a proliferation of big data distributed systems and cloud-streaming solutions such as Hadoop MapReduce [15], Apache Storm [16], S4 [17], and Apache Spark [18]. These systems support big amounts of data and allow fast distributed processing, therefore various indexing methods [1,7–9] are built on the top of it to inherit the scalability and fault-tolerance features of the underlying frameworks [13].

## 3    Proposed Indexing Technique Outline

In this section we explain our proposed indexing structure model, the underlying concept and used techniques.

### 3.1    Index Structure

We design a multi-layer indexing structure based on selective similar velocities partitioning as a first step. We represent the top layer in an R-tree structure, and the bottom is grid of 2 dimensions on a distributed cluster of nodes, because Tree-like structures are suitable for large amounts of objects and provides fast search, and grid-based ones are appropriate for frequent updates and outperform indexing trees in terms of maintenance cost. The second step is building the region based index, defined as an R-tree with minimal number of MBRs for root and non leaf nodes with no overlap. Combined with a grid based structure for leaf nodes to index enclosed objects with a low cost and a simple representation. On the other hand since we are tracking objects movement and visited regions, we integrate a storage model for multi-attribute trajectory querying in a supplementary structure in the master node, described in a hash-map representation for simple and fast retrieval.

**CKNN Querying' Velocity Selective Region-Based Structure.** In this section we present the proposed selective indexing technique, which will be performed into 3 steps as follows:

1. *Objects analysis:* the system will analyze velocities to classify objects into two distinct classes fast objects and slow objects. The selective classification (As shown in Fig. 1) is according to the predefined equation that calculates the standard deviation $\sigma$ of objects velocity in real-time, based on the dispersion of speed values as defined in Eq. (2). This partitioning will save the cost of unnecessary access to objects remained in the same region, and reduce search time and split/merge cost.

Let $n$ be the number of objects at any given time t and $v_i$ the objects velocities vector:

$$S = \frac{\sum\limits_{i=1}^{n} v_i^2}{n} - \left(\frac{\sum\limits_{i=1}^{n} v_i}{n}\right)^2 \tag{1}$$

$$\sigma = \sqrt{S} \tag{2}$$

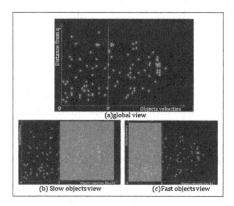

(a)global view

(b) Slow objects view          (c)Fast objects view

**Fig. 1.** Example of objects partitioning according to velocities into two views.

2. *Region-split:* after partitioning, we have two views the system will index objects of each one in a VS-TIMO index. The sub-space will be sliced into regions of interest: a minimalist number of vertical equal-sized splits according to the X axis. A split should satisfy the condition: min.max $\geq$ k where min(max) is the predefined minimum(maximum) number of objects per region to make sure that there are at least k nearest neighbors in the candidate split and avoid data skewness.
3. *Split indexing:* the indexing hybrid multi-layer structure is defined as a tree-like structure on the top layer and a grid-based one on the bottom.

The process is as follows: (1) a customized R-tree: a minimalist R-tree is created to index the root and the non-leaf nodes per splits, MBRs number is set to a minimum proper value to avoid overlapping. (2) a grid structure for leaf

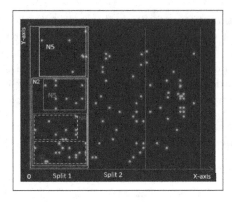

**Fig. 2.** Region-based splitting for indexing.

nodes: we divide the MBR space into M*M grid of equal size $\alpha$ squares (cells), based on a 2-D array that stores the set of objects within cells. Each cell C is denoted by C(pnt, i, j) corresponding to its row and column indices and the pointer to a list of buckets, referring to the enclosed objects meta-data. From the pointers in the bucket we access to information like the objects data, contained objects number and the next bucket. Given an object O at time t preliminary represented by the a triple tuple (id, (Ox, Oy), t) where Ox and Oy define its spatial position at time t, we can easily know in which cell it falls to start the kNN search by checking whether O' spatial components satisfy inequalities in (3) (Fig. 2):

Let $OC(t) = \{o_1, o_2, \ldots o_n\}$ be the set of moving objects that fall within the cell $C$ at any given time $t$,

$$O \in OC \text{ if and only if } \begin{cases} i * \alpha \leq Ox(t\,) \leq (i+1) * \alpha \\ \quad\quad \& \\ j * \alpha \leq Oy(t\,) \leq (j+1) * \alpha \end{cases} \tag{3}$$

The update algorithm in VS-TIMO is handled as a sequential companion of two processes: delete and insert, similar to the update process used for a moving objects database in [10], but adapted to our continuous context. To be the most responsive to the CKNN querying requirement the data layout is defined as shown in Fig. 3 we include time, velocity information and MO type in data representation thus the objects will be represented by the tuple *(id, (Ox, Oy), v, t, type)*. We also define a time interval $\Delta t$ for receiving new locations to update the index, to guarantee a compact data representation without loss of information on the one hand. And to have less space and calculations, and less but larger time units on the other hand, similarly to the proceedings of packing method used over trajectories databases for continuous tracking such as in [3].

**Multi-attribute Trajectory Query Key-Based Supplementary Structure.** Besides the described region-based structure for CKNN queries, we imple-

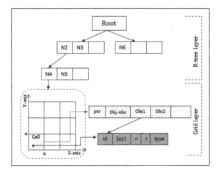

**Fig. 3.** The basic structure of the index.

ment a supplementary key-based structure on the top of VS-TIMO for users' multi-attribute trajectory retrieval, an emerging query type investigated in a set of proposals [5,12,14,19], it includes semantic attributes giving more querying possibilities for a better understanding of objects behaviour [3]; Since users tend to be of different types (for example taxi, person, or a bus) we integrate a hash map to store a key-value structure since hash maps are suitable for small amounts of data and support frequent changes. It stores in records users visited regions with other attributes that specifies some related data such as user type as follow: *(id, [trajectory], type)* . The hash map applies a straightforward search method that is a linear scan, hence the time complexity of a query is *O(m)*.

## 3.2   CKNN Query Processing

For a simple CKNN query processing we first don't make any assumption about objects movement pattern, and consider them in an euclidean space, we perform the query over the R-trees and combine the result sets at the end. The KNN query search method in conducted in an incremental way starting from the query issuer q in both views (low and high velocity views) following a filter-and-refine paradigm.we perform a minimum number of iterations to search the nearest objects, we detect the $\zeta$ candidate cells by identifying the nearest buckets that contain at least k neighbours all along a time interval $\Delta t$ such that $\zeta.\epsilon \geq k$ where $\epsilon$ is a minimum threshold. Then we take the distance $r$ between $q$ and the $k$-th nearest neighbour we take $q$ as a center and $r$ as a radius and draw the first iteration circle. Since we deal with continuously moving objects in real-time the results set of a query q sent at $t_0$ should be in coherence with data in response time $t_1$ ( $t_1 \geq t_0$). To achieve this we use a linear representation introduced in various approaches [4,5] to model the mobility as a linear function of time, and predict the position x of an object at any given time t denoted as $x(t)$ based on its position $x_{ref}$ at a time reference denoted as $t_{ref}$ and the velocity vector $v$ as defined in Eq. (4).

$$x(t) = x_{ref} + v(t - t_{ref}) : t_{ref} < t \qquad (4)$$

CKNN query search is performed over the grid cells to determine candidate objects easily as presented in Algorithm 1, and use Lemma 1 for objects selection.

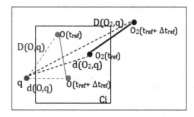

**Fig. 4.** Example of selecting objects candidates of a cell Ci.

---

**Algorithm 1** Candidate cells objects selection

**Input:** A set $MO$, grid cell $Ci$, a threshold number $\tau$ K, and a time interval $\Delta$ T

**Output:** A set of candidate objects $CCo$

1: **for** (each $Oi$ in $MO$) **do**
2:    compute $D(Oi,Ci)$;

3:    sort objects in ascending order of $D(Oi,Ci)$;
4: **end for**
5: dist-threshold ← The $\tau$ K-th object smallest $D(Oi,Ci)$;

6: **for** (each $Oi$ in $MO$) **do**
7:    compute $D(Oi,Ci)$;

     $(D(Oi,Ci) \leq$ dist-threshold) insert $Oi$ into $CCo$
8: **end for**
9: **return** $CCo$;

---

**Lemma 1.** *Given a grid candidate cells $Cs = \{Ci : 1 \leq i \leq m\}$. In Fig. 4 and for intuitive understanding we take a simple example of only one cell $Ci$; Let $Co$ be the enclosed objects set within cells $Cs$ and a CKNN query q within the time interval $[t_{ref}, t_{ref} + \Delta t]$ : $Co \subseteq Cs$. Let $PCo(t_{ref})$ and $PCo(t_{ref} + \Delta t)$ be the positions vector of objects in $Co$ at time $t_{ref}$ and $(t_{ref} + \Delta t)$ respectively when objects move according to the related velocities vector. If the position of object $O \in Co$ occurs in both $PCo(t_{ref} + \Delta t)$ and $PCo(t_{ref})$.i.e. the object still fills in the candidate cells all along the time interval, then we can say that it belongs to $CCo$ the query candidate objects set: $O \in CCo \implies O \in Co$. In other words the candidate objects set $CCo$ for a query is a subset of the enclosed objects in the whole grid candidate cells $Cs$ denoted as $Co : CCo \subseteq Co$.*

*Proof.* Let $O$ be an object within a candidate cell $Ci : O \in Ci$. As shown in Fig. 4, the positions $O(t_{ref})$ and $O(t_{ref}+\Delta t)$ fill in $Ci$ all along $[t_{ref}, t_{ref}+\Delta t]$. Because all positions along the line segment $\overline{(O(t_{ref} + \Delta t)O(t_{ref}))}$ are fully enclosed by $Ci$ during the time interval $[t_{ref}, t_{ref} + \Delta t]$ we have: (a) The minimal possible value of distance $d(O, q)$ between an object $O$ and the query issuer $q$ must be greater than or equal to the minimal possible value distance $d(Ci, q)$ between $q$ and cell $Ci$: $d(O, q) \geq d(Ci, q)$; and (b)The maximal possible distance value $D(O, q)$ between $O$ and $q$ must be less than or equal to the maximal possible distance value $D(O, Ci)$ between $O$ and $Ci$ : $D(O, q) \leq D(O, Ci)$. Proof by contradiction: Let's assume that a query candidate object is not enclosed in candidate cells $Ci$, suppose that an object $O'$ is a candidate object from the grid for query $q$ but does not belong to any of the candidate cells set $(O' \notin Cs)$ such that: $O' \in CCo$ & $O' \notin Co$. Intuitively there exist at least a number of $\tau$ K objects such that ($Ci$ is a candidate cell): $D(O, q) \leq D(Ci, q)$ & $d(O,q) \geq d(Ci, q)$, and whose maximal possible distances from $q$ $D(O, q)$ are less than the minimal distance $d(O', q)$ since it does not fill in Cs. Accordingly O' cannot be a candidate object for the query $q$ and does not belong to the set of query candidate objects $CCo$ which leads to a contradiction with the assumed statement $O O' \in CCo$. This proves that $O \in Co$ must be satisfied and cells candidates enclose the objects candidates set.

## 4   Conclusion

In this paper we have conducted a study on CKNN queries, an important research axis in nowadays world. We presented the research background and motivation to have an overview of the continuous querying field. we have highlighted the limitations of indexing approaches of state-of-the-art under distributed settings, such as data skeweness, search iterations unknown number for CKNN queries and the maintenance cost to support objects mobility. Then, we presented the model of the proposed indexing approach VS-TIMO; which introduces first a velocity objects distinction step followed by a region-based splitting. In order to maintain a compact data representation, we have defined a hybrid two-layered structure: R-tree+Grid. We have also introduced a processing technique similar to the trajectories packing method, and defined a CKNN query candidates selection method for scalable accurate querying. Finally, and with a view to enhance performances and parallelize processing tasks in a scalable manner; we envision to embed our index structure in an appropriate spatial platform to inherit the advantages of a scalable distributed processing system. Apache Spark is one of the most used distributed systems, that introduces the concept of partitioning and RDDs to store data, and seems to be suitable for our distributed parallel processing scenario.

**Acknowledgments.** This work was financially supported by the Natural Science Foundation of China (41571401).

# References

1. Yang, M., Ma, K., Yu, X.: An efficient index structure for distributed k-nearest neighbours query processing. Soft Comput. **22**, 1–12 (2018)
2. Tao, M.Y., Papadias, D., Sun, J.: The TPR*-tree: an optimized spatio-temporal access method for predictive queries. In: Proceedings of the 29th VLDB Conference, vol. 29, February 2013
3. Lee, J., Hong, B., Hong, J., Kim, C., Kim, C.W.: Optimal index partitioning of main-memory based TPR*-tree for real-time tactical moving objects. In: IEEE International Conference on Big Data and Smart Computing, January 2018
4. Jensen, C., Lin, D., Ooi, B.C.: Query and update efficient B-tree based indexing of moving objects. In: Proceedings of the 30th International Conference on Very Large Data Bases VLDB Endowment, October 2004
5. Parent, C., Spaccapietra, S., Renso, C., et al.: Semantic trajectories modeling and analysis. ACM Comput. Surv. **45**(4), article 42 (2013)
6. Tao, Y., Papadias, D., Shen, Q.: Continuous nearest neighbor search. In: International Conference on Very Large Databases VLDB, August 2002
7. Xiong, X., Mokbel, M. F., Aref, W.: SEA-CNN: scalable processing of continuous k-nearest neighbor queries in spatio-temporal databases. In: International Conference on Data Engineering (ICDE), pp. 643–654, May 2005
8. Yu, Z., Yu, X., Pu, K.Q., Liu, Y.: Scalable distributed processing of k nearest neighbor queries over moving objects. IEEE Trans. Knowl. Data Eng. **4347**(c), 1–14 (2015)
9. Zhang, F., Zheng, Y., Xu, D., Du, Z., Wang, Y., Liu, R.: Real-time spatial queries for moving objects using storm topology. In: The International Journal of Geo-Information, vol. 5, September 2016
10. Rslan, E., Hameed, H.A., Ezzat, E.: Spatial R-tree index based on grid division for query processing. Int. J. Database Manag. Syst. (IJDMS ) **9**(6), 25–36 (2017)
11. Fan, P., Li, G., Yuan, L., Li, Y.: Vague continuous K-nearest neighbor queries over moving objects with uncertain velocity in road networks. Syst. Inf. **37**(1), 13–32 (2012)
12. Zhang, C., Han, J., Shou, L., Lu, J., Porta, T.F.L.: Splitter: mining fine-grained sequential patterns in semantic trajectories. Proc. VLDB Endow. PVLDB **7**(9), 769–780 (2014)
13. Mahmood, A., Aref, W.G., Punni, S.: Spatio-temporal access methods: a survey (2010–2017). GeoInformatica **22**, 1–36 (2018)
14. Belhassena, A., HongZhi, W.: Distributed skyline trajectory query processing. In: Proceedings of the ACM Turing 50th Celebration Conference-China ACM TUR-C 2017, pp. 19–25. ACM, May 2017
15. Dittrich, J., Quiane-Ruiz, J.A.: Efficient big data processing in hadoop mapreduce. Proc. VLDB Endow. **5**(12), 2014–2015 (2012)
16. Toshniwal, A., Taneja, S., et al.: Storm@ Twitter. In: The International Conference on Management of Data (SIGMOD 2014), pp. 147–156, June 2014
17. Neumeyer, L., Robbins, B., Nair, A., Kesari, A.: S4: distributed stream computing platform. In: The 10th IEEE International Conference on Data Mining Workshops, pp. 170–177, December 2010
18. Zaharia, M., et al.: Apache spark: a unified engine for big data processing. Commun. ACM **59**(11), 56–65 (2016)
19. Xu, J., Guting, R.H.: MwgenG: a mini world generator. In: Proceedings of the IEEE 13th International Conference on MobileData Management (MDM 2012), pp. 258–267. IEEE, July 2012

# A New Bitcoin Address Association Method Using a Two-Level Learner Model

Tengyu Liu[1,2], Jingguo Ge[1,2(✉)], Yulei Wu[3], Bowei Dai[4], Liangxiong Li[1,2], Zhongjiang Yao[1,2], Jifei Wen[1,2], and Hongbin Shi[1,2]

[1] Institute of Information Engineering, Chinese Academy of Sciences, Beijing 100093, China
gejingguo@iie.ac.cn
[2] School of Cyber Security, University of Chinese Academy of Sciences, Beijing 100049, China
[3] College of Engineering, Mathematics and Physical Sciences, University of Exeter, Exeter EX4 4QF, UK
[4] Institute of Microelectronics of the Chinese Academy of Sciences, Beijing 100029, China

**Abstract.** Users in the Bitcoin system adopt a pseudonym-Bitcoin address as the transaction account, making Bitcoin address correlation analysis a challenging task. Under this circumstance, this paper provides a new Bitcoin address association scheme which makes address tracing possible in Bitcoin systems. The proposed scheme can be used to warn relevant institutions to study more secure encryption algorithms to protect users' privacy. Specifically, the important features of a Bitcoin address are extracted. After that, to reduce the computational complexity, we transform the clustering problem of addresses into a binary classification problem in which we integrate the features of two Bitcoin addresses. A novel two-level learner model is then built to analyze if the two Bitcoin addresses are belonging to the same user. Finally we cluster the addresses belonging to the same user accordingly. Extensive experimental results show that the proposed method outperforms the other address association schemes, which use deep learning models or heuristics, and can achieve an increase by 6%–20% in precision and by 10% improvement in recall.

**Keywords:** Bitcoin · Blockchain · Two-level learner · Bitcoin security

## 1 Introduction

Bitcoin is the first successful implementation of a digital currency that enables instant payments to anyone, anywhere in the world. The Bitcoin system is designed based on the idea of using a decentralized peer-to-peer network, rather than relying on central authorities. Blockchain is the basis of the Bitcoin system,

© Springer Nature Switzerland AG 2020
S. Wen et al. (Eds.): ICA3PP 2019, LNCS 11945, pp. 349–364, 2020.
https://doi.org/10.1007/978-3-030-38961-1_31

which provides a distributed infrastructure and an anonymous way of trading to guarantee the security and freedom of Bitcoin. There is no central server in the network. It uses distributed nodes to generate, update, and store data. Furthermore, it employs cryptography for data transmission and provides a secure and credible environment for Bitcoin transactions.

In details, when users use Bitcoin to trade with others, these transactions will be verified by the nodes in the blockchain network and then be recorded on the blockchain. To enhance privacy and security, users adopt pseudonyms-Bitcoin addresses [1] as their transaction accounts to send and receive bitcoins. The user's real identity will not be publicly bound to the transaction account. In principle, each user can have hundreds of different Bitcoin addresses.

This pseudo-anonymity nature has been improperly used by illegal activities, such as money laundering [2,3] and ransomware [4], to circumvent supervision. Recently, many researchers focus on Bitcoin de-anonymization to address this issue. De-anonymization makes Bitcoin address tracing possible and helps the regulatory system implement network security management. In addition, it can also warn relevant institutions to study more secure encryption algorithms to protect users' privacy.

Many heuristics and deep learning methods have been investigated to do Bitcoin de-anonymization. However, the performance of these methods were still not satisfactory. To make the Bitcoin de-anonymization more accurate and simple-to-implement, in this paper we resort to the important information of Bitcoin addresses based on the historical transactions and manage to associate the addresses. However, there exist many challenges because of the privacy issues: (1) The real identity of the owner of Bitcoin addresses involved in the transaction is unknown because of the anonymity. Thus, few features of Bitcoin addresses can be obtained to analyze. (2) Users may generate fresh Bitcoin addresses for each transaction. (3) It is difficult to apply the clustering analysis of Bitcoin addresses to associate them, because of the large number of users and massive categories to be classified.

To this end, we propose a new Bitcoin de-anonymization method, Bitcoin address association, in which we combine Bitcoin addresses in pairs and then use a two-level learner to determine whether two Bitcoin addresses belong to the same user. The main contributions of this paper can be summarized as follows:

- The large number of clustering categories increases the complexity of clustering algorithms. To tackle this problem, we transform the clustering problem into a binary classification problem to categorize the address pairs.
- In order to gain better classification results, we propose a new model, which is a two-level learner to classify Bitcoin address pairs. XGBoost, LightGBM and Gradient boosting decision tree (GBDT) are used in the first-level learner because they can handle all kinds of features well (see Sect. 3.3 for details). Deep neural network (DNN) is then employed in the second-level learner due to its excellent performance on classification (details in Sect. 3.3).

- Extensive experimental results are conducted to compare the performance of our proposed model with other existing methods. The results demonstrate that our model outperforms the existing methods, with almost 6%–10% improvement in precision and almost 20% improvement in recall.
- We further analyze the effect of Bitcoin address combination orders on the performance of the model. The results show that the combination order of two groups of addresses has little effect on the model's performance. It is therefore not necessary to consider the combination order of the two addresses when using our model for address association.

The rest of this paper is organized as follows. Section 2 gives a minimalistic introduction about related work on Bitcoin address de-anonymization. Section 3 presents the main method of how to associate two Bitcoin addresses. Section 4 shows the experimental results and carries out the performance analysis. Finally, we draw conclusions in Sect. 5.

## 2   Related Work

With the continuous development and maturity of the Bitcoin system, the number of Bitcoin users is increasing. Researchers began to study the de-anonymization of Bitcoin addresses mainly through three ways. One is clustering analysis on the basis of heuristics, another is analyzing the structure and characteristics of the underlying distributed system, and the last one is based on deep learning methods.

**Clustering Analysis.** In [5], the authors proposed two heuristic methods. One is "multi-input" heuristic, which is the most effective and simplest method. This method assumes that all input addresses participating in the same transaction belong to the same user. The second is "shadow" address. Assuming that a user seldom trades with two different users, the Bitcoin address storing the change resulted from a transaction also belongs to the input user of the transaction. The authors can easily map any Bitcoin addresses and users through the above two heuristic analysis of large transactions in the Bitcoin trading network. Fleder et al. [6] crawled the Bitcoin addresses and transaction information fragments from web forums. They then tracked users' whereabouts and conducted clustering analysis, so as to associate transactions and users, and achieve the purpose of de-anonymity. The authors in [7] introduced an efficient automatic cluster approach which used off-chain information as votes for address separation, and then considered it together with blockchain information obtained during the clustering model construction step. In [8], Nick combined the above two heuristics with other heuristics, and the mean recall for address mapping is approximately 0.709.

**Using Distributed Networks.** Several researches are performed to track the source of Bitcoin addresses by observing and analyzing the structure and characteristics of Bitcoin's underlying distributed networks. In [9], the authors used the open trading history of Bitcoin to establish the transaction network between addresses and between users. They combined these structures with external information and techniques such as context discovery and flow analysis to investigate an alleged theft of Bitcoin. The studies in [10] and [11] tracked user's identity information with distributed network characteristics and demonstrated the power of data mining technology in the de-anonymization of Bitcoin addresses. Sybil Attack in [12] and Fake Node Attack in [13] used the users' IP addresses to perform de-anonymization analysis. Mastan [14] proposed a new approach to link the sessions of unreachable Bitcoin nodes, based on the organization of the block-requests made by the nodes in a Bitcoin session graph with a precision of 0.90 and a recall of 0.71.

**Deep Learning Methods.** In [15], Shao, Li and Chen designed a pipeline for Bitcoin address featuring that converts raw address features into a primary address vector. Then they employed a deep learning system and $k$-NearestNeighbor algorithm that realize Bitcoin addresses clustering progressively whose precision achieves 0.766 and recall is 0.836.

Although many de-anonymization technologies for tracing illegal activities and strengthening supervision have been proposed, their performance is still not satisfactory and the risks still exist.

## 3   The Proposed Method

The main workflow of this article is as follows. We extract each Bitcoin address' base features (in Sect. 3.1) and pre-process the samples of Bitcoin addresses (in Sect. 3.2). We then use the GBDT, XGBoost and LightGBM mechanisms and the Function L to learn the new features, which are then fed into a DNN network (in Sect. 3.3) for further analysis. Finally, we cluster the addresses based on the relationship between them.

### 3.1   Basic Feature Analysis and Extraction

We analogize the transactions between Bitcoin addresses to those between users on e-commerce platforms. The user characteristics obtained from e-commerce platforms are mainly based on the users' behaviors, including transaction time, transaction amount, transaction times and other characteristics [16]. The transactions between Bitcoin addresses can be treated as a network, which we call a Bitcoin transaction network. Bitcoin addresses are the network nodes, and the transaction amount flows are the network links. When the address acts as a money receiver, the number of links pointing to it is called in-degree. Out-degree is the contrary. We analyze the in-degree and out-degree distributions of Bitcoin addresses in this network as shown in Fig. 1, which illustrates the number of Bitcoin addresses of in-degree and out-degree. From Fig. 1, we can find that there are a large number of small degree nodes in the network, and also a considerable number of high degree nodes (Hub nodes). The degree can

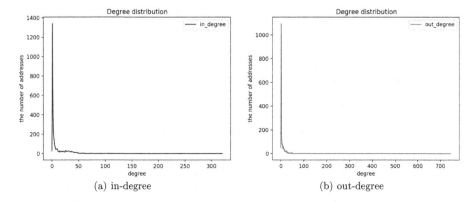

**Fig. 1.** The degree distribution of a Bitcoin transaction network: (a) represents the in-degree distribution of Bitcoin addresses which act as a money receiver, and (b) represents the out-degree distribution of Bitcoin addresses which act as a money sender.

therefore be considered as a key feature to distinguish the addresses of different users.

To enhance the security, one user may create multiple Bitcoin addresses and use different addresses for transactions. Many addresses are used only for a short time, such as "shadow address" in [5]. Different users may use Bitcoin addresses differently, so we consider the characteristics related to the lifetime of a Bitcoin address. 18 features of a Bitcoin address are exacted according to [17], which are shown in Table 1.

### 3.2 Clustering to Binary-Classification Transformation

A straightforward way to identify which addresses belong to the same user is addresses clustering analysis [18]. Each user represents a category. Bitcoin addresses belonging to the same user are classified into the same category. When the number of users go up to a certain scale, it increases the difficulty and complexity of clustering algorithms and reduces the accuracy of clustering results.

In view of the problem that there are a huge number of categories to be classified in the address association problem, we combine two Bitcoin addresses into a new sample, instead of directly using a single address as a classification sample according to the suggestion from social network account association [19]. The purpose of our approach is to perform the classification in Bitcoin address pairs. In this way, we transform a clustering problem into a binary classification problem to categorize the address pairs. The experiments in Sect. 4 illustrate this new method can alleviate the difficulty in the clustering problem. The combination method is as follows.

**Table 1.** The features and its interpretation of each bitcoin address

| Feature | Interpretation |
| --- | --- |
| Lifetime | The lifetime of each address |
| activity_days | The number of days that the address has participated in at least one transaction |
| max_trans_per_day | The maximum number of daily transactions of each address |
| total_received | The values of each address received |
| total_sent | The values sent out by each address |
| avg_val | The average of the values transformed from/to each address |
| std_val | The standard deviation of the values transformed from/to each address |
| in_gini | The Gini coefficient of the values transformed to the address |
| out_gini | The Gini coefficient of the values transformed from the address |
| in_trans_num | The number of transactions which the address acts as the input address |
| out_trans_num | The number of transactions which the address acts as the output address |
| ratio_btw_in_out | The ratio between in_trans_num and out_trans_num |
| in_digree | The number of addresses which have transferred money to the address |
| out_digree | The number of addresses which have received money from the address |
| max_time_diff | The maximum delay between the time when the address has received money and the time it has sent out to some others |
| min_time_diff | The minimum delay between the time when the address has received money and the time it has sent out to some others |
| avg_time_diff | The average delay between the time when the address has received money and the time it has sent out to some others |
| balance_two_days | The maximum difference between the balance of the address in two consecutive days |

We use $\boldsymbol{a_p} = (x_1, x_2, \ldots, x_i, \ldots, x_n)$ representing the address $p$'s eigenvector, and use $\boldsymbol{a_q} = (y_1, y_2, \ldots, y_i, \ldots, y_n)$ representing the address $q$'s eigenvector. $x_i$ and $y_i$ denote the $i$-dimensional feature of $\boldsymbol{a_p}$ and $\boldsymbol{a_q}$, respectively. Given the addresses $p$ and $q$, we can construct the address pair $(\boldsymbol{a_p}, \boldsymbol{a_q}) = (x_1, x_2, \ldots, x_i, \ldots, x_n, y_1, y_2, \ldots, y_i, \ldots, y_n)$. Then the relationship between addresses can be determined by a binary classification problem.

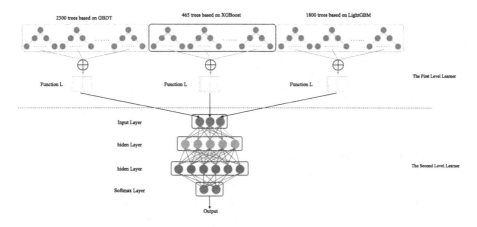

**Fig. 2.** The Two-level Learner Architecture: GBDT, XGBoost, LightGBM mechanism and Function L are used to perform the classification work in the first-level learner. Then the results are fed into the second-level learner which includes a DNN model to output the final results.

### 3.3 Model Stacking Architecture: A Two-Level Learner

Deep learning models [20] have been successfully applied in many fields and have performed well in classification problems. The structure of these models can be adjusted according to the experimental results.

Tree boosting is an effective and widely used machine learning method, due to its strong generalization ability and the capability of well-handling all kinds of features. Gradient boosting decision tree (GBDT) [21] is a typical tree boosting model, which can improve the accuracy of the final classifier by reducing the deviation in the training process. The experimental results in Sect. 4.3 show that it performs better than DNN [20] in classification problems. XGBoost [22] is a scalable end-to-end tree boosting system, which is widely used by data scientists to achieve state-of-the-art results on many machine learning challenges. Especially, it can automatically use the multi-threads of a CPU to parallelize the computation tasks and improve the accuracy on the basis of GBDT. LightGBM [23] is an efficient tree boosting model with higher computational speed and less memory consumption than GBDT and XGBoost. The experiments in Sect. 4.3 show that the three models perform well in the classification problems of Bitcoin addresses.

Stacking is a model ensembling technique used to combine information from multiple predictive models to generate a new model. Often, the stacked model (also called two-level learner) outperforms each of the individual models due to its smoothing nature, and the ability to highlight each base model on its best performance cases [24].

The model stacking is therefore employed in this paper to construct a two-level learner. The first-level learner is trained by the initial dataset, and the output is regarded as the input features of the second-level learner; the labels of the initial samples are still regarded as the labels of the second-level learner. The above three tree boosting models are considered as the first-level learner to get preliminary results. A DNN model is adopted in the second-level learner; its input features obey binomial distribution, and each input feature has the dimensions with the same order of magnitude. The experiments in Sect. 4.3 demonstrate that the use of DNN as the second-level learner outperforms that only using DNN as a classifier. The model architecture is shown in Fig. 2.

**The First-Level Learner.** As Fig. 2 shows, the first-level learner employs three tree models using the principle of GBDT, XGBoost, and LightGBM. At first, 2300 trees are built with the maximum depth of 8, which are based on GBDT. Then 465 trees are constructed using XGBoost mechanism; the maximum depth of the tree model is 15. Lastly we use the LightGBM mechanism to generate 1800 trees, and each tree has at most 34 leaves. These hyper-parameters are obtained experimentally. Because the outputs of the three models are decimal in $[0, 1]$, we employ a **function L:** $y = \begin{cases} 0 & 0 \le x \le 0.5 \\ 1 & 0.5 < x \le 1 \end{cases}$ to transform the output of each model to 0 or 1, where $x$ stands for the outputs of the three models, and $y$ stands for the final results of the first-level learner. We then get the three corresponding outputs marked as $(O_1, O_2, O_3)$.

**The Second-Level Learner.** The second-level learner is based on DNN, which consists of three fully connected layers. As for the design of the first layer, we adopt five units and the input is the output, $(O_1, O_2, O_3)$, of the first-level learner. ReLU [25] is employed as the activation function. The output is sent to the second layer, which has six units and also adopts ReLU as the activation function. Then, we take the output of the second layer to the last layer with two units. At this layer, we employ Softmax [26] as the activation function. The number of units in each layer are confirmed through experiments. The details of DNN is summarized in Table 2.

Table 2. The details of deep neural network part

| Layer | Input | Output | Activation function |
|-------|-------|--------|---------------------|
| First | 3*1 | 5*1 | ReLU |
| Second | 5*1 | 6*1 | ReLU |
| Third | 6*1 | 2*1 | Softmax |

# 4  Experimental Results and Analysis

## 4.1  Dataset

The dataset in [27,28] which contains the transaction history in Mt.Gox is used in this study. This data set records the user's access to Bitcoin on the platform.

The user uses WalletID as the identity on the Mt.Gox platform. When the user needs to send or receive bitcoin, he only needs to initiate a request to Mt.Gox, and Mt.Gox helps the user complete the transaction and will record (WalletID, Entry, Date, Amount) on the platform. Each row in the dataset corresponds to a transaction of a user, and the transaction is simultaneously recorded on the blockchain with the bitcoin addresses of the user or Mt.Gox as input or output. According to the transaction time and amount, each row in the datasets correspond a transaction record on the blockchain. Some WalletIDs appear multiple times in the dataset, representing the user's multiple bitcoin exchanges. Table 3 shows a segment of the data recorded on 15 June 2013. There are 1048196 users in the dataset. So, clustering is difficult.

**Table 3.** Partial transaction records on MtĠox.

| Wallet | Entry | Date | Operation | Amount |
|---|---|---|---|---|
| 292938a9-ea37-4d58-a6c2-a7774159dbf7 | 7e1db835-a3f0-4527-a804-dd47e5a5e59c | 2013/6/15 23:48 | Withdraw | −0.9318619 |
| 07df3a31-4bfe-4178-a05c-5daab4e96575 | 6603e225-cb52-45e9-8190-9ac8dc74048f | 2013/6/15 23:49 | Withdraw | −2.622 |
| 2720c9d5-add9-4319-aa4e-ffd3a0ef3e48 | 853b7efe-b3dc-4cc5-a779-c23c6ab759a8 | 2013/6/16 0:04 | Deposit | 0.01298472 |

How to map user-address is described as follows [29]. Let T_trans represent the transaction time recorded by Mt.Gox, and T_block denote the block creation time. Let B be the block height. When the user deposits money into Mt.Gox at time T_trans, we find the block B whose creation time is T_trans. Due to the delay of the transaction record in the blockchain, we traverse block B~B+36 to find the transaction whose amount uniquely matches to the amount recorded by Mt.Gox. If we find it, the output addresses of the transaction recorded on the blockchain are associated with the current user [9]. When the user withdraws the money from Mt.Gox, we traverse block B-6-36~B-6-1 (A transaction is confirmed after 6 new blocks generated [30]) to map the input addresses with the user. We choose the data from 1 May 2013 to 31 July 2013. Finally, 7945 unique (walletID, address) pairs are found. The partial results are shown in Table 4.

**Table 4.** Partial Results of (walletID, address) Pairs: Some walletIDs in MtĠox and Bitcoin addresses they used in blockchain.

| walletID | Address |
|---|---|
| 162aa442-0771-4bf7-a917-44b13506c139 | 1NYTFydxJFVEosz8M4KMQwWgmwPCTo6uVM |
| 162aa442-0771-4bf7-a917-44b13506c139 | 1HwjJsntuJ8EU1r3xa3yZ8unFBmdR4BnS7 |
| 2663b417-a49b-4e34-a102-17f6cb885bda | 1KmN99HPiYgVfvD1v648cccBbEkuRZgt19 |
| 203354fb-663b-4ee3-85e9-7d85c357927b | 1FHQV8uGggGBnsE6XggP3c4iJpdjpKcE5u |
| 203354fb-663b-4ee3-85e9-7d85c357927b | 18C7SGRMNuCrmKypEnDzyi92QoyBzvwkMV |

We arbitrarily extract two addresses from the above results to form address pairs marked as $(a_p, a_q)$. If $a_p$ and $a_q$ have the same walletID, the label of $(a_p, a_q)$ is 1. If not, the label is 0. There are then 5496694 negative samples and 13853 positive samples. Among them, the number of available samples of negative cases is much more than that of positive cases. In order to ensure the balance of the number of samples and improve model accuracy, the available samples of negative cases are randomly screened, so that the number of positive and negative cases are basically the same [17].

We use the API in [31] to find the transactions in which the Bitcoin addresses participate, and calculate the features' value of each unique address listed in Sect. 3.1. Then we combine the features of the two addresses in address pairs mentioned in Sect. 3.2, with the form: $(a_p, a_q, label) = (x_1, x_2, \ldots, x_i, \ldots, x_n, y_1, y_2, \ldots, y_i, \ldots, y_n, l)$, where $l$ represents the label. Then, the shape of the final samples is $(27670, 37)$.

**Table 5.** The evaluation scores of machine learning models

| Metrics | DNN | GBDT | XGBoost | LightGBM | Two-level learner |
|---------|-----|------|---------|----------|-------------------|
| Time (μs) | 368731 | 583201 | 497931 | 194901 | 1276033 |
| Precision | 0.5070 | 0.7940 | 0.7951 | 0.7870 | **0.9603** |
| Recall | 0.4763 | 0.8047 | 0.7855 | 0.7755 | **0.9570** |
| F1 | 0.4921 | 0.7993 | 0.7903 | 0.7812 | **0.9586** |

### 4.2   Model Training

For all the models, we divide 80% of the samples into the training dataset and 20% as the testing dataset. 4-fold cross validation is employed in the training dataset [32]. The size of verification datasets is kept consistent with that of testing datasets for the purpose of gaining the best results. The verification datasets are used to verify whether the hyper-parameters of the model are tuned to be optimal. The optimal hyper-parameters and training datasets are used to train the model again, and then the model is utilized on the testing datasets.

After we get the samples in Sect. 4.1, we send them to the first-level learner. To prevent overfitting, in GBDT model, every iteration adopts 90% samples and considers up to 35 features when looking for the best split. For each tree in the GBDT model, the minimum number of samples required to split an internal node is 88. In XGBoost model, we extract 90% of the samples and 70% of the features randomly to train each boosting tree. In LightGBM model, we choose 90% of the samples in every 5 iteration and 60% of the features randomly in each iteration.

The DNN in the second-level learner uses Adam [33] as the optimizer. The purpose is to minimize the sum of categorical cross-entropy loss. The class labels are encoded as a one-hot vector. Finally, the learning procedure stops after around 35 epochs.

### 4.3   Experimental Results

**Feature Analysis.** A heat-map is employed to represent the Pearson correlation [34] between features. The feature_1 and feature_2 are used to represent the address $i$'s features and address $j$'s features, respectively, according to Sect. 4.1. As Fig. 3 shows, most Pearson correlation coefficient is around 0, which explains there are not too many features strongly correlated with each other. This is good from the point of view of feeding these features into our learning model, because this means that there is not much redundant or superfluous data in our datasets, and each feature carries some unique information. For example, the Pearson correlation coefficient between in_degree_1 and avg_val_1 is 0.01; they are therefore both important to our classification model.

The feature importance ranking in each basic tree boosting model is calculated by how many times a feature is used to separate decision trees. It is shown in Fig. 4. We combine addresses in the order $(a_i, a_j)$, and the feature importance ranking of the three models is shown in Fig. 4(a), (b) and (c). As shown in the figures, these three models have the similar feature importance ranking. The features which have higher scores are related to time, value, amount and degree of the users' transaction. The result is consistent with the feature analysis in Sect. 3.1. We can then conclude that the addresses belonging to a user have the similar transaction characteristics, and we can safely utilize these features to correlate Bitcoin addresses.

**Address Clustering.** According to the experimental results of the classification, the addresses belonging to one user can be aggregated together. In the testing dataset, we get 1299 groups of addresses. The addresses in each group belong to the same user. Figure 5 shows the distribution of the number of addresses owned by the user. Most users have a small number of addresses. It illustrates that if we cluster the addresses directly, there are few samples in each category, and the information in each category is not sufficient. Clustering directly will thus be a complex task.

**Performance Analysis.** Three common and widely-used metrics, i.e., precision, recall and F1-measure are adopted to validate our proposed model. These metrics are formulated as follows: $precision = \frac{TP}{TP+FP}$, $recall = \frac{TP}{TP+FN}$, and $F_1 = \frac{2TP}{2TP+FP+FN}$, where $TP$ and $FP$ represent the number of items correctly and incorrectly labelled as belonging to the positive class, respectively. $FN$ is the number of items incorrectly labelled as belonging to the negative class.

We use the unit of millisecond to measure the time of the model spending from training to prediction. It reflects the operational efficiency of each model. The two-level learner's time is the sum of the time spent by all the models involved in the two-level learners.

The different evaluation scores of machine learning models are shown in Table 5. GBDT, XGBoost and LightGBM as tree boosting models can handle all kinds of features well and have strong generalization ability (see the analysis

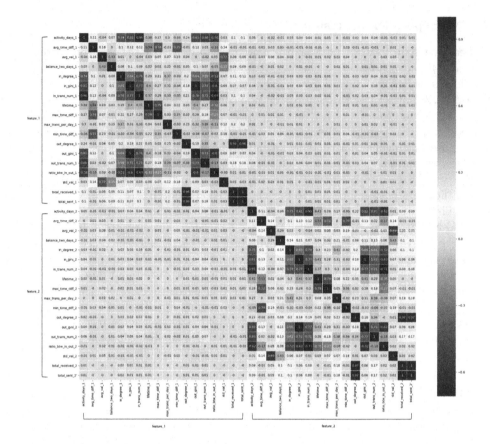

**Fig. 3.** Pearson Correlation Coefficient of Features: 1 denotes the total positive linear correlation, 0 means no linear correlation, and −1 represents the total negative linear correlation.

in Sect. 3.3). They perform better than the DNN model in precision, recall and $F_1$ scores. We can see that GBDT, XGBoost and LightGBM get similar precision, recall and $F_1$ scores. Because of the small number of features and samples, XGBoost and LightGBM cannot significantly improve precision and show advantages compared with GBDT. However, LightGBM is superior to XGBoost and GBDT in running speed due to its parallel optimization (see the discussions in Sect. 3.3). As to the two-level learner, we gain about 18% higher scores compared to other single models due to its smoothing nature and the ability to highlight each base model on its best performance cases (see Sect. 3.3 for details). It proves the effectiveness of adding the DNN layer after the three basic models and building the two-level learner model for Bitcoin address association.

In comparison with the existing works which use deep learning method in Sect. 2, where the precision is 0.766 and recall is 0.836, our model outperforms it by almost 20% in precision and 10% in recall. When using the heuristic methods

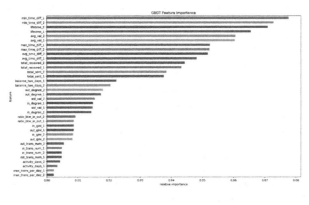

(a) GBDT

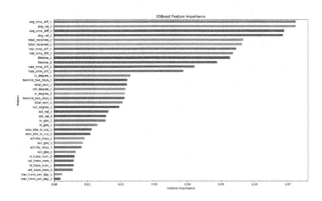

(b) XGBoost

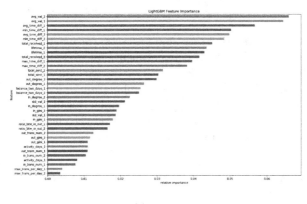

(c) LightGBM

**Fig. 4.** Feature importance ranking of GBDT, XGBoost and LightGBM

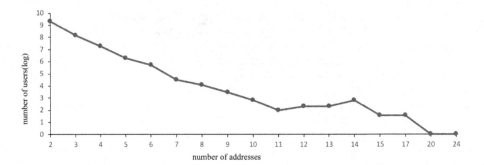

**Fig. 5.** The logistic function distribution of the number of addresses owned by the user: most users have a small number of addresses.

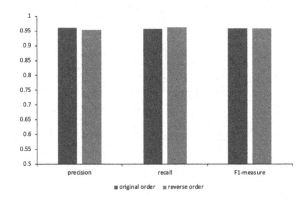

**Fig. 6.** The evaluation scores of different address combination orders: the blue bars represent original order and the red bars represent the reverse order. (Color figure online)

in Sect. 2, the mean recall is 0.709, our model outperforms it by almost 20% in recall.

For training sets, validation sets, and test sets, we generally use a consistent feature order as input. But in this experiment, we exchange the order of the two input addresses in the test datasets, that is, exchange the features' order. The evaluation scores of different address combination order are shown in Fig. 6. It shows that different orders result in the similar results. It illustrates that different orders do not have a significant impact on the classification of address pairs, which simplifies our address association work.

# 5   Conclusion

This paper has provided a Bitcoin address association method to perform de-anonymization, which transforms the clustering problem into a binary classification problem. The main idea of the method is to check if the two addresses belong to the same user and then cluster the addresses based on this check. The proposed method has combined both boosting tree models and deep learning models into a two-level learner to perform the classification. XGBoost, Light-GBM and GBDT have been utilized as the first-level learner, and a three-layer deep neural network has been adopted as the second-level learner. By performance comparison, our method has performed more excellently than the ones which only employ one boosting method, such as XGBoost, LightGBM and GBDT, or only employs a deep neural network model, in terms of higher precision, recall and F1-measure scores. The research results in this paper can offer the suggestions and references for the investigation and tracking of illegal activities in Bitcoin systems and guide the blockchain system to study more secure and reliable encryption mechanisms.

# References

1. ShenTu, Q.C., Yu, J.P.: Research on anonymization and de-anonymization in the bitcoin system. arXiv preprint arXiv:1510.07782 (2015)
2. Brenig, C., Accorsi, R., Müller, G.: Economic analysis of cryptocurrency backed money laundering. In: ECIS (2015)
3. Fanusie, Y., Robinson, T.: Bitcoin laundering: an analysis of illicit flows into digital currency services. Center on Sanctions & Illicit Finance memorandum, January 2018
4. Liao, K., Zhao, Z., et al.: Behind closed doors: measurement and analysis of CryptoLocker ransoms in bitcoin. In: 2016 APWG Symposium on Electronic Crime Research (eCrime), pp. 1–13. IEEE (2016)
5. Meiklejohn, S., Pomarole, M., et al.: A fistful of bitcoins: characterizing payments among men with no name. In: Proceedings of the 2013 Conference on Internet Measurement Conference, pp. 127–140. ACM (2013)
6. Fleder, M., Kester, M.S., Pillai, S.: Bitcoin transaction graph analysis. arXiv preprint arXiv:1502.01657 (2015)
7. Ermilov, D., Panov, M., Yanovich, Y.: Automatic bitcoin address clustering. In: 2017 16th IEEE International Conference on Machine Learning and Applications (ICMLA), pp. 461–466. IEEE (2017)
8. Nick, J.D.: Data-driven de-anonymization in bitcoin. Master's thesis, ETH-Zürich (2015)
9. Reid, F., Harrigan, M.: An analysis of anonymity in the bitcoin system. In: Altshuler, Y., Elovici, Y., Cremers, A., Aharony, N., Pentland, A. (eds.) Security and privacy in social networks, pp. 197–223. Springer, New York (2013). https://doi.org/10.1007/978-1-4614-4139-7_10
10. Ron, D., Shamir, A.: How did dread pirate roberts acquire and protect his bitcoin wealth? In: Böhme, R., Brenner, M., Moore, T., Smith, M. (eds.) FC 2014. LNCS, vol. 8438, pp. 3–15. Springer, Heidelberg (2014). https://doi.org/10.1007/978-3-662-44774-1_1
11. Ron, D., Shamir, A.: Quantitative analysis of the full bitcoin transaction graph. In: Sadeghi, A.-R. (ed.) FC 2013. LNCS, vol. 7859, pp. 6–24. Springer, Heidelberg (2013). https://doi.org/10.1007/978-3-642-39884-1_2
12. Kaminsky, D.: Black ops of TCP/IP. Black Hat USA, p. 44 (2011)
13. Biryukov, A., Pustogarov, I.: Bitcoin over tor isn't a good idea. In: 2015 IEEE Symposium on Security and Privacy, pp. 122–134. IEEE (2015)

14. Mastan, I.D., Paul, S.: A new approach to deanonymization of *unreachable* bitcoin nodes. In: Capkun, S., Chow, S.S.M. (eds.) CANS 2017. LNCS, vol. 11261, pp. 277–298. Springer, Cham (2018). https://doi.org/10.1007/978-3-030-02641-7_13

15. Shao, W., Li, H., Chen, M., Jia, C., Liu, C., Wang, Z.: Identifying bitcoin users using deep neural network. In: Vaidya, J., Li, J. (eds.) ICA3PP 2018. LNCS, vol. 11337, pp. 178–192. Springer, Cham (2018). https://doi.org/10.1007/978-3-030-05063-4_15

16. Sanjaya, C., et al.: Revenue prediction using artificial neural network. In: 2010 Second International Conference on Advances in Computing, Control, and Telecommunication Technologies, pp. 97–99. IEEE (2010)

17. Bartoletti, M., et al.: Data mining for detecting Bitcoin Ponzi schemes. In: 2018 Crypto Valley Conference on Blockchain Technology (CVCBT), pp. 75–84. IEEE (2018)

18. Ghahramani, Z.: Unsupervised learning. In: Bousquet, O., von Luxburg, U., Rätsch, G. (eds.) ML -2003. LNCS (LNAI), vol. 3176, pp. 72–112. Springer, Heidelberg (2004). https://doi.org/10.1007/978-3-540-28650-9_5

19. Fan, X., Hongbo, X., Liang, Y.: A sock-puppet relation detection method on social network. J. Chin. Inf. Process. **28**(6), 162–168 (2014)

20. LeCun, Y., Bengio, Y.: Deep learning. Nature **521**(7553), 436 (2015)

21. Friedman, J.H.: Greedy function approximation: a gradient boosting machine. Ann. Stat. **29**(5), 1189–1232 (2001)

22. Chen, T., Guestrin, C.: XGBoost: a scalable tree boosting system. In: Proceedings of the 22nd ACM SIGKDD International Conference on Knowledge Discovery and Data Mining, pp. 785–794. ACM (2016)

23. Ke, G., Meng, Q., et al.: LightGBM: a highly efficient gradient boosting decision tree. In: Advances in Neural Information Processing Systems, pp. 3146–3154 (2017)

24. A guide to model stacking in practice (2016). http://blog.kaggle.com/2016/12/27/a-kagglers-guide-to-model-stacking-in-practice

25. Nair, V., Hinton, G.E.: Rectified linear units improve restricted Boltzmann machines. In: Proceedings of the 27th International Conference on Machine Learning (ICML-10), pp. 807–814 (2010)

26. Liu, W., Wen, Y., Yu, Z., Yang, M.: Large-margin softmax loss for convolutional neural networks. In: ICML, vol. 2 (2016)

27. mtgox2014leak. https://www.reddit.com/r/mtgoxAddresses/wiki/mtgox2014leak (2014)

28. Chen, W., Wu, J., Zheng, Z., Chen, C., Zhou, Y.: Market manipulation of bitcoin: evidence from mining the Mt. Gox transaction network. In: IEEE INFOCOM 2019-IEEE Conference on Computer Communications, pp. 964–972. IEEE (2019)

29. Xing, Y., Li, X., et al.: Research on de-anonymization techniques of bitcoin trading network. A Thesis Submitted to Southeast University For the Academic Degree of Master of Engineering, China (2017)

30. Nakamoto, S.: Bitcoin: a peer-to-peer electronic cash system. Consulted (2008)

31. Blockchain data API. https://www.blockchain.com/zh/api/blockchain_api (2017)

32. Arlot, S., Celisse, A., et al.: A survey of cross-validation procedures for model selection. Stat. Surv. **4**, 40–79 (2010)

33. Kingma, D.P., Ba, J.: Adam: a method for stochastic optimization. Comput. Sci. (2014)

34. Pearson correlation coefficient. https://en.wikipedia.org/wiki/Pearson_correlation_coefficien (2019)

# Fog Computing Based Traffic and Car Parking Intelligent System

Walaa Alajali[1][(✉)] ⓘ, Shang Gao[1] ⓘ, and Abdulrahman D. Alhusaynat[2] ⓘ

[1] Deakin University, Melbourne 3217, Australia
{wkalajal,shang.gao}@deakin.edu.au
[2] Thi-Qar University, Nasiriyah 64001, Iraq
rahman_dakhil@yahoo.com

**Abstract.** Internet of Things (IoT) has attracted the attention of researchers from both industry and academia. Smart city, as one of the IoT applications, includes several sub-applications, such as intelligent transportation system (ITS), smart car parking and smart grid. Focusing on traffic flow management and car parking systems because of their correlation, this paper aims to provide a framework solution to both systems using online detection and prediction based on fog computing. Online event detection plays a vital role in traffic flow management, as circumstances, such as social events and congestion resulting from accidents and roadworks, affect traffic flow and parking availability. We developed an online prediction model using an incremental decision tree and distributed the prediction process on fog nodes at each intersection traffic light responsible for a connecting road. It effectively reduces the load on the communication network, as the data is processed, and the decision is made locally, with low storage requirements. The spatially correlated fog nodes can communicate if necessary to take action for an emergency. The experiments were conducted using the Melbourne city open data.

**Keywords:** Traffic flow · Prediction · ITS · Smart car parking

## 1 Introduction

In the past few years, many cities around the world are experiencing population growth. The United Nations estimates that 70% of the world's population will live in cities by 2050; in Australia, 90% of the population currently live in cities [7,23]. Consequently, the number of vehicles in cities has also increased. For example, in the United States (US) in 2013, there were 798 vehicles per 1,000 people [37]. The increase in the urban populations has led to many problems, including noise, environmental pollution and overcrowding, wasting time and money and adversely affecting people's mobility [32]. Intelligent transportation systems (ITS) have been developed in several cities [28], integrating various technologies, such as electronic information, artificial intelligence, geographical

S. Wen et al. (Eds.): ICA3PP 2019, LNCS 11945, pp. 365–380, 2020.
https://doi.org/10.1007/978-3-030-38961-1_32

information and global positioning systems [35,50]. An ITS is an effective way for smart cities to solve the transportation problems as a result of the significant increase in traffic levels in the urban areas. Managing traffic flow and reducing congestion are among the primary goals of ITS [18], as traffic congestion is a critical issue in most cities around the world [10,31,43].

Congestion refers to the traffic state when the number of vehicles using the road increases, impeding traffic flow. Its main signs are reduced speeds and increased trip times. Based on time and location, the definition of congestion may vary. For example, congestion differs on highways and arterial roads. In general, congestion is deemed to have occurred when a vehicle's average speed is below an identified threshold [37]. In Australia (according to the Bureau of Transportation Regional Economics), Melbourne, Sydney and Brisbane are congested cities, and an estimated 20.4 billion AUD is required to reduce congestion [7,40]. In addition, in 2015, the Texas A&M Transportation Institute released a report with a Mobility Scorecard showing that in the US in 2014, around 6.9 billion hours were wasted due to traffic congestion [38]. In China, traffic congestion causes $1 billion in economic losses for busy cities every day [27]. Air pollution is another adverse impact of traffic congestion [10]. Detecting and managing traffic congestion will result in improved environmental conditions and will solve the economic problems that are caused by congestion [33].

One solution to congestion is to expand the road network; another is to develop intelligent and efficient transportation management systems by installing sensors on roads and in cars (in addition to existing cameras) to collect data, which can then be analysed to better understand the characteristics of traffic, such as speed and flow. The first solution is difficult to implement in many cities due to financial concerns and the city's structure, so most efforts have concentrated on devising an efficient management system to solve traffic problems, including congestion [29]. The efficiency of ITS depends on how accurately it analyses and models transportation data [14]. Using multiple data sources to study traffic in an ITS has several benefits. Prediction model is significant for increasing the efficiency of the system and improving the services provided to users. It is also challenging, as traffic is dynamic and certain events can affect the accuracy of the forecast if the system is not updated in real time to ensure that appropriate decisions are made. Moreover, because sensors produce data continuously, this generates a substantial amount of data and increases the load of the network. To address this issue, in this paper, a distributive processing concept has been developed to process data locally and evaluate the response in real-time.

On-street car parking is another cause of traffic congestion. A driver starts searching for parking when approaching the destination and tries to find an available lot [1,8]. An estimated 30% of congestion in cities' internal roads results from drivers searching for a parking space, which consumes an average of 8 min [13]. Therefore, a smart car-park and traffic-management system will reduce congestion, providing accurate services to users and increasing the quality of smart cities.

Many cities have developed on-street smart car-park systems, including San Francisco (SFpark), Los Angeles (LA Express Park) and Melbourne [30]. One proposed solution to this problem involves providing users with real-time information on available parking spaces, but this solution alone is insufficient, as a driver may get to a space and find it occupied. Therefore, determining the number of available parking spaces plays an essential role by providing advance information about the parking situation so users can make appropriate decisions. These two parts of ITS are related to each other, where on-street car-park system causes increase in the traffic on the roads. In addition, road traffic conditions also affect the car-park system, as accidents and roadworks make finding suitable parking more difficult.

Fog computing has emerged as a way to moving processing to a network's edges (devices) by distributing a computing task among several devices (called nodes) at the edges. In recent years, several studies have investigated the advantages of applying fog computing in ITS (particularly in the case of VANETs), primarily to reduce the load on the communication network by not transferring all the data to the cloud for processing. This enhances the capacity for real-time processing and for making decisions locally and more quickly, especially in response to accidents and events. Finally, fog computing architecture is distributed, which reduces storage requirements by processing the data in a stream at each node for short-term decisions. A selected part of the data or the final decision can be transferred to the cloud for long-term planning, depending on what has been displayed.

In this paper, we offer a framework for traffic-flow management and an on-street smart car-park system based on a fog computing approach. We propose a computing network comprising nodes installed at each intersection with the ability to process data from each road segment belonging to the intersections. The main goal of this model is to detect congestion caused by accidents and roadworks and then, based on the current situation, to predict the traffic in the next time interval. This solution also includes a model for a parking system that provides the available car-park lots that are accessible from an intersection after detecting variables such as social and sporting events, taking traffic conditions into consideration and predicting the car-park lots available in the next time interval. The major contributions of this study comprise:

- Introducing a framework for traffic flow and car-park prediction based on fog computing to provide scalable real-time processing.
- Proposing a new, streaming-based concept drift detection to identify an abnormal situation in real time.
- Using incremental learning to update the model for newly arrived examples.

The remaining part of this paper is arranged as follows: Sect. 2 focuses on the related work. Section 3 presents the proposed framework, and Sect. 4 provides a discussion of the proposed prediction approach. The experiments, including datasets, outcomes and a comparative analysis of the approaches, are provided in Sect. 5. In the final Section VI, the conclusion and suggestions for further related research are presented.

## 2   Related Work

Short-term traffic prediction models can be classified by learning method as non-incremental learning models and incremental learning models [44]. In non-incremental learning models, the whole dataset is available and can be used for training and testing the model. Many models stand under this umbrella, some using parametric methods; the most popular of these models are auto-regressive integrated moving average (ARIMA) [22,30,34,42] and support vector regression (SVR) [9,49]. Non-parametric approaches include k-Nearest Neighbours (kNN) [15,18,53]. Deep learning is also used in [33]. Decision trees are the most understandable method and are used in various ways, such as gradient boosting trees (GBRT), random forest (RF) and extreme gradient boosting trees (XGboost) [3].

Traffic is dynamic [3,5] and a non-recurrent event on the street can affect traffic prediction accuracy. Therefore, online learning is essential to deal with streaming data generated by sensors on the roads and to update the model incrementally if a change appears. A few efforts have been made in this area: online SVR is used in [13] to predict traffic, and this approach was improved in [26] by adding weights for the incoming instances. Fast incremental model trees with drift detection (FIMT-DD) [25] is used in [4,41] to predict traffic. According to [51] *a social event is an occurrence that involves lots of people and is accompanied by an obvious rise in the human flow.* Social events involve large numbers of people [2,51], which affects traffic, causes congestion and changes the availability of parking in cities. Events must be taken into consideration when predicting traffic or available car parking. Although some events are planned, their time may change, and some events may occur suddenly. Also, some events do not specify in advance the time of their completion, so autonomously detecting these events is beneficial to users and managers.

In terms of concept drift detection, several approaches have been suggested in the literature. They can be classified into sequential analysis-based, data distributed-based and learner output-based [45].

In sequential analysis based methods, the probability distribution of the incoming data is determined by comparing the current distribution and the previous one, to see if it is significantly different. Page-Hinkley test [36] is the example of this category. However, the drawback is it is dealing with univariate time series. Its use is limited to multivariate time series. In our proposed External-DD method, we modify the sequential model to deal with multivariate time series.

The second type is the distributed-based model. It requires storing raw data in two windows and using a hypothesis test to compare the distribution in two windows and detect drift. The main drawback is that it requires identifying the length of the window.

Finally, the linear output-based detection method depends on building a classifier model and monitoring the error if the error increases significantly. DMM, RDDM are examples. However, it updates the model after the error number is more than a specific threshold. In that case, still, there is a possibility to produce errors especially if the drift is not sudden but incremental as in the case of congestion. In this paper, we combine FIMTDD prediction with External-DD to reduce the error caused if both of the sudden and incremental drift happen.

# 3  Proposed Framework

## 3.1  Preliminaries

The framework suggested in this paper is to enhance the prediction of traffic flow and improve a smart car-park system. Prediction based on multi-source data is important. Indeed, enhancing the services provided by a smart car-park system can reduce congestion, as unorganised searching for parking space increases traffic. Social and unexpected events also affect the accuracy of the system, and congestion due to accidents and roadworks can affect the car-park system by increasing prediction errors. Therefore, providing a framework for both systems is useful. The framework is described in Fig. 2, and the processing sequence is described in Fig. 3.

**Incremental Learning.** Data-stream processing plays an important role in the real-time collection, integration, and analysis of scalable and continuous data produced by IoT devices. Data-stream processing bridges the gap between the traditional batch processing approach and processing high-rate data. The other advantage of stream processing is its suitability in a dynamic environment, as it updates the query result incrementally [12]. According to [19], a data stream is an unlimited sequence of objects that flows continuously at a high rate of speed, and the distribution of the data may evolve over time. A stream-mining approach is a set of techniques used to extract information in real-time processing of streaming data [6]. A finite set of training data and a static model are used in standard data-mining approaches.

The situation is different for ITS data, however, as the data collection in a non-stationary environment changes over time in a dynamic way. Furthermore, the data will remain unused until the next update is performed, so the model needs to retrain, making a static model unsuitable. In addition, the data must be processed upon arrival, otherwise the value of the data will be wasted. According to IoT data features, the following requirements are fundamental in designing methods to analyse IoT data [17].

- The model should be maintained online and should incorporate data on the fly;
- The model should use an unbounded training dataset; the length of the dataset is infinite;
- A dynamic model should be used to detect and adapt to changes in the environment.

**Concept Drift Detection.** According to [20], "Concept drift primarily refers to an online supervised learning scenario when the relation between the input data and the target variable changes over time". In ITS concept drift caused by accidents, roadworks, and social events. Social event refers to the observed increase in the number of people at a particular place and time. The analysis and monitoring of these events are of great importance in determining their impact on the rest of the essential facilities in cities, including their effects on traffic and the parking management system [51]. Previously, the identification of these events was made in traditional ways, which required human effort to count any noticeable increase in the number of people. Now, with the evolution of technology, especially IoT applications in smart cities, roads are equipped with sensors and cameras to monitor traffic and pedestrians, through which it is possible to obtain data to judge their impact on the rest of a city's systems. Figure 1 illustrates the yearly average pedestrian volume of various streets in the Melbourne CBD. We aim to automatically detect social events and the congestion caused by accidents and roadworks using online concept drift detection and then predict the short-term traffic flow and car-park availability using locally based fog computing. Algorithm 1 is another description of the proposed framework.

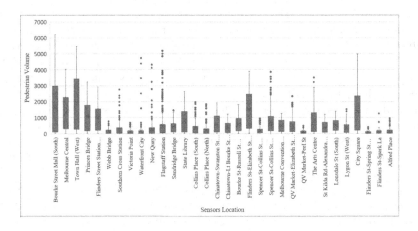

**Fig. 1.** Pedestrian volume all sensors 2014

**Fog Computing.** Recently Fog Computing has been proposed as a distributed solution to ITS applications. According to the survey [24] that presented the definition of fog computing, characteristics and applications, fog computing entails the integration of both cloud centres and network edge devices in a seamless manner. It is an effective solution for surmounting these limitations. Fog computing is a geographically distributed computing architecture using several heterogeneous devices at a network's edge that are connected ubiquitously to deliver elastic storage and computation services [47]. Importantly, the most noticeable facet of fog computing remains the extension of cloud service into the network's edge. This, in turn, makes computation, communication, storage and control closer to users by enabling the pooling of local resources.

Fog paradigm is capable of adequately addressing the real-time demands of applications that are latency-sensitive and remove bandwidth bottlenecks in case of video processing and VANET as examples. The architecture adds a layer

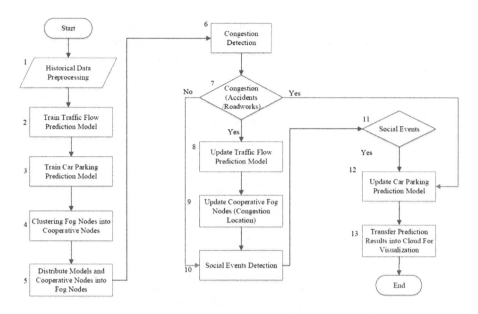

**Fig. 2.** Flow chart, step (1–5) on cloud and step (6–13) on fog node

between end devices and cloud to tackle the underlying problems affecting high security and reliability, low latency, high performance, interoperability and also mobility [39, 48]. This platform consists of several fog nodes that are inclusive of the number of devices as well as management systems, even encompassing a few virtualise data centres that are edge-centric [52]. Fog nodes can store the data that is created by edge devices as well as sensors. By collaborating with the conventional model of cloud computing, fog computing plays a more effective role in its utilisation as a green computing platform, something that it also helps cloud computing to accomplish [21, 46].

### 3.2 Framework Description

The framework consists of two components: cloud and fog nodes. The main steps of the process are done at the fog level, as shown in Fig. 3. The online concept drift detection and online prediction model are supposed to be installed in each node since they do not require storing data, making it efficient in time. In a fog node, the following steps are performed. First, the incoming data from sensors is detected for unusual events using External-DD method. The output of this method is used with other features as input to FIMTDD. The same process can be used for the car park system. The aim to detect events using the same detection method is detecting pedestrian volume. The output of detection from traffic and pedestrians can be used as a feature for FIMTDD to predict the car park availability in next time interval. Figure 2 and Algorithm 1 demonstrate the steps.

The second component is the cloud, instead of transferring all the data to the cloud for processing, especially if the data has to be transmitted from VANET. Statistics show aggregated data by day, week or month is transferred to the cloud. The reason is for the long term prediction for planning and to have a full system overview. In addition, the other feature is that the nodes can be spatially clustered as a group, so they can exchange information if required, as shown in Fig. 4. For the node clustering process, any common clustering method can be used, for instance, k-mean based on spatial features.

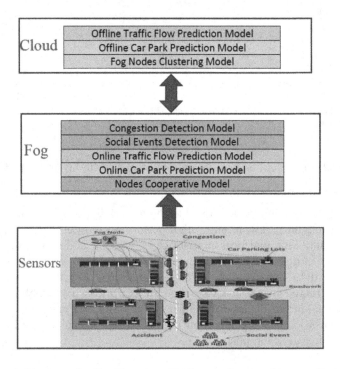

**Fig. 3.** Framework: distributed traffic flow and car parking prediction

### 3.3   Proposed Prediction Method

In this paper, we propose an incremental learning for prediction method called FIMT-DD [25] with an internal and external concept drift detection. This method has been used for the prediction of streaming data. It includes an internal concept drift detection method called the Page-Hinckley change detection test [36]. If the model error exceeds a defined threshold, the model will be updated. However, there remains a possibility of producing errors until the error exceeds the threshold. This paper proposes a simple mathematical concept drift model known as External-DD, as described in Algorithm 2. This method can be used to detect abnormal events such as accidents, roadworks and social events that could

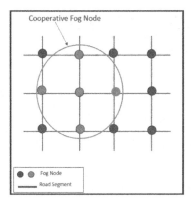

**Fig. 4.** Cooperative fog nodes

lead to changes in the prediction accuracy. It does not require saving the actual values, only the mean of the inputs, making it efficient in terms of time and memory. We used a 2D array (2, 97) to store the mean values, as shown in Fig. 5 below. Time complexity is $O(1)$. The threshold is set as $0.75 \times mean[i,j]$. The first column day indicates if it is a weekday or weekend and the other columns are for time, where each index represents 15 min time interval with total number 96 for a day. The input time value is compared with the mean value at time t after specifying if the day is weekday or weekend. If the difference between the mean and the input is greater than a threshold, the output is abnormal (event or congestion). Otherwise, the output is normal and a new mean is calculated. As such, we deal with a multivariate time series problem in concept drift which is the main problem for sequential concept drift model [45].

---

**Algorithm 1.** Hybrid TFCP Prediction (Fog Node Level)

---

**Input:** Three streaming data: Traffic Volume $Tv_t$, Car Park availability $CP_t$, Pedestrians Data $Pd_t$. /* t is the time */
**Output:** Traffic Flow and Car Park Availability for next 15 min.
**Procedure:**
1: For each $Tv_t$ do;
2:    Apply Concept Drift Detection Model for Congestion;
3:    Update Traffic Prediction Model (FIMT-DD);
4:    Update Car Park Prediction Model (FIMT-DD);
5:    Update Cooperative Nodes Status;
6: For each $Pd_t$ do;
7:    Apply Concept Drift Detection for Social Event;
8:    **IF** Event = True **Then**;
9:    Update Car Park model;
10: Transmit the Result.

---

---

**Algorithm 2.** Concept Drift Detection( External-DD)

---

**Input:** Data stream $X$, at time $t$ and date $d$

**Output:** Result; Normal $N$; Concept drift $(C)$ at time $t$; /*concept drift representing an event such as accident*/

**Procedure:**

1: Create $M(2 \times n)$ matrix; /* where $M_{i,t} = Mean(X_t), t \in \{1, t_{current}\}$; $96 \times 15$ min time slot in a day */

2: **For** each $X_t$ do

3:    From d find M[i,1]; /* 0 for weekday and 1 for weekend */

4:    **IF** $|(X_t - M[i,t]| >= C_{thr}$;/* Where $C_{thr}$ is concept drift threshold */

5:      Result= C;

6:    **Else**

7:      Result=N;

8:    Update $M[i,t]$ /* calculate new mean*/

9:    Feed Result as a feature in input features to Hybrid TFCP Prediction.

---

15 min time slot for 24 hours

| | | | | | | | |
|---|---|---|---|---|---|---|---|
| Weekday | 0 | $M_{t1}$ | $M_{t2}$ | $M_{t3}$ | $M_{t4}$ | .......... | $M_{t96}$ |
| Weekend | 1 | $M_{t1}$ | $M_{t2}$ | $M_{t3}$ | $M_{t4}$ | | $M_{t96}$ |

$M_{tj}$=mean($X_j$), j={$t_1$,..,$t_i$}

**Fig. 5.** 2D array to store time mean values

## 4 Experiments

Experiments were performed for evaluation of the performance of the above proposed approach. The real-world datasets were employed to evaluate the algorithms compared to state-of-the-art. The development and the tests were done on the Massive Online analysis software for streaming data processing (MOA) [11] and the Eclipse environment.

### 4.1 Dataset

We employed three datasets from various domains (on-street car parking data, pedestrian data and car traffic data) in Melbourne City, Australia, as shown in Table 1 [16]. For the on-street car parking dataset, the data was obtained by 4,600 in-ground sensors located in various streets in the Melbourne CBD. The data for the pedestrian dataset was from 36 sensors of the 44 in current use, which are installed in strategic streets in the Melbourne CBD to monitor pedestrian activities and analyse their changes and developments over time. The sensors count the number of pedestrians who pass the installed locations. The intersection traffic volume dataset covers more than 4,598 traffic signals throughout Melbourne and some suburban areas. The sensors installed at these traffic signals provide real-time traffic volume at the intersections. For each sensor, the data stream is aggregated into 96 bins for each day, in 15-minute time intervals.

**Table 1.** Description of Datasets

| Dataset | Type | Time interval |
|---|---|---|
| Traffic volume | Public | 15 min |
| Car parking | Public | 15 min |
| Pedestrian | Public | 15 min |

## 4.2 Discussion and Evaluation of Results

Two performance metrics, mean absolute error (MAE) and root mean square error (RMSE), were used to record the accuracy of the prediction methods and the advantages of internal and external concept drift methods. The most accurate model was proved to be with the smallest MAE and RMSE values, as described in following equations, where $\bar{y}_i$ is the predicted value and $y_i$ is the real value.

$$MAE = \sum_{i=1}^{N} |\bar{y}_i - y_i|/N \tag{1}$$

$$RMSE = \sqrt{\frac{1}{N} \sum_{i=1}^{N} (\bar{y}_i - y_i)^2} \tag{2}$$

Our prediction model comprises two parts: the first is the external concept drift detection model, to identify the unusual events or congestion namely, External- DD. The output of this part is either a normal or abnormal input. This feature along with other features such as time, day of the week, weekday/weekend, peak/off peak, were then used in incremental prediction model FIMTDD.

The results in Fig. 6 show that the accuracy of the FIMTDD prediction method increased after using an external concept drift to detect congestion caused by accidents and roadworks. Table 2 demonstrates the prediction error, MAE equal to 0.641 using FIMTDD only, however, this value is reduced to 0.606 after using FIMTDD and External-DD. The second metric is the RMSE, with the value equal to 0.822 using FIMTDD and also reduced to 0.79 using our method. Comparing the result with that of FIMT, the latter ignores the detected drift and does not update the model, so the error is increased with MAE equal to 0.686 and after using External-DD the error is reduced to 0.655 and RMSE records 0.89 and 0.859 respectively. Thus, FIMTDD with internal and external concept drift detection outperforms FIMT and ORTO. ORTO is a regression method for streaming data used in our experiments. It records the highest error and Table 3 shows the evaluation of FIMTDD with External-DD, suggesting efficiency in time and space.

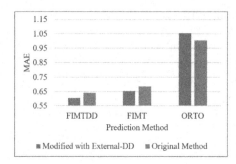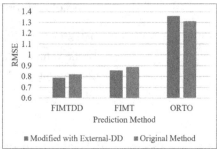

**Fig. 6.** Comparison of the results using MAE and MSE

**Table 2.** Results

| Method | MAE | RMSE | Drift Detection | |
|---|---|---|---|---|
| | | | Internal | External |
| Proposed Method | **0.606** | **0.79** | ✓ | ✓ |
| FIMTDD | 0.641 | 0.822 | ✓ | ✗ |
| External-DD &FIMT | 0.655 | 0.859 | ✗ | ✓ |
| FIMT | 0.6869 | 0.89 | ✗ | ✗ |
| External-DD &ORTO | 1.053 | 1.36 | ✗ | ✓ |
| ORTO | 1.003 | 1.31 | ✗ | ✗ |

**Table 3.** Description and evaluation of the results using proposed method

| Learning evaluation instances | Evaluation time (CPU seconds) | Model cost (RAM hours) | Classified instances | Mean absolute error | Root mean squared error |
|---|---|---|---|---|---|
| 100 | 0.015625 | 0 | 100 | 1.115124771 | 1.477465394 |
| 200 | 0.03125 | 0 | 100 | 0.946889883 | 1.255662272 |
| 300 | 0.03125 | 0 | 100 | 1.128056522 | 1.401679781 |
| 400 | 0.046875 | 0 | 100 | 0.800534942 | 1.18219432 |
| 500 | 0.046875 | 0 | 100 | 0.719083747 | 1.118842788 |
| 600 | 0.046875 | 0 | 100 | 0.327865222 | 0.459055334 |
| 700 | 0.0625 | 0 | 100 | 0.609368106 | 0.830886628 |
| 800 | 0.0625 | 0 | 100 | 0.683963244 | 0.924470448 |
| 900 | 0.0625 | 0 | 100 | 0.478814355 | 0.665715537 |
| 1000 | 0.078125 | 0 | 100 | 0.436850643 | 0.608490533 |
| 1100 | 0.078125 | 0 | 100 | 0.551151138 | 0.769255381 |
| 1200 | 0.09375 | 0 | 100 | 0.493650102 | 0.722171056 |
| 1300 | 0.09375 | 0 | 100 | 0.411204979 | 0.508820137 |
| 1400 | 0.09375 | 0 | 100 | 0.473706375 | 0.825236894 |
| 1500 | 0.09375 | 0 | 100 | 0.539935928 | 0.741406081 |
| 1600 | 0.09375 | 0 | 100 | 0.292022692 | 0.369561128 |
| 1700 | 0.09375 | 0 | 100 | 0.412347144 | 0.609504873 |
| 1800 | 0.09375 | 0 | 100 | 0.393053805 | 0.523699751 |
| 1900 | 0.109375 | 0 | 100 | 0.362047581 | 0.456501883 |
| 2000 | 0.109375 | 0 | 100 | 0.470569252 | 0.732041187 |
| 2100 | 0.109375 | 0 | 100 | 1.127129006 | 1.378777728 |

# 5    Conclusion

This paper investigated traffic prediction in ITS and its importance in improving services. Three concerns were addressed. First, traffic is dynamic because of potential unforeseen changes that may occur due to accidents, roadworks or presence of social events. A prediction model must deal with this issue. We suggested an upward prediction model that dealt with the concept drift resulting from a non-stationary environment by adding an external method to the prediction model used, to increase efficiency. A second concern was how to deal with the size of the data. We proposed a fog computing-based processing method, in which each node represents one intersection and the data is processed locally. This also adds the possibility of benefiting from the collaborating nodes, classified by location in transferring alert messages. The last concern was addressed by providing a framework for both systems due to the correlation between them. In future work, we aim to apply the proposed method in the detection of social events and highlight its advantage in a smart car-park system.

# References

1. Ahangari, S., Chavis, C., Jeihani, M., Moghaddam, Z.R.: Quantifying the impact of on-street parking information on congestion mitigation using a driving simulator. Technical report (2018)
2. Alajali, W., Wen, S., Zhou, W.: On-street car parking prediction in smart city: a multi-source data analysis in sensor-cloud environment. In: Wang, G., Atiquzzaman, M., Yan, Z., Choo, K.-K.R. (eds.) SpaCCS 2017. LNCS, vol. 10658, pp. 641–652. Springer, Cham (2017). https://doi.org/10.1007/978-3-319-72395-2_58
3. Alajali, W., Zhou, W., Wen, S.: Traffic flow prediction for road intersection safety. In: 2018 IEEE SmartWorld, Ubiquitous Intelligence and Computing, Advanced and Trusted Computing, Scalable Computing and Communications, Cloud and Big Data Computing, Internet of People and Smart City Innovation (SmartWorld/SCALCOM/UIC/ATC/CBDCom/IOP/SCI), pp. 812–820. IEEE (2018)
4. Alajali, W., Zhou, W., Wen, S., Wang, Y.: Intersection traffic prediction using decision tree models. Symmetry **10**(9), 386 (2018)
5. Anantharam, P., Thirunarayan, K., Marupudi, S., Sheth, A.P., Banerjee, T.: Understanding city traffic dynamics utilizing sensor and textual observations. In: AAAI, pp. 3793–3799 (2016)
6. Ángel, A.M., Bartolo, G.J., Ernestina, M.: Predicting recurring concepts on datastreams by means of a meta-model and a fuzzy similarity function. Expert Syst. Appl. **46**, 87–105 (2016)
7. Anwar, T.: Spatial partitioning of road traffic networks and their temporal evolution. Ph.D. thesis, Swinburne University of Technology (2017)
8. Arnott, R., Inci, E.: An integrated model of downtown parking and traffic congestion. J. Urban Econ. **60**(3), 418–442 (2006)
9. Asif, M.T., et al.: Spatiotemporal patterns in large-scale traffic speed prediction. IEEE Trans. Intell. Transp. Syst. **15**(2), 794–804 (2014)
10. Backfrieder, C., Ostermayer, G., Mecklenbräuker, C.F.: Increased traffic flow through node-based bottleneck prediction and V2X communication. IEEE Trans. Intell. Transp. Syst. **18**(2), 349–363 (2017)

11. Bifet, A., Holmes, G., Kirkby, R., Pfahringer, B.: MOA: massive online analysis. J. Mach. Learn. Res. **11**(May), 1601–1604 (2010)
12. Buyya, R., Dastjerdi, A.V.: Internet of Things: Principles and paradigms. Elsevier, Amsterdam (2016)
13. Castro-Neto, M., Jeong, Y.S., Jeong, M.K., Han, L.D.: Online-SVR for short-term traffic flow prediction under typical and atypical traffic conditions. Expert Syst. Appl. **36**(3), 6164–6173 (2009)
14. Chen, D.: Research on traffic flow prediction in the big data environment based on the improved RBF neural network. IEEE Trans. Ind. Inform. **13**, 2000–2008 (2017)
15. Clark, S.: Traffic prediction using multivariate nonparametric regression. J. Transp. Eng. **129**(2), 161–168 (2003)
16. City of Melbourne Data (2017). https://data.melbourne.vic.gov.au/. Accessed March 2017
17. De Francisci Morales, G., Bifet, A., Khan, L., Gama, J., Fan, W.: IoT big data stream mining. In: Proceedings of the 22nd ACM SIGKDD International Conference on Knowledge Discovery and Data Mining, pp. 2119–2120. ACM (2016)
18. Dell'Acqua, P., Bellotti, F., Berta, R., De Gloria, A.: Time-aware multivariate nearest neighbor regression methods for traffic flow prediction. IEEE Trans. Intell. Transp. Syst. **16**(6), 3393–3402 (2015)
19. Faria, E.R., Gonçalves, I.J., de Carvalho, A.C., Gama, J.: Novelty detection in data streams. Artif. Intell. Rev. **45**(2), 235–269 (2016)
20. Gama, J., Žliobaitė, I., Bifet, A., Pechenizkiy, M., Bouchachia, A.: A survey on concept drift adaptation. ACM Comput. Surv. (CSUR) **46**(4), 44 (2014)
21. Hajibaba, M., Gorgin, S.: A review on modern distributed computing paradigms: cloud computing, jungle computing and fog computing. J. Comput. Inform. Technol. **22**(2), 69–84 (2014)
22. Hamed, M.M., Al-Masaeid, H.R., Said, Z.M.B.: Short-term prediction of traffic volume in urban arterials. J. Transp. Eng. **121**(3), 249–254 (1995)
23. Henry, K.: To build or not to build: infrastructure challenges in the years ahead and the role of the government: address to the conference on the economics of infrastructure in a globalised world: issues, lessons and future challenges (2010)
24. Hu, P., Dhelim, S., Ning, H., Qiu, T.: Survey on fog computing: architecture, key technologies, applications and open issues. J. Netw. Comput. Appl. **98**, 27–42 (2017)
25. Ikonomovska, E., Gama, J., Džeroski, S.: Learning model trees from evolving data streams. Data Min. Knowl. Disc. **23**(1), 128–168 (2011)
26. Jeong, Y.S., Byon, Y.J., Castro-Neto, M.M., Easa, S.M.: Supervised weighting-online learning algorithm for short-term traffic flow prediction. IEEE Trans. Intell. Transp. Syst. **14**(4), 1700–1707 (2013)
27. Jia, R., Jiang, P., Liu, L., Cui, L., Shi, Y.: Data driven congestion trends prediction of urban transportation. IEEE Internet Things J. **5**, 581–591 (2017)
28. Kim, Y.J., Hong, J.S., et al.: Urban traffic flow prediction system using a multifactor pattern recognition model. IEEE Trans. Intell. Transp. Syst. **16**(5), 2744–2755 (2015)
29. Lana, I., Del Ser, J., Velez, M., Vlahogianni, E.I.: Road traffic forecasting: recent advances and new challenges. IEEE Intell. Transp. Syst. Mag. **10**(2), 93–109 (2018)
30. Lee, S., Fambro, D.: Application of subset autoregressive integrated moving average model for short-term freeway traffic volume forecasting. Transp. Res. Rec.: J. Transp. Res. Board **1678**, 179–188 (1999)

31. Levy, J.I., Buonocore, J.J., Von Stackelberg, K.: Evaluation of the public health impacts of traffic congestion: a health risk assessment. Environ. Health **9**(1), 65 (2010)
32. Lindley, J.A.: Urban freeway congestion: quantification of the problem and effectiveness of potential solutions. ITE J. **57**(1), 27–32 (1987)
33. Lopez-Garcia, P., Onieva, E., Osaba, E., Masegosa, A.D., Perallos, A.: A hybrid method for short-term traffic congestion forecasting using genetic algorithms and cross entropy. IEEE Trans. Intell. Transp. Syst. **17**(2), 557–569 (2016)
34. Min, X., Hu, J., Zhang, Z.: Urban traffic network modeling and short-term traffic flow forecasting based on GSTARIMA model. In: 2010 13th International IEEE Conference on Intelligent Transportation Systems (ITSC), pp. 1535–1540. IEEE (2010)
35. Mitrovic, N., Asif, M.T., Dauwels, J., Jaillet, P.: Low-dimensional models for compressed sensing and prediction of large-scale traffic data. IEEE Trans. Intell. Transp. Syst. **16**(5), 2949–2954 (2015)
36. Mouss, H., Mouss, D., Mouss, N., Sefouhi, L.: Test of page-hinckley, an approach for fault detection in an agro-alimentary production system. In: 2004 5th Asian Control Conference (IEEE Cat. No. 04EX904), vol. 2, pp. 815–818. IEEE (2004)
37. Networking, T.D.: Connected vehicular transportation. IEEE Veh. Technol. Mag. **12**, 42–54 (2017)
38. Schrank, D., Eisele, B., Lomax, T., Bak, J.: Urban mobility scorecard. Texas A&M Transportation Institute and the Texas A&M University System (2015)
39. Stojmenovic, I., Wen, S.: The fog computing paradigm: scenarios and security issues. In: 2014 Federated Conference on Computer Science and Information Systems (FedCSIS), pp. 1–8. IEEE (2014)
40. Bureau of Transport and Regional Economics: Estimating urban traffic and congestion cost trends for Australian cities (2007)
41. Wibisono, A., Jatmiko, W., Wisesa, H.A., Hardjono, B., Mursanto, P.: Traffic big data prediction and visualization using fast incremental model trees-drift detection (FIMT-DD). Knowl.-Based Syst. **93**, 33–46 (2016)
42. Williams, B.M., Hoel, L.A.: Modeling and forecasting vehicular traffic flow as a seasonal arima process: theoretical basis and empirical results. J. Transp. Eng. **129**(6), 664–672 (2003)
43. Wongcharoen, S., Senivongse, T.: Twitter analysis of road traffic congestion severity estimation. In: 2016 13th International Joint Conference on Computer Science and Software Engineering (JCSSE), pp. 1–6. IEEE (2016)
44. Xiao, J., Xiao, Z., Wang, D., Bai, J., Havyarimana, V., Zeng, F.: Short-term traffic volume prediction by ensemble learning in concept drifting environments. Knowl.-Based Syst. **164**, 213–225 (2019)
45. Yang, Z., Al-Dahidi, S., Baraldi, P., Zio, E., Montelatici, L.: A novel concept drift detection method for incremental learning in nonstationary environments. IEEE Trans. Neural Netw. Learn. Syst. (2019)
46. Yannuzzi, M., Milito, R., Serral-Gracià, R., Montero, D., Nemirovsky, M.: Key ingredients in an IoT recipe: fog computing, cloud computing, and more fog computing. In: 2014 IEEE 19th International Workshop on Computer Aided Modeling and Design of Communication Links and Networks (CAMAD), pp. 325–329. IEEE (2014)
47. Yi, S., Hao, Z., Qin, Z., Li, Q.: Fog computing: platform and applications. In: 2015 Third IEEE Workshop on Hot Topics in Web Systems and Technologies (HotWeb), pp. 73–78. IEEE (2015)

48. Yi, S., Li, C., Li, Q.: A survey of fog computing: concepts, applications and issues. In: Proceedings of the 2015 Workshop on Mobile Big Data, pp. 37–42. ACM (2015)
49. Zhan, H., Gomes, G., Li, X.S., Madduri, K., Sim, A., Wu, K.: Consensus ensemble system for traffic flow prediction. IEEE Trans. Intell. Transp. Syst. **19**, 3903–3914 (2018)
50. Zhang, J., Wang, F.Y., Wang, K., Lin, W.H., Xu, X., Chen, C.: Data-driven intelligent transportation systems: a survey. IEEE Trans. Intell. Transp. Syst. **12**(4), 1624–1639 (2011)
51. Zhang, W., Qi, G., Pan, G., Lu, H., Li, S., Wu, Z.: City-scale social event detection and evaluation with taxi traces. ACM Trans. Intell. Syst. Technol. (TIST) **6**(3), 40 (2015)
52. Zhang, Y., Niyato, D., Wang, P., Kim, D.I.: Optimal energy management policy of mobile energy gateway. IEEE Trans. Veh. Technol. **65**(5), 3685–3699 (2016)
53. Zhao, J., Sun, S.: High-order gaussian process dynamical models for traffic flow prediction. IEEE Trans. Intell. Transp. Syst. **17**(7), 2014–2019 (2016)

# Service Dependability and Security

# Semi-supervised Deep Learning for Network Anomaly Detection

Yuanyuan Sun[1,2,3(✉)], Lili Guo[3(✉)], Ye Li[3], Lele Xu[3],
and Yongming Wang[2]

[1] School of Cyber Security, University of Chinese Academy of Sciences,
Beijing, China
[2] Institute of Information Engineering, Chinese Academy of Sciences,
Beijing, China
[3] Key Laboratory of Space Utilization, Technology and Engineering Center
for Space Utilization, Chinese Academy of Sciences, Beijing, China
{sunyuanyuan, guolili}@csu.ac.cn

**Abstract.** Deep learning promotes the fields of image processing, machine translation and natural language processing etc. It also can be used in network anomaly detection. In practice, it is not hard to obtain normal instances. However, it is always difficult to label anomalous instances. Semi-supervised learning can be utilized to resolve this problem. In this paper, we make a comprehensive study of semi-supervised deep learning techniques for network anomaly detection. Three kinds of deep learning techniques including GAN (Generative Adversarial networks), Auto-encoder and LSTM (Long Short-Term Memory) are studied on the latest network traffic dataset of CICIDS2017. Five deep architectures based on semi-supervised learning are designed, including BiGAN, regular GAN, WGAN, Auto-encoder and LSTM. Seven schemes of semi-supervised deep learning for anomaly detection are proposed according to different functions of anomaly score. Grid search is utilized to find the threshold of anomaly detection. Two traditional schemes of machine learning are also adopted to compare performance. There are altogether nine schemes of anomaly detection for CICIDS2017. From results of the experiment for network anomaly detection, it can be found that Auto-encoder outperforms LSTM and the three kinds of GAN. BiGAN and LSTM are both better than WGAN and regular GAN. All the seven schemes of semi-supervised deep learning for anomaly detection outperform the two traditional schemes. The work and results in this paper are meaningful on the application of semi-supervised deep learning for network anomaly detection.

**Keywords:** Deep learning · Network anomaly detection · Auto-encoder · LSTM · BiGAN · Regular GAN · WGAN

## 1 Introduction

With the development of science and technology, network scale is increasing rapidly. The types of network are manifold, including Internet, mobile network, IoT (Internet of things), WSN (wireless sensor network), and industrial control network etc. There is a

S. Wen et al. (Eds.): ICA3PP 2019, LNCS 11945, pp. 383–390, 2020.
https://doi.org/10.1007/978-3-030-38961-1_33

pressing problem. How to guarantee the safety and reliability of network? The defects of network and various malicious attacks can lead to network anomaly.

Anomaly detection is to separate abnormal behaviors and characteristics from normal instances by specific methods. Anomaly detection can be widely used in various fields of intrusion detection, sensor networks, and fraud detection etc. Machine learning techniques have been used for anomaly detection for last two decades. As we known, there are three types of learning in machine learning, which are supervised learning, unsupervised learning, and semi-supervised learning.

In supervised learning, labels of all the samples are needed [1]. There are two problems for anomaly detection by supervised learning. One problem is that acquiring labels of instances is labor-intensive and time consuming. The other problem is that anomaly samples are always not easy to acquire. In most circumstance, only normal samples can be acquired.

In unsupervised learning, labels are not needed [2]. The typical example for unsupervised learning is clustering. Classical algorithms of clustering include K-means, K-medoids, CLARANS, DBSCAN, BIRCH, and CURE. There are no labels in the process of training by unsupervised learning, while the results of learning are always unsatisfactory.

In semi-supervised learning, a small number of labeled samples are utilized to train a learning model and the model can be used to predict the unlabeled samples [3]. For network anomaly detection, in most cases, normal labels can be acquired. But anomalous activities change constantly. It is a hard task to label the unknown anomalous activities. In this situation, semi-supervised learning is a good option for network anomaly detection.

It is well known that deep learning is an emerging branch of machine learning in recent years. Deep learning techniques power the domains of image processing, natural language processing, network information recommendation and machine translation etc. It is a promising branch to replace traditional machine learning techniques [4]. In terms of deep learning, there are representative techniques including CNN (Convolutional Neural Networks) [5], LSTM (Long Short-Term Memory) [6], Auto-encoder (AE) [7], and GAN (Generative Adversarial Networks) [8]. While most deep learning methods are supervised learning which still need the labels of normal and abnormal instances.

In [9], some kinds of traditional supervised learning techniques are adopted for an excellent network traffic dataset of CICIDS2017. In this paper, we make a comprehensive study of semi-supervised deep learning techniques for network anomaly detection. Seven anomaly detection schemes of semi-supervised deep learning are proposed for CICIDS2017.

The contributions of this paper are as follows:

– Nine schemes including seven schemes of semi-supervised deep learning and two traditional schemes are compared for network anomaly detection on the dataset of CICIDS 2017. To the best of our knowledge, semi-supervised deep learning techniques are first applied on analyzing the latest dataset of CICIDS2017. The work and results in this paper are meaningful on the application of semi-supervised deep learning for network anomaly detection [10].

- We study five deep architectures including BiGAN [11], regular GAN [12], WGAN [13], Auto-encoder [14, 15] and LSTM [16, 17] for semi-supervised anomaly detection. By different anomaly score functions, seven semi-supervised deep learning schemes for anomaly detection are proposed for CICIDS2017. Other two traditional schemes including one class SVM and K-means are utilized in the experiment for comparing performance.
- It is quite difficult to train the model of GAN. BiGAN, regular GAN, and WGAN are designed and successfully trained on CICIDS2017. Batch normalization and feature matching techniques are adopted in order to get more stable model of GAN.
- The deep architectures of Auto-encoder and LSTM are designed for the dataset of CICIDS2017. Two functions of anomaly score based on MAE and MSE for Auto-encoder and LSTM are utilized, respectively.

## 2   Experimental Methodology

In this section, we will introduce the methods used in the experiment. The experiment is done on GPU TITAN X. The runtime environment includes python3.7, Keras, and Tensorflow.

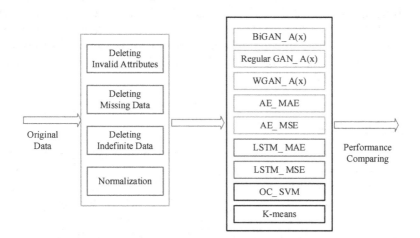

**Fig. 1.** The flow chart of experiment

The flow chart of experiment is shown in Fig. 1. Firstly, original data is processed by deleting invalid, missing and indefinite data. After normalization, data is detected by nine anomaly schemes proposed in the experiment. The nine schemes are as follows:

- BiGAN_A(x); Regular GAN_A(x); WGAN_A(x);
- AE_MAE; AE_MSE;
- LSTM_MAE; LSTM_MSE;
- OC-SVM, K-means.

For the first seven names of semi-supervised deep learning schemes, the left part of underscore represents the deep architecture and the right part of underscore represents the function of anomaly score.

A(x) represents the function of anomaly score which is the same with that in [18]; MAE means that the function of anomaly score is mean absolute error; MSE means that function of anomaly score is mean squared error. Grid search is utilized to find the threshold of anomaly detection. If the anomaly score of instance is higher than the threshold, the instance is anomalous. If the anomaly score of instance is lower than the threshold, the instance is normal.

The results of nine anomaly detection schemes are compared by metrics of precision, recall and F1-measure.

## 2.1 Dataset

In this paper, we choose the dataset of CICIDS2017 [9, 19] to conduct the experiment. CICIDS2017 is alike to the data of real-world. It includes the most up-to-date common attack types and normal activities. The dataset is labeled by CICFlowMeter. The dataset includes data of five days from Monday to Friday. There are only normal activities in Monday. From Tuesday to Friday, besides normal activities, many types of attacks occur, such as Botnet, DDoS, DoS, Heartbleed, Brute Force FTP, Web Attack, Brute Force SSH, and Infiltration. The data of Wednesday are selected for this experiment.

## 2.2 Data Processing

The dataset of CICIDS2017 is CSV file. We find that, there is one repetitive attribute and some invalid attributes in which all the values are 0. These attributes are deleted directly by the software of Excel. At last, there are 67 attributes left.

Before normalization, the missing instances are deleted for minimizing the uncertainty, other than using '0', '1' or mean of attributes to replace missing value. This is done by Pandas function "dropna". After this step, the infinite instances also have to be deleted. Otherwise, normalization could not be accomplished.

We use Z-score method to normalize the data to interval [−1, 1]. The values of attributes are subject to standard normal distribution.

$$z = \frac{x_i - \mu}{\delta} \tag{1}$$

## 3   Results of the Experiment

In the data of Wednesday from CICIDS2017, 80% of normal instances are used for training. The rest instances are used for validation set and test set. The ratio of instances for validation set and test set is 1:1. The metrics of precision, recall, and F1-measure are adopted in the experiment. The results are based on test set, which are shown in Table 1.

**Table 1.** The performance of nine schemes for network anomaly detection

| Scheme | Precision | Recall | F1-measure |
|---|---|---|---|
| BiGAN_A(x) | 0.845 | 0.999 | 0.915 |
| WGAN_A(x) | 0.786 | 0.999 | 0.879 |
| Regular GAN_A(x) | 0.762 | 0.993 | 0.863 |
| AE-MAE | 0.890 | 0.996 | 0.940 |
| AE-MSE | 0.866 | 0.997 | 0.927 |
| LSTM-MAE | 0.820 | 0.993 | 0.898 |
| LSTM-MSE | 0.813 | 0.977 | 0.887 |
| K-means | 0.410 | 0.310 | 0.350 |
| OC-SVM | 0.880 | 0.790 | 0.810 |

It can be compared in the three anomaly detection schemes for CICIDS2017 by three kinds of GAN. In BiGAN_A(x), Regular GAN_A(x) and WGAN_A(x), it is clearly that BiGAN_A(x) has the best performance. Its metrics of precision, recall, and F1-measure are 0.845, 0.999 and 0.915 respectively. The precision and F1-measure of BiGAN are higher than those of Regular GAN and WGAN. The recall of BiGAN is equal to that of WGAN and higher than that of Regular GAN. As for WGAN, its precision, recall and F1-measure are all higher than those of Regular GAN.

By the comparison, the performance of BiGAN is the best one in the three schemes of GAN. WGAN is better than regular GAN. The training loss of BiGAN can be seen in Fig. 2, the loss curves of encoder, generator and discriminator converged at about the epoch of 15000.

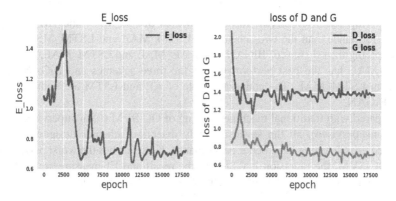

**Fig. 2.** Loss curves of encoder, generator, and discriminator for BiGAN

The two anomaly detection schemes of Auto-encoder (AE) adopt MSE and MAE as the functions of anomaly score respectively. The MSE and MAE curves of Auto-encoder are illustrated in Fig. 3. During the training of Auto-encoder, MSE is selected as loss function. The curve of MSE is also the curve of loss. It is shown that from the MSE curve in the left part of Fig. 3, at about the epoch of 12, the model of Auto-encoder becomes stable. We can see the results of anomaly detection by schemes of AE_MAE and AE_MSE from Table 1. The performance of AE_MAE is better than the performance of AE_MSE. For AE-MAE, the precision and F1-measure are both higher than those of AE_MSE. Especially F1-Measure is 0.940 for AE_MAE and 0.927 for AE_MSE. Though the recall of AE_MSE is higher than that of AE_MAE, we can still judge that the scheme of AE_MAE is better than the scheme of AE_MSE. Here, F1-Measure is the comprehensive metric of recall and precision. When there is a conflict between recall and precision, F1-measure can play an important role.

The two anomaly detection schemes of LSTM also adopt MAE and MSE as the functions of anomaly score. From Table 1, it is obvious that LSTM_MAE has better performance than LSTM_MSE. The precision, recall and F1-measure of LSTM_MAE are all higher than those of LSTM_ MSE. The MAE and MSE curves of LSTM are illustrated in Fig. 4. During the training of LSTM, MAE is selected as loss function. The curve of MAE is also the curve of loss. From the loss curve in the left part of Fig. 4, it can be seen that the model of LSTM converged at about the epoch of 10.

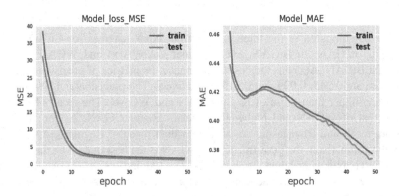

**Fig. 3.** MSE and MAE curves of Auto-encoder

Comparing AE with LSTM, the two schemes of AE_MAE and AE_MSE have better performance than the two schemes of LSTM_MAE and LSTM_MSE.

Comparing the schemes of BiGAN_A(x), AE_MAE and LSTM_MAE, we can find that AE_MAE has the best performance in the three schemes and BiGAN_A(x) is better than LSTM_MAE. The converging rate of AE and LSTM are both faster than that of BiGAN.

Compared with traditional machine learning of OC-SVM (semi-supervised) and K-means (unsupervised), all the seven anomaly detection schemes designed by semi-supervised deep learning have better performance. K-means has the worst performance. As for OC-SVM, the precision is 0.88, but the recall and F1-measure are both lower compared with the seven schemes of semi-supervised deep learning.

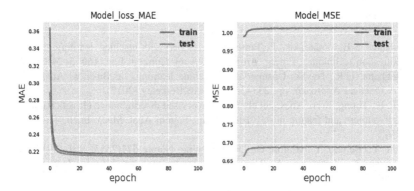

**Fig. 4.** MSE and MAE curves of LSTM

It is obvious that from the experiment, the anomaly detection scheme of Auto-encoder has best performance. The scheme of BiGAN is better than LSTM. The scheme of LSTM is better than the two schemes of Regular GAN and WGAN. The schemes of Regular GAN and WGAN are not good enough in the seven schemes of semi-supervised deep learning, but they are still better than OC-SVM and K-means.

## 4 Conclusion

In this paper, we make a comprehensive study of semi-unsupervised deep learning techniques for network anomaly detection. The network traffic dataset of CICIDS2017, which contains the most up-to-date attack types, is selected in the experiment. Five architectures of BiGAN, regular GAN, WGAN, Auto-encoder and LSTM are designed for network anomaly detection. Nine schemes including seven semi-supervised deep learning schemes and two traditional schemes for network anomaly detection are compared in the experiment. A(x) is selected as anomaly score function for BiGAN, regular GAN and WGAN. MAE and MSE are selected as anomaly score functions for Auto-encoder and LSTM. Grid search is utilized to determine the threshold of anomaly detection. From the experiment, we find that, two schemes of Auto-encoder perform best in the nine schemes. The scheme of BiGAN is better than the two schemes of LSTM. The two schemes of LSTM are both better than the schemes of WGAN and regular GAN. The scheme of selecting MAE as anomaly score has better performance than the scheme of selecting MSE. All the seven schemes of semi-unsupervised deep learning outperform the other two schemes of traditional machine learning. The work and results in this paper are meaningful on applying semi-supervised deep learning for network anomaly detection.

**Acknowledgement.** This work is supported by the National Natural Science Foundation of China (No. 61901454), and the Foundation of key Laboratory of Space Utilization, Technology and Engineering Center for Space utilization Chinese Academy of Sciences (No. CSU-QZKT-2018-08).

# References

1. Kotsiantis, S.B., Zaharakis, I., Pintelas, P.: Supervised machine learning: a review of classification techniques. Emerg. Artif. Intell. Appl. Comput. Eng. **160**, 3–24 (2007)
2. Hodeghatta, U.R., Nayak: Unsupervised machine learning. In: Business Analytics Using R - A Practical Approach, pp. 233–255. Apress, Berkeley (2017)
3. Adeli, E., Thung, K.H., An, L., et al.: Semi-supervised discriminative classification robust to sample-outliers and feature-noises. IEEE Trans. Pattern Anal. Mach. Intell. **41**(2), 515–522 (2019)
4. Lecun, Y., Bengio, Y., Hinton, G.: Deep learning. Nature **521**(7553), 436 (2015)
5. Simonyan, K., Zisserman, A.: Very deep convolutional networks for large-scale image recognition. Comput. Sci. (2014)
6. Tai, K.S., Socher, R., Manning, C.D.: Improved semantic representations from tree-structured long short-term memory networks. Comput. Sci. **5**(1), 36 (2015)
7. Chandar, A.P.S., Lauly, S., Larochelle, H., et al.: An autoencoder approach to learning bilingual word representations In: International Conference on Neural Information Processing Systems (2014)
8. Goodfellow, I.J., Pouget-Abadie, J., Mirza, M., et al.: Generative adversarial nets In: International Conference on Neural Information Processing Systems (2014)
9. Sharafaldin, I., Lashkari, A.H., Ghorbani, A.A.: Toward generating a new intrusion detection dataset and intrusion traffic characterization. In: 4th International Conference on Information Systems Security and Privacy (ICISSP), Portugal, January 2018
10. Springenberg, J.T.: Unsupervised and semi-supervised learning with categorical generative adversarial networks. Comput. Sci. (2015)
11. Donahue, J., Krähenbühl, P., Darrell, T.: Adversarial feature learning. arXiv preprint arXiv: 1605.09782 (2016)
12. Goodfellow, I.J., et al.: Generative adversarial nets. In: International Conference on Neural Information Processing Systems (2014)
13. Arjovsky, M., Chintala, S., Bottou, L.: Wasserstein GAN. arXiv preprint arXiv:1701.07875 (2017)
14. Zhang, J., Wang, H., Yang, H.: Dimension reduction method of high resolution range profile based on Autoencoder. J. Pla Univ. Sci. Technol. (2016)
15. Sakurada, M., Yairi, T.: Anomaly detection using autoencoders with nonlinear dimensionality reduction. In: Mlsda Workshop on Machine Learning for Sensory Data Analysis (2014)
16. Hochreiter, S., Schmidhuber, J.: Long short-term memory. Neural Comput. **9**(8), 1735–1780 (1997)
17. Jason Brownlee Blog. https://machinelearningmastery.com/convert-time-series-upervised-learning-problem-python/. Accessed 25 June 2019
18. Zenati, H., Foo, C.S., Lecouat, B., et al.: Efficient gan-based anomaly detection. arXiv preprint arXiv:1802.06222 (2018)
19. UNB. https://www.unb.ca/cic/datasets/index.html. Accessed 25 June 2019

# A Vulnerability Assessment Method for Network System Based on Cooperative Game Theory

Chenjian Duan, Zhen Wang$^{(\boxtimes)}$, Hong Ding, Mengting Jiang, Yizhi Ren, and Ting Wu

School of Cyberspace, Hangzhou Dianzi University, Hangzhou, China
duanchenjian1018@gmail.com, {wangzhen,dinghong,
renyizhi,wuting}@hdu.edu.cn, jiangmt9706@gmail.com

**Abstract.** It is very important for administrators to understand the severity of vulnerabilities in network systems. Although many systems such as CVSS can evaluate individual vulnerabilities, they do not take into account the specific environment, so the results are not helpful. In our paper, we construct a vulnerability dependency graph by modeling the complex dependencies between vulnerabilities, and introduce the Shapley value in the cooperative game. We consider an attack path as a cooperation between the vulnerability nodes, and use Access Complexity as the attack cost of each node, define the characteristic function in the cooperative. Finally, according to the Shapley value of each node, all the vulnerabilities are ranked, and the administrator can patch the high-rank vulnerabilities with the limited security resources. Our experimental results demonstrate that show that our method can more effectively assess the severity of vulnerabilities in specific environments.

**Keywords:** Vulnerability ranking · Shapley value · Vulnerability Dependency Graph

## 1 Introduction

Enterprise networks are essential for companies, government agencies, and universities, but as the size of the network grows, people are beginning to pay attention to security issues. Vulnerability is a key factor affecting network security. In general, the administrator will analyze the vulnerabilities in the network through vulnerability scanning tools, and choose some to patch. The practice has shown that although some isolated vulnerabilities have low impact, they are often related. If this relation is exploited by attackers through the network,

---

This work was supported by the Natural Science Foundation of Zhejiang Province (Grant No. LY18F020017, LY18F030007 and LQY19G030001), National Natural Science Foundation of China (Grant No. 61872120) and key technologies, system and application of Cyberspace Big Search, Major project of Zhejiang Lab (Grant No. 2019DH0ZX01).

© Springer Nature Switzerland AG 2020
S. Wen et al. (Eds.): ICA3PP 2019, LNCS 11945, pp. 391–398, 2020.
https://doi.org/10.1007/978-3-030-38961-1_34

the probability of a successful attack will be greatly improved. Therefore, it is worthwhile to study the vulnerability assessment in consideration of the specific network environment.

The vulnerability assessment of network systems presents significant challenges. First of all, because the real network is large in scale and has many vulnerabilities, analyzing the dependencies will become very complicated. Second, the same vulnerability can occur on different attack paths, and it plays different roles in different attack paths, and the benefits of each path are different, which makes the problem more complicated. The problem in some works is that their assessment method is not designed for vulnerability assessment, nor does it give a list of vulnerability rankings.

The existing vulnerability assessment methods are mainly CVSS [5] which is a published standard used by organizations worldwide. Therefore, we use the Vulnerability Dependency Graph to model the vulnerability dependencies in the network, then use the metrics of CVSS as the weight of the nodes, and introduce the idea of marginal contribution in the cooperative game. In this model, we calculate the Shapley value of each node and rank the vulnerabilities according to it. The results show that our method performs better than other papers.

## 2   Related Work

There are a lot of works based on CVSS and attack graph to assess the network environment. Miura-Ko et al. [6] raised a new scheme called SecureRank for prioritizing vulnerabilities. SecureRank takes into account the network topology and potential node interactions in calculating their relative risk and prioritizes vulnerabilities. Sawilla et al. [1] proposed AssetRank, a generalization of Google's PageRank algorithm. AssetRank addresses the unique semantics of dependency attack graphs to compute metrics. Homer et al. [2] use the existing work in attack graphs and probabilistic reasoning to generate a reliable quantitative risk model, and this model can handle cycles correctly. Li et al. [3] proposed an algorithm named NodeRank in state enumeration attack graphs, which not only the network topology relationship is put into consideration, but also they consider the effects of nodes' intrinsic attributes. Wang [4] et al. use Bayesian Attack Graph to model the attack events and the vulnerabilities. Their approach named HTV can identify the vulnerabilities in the system correctly and availably.

The attack graph is a common tool for us to analyze the security of the network system. Li et al. [7] based on system vulnerability data, system configuration data, and vulnerability scanner results defined exploitation graphs. Yang et al. [8] propose a new methodology called DBRank which prioritizes vulnerabilities based on the diffusibility and benefit of vulnerability exploitation. Jiang et al. [9] score and rank vulnerabilities in SOA. Their framework, VRank, which takes into account the contexts of the services having this vulnerability. Zhuang et al. [10] proposed a practical framework, NCVS, that offers an automatic and contextual scoring mechanism to assess the severity of vulnerabilities for cloud service.

Another line of research is about cooperative game theory. Shapley proposed to evaluate the role played by individual players in a coalitional game by comparing their marginal contributions to every possible coalition [11]. Up to now, Shapley values have been used in research in various fields. Michalak et al. [12] applied the Shapley value to find key players in terrorist networks. Szczepański et al. proposed algorithms for calculating the betweenness centrality based on the Shapley value in 2016 years of work [13].

# 3    Model

## 3.1    Network Environment

In a network system, all devices are represented as *Device*, and one of the devices is denoted by $d$, e.g. $d \in Device$. For each vulnerability, it can be evaluated by the existing standard like CVSS. The CVSS score is represented by $Impact(v)$. There is an exploitability metrics named *Access Complexity* in the standard of CVSS. This metric describes the conditions beyond the attacker's control that must exist in order to exploit the vulnerability, so we can use $AC(v)$ to represent the cost of an attacker exploiting a vulnerability.

## 3.2    Vulnerability Dependency Graph

We model the complex dependencies between vulnerabilities as vulnerability dependency graph and then analyze it. VDG is an abstract concept based on attack graph [7]. There are dependencies between vulnerabilities in VDG.

In our implementation, the vulnerability dependency graph is modeled as a direct graph $G = (V, E)$, and we use $V$ to denote the set of vulnerabilities and the total number of vulnerabilities is $N = |V|$. We use a set of edges $E$ to denote the relationship between vulnerabilities. As mentioned early, VDG is a directed graph, so we denote the in-degree and out-degree of a vertex $v$ as $d_{in}(v)$, and $d_{out}(v)$, respectively. Then for a vertex $v$, we denote $pre(v)$ as the set of predecessors, similarly, $post(v)$ is the set of successors. We define a path $p$ which is an ordered set of vertices beginning with the vertex which $d_{in}(v) = 0$, and we can use algorithm to obtain all the path sets $P$, e.g. $p \in P$. We assume that there is no loop in the graph. In order to facilitate the analysis of the boundary nodes, add a virtual node $S$ to represent the external attacker.

## 3.3    Shapley Value Based on VDG

With the dependency relationship in the vulnerability dependency graph, we use the idea of the cooperative game to regard each vulnerability node as an agent and regard an attacker's attack as a cooperation between these agents. We apply one of the fundamental concepts named the Shapley value in cooperative game theory to rank these vulnerabilities. Specifically, Shapley value is proposed to evaluate the role played by individual players in a coalitional game by comparing their marginal contributions to every possible coalition [12].

In order to formalize this concept, let $\pi \in \prod(V)$ denote a permutation of players in $V$, let $X_i^\pi$ denote the coalition made of all predecessors of node $v_i$ in

$\pi$. We denote the location of $v_j$ in $\pi$ by $\pi(j)$. In the vulnerability dependency graph context, we define every single coalition denoted by $c$ is an attack path $p$, and the Shapley value of $v_i$ denoted $SV_i(F)$ is then defined as the average marginal contribution of $v_i$ to coalition $X_i^\pi$ over all $p \in P$:

$$SV_i(F) = \frac{1}{N!} \sum_{c \in P} [F(X_i^c \cup v_i) - F(X_i^c)] \tag{1}$$

where $F$ is characteristic function, in coalitional games this function, $F : 2^V \to \mathbb{R}$, assigns to every coalition $c$ a numerical value representing its performance. And $F(X_i^\pi \cup v_i) - F(X_i^\pi)$ is the characteristic function difference before and after adding the node $v_i$, that is the marginal contribution.

### 3.4   Weighted Shapley Value

In the process of calculating the Shapley value, if we consider the importance of different individuals players, this generalization of the Shapley value is called weighted Shapley value. Castro et al. [14] define this concept. Let $w$ be a weight vector, where $w_i > 0$ represents the weight of the player $i$. Given $w$ and a permutation $\pi$, a probability function that represents the probability of the different orders in $\prod(N)$ can be defined as follows:

$$Pro_w(\pi) = \prod_{k=1}^{n} \frac{w_{\pi(k)}}{\sum_{l=1}^{k} w_{\pi(l)}} \tag{2}$$

According to Eq. 2, we consider *Access Complexity* of each vulnerability node as a weight vector. Finally, the weighted Shapley value of each vulnerability node $v_i$ is defined as:

$$SV_i(F) = \sum_{c \in P} \prod_{k=1}^{|c|} \frac{AC_{c(k)}}{\sum_{l=1}^{k} AC_{c(l)}} [F(X_i^c \cup v_i) - F(X_i^c)] \tag{3}$$

In general, the characteristic function represents the performance of the coalitional games, but in our paper, the performance of a game can also be called the utility of the attack. In the previous section, we converted the cost of vulnerability which is exploited by attackers into the weight of the node and calculated the probability of each path in the weighted Shapley values. For any coalition $c$, we define its characteristic function as follows:

$$F(c) = \sum_{v \in c} Impact(v) * length(v) \tag{4}$$

where $length(v)$ is the shortest path length from the virtual node $S$ to the node $v$.

## 4   Algorithm

We use Algorithm 1 to assess the vulnerabilities. After calculating the Shapley value of each vulnerability node, we rank the vulnerabilities based on the value.

**Algorithm 1.** Assess algorithm

---

**Input:** the VDG
**Output:** the vulnerability ranking list
1: $P \leftarrow$ EnumerationPaths();
2: $w \leftarrow$ CalcuateWeight();
3: list $\leftarrow$ CalcuateShapleyValue($P$, $w$);
4: **return** list

---

**Algorithm 2.** EnumerationPaths

---

**Input:** the VDG
**Output:** all the path sets $P$
$findpath$(start, end, visited[], Stack $s$)
$EnumerationPaths()$
1: Initialize visited[$N$+1]$\leftarrow$0, s$\leftarrow$ empty stack;
2: **for** $v \in V$ **do**
3:     **if** $v$ is target node **then**
4:         findpath($N$, $v$, visited, $s$)
5:         visited[$N$+1]$\leftarrow$0
6:         **while** $s$ not empty **do**
7:             pop $s$
8: **return** $P$

---

In Algorithm 1, all paths are obtained by Algorithm 2, then the weight of each path is calculated according to Eq. 2, the Shapley value of each node is calculated according to Eq. 3, and finally, the rank result is returned.

The function $findpath$ is a recursive function for finding a path from $start$ to $end$, and target node in the line 3 of function $EnumerationPaths$ indicates a node where $d_{in}(v) = 0$ without considering the virtual node $S$ in the pseudo-code of Algorithm 2.

## 5   Result

We use an example networked system in Fig. 1 to show our model and result. We used the same experimental environment as the paper [15] to compare experiments and ensure the objectivity of the results. There are a total of six devices including WEB server($v_2$,$v_3$), DNS server($v_1$), FTP server($v_{63}$), database server($v_7$), Windows host($v_5$), and Linux host($v_4$).

According to the impact and *Access Complexity* of nodes and Eq. 3, we can calculate the Shapley values of each node in Fig. 2, the results are shown in Table 1.

As can be seen from Table 1, the Shapley values of the nodes $v_1$, $v_4$, $v_3$ are higher than others, so they are more harmful in the network. Because $v_1$ is one of the starting nodes in this network, this node will be a breakthrough for successful attacks. And $v_2$, which is also the starting node, has only one successor node, so

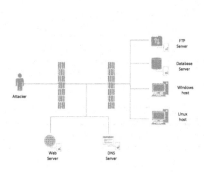

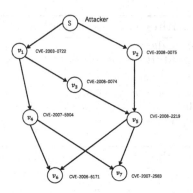

**Fig. 1.** Experimental network

**Fig. 2.** Vulnerability dependency graphs of experimental network

the harm is lower than $v_1$. As for $v_3$ and $v_4$, they are more harmful than other nodes because they are the first successor to the starting node $v_1$ and serve as a bridge in multiple attack paths.

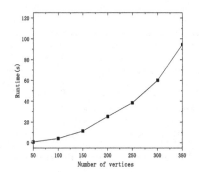

**Fig. 3.** The time to compute Shapley values at different network sizes

**Table 1.** Shapley values of vulnerability nodes

| Node | Shapley value |
|------|---------------|
| $v_1$ | 17.7463 |
| $v_2$ | 8.77885 |
| $v_3$ | 11.1209 |
| $v_4$ | 13.6318 |
| $v_5$ | 9.89147 |
| $v_6$ | 8.55995 |
| $v_7$ | 4.56531 |

## 5.1   Scalability Analysis

Unfortunately, the problem of computing the Shapley value is an NP-complete problem [16]. To illustrate this, we conduct experiments on synthetic networks as follows.

We use the Scale-Free network generated by Barabási-Albert model [17] to simulate large-scale VDG. Then we use the JSON Feed CVE-2019 published in [18]. Finally, we associate each node with a vulnerability randomly.

We perform all the experiments on a PC with 1.60 GHz double core CPU and 8.00 GB memory. The time of enumerating all paths are also taken into account. We get the average time over 50 samples for each size network. The result is shown in Fig. 3.

It can be seen from Fig. 3 that the calculation time is exponential, in this case, the vulnerability dependency graph becomes very large and difficult to analyze, so we only consider the effectiveness of this method in small-scale networks.

## 5.2   Solution Quality Analysis

CVSS provides a way to produce a numerical score reflecting its severity. Columns 2 and 3 of the Table 2 gives a comparison of our method and CVSS.

**Table 2.** A vulnerability rank list compared to CVSS

| Rank | Our method | CVSS | BM |
|---|---|---|---|
| 1 | CVE-2003-0722 | CVE-2003-0722 | CVE-2008-2219 |
| 2 | CVE-2007-5904 | CVE-2008-0075 | CVE-2007-5904 |
| 3 | CVE-2008-0074 | CVE-2006-6171 | CVE-2003-0722 |
| 4 | CVE-2008-2219 | CVE-2008-0074 | CVE-2008-0075 |
| 5 | CVE-2008-0075 | CVE-2007-5904 | CVE-2008-0074 |
| 6 | CVE-2006-6171 | CVE-2008-2219 | |
| 7 | CVE-2007-2583 | CVE-2007-2583 | |

In the CVSS standard, the impact of CVE-2006-6761 is greater than that of CVE-2007-5904, but in the experimental network, CVE-2007-5904 is the predecessor of CVE-2006-6761. If the attacker fails to exploit CVE-2007-5904, it is impossible to exploit CVE-2006-6761, so CVE-2007-5904 is more severe than CVE-2006-6761, this phenomenon can be reflected in our results. So our method is more effective than CVSS while considering the network environment.

Our paper uses the same example of the work of Jia et al. [15]. We call their method the modified betweenness (MB). Columns 2 and 4 of the Table 2 gives a comparison of our method and MB.

It can be seen that among the top three vulnerabilities, only one of our results is different from BM. According to the definition of Shapley value, if a node can provide more a marginal contribution in cooperation, its Shapley value will be higher. And node $v_3$ acts as a bridge in the network, so that the attacker has more choices from the starting node $S$ to $v_5$. Our method is more effective than BM in analyzing the effects of the same node on different paths.

## 6   Conclusions and Future Work

This paper proposes a vulnerability assessment method based on Shapley value, which combines the concept of cooperative game with CVSS standard. Our method fills in the gaps in which CVSS does not consider where the vulnerability is located, and compared with other methods, we find that our method is more reasonable and accurate.

However, it is still very difficult to analyze large-scale vulnerability dependency graphs. The problem of computing the Shapley value is an NP-complete problem, and we have not given a solution algorithm. We will focus on the problem in future work.

# References

1. Sawilla, R.E., Ou, X.: Identifying critical attack assets in dependency attack graphs. In: Jajodia, S., Lopez, J. (eds.) ESORICS 2008. LNCS, vol. 5283, pp. 18–34. Springer, Heidelberg (2008). https://doi.org/10.1007/978-3-540-88313-5_2
2. Homer, J., Ou, X., Schmidt, D.: A sound and practical approach to quantifying security risk in enterprise networks. Kansas State University Technical report 1–15 (2009)
3. Li, P., Qiu, X.: NodeRank: an algorithm to assess state enumeration attack graphs. In: 2012 8th International Conference on Wireless Communications, Networking and Mobile Computing, pp. 1-5. IEEE, Shanghai (2012)
4. Wang, H., Chen, F.W., Wang, Y.F.: An approach of security risk evaluation based on the Bayesian attack graph. Open Cybern. Syst. J. 9, 953–960 (2015)
5. Mell, P., Scarfone, K., Romanosky, S.: A complete guide to the common vulnerability scoring system version 2.0. FIRST-Forum of Incident Response and Security Teams 1:23 (2007)
6. Miura-Ko, R.A., Bambos, N.: SecureRank: a risk-based vulnerability management scheme for computing infrastructures. In: IEEE International Conference on Communications, pp. 1455–1460. IEEE, Scotland (2007)
7. Li, W., Vaughn, R.B., Dandass, Y.S.: An approach to model network exploitations using exploitation graphs. Simulation 82(8), 523–541 (2006)
8. Yang, X., Shunhong, S., Yuliang, L.: Vulnerability ranking based on exploitation and defense graph. In: 2010 International Conference on Information, Networking and Automation, pp. V1-163–V1-167. IEEE, Kunming (2010)
9. Jiang, J., Ding, L., Zhai, E., et al.: VRank: a context-aware approach to vulnerability scoring and ranking in SOA. In: 2012 IEEE Sixth International Conference on Software Security and Reliability, pp. 61–70, IEEE, Maryland (2012)
10. Zhuang, H., Pydde, F.: A non-intrusive and context-based vulnerability scoring framework for cloud services. arXiv:1611.07383 (2016)
11. Shapley, L.S.: A value for n-person games. Contrib. Theory Games 2(28), 307–317 (1953)
12. Michalak, T.P., Rahwan, T., Szczepanski, P.L., et al.: Computational analysis of connectivity games with applications to the investigation of terrorist networks. In: Twenty-Third International Joint Conference on Artificial Intelligence, pp. 293–301. AAAI, Beijing (2013)
13. Szczepański, P.L., Michalak, T.P., Rahwan, T.: Efficient algorithms for game-theoretic betweenness centrality. Artif. Intell. 231, 39–63 (2016)
14. Castro, J., Gómez, D., Tejada, J.: Polynomial calculation of the Shapley value based on sampling. Comput. Oper. Res. 36(5), 1726–1730 (2009)
15. Jia, W., Feng, D.G., Lian, Y.F.: Network-vulnerability evaluation method based on network centrality. J. Grad. Univ. Chin. Acad. Sci. 9(4), 529–535 (2012)
16. Deng, X., Papadimitriou, C.H.: On the complexity of cooperative solution concepts. Math. Oper. Res. 19(2), 257–66 (1994)
17. Barabási, A.L., Albert, R.: Emergence of scaling in random networks. Science 286(5439), 509–512 (1999)
18. NVD Home (2019). https://nvd.nist.gov/vuln/data-feeds

# Enhancing Model Performance for Fraud Detection by Feature Engineering and Compact Unified Expressions

Ikram Ul Haq[⊠], Iqbal Gondal, and Peter Vamplew

ICSL, School of Science, Engineering and Information Technology,
PO Box 663, Ballarat, VIC 3353, Australia
ikramulhaq@students.federation.edu.au,
{iqbal.gondal,p.vamplew}@federation.edu.au

**Abstract.** The performance of machine learning models can be improved in a variety of ways including segmentation, treating missing and outlier values, feature engineering, feature selection, multiple algorithms, algorithm tuning/ compactness and ensemble methods. Feature engineering and compactness of the model can have a significant impact on the algorithm's performance but usually requires detailed domain knowledge. Accuracy and compactness of machine learning models are equally important for optimal memory and storage needs. The research in this paper focuses on feature engineering and compactness of rulesets. Compactness of the ruleset can make the algorithm more efficient and derivation of new features makes the dataset high dimensional potentially resulting in higher accuracy. We have developed a technique to enhance model's performance with feature engineering and compact unified expressions for dataset of unknown domain using profile models approach. Classification accuracy is compared using well-known classifiers (Decision Tree, Ripple Down Rule and RandomForest). This technique is applied on fraud analysis bank dataset and multiple synthetic bank datasets. Empirical evaluation has shown that not only the ruleset size of training and prediction dataset is reduced but performance is also improved in other performance metrics including classification accuracy. In this paper, the transformed data is used for the experimental validation and development of fraud detection technique, but it can be used in other domains as well especially for scalable and distributed systems.

**Keywords:** Model performance · Fraud detection · Unified expressions · Feature engineering · Categorical data · Compactness · Ruleset · Situated profiles · RDR

## 1 Introduction

The accuracy of a machine learning model can be boosted with the use of various methods such as segmentation [1], adding more data, treating missing [2] and outlier values, feature engineering (FE) [3–5], feature selection, multiple algorithms, algorithm tuning and ensemble methods. Particularly, feature engineering helps to extract more

© Springer Nature Switzerland AG 2020
S. Wen et al. (Eds.): ICA3PP 2019, LNCS 11945, pp. 399–409, 2020.
https://doi.org/10.1007/978-3-030-38961-1_35

information from existing data by deriving new features from existing features. It helps to unleash the hidden relationships in a dataset. Derived features may help in explaining the variance in the training data more accurately and result in higher accuracy. FE could be done using indicator variables, features interaction, feature representation by extracting information from the existing features, transforming categorical to numeric features, by creating dummy features or by using external data. Feature representation can be mainly applied to categorical attributes. In this paper, we have focused on feature representation with minimum knowledge of the domain of an external dataset. One of the challenges in FE is to determine if FE can be applied on a particular feature and whether it could be applied via contextual expressions or via external sources, while another challenge is that data become high dimensional as new features are derived from existing features. We have developed a Feature Engineering and Compact Unified Expressions (FECUE) technique to improve model performance with feature engineering with minimal prior knowledge of the domain of the dataset coupled with compacting the ruleset and dataset with unified expressions using a model-based approach. Performance is measured using three well-known classifiers (Decision Tree [6], Ripple Down Rules(RDR) [7] and RandomForest [8]). The proposed technique is applied to bank datasets. The empirical evaluation has shown that model's performance has improved while training and prediction model sizes have also been reduced. Main contributions are listed below:

- Study of feature engineering and unified expressions to improve fraud analysis.
- Development of feature engineering technique using custom and configurable situated profile models (SPM) when the domain of a dataset is not known in advance.
- Empirical evaluation of the developed technique with multiple datasets.
- Ruleset compactness using contextual expressions and situated profile models.
- Evaluating performance in terms of standard performance metrics including classification accuracy, precision, recall, f-measure, time and ruleset size.

## 2  Related Work

Some of the known methods of improving model performance are highlighted below:

- Segmentation [1] by dividing the population into several groups.
- Adding more data to produce more accurate models and treating missing [2] and outlier values.
- Feature Engineering [3–5], extracting more information from existing features.
- Feature selection by finding and the most important subset of features.
- Multiple algorithms by applying a relevant model to see better suitability of models for a particular domain.
- Algorithm tuning by finding optimum parameter values used in the algorithm.

Our research focuses on feature engineering which is being used in different domains to improve model performance. In [3] authors have conducted an educational data mining study; and evaluated feature engineering for KDD Cup 2010 by training the model from students' past behavior and then predicting future performance. Authors in [4] have designed an information extraction technique using feature engineering with a combination of rule-based and machine learning methods. This technique is applied on narrative clinical discharge summaries. Turner et al. [5] proposed the concepts of FE and evaluated its impact on the software development life cycle. They proposed their research as the first step towards the development of feature engineering and its relationship to other domains. One text classification feature engineering technique is developed by [9], which is ontology guided. This technique utilizes the domain knowledge encoded in the taxonomical structure of the Medical Language System with the help of context-dependent relatedness between pairs of concepts.

These developed techniques have a variety of limitations and are either domain or context-specific. They do not discuss the problem or the solution of the increase of data dimension with the application of FE. Also, the performance impact in terms of either of the classification accuracy, time and model's size is not discussed. FE via external sources is also not used in these techniques. Considering these limitations, we have proposed an innovative technique which improves model performance over a variety of performance metrics. The proposed technique is a situated profile model-based, domain independent FE technique using compact unified expressions.

## 3    Methodology

Out of various methods available for improving model accuracy, research in this paper focuses on feature engineering and compression of ruleset of the training model. One of the challenges was to identify appropriate FE methods for individual attributes, ideally requiring minimal domain knowledge. Another challenge was the compactness of the ruleset. Four situated profiles models (SPM) are developed and used in this technique to predict features, which type of FE to use and how to apply the ruleset compactness. SPMs are explained in Sect. 3.1. Situated profile models make the technique more generic for different datasets. Consider the nomenclature of a typical bank transaction log as explained by Maruatona [10] Table 7-1.

Categorical attributes represent a type of data which may be divided into groups. Typically, a categorical attribute represents discrete values and have no concept of ordering the values of that attribute. From Maruatona [10] Table 7-1, some of the fields can be used for feature extraction. The developed technique is divided into two parts, feature representation and compactness of the ruleset. A situated profile (SP) [11] defines values relative to the situations, so these are only applied in situations for which they are valid. A situated profile could help in intelligence extraction efficiently. In RDR, the modelling is also based on SPs [10], as it describes every attribute for a particular case. The developed technique is explained in more detail in Sect. 3.4.

### 3.1    Feature Engineering Techniques for Bank Dataset

Many classification algorithms do not use attributes like Event-time, IP Address and Browser string as these type of attributes are ignored in the feature selection process. Feature engineering [12] is a critical and underexplored aspect of building high-quality knowledge base construction systems and is an understudied problem relative to its importance, especially in fraud detection. One way of FE is extracting information from the existing features, while another way is by using external data sources with some application program interface (APIs) or source like geocoding and demographics. In this paper, we have also applied FE with external data sources.

If we derive new attributes from existing attributes and train the model, we can see that the new attributes are used by the classifier. The newly derived features either can be numeric or can be easily transformed to numeric attributes. Numeric features give better performance in machine learning algorithms. Similarly, clustering algorithms work effectively on the data where all attributes are either numeric or categorical data, as compared to mixed data types [13]. [14] also proved higher classification accuracy with numeric data opposed to mixed datasets. In bank dataset, more attributes can be derived from Event-time, e.g. hour, day, month, year, day-of-week, holiday and weekend-flag. Browser string attribute may further produce attributes like O.S, browser and device identifiers. New attributes derived from an IP Address value could be either four segments separated by token character or location-based attributes. External data sources are available which provide geographic information of an IP Address. These newly derived attributes could also be helpful in identifying suspected transactions in terms of fraud. For example, if event hour is not in normal time, or if it is a holiday or weekend or if the location of the IP Address is different from the actual user's location, then there is higher chance of a potential fraud. Same applies with the attributes derived from Browser string attribute. Different SPMs are formed to aid this method be generic and domain-independent.

### 3.2    Situated Profile Models

A number of situated profiles models (SPM) were developed to process features and for the ruleset compactness. These models are used for banking dataset, but could also be modified for a specific dataset. Table 1 SPM is a set of tokenizer characters and their applicability to attributes, while Table 2 explains different measures to predict an attribute based on the type and category. With Table 3 FE could be categorized if it can be done via contextual expressions. E.g. extracting day-of-week from date field or getting geocoding and demographic information from an IP Address.

Below table is a sample list of UEL operators, which can be replaced with simple mathematical operator to achieve compactness in UEL ruleset.

**Table 1.** Tokenizer character model sample

| Token character | Category | Attribute index |
|---|---|---|
| . | Include | 2, 6 |
| _ | Include | 3, 5, 4 |
| ; | Include | 5 |
| , | Skip | all |
| ) | Skip | 5 |

**Table 2.** Feature prediction model sample

| Type | Category | Possible values |
|---|---|---|
| Attribute data type | Comparison | String, date, amount, integer |
| Tokenizer | Boolean exists | Yes/no |
| Tokenizer | Find | Ref: Table 1 |
| Tokenizer | Count | 1, 2, 3 |
| Attribute | Length | 0–100 |

**Table 3.** FE type model sample

| FE source | Attribute index |
|---|---|
| Contextual expressions | 3 |
| Contextual expressions | 4 |
| Contextual expressions | 5 |

**Table 4.** Rules compression model sample

| UEL operator | Simple operator | Types |
|---|---|---|
| Between | >= | Integer, amount |
| Between | <= | Integer, amount |
| Like/in | = | String |
| Not between | NA | Integer, amount |
| Not In | NA | String |

### 3.3    Challenges and Tokenizing a Feature Value

One of the challenges in FE is how to evaluate which information or features could be extracted from a particular feature, which already exists in the dataset. It cannot be done without domain knowledge or at-least heuristic approach needs to be applied based on the data type. Without domain knowledge of fraud dataset, how we will know that browser OSVer, O.S, Ver and device features can be extracted from raw Browser string. Heuristically, we know that hour, day, month, day-of-week, holiday and weekday flag information can be extracted from a date-time feature and that an IP Address contains geolocation data, which can be extracted by some external APIs.

A new way of FE is introduced in this paper, which can extract information from existing features with a minimum domain knowledge of the dataset. Four situated profile models (SPM) (Tables 1, 2, 3 and 4) are developed in this technique to predict a feature and to decide the source of feature engineering. This way is explained in Algorithm 1 and in Sect. 3.6 with a rule-based approach. By using this algorithm and the suggested rule-based approach, information can be extracted by tokenizing a feature value with non-alphanumeric characters. E.g comma, space, bracket, colon and semi-colon, Table 1 is configurable to update tokenizer characters with respect to attributes. From a sample date-time value "15/10/2018 23:55:10" six numeric attributes can be extracted by using Algorithm 1, which are "15 10 2018 23 55 10". A classifier doesn't need to know which value is an hour, day, month or a year. Similarly from a sample Browser string value "Mozilla/5.0 (iPad; CPU OS 3_2_1 like Mac OS X; en-us) AppleWebKit/531.21 (KHTML, like Gecko) Mobile", O.S, browser and device identifiers can be extracted. Although the contents of a Browser string will slightly vary based on the browser and the underlying operating system, but once the system knows that it is a Browser string field it can further extract these attributes. A ruleset can be further developed to extract browser name, operating system and the versions, as Browser string contents may vary based on the browser and the O.S. These newly extracted attributes are a combination of categorical and numeric attributes. But the extracted categorical attribute can also be converted to numeric attribute, which was not possible with the original attribute value of Browser string. Various SPMs are developed in this technique for bank dataset, but may also be customized for a particular dataset.

### 3.4    Algorithms

The developed technique is based on feature engineering and compactness of ruleset for the model. Feature engineering is explained in Algorithm-1, while ruleset compactness is explained in Algorithm 2. Tokenizer characters are maintained in situated profiles for every attribute, as a particular character could be a tokenizer character for one attribute, but not valid for other attributes.

## Algorithm-1.

**Input:** Instance from a dataset. **Output:** Instance with addition of new features with feature engineering.

#Load Source data and perform data cleaning. Do feature selection and filter categorical features and other features having tokenizer characters.

1. Process instances.

2. Process each Feature

3. IF Feature (Is Categorical) or (Having tokenizer characters)

i. Categorise the feature based on Table-1 and Table-2 (explained in more detail in section 3.6.)

ii. For each feature transform and extract new features with FE.

iii. Tokenize / Split with Tokenizer characters from Situated Profiles using Table-1 and Table-2

FOR Feature 1 to n LOOP

    IF NEW Tokenizer THEN Update Situated Profiles

    # Situated profiles will manage collection of tokenizer characters on attribute level.

    ELSE IF Tokenizer THEN NewFeatures = ExtractFeatures(feature)

    #Extract feature with the token

    NEXTVALUE

ENDLOOP

4. IF (more features in the row) Goto step-2

#Extract features from complete Row from Step 2-4, IF (more Row) Goto Step-1 ELSE FINISH

## Algorithm-2.

**Input:** A unified expression format rule from a ruleset. **Output:** A compact unified expression format rule.

#Load Ruleset.

1. Process each rule in the ruleset and compact the ruleset using fCompact function (1).

2. Process each expression in the rule.

3. IF (Expression is >= or <=) Process current rule and update UEL Rule 3.a

    #Update UEL Rule with BETWEEN operator

    ELSE if (Expression is ==)

    #Process current rule and update UEL Rule 3.a. Update UEL Rule with UEL operators as Table-4

    ELSE SKIP

    ENDIF

    3.a Update Unified Expression Rule (UEL)

#Update with appropriate UEL operator (BETWEEN, IN, NOT IN, LIKE, NOT LIKE) as explained in Table-4 and in section 3.5

4.IF (more expression) Goto step-2

#Process expressions from complete Rule from Step 2-4. IF (more Rules) Goto Step-1 ELSE FINISH

## 3.5    Unified Expressions Language

In this paper, we have considered rule-based classifiers. One of the well-known classifiers is RDR. We have suggested ruleset compactness in RDR using unified expressions using SPMs. Unified Expressions Language (UEL) can evaluate mathematical expressions with a lot of operators and enables dynamic scripting feature. Some of the advantages of UEL is that it supports more than 30 different operators; and expressions can also invoke functions, which can help in getting external data for feature engineering. For example, extracting geolocation data in bank dataset. Rule-based classifiers use only limited operators. However, using UEL many more operators can be used e.g. IN and LIKE Operators. In FE, features interaction can be achieved by dynamically evaluating expressions using Add, Subtract, Multiply and Divide operators instead of creating new features in the prediction phase. FE with feature interaction will be only needed for training the model. Authors in [14] have highlighted the importance of compactness of the prediction model and demonstrated that a compact prediction model is more efficient. The UEL expression will help in ruleset compactness and will improve performance in terms of the time taken for model prediction.

Algorithm 2 explains compactness with Expression Language using a configurable situated profile model (Table 4). This model uses a relevant UEL operator which can be used based on simple operator and attribute type. Ruleset compactness with unified expressions is explained below:

Rule-1: 'Source_Acc'='Personal' and 'Country'='AU' and Browser='MOZ-5Win' THEN FRAUD

Rule-2: 'Source_Acc'='Personal' and 'Country'='AU' and Browser='MOZ-5Lin' THEN FRAUD

Compressed Rule: (Using IN Operator)

'Source_Acc'='Personal' and 'Country'='AU' and Browser IN ('MOZ-5Lin', 'MOZ-5Win') THEN FRAUD

Other Operator could be BETWEEN for numeric features and LIKE for categorical features.

Compactness of an expression is explained with below equation.

$$R^{comp} = f\ Compact \int_{i}^{nY} (expSet)m \neq null \tag{1}$$

Where expSet is a set of expressions from RDR ruleset and $R^{comp}$ is a compact rule set with unified expressions and fCompact is a function to compact an RDR ruleset which compacts simple mathematical expressions from 1 to n from SPM Table 4 on $i^{th}$ rule index having m value which is non-null.

**Contextual Expressions**

Unified expressions can be used to get further useful information from the existing attributes through external sources, e.g. getting geocoding and demographic information from IP Address in bank dataset. Which can help in making further decisions related to fraudulent transactions and will improve model accuracy as well. To make it generic which attributes needs FE from an external source, a situated profile model Table 3 is developed and used in this technique. This model decides FE based on the attributes, which is predicted from two other models Tables 1 and 2. E.g. Get country information from IP Address may help in detecting suspected tunnel sites usage. We can add a rule when IP Address and user's actual country are different.

Rule: 'Source_Acc' == 'Personal' and 'UserCountry' <> 'IPCountry' THEN FRAUD

### 3.6 Constructing a Feature

Extracting features from the existing feature is a challenging task, especially without knowing the domain of the dataset. However, if we know the feature name in a particular dataset, it will help in extracting more features from this feature. Considering commonly used data types explained by [15], [16] and adding some further measures of feature content length and presence of the token character, a rule-based approach is developed to predict a feature name. To make the technique more generic, four situated profile models are developed and used in this technique. See a ruleset example.

```
Rule-1: DataType='String' and Count(Token_Character='.')=3 THEN IPAddress
Rule-2: DataType ='String' and Token_Character==';' THEN BrowserString
Rule-3: DataType ='String' and (No_Token_Character or Token_Character='_')
THEN SourceAccount
Rule-1, 2 and 3 can also be represented as:
    DataType ='String'
        Count (Token_Character='.') = 3 THEN IPAddress
        Token_Character=';' THEN BrowserString
        (No_Token_Character or Token_Character='_') THEN SourceAccount
```

Comparison with attribute types and checking the existence of a particular and using other measures of length or count is used from the SPMs explained in Sect. 3.1

## 4 Results

Empirical evaluation was done for both original and the dataset produced by FECUE technique. Performance was measured with a variety of performance metrics including classification accuracy, precision, recall, f-measure, time and ruleset compactness.

$$\text{Accuracy} = \frac{TP + TN}{TP + FP + FN + TN} \tag{2}$$

$$\text{Precision} = \frac{TP}{TP + FP} \tag{3}$$

$$\text{Recall} = \frac{TP}{TP + FN} \tag{4}$$

$$\text{F-measure} = \frac{2 * (\text{Recall} * \text{Precision})}{(\text{Recall} + \text{Precision})} \tag{5}$$

Where TP are correctly predicted positive and TN are correctly predicted negative values, FP when actual class is no and predicted class is yes and FN when actual class is yes but predicted class is no.

### 4.1 Bank Datasets

Various performance metrics with three well-known classifiers has been compared for the original datasets and corresponding datasets with derived attributes after feature engineering using FECUE. The results in Tables 5 and 6 show that there is an improvement in performance metric results. In this study, 30% and 70% split is done for training and testing datasets. Average measurement was calculated for various dataset sizes ranging from small to large datasets and for multiple simulation runs for each classifier. RIDOR is RDR and J48 is decision tree implementation in WEKA.

**Table 5.** Performance with reference bank dataset

| Classifier | Accuracy | Precision | Recall | F-Measure | Time | Ruleset |
|---|---|---|---|---|---|---|
| RIDOR | 3.96% | 1.85% | 4.05% | 4.05% | 58.06% | 26.09% |
| C45/J48 | 0.32% | −0.10% | 0.00% | 0.00% | 50.00% | −10.67% |
| R. forests | 49.39% | 91.49% | 33.68% | 97.39% | −8.33% | |

**Table 6.** Performance with Synthetic Bank dataset

| Classifier | Accuracy | Precision | Recall | F-Measure | Time | Ruleset |
|---|---|---|---|---|---|---|
| RIDOR | 6.75% | 7.34% | 6.75% | 7.91% | 165.32% | 50.32% |
| C45/J48 | 2.64% | 5.87% | 6.37% | 2.53% | 108.41% | 15.53% |
| R. forests | 50.58% | 52.42% | 50.58% | 119.64% | 20.26% | |

Above tables shows that there is an overall improvement (original and corresponding datasets after FE with FECUE) in all performance metrics with both bank's datasets.

## 5 Conclusion

Model performance can be improved in a variety of ways including segmentation, treating missing and outlier values, feature engineering, feature selection, multiple algorithms, algorithm tuning and ensemble methods. This paper has presented model accuracy and compactness technique (FECUE), and it is observed that derivation of new features makes the dataset high dimensional. The developed technique has enhanced the model's performance with feature engineering (when the domain of a dataset is not known in advance), with the use of external sources and compact unified expressions. Multiple situated profile models (SPM) are used to make the technique more generic so that it is applicable on multiple datasets and domains. Performance in terms of classification accuracy, precision, recall, f-measure, time and ruleset compactness is compared using three well-known classifiers. FECUE has been applied on reference bank dataset and multiple synthetic bank datasets. The empirical evaluation has shown that not only the ruleset in training and prediction model are reduced but the performance improvement is also observed in other standard performance metrics. The developed technique is mainly applied in fraud detection area, but it can be used in other domains as well. One of the future works would be to test this technique on a variety of datasets especially with high dimensional data.

# References

1. Bijak, K., Thomas, L.C.: Does segmentation always improve model performance in credit scoring? Expert Syst. Appl. **39**(3), 2433–2442 (2012)
2. Xiaofeng, Z., Shichao, Z., Zhi, J., Zili, Z., Zhuoming, X.: Missing value estimation for mixed-attribute data sets. IEEE Trans. Knowl. Data Eng. **23**(1), 110–121 (2011)
3. Yu, H.-F., et al.: Feature Engineering and Classifier Ensemble for KDD Cup 2010 (2010)
4. Xu, Y., Hong, K., Tsujii, J., Chang, E.I.-C.: Feature engineering combined with machine learning and rule-based methods for structured information extraction from narrative clinical discharge summaries. J.A.M.I.A **19**(5), 824–832 (2012)
5. Turner, C.R., Fuggetta, A., Lavazza, L., Wolf, A.L.: A conceptual basis for feature engineering. J. Syst. Softw. **49**(1), 3–15 (1999)
6. Quinlan, J.R.: C4.5: Programs for Machine Learning. Morgan Kaufmann (1993)
7. Compton, P., Jansen, R.: Knowledge in context: a strategy for expert system maintenance. In: Barter, C.J., Brooks, Michael J. (eds.) AI 1988. LNCS, vol. 406, pp. 292–306. Springer, Heidelberg (1990). https://doi.org/10.1007/3-540-52062-7_86
8. Breiman, L.: Random forests. Mach. Learn. **45**, 5–32 (2001)
9. Garla, V.N., Brandt, C.: Ontology-guided feature engineering for clinical text classification. J. Biomed. Inform. **45**(5), 992–998 (2012)
10. Maruatona, O.O.: Internet Banking Fraud Detection Using Prudent Analysis. University of Ballarat, Ballarat (2013)
11. Vastenburg, M.H.: SitMod: A Tool for Modeling and Communicating Situations (2004)
12. Ré, C., et al.: Feature Engineering for Knowledge Base Construction (2014)
13. Shih, M.-Y., Jheng, J.-W., Lai, L.-F.: A two-step method for clustering mixed categroical. Tamkang J. Sci. Eng. **13**(1), 11–19 (2010)
14. Ul Haq, I., Gondal, I., Vamplew, P., Brown, S.: Categorical features transformation with compact one-hot encoder for fraud detection in distributed environment. In: The 16th Australasian Data Mining Conference, Bathurst NSW, Australia (2018)
15. Witten, I.H., Eibe, F.: Data Mining: Practical Machine Learning Tools and Techniques, 2nd edn. edited by Jim Gray, M.R. Morgan Kaufmann (2005)
16. Durrant, B.: An ARFF (Attribute-Relation File Format). University of Waikato. https://waikato.github.io/weka-wiki/arff_stable/. Accessed 9 Nov 2018

# Network Intrusion Detection Framework Based on Embedded Tree Model

Jieying Zhou[✉], Pengfei He, Rongfa Qiu, and Weigang Wu

School of Data and Computer Science, Sun Yat-sen University,
Guangzhou, China
{isszjy,wuweig}@mail.sysu.edu.cn,
{hepf3,qiurf5}@mail2.sysu.edu.cn

**Abstract.** Network intrusion detection system plays a vital role in network security protections that could be used to protect personal privacy and property security so as to protect users from attackers. However, there are a few samples of attack types with various characteristics. To solve the problem of class-imbalance in network security and correctly detect the attack, this paper proposes a network intrusion detection framework: random forest and gradient boosting decision tree (RF-GBDT). Random forest model is used for feature transformation and gradient boosting decision tree model is used for classification. RF-GBDT was used on the UNSW-NB15 dataset in which only 8 features were selected for training and a large number of irrelevant features were deleted. RF-GBDT not only reduced the training time but also improved the detection rate. The experiment result shows that RF-GBDT model has a higher detection rate and lower false alarm rate compared with other relative algorithms.

**Keywords:** Network intrusion detection system · UNSW-NB15 dataset · Class-imbalance

## 1 Introduction

With the rapid development of computer application, network intrusion detection has become an important barrier to ensure the computer security. The intrusion detection is a security mechanism. It monitors and filters the network behaviors by analyzing the data such as the host audit data and the network flow data, then identifies the abnormal access in the network communication and notice the administrator in time. By doing all this, it achieves the purpose of protecting network information security.

Intrusion detection system (IDS) can be divided into three modules, which showed by Fig. 1. There are data collection module, intrusion detection module and response module. The data collection module collects data from system logs, network data flow, host audit data and etc. These data collected will be sent to the intrusion detection module. Then the intrusion detection module will conduct data processing and modeling analysis on these data, which helps determine whether the behaviors are aggressive or what type of attacks it belongs to. This module is the kernel of intrusion detection system, which directly affects the performance of the system. Finally, the

© Springer Nature Switzerland AG 2020
S. Wen et al. (Eds.): ICA3PP 2019, LNCS 11945, pp. 410–417, 2020.
https://doi.org/10.1007/978-3-030-38961-1_36

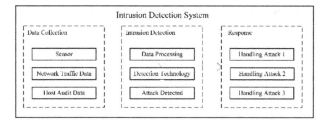

**Fig. 1.** Intrusion detection system

response module receives the attacking data detected by intrusion detection module and takes some actions according to what type the data belongs to.

The network intrusion detection mainly faces the following several problems: (1) diversity: there are many types of attack. So, in the detection, so intrusion detection is regarded as a multi-classification problem rather than a binary classification problem; (2) class-imbalance: there are small amounts of samples belonging to attacking types.

To solve the multi-classification and class-imbalance problem, this paper proposes the RF-GBDT model framework which has a high accuracy of detection and good generalization performance.

The rest of the paper is organized as follows. Section 2 provides a summary of the previous work in intrusion detection. Section 3 presents the theory of grandient boosting decision tree. Section 4 shows the proposed model framework. Section 5 presents the UNSWNB-15 dataset and evaluation. Section 6 discusses experimental results. The conclusion is drawn in Sect. 7.

## 2    Related Works

In the recent years, the machine learning algorithms are widely used to solve the network intrusion detection problem. Aiming at the class-imbalance problem in network intrusion detection, people mainly study in the following two aspects: the methods based on data level [1] and the methods based on algorithm level [6, 7].

The methods based on data level use data sampling technology, and the train the model. In the recent years, the classification method based on sampling algorithm has become the research hotspot [8, 9]. Chawla [2] proposed a boosting method based on Synthetic Minority Over-sampling Technique (SMOTE), which used SMOTE to up-sample in every iteration. However, the methods based on sampling have changed the distribution of data. Majority of machine learning algorithms are established on the hypothesis of the training data and testing data having the same data distributions.

The methods based on algorithm level promote the classification performance mainly by improving the process of training and adopting several integrated strategies, such as adopting the feature selection technologies and classifier integrated technologies. The integrated classifier combines multiple weak classifiers to improve the performance of multi-classification through majority voting, boosting or bagging. It can effectively avoid the problem of consumption of resource and the bias of classification that the single classifier will cause. By the way, it can also improve the classifier performance of the detection models and reduce the variance which can prevent overfitting.

## 3  Theory of Gradient Boosting Decision Tree

The GBDT model [5] is an additive model:

$$F(x) = \sum_{t=1}^{T} \alpha_t h_t(x) \tag{1}$$

where $x$ is the input sample, $h_t(x)$ is Classification and Regression Trees (CART), and $T$ is the number of trees, $\alpha_t$ is the weight of the $t^{th}$ tree.

GBDT uses the forward distribution algorithm. At first, it selects a constant $F_0(x)$ as the initial value of model. And the model of the $m^{th}$ step is:

$$F_m(x) = F_{m-1}(x) + \alpha_m h_m(x) \tag{2}$$

$F_{m-1}(x)$ is the current model. $h_m(x)$ is computed by the minimizing loss function:

$$h_m = \arg\min_h \sum_{i=1}^{N} L(y_i, F_{m-1}(x_i) + h(x_i)) \tag{3}$$

$N$ is the number of samples.

GBDT uses the gradient descent method to compute the optimal model, which regards the negative gradient value of loss function on the current model $F_{m-1}(x)$ as the direction of gradient descent:

$$F_m(x) = F_{m-1} - \alpha_m \sum_{i=1}^{N} \nabla_F L(y, F_{m-1}(x_i)) \tag{4}$$

$\alpha_m$ is computed by the line search:

$$\alpha_m = \arg\min_{\alpha} \sum_{i=1}^{N} L(y_i, F_{m-1}(x_i)) - \alpha \frac{\partial L(y_i, F_{m-1}(x_i))}{\partial F_{m-1}(x_i)} \tag{5}$$

The regularization of GBDT can be controlled by setting the learning rate:

$$F_m(x) = F_{m-1}(x) + v\alpha_m h_m(x) \tag{6}$$

$v$ is the learning rate. The smaller it is, the more CART we need and we will have the less error. But it will increase the training time. Therefore, we need to control learning rate and the amount of CART at the same time to confirm a model having high velocity and accuracy.

## 4 The Proposed Framework

RF-GBDT is constituted of three parts which are feature selection, feature transformation and the classifier. Firstly, the training data is trained by GBDT to get the features which are sorted by feature importance from high to low. Then Recursive Feature Elimination method is used to select the optimal feature subset. Next, Random Forest model is used to train the feature subset. And this paper uses the indexes of the leaves which the samples fall on as the final input of the classifier. If there are $m$ samples and the Random Forest model has $n$ trees, then the size of transformed data is $m \times n$. At last, this paper use GBDT model to train and predict on transformed features. The overall structure of the model framework is shown by Fig. 2.

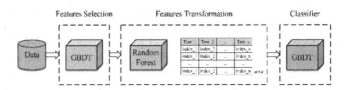

**Fig. 2.** The proposed model framework

The first part is feature selection. This paper proposes a Recursive Feature Elimination method based on GBDT (GBDT-RFE), which belongs to one kind of wrapper feature selection algorithms. The algorithm uses GBDT model for multi-round training and records the loss value in each round of training. Then it eliminates the feature with the least importance, and then carries on the next round of training based on the new feature set until the feature set is eliminated completely. The workflow is given in Algorithm 1.

```
Algorithm 1 GBDT-RFE
1: Input: X, y
2: Output: F
3: while features_list:
4:     Train GBDT (X, y) on features_list:
5:     Record Loss:
6:     Get ranked_feature_importances;
7:     del ranked_feature_importances [-1];
8:     features_list = ranked_feature_importances;
9: end while
10: F = argminLoss
        F
```

The second part is feature transformation. He [3] has proposed the decision tree feature transformation method. On the basis of that method, this paper uses Random Forest model to generate the embedded leaf features, as shown in Fig. 3. For example,

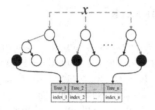

**Fig. 3.** New features generated by leaf index

while sample $x$ traversing all trees, it will firstly pass the first tree and fall on the leaf having index of *index_1*. Then it will pass the second tree and fall on the leaf having index of *index_2*. This process stops when the sample passes the last tree and falls on the leaf having index of *index_n*. All these indexes of leaves will combine to generate the embedded leaf features. The paper [3] use One-Hot Encoding after feature transformation. But this paper directly uses indexes of leaves as features rather than encode. If there are $n$ trees in Random Forest model and $m$ samples, then the shape of the embedded leaf features will be $m \times n$.

The third part is the classifier. GBDT model is used to train on the training dataset with size of m × n. And then predict on the testing dataset. The cross-validation method is used to select parameters such as the number of trees and learning rate.

## 5 Dataset and Evaluation

### 5.1 UNSW-NB15 Dataset

In 2015, Nour and Slay proposed the UNSW-NB15 dataset [4]. As a new benchmark dataset in the field of intrusion detection, UNSW-NB15 dataset can correctly reflect today's diverse attack types and complex network conditions.

There are 2,540,044 samples in the UNSW-NB15 dataset, including 49 features and 10 classes. They are normal samples "Normal" and nine attack types: "Fuzzers", "Analysis", "Backdoors", "DoS", "Exploits", "Generic", "Reconnaissance", "Shellcode" and "Worms". The UNSW-NB15 dataset [4] has a subset version, with 175,341 samples in the training dataset and 82,332 samples in the test dataset, including 41 features.

### 5.2 Testing Evaluation and Performance Measures

The flow data in network security field has the characteristic of unbalanced distributed samples. So, in order to reflect the true effect of model, the proper evaluation measures should be selected. For performance estimation, precision, detection rate, false alarm rate and F1-score performance measures are taken. Confusion matrix is used and shown in Table 1.

**Table 1.** Confusion matrix

| True label | Predicted label | |
| --- | --- | --- |
| | 1. Attack | 2. Normal |
| Attack | True Positive (TP) | False Negative (FN) |
| Normal | False Positive (FP) | True Negative (TN) |

- Detection Rate (DR): the ratio of the abnormal samples correctly predicted in all abnormal samples;
- False Alarm Rate (FAR): the ratio of the normal samples mistakenly predicted abnormal;
- Precision: the ratio of the abnormal samples correctly predicted in the samples predicted abnormal;
- F1-score: the harmonic mean of recall rate and precision.

$$DR = \frac{\sum_{i=1}^{C} TP_i}{\sum_{i=1}^{C} (TP_i + FN_i)} \tag{7}$$

$$FAR = \frac{\sum_{i=1}^{C} FP_i}{\sum_{i=1}^{C} (FP_i + TN_i)} \tag{8}$$

$$Precision = \frac{\sum_{i=1}^{C} TP_i}{\sum_{i=1}^{C} (TP_i + FP_i)} \tag{9}$$

$$F_1 score = \frac{2 \times Precision \times Recall}{Precision + Recall} \tag{10}$$

$C$ is the amount of types.

## 6 Experiments and Results

The hardware configuration of experiment is the 2.5 GHz Intel Core processor with 8 GBs of RAM and 64-bit Windows10 operating system. This experiment uses Python on Anaconda Platform by calling the Scikit-learn Toolkit. RF-GBDT model framework are also compared to the following four algorithms such as Adaboost, Random Forest, K-Nearest Neighbor and Logistic Regression.

## 6.1 Result Comparison

Table 2 shows the performance with 10-fold cross-validation on training dataset of K-NN, AdaBoost, LR, RF-GBDT, DNN [10] and RICSA-KELM [11]. In the table, N/A indicates that the result doesn't exist or can't do comparison because of different evaluating methods. RF-GBDT has a detection rate of 83.78%, a false alarm rate of 1.8% and the F1 score is 83.78%.

**Table 2.** Comparisons of results

| Method | DR (%) | FAR (%) | F1Score (%) |
|---|---|---|---|
| K-NN | 64.35 ± 1.05 | 3.96 ± 0.12 | 64.35 ± 1.05 |
| AdaBoost | 73.85 ± 0.82 | 2.91 ± 0.09 | 73.85 ± 0.82 |
| LR | 63.16 ± 0.84 | 4.09 ± 0.84 | 63.16 ± 0.84 |
| DNN | 80 | N/A | 76 |
| RICSA-KELM | N/A | 2.12 | N/A |
| RF-GBDT | **83.78 ± 0.91** | **1.80 ± 0.91** | **83.78 ± 0.91** |

As a multi-class model, RF-GBDT also has a high detection rate on each class. Table 3 shows the detection performance on each class of four algorithms such as K-NN, AdaBoost, LR and RF-GBDT. Although there ara a few samples of minority class, the detection rate is over 84%, such as "Worns", "Reconnaissance", "Shellcode" and "Generic".

**Table 3.** Results of detection rate

| Type | Rate (%) | DR (%) | | | |
|---|---|---|---|---|---|
| | | K-NN | LR | Ada-Boost | RF-GBDT |
| Analysis | 1.85 | 52.71 | 25.74 | 52.43 | **70.52** |
| Backdoor | 1.84 | 14.23 | 3.66 | 53.65 | **72.65** |
| DoS | 3.88 | 7.78 | 1.39 | 21.98 | **35.47** |
| Exploits | 12.53 | 60.50 | 60.29 | 62.71 | **76.31** |
| Fuzzers | 9.33 | 39.02 | 33.12 | 54.00 | **63.50** |
| Generic | 9.73 | 56.31 | 40.51 | 67.90 | **84.27** |
| Normal | 43.79 | 85.57 | 87.58 | 86.34 | **92.38** |
| Reconn. | 8.69 | 56.19 | 71.71 | 81.68 | **92.89** |
| Shellcode | 7.51 | 48.49 | 43.36 | 83.72 | **91.23** |
| Worms | 0.87 | 14.62 | 0.00 | 41.54 | **82.31** |

# 7   Conclusion

In this paper, RF-GBDT model is proposed to solve the problem of multi-classification of class-imbalance in network intrusion detection. The model framework consists of three parts: feature selection, feature transformation and classifier.

Selecting features by the feature importance of GBDT and deleting the irrelevant features, can not only reduce training time cost and reduce the amount of calculation, but increase the detection rate of model.

Experiments show that RF-GBDT has the characteristics of high detection rate and low false alarm rate on UNSW-NB15 data set. The results showed that the detection rate was 83.78%, the false alarm rate was 1.8%, the F1 score was 83.78%, and the ROC AUC was 98.31%

The RF-GBDT model can accurately detect the attack types in the network flow data, especially the attack types with litter samples. For example, "Worns", "Shellcode", "Reconnaissance" and "Generic" have a few samples, but the detection rate is more than 84%. Therefore, RF-GBDT has an obvious advantage on solving the multi-class problem with class-imbalanced data in network intrusion detection.

**Acknowledgment.** This work is supported by the National Key R&D Program of China (2018YFB0203803), the National Natural Science Foundation of China (U1801266), and the Program of Science and Technology of Guangdong (2015A010103007).

# References

1. Garca, S., Derrac, J., Triguero, I., et al.: Evolutionary-based selection of generalized instances for imbalanced classification. Knowl.-Based Syst. **25**(1), 3–12 (2012)
2. Chawla, N.V., Lazarevic, A., Hall, L.O., Bowyer, K.W.: SMOTEBoost: improving prediction of the minority class in boosting. In: Lavrač, N., Gamberger, D., Todorovski, L., Blockeel, H. (eds.) PKDD 2003. LNCS (LNAI), vol. 2838, pp. 107–119. Springer, Heidelberg (2003). https://doi.org/10.1007/978-3-540-39804-2_12
3. He, X., Pan, J., Jin, O., et al.: Practical lessons from predicting clicks on ads at Facebook. In: Proceedings of the Eighth International Workshop on Data Mining for Online Advertising, pp. 1–9. ACM (2014)
4. Moustafa, N., Slay, J.: UNSW-NB15: a comprehensive data set for network intrusion detection systems (UNSW-NB15 network data set). In: 2015 Military Communications and Information Systems Conference (MilCIS), pp. 1–6. IEEE (2015)
5. Friedman, J.H.: Greedy function approximation: a gradient boosting machine. Ann. Stat. **29**(5), 1189–1232 (2001)
6. Sun, Z., Song, Q., Zhu, X., et al.: A novel ensemble method for classifying imbalanced data. Pattern Recognit. **48**(5), 1623–1637 (2015)
7. Zhang, Z., Krawczyk, B., Garcìa, S., et al.: Empowering One-vs-One Decomposition with Ensemble Learning for Multi-Class Imbalanced Data. Elsevier Science Publishers B.V. (2016)
8. Sain, H., Purnami, S.W.: Combine sampling support vector machine for imbalanced data classification. Procedia Comput. Sci. **72**(Complete), 59–66 (2015)
9. Jian, C., Gao, J., Ao, Y.: A New Sampling Method for Classifying Imbalanced Data Based on Support Vector Machine Ensemble. Elsevier Science Publishers B.V. (2016)
10. Cai, H., Wang, Q.: Research on intrusion detection technology based on deep learning. In: Network Security Technology and Application (2017)
11. Chao, M.: A parallel intrusion detection method based on relieff and improving crow search optimization. Comput. Appl. Res. **11**, 1–3 (2019)

# Generative Adversarial Nets Enhanced Continual Data Release Using Differential Privacy

Stella Ho[1,2(✉)], Youyang Qu[1], Longxing Gao[1], Jianxin Li[1], and Yong Xiang[1]

[1] Deakin University, 221 Burwood Highway, Burwood, VIC 3125, Australia
{hoste,y.qu,longxiang.gao,jianxin.li,yong.xiang}@deakin.edu.au
[2] Cyber Security Cooperative Research Centre, Joondalup, WA 6027, Australia

**Abstract.** In the era of big data, increasing massive volume of data is generated and published consecutively for both research and commercial purposes. The potential value of sensitive information also attracts interest from adversaries and thereby arises public concern. Current research mostly focuses on privacy-preserving data release in a statistic manner rather than taking the dynamics and correlation of context into consideration. Motivated by this, a novel idea is proposed by combining differential privacy and generative adversarial nets. Generative adversarial nets and its extensions are used to generate a synthetic data set with indistinguishable statistic features while differential privacy guarantees a trade-off between the privacy protection and data utility. Extensive simulation results on real-world data set testify the superiority of the proposed model in terms of privacy protection and improved data utility.

**Keywords:** Differential privacy · Generative adversarial nets · Continue data

## 1 Introduction

Nowadays, with the explosive growth of data in terms of volume, velocity and variety and rapid development of the Internet-of-Things (IoT), massive volume of streaming data that generated by individuals are being collected and analyzed due to its great potential for both research and commercial uses [15]. Streaming data refers to data that is updating and evolving through time, such as social media data, online game data and search engineer data [3]. Thereby, those continually updating data that contains personally identifiable information requires strong privacy protection in case of personally identifiable information leakage and violation of one's privacy [3,6].

In the context of raised data privacy concern in the era of big data, differential privacy as a wide-utilized privacy guarantee aims to address the issue regarding privacy-preserving data analysis [5,6]. As one of the effective privacy-preserving techniques [21], research on differential privacy that focuses on

S. Wen et al. (Eds.): ICA3PP 2019, LNCS 11945, pp. 418–426, 2020.
https://doi.org/10.1007/978-3-030-38961-1_37

privacy-preserving data publishing by taking the dynamics and correlation of context into consideration is yet to be well-developed [2,7]. In order to cope with explosive data traffic and public arising security concern of compromising individuals' privacy by streaming data, differentially private algorithms require to provide strong privacy guarantees not only for static data but also for data that is varying or evolving through time. Therefore, a study named differential privacy under continual observation initiated by Dwork et al., is used to address such issue due to the urgent demand for streaming data privacy preservation [6].

In differentially private mechanism, algorithms are desired to characterize data distribution so as to not affect the output [5,7,9]. The demand of learning true data distribution and balancing the level of privacy protection and data utility [13,18] can be satisfied by utilizing deep generative models, especially by using generative adversarial nets (GANs) [8,14,20].

In deep generative models that employed for synthesizing new data via capturing the underlying distribution [1], GANs has drawn a lot of attentions recently showing its significant performance on generating realistic images [8,11,16]. In GANs, the objective is to drive two models to contest with each other in order to improve their method until indistinguishable results are produced by the generative model. Such that, the adversarial modelling framework of GAN refers to two-player game [8,10,11]. In state-of-the-art, triple generative adversarial nets (triple-GANs) model introduces a classifier for classification to solve the problems that existed in two-player adversarial model, as mentioned in Sect. 3.1. Therefore, the updated three-player model has shown a novel perspective of solving problems by not limiting two players in the modelling framework for better performance on generating realistic data-label pairs [11].

The three-player adversarial model has never been considered for the use of privacy preserving on continual data release. In the differential private mechanism, an objective is to produce approximately the same data set compared to the given data for privacy protection [5–7], which can apply the framework that utilized in triple-GANs. In this paper, we present an adversarial generative model for privacy protection on continual data release in the adversarial modelling framework of three players. The novel idea is introducing an identifier into the system for labelling the generated samples whether they fulfill the demand of differential privacy. The modelling details are illustrated in Sect. 3. The evaluation results on the proposed model are shown in Sect. 4.

## 2    Related Work

Recent years, due to the explosive data traffic in big data era, the methods of handling static data cannot be fully adapted to dealing with data that is varying with respect to time. In the field of privacy protection, the study of differential privacy under continual observation was first proposed by Dwork et al. to address such problems. The concept of pan-privacy and user-level/event-level privacy was also proposed for the privacy preserving on continual data release. A differentially private continual counter with error $O(\frac{1}{\epsilon} \cdot (\log t)^{1.5})$ where $t$ refers to the number of time steps [3,6]. Later, a model called $\epsilon-$deferentially private continual counter with poly-log error with a guaranteed error $O(\frac{1}{\epsilon} \cdot (\log t)^{1.5} \cdot \log \frac{1}{\delta})$ was established by Chan, Shi and Song [3]. The model that outputs at every time step, not only has a considerably small error with respect to time but achieves time unboundedness in its mechanism to work functionally with respect to data utility and privacy guarantee for continual data publishing based on the theoretical results [3].

Recently, several studies have focused on the privacy guarantee of differential privacy on correlated data. The temporal correlation of continuously generated data can be obtained by adversaries and have a high risk of compromising private information, which significantly lower the level of privacy protection [2,12]. Based on [2], it shows that employing differential privacy in the Laplace mechanism as primitives for data privacy preserving of continual data release at each time slot (i.e., event-level privacy) can lead to temporal correlations of data due to the nature of dynamic data, which result in potential privacy leakage.

Due to GANs' outstanding ability of estimating the underlying distribution and generating realistic samples that can be employed for data privacy preservation, the idea of combining GANs and differential privacy has shown in [20] and proven its feasibility. A differentially private GAN, namely GANobfuscator, that proposed for mitigating the privacy leakage under GANs, successfully achieves $(\epsilon, \sigma)-$differential privacy under GANs through introducing controlled noise to gradients during the training procedure. GANobfuscator shows its great performance on generating synthetic samples for arbitrary analysis tasks with a strong privacy guarantee for training data [20]. However, the model is only used for privacy preservation in a static manner.

## 3   System Modelling

The proposed model, namely DP-GAN model, is illustrated in this section.

### 3.1   Preliminary

**Differential Privacy.** The definition of Differential Privacy (DP) refers to a stringent mathematical interpretation of privacy in the context of privacy-preserving data analysis [7]. A differentially private mechanism serves on a database or a data set, which holds the collections of individuals' records. Each row in the database or the data set represents a record that contains an individual's data. A differential privacy algorithm ensures presence or absence of an individual will not have a significant impact on the output of the algorithm. The mathematical interpretation in differentially private mechanisms focus on the probability of a given output and its variation by adding or removing of any row in the given database or the given data set. Let $(D, D')$ be a pair of database, where one is a subset of the other by missing one row compared to the larger database, implying two database only varying in one row [5,6].

**Definition 1.** [5] *A randomized function* $\mathcal{K}$ *gives* $\varepsilon$-*differential privacy if for all data sets* $D$ *and* $D'$ *differing in at most one row, and all* $S \subseteq Range(\mathcal{K})$,

$$Pr[\mathcal{K}(D) \in S] \leq \exp(\varepsilon) \times Pr[\mathcal{K}(D') \in S], \qquad (1)$$

*where the probability is over the coin flips of* $\mathcal{K}$.

For appropriate privacy loss $\varepsilon$, a mechanism $\mathcal{K}$ satisfying this definition by generating approximately the same results with or without the presence of any individual's data in the data set [5].

**Pan-Privacy.** Pan-Privacy can be defined as the internal state of an algorithm in a differential private mechanism satisfied the constraint of differential privacy in the context of private preserving for data analysis on a streaming of data [7]. That is, differential privacy preservation maintains even if the database experiences an intrusion once and the internal state of the algorithm is exposed to the adversary. Hereby, a pan-private algorithm is capable of hiding its internal state evolving process from one intrusion which can occur at an unpredictable time so as to private-preserve on continuous data releasing by prevent the internal state and output sequence leakage. Pan-Privacy can be described as the orthogonal presentation of differential privacy preserving under continual observation [3,6,7].

**Triple-GANs.** Triple-GANs, proposed by Chongxuan et al., show its outstanding performance on generating indistinguishable samples by updating two-player game framework of GANs in general into three-players formulation [11] to address the issue caused by the restriction of existing two-player generative models in semi-supervised learning, i.e., discriminator in two-player formulation is incapable of playing two crucial roles at the same time—competing with generator by identifying fake sample and acting as a classifier for label prediction of unlabelled real samples [11,17,19].

In deep generative model, the adversarial generative framework with two player forms a minimax game, the training procedure is shown as:

$$\min_G \max_D \tilde{U}(D, G) = E_{x \sim p(x)}[\log D(x)] + E_{z \sim p_z(z)}[\log(1 - D(G(Z)))] \qquad (2)$$

where $p_z(z)$ is a simple distribution(e.g., uniform and standard norm), which is the random noise $z$ that taken by the generator $G$ drawn from [8,11].

From two-player adversarial modeling framework that only consists of generator and discriminator, Triple-GANs upgraded into three-player game framework by introducing a classifier and highlighting classification and class-conditional generation through conditional distributions and joint distribution, where both the classifier $C$ and the generator $G$ are expected their outputs to converge to the real data distribution by characterizing their conditional distributions, which is $p_c(y|x) \approx p(y|x)$ for the classifier and $p_g(x|y) \approx p(x|y)$ for the generator. The discriminator $D$ only focus on distinguishing fake sample pairs from the true data distribution $p(x, y)$ [11]. The training process in Triple-GANs is described as a three-player minimax game:

$$\min_{C,G} \max_D \tilde{U}(C, G, D) = E_{(x,y)\sim p(x,y)}[\log D(x,y)] + \alpha E_{(x,y)\sim p_c(x,y)}[\log(1 - D(x,y))]$$

$$+ (1 - \alpha)E_{(x,y)\sim p_g(x,y)}[\log(1 - D(G(y,z),y))] + \mathcal{R}_{\mathcal{L}},$$

(3)

where $\alpha \in (0, 1)$ is a constant that implies the relative weighting of generation and classification in the Triple-GANs system [11], and $\mathcal{R}_{\mathcal{L}} = E_{(x,y)\sim p(x,y)}$ $[-\log p_c(y|x)]$ denotes the standard supervised loss to $C$ to address the global optimum of the equilibrium, i.e., $p(x, y) = p_g(x, y) = p_c(x, y)$.

### 3.2 Differential Privacy Identifier

Inspired by three-player game framework, the proposed differential privacy algorithm that utilized GANs generative model can be seen as semi-supervised or unsupervised learning model with classification and class-conditional generation work done inside the adversarial modelling framework in the purpose of learning true data distribution while providing a strong differential privacy guarantee by introducing a differential privacy identifier.

The proposed generative model consists of three components: (1) a differential privacy identifier $I$ to identify whether the generated data set fulfill the stringent requirement of differential privacy; (2) a generator $G$ for characterizing the true data distribution and data-label pairs generation; and (3) a discriminator $D$ to distinguish a received sample-label pair from whether the generator or the true data-label pairs, produced by the identifier, that draw from realistic data distribution and contain correct label information regarding differentially privacy preservation. In the three-player game, by competing with both the discriminator and the differential privacy identifier, the generator are expected its outputs to not only converge to the true data distribution, but also have correct labels with respect to the level of privacy protection, especially the labels that indicate a strong privacy guarantee in a differentially private mechanism under continual observation (Fig. 1).

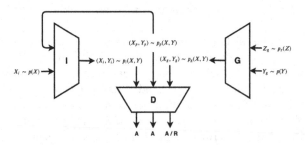

**Fig. 1.** "A" and "R" stand for acceptance and rejection in $D$ correspondingly.

In the proposed three-player game framework, differential privacy identifier $I$ produce a label y for a given data $x$ that sampled from true data distribution $p(x)$ by examining $x$ through differentially private mechanisms. Under given $x$ for label $y$, the conditional distribution refers to $p_i(y|x)$. Hereby, the data-label pair can be expressed as a data that draw from the joint distribution $p_i(x, y) = p(x)p_i(y|x)$. Similarly, generator $G$ generates a pseudo data $x$ for a given label $y$ that draw from label distribution $p(y)$. The conditional distribution for $x$ under label $y$ is $p_g(x|y)$ while the joint contribution that applied in generator represents as $p_g(x, y) = p(y)p_g(x|y)$. In order to learn generator's distribution over input data $x$ given label $y$, $G$ receives an input noise variable $z$ that sampled from a simple distribution $p_z(z)$ (e.g., uniform or standard normal), hereby $x$ can be defined mathematically by $x = G(y, z)$. After classification and generation process through $I$ and $G$ respectively, the synthetic data-label pairs $(x, y)$ are sent to discriminator $D$ for identifying whether the given data-label pairs are from true data distribution or the generator $G$. Seeing that the proposed model is a time-related adversarial generative model as described in Sect. 3.3, $p_p(x, y)$ is introduced and denotes the joint distribution of synthetic data-label pairs at the previous time slots. The formulation of modified three-player adversarial model that stated as above is represented as a minimax equation,

$$\min_{I,G} \max_D \tilde{U}(I, G, D) = E_{(x,y)\sim p(x,y)}[\log D(x,y)] + \alpha E_{(x,y)\sim p_i(x,y)}[\log(1 - D(x,y))]$$

$$+ E_{(x,y)\sim p_g(x,y)}[\log(1 - D(G(y,z),y))] + \mathcal{R}_{\mathcal{L}}, \tag{4}$$

where $\alpha \in (0, 1)$ is a constant that shows the importance of fulfilling the stringent requirement of differential privacy in the game, implying the relaxations of differential privacy is applicable in Eq. 4 in order to improving the degree of data utility. $\mathcal{R}_{\mathcal{L}}$ mentioned in Sect. 3.1 is to address the unique global optimum issues regarding the equilibrium $p(x, y) = p_g(x, y) = p_i(x, y)$. Additionally, seeing the privacy requirements for label information in Sect. 3.3, our model does not have the issue regarding insufficiency of label information in semi-supervised learning. Pseudo discriminative loss is not considered in the training procedure of our model.

### 3.3   Differential Privacy Under Continual Observation

Applying differential privacy on a streaming of data that continues to change and evolve through time refers to differential privacy under continual observation. The continual observation algorithms can be seen as applying strong privacy protection in a differentially private mechanism at discrete time intervals [6].

To define privacy under continual observation, Dwork et al. proposed the concept of adjacency with respect to time in [6]. The definition of adjacency can be expressed as follows: $Adj(S, S')$ if and only if $\exists x, x' \in X$ and $\exists T \subseteq [|S|]$, such that $S|_{T:x \rightarrow x'} = S'$, i.e., $S$, a stream prefix, is defined to be adjacent to $S'$, another stream prefix with a different length compared to $S$ if and only if there exist $x$, $x'$ from $X$, the universe of possible input symbols so that some of the occurrences of $x$ in $S$ is replaced with $x$ at $T$, a set of indices in the $S$, and then form $S'$ [6].

Our model utilizes the definition of adjacency in the differential privacy identifier for streaming data. The proposed model divides streaming data into multiple time slots for data privacy preserving. The data at each time slot are processed respectively, meaning our generative model treats the streaming data separately with respect to time. At each time unit, the generated data are required to not only can be identified as differential private-preserving data and also be the adjacent to the generated data that passed the test from the differential privacy identifier at the last time unit.

In differential privacy identifier $I$, data is processed in differentially private mechanisms by each time unit. At given time slot $t$, the input data is categorized by binary classification to examine whether the input data meet the demand of differential private preserving under continual observation. In our model, we altered Eq. 1 to be dependent on time, the mathematical interpretation of differential privacy with respected to time as follows:

$$Pr[\mathcal{K}(D) \in S] \leq \exp(\varepsilon(t)) \times Pr[\mathcal{K}(D') \in S], \quad (5)$$

At each time unit, the input data that provide a strong differential privacy guarantee should not only satisfy the differential privacy in Eq. 5 but also be identified as adjacency to the input data at previous time unit. The input data that failed to any the two conditions stated as above is classified into unsatisfactory data for privacy protection. Namely, the identifier classifies data according to the two strict requirements for differential privacy under continual observations, i.e., the definition of differential privacy and the definition of adjacency for pan-privacy.

## 4    Performance Evaluation

The experimental results of the proposed model are illustrated in this section by using real-world data. First, the preliminaries of the data set are briefly described. Then, we conduct performance evaluations of the proposed model in terms of privacy protection. The evaluation results of DP-GAN model shows its superiority in comparison of two conventional differentially private mechanisms, i.e., Laplace mechanism and Gaussian mechanism.

The applied data set for performance evaluation is *Iris Data Set* from UCI Machine Learning Repository [4]. The correlations of this data set is as Fig. 2. In the following subsections, for testifying the effectiveness and feasibility of DP-GAN model, Laplace mechanism and Gaussian mechanism are employed for comparisons. The level of data utility is evaluated by Root Mean Square Error (RMSE) between the true data and the synthetic data.

Refers to Fig. 3, it is evident that the level of data utility is raised along with the increase of the value of $\epsilon$. Namely, the privacy level decreases during this procedure seeing that the value of $\epsilon$ can be interpreted as privacy loss. Thus, it can be said that the proposed model can offer higher privacy-preserving level where data utility is a constant.

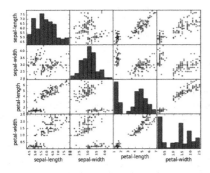
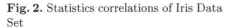
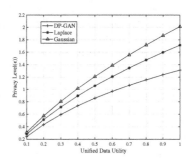

**Fig. 2.** Statistics correlations of Iris Data Set

**Fig. 3.** Privacy level comparison

## 5 Summary and Future Work

In this paper, we present our adversarial generative model that employ theoretical game framework with three players, i.e., a generator, a discriminator and a differential privacy identifier, to provide a strong privacy protection in differentially private mechanisms on continual data publishing through semi-supervised learning process. Also, we shows extensive simulation results on real-world data set for evaluation our proposed model. The outcome indicates the performance of our model is significantly great in terms of privacy protection.

In the future work, we plan to manage the resource allocation of the two games to improve the efficiency and lessen the requirements of computational power. In addition, we intend to use generative adversarial nets to further optimize the privacy protection by deriving the minimum privacy budget.

**Acknowledgements.** The work has been supported by the Cyber Security Research Centre Limited whose activities are partially funded by the Australian Government's Cooperative Research Centres Programme.

## References

1. Audebert, N., Le Saux, B., Lefevre, S.: Generative adversarial networks for realistic synthesis of hyperspectral samples. In: IGARSS 2018–2018 IEEE International Geoscience and Remote Sensing Symposium, pp. 4359–4362 (2018)
2. Cao, Y., Yoshikawa, M., Xiao, Y., Xiong, L.: Quantifying differential privacy in continuous data release under temporal correlations. IEEE Trans. Knowl. Data Eng. **9**, 1281–1295 (2018)
3. Chan, T.H.H., Shi, E., Song, D.: Private and continual release of statistics. ACM Trans. Inf. Syst. Secur. (TISSEC) **14**(3), 26 (2011)
4. Dheeru, D., Karra Taniskidou, E.: UCI machine learning repository (2017). http:// archive.ics.uci.edu/ml
5. Dwork, C.: Differential privacy. In: Bugliesi, M., Preneel, B., Sassone, V., Wegener, I. (eds.) ICALP 2006. LNCS, vol. 4052, pp. 1–12. Springer, Heidelberg (2006). https://doi.org/10.1007/11787006_1

6. Dwork, C., Naor, M., Pitassi, T., Rothblum, G.N.: Differential privacy under continual observation. In: Proceedings of the Forty-Second ACM Symposium on Theory of Computing, pp. 715–724. ACM (2010)
7. Dwork, C., Roth, A., et al.: The algorithmic foundations of differential privacy. Found. Trends® Theor. Comput. Sci. **9**(3–4), 211–407 (2014)
8. Goodfellow, I.J., et al.: Generative adversarial nets. In: Advances in Neural Information Processing Systems 27: Annual Conference on Neural Information Processing Systems 2014, 8–13 December 2014, Montreal, Quebec, Canada, pp. 2672–2680 (2014)
9. Inan, A., Gursoy, M.E., Saygin, Y.: Sensitivity analysis for non-interactive differential privacy: bounds and efficient algorithms. IEEE Trans. Dependable Secur. Comput. (2018)
10. Kawai, Y., Seo, M., Chen, Y.: Automatic generation of facial expression using generative adversarial nets. In: 2018 IEEE 7th Global Conference on Consumer Electronics (GCCE), pp. 278–280 (2018)
11. Li, C., Xu, T., Zhu, J., Zhang, B.: Triple generative adversarial nets. In: Advances in Neural Information Processing Systems 30: Annual Conference on Neural Information Processing Systems 2017, 4–9 December 2017, Long Beach, CA, USA, pp. 4091–4101 (2017)
12. Ou, L., Qin, Z., Liao, S., Hong, Y., Jia, X.: Releasing correlated trajectories: towards high utility and optimal differential privacy. IEEE Trans. Dependable Secur. Comput. (2018)
13. Qu, Y., Yu, S., Gao, L., Zhou, W., Peng, S.: A hybrid privacy protection scheme in cyber-physical social networks. IEEE Trans. Comput. Soc. Syst. **5**(3), 773–784 (2018)
14. Qu, Y., Yu, S., Zhang, J., Binh, H.T.T., Gao, L., Zhou, W.: GAN-DP: generative adversarial net driven differentially privacy-preserving big data publishing. In: ICC 2019–2019 IEEE International Conference on Communications (ICC), pp. 1–6. IEEE (2019)
15. Qu, Y., Yu, S., Zhou, W., Peng, S., Wang, G., Xiao, K.: Privacy of things: emerging challenges and opportunities in wireless Internet of Things. IEEE Wireless Commun. **25**(6), 91–97 (2018)
16. Radford, A., Metz, L., Chintala, S.: Unsupervised representation learning with deep convolutional generative adversarial networks. arXiv preprint arXiv:1511.06434 (2015)
17. Salimans, T., Goodfellow, I., Zaremba, W., Cheung, V., Radford, A., Chen, X.: Improved techniques for training GANs. In: Advances in Neural Information Processing Systems, pp. 2234–2242 (2016)
18. Soria-Comas, J., Domingo-Ferrer, J., Sánchez, D., Megías, D.: Individual differential privacy: a utility-preserving formulation of differential privacy guarantees. IEEE Trans. Inf. Forensics Secur. **12**(6), 1418–1429 (2017)
19. Springenberg, J.T.: Unsupervised and semi-supervised learning with categorical generative adversarial networks. arXiv preprint arXiv:1511.06390 (2015)
20. Xu, C., Ren, J., Zhang, D., Zhang, Y., Qin, Z., Ren, K.: GANobfuscator: mitigating information leakage under GAN via differential privacy. IEEE Trans. Inf. Forensics Secur. **14**(9), 2358–2371 (2019)
21. Xu, C., Ren, J., Zhang, Y., Qin, Z., Ren, K.: DPPro: differentially private high-dimensional data release via random projection. IEEE Trans. Inf. Forensics Secur. **12**(12), 3081–3093 (2017)

# Data Poisoning Attacks on Graph Convolutional Matrix Completion

Qi Zhou, Yizhi Ren$^{(\boxtimes)}$, Tianyu Xia, Lifeng Yuan, and Linqiang Chen

School of Cyberspace, Hangzhou Dianzi University, Hangzhou 310018, China
isq.zhou@gmail.com, {renyz,15084234,yuanlifeng,clq}@hdu.edu.cn

**Abstract.** Recommender systems have been widely adopted in many web services. As the performance of the recommender system will directly affect the profitability of the business, driving bad merchants to boost revenue for themselves by conducting adversarial attacks to compromise the effectiveness of such systems. Several studies have shown that recommender systems are vulnerable to adversarial attacks, e.g. data poisoning attack. Since different recommender systems adopt different algorithms, existing attacks are designed for specific systems. In recent years, with the development of graph deep learning, recommender systems have been also starting to use new methods, like graph convolutional networks. More recently, graph convolutional networks have also been found to be affected by poisoning attacks. However, characteristics of data sources in recommender systems, such as heterogeneity of nodes and edges, will bring challenge to solve attack problem. To overcome this challenge, in this paper, we propose data poisoning attacks on graph convolutional matrix completion (GCMC) recommender system by adding fake users. The key point of the method is to make fake users mimicrking rating behavior of normal users, then pass the information of thier rating behaviors towards the target item back to related normal users, attempting to interfere with the prediction of the recommender system. Futhermore, on two real-world datasets ML-100K and Flixster, the results show that our method significantly overmatches three baseline methods: (i) random attack, (ii) popular item based attack, (iii) and mimicry with random scores based attack.

**Keywords:** Poisoning attack · Recommender system · Graph auto-encode

## 1 Introduction

Recommender systems have been widely used in web services and E-commerce to help users to screen out valuable information from miscellaneous choice. The

This work was supported by the Natural Science Foundation of Zhejiang Province (Grant No. LY18F020017, LY18F030007, and LQY19G030001, National Natural Science Foundation of China (Grant No. 61872120) and Key Technologies, System and Application of Cyberspace Big Search, Major project of Zhejiang Lab (Grant No. 2019DH0ZX01)).

S. Wen et al. (Eds.): ICA3PP 2019, LNCS 11945, pp. 427–439, 2020.
https://doi.org/10.1007/978-3-030-38961-1_38

recommender system first collects the user's historical behavior data to obtain the user-rating matrix through the pre-processing method, and then uses the relevant recommendation technology in the machine learning field to form a personalized recommendation for the user. Existing approaches include content-based methods [1], graph-based methods [2] and factoriztion-based methods [3]. As the most popular method, collaborative filtering (CF) models solve the matrix completion task. The key idea is considering the historical collective interaction information to make predictions.

As the recommendation system plays an increasingly important role in current web services, their vulnerability to malicious attacks is exposed as well. *Poisoning attack* is one of the most common attacks, which injects fake users to a recommonder system with crafted rating behaviors to reduce the effectiveness of the system. For instance, a merchant may hire the black market to promote the rating score and exposure rate of his merchandise. In recent years, several studies proposed poisoning attacks to recommender systems [4,5]. Li et al. [4] proposed poisoning attacks to factorization-based recommender systems. Fang et al. [5] proposed poisoning attacks to graph-based recommender systems. These attack methods are designed for specific systems respectively.

Recently, researchers focus on the application of *graph convolution networks* (GCN) in recommendation system [6,7]. GCN models capture structured features through information propagation between nodes, eliminating the need of feature engineering. However several studies revealed that GCN models were also vulnerable in face of data poisoning attacks [8–10]. Literature [8,9] first proposed adversarial attacks against GCN models at different knowledge levels of the attacker respectively, Chen et al. [10] proposed data poisoning attacks on graph convolutional auto-encoder under the task of link prediction. However, above attack methods are for the general case and most existing attacks are performed on homogenous graph. Since the special data characteristics in recommendation task, such as heterogeneity of nodes and edges, implement data poisoning attacks on such GCN based system will be more challenging.

In this work, we focus on designing data poisoning attacks for recommender system using graph convolutional networks, more specifically, *Graph Convolutional Matrix Completion* (GCMC) [11]. GCMC uses a graph auto-encoder to achieve the edge-type specific information passing on a user-item bipartite graph, information only be passed along the edge of the same rating type. To get the user embeddings and item embeddings, the output of graph convolution layer will be sent to a dense layer. Finally, with a bilinear decoder, links in the bipartite interaction graph will be reconstructed to achieve the predicting process.

This paper makes the following key contributions. First, in order to evaluate the threats faced by recommender systems, we propose a special data poisoning attack on graph auto-encoder recommender system GCMC. According to the principle of information propagation through specific rating types in GCMC, The attack method is based on mimicry related normal users, inducing related normal users giving high rating scores to target items. Then achieve the goal of fooling recommender system. Second, we conduct several evaluation of our method on different real-world datasets, and the results show that (i) our attack method is overmatch other baseline methods, 5–8 times in best cases. (ii) For cold start, our method can significantly increase the rating scores.

## 2 Related Work

### 2.1 Recommender System

Recommender systems have been widely deployed in web services. As the most common method, collaborative filtering can be divided into user-based CF, item-based CF and model-based CF. User-based CF is based on the assumption that similar users may have the same preferences [12], therefore system first calculate the users similarity, ratings by more similar users contribute more to predicting the item rating. Item-based CF approaches [13] apply the same idea but use similarity between items instead of users. Similarly, ratings from more similar items contribute more. Model-based CF uses machine learning methods to solve the recommendation problem, such as matrix factorization [3] and graph-based model [14].

### 2.2 Graph Convolutional Network

In recent years, with the development of deep learning, deep learning methods are gradually applied to graph data, i.e. graph convolutional networks (GCN) [15]. The key idea of GCN is to aggregate information from neighbor nodes in the graph to capture structured features. Due to the good nature of GCN, it is gradually applied in the recommendation field [6,7,11,19,20]. Most relevant work to this paper is GCMC [11], it proposed a novel recommendation method using graph convolutional auto-encoder. The first order convolution of user nodes and item nodes is carried out to obtain respective embeddings. Then a decoder is used for link prediction.

### 2.3 Adversarial Attacks to Recommender Systems

In recommender system, a mainstream attack aims to spoof a recommender system so that the target item is recommended to more users or fewer users. Specially, poisoning attacks [4,5,16,18] implement attacks by injecting fake users with crafted fake rating scores to the system such that a biased model is learnt from contaminated dataset. E.g., In random attacks [16], given the number of fake users an attacker can inject and the number of items that each fake user can rate, the attacker randomly selects the items to be filled and the rating scores from a normal distribution.

More recent poisoning attacks [4,5] generate fake rating scores for specific recommender system by respective optimization algorithm. Specially, Li et al. [4] proposed poisoning attacks to matrix-factorization-based recommender systems. Fang et al. [5] proposed poisoning attacks to graph-based recommender systems.

For each fake user, attack algorithm selects filler items by computing gradient. However, above methods are proposed corresponding to specific recommender systems.

### 2.4    Adversarial Attacks to GCN

In this work, we focus on attacking graph convolutional auto-encoder model. There are some similar work about adversarial attacks against GCN models [8–10]. Zügner et al. [8] proposed the first adversarial attacks against GCN and defined some metrics about attacking neural networks for graph data. In their settings, the attacker have full knowledge of the model. Dai et al. [9] proposed a reinforcement learning based attack method, and studied the method and effect of corresponding attack under different knowledge level of the attacker. Chen et al. [10] studied data poisoning attacks on graph convoluntional auto-encoder model under the task of link prediction. The data and downstream task in the model mentioned above are different from those of the recommender systems, which will cause the definition of the attack problem to be different.

## 3    Problem Definition and Preliminary

### 3.1    Victim Model

In this work, we choose GCMC as the victim model. GCMC contains **a graph convolutional encoder** and **a bilinear decoder**. Given a rating matrix $M$ of shape $N_u \times N_v$, where $N_u$ is the number of users and $N_v$ is the number of items. Entries $M_{ob}$ denotes the observed ratings and $M_{un}$ denotes the unobserved ratings. Notice that $M_{ob} \cup M_{un} = M$. GCMC model uses $M_{ob}$ as the training set and then predict the unobserved ratings (Fig. 1). The whole model is shown in Fig. 1.

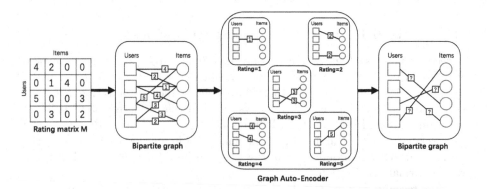

**Fig. 1.** GCMC model

**Graph Convolutional Encoder.** Consists of two parts: (1) a graph convolution layer and (2) a dense layer. By assigning a specific transformation for each rating level in rating matrix, user nodes and item nodes will be represented as node embedding. Then the final embedding can be reached with a dense layer Eq. (3).

$$\mu_{j \to i, r} = \frac{1}{c_{ij}} W_r x_j \qquad (1)$$

$$h_i = \sigma[accum(\sum_{j \in \mathcal{N}_{i,1}} \mu_{j \to i,1}, ..., \sum_{j \in \mathcal{N}_{i,R}} \mu_{j \to i,R})] \qquad (2)$$

The convolution process is shown in Eqs. (1) and (2), Eq. (1) shows the information passing from node $j$ to node $i$ along the edge type $r \in R$, where $c_{ij}$ is a normalization constant, $W_r$ is an edge-type specific parameter matrix and $x_j$ is the feature vector of node $j$. Equation (2) shows the aggregation process, where $\mathcal{N}_{i,r}$ denotes the neighbors of node $i$ under a specific edge-type $r$. And $accum(\cdot)$ denotes an accumulation operation such as $stack(\cdot)$ or $sum(\cdot)$.

$$\mu_i = \sigma(W h_i) \qquad (3)$$

Equation (3) shows the dense layer, where $W$ is the parameter matrix. After that, encoder can get both user embeddings and item embeddings.

**Bilinear Decoder.** Reconstructs links in the bipartite interaction graph, and treat each rating level as a separate class. $\breve{M}_{ij}$ is defined as the reconstructed rating score that user $i$ gives to item $j$, the decoder produces a probability distribution over possible rating levels, using a bilinear operation and then applying a softmax function Eq. (4).

$$p(\breve{M}_{ij} = r) = \frac{e^{u_i^T Q_r v_j}}{\sum_{s \in R} e^{u_i^T Q_s v_j}} \qquad (4)$$

where $Q_r$ a trainable parameter matrix of shape $E \times E$, and $E$ is the dimensionality of user(item) embedding $u_i(v_j)$. The predicted rating is computed as

$$\breve{M}_{ij} = g(u_i, v_j) = \mathbb{E}_{p(\breve{M}_{ij}=r)}[r] = \sum_{r \in R} r p(\breve{M}_{ij} = r) \qquad (5)$$

### 3.2   Threat Model

**Attacker's Goal.** In our *single attack* setting, the attacker's goal is to design a data poisoning attack to promote the ratings of a certain item. For example, the seller may try to promote his own products by buying fake users and ratings from the balck market. It means that the attacker can construct data poisoning attack by adding fake users and controlling these fake users' rating behaviors.

**Attacker's Knowledge and Capability.** In our attack scenarios, we assume the attacker has full knowledge of victim model, including dataset and model architecture. Corresponding attack scenario setting in *adversarial machine learning* is called *white-box attack*. Although this assumption is somewhat unrealistic, we can evaluate the security of the victim model in the worst case.

**Limitation.** We assume that the attacker has limited resource. Consider that creating fake users costs money, it can be seen as money resources. Besides, the attacker need to make sure their attacks are undetectable. We regard this limitation as a tiny change of the original graph, by limiting the number of items that per fake user can rate.

### 3.3 Attack Model

Given the original rating matrix $M$, the victim model $f$ and target item $v_t \in V$. In this work, we focus on GCMC, thus $f$ can be regarded as the GCMC model. Let $r_{u'}$ be the rating score vector of a fake user $u' \in U_{fake}$, where $r_{uv}$ is the rating score that the fake user $u$ gives to the item $v$. Noticed that the number of fake user $|U_{fake}|$ and the number of items that a fake user can rate, which can be represented as $|r_{u'}|$, are limited. We consider a rating score is in the set of integers $\{1, 2, ..., r_{max}\}$, where $r_{max}$ is the maximum rating score value. Our goal is to find the rating score vector for each fake user which maximizes the total predicted rating score of the target item $\check{M}_{v_t}$. Mathematically, the attack problem can be formally described as the following optimization problem:

$$\max \check{M}_{v_t} = \sum_{u \in \mathcal{N}_{v_t}} \check{M}_{u,v_t} \tag{6}$$

$$\text{s.t. } |U_{fake}| \leq N, \ |r_{u'}| \leq n, \ r_{u'v} \in \{1, 2, ..., r_{max}\} \tag{7}$$

Where $\check{M}_{v_t}$ is the total predicted ratings of target item $v_t$, which is the sum of predicted ratings to target item by each neighbor user $u \in \mathcal{N}_{v_t}$ in test dataset. Notice that fake users $|U_{fake}|$ are not included in the test dataset, which means $u \notin U_{fake}$. $N$ and $n$ are the restrictions on the number of fake users and ratings.

## 4    Poisoning Attacks

In this section, we will introduce our data poisoning attack algorithm. Before that, We review how GCMC works. Specifically, the key operation of graph convolutional layer. For both user nodes and item nodes, information is propagated through the special rating-type edges, see formula (2). Thus the same rating behavior from a user to an item will work on both the user embedding and item embedding, which means we can implement attacks by poisoning target item embedding and corresponding user embeddings. However, we have no access to manipulate the rating behavior of normal users but only fake users whose rating behaviors can not influence above embeddings directly. Beyond that, unlike the 2nd-order graph convolution on homogeneous graph, GCMC uses 1st-order graph convolution for two types of nodes on heterogeneous bipartite graph. It is hard to evaluate the equivalency or similarity of this two different methods, meaning that the indirect effects by 2-hop neighbors are also ambiguous. These make it difficult to solve the above optimization problem. To tackle our data poisoning attack, we use a heuristic method to approximate the solution.

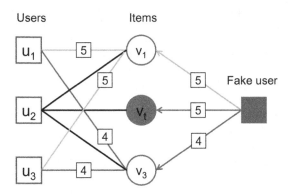

**Fig. 2.** Illustration

The key idea of our method is that under the constraint of the limited ability of the attacker, we try to make fake users to mimic the normal users as much as possible, and guide the normal users to rate the target item with a higher score. First, given a target item $v_t$, we find the users to predict as candidate sets, denoted by $\mathcal{U}_{v_t}$, as nodes $u_1$ and $u_3$ shown in Fig. 2. Then sort in descending order according to co-occurrence frequency of (*item, rating*) pair from the collection of users' rating behaviors in $\mathcal{U}_{v_t}$, the sorted rating behavior set is denoted by $C = (V, R)$. For each fake user, we give $v_t$ the highest score and then select top-$(n-1)$ (*item, rating*) pair from $C$ as remaining items and corresponding rating scores. For example, if top-1 pair is $(v_1, 5.0)$, then each fake user will give a rating score 5.0 to item $v_1$. Since co-occurrence frequency of (*item, rating*) pair represents how many related normal users have the same rating behavior. The intuition behind this step is that, by rating items with higher co-occurrence frequency, fake users will be more likely to simulate more normal users who are related to the target item. After that, the high score rating behavior from fake users to target item may be excepted to contaminate related user embeddings. The algorithm is giving as follows.

---

**Algorithm 1.** Mimicry related users

---

**Input:** Rating matrix $M$, target item $v_t$, fake user number limitation $N$, fake
  user rating number limitation $n$
**Output:** Poisoning rating matrix $M'$
$M' \leftarrow M$;
Get user candidate set $\mathcal{U}_{v_t}$;
Sort $(item, \ rating)$ pair in descending order by co-occurrence frequency, get
sorted rating behavior candidate set $C = (V, \ R)$;
**for** $i = 1, 2, ..., N$ **do**
$\quad$ Create fake user node $u'_i$;
$\quad$ **for** $j = 1, 2, ..., n$ **do**
$\quad\quad$ **if** $j = 1$ **then**
$\quad\quad\quad$ $r_{u'_i v_t} = r_{max}$
$\quad\quad$ **else**
$\quad\quad\quad$ $v_j = V[j-1]$
$\quad\quad\quad$ $r_{u'_i v_j} = R[j-1]$
$\quad\quad$ **end**
$\quad$ **end**
$\quad$ add $r_{u'}$ to $M'$
**end**
return $M'$

---

## 5   Experiments

In this section, we compare our data poisoning attacks on with three baseline
attack methods. And show advantages of our method, then we will give the
analysis about the results and explain why our method is effective.

### 5.1   Experiments Setup

**Datasets.** We choose two real-world datasets, which are very common in recommendation systems and data mining tasks. We use the same datasets as in
literature [11].

*MovieLens 100K.* This dataset consists of 943 users, 1682 items and 100,000
ratings.

*Flixster.* This dataset consists of 3000 users, 3000 items and 26,173 ratings.
$\quad$ We follow the definition of sparsity proposed in literature [5].

$$Sparsity = 1 - \frac{\text{number of ratings}}{\text{number of users} \times \text{number of items}} \quad (8)$$

In the following experiments, we will show the influence of data sparsity on the
attack results. The statistic analysis of these two datasets is given in Table 1.

**Baseline.** We compare our attack method with three baseline methods. In all
these attacks, an attacker injects $N$ fake users to the recommender system. Each

**Table 1.** Datasets statistic analysis

| Dataset | Users | Items | Ratings | Sparsity | Ratings |
|---------|-------|-------|---------|----------|---------|
| MovieLens 100K (ML-100K) | 943 | 1682 | 100,000 | 93.67% | $1, 2, \ldots, 5$ |
| Flixster | 3000 | 3000 | 26,173 | 99.71% | $1, 2, \ldots, 10$ |

fake user rates $n$ items, gives the maximum rating score to the target item and gives respective ratings scores to the remaining $n - 1$ items. The main difference between those attack methods is how to select remaining items and generate corresponding ratings scores to these items.

*Random Attack (RA)* [16]: RA uniformly selects remaining items from the item list at random. For each selected remaining item, RA randomly generates rating scores from the distribution of ratings occurred in training data.

*Popular-item based Attack (PA):* Different from RA, this attack selects the top $n - 1$ most popular items in training data as remaining items. We define popular as the total rating scores of an item. For each selected item, rating scores is generated in the same way as RA.

*Mimicry with Random scores Attack (MRA):* This attack considers having fake users to mimic the users related with the target item. MRA selects the top $n - 1$ most popular 2nd-order neighbor items of the target item, trying to mimic as many related users as possible. Rating scores generation is the same as RA.

**Target Item.** In ML-100K dataset, we divide items into three categories: Popular, Medium and unpopular. In Flixster, considering that the dataset is sparse, we only divide items into popular and unpopular.

**Parameter Setting.** There are two main parameters in our experiment, the number of fake users $N$ and the number of items that a fake user can rate $n$. In order to make the disturbance as small as possible, we set $N = \lambda \cdot$ number of users, $\lambda$ is the inject size of fake users (e.g. $\lambda = 1\%$, $2\%$ and $3\%$), and $n = 10$. In order to avoid deviation, we choose $T = 20$ target items for each dataset. For each target item, we conduct the experiment three times to take the average of predicted results of the model.

**Evaluation.** To evaluate the performance of different attack methods, we use the average improvement ratio of predicted rating scores as metric. The formula as follows.

$$\text{Improvement Rate} = \frac{1}{T} \sum_T \frac{\check{M}_{v_t} - M_{v_t}}{M_{v_t}} \tag{9}$$

## 5.2   Results

In this part, we will show our attack method results in different experiments parameter settings on several real-world datasets, and show how our method is more competitive than baseline methods.

**Table 2.** Improvement rate of predicted ratings for different attack methods with different inject size

| Dataset | Attack | Inject size $\lambda$ | | | | | | | | |
|---|---|---|---|---|---|---|---|---|---|---|
| | | Popular items | | | Moderate items | | | Unpopular items | | |
| | | 0.01 | 0.02 | 0.03 | 0.01 | 0.02 | 0.03 | 0.01 | 0.02 | 0.03 |
| ML-100K | RA | 0.00551 | 0.00857 | 0.00900 | 0.02515 | 0.04735 | 0.06606 | 0.06189 | 0.11240 | 0.16716 |
| | PA | 0.01240 | 0.01299 | 0.01505 | 0.05961 | 0.08761 | 0.12553 | 0.01251 | 0.03548 | 0.07424 |
| | MRA | 0.01470 | 0.01183 | 0.01311 | 0.02889 | 0.04126 | 0.07941 | 0.05869 | 0.10854 | 0.17710 |
| | **Ours** | **0.01692** | **0.02695** | **0.03459** | **0.12353** | **0.15396** | **0.19085** | **0.86140** | **0.84865** | **0.86380** |
| Flixster | RA | 0.09545 | 0.13409 | **0.15266** | – | – | – | 0.43948 | 0.43946 | 0.40501 |
| | PA | 0.09886 | 0.12647 | 0.13258 | – | – | – | 0.28090 | 0.41489 | 0.46656 |
| | MRA | 0.09876 | 0.12914 | 0.142068 | – | – | – | 0.31826 | 0.37060 | 0.37506 |
| | **Ours** | **0.11991** | **0.14381** | 0.14768 | – | – | – | **1.22288** | **1.29889** | **1.38286** |

**Overall Comparison.** Table 2 shows the improvement rate of predicted ratings for different attack methods under different inject size. In detail, inject size on each dataset are 1%, 2% and 3%. In ML-100K dataset, we divide the target items into popular, moderate and unpopular. The criteria for classifying categories is that, a target item is regarded as popular if its total predicted rating score is higher than 50 under the circumstance that the victim model is not contaminated. Likewise, lower than 20 is unpopular, between 20 and 50 is moderate. In Flixster dataset, due to the sparsity of ratings, we only divide them into two categories, popular with total rating score higher than 20 and unpopular with total rating score lower than 5. As we can see in each column, in all datasets, our attack method is significantly better than other baseline approaches for different categories of target items and inject size. Besides, comparing different item types under the same inject size, we can see as the popularity of target items decreases, the improvement rate of each method increases. That is because for an popular item, it will be influenced by more neighbors, which will weaken the impact of the attack. In addition, the evaluation metric itself also limits the numerical values toward popular target items.

**Compare with Different Methods.** In ML-100K dataset, our method is slightly better than the others for popular items, but none of the attack methods work particularly well. In other two item type settings, the result of our method is several times that of the others. For moderate items, our method is almost twice as many as the others. And for unpopular items, our method turns out to be about seven times better than the others. Contrary to popular items, moderate and unpopular items have less connection with other nodes, and this

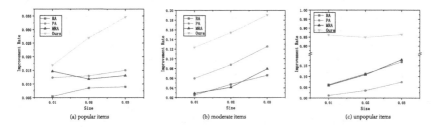

**Fig. 3.** Improvement rate varies with the size of inject size for each target item type on ML-100K

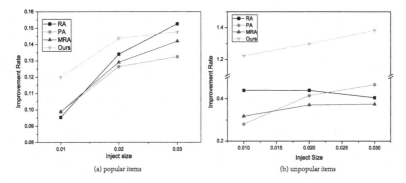

**Fig. 4.** Improvement rate varies with the size of inject size for each target item type on Flixster

makes the impact of the attack more significant. Things are almost the same on Flixster dataset. For popular items, all four methods performed similarly. For un-popular items, baseline methods are almost equaled but slightly difference with different inject size. Our method is also 3–4 times better than baseline methods. Noticed that MRA is the ablation of our method, the main difference is giving rating scores to remaining items. The results show that MRA has no advantage over the other two baseline methods. We regard this phenomenon as the basis for verifying the validity of our method. Since the information propagation of GCMC model is based on specific rating type edge, this may result in not all rating behavior information being passed to the neighbor nodes, but only those behaviors with consistent ratings.

**Impact of Inject Size.** As is shown in Figs. 3 and 4, with the growth of inject size, the influence of fake users to the target item also increases, leading the overall trend of improvement rate to increase. For RA and PA method, it is more likely to hit the remaining items and appropriate rating behaviors due to the randomness of attack algorithm. For MRA, since the remaining items are the same as our proposed method, so there will be no more contingency for nodes. But rating behaviors still have more chance to impact the target items. Notice that on ML-100K dataset, for unpopular items, the effect of our method is no longer increased as inject size increasing, but it is maintained at around 0.85. This may due to the effectiveness of the method has reached the upper limit.

## 6    Conclusion

In this work, we propose the data poisoning attacks on GCMC recommender system. We show that the recommender system based on graph convolutional network is vulnerable in the face of adversarial attacks. Although the GCMC model restricts the propagation of information on bipartite graphs by distinguishing different types of ratings, which makes the attack more difficult. We propose attack method based on mimicking the rating behaviors of related users. And by selecting the behaviors that are the highest frequency of co-occurrence, fake users can share their high-score rating behaviors towards the target item to the relevant users as much as possible. We verify the effectiveness of our method by experiments on several real-world datasets, and the results show that our method is more effective than baseline method in different experimental settings. However, our method assumes the attacker has full knowledge, which may be unrealistic. We leave more strict limits on the knowledge of attacker for future work.

Interesting extensions of this work includes (i) studying data poisoning attacks against graph convolutional recommender systems in more practical scenarios, (ii) exploring how to use gradient-based method to solve the optimization problem in this paper, (iii) studying to improve the robustness of graph convolutional recommender system in the face of attack.

## References

1. Pazzani, M.J., Billsus, D.: Content-based recommendation systems. In: Brusilovsky, P., Kobsa, A., Nejdl, W. (eds.) The Adaptive Web. LNCS, vol. 4321, pp. 325–341. Springer, Heidelberg (2007). https://doi.org/10.1007/978-3-540-72079-9_10
2. Pirotte, A., Renders, J.M., Saerens, M.: Random-walk computation of similarities between nodes of a graph with application to collaborative recommendation. IEEE Trans. Knowl. Data Eng. **19**, 355–369 (2007)
3. Koren, Y., Bell, R., Volinsky, C.: Matrix factorization techniques for recommender systems. Computer **39**(8), 30–37 (2009)
4. Li, B., Wang, Y., Singh, A., et al.: Data poisoning attacks on factorization-based collaborative filtering. Advances in Neural Information Processing Systems, pp. 1885–1893 (2016)
5. Fang, M., Yang, G., Gong, N.Z., et al.: Poisoning attacks to graph-based recommender systems. In: Proceedings of the 34th Annual Computer Security Applications Conference, pp. 381–392. ACM (2018)
6. Ying, R., He, R., Chen, K., et al.: Graph convolutional neural networks for web-scale recommender systems. In: Proceedings of the 24th ACM SIGKDD International Conference on Knowledge Discovery & Data Mining, pp. 974–983. ACM (2018)

7. Wu, S., Tang, Y., Zhu, Y., et al.: Session-based recommendation with graph neural networks. arXiv preprint arXiv:1811.00855 (2018)
8. Zügner, D., Akbarnejad, A., Gnnemann, S.: Adversarial attacks on neural networks for graph data. In: Proceedings of the 24th ACM SIGKDD International Conference on Knowledge Discovery & Data Mining, pp. 2847–2856. ACM (2018)
9. Dai, H., Li, H., Tian, T., et al.: Adversarial attack on graph structured data. In: Proceedings of the 35th International Conference on Machine Learning, PMLR, 80 (2018)
10. Chen, J., Shi, Z., Wu, Y., et al.: Link prediction adversarial attack. arXiv preprint arXiv:1810.01110 (2018)
11. Berg, R., Kipf, T.N., Welling, M.: Graph convolutional matrix completion. arXiv preprint arXiv:1706.02263 (2017)
12. Breese, J.S., Heckerman, D., Kadie, C.: Empirical analysis of predictive algorithms for collaborative filtering. In: Proceedings of the Fourteenth conference on Uncertainty in artificial intelligence. Morgan Kaufmann Publishers Inc., pp. 43–52 (1998)
13. Deshpande, M., Karypis, G.: Item-based top-n recommendation algorithms. ACM Trans. Inf. Syst. (TOIS) **22**(1), 143–177 (2004)
14. Pirotte, A., Renders, J.M., Saerens, M.: Random-walk computation of similarities between nodes of a graph with application to collaborative recommendation. IEEE Trans. Knowl. Data Eng. **3**, 355–369 (2007)
15. Kipf, T.N., Welling, M.: Semi-supervised classification with graph convolutional networks. arXiv preprint arXiv:1609.02907 (2016)
16. Lam, S.K., Riedl, J.: Shilling recommender systems for fun and profit. In: Proceedings of the 13th international conference on World Wide Web, pp. 393–402. ACM (2004)
17. Mobasher, B., Burke, R., Bhaumik, R., et al.: Toward trustworthy recommender systems: an analysis of attack models and algorithm robustness. ACM Trans. Internet Technol. (TOIT) **7**(4), 23 (2007)
18. O'Mahony, M., Hurley, N., Kushmerick, N., et al.: Collaborative recommendation: a robustness analysis. ACM Trans. Internet Technol. (TOIT) **4**(4), 344–377 (2004)
19. Fan, W., Ma, Y., Li, Q., et al.: Graph neural networks for social recommendation. In: The World Wide Web Conference, pp. 417–426. ACM (2019)
20. Song, W., Xiao, Z., Wang, Y., et al.: Session-based social recommendation via dynamic graph attention networks. In: Proceedings of the Twelfth ACM International Conference on Web Search and Data Mining, pp. 555–563. ACM (2019)

# Secure Data Deduplication with Resistance to Side-Channel Attacks via Fog Computing

Fuyou Zhang, Saiyu Qi$^{(\boxtimes)}$, Haoran Yuan, and Meng Zhang

School of Cyber Engineering, Xidian University, Xi'an, Shaanxi, China
fuyouzhang@yeah.net, syqi@connect.ust.hk, hryuan1@163.com,
zhangmeng1575431@163.com

**Abstract.** Deduplication could greatly save the storage overhead of cloud server by eliminating duplicated data and retaining one copy. In order to ensure the data privacy, many researchers try to make deduplication feasible in ciphertext. A typical scheme is message-locked encryption (MLE) which takes cryptographic hash value of message as encryption key. However, MLE is vulnerable to side-channel attacks. To our knowledge, the existing schemes try to mitigate these attacks with either security drawbacks or expensive overhead. In this paper, we propose two new techniques to solve two typical side-channel attacks named probe attack and key-cache attack via fog computing with new security and efficiency tradeoffs. Built on the new techniques, we propose a secure data deduplication system in fog computing environment. Our evaluation shows that our system has better performance compared with previous works.

**Keywords:** Message-locked encryption · Deduplication · Fog computing · Side-channel attack

## 1 Introduction

Cloud computing is a new computation paradigm which was proposed by Google in 2006. It adopts a centralized computation mode to mitigate the resource-constrained burden of personal devices [7]. Many users incline to store data on the cloud server instead of local devices. According to the analysis report of International Data Corporation (IDC), the volume of data in the world is expected to reach 40 trillion gigabytes in 2020 [20,37]. Another survey of IDC indicates that 75% data items are duplicated, since the redundant data wastes numerous storage resources [17,27].

In order to ensure the data privacy, end users usually encrypt data before uploading to the cloud server [6,7,34]. However, traditional encryption schemes do not support data deduplication [32,36]. Specifically, same data items will be encrypted into different ciphertexts since different users might select distinct encryption keys. In response, MLE [2] provides a viable option to ensure data

© Springer Nature Switzerland AG 2020
S. Wen et al. (Eds.): ICA3PP 2019, LNCS 11945, pp. 440–455, 2020.
https://doi.org/10.1007/978-3-030-38961-1_39

confidentiality while enabling data deduplication. It encrypts a data item with a MLE key, which is derived from a cryptographic hash value of the data item content. In this way, MLE ensures that data items with the same content will be encrypted into same ciphertexts.

However, MLE is vulnerable to two side-channel attacks: Probe attack [13] and key-cache attack [21]. Through probe attack, the adversary can infer the existence of certain data items of other users by checking whether these data items have been deduplicated at the cloud side. Through key-cache attack, the adversary can decrypt data items by caching MLE keys. Both attacks leak the privacy of user data. Although several works [12,21] have been proposed to solve these attacks, these works either suffer security drawbacks or incur expensive overhead. For example, one solution to prevent key-cache attack is to select a new MLE key to re-encrypt the whole data item. However, this solution requires to download/upload a large volume of data, incurring high communication overhead. In order to solve above questions, another solution is REED [21], which supports lightweight rekeying by re-encrypting only a little part of a data item named stub. However, REED does not completely mitigate the key cache attack since the adversary can still access the rekeyed data item by retaining the stub.

Based on above questions, we design a new secure deduplication storage system to prevent the two side-channel attacks with new security and efficiency tradeoffs. Our system leverages fog nodes [5] to execute some critical tasks in data deduplication process. Compared with cloud server, fog nodes have higher credibility since they only afford simple tasks and thus suffer less software attacks [1,26,28,30,33]. As a result, many cloud-based secure data storage systems rely on fog nodes to afford critical tasks [15,16,19,31,35]. Different with these works, we design two new techniques to mitigate side-channel attacks in the context of secure data deduplication via fog computing. This paper makes the following contributions.

- We design a random position re-encryption technique to prevent key-cache attack. Our technique combines message-locked encryption and convergent All-or-nothing transform (CAONT) [22], which allows a fog node to re-encrypt a random position of a data item in a rekeying operation. Compared with previous works, our technique mitigates key-cache attack while avoiding re-encrypting large volume of data.
- We design a two-stage deduplication technique to prevent probe attack. Our technique deduplicates data items through two stages. The first stage is conducted between a user and a fog node, and the second stage is conducted between a fog node and the cloud server. Both stages adopt client-side deduplication. Compared with previous works, our technique mitigates probe attack while avoiding high communication overhead incurred by server-side deduplication and high computation overhead incurred by client-side deduplication.
- We implement our system on a local PC and a commercial cloud server. We evaluate some critical parameters in our experiment to prove the efficiency of

our system. The result shows that our system ensures high speed of encryption and data access. Our system also supports efficient data deduplication.

This paper is organized as follows. In Sect. 2, we introduce some related backgrounds about our system. In Sect. 3, we elaborate the system details of our scheme. The security analysis of our system is presented in Sect. 4. The implementation and evaluation have been given in Sect. 5 and Sect. 6, respectively. In Sect. 7, we introduce related works. Finally, we draw conclusions in Sect. 8.

## 2   Background

In this section, we describe the architecture of fog storage systems and cryptographic primitives. Besides, we define the threat model of our system.

### 2.1   System Overview

Fog computing is considered as an extension of the cloud computing to the edge of the internet [19]. It always provides computation, storage, and networking services to undertake the cloud computation task. Our system model consists of a cloud server, a set of fog nodes, a set of users and a key manager. Different with traditional cloud based on system model, our model considers the deployment of fog nodes. Fog computing is a combination of the Internet of Things (IoT) and cloud computing [5,14]. Our system architecture is shown in Fig. 1.

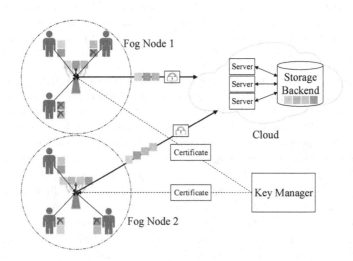

**Fig. 1.** System model (Wireless base stations represent fog nodes. Each fog node manages three users. Key manager sends certificates to fog nodes and users. The small square denotes user data item. Different color express different data contents).

**Cloud Server:** The cloud server is a centralized service which provides long-term data storage and retrieval services. It stores deduplicated data items and maintains a global data index.

**Fog Node:** A fog node is a distributed entity which acts as a proxy for the cloud server by providing storage services and lightweight computation tasks at the network edge. The fog node is connected to the cloud server and manages data ownership of users within certain area.

**User:** A user uploads/downloads the data items to/from the cloud server. In a data uploading/downloading operation, the user selects a local fog node as a proxy to forward the data item between it and the cloud server. According to the uploading order of a data item, a user can be classified as an initial uploader and a subsequent uploader.

**Key Manager:** Our system deploys a key manager to provide certificates for fog nodes and users. The certificate is used to distinguish users and fog nodes in data uploading/downloading operations. Specifically, a user needs to prove its role to a fog node and a fog node needs to prove its role to the cloud sever.

### 2.2   Threat Model and Design Goals

We consider a malicious cloud server that aims to explore the contents of user data items. As a result, data items of users should be encrypted before uploading to the cloud server. Also, we consider malicious users that aim to explore the contents of data items outside their access authorities. MLE is an important tool to reconcile encryption and deduplication. However, the usage of MLE also opens a door for malicious users to launch side-channel attacks since it is a deterministic encryption.

We mainly consider two side-channel attacks namely probe attack and key-cache attack, which can be used to explore the contents of data items as demonstrated by previous works [12,13]. For key-cache attack, once a user deletes his data item, the user should have no authority to access the data item anymore. However, the user can cache the MLE key of the data item to continuously access it. For probe attack, a malicious user could infer the existence of data items of other users by checking whether these data items have been deduplicated by the cloud server. In a typical storage system with deduplication, a user first sends a hash value of his data item to the cloud server to check if the hash value has already existed in its storage. If yes, the user does not need to upload the data item anymore. However, a malicious user can upload the hash values of interested data items, too. If the cloud server does not require the user to upload these data items, the user knows that these data items have already existed in the cloud server.

Our threat model makes the following assumptions. We assume that the key manager could contact with fog nodes and users in a security channel which means the malicious user cannot launch man-in-the-middle attack. Besides, each user and key manager cannot infer the fingerprint information and learn the message content. We also assume that any adversaries don't have abilities to compromise or gain access to the key manager.

To this end, our system focuses on the following main security goals. First, our system ensures confidentiality, such that data items are kept secret against malicious server and users. In addition, our system could prevent the probe attack and the key-cache attack.

### 2.3  Cryptography Primitives

In this section, we first introduce the basic definition and properties of message-locked encryption and then we present the definition of convergent All-or-nothing-transform and broadcast encryption.

**Message-Locked Encryption:** Message-locked encryption is a symmetric encryption scheme in which the key used for encryption and decryption is derived from the message itself. $\text{Enc}^{\text{MLE}}(\cdot)$ algorithm has such steps. First, $\text{KenGen}^{\text{MLE}}(\cdot)$ calculates cryptographic hash value of plaintext $M$ and generates the symmetric key $K_{\text{MLE}}$. Then, $\text{Enc}^{\text{AES}}(M, K_{\text{MLE}}) \rightarrow C_1$. Decryption algorithm is similar as encryption algorithm. It takes $C_1$ and $K_{\text{MLE}}$ as input and outputs plaintext $M$: $\text{Dec}^{\text{MLE}}(C_1, K_{\text{MLE}}) \rightarrow M$.

**Convergent All-or-Nothing Transform:** CAONT is derived from All-or-nothing transform (AONT) which was proposed by Rivest in 1997. $\text{Enc}^{\text{AONT}}(\cdot)$ algorithm [29] needs two inputs: data $M$ and a random key $K_{\text{AONT}}$. And it generates AONT ciphertext $C_{\text{AONT}}$ which could define property that one must decrypt the entire ciphertext before one can determine even one message block. CAONT is the same as AONT which replaces the random key to a hash value $h$. Encryption and decryption algorithms could be denoted as follows respectively: $\text{Enc}^{\text{CAONT}}(M, h) \rightarrow (C_2, t)$, $\text{Dec}^{\text{CAONT}}(C_2, t) \rightarrow M$.

**Broadcast Encryption:** A broadcast encryption system is made up of the following randomized algorithms [4]. $\text{Setup}^{\text{BE}}(\cdot)$ algorithm takes the number of receivers $n$ as input, and it outputs $n$ private keys $d_1, d_2, ..., d_n$ and a public key $PK$. Encryption algorithm takes a subset $S \subseteq \{1, ..., n\}$ and a public key $PK$ as input. It outputs a pair $(Hdr, K)$ where $Hdr$ is called the header and $K$ is a message encryption key. $C_{\text{M}}$ is the ciphertext of $M$ under the symmetric key $K$. For simplicity, the encryption and decryption algorithms could be expressed as follows respectively: $\text{Enc}^{\text{BE}}(M, S, PK) \rightarrow (C_{\text{M}}, Hdr, S)$, $\text{Dec}^{\text{BE}}(S, d_i, Hdr, C_{\text{M}}) \rightarrow M$.

## 3   System Design

### 3.1  Design Philosophy

As mentioned above, a secure data deduplication system that adopts MLE always suffered two kinds of side-channel attack: key-cache attack and probe attack. Accordingly, we design two new techniques: random position re-encryption and two-stage deduplication to mitigate the two attacks via fog nodes.

**Random Position Re-encryption:** In key-cache attack, a revoked user could cache old MLE keys to decrypt the encrypted data items. One solution [19, 25] is to re-encrypt the data item by a new MLE key to make the revoked key invalid. This procedure, however, needs to download/upload a large volume of data and distribute the new MLE key to other users, incurring high communication overhead. Another solution is REED [21], which supports lightweight rekeying by re-encrypting a small part of data item. REED adopts two-layers encryption to protect user data. The first layer uses MLE and the second layer uses CAONT. A small part of CAONT ciphertext named stub is encrypted by a data key. The usage of CAONT ensures that the data item cannot be decrypted unless the stub can be decrypted. In a rekeying operation, REED uses a new data key to re-encrypt the stub. Since a revoked user cannot access the new data key, he cannot access the data even he caches the MLE key. However, since the location of the stub is fixed, a revoked user that has accessed the data item can just retain the stub to continuously access the rekeyed data item.

We design a random position re-encryption technique to prevent the key-cache attack as shown in Fig. 2. Similar with REED, we also adopt two-layers encryption to protect a data item. Unlike REED, we always select a random part of the data item as the new stub in each rekeying operation. We rely on fog nodes to conduct rekeying operations and to record the current location of the current stub. When an authorized user needs to access the data item, the fog node sends the location and the data key to the user to decrypt the data item. The random position re-encryption technique could prevent key-cache attack while avoiding re-encryption of the whole data item. Importantly, a revoked user that retains the old stub cannot access the data item anymore since the position of the stub is changed after each re-keying operation.

**Two-Stage Deduplication:** To prevent the probe attack, one solution is to adopt server-side deduplication, in which users directly upload their data items to the cloud sever without any check, and the cloud server conducts the deduplication tasks. This solution prevents a malicious user from checking the existence of data items of other users, but incurs high communication overhead since same data items can be submitted multiple times. Another solution is to adopt client-side deduplication with a proof of ownership (PoWs) protocol [12], in which a user proves to the cloud server that he owns the data item. However, this solution incurs high computation overhead at user side. For example, when the bandwidth is 100 Mbps, PoWs might consumes 20% uploading time [12].

**Table 1.** Local data index

| Data tag | Data item | Data owner |
| --- | --- | --- |
| dcfc5f6e22231a2a | data0596 | UserA UserF |
| 4c15796c7aec02b8 | data4368 | UserB UserG UserJ |
| b0b2676cb488c75c | data5968 | UserA UserG UserK |

We design a two-stage deduplication technique to prevent probe attack. Our technique requires that the cloud server maintains a global data index and each fog node maintains a local data index to keep track of which data items have been stored and deduplicated. The local data index records data tags, corresponding data items and data owners, an example of a local data index is shown in Table 1. The global data index just records data tags and corresponding data items. The two deduplication stages are implemented as follows. In the first stage, a user uploads his data item to a local fog node. Before uploading the data item, the user uploads the tag of data item to the fog node. With the tag value, the fog node checks its local data index to decide if the data item has been uploaded before. If yes, the user does not need to upload the data item. In the second stage, the fog node further uploads the tag of the data item to the cloud server. The cloud server uses its global data index to decide whether the data item has been uploaded before. If yes, the fog node does not need to upload the data item. Figure 3 shows the details of the two-stage deduplication technique. In this way, our system achieves a similar communication efficiency with client-side deduplication while constraining a malicious user to be able to only check the existence of data items of users managed by a certain fog node.

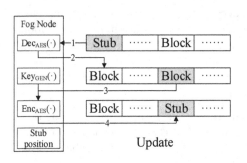

**Fig. 2.** Random position re-encryption     **Fig. 3.** Two stage deduplication

## 3.2   Encryption Scheme

Our encryption scheme uses and adjusts CAONT. We use CAONT to protect MLE ciphertext rather than plaintext. The encryption procedure to encrypt a data item $M$ is shown in Fig. 4. The first step is executed at user side, the specific process is as follows. First, $M$ is encrypted by the message-locked encryption: $\text{Enc}_{\text{MLE}}(M, K_{\text{MLE}}) \rightarrow C_1$. The user retains the MLE key $K_{\text{MLE}}$. Second, the user encrypts $C_1$ by CAONT. Specifically, the user calculates the hash function over $C_1$ and gets a hash value $h$. Let $h$ XOR with $C_1$ to get a ciphertext $C_2$. Let the hash value of $C_2$ XOR $h$ to get a tail part $t$. $C_2$ and $t$ form a CAONT package.

The second step is executed at fog node side, the process is shown in Fig. 5. First, the fog node randomly selects a 256 bits long part $p$ from the CAONT package. Second, the fog node creates a data key $K_F$. Third, the fog node symmetrically encrypts $p$: $\text{Enc}_{\text{AES}}(p, K_F) \rightarrow C_{\text{stub}}$. Fourth, the fog node treats the rest of the CAONT package as a rest package $C_{\text{rest}}$. Finally, the fog node uploads $C_{\text{rest}}$ and $C_{\text{stub}}$ to the cloud server.

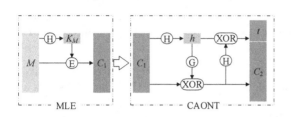

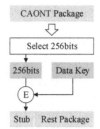

**Fig. 4.** Encryption at user side

**Fig. 5.** Encryption at fog node side

### 3.3   Design Details

Our deduplication system supports three types of operations: data upload, data download, and authority update. Each operation involves a user, a fog node and the cloud server. The fog node needs to check the certificate of the user to confirm its role and the cloud server needs to check the certificate of the fog node to confirm its role. The detailed processes are as follows.

**Data Upload:** The specific steps of data upload are as follows.

- When a user wants to upload a data item $M$, the user sends a data tag *tag* of $M$ to a local fog node. The fog node then checks if its local data index contains *tag*. If yes, the user does not need to upload $M$ anymore and the fog node updates its local data index to add the user as an owner of $M$. Otherwise, the user is considered as an initial uploader and the fog node asks user to upload $M$.
- If the user is required to upload $M$. $M$ needs to be encrypted before uploading. First, the user executes encryption operation: $\text{Enc}^{\text{MLE}}(M, K_{\text{MLE}}) \rightarrow C_1$. The user next encrypts $C_1$ by CAONT to get a CAONT package: $\text{Enc}^{\text{CAONT}}(C_1, h) \rightarrow (C_2, t)$. After that, the user sends $C_2$, $t$ and *tag* to the fog node.
- The fog node adds a new encryption layer. As shown in Fig. 5, the fog node creates a data key $K_F$ to encrypt a randomly selected part of $C_2$ and gets $C_{\text{rest}}$ and $C_{\text{stub}}$ respectively: $\text{Enc}^{\text{AES}}(p, K_F) \rightarrow C_{\text{stub}}$. Finally, the fog node preserves $C_{\text{rest}}$, $C_{\text{stub}}$ and sends *tag* to the cloud server. Besides, the fog node also inserts *tag* into its local data index and adds the user as an owner of the data item.
- Upon receiving *tag*, the cloud server checks if its global data index contains *tag*. If yes, the fog node does not need to upload $M$ anymore. Otherwise, the fog node uploads $C_{\text{rest}}$, $C_{\text{stub}}$ to the cloud server. In this case, the cloud server adds $C_{\text{rest}}$, $C_{\text{stub}}$ and *tag* to its global data index.

**Data Download:** When a user wants to download a data item $M$, he sends a download request $tag$ to a local fog node. The fog node checks the user's authority in its local data index. If the user is not a legal data owner of $M$, the fog node will refuse the request. Otherwise, the fog node forwards the download request to the cloud server and gets the $C_{\text{rest}}$ and $C_{\text{stub}}$. After that, the fog node returns $K_F$ and the position of stub to the user. First, the user decrypts $C_{\text{stub}}$: $\text{Dec}^{\text{AES}}(C_{\text{stub}}, K_F) \to (C_2, t)$. Second, the user decrypts $C_2$ and $t$: $\text{Dec}^{\text{CAONT}}(C_2, t) \to C_1$. Finally, the user decrypts $C_1$ and gets $M$: $\text{Dec}^{\text{MLE}}(C_1, K_{\text{MLE}}) \to M$.

**Authority Update:** When a user deletes a data item $M$, its access authority about $M$ is revoked by the responsible fog node. Suppose that the user sends a data deletion request to the fog node. The fog node first downloads $C_{\text{stub}}$ and executes $\text{Dec}^{\text{AES}}(C_{\text{stub}}, K_F) \to p$. The fog node also randomly selects a new part $p_2$ of $C_{\text{rest}}$ and creates a new data key $K_{F2}$. Then the fog node symmetrically encrypts $p_2$ by $K_{F2}$ to get a new stub $C_{\text{stub2}}$: $\text{Enc}^{\text{AES}}(p_2, K_{F2}) \to C_{\text{stub2}}$. Finally, the fog node uploads $C_{\text{rest2}}$ and $C_{\text{stub2}}$ for the cloud server to update the old $C_{\text{rest}}$ and $C_{\text{stub}}$. In addition, the fog node updates its local data index and deletes the ownership of the user about $M$. Meanwhile $K_{F2}$ and the position of $C_{\text{stub2}}$ are recorded in the fog node.

After updating the ciphertext, there is only one fog node knowing $K_{F2}$ and the position of the $C_{\text{stub2}}$. If other fog nodes need to download $M$, they cannot decrypt it anymore since $C_{\text{stub2}}$ and $K_{F2}$ are not equal to $C_{\text{stub}}$ and $K_F$ respectively. Thus, the fog node who reserves $K_{F2}$ needs to distribute the position of $C_{\text{stub2}}$ and $K_{F2}$ to other fog nodes. Our system uses the broadcast encryption [10] to do so. The fog node encrypts $K_{F2}$ and the position of $C_{\text{stub2}}$: $\text{Enc}^{\text{BE}}(M_{\text{BE}}, S, PK) \to (C_M, S, Hdr)$ where $M_{\text{BE}}$ contains $K_{F2}$ and the position of $C_{\text{stub2}}$, $C_M$ is the ciphertext of $M_{\text{BE}}$. The fog node sends $C_M, S$ and $Hdr$ to the cloud server. Other fog nodes who want to download $M$ need to decrypt $C_M$ by their private keys: $\text{Dec}^{\text{BE}}(C_M, S, d_i, Hdr) \to M_{\text{BE}}$.

## 4    Security Analysis

We now analyze the security of our system. We assume that our system has $n$ fog nodes. Each fog node locates in the area and manages $l_i$ users.

**Key-Cache Attack:** We elaborate the key-cache attack firstly. We consider two different abilities of a malicious revoked user. First of all, we consider that the malicious user cannot compromise the cloud server. We assume that the $l_k$th user in the $l$th fog node has been revoked. If the user intends to get data $C_{\text{stub}}$ and $C_{\text{rest}}$ through the $l$th fog node, the $l$th fog node will check his authority and reject the request. In this way, the malicious user cannot obtain $C_{\text{stub}}$ and $C_{\text{rest}}$. Thus, he cannot get $M$, either. Second, we consider that the user could compromise the cloud server to get the updated $C_{\text{stub2}}$ and $C_{\text{rest2}}$. In addition, the malicious user could also cache $K_{\text{MLE}}$, $K_F$ and $C_{\text{stub}}$ in the latest data download operation. Due to our random re-encryption technique, the fog node selects a random position of $C_{\text{rest}}$ and encrypts it by a new data key $K_{\text{F2}}$ in an authority update operation. As a result, the user cannot use either the old data key $K_{\text{F}}$ or the old stub $C_{\text{stub}}$ to recover the CAONT package. Without the CAONT package, the user cannot further use the $K_{\text{MLE}}$ to decrypt it to recover $M$.

**Probe Attack:** We next discuss the probe attack that a malicious user can check the existence of a data item $M$ by using its data tag $tag$ in a data upload operation. We still assume that the $l_k$th user belonging to the $l$th fog node is the malicious user. Due to our two-stage deduplication technique, the user needs to submit $tag$ to the $l$th fog node. The $l$th fog node checks $tag$ and replies the check result to the user. Since a fog node only manages a small number of users, the user can only probe if the data items of these users are existed in the fog node. As a result, our system constrains the user to launch probe attack in the certain area.

## 5 Implementation

In this section, we discuss the implementation details of our system. We implement our system in Python 3.5. We use pyCrypto library [23] and pyCryptodome library [9] to realize related cryptographic primitives. In our system, we divide a whole data into different chunks. Each chunk is encrypted by the message-locked encryption and CAONT. The MLE key is generated from its corresponding chunk. We calculate SHA-256 hash function value for the plaintext as MLE key and record it into a text file which is preserved by the user. Meanwhile, we use 'scp' protocol measuring upload and download performance.

The previous works point out the key management overhead of convergent encryption (e.g., especially fine-grained block-level deduplication) becomes more prominent [20]. For example, if we store 1 TB of data with all unique blocks of size 4 KB each, and that each convergent key is the hash value of SHA-256. Then the total size of keys is 8 GB. The key storage occupies a lot of overhead. Thus, to optimize the efficiency of our system, we implement the upload performance, encryption performance, update performance and deduplication performance in different chunk sizes. And we conclude which size of block could maintain the maximum storage efficiency.

## 6 Evaluation

We evaluate the encryption efficiency on the local PC, which is equipped with a Intel Xeon E5-1630 v3 3.5 GHz CPU and 16 GB RAM. We cascade three physical machines as fog nodes. Each of them has the same performance index as ibid. In addition, we deploy the cloud server based on the aliyun cloud with 1 CPU and 2 GB RAM, and the bandwidth is 2 MB.

## 6.1  Real-World Data

First, we utilize real-world data to evaluate the encryption, decryption, update and upload performance. We select the data which is suitable for experiments from Aliyun Tianchi dataset. We select different dataset sizes to accomplish our experiment which ensure our experiment more credible. For example, we evaluate encryption performance in 10 MB, 50 MB, 100 MB, 500 MB, and 1 GB, respectively. We divide the total file size by the time to get the encryption rate. The experimental diagram shows the average encryption speed.

**Experiment 1.1 (Encryption Performance):** We first measure the performance of encryption and decryption. What needs to be emphasized is that the test results include two steps: message-locked encryption and CAONT operation. Both of them are executed at the user side. Besides, the test range doesn't contain re-encryption process in the fog node.

Figure 6 shows the encryption and decryption performance at the user side. We observe that when the average chunk size increases, the encryption speed also gets the corresponding improvement. For example, when the chunk size is 32 KB, the encryption speed reaches 37.5 MB/s.

**Experiment 1.2 (Re-encryption Performance):** We measure the decryption performance and re-encryption performance in fog node. It is worth noting that this operation not only contains encryption process but also includes file reading and writing operations.

Figure 7 shows the result of re-encryption performance. As mentioned above, the chunk size is large, fewer chunks need to be processed, and the performance has been improved. We observe the re-encryption speed reaches 104.77 MB/s when the chunk size is 32 bits. The reason why re-encryption performance better than encryption performance at user side is that re-encryption operation just needs to encrypt 256 bits part rather than the whole chunk. We observe that the decryption operation has the higher performance than encryption operation. Because re-encryption operation contains file reading and writing.

**Experiment 1.3 (Update Performance):** We measure the update performance in fog node. To prove the advantage of our system, we compare our system with traditional update and REED.

Figure 8 shows the result of update speed. We observe that our system is much better than the traditional update but lower than REED. As Fig. 8 shows, when the chunk size is 16 KB, the update speed of our system could reach 52 MB/s and traditional update speed merely reaches 37 MB/s. But REED reaches up to 120 MB/s. The reason is that REED just needs to encrypt and decrypt a little stable part of data. Our system wastes some time in file reading and writing. Although we have the performance disadvantages compared with REED, our system still achieves satisfactory result.

**Experiment 1.4 (Upload Performance):** We measure upload performance at the user side. In order to ensure maximum upload efficiency, we set the chunk size into 4 KB, 8 KB, 16 KB, 32 KB, 64 KB, respectively. Theoretically, in 2 MB bandwidth network, the maximum upload speed is 256 KB/s. As shown in Fig. 9, when the chunk size reaches 16 KB, the upload speed could achieve 228 KB/s closing to the theoretical maximum speed. Henceforth, the upload speed does not improve with growing of chunk size.

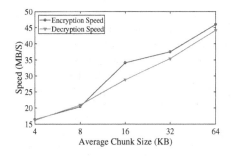

**Fig. 6.** Encryption performance

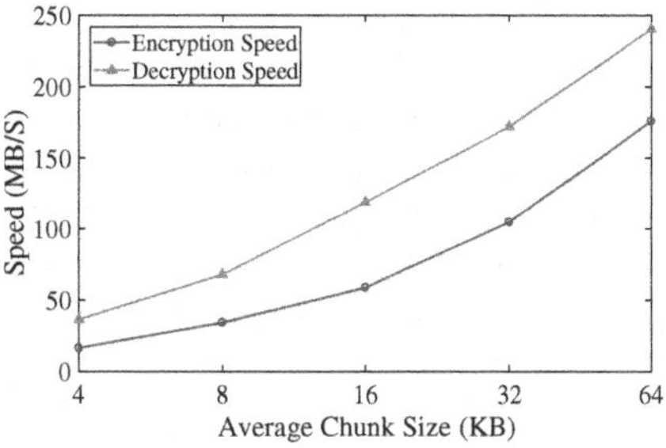

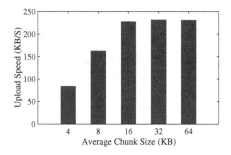

**Fig. 7.** Re-encryption performance

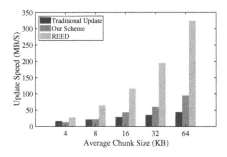

**Fig. 8.** Update performance

**Fig. 9.** Upload performance

## 6.2 Synthetic Data

We evaluate upload performance in synthetic data. We set different deduplication rate to measure the upload speed and deduplication performance.

**Experiment 2.1 (Upload Performance):** Unlike experiment 1.4, we measure the upload performance in different deduplication rates. We set the deduplication rate as 10%, 20%, 30%, 40%, 50% and 60%. Meanwhile, we set the chunk size in 32 KB which ensures the maximum upload speed in 2 MB bandwidth. As shown in Fig. 10, the upload speed increases with the raising of deduplication rate. Because our system uses client-side deduplication, the duplicate files need not be uploaded again. Therefore, when the deduplication rate reaches 10%, the upload speed is 260 KB/s which is higher than the theoretical maximum speed.

In addition, we could find that our system has much upload performance advantages than REED and traditional re-encryption, especially when deduplication rate is high. As shown in Fig. 10, when the deduplication rate is 50%, the upload speed of our system could reach 460 KB/s, which is twice as much as REED. In addition, unlike PoWs [12], our system doesn't need user to do extra calculation which saves their computation resource. We also find that REED is slower than traditional encryption. Because REED separates each chunk into two parts: stub and trimmed package. And 'scp' protocol always processes data one by one. More data parts will waste more time.

**Experiment 2.2 (Deduplication Performance):** We generate 12 GB synthetic file to measure the deduplication performance. The dataset has 75% duplicated chunks. We measure the deduplication rate changes of file number and file size over time. As shown in Fig. 11, file is gradually decreasing over time. After 120 min deduplicating, the file size reaches 3 GB which means most duplicated chunks have been eliminated.

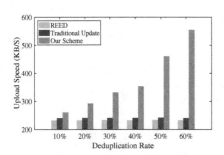
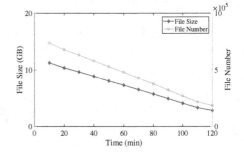

**Fig. 10.** Upload performance                   **Fig. 11.** Deduplication performance

## 7   Related Work

Deduplication which stores only one copy of identical messages and eliminates redundant data could achieve storage efficiency. Each Message is identified by a fingerprint and computed as a cryptographic hash value of the message content. We assume that two messages are identical if their fingerprints are identical [3]. However, deduplication in ciphertext is infeasible, because the same content might be encrypted into different ciphertext.

Message-locked encryption [2] is a cryptographic primitive that provides confidentiality guarantee for deduplication storage. It is a symmetric encryption and MLE key is derived from message itself. Thus, MLE ensures the same plaintext will be encrypted as the same ciphertext which makes deduplication plausible. A special case of MLE is convergent encryption (CE) [8], which directly uses the message's fingerprint as the MLE key.

However, MLE (e.g., including CE) is inherently vulnerable to brute-force attacks, and achieves security only for unpredictable messages [2]. To address the unpredictability assumption, DupLESS [18] uses oblivious pseudo-random function (OPRF) [11] to implement server-aided MLE. But it needs a trusted third party. Liu et al. [24] propose a password authenticated key exchange protocol for MLE key generation. Li et al. [21] propose a deduplication scheme which supports lightweight update. Koo et al. [19] design a deduplication system based on fog computing environment, but the update process wastes too much calculation resources.

# 8  Conclusion

We present a new encrypted data deduplication system in fog computing environment that aims for secure and lightweight rekeying. The core idea is re-encrypting a part of CAONT package. We propose a two-stage deduplication which saves communication overhead and resists probe attack. In addition, when the data needs to be updated, we propose random position re-encryption technique to prevent key-cache attack. Furthermore, we show the confidentiality and security of our system under our security definitions. Last, we implement our system to show its performance and storage effectiveness.

**Acknowledgement.** We acknowledge the support from National Natural Science Foundation of China (No. 61602363), China Postdoctoral Science Foundation (No. 2016M590927), National Cryptography Development Fund (No. MMJJ20180110) and Graduate Innovation Foundation, School of Cyber Engineering, Xidian University (No. 20109194858).

# References

1. Ahmad, M., Amin, M.B., Hussain, S., Kang, B.H., Cheong, T., Lee, S.: Health fog: a novel framework for health and wellness applications. J. Supercomput. **72**(10), 3677–3695 (2016). https://doi.org/10.1007/s11227-016-1634-x
2. Bellare, M., Keelveedhi, S., Ristenpart, T.: Message-locked encryption and secure deduplication. In: Johansson, T., Nguyen, P.Q. (eds.) EUROCRYPT 2013. LNCS, vol. 7881, pp. 296–312. Springer, Heidelberg (2013). https://doi.org/10.1007/978-3-642-38348-9_18
3. Black, J.: Compare-by-hash: a reasoned analysis. In: Proceedings of the 2006 USENIX Annual Technical Conference, Boston, MA, USA, 30 May–3 June 2006, pp. 85–90 (2006). http://www.usenix.org/events/usenix06/tech/black.html
4. Boneh, D., Gentry, C., Waters, B.: Collusion resistant broadcast encryption with short ciphertexts and private keys. In: Shoup, V. (ed.) CRYPTO 2005. LNCS, vol. 3621, pp. 258–275. Springer, Heidelberg (2005). https://doi.org/10.1007/11535218_16
5. Bonomi, F., Milito, R.A., Zhu, J., Addepalli, S.: Fog computing and its role in the Internet of Things. In: Proceedings of the First Edition of the MCC Workshop on Mobile Cloud Computing, MCC@SIGCOMM 2012, Helsinki, Finland, 17 August 2012, pp. 13–16 (2012). https://doi.org/10.1145/2342509.2342513
6. Chen, X., Li, J., Huang, X., Ma, J., Lou, W.: New publicly verifiable databases with efficient updates. IEEE Trans. Dependable Sec. Comput. **12**(5), 546–556 (2015). https://doi.org/10.1109/TDSC.2014.2366471

7. Chen, X., Li, J., Ma, J., Tang, Q., Lou, W.: New algorithms for secure outsourcing of modular exponentiations. IEEE Trans. Parallel Distrib. Syst. **25**(9), 2386–2396 (2014). https://doi.org/10.1109/TPDS.2013.180
8. Douceur, J.R., Adya, A., Bolosky, W.J., Simon, D., Theimer, M.: Reclaiming space from duplicate files in a serverless distributed file system. In: ICDCS, pp. 617–624 (2002). https://doi.org/10.1109/ICDCS.2002.1022312
9. Eijs, H.: Pycryptodome-the Python cryptography toolkit. https://pypi.org/project/pycryptodome/ (2019)
10. Fiat, A., Naor, M.: Broadcast encryption. In: Stinson, D.R. (ed.) CRYPTO 1993. LNCS, vol. 773, pp. 480–491. Springer, Heidelberg (1994). https://doi.org/10.1007/3-540-48329-2_40
11. Goldwasser, S., Bellare, M.: Lecture notes on cryptography. Summer course "Cryptography and computer security" at MIT 1999 (1996)
12. Halevi, S., Harnik, D., Pinkas, B., Shulman-Peleg, A.: Proofs of ownership in remote storage systems. In: Proceedings of the 18th ACM Conference on Computer and Communications Security, CCS 2011, Chicago, Illinois, USA, 17–21 October 2011, pp. 491–500 (2011). https://doi.org/10.1145/2046707.2046765
13. Harnik, D., Pinkas, B., Shulman-Peleg, A.: Side channels in cloud services: deduplication in cloud storage. IEEE Secur. Priv. **8**(6), 40–47 (2010). https://doi.org/10.1109/MSP.2010.187
14. Hong, K., Lillethun, D.J., Ramachandran, U., Ottenwälder, B., Koldehofe, B.: Mobile fog: a programming model for large-scale applications on the Internet of Things. In: Proceedings of the Second ACM SIGCOMM Workshop on Mobile Cloud Computing, MCC@SIGCOMM 2013, Hong Kong, China, 16 August 2013, pp. 15–20 (2013). https://doi.org/10.1145/2491266.2491270
15. Huang, H., Chen, X., Wu, Q., Huang, X., Shen, J.: Bitcoin-based fair payments for outsourcing computations of fog devices. Future Gener. Comp. Syst. **78**, 850–858 (2018)
16. Huang, Q., Yang, Y., Wang, L.: Correction to "secure data access control with ciphertext update and computation outsourcing in fog computing for Internet of Things". IEEE Access **6**, 17245 (2018)
17. Jiang, T., Chen, X., Wu, Q., Ma, J., Susilo, W., Lou, W.: Secure and efficient cloud data deduplication with randomized tag. IEEE Trans. Inf. Forensics Secur. **12**(3), 532–543 (2017)
18. Keelveedhi, S., Bellare, M., Ristenpart, T.: DupLESS: server-aided encryption for deduplicated storage. In: Presented as part of the 22nd USENIX Security Symposium (USENIX Security 2013), pp. 179–194. USENIX, Washington, D.C. (2013). https://www.usenix.org/conference/usenixsecurity13/technical-sessions/presentation/bellare
19. Koo, D., Hur, J.: Privacy-preserving deduplication of encrypted data with dynamic ownership management in fog computing. Future Gener. Comp. Syst. **78**, 739–752 (2018)
20. Li, J., Chen, X., Li, M., Li, J., Lee, P.P.C., Lou, W.: Secure deduplication with efficient and reliable convergent key management. IEEE Trans. Parallel Distrib. Syst. **25**(6), 1615–1625 (2014). https://doi.org/10.1109/TPDS.2013.284
21. Li, J., Qin, C., Lee, P.P.C., Li, J.: Rekeying for encrypted deduplication storage. In: 46th Annual IEEE/IFIP International Conference on Dependable Systems and Networks, DSN 2016, Toulouse, France, 28 June–1 July 2016, pp. 618–629 (2016). https://doi.org/10.1109/DSN.2016.62

22. Li, M., Qin, C., Lee, P.P.C.: CDStore: toward reliable, secure, and cost-efficient cloud storage via convergent dispersal. In: USENIX Annual Technical Conference, pp. 111–124. USENIX Association (2015)
23. Litzenberger, D.C.: Pycrypto-the Python cryptography toolkit. https://www.dlitz. net/software/pycrypto (2016)
24. Liu, J., Asokan, N., Pinkas, B.: Secure deduplication of encrypted data without additional independent servers. In: Proceedings of the 22nd ACM SIGSAC Conference on Computer and Communications Security, Denver, CO, USA, 12–16 October 2015, pp. 874–885 (2015). https://doi.org/10.1145/2810103.2813623
25. Liu, J., Duan, L., Li, Y., Asokan, N.: Secure deduplication of encrypted data: refined model and new constructions. In: Smart, N.P. (ed.) CT-RSA 2018. LNCS, vol. 10808, pp. 374–393. Springer, Cham (2018). https://doi.org/10.1007/978-3-319-76953-0_20
26. Madsen, H., Burtschy, B., Albeanu, G., Popentiu-Vladicescu, F.: Reliability in the utility computing era: Towards reliable fog computing. In: 2013 20th International Conference on Systems, Signals and Image Processing (IWSSIP), pp. 43–46, July 2013. https://doi.org/10.1109/IWSSIP.2013.6623445
27. Miao, M., Wang, J., Li, H., Chen, X.: Secure multi-server-aided data deduplication in cloud computing. Pervasive Mob. Comput. **24**, 129–137 (2015)
28. Ni, J., Lin, X., Zhang, K., Yu, Y.: Secure and deduplicated spatial crowdsourcing: a fog-based approach. In: 2016 IEEE Global Communications Conference (GLOBE-COM), pp. 1–6, December 2016. https://doi.org/10.1109/GLOCOM.2016.7842248
29. Rivest, R.L.: All-or-nothing encryption and the package transform. In: Fast Software Encryption, 4th International Workshop, FSE 1997, Haifa, Israel, 20–22 January 1997, Proceedings, pp. 210–218 (1997). https://doi.org/10.1007/BFb0052348
30. Stojmenovic, I., Wen, S.: The fog computing paradigm: scenarios and security issues. In: Proceedings of the 2014 Federated Conference on Computer Science and Information Systems, Warsaw, Poland, 7–10 September 2014, pp. 1–8 (2014). https://doi.org/10.15439/2014F503
31. Vaquero, L.M., Rodero-Merino, L.: Finding your way in the fog: towards a comprehensive definition of fog computing. ACM SIGCOMM Comput. Commun. Rev. **44**(5), 27–32 (2014)
32. Wang, J., Chen, X., Li, J., Kluczniak, K., Kutylowski, M.: TrDUP: enhancing secure data deduplication with user traceability in cloud computing. IJWGS **13**(3), 270–289 (2017)
33. Wang, Y., Uehara, T., Sasaki, R.: Fog computing: issues and challenges in security and forensics. In: 39th Annual Computer Software and Applications Conference, COMPSAC Workshops 2015, Taichung, Taiwan, 1–5 July 2015, pp. 53–59 (2015). https://doi.org/10.1109/COMPSAC.2015.173
34. Xiang, Y., Bertino, E., Kutylowski, M.: Security and privacy in social networks. Concurr. Comput.: Pract. Exp. **29**(7) (2017)
35. Yu, Z., Au, M.H., Xu, Q., Yang, R., Han, J.: Towards leakage-resilient fine-grained access control in fog computing. Future Generation Comp. Syst. **78**, 763–777 (2018). https://doi.org/10.1016/j.future.2017.01.025
36. Yuan, H., Chen, X., Jiang, T., Zhang, X., Yan, Z., Xiang, Y.: DedupDUM: secure and scalable data deduplication with dynamic user management. Inf. Sci. **456**, 159–173 (2018)
37. Zhang, X., Jiang, T., Li, K., Castiglione, A., Chen, X.: New publicly verifiable computation for batch matrix multiplication. Inf. Sci. **479**, 664–678 (2019)

# Practical IDS on In-vehicle Network Against Diversified Attack Models

Junchao Xiao[1,2], Hao Wu[4], Xiangxue Li[2,3(✉)], and Yuan Linghu[2]

[1] School of Systems Science and Engineering, Sun Yat-Sen University,
Guangzhou, China
[2] School of Software Engineering, East China Normal University, Shanghai, China
xxli@cs.ecnu.edu.cn
[3] Westone Cryptologic Research Center, Beijing, China
[4] CNCERT/CC, Beijing, China
wuhao@cert.org.cn

**Abstract.** A vehicle bus is a specialized internal communication network that interconnects components inside a vehicle. The Controller Area Network (CAN bus), a robust vehicle bus standard, allows microcontrollers and devices to communicate with each other. The community has seen many security breach examples that exploit CAN functionalities and other in-vehicle flaws. Intrusion detection systems (IDSs) on in-vehicle network are advantageous in monitoring CAN traffic and suspicious activities. Whereas, existing IDSs on in-vehicle network only support one or two attack models, and identifying abnormal in-vehicle CAN traffic against diversified attack models with better performance is more expected as can be then implemented practically. In this paper, we propose an intrusion detection system that can detect many different attacks. The method analyzes the CAN traffic generated by the in-vehicle network in real time and identifies the abnormal state of the vehicle practically. Our proposal fuses the autoencoder trick to the SVM model. More precisely, we introduce to the system an autoencoder that learns to compress CAN traffic data into extracted features (which can be uncompressed to closely match the original data). Then, the support vector machine is trained on the features to detect abnormal traffic. We show detailed model parameter configuration by adopting several concrete attacks. Experimental results demonstrate better detection performance (than existing proposals).

**Keywords:** In-vehicle network · Intrusion detection systems · Autoencoder

## 1 Introduction

Vehicles have become a ubiquitous means of transportation in modern society, and the development of information technology has made the communication components inside the vehicles more complicated. A variety of electronic products are rapidly applied in vehicles. For example, the GPS units provide vehicle

© Springer Nature Switzerland AG 2020
S. Wen et al. (Eds.): ICA3PP 2019, LNCS 11945, pp. 456–466, 2020.
https://doi.org/10.1007/978-3-030-38961-1_40

location information. The vehicle sound equipment is connected to the phone for functions such as calling and play music. The Internet of Vehicles is designed to meet the communication of various devices in the vehicles. However, unregulated communication between electronic devices exposes in-vehicle network to security threats [1]. Recent work show that attackers can take advantage of the vulnerabilities of in-vehicle network to manipulate vehicles [2].

For now, in-vehicle network communication mainly relies on the Controller Area Network (CAN) bus. The CAN bus, a robust vehicle bus standard, allows microcontrollers and devices to communicate with each other. There are many Electronic Control Units (ECUs) in the vehicles, and the ECU sends instructions through the CAN bus to handle specific tasks. The CAN protocol was invented in the 1980s for the internal communication without considering cybersecurity threat. For example, the CAN bus uses Carrier Sense Multiple Access with Collision Detection (CSMA/CD) to transmit instructions. All ECUs can receive instructions transmitted on the bus. As long as one ECU is controlled by attackers, all ECUs confront potential security threats [3].

There are currently two methods for reinforcing CAN protocol security. One line is to design Message Authentication Code (MAC), and another method installs an intrusion detection system on the vehicles [4]. The MAC has higher reliability but needs to occupy the data fields in the CAN packets. The data fields in the CAN packets are only 8 bytes and the maximum transmission rate is only 1 M/s, which cannot meet the MAC requirements. Therefore, the application of the MAC is bound to change the existing vehicle structure and increase product costs, which is currently difficult to widely popularize. Another line is to construct intrusion detection system (IDS) [5], which does not need to change the structure of the vehicles and can be installed in the vehicle's gateway to detect malicious behavior in real time. Therefore, the IDS is more suitable for improving the security of in-vehicle network.

In the paper, we propose a practical IDS of in-vehicle network to extract features using an autoencoder and solicit support vector machines (SVM) for anomaly detection. The proposed detection model provides better performance and can monitor the status of the in-vehicle network in real time. Our contributions include the following.

1. We present a novel detection model that can effectively detect multiple attacks. In our model, each CAN bus instruction is encoded by the autoencoder into a number which can represent the instruction itself. The SVM analyzes the features of these numbers to identify abnormal instructions.
2. Unlike most detection models that can only detect one single attack [6], our method can detect a variety of attack models. We mention that IDS of in-vehicle network should support the capability of detecting multiple attacks. If the detection model can only detect one single attack model, then security engineers have to accumulate multiple models to effectively detect the abnormal state of the in-vehicle network. This greatly increases the computational cost and is clearly not suitable for practical application in vehicles. Our proposal effectively fills the gap.

3. A detection model based on time series prediction needs to separately train the model by instruction ID. However, the CAN protocol is a combination of multiple instructions, and one single ID prediction may lose some important information. The numbers we get by the autoencoder are grouped in chronological order per 100 as the input data of the SVM. This effectively solves the problem of training multiple time series prediction models.
4. Oue model is compared to the other two methods to demonstrate its detection superiority. One method is to use 100 instructions directly as input data to the SVM. This method does not use an autoencoder to extract numbers. Other method is to use LSTM, a time series prediction model, to train only one model without separate ID for training.

## 2  Related Work

Machine learning methods are widely used in IDS. Larson et al. present a CAN bus network attack detection method based on security rules [7]. The method calls the object dictionary of CANopen protocol, uses protocol-level security rules to detect illegal ECU behavior, and provides a set of example security rules [7]. Wang et al. propose a time series prediction model that trains different types of instructions in the CAN protocol separately and finally combines them [8]. Müter et al. detect attacks by calculating the entropy of normal traffic and abnormal traffic on the CAN bus [5]. Hu et al. use the SVM model to detect the abnormal states of vehicles [9]. However, these methods have inherent drawbacks. For example, in [8], the instruction IDs of the CAN bus are separated, resulting in the loss of part of the information. In [5], the entropy-based method can only perform preliminary statistics and detect a part of the attack methods, which has no effect on most attack methods. In [9], directly detecting the abnormal state of the vehicle with SVM requires excessive computational resources, and it is difficult to ensure real-time monitoring. Our proposed IDS of in-vehicle network can defeat these obstacles.

## 3  Can Bus Data and Attack Models

An attacker can operate the vehicle ECU remotely or physically. Access points include but are not limited to Bluetooth, OBD_II, Wi-Fi, physical access, and USB ports, as shown in Fig. 2. In the paper, we suppose two types of attacks, one by directly modifying the vehicle ECU nodes and the other by injecting unauthenticated information directly into the in-vehicle network. First, we assume that the attacker can physically control and modify the target ECU node in the vehicles, so that the target ECU node stops transmitting information and allows the impersonating node to replace the sending information, causing the malicious node to replace the target ECU node. For the overall information, the system has not changed. We call this attack as impersonation attack. We may also assume that the attacker can not control the target ECU node, but inject malicious information to induce a fault or remotely maneuver the vehicle which lead to two types of injection attacks-DoS attack and fuzzy attack [10].

Our CAN bus data is sourced from http://ocslab.hksecurity.net/Data set/CAN-intrusion-dataset. It is constructed by logging CAN traffic via the OBD-II port from a real vehicle of KIA SOUL while message injection attacks are performing. We view the data as the following four sets:

- Attack-free: Capture the CAN bus data under normal conditions once the vehicle starts.
- Dos attack: The attacker periodically injects high-priority CAN bus instructions so that legitimate instructions do not respond in time. One quintessential trick is to inject the highest priority instruction with ID 0000.
- Fuzzy attack: The attacker randomly sends instructions to cause the vehicle to perform unexpected behavior. In order to implement a fuzzy attack, the attacker needs to find specific information about the vehicle and an instruction ID that can produce unexpected behavior. Unlike the Dos attack, it paralyzes functions of a vehicle rather than delaying normal messages via occupancy of the bus.
- Impersonation attack: After an adversary attacks an ECU causing it to lose the capacity of work, he inserts his ECU for a specific purpose. The inserted ECU is disguised as this ECU that stops working, and can periodically reply to the remote frame [10, 11]. Refer to [10] for the construction and distribution characteristics of the attack data set.

The attack-free data set takes the first 2300000 instructions. The Dos attack, fuzzy attack, and impersonation attack data sets take the first 590000 instructions. All data sets are normalized (from 0 to 1) by the following formula: $x_{normali} = \frac{x-min}{max-min}$, where $x$ represents data that needs to be normalized, and $min$ (and $max$) represents the minimum (maximum, respectively) value of an attribute in the dataset.

# 4 Autoencoder and Feature Extraction

We apply the classic autoencoder [12]. The autoencoder is forced to learn the features of the input data by setting the number of the hidden layer's units less than the input layer's units and using the smaller variables to store the information as much as possible. If the activation function is linear, the autoencoder is similar to the principal component analysis (PCA), with the function of a low-dimensional linear representation, and then decoded to produce output data similar to the input data, $x \approx \overline{x}$. The autoencoder consists of two functions: the first one is the encoding function $f_{\theta_1}(x)$, which maps the high-dimensional input data to the latent feature $h$, where $x$ denotes the input data, and $\theta_1 = \{w^{(1)}, b^{(1)}\}$ denotes transition and bias parameter, respectively. The second one is the decoding function $g_{\theta_2}(f_{\theta_1}(x))$, which maps the latent feature to the reconstructed input data $\overline{x}$ and the parameter $\theta_2 = \{w^{(2)}, b^{(2)}\}$. It should be noted that the mapping rules of the encoding function and the decoding function are respectively $f : R^N \to R^K$ and $g : R^K \to R^N$, where $N$ is the dimension of the input data, and $K$ is the number of units in the hidden layer.

Specifically, the encoding function maps the input vector $x \in R^N$, which can be expressed as $h = \delta(w_1 x + b_1)$, where $\delta(\cdot)$ is the activation function, $w_1$ is $N \times K$ matrix, and $b_1$ is the bias vector. In other words, the decoding function maps the latent features to the output reconstructed data, which can be expressed as $\overline{x} = \delta(w_2 x + b_2)$. Herein, $w_2$ is a $K \times N$ matrix and $b_2$ is the bias vector. Each input sample data $x^i$ maps to the latent feature $h^i$, and is then reconstructed to $\overline{x}$. The parameter $\theta = \{\theta_1, \theta_2\}$ of autoencoder is determined by the loss function that can minimize the average reconstruction error. The loss function is expressed by the cross-entropy as follows:

$$L_H(x, \overline{x}) = -\sum_{i=1}^{n} x_i log\overline{x_i} + (1 - x_i)log(1 - \overline{x_i}). \qquad (1)$$

We define the stacked autoencoder with three hidden layers, one input layer, and one output layer. The number of neurons per layer and activation function are shown in Table 1, and each layer is fully connected. The neurons number of input and output layer neurons are defined as 11 as the vector length of input data is 11. The activation functions of the third hidden layer and the output layer are set to Sigmoid in order to renormalize the encoded data (from 0 to 1). The structure of the autoencoder is shown in Fig. 1.

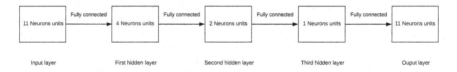

**Fig. 1.** The structure of the autoencoder.

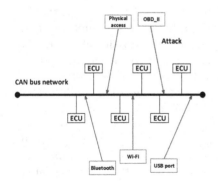

**Fig. 2.** Attack models.

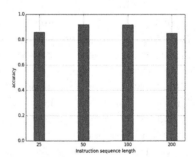

**Fig. 3.** The effect of instruction sequence length on the detection performance.

We employ keras to implement a standard stacked autoencoder. The optimization algorithm adopts an improved version of the stochastic gradient descent (SGD) so that the learning rate can be updated automatically according to different learning environments and the convergence rate can also be speed up simultaneously. Mini-batch is defined as 128. The model converges after iterations of 500.

We use the first 500000 instructions in the attack-free data set to train the autoencoder. The reason for using the attack-free dataset is to make the model better understand the data distribution characteristics under normal conditions. We view the output data from the third layer of the autoencoder as the CAN bus instruction feature data are extracted from the autoencoder. An instruction originally conveys 11 attributes, and is now reduced to 1 attribute by the autoencoder encoding feature, which greatly degrades the data set sample dimension. The obtained features are different from those produced by principal component analysis (PCA) [13]. In general, PCA extracts the main components of the data and loses some of the information in the process. However, the autoencoder abstracts data in a way that minimizes information loss, and the abstracted data can better reflect the original data. The feature largely filters out irrelevant information that contributes less to the classification, resulting in better performance of the IDS. Unless otherwise stated, the instructions in the following represent instructions that are already handled by the autoencoder.

## 5    Support Vector Machine Model and Abnormal State Detection

In machine learning, support vector machines (SVMs) are supervised learning models with associated learning algorithms that analyze data used for classification and regression analysis. In our context, the SVM attempts to find a classification hyperplane to detect the abnormal state of the vehicle. We refer to [14] for more details on SVM.

Given training vectors $x_i \in R^p$ $i = 1, ..., n$, in two classes, and a vector $y \in \{1, -1\}^n$, SVM mainly performs the calculation of the following tasks:

$$\min_{w,b,\zeta} \frac{1}{2} w^T w + C \sum_{i=1}^{n} \zeta_i$$
$$s.t.\ y_i(w^T \phi(x_i) + b) \geq 1 - \zeta_i$$
$$\zeta_i \geq 0, i = 1, ..., n$$

(2)

Its dual is

$$\min_{\alpha} \frac{1}{2} \alpha^T Q \alpha - e^T \alpha$$
$$s.t.\ y^T \alpha = 0$$
$$0 \leq \alpha_i \leq C, i = 1, ..., n$$

(3)

where $C > 0$ is an $n \times n$ positive semidefinite matrix, $Q_{ij} \equiv y_i y_j K(x_i, x_j)$, $K(x_i, x_j) = e^{\gamma(\|x-y\|^2)}$ is the kernel. Training vectors are implicitly mapped into a higher (maybe infinite) dimensional space by the function $K(x_i, x_j)$.

The decision function is: $sgn(\sum_{i=1}^{n} y_i \alpha_i K(x_i, x_j) + \rho)$.

## 6    Model Parameter Configuration

The instruction sequence is used as the input data of the SVM, and its length affects the final detection results. If the length of the instruction sequence is too short, then the computational consumption per second could be too large. In addition, it is not easy to find the statistical distribution characteristics of the data, resulting in unexpected detection accuracy. Conversely, due to errors in the instructions of the autoencoder transition, the accumulated error will affect the detection accuracy if the length of instruction sequence is too long. Therefore, we need to determine the appropriate length of the instruction sequence. We separately take the length of 25, 50, 100, 200 as the input sequence of the SVM to find the appropriate sequence length. The experimental results are shown in Fig. 3.

We take the instructions of 500000 to 2000000 in the attack-free dataset, and Dos attack, fuzzy attack, and impersonation attack data set from 0 to 500000 instructions as sample data using the 10-cross-validation method for experiments. The main method is to split the attack-free, Dos attack, fuzzy attack, impersonation attack data sets into 10 groups, and 9 groups as the training set and 1 group as the test set. We train the SVM model 10 times, and each group is used as a test set to produce 10 detection results. Figure 3 shows the average detection results.

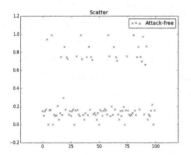 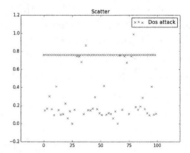

**Fig. 4.** Attack-free data distribution.          **Fig. 5.** Dos attack data distribution.

As shown in Fig. 3, a sequence length of 25 as the input sample does not obtain good results, as too little data volume makes it impossible for SVM to find a relatively good hyperplane in detecting the abnormal state. The sequence of length 200 has a relatively large volume of data, which amplifies the autoencoder transform error and obtains relatively unsatisfactory result. The sequence length of 50 obtains a slightly better result than the sequence length of 100, but the detection of a short sequence results in a multiplication of the computational requirements. In order to match the computational power of the in-vehicle gateway, we select a sequence length of 100 as the input sample of the SVM.

We randomly select consecutive 100 instructions in the four data sets to check their data distribution characteristics. Figure 4 indicates the distribution of data under normal conditions. Figure 5 shows the data distribution under Dos attack. It can be seen that most of the data gather around 0.78, and the data distribution is quite different from the attack-free data.

Figure 6 shows the data distribution under fuzzy attack, with significantly more data near 0 compared to attack-free data. Figure 7 shows the data distribution under impersonation attack, with more data between 0.8 and 1 compared to attack-free. Through the above analysis, it can be seen that the instruction features extracted by the autoencoder have a tendency to distinguish the normal state data and the abnormal state data. Better detection performance can be obtained by further processing of the SVM model.

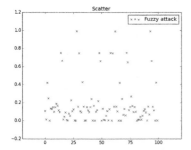 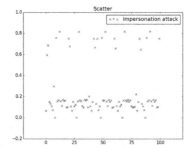

**Fig. 6.** Fuzzy attack data distribution.     **Fig. 7.** Impersonation attack data distribution.

# 7  Experiments and Evaluations

The training process of the SVM is shown in Fig. 8. We take the instructions of 500000 to 2000000 in the attack-free dataset, and Dos attack, fuzzy attack, and the impersonation attack data set from 0 to 500000 instructions as training set.

Simultaneously, we take the instructions of 2000000 to 2270000 in the attack-free data set, and DoS attack, fuzzy attack, and impersonation attack data set from 500000 to 590000 instructions as test set. The IDS performance evaluation and testing steps are shown in Fig. 9.

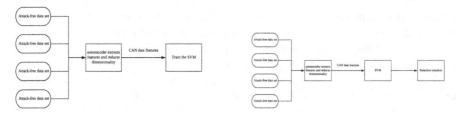

**Fig. 8.** SVM training process.

**Fig. 9.** The IDS performance evaluation and testing steps.

We introduce as reference models the LSTM model [15] and the original instruction SVM model that does not extract instruction features from the autoencoder. At present, the LSTM detection predicted by time series is separating out common instruction ID for training and detection, and an LSTM model can only detect common instruction ID. Since our detection model takes the generated input sequence in the order of instruction generation, we do not train the LSTM model based on the same instruction ID, but only train one LSTM model in the order of instruction generation. The reason why LSTM is used as a comparative experiment is that LSTM is a classic time series prediction algorithm, and many improved algorithms use it as a prototype. The experimental results are shown in Table 1, Fig. 10.

**Table 1.** Detection accuracy comparison.

| Detection model | Accuracy |
|---|---|
| Autoencoder + SVM | 0.970 |
| SVM | 0.927 |
| LSTM | 0.500 |

Figure 10 shows the ROC curves of the three detection models that visually explain their detection performance. The True Positive Rate (TPR) is the ratio of finding the abnormal state correctly. The False Positive Rate (FPR) indicates the rate that the detected model misjudges the abnormal state for actual normal state. The area under the ROC curve represents the Area Under Curve (AUC) value, and the greater the AUC, the better the performance of the model.

As shown in Fig. 11, compared with the other two methods, our proposed IDS provides the best performance. The SVM model that does not extract features from the autoencoder has a slightly lower performance. The input sequence length of the SVM model is 1100, while that of the autoencoder + SVM model is only 100. The experimental results show that the proposed detection model not only improves the performance of the model but also reduces the computational complexity after shrinking the instruction attributes. The LSTM model basically has no detection capability, as the instruction data of different IDs are too different, so that the LSTM model cannot find out the existing rules in a limited time period.

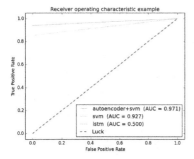

**Fig. 10.** Detection ROC curve comparison.

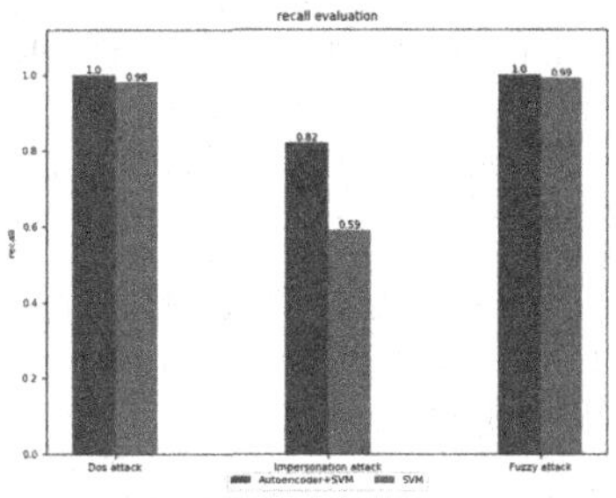

**Fig. 11.** Recall rates for different attack data.

We also calculate the recall rates of SVM and autoencoder+SVM under different attack types. The recall rate represents the rate at which the attack data is correctly detected. The recall of each attack is shown in Fig. 11. It can be seen that for the detection of impersonation attack, the autoencoder+SVM method has better performance.

To evaluate the performance of IDS for unknown attack data, we remove the impersonation attack data in the training set and only the impersonation attack data is available in the test set. We choose the impersonation attack as the unknown attack because the features of the impersonation attack are similar to those of the attack-free and are more difficult to detect. The experimental results show that the SVM and autoencoder + SVM for impersonation attack detection precision is 1.

The FPR is one of the important evaluation indicators for vehicle IDS. If the false alarm is too frequent for a driving vehicle, it will affect the driver's driving experience and might cause traffic accident. The FPR of SVM is 0.13 and the FPR of autoencoder+SVM is 0.06.

**Acknowledgement.** The paper is supported by the National Natural Science Foundation of China (Grant Nos. 61572192, 61971192) and the National Cryptography Development Fund (Grant No. MMJJ20180106).

# References

1. Woo, S., Jo, H.J., Lee, D.H.: A practical wireless attack on the connected car and security protocol for in-vehicle CAN. IEEE Trans. Intell. Transp. Syst. **16**(2), 993–1006 (2015)
2. Foster, I., Prudhomme, A., et al.: Fast and vulnerable: a story of telematic failures. In: USENIX Workshop on Offensive Technologies (2015)

3. Li, X., Yu, Y., Sun, G., et al.: Connected vehicles' security from the perspective of the in-vehicle network. IEEE Netw. **32**(2), 58–63 (2018)
4. Groza, B., Murvay, S.: Efficient protocols for secure broadcast in controller area networks. IEEE Trans. Ind. Inform. **9**(4), 2034–2042 (2013)
5. Muter, M., Asaj, N.: Entropy-based anomaly detection for in-vehicle networks. In: Proceedings of IEEE Intelligent Vehicles Symposium (IV), June, pp. 1110–1115 (2011)
6. Ji, H., Wang, Y., Qin, H., Wang, Y., Li, H.: Comparative performance evaluation of intrusion detection methods for in-vehicle networks. IEEE Access **6**, 37523–37532 (2018)
7. Larson, U.E., Nilsson, D.K., Jonsson, E.: An approach to specification-based attack detection for in-vehicle networks. In: 2008 Intelligent Vehicles Symposium, pp. 220–225. IEEE (2008)
8. Wang, C., Zhao, Z., Gong, L., et al.: A distributed anomaly detection system for in-vehicle network using HTM. IEEE Access **6**(99), 9091–9098 (2018)
9. Hu, W., Liao, Y., Vemuri, V.R.: Robust anomaly detection using support vector machines. In: Proceedings of International Conference on Machine Learning, pp. 282–289 (2003)
10. Lee, H., Jeong, S.H., Kim, H.K.: OTIDS: a novel intrusion detection system for in-vehicle network by using remote frame. PST (Privacy, Security and Trust) (2017)
11. Cho, K.-T., Shin, K.G.: Fingerprinting electronic control units for vehicle intrusion detection. In: Proceedings of USENIX (2016)
12. Cozzolino, D., Verdoliva, L.: Single-image splicing localization through autoencoder-based anomaly detection. IEEE International Workshop on Information Forensics and Security. IEEE (2017)
13. Abdi, H., Williams, L.J.: Principal component analysis. Wiley Interdiscip. Rev. Comput. Stat. **2**(4), 433–459 (2010)
14. Chang, C.C., Lin, C.J.: LIBSVM: a library for support vector machines. ACM Trans. Intell. Syst. Technol. (TIST) **2**(3), 1–27 (2011)
15. Hochreiter, S., Schmidhuber, J.: Long short-term memory. Neural Comput. **9**(8), 1735–1780 (1997)

# Ultragloves: Lowcost Finger-Level Interaction System for VR-Applications Based on Ultrasonic Movement Tracking

Si Li, Yanchao Zhao[✉], and Chengyong Liu

Nanjing University of Aeronautics and Astronautics, Nanjing, China
{lisinuaa1994,yczhao}@nuaa.edu.com

**Abstract.** The VR technology is undergoing a series of evolution, including not only the higher video quality but also more fluent human-computer interaction operations. Current solution mainly use camera-based or specified equipment-based hand gesture recognition, where the former suffers from low accuracy and limited interaction capacity while the latter suffers from expensive cost. In this paper, we propose Ultra-gloves, which is a high accurate finger-level and low cost gesture inter-action system for VR and smartphones. Specifically, it is enabled by the gloves implanting multiple microphones with which the ultrasonic signal are played. In the VR or mobile devices, the hand gestures are rebuilt with the recorded FMCW-signal and the hand motion model. Further-more, to improve the accuracy and calculation speed, we propose a par-allel processing algorithm for the signals, which could greatly accelerate our solution to real time manner even in COTS smartphones. The real implementation based experiments shows that, Ultragloves could capture the gestures in the accuracy of 2 cm while the parallel algorithm could accelerate the solution with about 3 times.

**Keywords:** Ultrasonic tracking · Multi-finger gesture capture · System implementation

## 1 Introduction

### 1.1 Background

VR (Virtual Reality) is envisioned to be the dominant displaying and interaction technology that bridge the gap between virtual and real world. It basically generates an interactive 3D dynamic view and can provide the user with an immerse experience where user can have a virtual space view, auditory, tactile and other sensory simulation in a full range of embedded virtual environment. VR is not only applied to the field of human-computer interaction, but also has a large number of applications for CAD, technical education and training institution.

Despite its bright future and fast development in the high quality display, the interaction technology in VR is still facing challenges such as low recognition accuracy and high recognition cost in gesture recognition, which greatly increase the device expense and compromise the user experience to a large extent (Fig. 1).

© Springer Nature Switzerland AG 2020
S. Wen et al. (Eds.): ICA3PP 2019, LNCS 11945, pp. 467–481, 2020.
https://doi.org/10.1007/978-3-030-38961-1_41

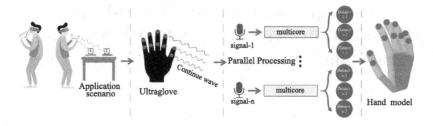

**Fig. 1.** VR glove usage scenario and the processing flow of the system.

## 1.2   Limitation of State-of-the-Art

Both research and industry community endeavored to improve the interaction experience for VR systems, the solutions of which can be mainly divided into two categories: device-free and device-wearing.

**Device-Free Method.** Among the devices-free methods, the techniques are mainly divided into visual-based and acoustic signal-based solutions.

In vision-based recognition, such as Leap Motion [3], through the parallax, the depth of the space is obtained. Leap Motion uses this principle to identify the movement of the hand within the range by capturing the difference between the images captured by the two cameras. Recently, in [6], the authors use visual and acoustic signals together to make it possible to detect inputs in the air. However, vision based solution suffers from limited recognition range and poor performance in multi-finger tracking scenario.

Another tack of solution uses acoustic signal to capture the hand movements, such as EchoTrack [1], which utilizes the motion of the palm near the echo recognition device; LLAP [9]and Strata [11] measure the phase change caused by the motion of the hand on the transmitted sine wave; FingerIO [5] measures the single-finger motion near the device by the echo of the OFDM signal; UltraGesture [4] uses sound signal driven model training to detect gestures; QGesture [10] and Venkatnarayan's research [8] use the wifi to measures the Gesture.

Although device-free solutions have advantage in the interaction experience, none of them can achieve fine-grained multi-finger identification, as the movement of the multi-finger is too complicated with multiple degrees of freedom while the reflection of signal suffers from multipath effect and thus hinder to identification of multiple objects.

**Device-Wearing Method:** The main focus of this track is on how to accurately measure the movement of the multiple components of the wearing devices so as to extract the gesture in a fine-grained manner.

Typically, Dison Technologies' Synertial [7] finger-capture system captures the movement of the wearer's hand through an inertial measurement unit (IMU) on the glove; 5DT magnetic resonance imaging(MRI) gloves measure the movement of the glove through magnetic resonance techniques. The SoundTrak [12] study uses a ring with a mini-speaker to acquire the motion of the ring by measuring the phase of the motion through the matrix of the microphone; In [7], expensive IMUs and complex initialization were used. However, MRI not only requires expensive equipment, but also requires a magnetic resonance environment. In [12], the device is still not applicable to multi-finger scenarios.

### 1.3   Challenges

In summary of the existing work, gesture recognition in the VR field mainly has challenges such as high recognition cost, large amount of calculation, and lack of multi-finger movement capture support in mobile devices. Specifically, they could be summarized as follows.

- **High wearable equipment cost:** VR technology often requires additional equipment which are specifically designed and often very expensive. Plus, the redundant equipment is cumbersome and inflexible, which seriously reduces the user experience.
- **Low recognition accuracy:** The positioning error of the low-cost device node is large, and it is impossible to accurately recognize the finger-level movement and reduce the user experience.
- **Lack of Multi-finger movement capture:** Multi-finger movement support is the milestone to enable fully VR interaction. How to capture multi-finger motion with portable equipment is indispensable and urgently.

### 1.4   Our Approach and Contributions

To tackle above challenges, we proposed a comprehensive solution consisting of following components.

- For issue of redundant and expensive equipment, we designed a glove implanting multiple speakers in the crucial position, with which the motion model of the finger could be obtained by measuring the motion trajectory of the nodes. This glove is very light and agile, thus rarely affects the user experience.
- Considering that the accuracy of inexpensive gyroscopes and accelerometer nodes is almost ad-hoc, we chose the microphone node as the measurement sensors. Inspired by the radar system, we chose the FMCW as the signal to measure the location of the microphone. It basically compares the frequency difference between the transmitted frequency modulated continuous wave and the received signal. Then, the distance from the microphone to the sound source is calculated by converting the frequency difference into a ToA.
- To achieve multi-finger motion capture in commercial mobile devices, we first draw the trajectory of a single node by periodically measuring the position of the node on the glove. Then we proposed an parallel algorithm to fully utilize the multicore nature of the mobile devices and accelerate the system to a realtime manner.

The contributions of our work could be summarized as follows:

– We propose and implement a prototype system for capturing finger-level gestures for VR and smartphone devices based on ultrasonic ranging techniques.
– Compared to the previous study, our system could achieve acceptable accuracy with only low end speaker nodes and low end mobile processors. With this, our technique shows great potential for severing as an substitute of the other interaction techniques, as the accuracy could be improved by upgrading the hardware without change our technical frameworks.
– Compared to current acoustic signal-based studies, our research manage to capture the multiple finger gestures. At the same time, to achieve both fine-grained and realtime interaction, we design a parallel algorithm to measure and process signals recorded from multiple speaker nodes in a realtime manner. This algorithm gives great potential for the future when we can increase the density of speaker nodes so as to capture finer-grained interaction gestures.

## 2   Modeling and Preliminaries

In this section, we introduce our basic positioning model and preliminaries.

### 2.1   Application Model

We envision that our system is mainly used for interactive assistance of VR systems, including smartphone VR systems and conventional VR systems. In smartphone VR system, a mobile phone generally has a plenty of speakers. By using these speakers to send inaudible ultrasound that does not affect the user, we can use Ultraglove to interact in the VR world. In conventional VR systems, ultrasound can also be played through the bypass of the device's speakers for the positioning of our system.

In these scenarios, we can enrich the user's interaction with the VR world without affecting the user experience.

### 2.2   One-Dimensional Positioning Model

To locate the position of the node in space, we first need to measure the one-dimensional distance from the node to the sound source.

The basic idea is as follows. The distance from the node to the source is measured using the FMCW(Frequency-Modulated Continuous Wave) technique. FMCW technology has mature applications and high-precision ranging in the radar field. In FMCW technology (as shown in Fig. 2), it changes its operating frequency during the measurement. The transmitted signal is shown as a solid line in Fig. 2. From the initial frequency $f_0$, the frequency increases with a slope $B/T$ over time. The received signal is shown in the dotted line in Fig. 2. At time $t_0$, the frequency of the transmitted signal is $f_2$, and the frequency of the received signal is $f_1$. The difference between the two frequencies is obtained by: $\triangle f = f_2 - f_1$. By dividing by the frequency change slope $B/T$, we can get:

$$\triangle t = \triangle f \times T/B = (f_2 - f_1) \times T/B, \tag{1}$$

where $\triangle t$ is the time from the sound source to the node. By multiplying the speed of sound, the distance from the source to the node can be obtained.

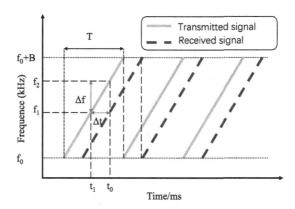

**Fig. 2.** The frequency curve of the original waveform and the recorded waveform of the FMCW.

## 2.3 Three-Dimensional Positioning Model

An one-dimensional distance alone cannot locate a node's position in space. In order to locate the position of the node in space, we use a plurality of sound sources for positioning, as shown in Fig. 3.

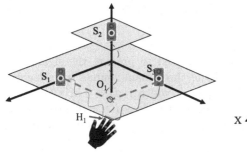

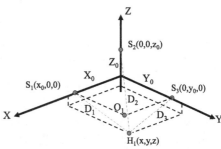

**Fig. 3.** Schematic diagram of a three-position model.

**Fig. 4.** Schematic diagram of measuring the node to sound in a three-dimensional model

For one speaker node on the Ultraglove, we can get its distance to all sources using the one-dimension technique. After obtaining the distance from the three sources to the node (taking three speakers as an example), we use three sound sources as the center of the sphere and the distance is the radius. Then we will get three balls and there are two intersections between the three balls. In order to select the correct position from the position in these two spaces. We choose an initial position in the space. At the beginning of the system, the node is in its initial position. We rely on the near and low speed criteria to select the location near the previous location as the correct location.

Assuming that the location of the sound source and the initial position of the node are known, then we model it as in Fig. 4. Here, $S_1, S_2$ and $S_3$ represent three sound sources, and $S_1(x_0, 0, 0)$, $S_2(0, 0, z_0)$, $S_3(0, y_0, 0)$, $O_1$ represents the coordinates of initial location. $H$ is a node, the coordinates are $(x, y, z)$, and the distance to three sound sources is $D_1, D_2, D_3$. The coordinates of the point $H$ can be obtained by solving the following equations:

$$\begin{cases} D_1 = \sqrt{(x - x_0)^2 + y^2 + z^2}, \\ D_2 = \sqrt{x^2 + y^2 + (z - z_0)^2}, \\ D_3 = \sqrt{x^2 + (y - y_0)^2 + z^2}. \end{cases} \tag{2}$$

After obtaining the coordinates of the node at the current time, we compare it to the coordinate position of the previous moment (or initial position). Then we can determine that the position with less movement is the correct position.

## 3   Technical Details

In this section, we will provide a detailed description of the specific implementation steps of the system.

### 3.1   System Overview

The basic idea of our prototype system is to measure the position of the nodes that fit on the surface of the hand in space. This represents the position of the hand in space. As shown in Fig. 5, the system mainly includes the following steps:

**Initialization:** In the initialization phase, we need to prepare the signal for FMCW ranging. The initial frequencies of the three main sources we selected were 18 kHz, 19.5 kHz and 21 kHz. The frequency width is $B = 1$ kHz and the sweep time is $T = 5000 points$. The sampling rate we use is $FS = 48$ kHz. Then we initialize the microphone, speaker and place the speaker in the initial position. And we place the node of the ultraglove at the initial position for 1 s.

**Recording the Signal:** After initializing the original waveform, speaker and microphone, we use multiple speakers to play the signals in the respective bands. Then we need to keep all the microphone nodes in the recording state and record all the microphones at the same time.

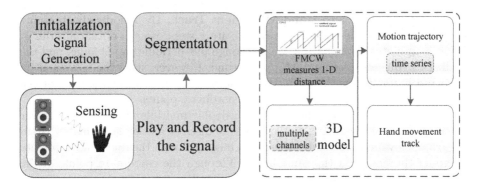

**Fig. 5.** System overview of ultraglove.

**Segmentation:** After recording each microphone separately, we cut the window of the recorded signal. The recorded signal contains a mix of signals from all speakers.

**Measuring 1-D Distance:** After adding the initial translation, we obtain the sinusoidal waveform of the frequency difference between the original signal and the received signal by multiplying the original signals of the different speakers. Then we perform a discrete Fourier transform on it and measure the position of the frequency response peak that is less than half the original signal bandwidth. The distance from the node to the source of the original signal is measured by measuring the frequency difference between the original signal and the received signal.

**Calculating 3D Position:** After obtaining the $1-D$ distance from the node to all sources, we can find the position of the node in space through the $3-D$ model. Although in the context of the three sound sources, the resolved coordinate points have two points that are spatially symmetric. But since the initial position is known, we can choose one of the solutions by the continuous motion trajectory.

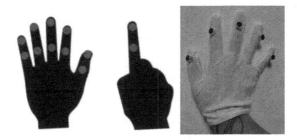

**Fig. 6.** A schematic diagram of the node on the glove, the red dot represents the microphone node on the glove. (Color figure online)

**Motion Trajectory and Hand Movement Track.** By periodically obtaining the coordinates of the node in the coordinate system, we can obtain a series of point coordinates arranged in time. Although our periodic gap is large, and the feedback frequency is around 9 Hz per second. However, since the movement of the hand is not too fast, we can still restore the movement trajectory of the human hand by connecting the discrete coordinate points.

According to the user's needs, we can deploy multiple nodes on each finger of the glove, as shown in Fig. 6. We know the discrete motion coordinate system of each node which constitutes the user's finger through the directed coordinate points of the nodes on the same finger. Through the coordinate points of the time series, we can fit the movement of the hand.

## 3.2   Distance Measurement

We modulate the signal sent by the FMCW as follows:

$$Y(t) = cos(2\pi \times f(t) \times t), \tag{3}$$

where $f(t)$ is expressed as follows:

$$f(t) = \frac{B}{2T} \times t + f_0. \tag{4}$$

Here $B$ is the frequency width; $T$ is the period length; $f_0$ is the initial frequency. The received signal is:

$$R(t) = \sum_{k=1}^{N} cos(2\pi \times f(t - \triangle t_k) \times t). \tag{5}$$

And this includes echoes of all multipaths. In order to calculate the distance, we need to get the $\triangle t$ first. According to the trigonometric formula, we multiply the received signal by the original signal:

$$I(t) = Y(t) \times R(t) = cos(2\pi \times f(t) \times t) \times \sum_{k=1}^{N} cos(2\pi \times f(t - \triangle t_k) \times t) \tag{6}$$

According to the following trigonometric function, we simplify one of the paths.

$$cos(\alpha) \times cos(\beta) = \frac{cos(\alpha + \beta) + cos(\alpha - \beta)}{2} \tag{7}$$

Since we are using high frequency signals, the high frequency part is filtered out, and the rest are as follows:

$$I(t) = \sum_{k=1}^{N} cos(2\pi \times t \times \triangle f_k). \tag{8}$$

Although $I$ include multipath, the energy of others is much smaller than the energy of the direct path. By measuring the frequency component of the most energy in $I$, $\triangle f$ can be obtained. Thereby we obtain fly time $\triangle t$. Then we multiply the speed of sound $v_s$ to get the distance $D$ from the source to the microphone node:

$$D = v_s \times \triangle t \tag{9}$$

We use the above method to find the distance. Therefore, we need to get the signal emitted at the current moment and the exact frequency at which the waveform needs to be solved.

If the system is integrated, it is not difficult to obtain the signal transmitted at the current moment. But our system is a separate system, we will use the following way to get the signal transmitted at the current time.

1. After turning on the system, user place the node at the initial position $D_{initial}$ where the distance is known in advance for a certain period of time.
2. After taking a signal $R$ of several period, the signal is multiplied by the original transmitted signal to obtain the signal $I$.
3. The Fourier transform is performed on the signal I to convert the original signal from the time domain to the frequency domain. In the 0 to B frequency band, we find the highest frequency $f_{peak}$.
4. We get the amount of translation $Mov_{initial} = \frac{f_p \times T}{B}$.

The period we chose was $T = 5000$ sample points. So the coverage is about 35 m. After solving this translation $Mov_{initial}$, we plus $T/4$ to get $Mov$. This will ensure that the measured $f_p$ will not jump from 0 to $B/2$ during the run.

To get the exact frequency, we use a discrete Fourier transform at the end of the zero. After we have obtained the exact frequency of $f_{peak}$, we obtain the distance by:

$$D = v_s \times (f_{peak} - Mov)/B \times T + D_{initial} \tag{10}$$

### 3.3   Parallel Acceleration for Multi-finger Signal

In our prototype system, simultaneous recording of multiple microphone nodes is required. So we also introduce parallel processing of signals recorded by multiple microphones.

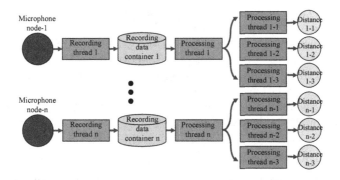

**Fig. 7.** Parallel processing frame schematic.

Our processing framework is shown in Fig. 7. We use multiple threads for parallel processing. Each microphone corresponds to one thread and this thread is only responsible for recording a corresponding microphone and storing the recorded signal in a corresponding container. There are also a large number of processing threads in the system. These processing threads are shown in the figure. In our processing, there is a rate matching algorithm, as shown in Algorithm 1. This algorithm ensures that the distance we are getting is the latest data.

---

**Algorithm 1.** Rate Matching algorithm

---

**Input:** *Stack* for storing the recording signal window. The latest data is at the top of
    the stack. The stack lock *flag*
**Output:** *Data* used to calculate distance.
 1: $last = list()$
 2: **while** $isempty(stack)$ **do**
 3:    wait for a short time.
 4: **end while**
 5: **while** $flag == False$ **do**
 6:    wait for a short time.
 7: **end while**
 8: $flag = False$ %add lock to the stack
 9: **if** $size(stack) >= 2$ **then**
10:    $now = pop(stack)$
11:    $last = pop(stack)$
12:    $clear(stack)$
13: **else if** $size(stack) == 1$ **then**
14:    $last = now$ %last *now* is the last window of signal
15:    $now = pop(stack)$
16: **end if**
17: $flag = True$ %unlock the stack
18: $Data = connect(last, now)$
19: **return** $Data$

---

The thread then adds the data to the pre-allocation value $Mov$ and assigns the data to the child thread of the thread. Each child thread calculates only the distance between one microphone node and one sound source.

Such a framework is mainly to match the calculation rate and the recording rate. Our prototype system not only supports large VR systems, but also supports smartphone VR systems. The computing power of smartphones is not strong and the calculation rate is slow. Therefore, we only consider the latest data for each calculation and then clear the old data. We use this method to reduce the amount of calculation.

## 3.4    3D Position Measurement

In order to get the position in the three-dimensional space. We need to know the distance between the node and the sound source at the same time. So we use *multi − channel FMCW*. We call the signal sent by each sound source a **channel**. Each channel uses a different frequency and does not intersect each other. We put the signals of all channels into the same spectrogram in Fig. 8. We choose B = 1000 Hz, $T$ = 5000 points. At a sampling rate of 48 kHz, the period is approximately 0.1 ms. The bright band between 18 kHz and 19 kHz on the map is emitted by a sound source. The period is 5000 point. After receiving the signal, by multiplying the original signal of each sound source, the spectrum is shown in Fig. 9.

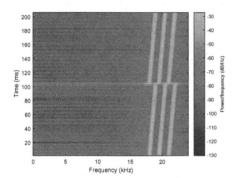

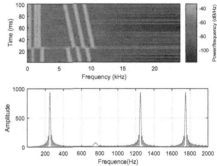

**Fig. 8.** Spectrogram of multiple sound sources. Put the frequencies of the three channels into the same spectrum.

**Fig. 9.** The figure above is the spectrum diagram, and the figure below is the diagram after the signal is Fourier transformed.

With a sound speed of 343 m/s, our measurement range is approximately 34.3 m. Since we added the $T/4$ offset value to the original signal, the measured peak value falls within the cell near B/4 when the user moves. The distance corresponding to B/4 is about 8 m. After the received signal $R$ is multiplied by a certain channel, as shown in the following figure of Fig. 9. Between 0 and 1000 Hz, we observed a large peak at about 260 Hz. After subtracting the initial translation value of 250 Hz, 10 Hz is obtained. By calculation, the distance is 35.7 cm. Further observation, we can find that there are two peaks between 1000 *and* 2000 Hz, which are the frequency difference between the original signal of the current channel and the received signal of the remaining channels. These signals do not interfere with our measurements. To reduce the amount of computation, we can choose not to filter. In the middle of 5000 *to* 10000 Hz, there are three bright bands that are inversely related to the original signal. The reason for these signals is that at a sampling rate of 48 kHz, the frequency of the original signal and the received signal is greater than the maximum frequency that can be sampled. The frequency is 48 kHz $− (\alpha + \beta)$ [2].

Therefore, we only need to multiply the original signal of each channel and measure the frequency corresponding to the peak value of $B/2$, so that the distance from the node to the sound source corresponding to the channel can be measured.

## 4  Experiment and Evaluation

### 4.1  Environment

**Hardware:** We validated our prototype system with five microphones and two separate speakers. Both the microphone and the speakers are connected to the PC via a USB sound card. The microphone uses an electric microphone. The USB sound card uses a *VENTION* sound card. The speaker is the *Philips SPA5270*.

**Environment Setting:** Our programming environment is python3.6.0 on windows 10. We also use the Pyaudio library in python. It allows simultaneous calls to multiple microphones and speakers. The experimental site was selected in a laboratory with a large table in the middle of $5 * 6\,m^2$. We choose the sound speed as $343\,m/s$.

### 4.2  Experiment Setup and Evaluation of Results

In this subsection, we will conduct a series of experiments to verify the feasibility and accuracy of our prototype system.

**Distance Measurement.** This experiment is mainly to verify the accuracy of distance measurement in our system. In this experiment, we only used one microphone and speaker. And the modulated signal is played using a single channel. We set $T = 5000, B = 1000, f_0 = 19500$.

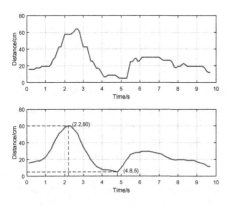

**Fig. 10.** The picture above is the distance measured by our approach. After passing the sliding window filtering algorithm, the following figure is obtained.

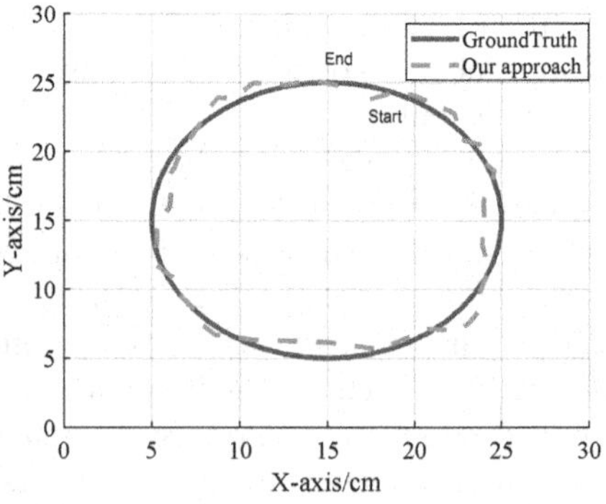

**Fig. 11.** The result of 2D trajectory tracking.

The results of the first experiment are shown in Fig. 10. The above figure shows the distance measured by our approach. It can be seen that the distance measurement is not smooth. Therefore, we perform sliding window filtering on the obtained distance. As we can see from the figure, the node starts at a position 15 cm from the sound source position. At 2.2 s, it moves to 60 cm. Then it closes to the sound source, 4 cm away from the sound source to the position of 5 cm. And it keeps away from the sound source and moves to 35 cm. Finally it moves to 20 cm.

**2D Trajectory Tracking.** Then, we set $T = 5000, B = 1000, f_0 = 18000, f_1 = 19500$ in order to further verify that our method can accurately obtain the position of the node on the $2D$ plane. In this experiment, we wore nodes and moved along the trajectory printed on the grid paper.

In the 2D tracking experiment, the results are shown in Fig. 11. According to the experimental requirements, we follow the circular motion prepared in advance. In the experiment, the maximum error measured was 2.5 cm.

**3D Position Measurement.** We then verified the accuracy and robustness of the 3D position measurement. Single-node and multi-speaker with multi-channels were used in the experiment. We set $T = 5000$, $B = 1000$, $f_0 = 18000$, $f_1 = 19500$, $f_2 = 21000$. And we place a single microphone node in a pre-known position in space, as shown in Fig. 12, comparing the measured value of that position with the location of the exact value. The nodes we prepared in advance have a total of $4 \times 4 \times 4$ points.(From 5 cm to 35 cm, one point per 10 cm).

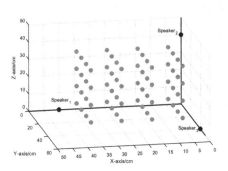

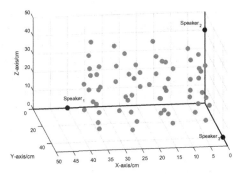

**Fig. 12.** The ground-truth of the third experiment.

**Fig. 13.** The measurement position of the third experiment.

In the third experiment, after we opened the playback system, we placed the node at the coordinate position of the previously known position as shown in Fig. 13. Then we start measuring the position. The maximum error of the coordinate position we measured from the real coordinate is 3.2 cm.

**3D Moving Tracking.** We further increase the difficulty of the experiment. We want to draw in the three-dimensional space around the sound source and track the motion of the nodes.

In our experiments, after playing the preset modulation sound on multiple speakers, the node is initialized, and then the wearing node moves in space. The results of the experiment are shown in Fig. 14. In the experiment, during the initialization phase, we wore the node at the initial coordinate position $(10, 10, 0)$. Then we drew a circle in the space. As shown in the figure, it can be clearly seen that a circular motion trajectory starting from the coordinate position $(10, 10, 0)$.

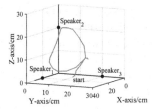

Fig. 14. The result of 3D trajectory tracking.

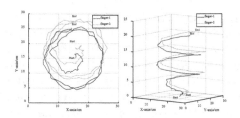

Fig. 15. The trajectory of two fingers moving in 3D space. The left figure is a top view and the right figure is a 3D view.

**Muti-node Tracking.** In order to verify the performance of multiple fingers gesture tracking and if it will affect our processing time and feedback frequency, we designed following experiments. We wear all the nodes, starting from the initial position and moving in space. Then, we compare the time spent on multi-node processing threads with the time spent on a single node. In the experiment, we wore a prototype system and moved in space. In order to calculate the average processing time, we collected runtime data for multiple experiments. The processing time of experiment is shown in Table. 1. In the experiment, the motion trajectory of the two nodes is shown in Fig. 15. From the top view, the experimenter wears a node to do the spiral motion. From the 3D view, the hand moves continuously upward in the Z-axis direction.

Table 1. Time to process with 5000 points duration.

| Initialization time | Processing time | | | |
|---|---|---|---|---|
| | Fast Fourier transform | Distance calculation | Moving average | Total |
| 2.2144 s | 8.40 ms | 1.22 ms | 0.35 ms | 9.97 ms |

Through statistics, the initialization phase takes about 2.2 s. After the initialization phase, our average processing time is stable at about 10 ms. We have 5,000 points per time. At a sampling rate of 48 kHz, it is approximately 104.2 ms. We can track the movement of the finger at the feedback frequency of 10 times per second by prioritizing the principle of processing the latest data. With our rate matching design, even on devices with insufficient computation, we will only sacrifice the recognition frequency without causing much delay.

# 5   Conclusion

In order to enrich the VR input method, we proposed wearable VR gloves, Ultra-graves. This is a prototype system capable of active acoustic sensing for multiple finger tracking based on ultrasonic signals. The prototype system consists of a plurality of microphone nodes with a plurality of sound sources and a processing unit. The system plays the FM signal of different frequency bands through the sound source and records the signal through the microphone node. At the same time, the processing unit performs parallel processing. We verified the feasibility and performance of the prototype system. In our experiments, the error was about 2 cm at rest. When moving, the error is about 3.2 cm. The frequency of measurement is approximately 5 times per second. Therefore, we believe that we can provide a low-overhead and low-cost VR interaction method for smart-phones. In the future, if higher sampling rates are supported with better devices, our accuracy will be further improved.

# References

1. Chen, H., Li, F., Wang, Y.: EchoTrack: acoustic device-free hand tracking on smart phones. In: Proceedings of IEEE INFOCOM, pp. 1–9. IEEE (2017)
2. Chen, Y., Wei, G., Liu, J., Yong, C.: Fine-grained ultrasound range finding for mobile devices: sensing way beyond the 24 khz limit of built-in microphones. In: Computer Communications Workshops (2017)
3. Leap: Leapmotion (2019). https://www.leapmotion.com/
4. Ling, K., Dai, H., Liu, Y., Liu, A.X.: Ultragesture: fine-grained gesture sensing and recognition. In: Proceedings of IEEE SECON. IEEE (2018)
5. Nandakumar, R., Iyer, V., Tan, D., Gollakota, S.: FingerIO: using active sonar for fine-grained finger tracking. In: Proceedings of the ACM CHI, pp. 1515–1525. ACM (2016)
6. Sun, K., Wang, W., X. Liu, A., Dai, H.: Depth aware finger tapping on virtual displays, pp. 283–295, June 2018. https://doi.org/10.1145/3210240.3210315
7. Discuz! Techology: Synertial (2019). http://www.disonde.com/
8. Venkatnarayan, R.H., Page, G., Shahzad, M.: Multi-user gesture recognition using WiFi. In: Proceedings of ACM Mobisys, pp. 401–413. ACM (2018)
9. Wang, W., Liu, A.X., Sun, K.: Device-free gesture tracking using acoustic signals. In: Proceedings of ACM Mobicom, pp. 82–94. ACM (2016)
10. Yu, N., Wang, W., Liu, A.X., Kong, L.: QGesture: quantifying gesture distance and direction with WiFi signals. Proc. ACM Ubicomp **2**(1), 51 (2018)
11. Yun, S., Chen, Y.C., Zheng, H., Qiu, L., Mao, W.: Strata: fine-grained acoustic-based device-free tracking. In: Proceedings of the 15th Annual International Conference on Mobile Systems, Applications, and Services, pp. 15–28. ACM (2017)
12. Zhang, C., et al.: Soundtrak: continuous 3D tracking of a finger using active acoustics. Proc. ACM Ubicomp **1**(2), 30 (2017)

# Adaptive Detection Method for Packet-In Message Injection Attack in SDN

Xinyu Zhan[1,2], Mingsong Chen[1,2], Shui Yu[3], and Yue Zhang[1,2(✉)]

[1] Shanghai Key Laboratory for Trustworthy Computing,
East China Normal University, Room 308, Math Building, No. 3663,
North Zhongshan Rd., Shanghai 200062, China
51174500153@stu.ecnu.edu.cn,
{mschen,yzhang}@sei.ecnu.edu.cn
[2] Software/Hardware Co-design Engineering Research Center of MOE,
East China Normal University, Shanghai, China
[3] School of Computer Science, University of Technology, Sydney, Australia
shuiyu@uts.edu.au

**Abstract.** Packet-In message injection attack is severe in Software Defined Network (SDN), which will cause a single point of failure of the centralized controller and the crash of the entire network. Nowadays, there are many detection methods for it, including entropy detection and so on. We propose an adaptive detection method to proactively defend against this attack. We establish a Poisson probability distribution detection model to find the attack and use the flow table filter to mitigate it. We also use the EWMA method to update the expectation value of the model to adapt the actual network conditions. Our method has no need to send additional packets to request the switch information. The experiment results show that there is 92% true positive rate in case of attack with random destination IP packets injected, and true positive rate is 98.2% under the attack with random source IP packets injected.

**Keywords:** Software-Defined Network · Packet-In message injection attack · Controller security · Adaptive detection

## 1 Introduction

Software-Defined Networking (SDN) is a new type of network architecture that provides flexible network management with the separation of the data and control planes. SDN is flexible, programmable and maintainable, so that it has been widely studied for its applications in backbone networks, data centers, enterprise networks, access networks, wireless networks, and etc. [1, 19]. The wide variety of use cases makes SDN security a serious concern [2, 17, 18]. Several main potential security issues have been presented in Kreutz's [3] research, such as attacks on the vulnerabilities in controllers, attacks on control plane communications, forged or faked traffic flows, and so on. After the controller is attacked, due to its centralized characteristics, it will cause different degrees of defects in the controller, switch, attacked host, and the entire network.

© Springer Nature Switzerland AG 2020
S. Wen et al. (Eds.): ICA3PP 2019, LNCS 11945, pp. 482–495, 2020.
https://doi.org/10.1007/978-3-030-38961-1_42

In the communication process of SDN, the OpenFlow protocol is responsible for maintaining communication through a secure channel. When a new flow occurs on the network, the switch cannot find a match entry in the flow table, and then sends a Packet-In message to query the controller for the related forwarding policy. The controller needs to process the Packet-In message sent by the switch. If the attacker injects a large amount of malicious data packets, the switch cannot find a match of these new packets, and then sends Packet-In messages to the controller, thereby triggering the Packet-In message injection attack. A massive injections of such new packets to the network will quickly choke the processing of the controllers [22]. If the attack continues, it will lead to topology spoofing, controller overload, etc., so that the controller can't handle normal packets. A single point of failure, even the embarrassment of the upper application probably occurs. Based on the analysis, the Packet-In message injection attack is concluded with two important features: burst large traffic rate and traffic rate deviation [5].

To solve to these problems, this paper proposes a Packet-In message Security Method (PSM) that combines the advantages of multiple methods to detect and defend against Packet-In message Injection attack.

The rest of the paper is structured as follows. Section 2 presents the related survey that includes the defenses for controller. Our proposed method to detect and mitigate the Packet-In message injection attack, is described, in detail, in Sect. 3. Section 4 shows the results of experiment of our proposed method. Finally, Sect. 5 concludes the paper with future enhancements.

## 2   Related Work

With the security defense research for controllers, detection and defense methods for various vulnerabilities and attacks are proposed. Packet-In message injection attack can be triggered by multiple vulnerabilities, such as topology spoofing, packet flooding, and DDoS attacks.

The controller has a host tracking function that relies on Packet-In messages and does not require any verification. The attacker sends a malicious message to the switch, spoofing the controller that the host has moved to a physical location which controlled by the attacker. It's called topology spoofing. To defend against this attack, the TopoGuard [6] tool was proposed to verify the authenticity of the host migration by verifying whether the Port Down signal is present. However, this method can't handle the case of a large number of Packet-In message flooding.

For packet flooding attacks, PacketChecker [7] is proposed. It's a defense strategy based on Packet-In message legality detection. According to the whitelist, the method can verify the legality of a Packet-In message and determine whether the message should be forwarded to the controller. However, this method does not consider the case where an attacker sends messages with forged IP address.

Among the current security problems, one of the most urgent and hardest security issues is Distributed Denial of Service (DDoS) [8, 10, 15]. It is easy to start, hard to defend and trace [4, 9]. Many attack detection methods have been proposed for DDoS attacks. Methods based on statistics and threshold detection have also been proposed.

For example, a method [16] calculates the entropy value of the target IP and sets the threshold to judge whether the stream characteristics in the network are normal, then achieves early detection of the attack. This method cannot simultaneously target topology attacks and flooding attacks, and does not propose related mitigation strategies. More targeted detection methods [11–13], are to collect the flow information and extract the relevant features, then calculate the corresponding entropy value and perform threshold detection. They also release the mitigation strategy. However, the methods usually require a large amount of information to be collected, which is expensive. With the development of machine learning, some methods combined with it also have certain reference. A method [5], which a controller uses Bayesian networks to classify switches to identify switches with potential attack risks. Use a special flow table when mitigating attacks. This method is susceptible to the complexity of the topology.

## 3  PSM Method

The specific module of the PSM method mainly consists of four parts: a model-building module, a message-filtering module, a detection module, and a defense module. In the PSM, firstly establish a probability detection model and set a threshold. The controller models the historical sample according to the number of Packet-In messages, combined with the set parameters, and then sets a whitelist filtering forged IP/MAC address message. The new Packet-In messages are detected based on the model. Finally, if the detection result is an attack, the statistical sorting method is used to find the attack source and defend, and the whitelist and the threshold are dynamically updated. Figure 1 is a specific detection flow chart of the PSM Method. The details of each part are described in Sects. 3.1, 3.2, 3.3 and 3.4.

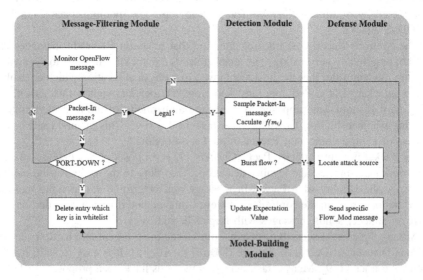

**Fig. 1.** Flowchart of the detection and defense in PSM method.

### 3.1   Model-Building Module

The model-building module is configured to establish a probability detection model based on historical data, set a threshold, and then receive new sampling data sent by the detecting module to update the model parameters. There is a lot of valuable information in the historical data [20, 21], and threshold detection is a common method in attack detection. Compared with the simple threshold detection method, this module can update the parameters according to the actual changes of the network, and has the advantage of self-adaptation.

First, the module obtains historical data of the normal flow, and sampling according to the set window size. Then the expectation of the sample is calculated by the moment estimation method. The calculation method is as shown in Eq. (1).

$$m = \sum\nolimits_{i=1}^{n} \frac{m_i}{n} \tag{1}$$

$m_i$ is the number of samples in the $i^{th}$ window, and $n$ is the number of sample windows. The sample mean calculated using the moment estimation method is the overall expectation.

Then this module establishes a probability detection model. In the actual network environment, the Packet-In message arrival rate satisfies the Poisson distribution [14] $P(\lambda)$, where $\lambda > 20$. The PSM is modeled using a Poisson distribution probability model. According to the nature of the Poisson distribution, we use probability formula of the normal distribution $N(\lambda, \lambda)$ to caculate. In this way, a Poisson distribution detection model under normal flow is established. Its probability density function is shown in Eq. (2).

$$f(m) = \frac{1}{\sqrt{2\pi m}} \tag{2}$$

Thirdly, the administer sets the threshold $\alpha$, the size of which ranges from 0 to 1. The specific value should be weighed according to the false alarm rate and the false negative rate. The smaller the network can bear the attack, the larger the value of $\alpha$ can be set.

However, in actual networks, the number of Packet-In messages may change. This change is softer and is not the same as a sudden change in a sudden attack, but it can still result in a change in parameters. Therefore, a self-learning mechanism for adaptively updating parameters is needed. Since the exponentially weighted moving average method can increase the weight of the latest sampled data, this method can be utilized to effectively adapt to such changes in the network. As shown in Fig. 1, after receiving the sampled value of the normal window, the exponential weighted moving average method is used to update the value of the expectation parameter. The specific calculation method is as shown in Eq. (3),

$$EWMA_t = \lambda y_t + (1 - \lambda)EWMA_{t-1}, t = 1, 2, 3, \ldots, n \tag{3}$$

where $EWMA_t$ is the estimated value at time $t$, $y_t$ is the measured value at time $t$, $\lambda$ is the fading factor, indicating the weighting coefficient of EWMA for historical measurements, and $n$ is the number of observations tracked by the EWMA model. The new normal sample value takes a greater weight to adaptively satisfy the soft flow changes in the network.

## 3.2  Message-Filtering Module

The message-filtering module mainly establishes a whitelist filtering forged IP/MAC message, and collects the attack source information of the defense module to update the whitelist in real time. The method of filtering forged IP/MAC messages can reduce the load on the controller, and defend against attacks such as topology explosions and broadcast storms. The way of the real-time update of the whitelist can also adapt to network changes and enhance security.

When a whitelist is established, the four fields of the DPID, IN_PORT, source IP address, and source MAC address combine into one entry, and the port fields DPID and IN_PORT are key values. Since the port field (DPID and IN_PORT fields) can uniquely identify a host, it is possible to determine a forged IP/MAC address message attack by recording the port field and the corresponding source IP and source MAC address. When a new host joins the network, the new packet sent by it does not match the flow entry in the switch, so the switch sends a Packet-In message to the controller. When a host leaves the network, the switch sends a Port-Status message to the controller and sets the PORT-DOWN field in the message to indicate that the host on the port leaves the network. We can use this mechanism to get real-time changes to the network and update the whitelist entries.

As shown in Fig. 1, the message-filtering module in the controller listens to and receives an OpenFlow message from the switch (here specifically referred to as a Packet-In message and a Port-Status message). The message-filtering module determines the type of the received message and performs the corresponding operation. The details of this filtering algorithm are shown as Table 1.

If the PORT-DOWN field of the Port-Status message is set, the host corresponding to the key value in the message leaves the network. In this case, the whitelist needs to be updated to delete related entries. When the source IP address and the source MAC address field in the Packet-In message are different from the existing corresponding entries, the message is invalid, and it is determined that a pseudo IP/MAC address message attack occurs, and the host corresponding to the message port is Attack the host. At this point, the key value of the attack source is sent to the sdefense module. Otherwise, the message is valid, there is no pseudo IP/MAC address message attack, and the message is forwarded to the detection module.

**Table 1.** Filtering algorithm to the OpenFlow messages

---

**Algorithm 1** Filtering the OpenFlow messages

---

```
Input: PS (Port-Status messages), PI (Packet-In messages), W (White-
list)
Output: D (send message to Detection Module), A (send to Defense Mod-
ule)
if PS then
  if PS.PORT_DOWN then
    del W [(PS.dpid, PS.in_port)]
    return
  end if
end if
if PI then
  if (PI.dpid, PI.in_port) ∈ W
    if W [(PI.dpid, PI.in_port)] ≠ (PI.srcIP, PI.srcMAC)
      return A (PI.dpid, PI.in_port)
    else
      W [(PI.dpid, PI.in_port)] = (PI.srcIP, PI.srcMAC)
      return D
    end if
  else
    W [(PI.dpid, PI.in_port)] = (PI.srcIP, PI.srcMAC)
    return D
  end if
end if
```

---

## 3.3 Detection Module

The detecting module is configured to, firstly sample the Packet-In message, then calculate the probability density and determine the attack detection result, and finally send the normal sampling data to the model-building module. The method of this module has a small amount of calculation and can quickly and sensitively obtain the detection result.

A large number of Packet-In message injections will inevitably lead to an abnormal distribution of the number of messages received by the controller, which will deviate from the Poisson distribution under normal conditions. Based on this principle, the Poisson distribution model under the normal network can be used for attack detection.

When a large number of Packet-In message injection attacks occur, the distribution of the number of messages received by the controller must be abnormal, which may cause a certain degree of deviation from the Poisson distribution under normal conditions. Based on this principle, the Poisson distribution model under the normal network can be used for attack detection.

In the detection module, firstly, the received normal Packet-In message is sampled by the window counter, and the probability density of the current window is calculated, and the specific calculation method is shown in the formula (4).

$$f(m_t) = \frac{1}{\sqrt{2\pi m}} e^{-\frac{(m_t - m)^2}{2m}} \tag{4}$$

Where $m_t$ is the number of Packet-In messages collected in the window, and m represents the expected parameters of the detection model.

The ratio $\frac{f(m_t)}{f(m)}$ of $f(m_t)$ to $f(m)$ is calculated, where $f(m)$ represents the probability density of the Poisson distribution detection model and is compared with the threshold $\alpha$. The details of the detection algorithm are shown as Table 2.

**Table 2.** Detection algorithm to the packet-in message injection attack

---

**Algorithm 2** Detection the Packet-In message Injection Attack

---

Input: r (the ratio of $\frac{f(m_t)}{f(m)}$), r_new (the ratio of $\frac{f(m_{t+1})}{f(m)}$), $\alpha$ (threshold)
Output: M (send sampling number to Model-Building Module), A (start up Defense Module)
if r < $\alpha$ then
  if r_new < $\alpha$ then
    return A
  else
    return M
else

  return M
end if

---

In the algorithm, if the ratio is not less than the threshold $\alpha$, there is no attack and the EWMA is used to update the desired parameters. But when the ratio is smaller than the threshold $\alpha$, it is determined to be a suspected attack, and the sample record of the window is saved. At the same time, the window counter is reset to zero, a new window is sampled and the ratio of probability density is calculated. If the ratio of the new window is not less than the threshold $\alpha$, it is determined that a burst flow occurs, there is no attack, and the number of samples is sent to the model building module for parameter update. Otherwise, the ratio of the new window is still less than the threshold $\alpha$, it is determined that an attack has occurred, and the defense module is started.

## 3.4  Defense Module

When the attack occurs, the defense module starts up. This module searches for the attack source and sends a defense message to the switch, and sends the attack source information to the message-filtering module. The method of sorting and finding attack sources can find multiple attack sources at one time and improve the defense effect.

First, the module separately counts the number of packet incoming messages sent by different source IPs, and sorts the number of incoming packets corresponding to each source IP address in descending order. After the corresponding highest number of entries are sequentially removed in this order, the remaining packets - the probability density of the total number of messages - are compared and compared with the detection model until the ratio of the number of remaining messages is not greater than the threshold. The key value of the switch port corresponding to the deleted entry is the attack source.

Then, the module records the key value of the attack source, constructs and sends a Flow_Mod message to the switch, and adds a flow table entry with the key value of the attack source as the matching keyword in the flow table, and the ACTION is DROP. When the packet port field received by the switch matches the attack source, a discard operation is performed to discard all the packets sent by the port. Because the attack source sends malicious packets in a large amount at a high speed, the idle time is set for the specific flow entry. After the attack stops, the switch automatically deletes the flow entry without affecting normal communication.

Finally, the module sends the key value of the attack source to the message-filtering module, deletes the related records in the whitelist, and updates the whitelist in real time to further defend against attacks on the port.

# 4  Experiments

In this section, we first describe the attack model and the configuration of the key parameters of our experiment. Then we examine the validity of our proposed method. Finally, we evaluate the effectiveness of mitigating Packet-In message injection attacks.

## 4.1  Attack Model

A schematic diagram of an attack model for the Packet-In injection attack, as shown in Fig. 2, including a controller, a switch network, an attacker, and a normal host.

The normal host sends normal packets to the switch for normal communication. An attacker could be an external attacker, or an attacker who invades an internal host, sending a large number of malicious packets, such as forged IP/MAC messages, or injecting malicious packets into the controller.

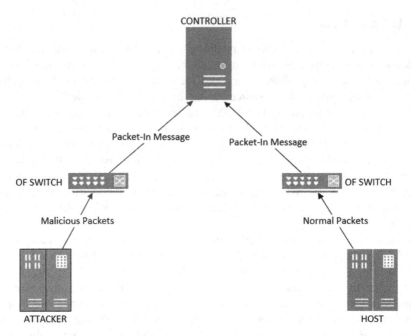

**Fig. 2.** Packet-In messages injection attack

The switch network receives a new packet from the host or the attacker and cannot find a match in the flow table. Therefore, the Packet-In message is sent to the controller to query the forwarding path.

The controller receives a large number of Packet-In packets and cannot determine its authenticity. It may send a large number of broadcasts to the non-existent port for pseudo IP/MAC messages, or add the forged host to the network topology, causing waste of resources and topology explosion. A large number of Packet-In packets are injected, which requires a lot of resources to process, resulting in reduced processing power of the controller, failure to process normal data packets, and even single-point failures, upper-layer application crashes, and network flaws.

### 4.2 Experiment Environment

The experimental development environment is Ubuntu 14.04, Python 2.7. We use Mininet [24] as a network simulation tool. The specific detection method runs on the Floodlight [25] controller, which is a Java-based open source controller. Among them, we use Scapy [23] to simulate injected malicious packets, which is a packet generation tool. In this Experiment, we use the Mininet to build a network including 3 layers, including 7 switches (s1–s7) and 8 hosts (h1–h8).

The other parameters used in our experiment are as shown in Table 3.

**Table 3.** Parameters used in the experiment

| Parameter | Definition | Value |
|---|---|---|
| α | Threshold | 0.000006 |
| p | Detection period | 4(s) |
| m | Expectation value | 90 |
| ATTACK1 | Random dst IP Packets | 100 |
| ATTACK2 | Random src IP Packets | 100 |

### 4.3    Results Analysis

As shown in Fig. 3, because of the process of the topology discovery, it occurs a burst flow in the beginning, then, a normal burst flow is shown. When they last for only 1 period, so our method dose not start the defense module.

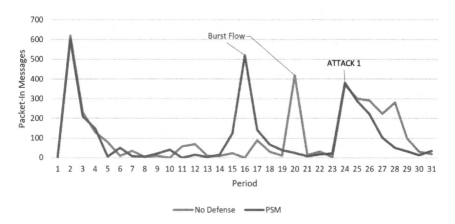

**Fig. 3.** Numbers of Packet-In messages of ATTACK1.

At the 24th period, we inject the malicious packets, which have random destination IP address. This attack triggers low speed Packet-In messages, but last for a long time. Our Method successfully detect the attack, and in the 26th period, the number of Packet-In messages starts to decrease to the normal level. But without our method, the attack still continues, wastes the resources of the controller.

Our method has 92% true positive rate, and 5% false alarm rate.

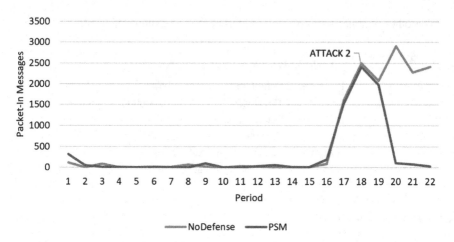

**Fig. 4.** Numbers of Packet-In messages of ATTACK2

As shown in Fig. 4, in case of ATTACK2, we send massive packets with the source IP of host3. It is clear that the Packet-In message number is rapidly increase. Our method detects the attack and starts up the defense module. Then it starts to decrease. The Packet-In messages are filtered by the defense entry in the flow table. Compared to that, the situation without our method still bear the severe attack, it may cause the controller crush.

In this situation, the method has 98.2% true positive rate, and the false alarm rate is 8.9% (Fig. 5).

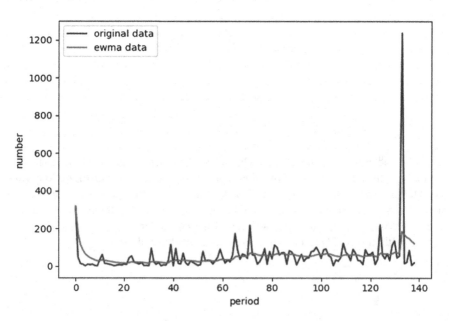

**Fig. 5.** The adaptive change trend of $m$ with EWMA

During the experiment, we use the scapy tool to simulate network packets. In the initial phase of the network, we do not inject other packets into the network, except for packets that were automatically generated when the SDN network was initialized. At the period of 24$^{th}$, we began to inject TCP, UDP, ICMP, and packets generated by the iperf and iperfudp commands into the network. As the number of data packets in the network grows, the change of $m$ with EWMA shows the smooth growth. EWMA can also reduce the impact of burst data on overall expectation value, makes the detection model more accurate.

**Table 4.** True position rate of different methods in different attacks

| Method (Attack) | True position rate |
|---|---|
| PacketChecker (Forged MAC Packets Attack) | 100% |
| PSM (Forged MAC Packets Attack) | 98.8% |
| PacketChecker (Forged IP Packets Attack) | 0% |
| PSM (Forged IP Packets Attack) | 96.4% |

PacketChecker is a lightweight Packet-In injection message detection mechanism with high efficiency, accuracy and low resource occupancy. However, this method can only detect the attack with forged MAC address packets and cannot detect attack with forged IP address packets.

As shown in Table 4, compared with the PacketChecker method, we can find that in the attack of forged MAC address packets, our method PSM has a true position rate of 98.8%, which is close to PacketChecker; With the attack of forged IP address packets, PacketChecker cannot detect malicious packets, but PSM has a true position rate of 96.4%. Forged IP address packet attacks can also trigger a large amount of Packet-In message injection. Obviously, PacketChecker is still poisoned by this attack.

# 5  Conclusion

In this paper, we propose an adaptive detection method for Packet-In injection attacks in SDN environment. The method is designed according to the Poisson distribution of the arrival of Packet-In messages. It enables the controller to obtain efficient detection and defense of Packet-in Injection attack results with low overhead. Due to the limitations of the experiment environment, the topological structures may have effects on threshold settings and experiment results.

In the future work, we will try to combine the programmability of the switch to relieve the defense pressure of the controller, and to improve the efficiency of detection and defense.

**Acknowledgments.** This paper is finished under the support of Key Program of the National Key Research and Development Program of China (No. 2018YFB2101300, No. 2018YFB2101301); Open Project Fund of Shanghai Key Lab for Trustworthy Computing (No. 07dz22304201607); National Nature Science Foundation of China (No. 61772034, No. 61872147); Natural Science Foundation of Anhui Province (1808085MF172).

# References

1. Cui, Y., et al.: SD-Anti-DDoS: fast and efficient DDoS defense in software-defined networks. J. Netw. Comput. Appl. **68**, 65–79 (2016)
2. Akhunzada, A., Ahmed, E., Gani, A.: Securing software defined networks: taxonomy, requirements, and open issues. IEEE Commun. Mag. **53**(4), 36–44 (2015)
3. Kreutz, D., Ramos, F., Verissimo, P.: Towards secure and dependable software-defined networks. In: Proceedings of the Second Workshop on Hot Topics in Software Defined Networking (HotSDNb12), pp. 55–60 (2013)
4. Yu, S., Tian, Y., Guo, S., Wu, D.O.: Can we beat DDoS attacks in clouds? IEEE Trans. Parallel Distrib. Syst. **25**(9), 2245–2254 (2014)
5. Gao, D., Liu, Z., Liu, Y., Heng, C., Ting, F., Chao, Z.H.: Defending against Packet-In messages flooding attack under SDN context. Soft. Comput. **22**(20), 6797–6809 (2018)
6. Hong, S., Xu, L., Wang, H., Gu, G.: Poisoning network visibility in software-defined networks: new attacks and countermeasures. In: Internet Society (2015)
7. Deng, S., Gao, X., Lu, Z., Gao, X.: Packet injection attack and its defense in software-defined networks. IEEE Trans. Inf. Forensics Secur. **13**(3), 695–705 (2018)
8. You, X., Feng, Y., Sakurai, K.: Packet in message based DDoS attack detection in SDN network using OpenFlow. In: International Symposium on Computing & Networking. IEEE Computer Society (2017)
9. Sunny, B., Krishan, K., Monika, S.: Discriminating flash events from DDoS attacks: a comprehensive review. Int. J. Netw. Secur. **19**(5), 734–741 (2017)
10. Shui, Y., Zhou, W., Guo, S., Guo, M.: A feasible IP traceback framework through dynamic deterministic packet marking. IEEE Trans. Comput. **65**(5), 1418–1427 (2016)
11. Kalkan, K., Altay, L., Gur, G., Alagoz, F.: JESS: joint entropy based DDoS defense scheme in SDN. IEEE J. Sel. Areas Commun. **36**, 2358–2372 (2018)
12. Kumar, P., Tripathi, M., Nehra, A., Conti, M., Lal, C.: Safety: early detection and mitigation of TCP SYN flood utilizing entropy in SDN. IEEE Trans. Netw. Serv. Manag. **15**, 1545–1559 (2018)
13. Yu, S., Zhou, W., Doss, R., Jia, W.: Traceback DDoS attacks using entropy variations. IEEE Trans. Parallel Distrib. Syst. **22**(3), 412–425 (2011)
14. La-Lin, J., Xia, P., Bing, X.: Performance evaluation of SDN controllers based on hybrid queuing model. Computer Engineering Science (2017)
15. Yu, S., Zhou, W., Jia, W., Guo, S., Xiang, Y., Tang, F.: Discrim-inating DDoS attacks from flash crowds using flow correlation coefficient. IEEE Trans. Parallel Distrib. Syst. **23**(6), 1073–1080 (2012)
16. Mousavi, S.M., St-Hilaire, M.: Early detection of DDoS attacks against SDN controllers. In: 2015 International Conference on Computing, Networking and Communications (ICNC). IEEE Computer Society (2015)
17. Yu, S., Guo, S., Stojmenovic, I.: Fool me if you can: mimicking attacks and anti-attacks in cyberspace. IEEE Trans. Comput. **64**(1), 139–151 (2015). (Spotlight paper of the issue)
18. Yu, S., Wang, G., Zhou, W.: Modeling malicious activities in cyber space. IEEE Netw. **29**(6), 83–87 (2015)
19. Feng, B., Zhang, H., Zhou, H., Yu, S.: Locator/identifier split networking: a promising future internet architecture. IEEE Commun. Surv. Tutor. **19**(4), 2927–2948 (2017). (Impact factor 17.2)
20. Chen, J., et al.: A parallel random forest algorithm for big data in a spark cloud computing environment. IEEE Trans. Parallel Distrib. Syst. **28**(4), 919–933 (2017)

21. Yu, S., Liu, M., Dou, W., Liu, X., Zhou, S.: Networking for big data: a survey. IEEE Commun. Surv. Tutor. **19**(1), 531–549 (2017). (Impact factor 17.2)
22. Mousavi, S.-M., St-Hilaire, M.: Early detection of DDoS attacks against SDN controllers. In: International Conference on Computing, Networking and Communications, pp. 77–81, (2015)
23. Scapy. http://www.secdev.org/projects/scapy/. Accessed 2018
24. Mininet. http://mininet.org/. Accessed 2019
25. Floodlight. http://www.projectfloodlight.org/. Accessed 2019

# PMRS: A Privacy-Preserving Multi-keyword Ranked Search over Encrypted Cloud Data

Jingjing Bao[1], Hua Dai[1,2(✉)], Maohu Yang[1], Xun Yi[3], Geng Yang[1,2], and Liang Liu[4]

[1] Nanjing University of Posts and Telecommunications, Nanjing 210023, China
jing874444051@163.com, {daihua,yangg}@njupt.edu.cn, yangmh1234@163.com
[2] Jiangsu Security and Intelligent Processing Lab of Big Data, Nanjing 210023, China
[3] Royal Melbourne Institute of Technology University, Melbourne 3001, Australia
xun.yi@rmit.edu.au
[4] Nanjing University of Aeronautics and Astronautics, Nanjing 210016, China
liangliu@nuaa.edu.cn

**Abstract.** In cloud computing, data owners outsource their data to clouds for saving cost of data storage and computation. However, while enjoying the benefits of cloud computing, users have to face the risk that sensitive outsourced data could be leaked. This paper proposes a privacy-preserving multi-keyword ranked search scheme over encrypted cloud data, which adopts a novel two-layer complete binary tree index structure. The upper layer index is used to filter the candidate documents while the lower layer index is used to prune those unqualified documents, and then the search result is efficiently determined. Security analysis is presented which indicates that the proposed scheme is capable of preserving the privacy of outsourced data. Experiment results show that the proposed scheme has good performance in terms of search time cost.

**Keywords:** Cloud computing · Searchable encryption · Multi-keyword search · Top-$k$ · Two-layer complete binary tree index

## 1 Introduction

With the maturity and popularity of cloud computing [1], more and more enterprises and individuals tend to outsource their storage, computing, and other resources to the cloud server provider (CSP) for easy access and cost savings. However, while enjoying the large-scale and efficient services provided by cloud server (CS), users are also at risk of leaking sensitive information of outsourced

Supported by the National Natural Science Foundation of China under the grant Nos. 61872197, 61972209, 61572263, 61672297 and 61872193; the Postdoctoral Science Foundation of China under the Grand No. 2019M651919; the Natural Research Foundation of NJUPT under the grand No. NY217119.

© Springer Nature Switzerland AG 2020
S. Wen et al. (Eds.): ICA3PP 2019, LNCS 11945, pp. 496–511, 2020.
https://doi.org/10.1007/978-3-030-38961-1_43

data. To decrease the risk of leaking sensitive information of the outsourced data, data encryption before outsourcing seems to be a feasible countermeasure. However, the encrypted data is hard and costly to be utilized to perform comprehensive computations, such as ranked search, etc. It is a challenge to perform an efficient ranked search over encrypted cloud data.

In the last decade, searchable encryptions (SE) [2] is an important way to realize the safety of keywords search in the cloud environment. At present, more mature solutions are based on TF-IDF algorithm and vector space model which abstracting keywords into points in multi-dimensional space. The correlation between documents and search keywords is described by a secure inner product between the vectors. Of course, it is obviously a waste of time and resources to simply calculate the safe inner product between the search vector and all the document vectors. To balance the privacy and practicability of the data, various solutions are provided [3–18].

Song et al. [3] proposed the first SE scheme. Although its security has been proved, no security model has been given. Goh et al. [4] first defined the security model of searchable symmetric encryption index and proposed a new scheme based on Bloom filter. Curtmola et al. [5] used inverted index for the first time to construct a searchable symmetric encryption scheme, which greatly improved the search efficiency. The above work is for the single keyword search, but in fact, multi-keyword search is more in line with the user's needs. In 2011, Cao et al. [6] first proposed the multi-keyword ranked search scheme (MRSE). In this scheme, documents are represented as vectors by using space vector model. Then, the vectors are encrypted by security KNN [7], [8] to solve the security problem of index. Later, Cao et al. [9] optimized the MRSE and adopted the TF-IDF algorithm to improve the accuracy of search, but the scheme is still linear time cost and inefficient. Xia et al. [10] proposed a multi-keyword search scheme based on balanced binary tree (DMRS), which can reduce the large inner product calculation by pruning function. In the same year, Chen et al. [11]proposed an efficient index scheme based on hierarchical clustering (MRSE-HCI), which can achieve linear time complexity with the exponential growth of the document set.

To improve the search efficiency, in this paper, we propose a privacy-preserving multi-keyword ranked search scheme (PMRS) over encrypted cloud data, which is based on the two-layer complete binary tree index (TCBT-index). The TCBT-index is composed of two layers of complete binary trees and it is the key structure for improving search efficiency. The upper layer has one complete binary tree and it is used to filter the candidate documents. While the lower layer has multiple complete binary trees whose roots are the nodes of upper layer tree, and they are used to prune those unqualified documents. Security analysis indicates that our proposed scheme PMRS is able to prevent the privacy of the search keywords, the outsourced encrypted documents, and index. Experiment results show that PMRS has good performance in terms of search time cost.

The contributions of this paper are: (1) We present the $\varepsilon$-posting partition vector model, which conceals the "hot words" and "cold words" into vectors. On the basis of such model, the two-layer complete binary tree index (TCBT-index) is proposed. (2) By adopting the TCBT-index, we propose an efficient privacy-preserving ranked search algorithm over encrypted cloud data. (3) We analyze the security and evaluate the search performance. The result shows that the proposed scheme can achieve efficient search while preserving data privacy.

## 2  Notations and Preliminaries

- $DS$ : The document set, $DS = \{d_1, d_2, ..., d_n\}$. $\widetilde{DS}$ is the encrypted form.
- $W$ : The keyword dictionary including $m$ keywords, $W = \{w_1, w_2, ..., w_m\}$.
- $Q$ : The set of search keywords, $Q = \{w_1, w_2, ..., w_q\}$.
- $DS(w_i)$ : The set of documents containing the keyword $w_i$.
- $VD_i$ : The document vector of $d_i$. $\widetilde{VD_i}$ is the encrypted form.
- $P_{i,j}$ : The posting corresponding to the document $d_j$ containing keyword $w_i$, $P_{i,j} = <id(d_j), VD_j>$. $\widetilde{P}_{i,j}$ is the encrypted form.
- $PL_i$ : The posting list corresponding to the keyword $w_i$.
- $\varepsilon$ : The posting list partition parameter.
- $PAR(PLi)$ : The set of $\varepsilon$-partitions after partitioning $PL_i$.
- $PL_{i,j}$ : The $j$th $\varepsilon$-partition of $PL_i$.
- $\mathcal{H}$ : The $\varepsilon$-partition set corresponding to all posting lists.
- $VP_{i,j}$ : The partition vector of $PL_{i,j}$ and $\widehat{VP}_{i,j}$ is its encrypted form.
- $PVS(w_i)$ : The partition vector set corresponding to the keyword $w_i$.
- $VF_Q$ : The query filter vector of $Q$ and $\widehat{VF}_Q$ is its encrypted form.
- $\Psi$ : The set of KPV-pairs (see Definition 6).
- $\Gamma$ : The set of PVP-pairs (see Definition 7).
- $\mathcal{U}$ : The plaintext two-layer complete binary tree index and $\widetilde{\mathcal{U}}$ is its encrypted form.
- $VQ$ : The vector of search keywords and $\widetilde{VQ}$ is its encrypted form.
- $TD$ : The trapdoor of a ranked search.

**Vector Space Model (VSM) and TF-IDF Model.** The VSM and TF-IDF can transform the processing of text into operations on vectors in vector space, and use spatial similarity to express relevance on text. The term frequency (TF) refers to the number of times a given keyword or term appears in documents, while the inverse document frequency (IDF) is equal to the total number of documents in the set divided by the number of documents containing a given keyword. VSM is used to convert a given document $d_i$ into a vector $VD_i$, or search keywords set $Q$ into vectors $VQ$, which are shown in the following formulas:

$$VD_i[j] = TF_{d_i,w_j} / \sqrt{\sum_{w_j \in d_i \wedge d_i \in DS} \left(TF_{d_i,w_j}\right)^2} \qquad (1)$$

$$VQ[j] = IDF_{w_j} / \sqrt{\sum_{w_j \in Q} \left(IDF_{w_j}\right)^2} \qquad (2)$$

where $TF_{d_i,w_j}$, and $IDF_{w_j}$ represent TF and IDF values of $w_j$, respectively. The relevance score between $d_i$ and $Q$ are calculated as:

$$Score(VD_i, VQ) = VD_i \cdot VQ = \sum_{j=1}^{m} VD_i[j] \times VQ[j] \qquad (3)$$

**Secure Inner Product Operation.** This scheme uses the secure inner product operation to calculate the inner product of two encrypted vectors without knowing the plaintext value. The basic idea of this is as follows. Assuming that $p$ and $q$ are two $n$-dimensional vectors and $M$ is a random $n \times n$-dimensional invertible matrix. $M$ is treated as the secure key. The encrypted form of $p$ and $q$ are denoted as $\widetilde{p}$ and $\widetilde{q}$ respectively, where $\widetilde{p} = pM^{-1}$ and $\widetilde{q} = qM^T$. Then we have $\widetilde{p} \cdot \widetilde{q} = (pM^{-1}) \cdot (qM^T) = pM^{-1}(qM^T)^T = pM^{-1}Mq = p \cdot q$, i.e. $\widetilde{p} \cdot \widetilde{q} = p \cdot q$. Therefore, we have that the inner product of two encrypted vectors equals the inner product of the corresponding two plaintext vectors.

**Inverted Index.** The inverted index is a kind of indexing method for quickly finding a list of documents containing a certain keyword [5]. As shown in Fig. 1, it consists of dictionary and inverted files. The dictionary represents a collection of all the keywords that have appeared in the files of $DS$. Each keyword corresponds to a posting list. The posting list records a list of all documents in which a specified keyword appears. The element of a posting list is called a posting which includes the information of a document.

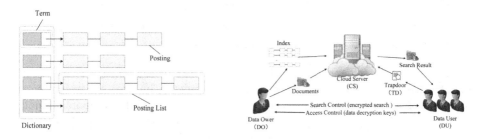

**Fig. 1.** Inverted Index          **Fig. 2.** System Model

# 3 Models and Problem Formulation

## 3.1 System Model

This paper adopts the same system model as [9–11]. As shown in Fig. 2, the system mainly includes three types of entities, namely Data Owner (DO), Data User (DU) and Cloud Server (CS). Their collaboration mode is as follows: (1) DO is in charge of preprocessing the original data and then uploading the preprocessed data (encrypted documents and indexes) to CS. (2) CS provides data storage and computation service to DU. When a search request is submitted,

CS performs the search and returns the corresponding result. (3) DU generates a trapdoor corresponding to the queried keywords and then submits it to CS as a search command. After that, DU waits the final search result.

### 3.2  Problem Statement

Given a set of $q$ queried keywords $Q = \{w_1, w_2, ..., w_q\}$, a multi-keyword ranked search is to retrieve the $k$ ranked documents that have the $k$ highest relevance scores to $Q$. Formally, we define the multi-keyword ranked search as $Query = (DS, Q, k)$ where $k$ is the number of requested documents and $k << n$ generally. For simplicity, we use $Q$ to represent a search.

The threat model in this paper adopts the same "honest but curious" model as in [9–11]. It assumes that CS follows the pre-established protocols to perform ranked searches honestly, but it is curious of sniff the private information from the outsourced data through analysis and deduction.

In order to achieve the multi-keyword ranked search over encrypted data, it is inevitable for CS to know some information such as search mode and results. Here, we adopt the same security definition as Curtmola et al. [5] for security analysis in our proposed scheme, which has three aspects as follows:

- History: Each interaction between DU and CS becomes a history, including the corresponding documents, index, and search keywords.
- View: The contents can be seen for CS, including the encrypted document set, index, and trapdoor.
- Trace: The sensitive information can be obtained by CS, such as search results and mode. In this scheme, CS can generate a search mode by recording the trapdoor uploaded by DU and the search results.

The proposed scheme PMRS should satisfy two goals. First, the contents directly seen by CS only include encrypted documents, indexes and trapdoors. It means that the confidentiality of documents, indexes, and trapdoors cannot be leaked. Second, PMRS is designed to ensure the comprehensive performance of search efficiency, that is, the search process should be concise and efficient.

## 4  $\varepsilon$-Posting Partition Vector Model

To describe the multi-keyword ranked search scheme, we redefine the posting and posting list.

**Definition 1.** *Posting and Posting List.* Given a keyword $w_i$, a posting of $w_i$ is a document access structure which is denoted as a pair $(docID, docVec)$, where $docID$ points to a document having $w_i$ and $docVec$ is the vector of the document. Let $DS(w_i) = \{d_1, d_2, ..., d_g\}$, the posting list of $w_i$ is denoted as $PL_i = \{P_{i,1}, P_{i,2}, ..., P_{i,g}\}$, where $P_{i,j}$ is one of the postings of $w_i$.

**Definition 2.** $\varepsilon$-*partition.* Given a partition parameter $\varepsilon$ and a keyword $w_i$, the posting list $PL_i$ of $w_i$ is randomly divided into $t_i$ partitions by $\varepsilon$. A generated partition is denoted as a $\varepsilon$-partition that is a sub-posting list of $PL_i$. The set of generated partitions is $PAR(PL_i) = \{PL_{i,1}, PL_{i,2}, \ldots, PL_{i,t_i}\}$ which satisfies Eqs. (4) and (5).

$$|PL_{i,1}| = |PL_{i,2}| = \ldots = |PL_{i,t_i-1}| = \varepsilon, 1 \le |PL_{i,t_i}| \le \varepsilon \tag{4}$$

$$PL_i = PL_{i,1} \cup PL_{i,2} \cup \ldots \cup PL_{i,t_i}, \tag{5}$$

where $t_i = \left\lceil \frac{|PL_i|}{\varepsilon} \right\rceil$, $|X|$ is the number of items in $X$.

**Definition 3.** $\varepsilon$-*partition dictionary.* For the keyword dictionary $W$, the $\varepsilon$-partition dictionary, denoted as $\mathcal{H}$, is the list of $\varepsilon$-partitions that are generated by partitioning the posting lists of $w_1$, $w_2$,... and $w_m$ according to Definition 2.

$$\mathcal{H} = PAR(PL_1) \cup PAR(PL_2) \cup \ldots \cup PAR(PL_m) \tag{6}$$

According to Definitions 1 and 2, a given keyword corresponds to a posting list and multiple $\varepsilon$-partitions. Definition 3 indicats that, for the dictionary $W = \{w_1, w_2, \ldots, w_m\}$, there are $\sum_{i=1}^{m} |PAR(PL_i)|$ generated $\varepsilon$-partitions in $\mathcal{H}$, i.e. $|\mathcal{H}| = \sum_{i=1}^{m} |PAR(PL_i)|$. We use $\mathcal{H}[i]$ to represent the $i$th $\varepsilon$-partition of $\mathcal{H}$.

**Definition 4.** *Partition Vector.* Given an $\varepsilon$-partition $PL_{i,j} \in \mathcal{H}$, the partition vector of $PL_{i,j}$ is denoted as $VP_{i,j}$ which is a $|\mathcal{H}|$-dimensional bit vector. The calculation of $VP_{i,j}$ is given in Eq. (7) where $v = 1, 2, \ldots, |\mathcal{H}|$.

$$VP_{i,j}[v] = \begin{cases} 1 & \mathcal{H}[v] = PL_{i,j} \\ 0 & \text{otherwise} \end{cases} \tag{7}$$

According to Definition 4, the dimensions of a partition vector equal to the total number of the posting partitions of all keywords in the dictionary and only one bit of a partition vector is 1 while others are all 0.

**Definition 5.** *Query Filter Vector.* Given a query $Q$ with multiple keywords, the query filter vector of $Q$ is denoted as $VF_Q$ which is also a $|\mathcal{H}|$-dimensional bit vector. The calculation of $VF_Q$ is given in Eq.(8) where $v = 1, 2, \ldots, |\mathcal{H}|$.

$$VF_Q[v] = \begin{cases} 1 & \exists PL_{i,j} (w_i \in Q \wedge PL_{i,j} \in PAR(PL_i) \wedge \mathcal{H}[v] = PL_{i,j}) \\ 0 & \text{otherwise} \end{cases} \tag{8}$$

**Definition 6.** *Keyword and Partition Vectors Pair (KPV-pair).* Given a keyword $w_i \in W$, the KPV-pair for $w_i$ is denoted as $< w_i, PVS(w_i) >$. Here, $PVS(w_i) = \{VP_{i,1}, VP_{i,2}, \ldots, VP_{i,t_i}\}$ is the set of vectors of the $\varepsilon$-partitions corresponding to $w_i$, where $t_i$ is the number of the $\varepsilon$-partitions. We denote $\Psi$ as the set of KPV-pairs for all keywords in $W$, then we have

$$\Psi = \{< w_i, PVS(w_i) > | w_i \in W\}. \tag{9}$$

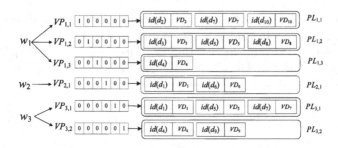

**Fig. 3.** An Example of $\varepsilon$-Posting Partition Vector Model

**Definition 7.** *Partition Vector and $\varepsilon$-Partition Pair (PVP-pair).* A PVP-pair is composed of a partition vector and its corresponding $\varepsilon$-partition. We denote $\Gamma$ as the set of PVP-pairs for all $\varepsilon$-Partitions in $\mathcal{H}$, then we have

$$\Gamma = \{< VP_{i,j}, PL_{i,j} > | PL_{i,j} \in \mathcal{H}\}. \tag{10}$$

We take an example to explain the above definitions. We assume $DS = \{d_i | i = 1, .., 10\}$ and $W = \{w_1, w_2, w_3\}$. According to Definition 1, $DS(w_1) = \{d_2, d_3, d_4, d_5, d_7, d_8, d_{10}\}$ and $PL_1 = \{P_{1,2}, P_{1,3}, P_{1,4}, P_{1,5}, P_{1,7}, P_{1,8}, P_{1,10}\}$. Then, $PL_1$ is randomly divided into three sublists by using partition parameter $\varepsilon = 3$, namely $PL_{1,1} = \{P_{1,2}, P_{1,7}, P_{1,10}\}$, $PL_{1,2} = \{P_{1,3}, P_{1,5}, P_{1,8}\}$ and $PL_{1,3} = \{P_{1,4}\}$. All $\varepsilon$-partitions generated by $PL_1$ are recorded as $PAR(PL_1) = \{PL_{1,1}, PL_{1,2}, PL_{1,3}\}$. Similarly, the posting lists of $w_2$ and $w_3$ are treated the same way to generate $\varepsilon$-partition dictionary $\mathcal{H} = \{PL_{1,1}, PL_{1,2}, PL_{1,3}, PL_{2,1}, PL_{3,1}, PL_{3,2}\}$. For each $\varepsilon$-partition $P_{i,j} \in \mathcal{H}$, the corresponding partition vector following Definition 4 is shown in Fig. 3. On the basis of partition vectors, KPV-pairs and PVP-pairs are generated according to Definitions 6 and 7 and they are also shown in Fig. 3. Assuming that the search keywords are $Q = \{w_1, w_3\}$, the corresponding query filter vector is $VF_Q = (1, 1, 1, 0, 1, 1)$ according to Definition 5.

## 5    Two-Layer Complete Binary Tree Index

The two-layer complete binary tree index (TCBT-index) is composed of an upper-layer complete binary tree and multiple lower-layer complete binary trees. According to [19], we know that an array is an appropriate structure to store a complete binary tree. Assuming that the array $\mathcal{A}$ represents a complete binary tree, then we have that: for the node $\mathcal{A}[i]$, if $i = 1$, $\mathcal{A}[i]$ is the root; if $i > 1$, the parent node of $\mathcal{A}[i]$ is $\mathcal{A}[\lfloor i/2 \rfloor]$; if the left and right child nodes of $\mathcal{A}[i]$ are $\mathcal{A}[2i]$ and $\mathcal{A}[2i + 1]$ respectively if they exist.

Adopting the array-based complete binary tree description, we give the definitions of the upper-layer and lower-layer complete binary trees of the TCBT-index.

**Definition 8.** *Upper-layer Complete Binary Tree (UCB-tree).* Assuming that the corresponding array of the UCB-tree is $\mathcal{U}[1, 2, ..., |\mathcal{H}|]$. A node of the

UCB-tree, $\mathcal{U}[i]$, represents an $\varepsilon$-partition of $\mathcal{H}$ and it is a three-element tuple: $\mathcal{U}[i] = <parVec, pruVec, \mathcal{L}>$, where $\mathcal{U}[i].parVec$ is the partition vector, $\mathcal{U}[i].pruVec$ is the pruning vector, and $\mathcal{U}[i].\mathcal{L}$ points to the lower-layer complete binary tree corresponding to the $\varepsilon$-partition.

**Definition 9.** *Lower-layer Complete Binary Tree (LCB-tree).* The lower layer of the TCBT-index consists of multiple LCB-trees. Each LCB-tree represents an $\varepsilon$-partition and corresponds to a node of UCB-tree. Assuming that a LCB-tree corresponds to the $\varepsilon$-partition $\mathcal{H}[i]$ and its corresponding array is $\mathcal{L}[1, 2, ..., |\mathcal{H}[i]|]$ where $|\mathcal{H}[i]|$ is the number of postings in $\mathcal{H}[i]$. A node of the LCB-tree, $\mathcal{L}[j]$, is also a three-element tuple: $\mathcal{L}[j] = <docVec, pruVec, docID>$, where $\mathcal{L}[j].docVec$ and $\mathcal{L}[j].pruVec$ are the document vector and pruning vector, and $\mathcal{L}[j].docID$ is the identity of a document.

According to Definitions 8 and 9, the LCB-trees are embedded in the UCB-tree and the latter tree can be treated as the entrance of the TCBT-index. Therefore, for simplicity, we use $\mathcal{U}$ to represents the TCBT-index.

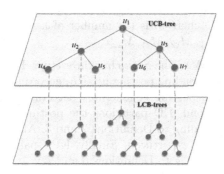

**Fig. 4.** An example of TCBT-index

Figure 4 shows an example of TCBT-index. There are an UCB-tree in the upper layer while several LCB-trees in the lower layer. Each node of the UCB-tree corresponds to a LCB-tree.

# 6   The Proposed Scheme

## 6.1   Framework

We first present the framework of the proposed scheme PMRS as shown in Fig. 5. It consists of six modules: *GenKey, Setup, BuildIndex, Encrypt, GenTrapdoor,* and *Search*.

- *Genkey*: DO generates a set of keys $K$ for encryption and share it with DU.
- *Setup*: DO preprocesses the original document set $DS$ to generate the KPV-pair set $\Psi$ and the PVP-pair set $\Gamma$.
- *BuildIndex*: DO constructs the plaintext TCBT-index $\mathcal{U}$.
- *Encrypt*: DO uses $K$ to encrypt $DS$ and $\mathcal{U}$ into $\widetilde{DS}$ and $\widetilde{\mathcal{U}}$, respectively.
- *GenTrapdoor*: DU generates the trapdoor for the search keywords.
- *Search*: CS performs the ranked search by using the trapdoor and $\widetilde{\mathcal{U}}$, and then returns the search result to DU.

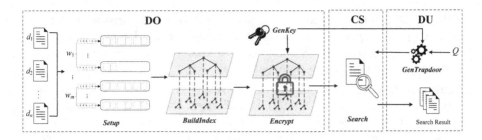

**Fig. 5.** Framework of PMRS

## 6.2   Data Preprocessing and Outsourcing

$K \leftarrow GenKey$ $(1^\lambda, u, m)$: DO generates a secret key set $K$, including a random $\lambda$-bit key $sk$, a random $|\mathcal{H}|$-bit vector $S_1$, a random $m$-bit vector $S_2$, two $|\mathcal{H}| \times |\mathcal{H}|$ invertible matrices $M_{11}$ and $M_{12}$, and two $m \times m$ invertible matrices $M_{21}$ and $M_{22}$, where $m$ is the number of keywords. Thus, we denote that $K = (sk, S_1, S_2, M_{11}, M_{12}, M_{21}, M_{22})$.

$(\Psi, \Gamma) \leftarrow Setup$ $(DS, W)$: The *Setup* has three steps. First, for each $w_i \in W$, DO constructs the corresponding posting list $PL_i$ according to Definition 1. The document vector in each posting is generated according to VSM and TF-IDF model (see Sect. 2). Second, DO partitions the posting lists and generates $\varepsilon$-partition dictionary according to Definitions 2 and 3. After that, DO computes partition vectors for every $\varepsilon$-partition. Finally, KPV-pair set $\Psi$ and PVP-pair set $\Gamma$ are generated according to Definitions 6 and 7.

$\mathcal{U} \leftarrow BuildIndex$ $(\Gamma)$: Based on the definition of the two-layer complete binary tree index, we give the TCBT-index construction algorithm as in Algorithm 1. The DO can generate the TCBT-index $\mathcal{U}$ according to the PVP-pair set $\Gamma$. In Algorithm 1, lines 5-13 are the process of constructing the LCB-tree, and the rest are used to describe the construction process of the UCB-tree. The time complexity of Algorithm 1 is $O(|\Gamma| \times \varepsilon)$.

$(\widetilde{DS}, \widetilde{\mathcal{U}}) \leftarrow Encrypt$ $(DS, \mathcal{U}, K)$: The document set $DS$ and TCBT-index $\mathcal{U}$ are respectively encrypted into $\widetilde{DS}$ and $\widetilde{\mathcal{U}}$. The encryption of $DS$ is to use the key $sk$ to symmetrically encrypt each document in $DS$. While the encryption of the $\mathcal{U}$ has two steps. We give the vector encryption algorithm $EncVec(V, S, M_1, M_2, flag)$ as in Algorithm 2 according to the references [10,11]. First, using $\{S_1, M_{11}, M_{12}\}$ in $K$ to encrypt each node $\mathcal{U}[i]$ in the UCB-tree to generate the encrypted form $\widetilde{\mathcal{U}}[i]$. That is to say, $EncVec$ $(\mathcal{U}[i].parVec, S_1, M_{11}, M_{12}, true)$ and $EncVec(\mathcal{U}[i].pruVec, S_1, M_{11}, M_{12}, true)$ are executed respectively. Then, for each node $\widetilde{\mathcal{U}}[i].\mathcal{L}[j]$ in the LCB-tree corresponding to node $\widetilde{\mathcal{U}}[i]$, $EncVec(\widetilde{\mathcal{U}}[i].\mathcal{L}[j].docVec, S_2, M_{21}, M_{22}, true)$ and $EncVec(\widetilde{\mathcal{U}}[i].\mathcal{L}[j].pruVec, S_2, M_{21}, M_{22}, true)$ are used to encrypt this node into $\widetilde{\mathcal{U}}[i].\widetilde{\mathcal{L}}[j]$.

---

**Algorithm 1.** $BuildIndex(\Gamma)$

---

1   $i = 1$;
2   Create an array $\mathcal{U}[1, 2, ..., |\Gamma|]$ to store UCB-tree nodes;
3   **for** *each $< v, p > \in \Gamma$ where $p$ is an $\varepsilon$-partition and $v$ is its partition vector* **do**
4      $\mathcal{U}[i].parVec = \mathcal{U}[i].pruVec = v$;
5      $j = 1$;
6      Create an array $\mathcal{U}[i].\mathcal{L}[1, 2, ..., |p|]$ to store LCB-tree nodes;
7      **for** *each $< id(d_x), VD_x > \in p$* **do**
8         $\mathcal{U}[i].\mathcal{L}[j].docVec = \mathcal{U}[i].\mathcal{L}[j].pruVec = VD_x$;
9         $j + +$;
10      **end**
11      **for** $j = \lceil (|p| - 1)/2 \rceil$ *to* 1 **do**
12         $\mathcal{U}[i].\mathcal{L}[j].pruVec = max\{\mathcal{U}[i].\mathcal{L}[j].docVec, \mathcal{U}[i].\mathcal{L}[2j].pruVec, \mathcal{U}[i].\mathcal{L}[2j + 1].pruVec\}$;
13      **end**
14      $i + +$;
15   **end**
16   **for** $i = \lceil (|\Gamma| - 1)/2 \rceil$ *to* 1 **do**
17      $\mathcal{U}[i].pruVec = max\{\mathcal{U}[i].parVec, \mathcal{U}[2i].pruVec, \mathcal{U}[2i + 1].pruVec\}$;
18   **end**
19   **return** $\mathcal{U}$

---

**Algorithm 2.** $EncVec(V, S, M_1, M_2, flag)$

---

1   Split $V$ into two random vectors $V'$ and $V''$ according to the vector $S$;
2   **if** *flag=true* **then**
3      $\begin{cases} V'[j] = V''[j] = V[j] & S[j] = 0 \\ V'[j] + V''[j] = V[j] & S[j] = 1 \end{cases}$;
4      **return** $\tilde{V} = \left( M_1^T V', M_2^T V'' \right)$
5   **end**
6   **else if** *flag=false* **then**
7      $\begin{cases} V'[j] + V''[j] = V[j] & S[j] = 0 \\ V'[j] = V''[j] = V[j] & S[j] = 1 \end{cases}$;
8      **return** $\tilde{V} = \left( M_1^{-1} V', M_2^{-1} V'' \right)$
9   **end**

---

### 6.3   Multi-keyword Ranked Search

$TD \leftarrow GenTrapdoor\ (K, Q, k, \Psi)$: The process of DU generating $TD = (\widetilde{VF_Q}, \widetilde{VQ}, k)$ based on search keywords $Q$ is as follows:

(1) The generation of $\widetilde{VF_Q}$. According to Definition 5, the query filter vector $VF_Q$ of $Q$ can be constructed by the KPV-pairs set $\Psi$. Then, the $VF_Q$ is reconstructed according to the follow formula, where $rand()$ is a random number generator. Finally, use the key $\{S_1, M_{11}, M_{12}\}$ in $K$ to generate the $VF's$ encrypted form $\widetilde{VF}$, ie $\widetilde{VF} = EncVec(VF, S_1, M_{11}, M_{12}, false)$.

$$VF_Q[i] = \begin{cases} rand() & VF_Q[i] = 1 \\ 0 & VF_Q[i] = 0 \end{cases} \tag{11}$$

(2) The generation of $\widetilde{VQ}$. The search vector $VQ$ is constructed according to $Q$. For any $w_i \in W$, if $w_i \in Q$, the IDF value of $w_i$ is stored in $VQ[i]$, otherwise the value of $VQ[i]$ is 0. Then use the key $\{S_2, M_{21}, M_{22}\}$ in $K$ to generate the $VQ's$ encrypted form $\widetilde{VQ}$, ie $\widetilde{VQ} = EncVec(VQ, S_2, M_{21}, M_{22}, false)$.

$RS \leftarrow Search\ (\widetilde{\mathcal{U}}, TD)$: CS receives the trapdoor $TD$, it performs the ranked search on the basis of the encrypted index $\widetilde{\mathcal{U}}$, and then returns the search results $RS$. The search process consists of three algorithms. Algorithm 3 is the entry to the search. Algorithm 4 and 5 are two recursive algorithms. In the UCB-tree, Algorithm 4 uses the query filter vector $TD.VF_Q$, the pruning vector $\widetilde{\mathcal{U}}[i].pruVec$, and the partition vector $\widetilde{\mathcal{U}}[i].parVec$ to determine whether the partition $\widetilde{\mathcal{U}}[i].\widetilde{\mathcal{L}}$ corresponding to the root of the current subtree $\widetilde{\mathcal{U}}[i]$ is related to the search keywords $Q$. If the current partition $\widetilde{\mathcal{U}}[i].\widetilde{\mathcal{L}}$ is related to the search keywords $Q$, Algorithm 5 is invoked. And the currently most relevant $k$ documents are found based on a "dynamic depth-first" algorithm using the search vector $TD.\widetilde{VQ}$, the pruning vector $\widetilde{\mathcal{U}}[i].\widetilde{\mathcal{L}}[j].pruVec$ and the document vector $\widetilde{\mathcal{U}}[i].\widetilde{\mathcal{L}}[j].docVec$ in the LCB-tree.

---

**Algorithm 3.** $Search(\widetilde{\mathcal{U}}, TD)$

---

1 Initial the query result $RS = null$;
2 $UpperSearch(\widetilde{\mathcal{U}}, 1, TD, RS)$;
3 **return** $RS$;

---

**Algorithm 4.** $UpperSearch(\widetilde{\mathcal{U}}, i, TD, RS)$

---

1 **if** $Score(\widetilde{\mathcal{U}}[i].pruVec, TD.\widetilde{VF_Q}) \neq 0$ **then**
2   **if** $Score(\widetilde{\mathcal{U}}[i].parVec, TD.\widetilde{VF_Q}) \neq 0$ **then**
3     | $LowerSearch(\widetilde{\mathcal{U}}[i].\widetilde{\mathcal{L}}, 1, TD, RS)$;
4   **end**
5   **if** $\widetilde{\mathcal{U}}[2i] \neq null$ **then**
6     | $UpperSearch(\widetilde{\mathcal{U}}, 2i, TD, RS)$;
7   **end**
8   **if** $\widetilde{\mathcal{U}}[2i + 1] \neq null$ **then**
9     | $UpperSearch(\widetilde{\mathcal{U}}, 2i + 1, TD, RS)$;
10   **end**
11 **end**

# 7    Security Analysis

In the PMRS scheme, we record history as $H = (\mathcal{U}, DS, Q)$, that is, the interaction between DU and CS only involves the index $\mathcal{U}$, the document set $DS$ and the search keywords $Q = \{w_1, w_2, ..., w_q\}$. The view is recorded as $V(H) = (\widetilde{\mathcal{U}}, \widetilde{DS}, TD)$, that is, CS only get the encrypted index $\widetilde{\mathcal{U}}$, the encrypted document set $\widetilde{DS}$ and trapdoor $TD$. The trace is recorded as $T(H) = \{T(w_1), T(w_2), ..., T(w_q)\}$, where $T(w_i) = \{(d_j, score_{i,j}), 1 \leq j \leq |DS|\}$, and $score_{i,j}$ is the relevance score of keywords $w_i$ and document $d_j$. Thus, the sensitive data that CS can obtain include the relevance score between $w_i$ and $d_j$. Specific security analysis are as follows:

**Theorem 1.** *This scheme is secure in the case of known ciphertext attack.*

*Proof.* Under the known ciphertext attack, $V(H)$ and $V(H')$ can be generated respectively according to two sets of history with the same trace. If CS cannot distinguish the two views, it can be proved that CS cannot obtain information other than the search result and the search mode, and the scheme is safe under the known ciphertext attack. Suppose the simulator can generate an $H'$ that CS cannot distinguish $V(H)$ from $V(H')$. The simulator needs to construct $\mathcal{U}'$, $DS'$, $Q'$ and the key $K' = (sk', S_1', S_2', M_{11}', M_{12}', M_{21}', M_{22}')$. Then $\widetilde{\mathcal{U}}'$, $\widetilde{DS}'$, and $TD'$ need to be constructed on this basis. Even if the constructed $V(H')$ has the same trace as the $V(H)$, the CS cannot distinguish between $\widetilde{DS}$ and $\widetilde{DS}'$ without the key $sk$. Besides, $\widetilde{\mathcal{U}}'$ and $TD'$ are encrypted by the secure KNN algorithm. The CS cannot recover the plaintext form of $\widetilde{\mathcal{U}}'$ and $TD'$ without the key $K$. In summary, CS cannot distinguish between $V(H)$ and $V(H')$, so the scheme is safe under the known ciphertext attack.

**Theorem 2.** *This scheme is secure in the case of known partial plaintext attack.*

*Proof.* Under the known partial plaintext attack, the trace that CS can obtain includes the plaintext form of partition vectors and document vectors in addition to the relevance score of the documents. Suppose that the number of plaintext partition vectors is $a$. The partition vector $VP_{i,j}$ is randomly divided into $VP'_{i,j}$ and $VP''_{i,j}$. Then according to the encryption method, two linear equations as shown in the following formula are obtained:

$$\begin{cases} VP'_{ij} \cdot M_{11}^{-1} = \widetilde{VP}'_{i,j} \\ VP''_{i,j} \cdot M_{12}^{-1} = \widetilde{VP}''_{i,j} \end{cases} \tag{12}$$

In this system of equations, $M_{11}$ and $M_{12}$ each contain $m^2$ unknown numbers, so there are $2m^2$ unknown numbers need to be solved in total. However, there are only $2|\mathcal{H}|$ equations in the equation system. Because the number of equations is far less than the number of unknown numbers, it is impossible to find the true values of $M_{11}$ and $M_{12}$. Similarly, it is difficult to obtain $M_{21}$ and $M_{22}$. Therefore, the scheme is safe under the known partial plaintext attack.

---

**Algorithm 5.** $LowerSearch(\widetilde{\mathcal{L}}, j, TD, RS)$

---

1  **if** $|RS| < TD.k$ **then**
2  │    Add the node $\widetilde{\mathcal{L}}[j]$ into $RS$;
3  **end**
4  **else**
5  │    **if** $Score(\widetilde{\mathcal{L}}[j].pruVec, TD.\widetilde{VQ}) > minScore(RS)$ **then**
6  │    │    // $minScore(RS)$ is the minimum value of the relevance score in the
   │    │    current result $RS$;
7  │    │    **if** $Score(\widetilde{\mathcal{L}}[j].docVec, TD.\widetilde{VQ}) > minScore(RS)$ **then**
8  │    │    │    Delete the node from $RS$, whose relevance score is the lowest;
9  │    │    │    Add the node $\widetilde{\mathcal{L}}[j]$ into $RS$;
10 │    │    **end**
11 │    │    **if** $\widetilde{\mathcal{L}}[2j] \neq null$ **then**
12 │    │    │    $LowerSearch(\widetilde{\mathcal{L}}, 2j, TD, RS)$
13 │    │    **end**
14 │    │    **if** $\widetilde{\mathcal{L}}[2j + 1] \neq null$ **then**
15 │    │    │    $LowerSearch(\widetilde{\mathcal{L}}, 2j + 1, TD, RS)$
16 │    │    **end**
17 │    **end**
18 **end**

---

**Theorem 3.** *This scheme has trapdoor unlinkability.*

*Proof.* In this scheme, the trapdoor $TD=(\widetilde{VF_Q}, \widetilde{VQ}, k)$. Each bit of $VF_Q$ is a randomly generated value, even if the value generated for the same search request is different, so it is unique. And $\widetilde{VF_Q}$ is the encrypted form of $VF_Q$ encrypted by matrix $M_{11}$ and $M_{12}$ and vector $S_1$. $VQ$ is also encrypted by matrix $M_{21}$ and $M_{22}$ and vector $S_2$, where $S_2$ can randomly divide $VQ$ into $VQ'$ and $VQ''$, and the divide results are different each time. So $\widetilde{VQ}$ is also unique, even if it's constructed from the same search keywords $Q$. Thus, the confidentiality of the TD can be well protected. However, the CS can track the relevance scores corresponding to the partition vector and the document vector to analyze whether the search keywords are the same. For this problem, several random values can be added to the $VF$ and $VQ$ to ensure that the relevance score is different, thereby ensuring trapdoor unlinkability.

## 8  Performance Evaluation

In this section, we implement the PMRS scheme proposed in this paper and compare the search time cost with DMSR scheme [10]. The test data set used in the experiment is New York Times Dataset [20]. The experimental hardware environment is 2.5 GHz 2 core CPU, 8G memory, 256G solid state hard disk; the software environment is 64 bit Windows 10 operating system and some Java development tools. Default parameters are $n = 50,000$, $|Q| = 5$, $k = 15$, and $\varepsilon = 500$ which are the number of documents, search keywords, search documents and partition parameter.

In the following experiments, we evaluate the time cost of searches where one of the parameters $n, k$, and $|Q|$ changes and the other parameters adopt the default values. The results are shown in Figs. 6, 7 and 8.

Figures 6, 7 and 8 all show that the proposed PMRS outperforms DMSR in the time cost of ranked searches, and the former saves at least 65.2% of the time cost compared with the latter. The reason is that both PMRS and DMSR are based on the tree structure, and the more documents there are, the more nodes there are in the tree. As the number of documents increases, so does the number of extracted keywords. Then, the dimension of document vector and search vector will also increase, which will consume more time to calculate relevance score. In PMRS, through the calculation of the query filter vector and pruning vector in the UCB-tree, a large number of documents unrelated to the search keyword can be quickly filtered out, and the "dynamic depth-first" algorithm is used in the LCB-tree to continuously prune to obtain search results. While DMSR only performs the pruning operation in the tree composed of the entire document set, which is not as efficient as PMRS. In addition, as the number of search keywords increases, the relevance score between the pruning vector and the search vector is higher, and the $minScore(RS)$'s pruning effect is less obvious. And as the number of search documents increases, the value of $minScore(RS)$ decreases, and the pruning effect is also not obvious. Thus, the efficiency of performing pruning on the entire tree in DMSR is less than that of PMRS, which first filter the candidate documents and then performs a "dynamic depth-first" search algorithm for them.

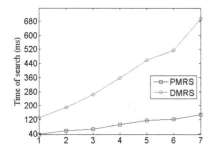

**Fig. 6.** The impact of $n$ ($\times 10^4$)

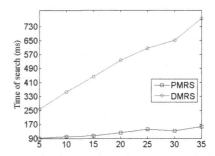

**Fig. 7.** The impact of $k$

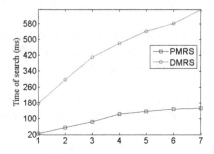

**Fig. 8.** The impact of $|Q|$

## 9  Conclusion

Under the "honest but curious" threat model, this paper proposes a privacy-preserving multi-keyword ranked search over encrypted cloud data scheme based on two-layer complete binary tree index. In this scheme, the two-layer complete binary tree index is designed, which can be used for efficient search. On the basis of such index, a "dynamic depth-first" search algorithms are proposed. Security analysis and experiment evaluation show that the PMRS scheme is privacy-preserving and efficient.

## References

1. Linthicum, D.S.: Approaching cloud computing performance. IEEE Cloud Comput. **5**(2), 33–36 (2018)
2. Bösch, C., Hartel, P.H., Jonker, W., Peter, A.: A survey of provably secure searchable encryption. ACM Comput. Surv. **47**(2), 18:1–18:51 (2014)
3. Song, D.X., Wagner, D.A., Perrig, A.: Practical techniques for searches on encrypted data. In: 2000 IEEE Symposium on Security and Privacy, Berkeley, California, USA, 14–17 May 2000, pp. 44–55 (2000)
4. Goh, E.: Secure indexes. IACR Cryptology ePrint Archive 2003, 216 (2003)
5. Curtmola, R., Garay, J.A., Kamara, S., Ostrovsky, R.: Searchable symmetric encryption: improved definitions and efficient constructions. J. Comput. Secur. **19**(5), 895–934 (2011)
6. Cao, N., Wang, C., Li, M., Ren, K., Lou, W.: Privacy-preserving multi-keyword ranked search over encrypted cloud data. In: 30th IEEE International Conference on Computer Communications, Joint Conference of the IEEE Computer and Communications Societies, INFOCOM 2011, 10–15 April 2011, pp. 829–837, Shanghai (2011)
7. Wong, W.K., Cheung, D.W., Kao, B., Mamoulis, N.: Secure KNN computation on encrypted databases. In: Proceedings of the ACM SIGMOD International Conference on Management of Data, SIGMOD 2009, Providence, Rhode Island, USA, 29 June–2 July 2009, pp. 139–152 (2009)
8. Cheruku, H., Subhashini, P.: Ranked search over encrypted cloud data in Azure using secure K-NN. In: Aggarwal, V.B., Bhatnagar, V., Mishra, D.K. (eds.) Big Data Analytics. AISC, vol. 654, pp. 341–350. Springer, Singapore (2018). https://doi.org/10.1007/978-981-10-6620-7_33

9. Cao, N., Wang, C., Li, M., Ren, K., Lou, W.: Privacy-preserving multi-keyword ranked search over encrypted cloud data. IEEE Trans. Parallel Distrib. Syst. **25**(1), 222–233 (2014)

10. Xia, Z., Wang, X., Sun, X., Wang, Q.: A secure and dynamic multi-keyword ranked search scheme over encrypted cloud data. IEEE Trans. Parallel Distrib. Syst. **27**(2), 340–352 (2016)

11. Chen, C., et al.: An efficient privacy-preserving ranked keyword search method. IEEE Trans. Parallel Distrib. Syst. **27**(4), 951–963 (2016)

12. Zhu, X., Liu, Q., Wang, G.: A novel verifiable and dynamic fuzzy keyword search scheme over encrypted data in cloud computing. In: 2016 IEEE Trustcom/BigDataSE/ISPA, Tianjin, China, 23–26 August 2016, pp. 845–851 (2016)

13. Deepa, N., Vijayakumar, P., Rawal, B.S., Balamurugan, B.: An extensive review and possible attack on the privacy preserving ranked multi-keyword search for multiple data owners in cloud computing. In: 2017 IEEE International Conference on Smart Cloud, SmartCloud 2017, New York City, NY, USA, 3–5 November 2017, pp. 149–154 (2017)

14. Guo, Z., Zhang, H., Sun, C., Wen, Q., Li, W.: Secure multi-keyword ranked search over encrypted cloud data for multiple data owners. J. Syst. Softw. **137**, 380–395 (2018)

15. Yao, X., Lin, Y., Liu, Q., Zhang, J.: Privacy-preserving search over encrypted personal health record in multi-source cloud. IEEE Access **6**, 3809–3823 (2018)

16. Fu, Z., Ren, K., Shu, J., Sun, X., Huang, F.: Enabling personalized search over encrypted outsourced data with efficiency improvement. IEEE Trans. Parallel Distrib. Syst. **27**(9), 2546–2559 (2016)

17. Sun, W., et al.: Verifiable privacy-preserving multi-keyword text search in the cloud supporting similarity-based ranking. IEEE Trans. Parallel Distrib. Syst. **25**(11), 3025–3035 (2014)

18. Wan, Z., Deng, R.H.: VPSearch: achieving verifiability for privacy-preserving multi-keyword search over encrypted cloud data. IEEE Trans. Dependable Secure Comput. **15**(6), 1083–1095 (2018)

19. Bulut, M.: ReducedCBT and superCBT, two new and improved complete binary tree structures, CoRR, vol. abs/1401.7741 (2014)

20. B.D.: New York times dataset [db/ol] (2018). http://developer.nytimes.com/docs

# Privacy-Preserving Fine-Grained Outsourcing PHR with Efficient Policy Updating

Zuobin Ying[1,2(✉)], Wenjie Jiang[1], Ximeng Liu[3,4], and Maode Ma[2]

[1] School of Computer Science and Technology, Anhui University, Hefei 230601, China
yingzb@ahu.edu.cn, wenjie941105@gmail.com
[2] School of Electrical and Electronic Engineering, Nanyang Technological University,
Singapore 639798, Singapore
EMDMa@ntu.edu.sg
[3] College of Mathematics and Computer Science, Fuzhou University,
Fuzhou 350108, China
snbnix@gmail.com
[4] Key Lab of Information Security of Network System (Fuzhou University),
Fuzhou 350108, Fujian, China

**Abstract.** Personal Health Record (PHR) is a novel way of managing individual health. With the help of cloud computing and Ciphertext-Policy Attribute-Based Encryption (CP-ABE), the fine-grained access control, as well as authenticated sharing, can be realized. However, the importance of policy preserving has emerged to be a major security risk in the PHR cloud sharing scenario. Besides, cloud-assist policy updating, as a practical functionality of the PHR system, should also be taken into consideration. In this paper, we present an integrated policy preserving PHR scheme with efficient policy updating. The scheme is proved to be selectively secure under $q$-BDHE assumption. The evaluation result indicates that our scheme can reach a balanced trade-off in terms of privacy and efficiency.

**Keywords:** Personal Health Record · Policy preserving · Policy updating · Attribute-Based Encryption

## 1 Introduction

Noncommunicable diseases (NCDs), including cardiovascular diseases, cancer, diabetes, chronic respiratory diseases, and mental disorders, have become the top health hazards worldwide. Collectively, cancer, diabetes, lung and heart diseases kill 41 million people annually, accounting for 71% of all deaths globally according to the statistics from the world health organization (WHO). NCDs tend to be of long duration and are the result of a combination of genetic, physiological, environmental and behavioral factors. Tobacco use, physical inactivity,

Supported by Anhui University.

S. Wen et al. (Eds.): ICA3PP 2019, LNCS 11945, pp. 512–520, 2020.
https://doi.org/10.1007/978-3-030-38961-1_44

the harmful use of alcohol and unhealthy diets all increase the risk of dying from an NCD. Detection, screening, and treatment of NCDs, as well as palliative care, are key components of the response to NCDs [1]. Although the advanced medical treatment can provide with more precise therapies. Whereas, NCDs require long-term treatment as well as daily self-monitoring and detailed life-style recording. Therefore, a patient-centric record management pattern became widely prevalent, namely, Personal Health Record (PHR). PHR is a derivation of Electronic Medical Record (EMR), which integrates the patient-generated health data (PGHD). Through applying of PHR, patients could update their daily health conditions and activities online after discharge, the medical institutions could also keep abreast of the patient's current situation and give appropriate advice remotely. Meanwhile, the development of cloud computing makes sharing more convenient. PHR owners could share their documents with other patients who have the same symptoms. Nowadays, more and more commercial medical service providers are committed to providing users with more convenient cloud outsourcing personal health record solutions.

However, since the PHR contains more health-related privacy information than any other documents. Thus, the PHR is more preferred to malicious users. For instance, in July 2018, Singapore Health Group (SingHealth) health data was hacked and 1.5 million people's personal information was illegally obtained. Obviously, without any effective protection, the outsourced PHR will suffer from various kinds of security threats in the cloud. In the past few years, Ciphertext-Policy Attribute-Based Encryption (CP-ABE) has been widely studied and adopted in the cloud outsourcing scenario [2–10], CP-ABE can realize one-to-many fine-grained access control by formulating the proper access policies. Users have the corresponding attributes are able to get access to the ciphertext. For example, if an outsourced file is encrypted by using the policy formed as (("School of Computer Science" AND "Graduate Student") OR "Supervisor"). It means that all of the graduate students from the school of computer science or the supervisors can correctly decrypt the file.

First of all, the policy in traditional CP-ABE schemes is in plaintext form. Imagine a scenario of new employees' physical examination. The results will be recorded in the PHR, encrypted by using CP-ABE and then uploaded to the cloud server. Assume that there exists a policy expressed like (("Cardiology" AND "Age < 25" AND "Male") OR "ICU"), then it can be inferred that this young man may have serious heart disease. Combined with some background knowledge, the specific individual can be identified easily. As we can see, privacy remains leaked while the ciphertext is protected well. Secondly, for ease of use, PHR owners prefer to store their medical records in the cloud, and it became cumbersome when the data owner wants to update the policy. The simplest way is to retrieve and decrypt the PHR ciphertext, re-encrypt it by using the new policy, and then outsource it to the cloud again. However, this approach will result in a significant computation and communication cost to the resource constraint end side. Thus, updating ciphertext in the cloud turns into an essential function of a mature cloud assistant PHR system. Motivate by solving these two problems in one integrated scheme. We present the Privacy-Preserving Policy Updating (3PU) scheme. Our main contribution can be summarized as follows:

(1) We are the first to consider solving the problem of policy updating under the premise of policy hidden. It is noteworthy that we are not just simply add these two functions together, we try to make a dynamic integration so as to realize usability in the outsourcing PHR environment.
(2) To enhance the security of previously proposed ACF scheme. We design a lightweight encryption mechanism, thus, only the authenticated users can recover the corresponding attributes.
(3) The scheme is constructed on the prime order group, we proved the security of our scheme, and the experiment result shows that our scheme has an equilibrium in both security and efficiency.

## 2    System Model

Our proposed scheme has five entities as shown in Fig. 1. We briefly describe the function of each entity.

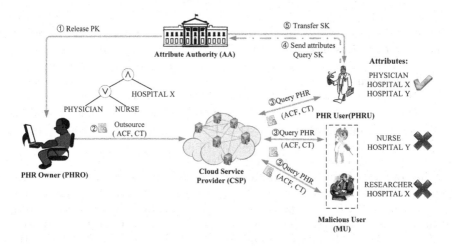

**Fig. 1.** Proposed 3PU system model and the different entities

(1) Attribute Authority (AA) is responsible to distribute attributes to the authenticated users. Besides, AA generates the public parameter and the master secret key. It also issues private keys to the PHR users according to their attributes.
(2) Cloud Service Provider CSP provides with resource infinit computing, mass storage and other services to the users. CSP will follow the instructions issued by all the users (include the malicious users).

(3) PHR Owner (PHRO) create their own health records, formulate the access policy to decide who has the right to access to their files.

(4) PHR User (PHRU) could be patients, clinicians, assurance company employees, health-care institutes, etc. PHRU can obtain the encrypted PHR from CSP. They can proceed decryption only when their own attributes could satisfy the access policy formulate by the PHRO.

(5) Malicious User (MU) in our proposed scheme have the polynomial time capacity. They always want to mine privacy from the PHR outsourcing to the cloud.

# 3 Privacy Preserving Policy Updating Scheme

Recently, we have proposed a new algorithm called ACF to protect the pivacy of the policy [8]. We improve the security of the attributes stored in the ACF in the ACF Create sub-routine, then in the ACF Updating algorithm, we will detail the privacy policy updating.

(1) Setup AA initials the Setup algorithm. Let $\mathbb{G}$ and $\mathbb{G}_T$ be two multiplicative cyclic groups of prime order $p$. $e : \mathbb{G} \times \mathbb{G} \rightarrow \mathbb{G}_T$ is a bilinear map. AA selects a generator $g \in \mathbb{G}$ and $N$ random group elements $h_1, \ldots, h_N \in \mathbb{G}$ associated with the $N$ attributes in the system. Select $\alpha, a \in \mathbb{Z}_p$ at random. Denote $L_{att}$ and $L_{rnum}$ to be the maximum bit length of the attributes as well as the maximum bit length of row numbers of the $LSSS$ matrix respectively. Let $H_f$ be the collision-resistant hash functions of generating fingerprint of an element. Let $H_e$ be the collision-resistant hash function which maps an element to an entry in the ACF buckets.

The public parameter is formed as:

$$\mathsf{pk} = \langle\; g, e(g,g)^{\alpha}, g^a, h_1, \ldots, h_N, L_{att}, L_{rnum}, H_f, H_e \;\rangle.$$

The master secret key is set to be $\mathsf{msk} = g^{\alpha}$.

(2) KeyGen To get access to the encrypted PHR in the cloud, users queries secret key from AA. AA distributes the attributes $\mathcal{S}$ according to the user's characteristic. AA also generates the corresponding secret key for the users. It takes the input as $\mathsf{pk}$, $\mathsf{msk}$ and $\mathcal{S}$, randomly chooses $t, m \in \mathbb{Z}_p$, then computes

$$E = g^{\alpha}g^{at}, I = g^t, K_1 = g^{\alpha}g^{am}, K_2 = g^{-m}, \{E_x = h_x^t\}_{x \in \mathcal{S}}.$$

Then the secret key is set to be:

$$\mathsf{sk} = \langle\; E, I, K_1, K_2, \{E_x\}_{x \in \mathcal{S}}, \mathcal{S} \;\rangle.$$

It should be noted that $K_1, K_2$ are the secret key component designed for the micro encryption algorithm to get the attributes stored in the ACF.

(3) **Encryption** This algorithm contains two subroutines, namely, **Enc** and **ACF Create**. At first, **Enc** takes as input the public parameter pk, the plaintext M and the access matrix $(\mathbb{M}, \rho)$. M is an $l \times n$ matrix by using function $\rho$ to map attributes to rows of M. This subroutine randomly chooses a vector $v = (s, y_2, \ldots, y_n) \in \mathbb{Z}_p^n$, where $y_2, \ldots, y_n$ are used to share the encryption secret $s$, a $d \in \mathbb{Z}_p$. For $i = 1$ to $l$, it calculates $\lambda_i = \mathbb{M}_i \cdot v$, in which $\mathbb{M}_i$ is the vector related to the $i$th row of M. In addition, the algorithm chooses random $r_1, \ldots, r_l \in \mathbb{Z}_p$. Then it outputs the ciphertext:

$$CT = \langle C = Me(g,g)^{\alpha s}, \mathsf{ACF}, C' = g^s, \overline{C}_1 = g^d,$$
$$\overline{C}_2 = g^{ad}, \{C_i = g^{a\lambda_i} h_{\rho(i)}^{-r_i}, D_i = g^{r_i}\}_{i=1,\ldots,l}\rangle. \tag{1}$$

Here, $\overline{C}_1, \overline{C}_2$ are also for the protection of attributes in the ACF.

The next subroutine is to embed the access policy into the ACF to realize policy hidden. In our previous work [8], we have proposed a prototype of ACF on the basis of the Cuckoo Filter. Here, we briefly introduce the main idea of the original ACF and then detailed the improvement we have made in this scheme.

**ACF Create** subroutine takes as input the access policy $(\mathbb{M}, \rho)$. It concatenate the attributes with the relevant row number in the access matrix M and generate a set of elements $\mathbb{S} = \{(i\|att_x)\}_{i\in[1,l]}$, in which the $i$-th row of the access matrix maps to the attribute $att_x = \rho(i)$. Afterwards, the algorithm create the ACF by taking the $\mathbb{S}$ as a input. When we have to insert a new element $x$ in the set $\mathbb{S}$ to the ACF, the algorithm first calculate the fingerprint $H_f(x)$ of $x$, which we denote as $F$. In the original ACF, the tuple stored in the ACF is formed as $\langle F, F \oplus x \rangle$. Although it can achieve the policy hidden function, however, the malicious user can launch a brute force violence crack to get the value of $x$. Therefore, we made some modifications, the new element stored in the ACF is formed as $\langle F, x \cdot e(g,g)^{ad} \rangle$. Later in the decryption phase, only the user who matches the fingerprint can get the valid value by initial the micro encryption algorithm.

(4) **Decryption** When the user wants to get access to the encrypted PHR in the cloud, the access control is triggered to check if the user has the proper attributes. Since the access policy is hidden in the ACF, an **ACF Check** subroutine will be initialed to find out which attributes of the user is in the access matrix.

**ACF Check** takes pk, the attribute set $\mathcal{S}$ of the user and $ACF$ as the input. For each attribute $att$ in the set $\mathcal{S}$, the algorithm first computes $H_e$ and $H_f$ to locate the candidate buckets. If either bucket have the fingerprint, then the algorithm obtains the element $x$ in the ACF as follows:

$$\theta = e(\overline{C}_1, K_1) \cdot e(\overline{C}_2, K_2) = e(g^d, g^\alpha g^{am}) \cdot e(g^{ad}, g^{-m}) = e(g,g)^{\alpha d}$$

The element $x$ can then be recovered by using $x = C_x \oplus \theta$, where $C_x = x \oplus e(g,g)^{\alpha d} \in \mathsf{ACF}$. Element $x$ is formed as $x = \{i\|att_x\}$. The algorithm

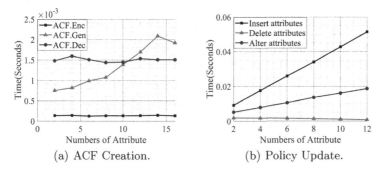

**Fig. 2.** Performance evaluation of policy hidden & policy updating

will automatically remove the zero bits to the left of the string $L_{att}$ to get the attribute $att_x$, then the same operation to obtain the row number $i$ from $L_{rnum}$. Otherwise, the attribute $att$ does not exit in the access policy if $att$ is not the same as $att_x$, Finally, the new attribute mapping function $\rho'$ will be reconstructed as: $\rho' = \{rnum, att\}_{att \in S}$. Then the row number in the matrix $\mathbb{M}$ will be determined. When the access policy $(\mathbb{M}, \rho')$ is obtained. The final decryption algorithm can proceed just the same as in the original CP-ABE scheme.

(5) **Update** The updating process consists of two parts, **UPKeyGen** and **CTUpdate**. The **UPKeyGen** is executed on local side, while **CTUpdate** is executed on the cloud.

Note that at the end of the **Encryption** procedure, we store the $\mathbb{E}_M$ as:

$$\mathbb{E}_M = ((\mathbb{M}, \rho), \boldsymbol{v} = (s, y_2, \ldots, y_n), e(g, g)^{\alpha d}).\qquad(2)$$

These components will be used to sustain the creation of new ciphertext and **ACF**.

**UPKeyGen** After generating the new access structure $(\mathbb{M}', \rho')$, the algorithm will compare $(\mathbb{M}, \rho)$ with $(\mathbb{M}', \rho')$ and generate the update access structure $(\widehat{\mathbb{M}}, \widehat{\rho})$.

For convenience of description. We divide the attributes that need to be updated into three sets $\mathbb{S}_{\text{insert}}$, $\mathbb{S}_{\text{delete}}$ and $\mathbb{S}_{\text{alter}}$. If the number of columns in $\widehat{\mathbb{M}}$ is more than the dimension of $\boldsymbol{v}$, the dimension of $\boldsymbol{v}$ need to be raised. For example, if the number of columns is $n + m$ and the $\boldsymbol{v} = (s, y_2, \ldots, y_n)$, then we set $\boldsymbol{w} = (s, y_2, \ldots, y_n, y_{n+1}, \ldots, y_{n+m})$ where $y_{n+1}, \ldots, y_{n+m} \in \mathbb{Z}_p$. In the following description, we assume that the number of columns in $\widehat{\mathbb{M}}$ is more than the dimension of $\boldsymbol{v}$. For $att \in \mathbb{S}_{\text{insert}}$, it calculates $\widehat{\lambda}_i = \boldsymbol{w} \cdot \widehat{\mathbb{M}}_i$, where $\widehat{\rho}(i) = att$, and $\widehat{\mathbb{M}}_i$ is the vector corresponding to the $i$th row of $\widehat{\mathbb{M}}$. Suppose there are $\mathcal{L}$ attributes in $\mathbb{S}_{\text{insert}}$. The algorithm chooses random $r'_1, \ldots, r'_{\mathcal{L}} \in \mathbb{Z}_p$ and computes $\mathsf{UK}_{\text{value}} =$

$$\langle\langle rnum, \widehat{C}_i = g^{a\widehat{\lambda}_i} h_{\widehat{\rho}(i)}^{-r'_j}, \widehat{D}_i = g^{r'_j}\rangle, \langle f, \mathsf{ACF}_f = (i\|att) \oplus e(g, g)^{\alpha d}\rangle\rangle\qquad(3)$$

where $rnum = i, f = H_f(att)$, $j \in 1, 2, \ldots, \mathcal{L}$. Then, the algorithm sets the $UK_{type} = $ "insert", and lets $UK_i = \langle UK_{type}, UK_{value} \rangle$.

For $att \in \mathbb{S}_{delete}$, the $UK_{value} = \langle rnum, f \rangle$, where $rnum \in \widehat{\rho}, f = H_f(att)$, and the $UK_{type} = $ "delete". The $UK_i$ is defined as $\langle UK_{type}, UK_{value} \rangle$.

For $att \in \mathbb{S}_{alter}$, computes

$$UK_i = \langle UK_{type} = \text{"alter"}, UK_{value} = \langle rnum, \widetilde{C}_i = g^{a\widetilde{\lambda}_i} \rangle \rangle \qquad (4)$$

where $\widetilde{\lambda}_i = \widehat{\mathbb{M}}_i \cdot \boldsymbol{w} - \mathbb{M}_i \cdot \boldsymbol{v}$, $rnum = i$.

At the end, the UPKeyGen sub-algorithm outputs the update keys as $\{UK_i | \widehat{\rho}(i) = att, att \in \mathbb{S}_{insert} \vee \mathbb{S}_{delete} \vee \mathbb{S}_{alter}\}$.

CTUpdate After receiving the update keys $\{UK_i\}$, the cloud server updates the ciphertext CT and ACF. If the $UK_{type} = $ "insert", the cloud updates the ciphertext based on the $\langle \widehat{C}_i, \widehat{D}_i, ACF_f \rangle \in UK_{value}$. If the $UK_{type} = $ "delete", the cloud deletes the components form the CT and ACF as per the $\langle rnum, f \rangle \in UK_{value}$, otherwise the cloud alters the ciphertext.

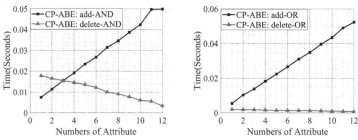

(a) Modification on AND structure. (b) Modification on AND structure.

**Fig. 3.** Performance evaluation of policy updating in the cloud

## 4   Performance Analysis

We deploy the experiment environment on the Ubuntu Linux Desktop 64-bit system with an Intel Core i7 CPU at 3.4 GHz and 8.00 GB RAM. The code utilizes the charm library version 0.50.2 and an asymmetric elliptic curve, where the base field size is 512-bit and the embedding degree is 2. All the experimental results are the mean of 100 trails.

Figure 2(a) presents the time cost of the generation of ACF as well as encryption, decryption of the micro encryption algorithm. The encryption and decryption phase in this micro algorithm is not related to the attributes. The decryption performs two pairing computations in $\mathbb{G}_T$ group, while the encryption performs one pairing computation in $\mathbb{G}_T$ group. Figure 2(b) shows the policy updating costs. The updating mode is divided into three types, delete, insert and alter.

Deletion only needs to find the corresponding element in the bucket of the ACF, the location tracing is realized by using traversing fingerprints in the access policy tree. Alter attribute in the ACF needs to locate the bucket first. Then replace the $C_x$ with the new element $C'_x$. It is correlated with the attributes. Insert new attributes is just the same as in the ACF creation phase. The cost will also be related to the number of attributes.

We also evaluate the computation cost of the policy updating in the cloud. Note that in this scheme, we concentrate on the monotone policy only. In this case, we can consider the following two structures in the policy updating, namely, the "AND" structure and the "OR" structure. Figure 3(a) and (b) express these two situations separately.

# 5   Conclusion

We proposed an outsourcing PHR scheme with policy preserving and cloud-assist policy updating. We are the first to consider the integration of both functions. Besides, to protect the security of the attributes in the ACF, we designed a lightweight tiny encryption algorithm. The performance evaluation, as well as security analysis, indicated that our scheme could achieve an ideal balance between efficiency and privacy.

**Acknowledgment.** This work is supported by the key project of Anhui provincial department of education (Grant No. KJ2018A0031), the National Natural Science Foundation of China under Grant Nos. U1804263 and 61702105.

# References

1. Heron, M.: Deaths: leading causes for 2016. Natl. Vital Stat. Rep. Cent. Dis. Control Prev. Natl. Cent. Health Stat. Natl. Vital Stat. Syst. **67**(6), 1 (2018)
2. Lin, G., Ying, C., Tan, S., et al.: ARP-CP-ABE: toward efficient, secure and flexible access control for personal health record systems. In: 2018 IEEE 16th International Conference on Dependable, Autonomic and Secure Computing, 16th International Conference on Pervasive Intelligence and Computing, 4th International Conference on Big Data Intelligence and Computing and Cyber Science and Technology Congress (DASC/PiCom/DataCom/CyberSciTech), pp. 54–61. IEEE (2018)
3. Liu, X., Xia, Y., Yang, W., et al.: Secure and efficient querying over personal health records in cloud computing. Neurocomputing **274**, 99–105 (2018)
4. Yeh, L.Y., Chiang, P.Y., Tsai, Y.L., et al.: Cloud-based fine-grained health information access control framework for lightweight IoT devices with dynamic auditing and attribute revocation. IEEE Trans. Cloud Comput. **6**(2), 532–544 (2018)
5. Xue, K., Chen, W., Li, W., et al.: Combining data owner-side and cloud-side access control for encrypted cloud storage. IEEE Trans. Inf. Forensics Secur. **13**(8), 2062–2074 (2018)
6. Li, J., Chen, X., Chow, S.S.M., et al.: Multi-authority fine-grained access control with accountability and its application in cloud. J. Netw. Comput. Appl. **112**, 89–96 (2018)

7. Zhang, Y., Zheng, D., Deng, R.H.: Security and privacy in smart health: efficient policy-hiding attribute-based access control. IEEE Internet Things J. **5**(3), 2130–2145 (2018)
8. Ying, Z., Wei, L., Li, Q., et al.: A lightweight policy preserving EHR Sharing scheme in the cloud. IEEE Access **6**, 53698–53708 (2018)
9. Ying, Z., Li, H., Ma, J., et al.: Adaptively secure ciphertext-policy attribute-based encryption with dynamic policy updating. Sci. China Inf. Sci. **59**(4), 042701 (2016)
10. Ying, Z., Jang, W., Cao, S., et al.: A lightweight cloud sharing PHR system with access policy updating. IEEE Access **6**, 64611–64621 (2018)

# Lightweight Outsourced Privacy-Preserving Heart Failure Prediction Based on GRU

Zuobin Ying[1,2(✉)], Shuanglong Cao[2], Peng Zhou[2], Shun Zhang[2], and Ximeng Liu[3,4]

[1] School of Electrical and Electronic Engineering, Nanyang Technological University, Singapore 639798, Singapore
[2] School of Computer Science and Technology, Anhui University, Hefei 230601, China
yingzb@ahu.edu.cn
[3] College of Mathematics and Computer Science, Fuzhou University, Fuzhou 350108, China
[4] Key Lab of Information Security of Network System (Fuzhou University), Fuzhou 350108, Fujian, China

**Abstract.** The medical service provider establishes a heart failure prediction model with deep learning technology to provide remote users with real-time and accurate heart failure prediction services. Remote users provide their health data to the health care provider for heart failure prediction through the network, thereby effectively avoiding the damage or death of vital organs of the patient due to the onset of acute heart failure. Obviously, sharing personal health data in the exposed data sharing environment would lead to serious privacy leakage. Therefore, in this paper, we propose a privacy-preserving heart failure prediction (PHFP) system based on Secure Multiparty Computation (SMC) and Gated Recurrent Unit (GRU). To meet the real-time requirements of the PHFP system, we designed a series of data interaction protocols based on additional secret sharing to achieve lightweight outsourcing computing. Through these protocols, we can protect the user's health data privacy while ensuring the efficiency of the heart failure prediction model. At the same time, to provide high-quality heart failure prediction services, we also use the new mathematical fitting method to directly construct the safety activation function, which reduces the number of calls to the security protocol and optimizes the accuracy and efficiency of the system. Besides, we built a security model and analyzed the security of the system. The experimental results show that PHFP takes into account the safety, accuracy, and efficiency in the application of heart failure prediction.

**Keywords:** Secure Multiparty Computation · Privacy-preserving · Heart failure prediction · Gated Recurrent Unit

© Springer Nature Switzerland AG 2020
S. Wen et al. (Eds.): ICA3PP 2019, LNCS 11945, pp. 521–536, 2020.
https://doi.org/10.1007/978-3-030-38961-1_45

# 1  Introduction

Heart Failure (HF) is a complex clinical symptom cluster and a severe stage of various heart diseases with high morbidity and mortality. According to the European Society of Cardiology (ESC), 26 million adults worldwide are diagnosed with heart failure, and 3.6 million people are newly diagnosed each year. About 20% heart failure patients die within one year after diagnosis, and about 50% decease in five years once have been diagnosed [1]. To effectively reduce the incidence and mortality of heart failure, early accurate prediction of heart attack episodes is indispensable. It is difficult for the traditional clinical methods to diagnose the occult acute heart failure at an early stage, so usually, the patient diagnosed after being admitted to the emergency department. If the essential organs of some patients not diagnosed in time, irreversible damage or death may occur [2]. Therefore, it is essential to provide an early and accurate heart failure prediction service. In recent years, with the development of deep learning, medical research institutions have trained high-precision heart failure prediction models by acquiring patient health data to provide users with high-quality heart failure prediction services. Among them, Edward Choi et al. [3] used a GRU neural network to establish a time series model the records related to EMR and realized much accurate prediction at an early stage of heart failure. Moreover, the Area Under the Curve (AUC) of the model reaches 0.777, compared to the traditional clinical diagnostic (correct rate of 0.513) has better accuracy. Using this result, healthcare providers can establish a heart failure pre-diagnosis model to provide real-time, accurate, and convenient heart failure prediction services to remote users. As a result, more and more users are providing their personal health data to medical service providers via the Internet for the purpose of obtaining real-time, accurate heart failure prediction services.

However, in the actual scenario, it is necessary to comprehensively consider the privacy of personal health data and the security of the heart failure prediction model provided by the medical service provider. In general, there are two ways to predict data sharing for heart failure. One way is that the user provides personal health data to the medical provider through the network, and then the medical service provider performs heart failure prediction and returns the result to the user locally. However, the medical service provider may disclose the user's health data during the process, which will lead to the leakage and abuse of the user's personal health data privacy. Another way is for the medical service provider to send the heart failure model to the user, and heart failure prediction is made by the user locally. Due to the high commercial value attached to this kind of prediction, the leakage of the model will bring economic losses to the medical service provider. Therefore, how to design a heart failure prediction system that can protect data privacy has become a vital issue to be solved. Furthermore, considering that the sudden onset of acute heart failure is often life-threatening and requires urgent rescue measures, this requires us to balance the timeliness and accuracy of the PHFP system.

In recent years, researchers have proposed a variety of technologies to protect medical privacy, such as anonymous technology and homomorphic encryption, in response to data privacy breaches in telemedicine scenarios. Nevertheless, anonymous technology [4] only protects the privacy of users to a certain extent, which makes it easy to lose valuable information, and then its prediction accuracy will be affected. And, studies have shown that anonymous techniques are not sufficient to resist re-identification attacks [5]. Moreover, current frameworks based on homomorphic encryption [6] are time-consuming and memory-intensive, and its computational overhead is enormous, which is not suitable for real-time heart failure prediction scenarios. None of the current work takes into account the balance between efficiency and precision. Therefore, when constructing a privacy-preventing heart failure prediction system, we must realize privacy protection in the premise of system accuracy and efficiency.

To achieve the above objectives, we propose a PHFP system. Our main contributions can be summarized as follows:

- We are first design a lightweight system to protect privacy data and service provider model parameters for the medical user heart failure prediction. The system is based on the addition of secret sharing technology in secure multiparty computing, which transfers intricate work to the edge server, reducing the cost of medical users. Moreover, the system avoids the interaction between the medical user terminal and the server, with the results that the overall efficiency of the system is improved.
- PHFP use a new mathematical method to directly construct the secure *Sigmoid* function and the *Tanh* function, which avoids the time overhead caused by the system calling too many security components during the running process. At the same time, system solve the problem of low function fitting precision within a specific interval caused by the local fitting of the Taylor series. Compared with the existing additive secret sharing scheme, our system has significantly improved in terms of computational overhead and precision.
- We conduct a comprehensive experimental evaluation to measure the performance of our program. The experimental results show that the system is superior to the previous work in terms of computational overhead, communication overhead, and computational accuracy while protecting the privacy of heart failure prediction data.

The remaining part of this paper is organized as follows. We formulate the problem and present the system model and security model in Sect. 2. In Sect. 3, the primitives about GRU and secure multiparty computation are briefly introduced followed by problem analysis and model presentation. Then the building blocks that support efficient, secure computation based on secret sharing techniques are provided in Sect. 4. On the basis of that, we propose the details of our system in Sect. 5. And Sects. 6 and 7 covers the theoretical analysis and experimental results respectively. Finally, related conclusion is stated Sect. 8.

## 2    Problem Formulation

In this section, we formalize the system model, security model and identify our design goal.

### 2.1    System Model

In our system model, we focus on how users with sensitive medical data can obtain accurate and privacy-preserving real-time heart failure prediction services from cloud service providers. Precisely, the system consists of five parts: (1) smart wearable device (SWD); (2) the Medical User (MU); (3) the Edge Servers (ESs); (4) the Medical Service Provider (MSP); (5) Trusted Third Party (TTP). As shown in Fig. 1.

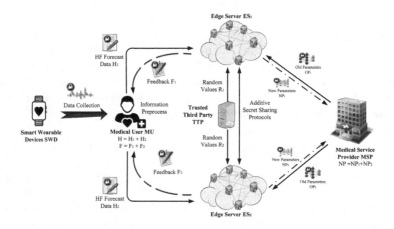

**Fig. 1.** System model under consideration

- SWD is used to collect various health data of healthy users. The collected data has a total of 279 feature dimensions, such as heart rate, blood pressure, body temperature, etc. And it sends the collected health data to the medical user.
- MU wants to know its future heart failure attack risk coefficient, and it will preprocess its heart failure data on the phone to form the heart failure eigenvector, which is randomly divided into the different secret values and sent to different ESs. Besides, MU was able to accept the feedback results from ESs and combines the feedback results to obtain the final correct prediction results of heart failure.

- ESs can be a cloud service provider that assists healthcare providers in collecting data related to heart failure prediction and training new data. At the same time, ESs can return the correct heart failure prediction result to the user, and promote the user to provide more data sets.
- MSP, such as pharmaceutical companies or hospitals, can provide real-time heart failure risk prediction services. Individually, with the help of ESs, MSP can obtain the latest training parameters of GRU recurrent neural network. Considering the benefits of ESs, MSP is also willing to commission ESs to effectively predict the risk of heart failure and return the results to MU.
- TTP is only responsible for generating random numbers, which means that TTP doesn't require a lot of computing power. It can be replaced by a light server or even a personal computer.

### 2.2  Security Model

In the security model, we use the standard semi-honest security model [7], which is also perceived as passive or honest-but-curious. In this security model, each edge server enforces the protocol as required by the contract. But out of curiosity, they can try to get as much information as they can from the data they receive and the data they process.

Also, we assume that the two edge servers $ES_1$ and $ES_2$ are independent of each other, and there is no collusion between them. This means that the data acquired by each of the edge servers will not be revealed. In this way, even if each edge server has durable computing power, they can only get some split medical or intermediate interaction data and model parameters. In other words, real raw medical data and model parameters cannot be recovered.

It is worth noting that TTP is merely responsible for generating random numbers, and it is honest and trustworthy. Last but not least, we also assume that medical users and service providers are honest and a secure channel for communication exists between the entities.

## 3  Preliminaries

### 3.1  Features of GRU

GRU neural network is a variant of Long Short Term Memory (LSTM), besides, GRU maintains the effect of LSTM while making the structure simpler. It's a very popular neural network. The mathematical expression is shown below:

$$z_t = \sigma(W_z \cdot [h_{t-1}, x_t] + b_z)$$
$$r_t = \sigma(W_r \cdot [h_{t-1}, x_t] + b_r)$$
$$\tilde{h}_t = \tanh(W_{\tilde{h}} \cdot [r_t \odot h_{t-1}, x_t] + b_{\tilde{h}})$$
$$h_t = z_t \odot h_{t-1} + (1 - z_t) \odot \tilde{h}_t$$

It is worth noting that in GRU, the value of hidden layer $h_{t-1}$ at time step $t-1$ and the input value at time step $t$ doesnt directly change the value of $h_t$.

The value of $h_t$ is determined by updating gate $z_t$, resetting gate $r_t$, and intermediate storage cell $\tilde{h}_t$. In short, reset gates allow the hidden layer to remove any information that is not useful for future prediction, while update gates determine how much information from the previous hidden layer should be retained by the current hidden layer.

### 3.2   Additive Secret Sharing Protocols

Secret sharing protocol is mainly used for secure multiply party computing (SMC) and privacy protection. The encryption protocol based on secret sharing has good performance and can be used to design an efficient privacy protection computing model [8]. The secret sharing protocol can be thought of as consisting of a large number of "components" through which we can build a larger and equally secure system.

**Lemma 1.** *If all the sub-protocols of a protocol are fully emulated, then the protocol is fully emulated* [9].

- *Random Bit Protocol.* The $RanBits(\cdot)$ protocol [10] can be thought of simply as a random number generator. It doesn't need any input to generate any bit sequence $(r_0, \cdots, r_l)$. At the same time, a hex random number $r$ can be calculated by

$$r = \sum_{i=0}^{l} r_i \cdot 2^i.$$

- *Secure Addition and Subtraction Protocol.* The $SecAdd(\cdot)$ and $SecSub(\cdot)$ protocol [10] can calculate $f(u,v) = u \pm v$. Since $u \pm v = (u_1 + u_2) \pm (u_1 + v_2) = (u_1 \pm v_1) + (u_2 \pm v_2)$, it's easy to see that the protocol can perform secure additions and subtractions locally without the need for interaction between servers. After the computation, each participating party will output $f_i = u_i \pm v_i$. Obviously, we have $f_1 + f_2 = u \pm v$.
- *Secure Multiplication Protocol.* The $SecMul(\cdot)$ protocol [10] is based on the Beaver's triplet [11]. Given an input binary group $(u, v)$, the protocol outputs another binary group $(f_1, f_2)$ to the two participants, where $f = f_1 + f_2 = u \cdot v$. In this process, a trusted third party is required to generate a random triple $(x, y, z)$ and $z = x \cdot y$. It is worth noting that the Participants will not be informed of each other's input.
- *Secure Comparison Protocol.* The $SecCmp(\cdot)$ protocol [10] can be achieved in the comparison of the size of two inputs $u$ and $v$, the input of both sides will not be leaked. And, if $u < v$, $SecCmp(\cdot)$ outputs 1, otherwise outputs 0.
- *Secure Vector Concatenation Protocol.* The $SecCon(\cdot)$ protocol [12] is to connect two short vectors into one long vector. That is (Table 1),

$$[\boldsymbol{u}, \boldsymbol{v}] = [(\boldsymbol{u_0}, \boldsymbol{u_1}, \boldsymbol{u_2}, ...), (\boldsymbol{v_0}, \boldsymbol{v_1}, \boldsymbol{v_2}, ...)]$$
$$= (\boldsymbol{u_0}, \boldsymbol{u_1}, \boldsymbol{u_2}, ..., \boldsymbol{v_0}, \boldsymbol{v_1}, \boldsymbol{v_2}, ...).$$

**Table 1.** Variables and their description

| Variables | Description |
|---|---|
| $z_t$ | The update gate at timestep $t$ |
| $r_t$ | The reset gate at timestep $t$ |
| $\widetilde{h}_t$ | The intermediate memory unit at timestep $t$ |
| $h_t$ | The hidden layer at timestep $t$ |
| $W$ | The weight matrix |
| $b$ | The bias term |
| $\odot$ | Hadamard product |
| $\sigma_{sec}(.)$ | Secure sigmoid function |
| $\tanh_{sec}(.)$ | Secure tanh function |
| $W_{ih}, W_{ix}$ | The split matrixes of $W_i$ |
| $\sigma$ | The sigmoid function |
| $\delta_t$ | The error vector at time |
| $\nabla$ | The symbol of gradient |

## 4 Secret Sharing Based Functions

### 4.1 Nonlinear Function Fitting Method

The GRU has at least one activation function deployed in each gate. These activation functions are nonlinear functions such as the *Sigmoid* function and the *Tanh* function. According to Lemma 1, we can use some of the security protocols mentioned in Sect. 3 to build a secure nonlinear function protocol. However, nonlinear functions need to include not only addition and multiplication operations, but also complex operations, such as exponents and reciprocals. It is impossible to construct safe nonlinear functions directly with the security protocols mentioned before. Therefore, we need to fit the nonlinear numbers in the GRU gates with polynomials that only contain multiplication and addition.

**Scheme I: Taylor Series** [13]. At present, a useful tool for solving nonlinear problems is the Taylor series. By using Taylor expansion multi-order approxima-

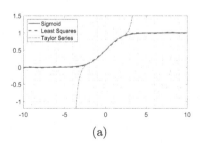
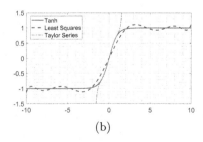

(a)                                    (b)

**Fig. 2.** Taylor series and least squares approximations for Sigmoid and Tanh function

tion, nonlinear problems can be linearised, which makes calculation and understanding more conveniently. The literature [12] uses the Taylor series to construct a secure exponential function with base e and Newton iteration method to build a secure reciprocal function. Then the security sigmoid and the secure tanh function are further built by invoking the security exponential and the reciprocal function, but this will make the overhead of the two edge servers more massive. Also, each invoke to the security function will result in a loss of precision, and too many invokes to the safety function will result in more loss of accuracy. It is not suitable for high-precision and high-efficiency scenarios like heart failure prediction. Hence, in our scheme I, we borrowed homomorphic encryption [14] to direct fit *Sigmoid* and *Tanh* using Taylor series directly and then build the addition secret sharing protocol.

**Scheme II: Least Square Method.** Although in scheme I, direct fitting of the sigmoid and tanh functions using Taylor series can reduce the overhead and precision loss of the edge server, this method still has a defect. As shown in the Fig. 2(a)–(b), the basic idea of the Taylor series is to approximate a function in the neighborhood of a point. For points that are not included in the neighborhood, the approximation error is much larger than the point contained within the area. To avoid the problem of the local fitting function in scheme I, we additionally consider the method of fitting the function by least squares [15], which finds the best function matching of the activation function by minimizing the sum of the squares of the errors. Its expression is as follows,

$$E_{min} = \sum_{i=1}^{n} (p(x_i) - y_i)^2.$$

Where $y_i$ is the function value of the activation function to be fitted, and $p(x_i)$ is the function value of the polynomial to be constructed. Next, we will give the process of fitting the *Sigmoid* function $\sigma(x)$ by the least squares method, as shown below.

1. Let the least squares fit the polynomial as follows,

$$p(x) = a_0 + a_1 x +, \cdots, +a_m x^m. \tag{1}$$

2. The expression of the sum of squares of deviations is as follows,

$$E = \sum_{i=1}^{n} (a_0 + a_1 x_i +, \cdots, +a_m x_i^m - \sigma(x_i))^2. \tag{2}$$

3. To find the $a_j$ value satisfying the minimum value of $E$, it is necessary to derive the partial derivative of Eq. (2) on the right side of $a_j$.

$$\frac{\partial E}{\partial a_j} = \sum_{i=1}^{n} 2 \cdot (a_0, a_1 x_i +, \cdots, a_m x_i^m - \sigma(x_i)) x_i^j. \tag{3}$$

4. By sorting out, we can get the following equations.

$$\begin{cases} na_0 + (\sum_{i=1}^n x_i)a_1 +, \cdots, +(\sum_{i=1}^n x_i^m)a_m = \sum_{i=1}^n y_i \\ (\sum_{i=1}^n x_i)a_0 + (\sum_{i=1}^n x_i^2)a_1 +, \cdots, +(\sum_{i=1}^n x_i^{m+1})a_m \\ \quad = \sum_{i=1}^n x_i \sigma(x_i) \\ \cdots\cdots \\ (\sum_{i=1}^n x_i^m)a_0 + (\sum_{i=1}^n x_i^{m+1})a_1 +, \cdots, +(\sum_{i=1}^n x_i^{2m})a_m \\ \quad = \sum_{i=1}^n x_i^m \sigma(x_i). \end{cases}$$

Finally, by solving the equations, we can get the values of $(a_1, a_2, \cdots, a_m)$ and get the least squares fit polynomial $p(x)$ of $\sigma(x)$. Therefore, $\sigma(x) \approx p(x)$.

### 4.2 Secure Sigmoid Function.

We use the least squares method to fit the Sigmoid function, Let $x$ denote the input, the polynomial of the least squares fitting of the sigmoid function is expressed as follows,

$$f(x) = \sigma(x) = \frac{1}{1 + e^{(-x)}} \approx \sum_{i=0}^{\infty} C_i x^i.$$

Where $C_i$ represents the coefficient of the least squares polynomial and $i$ represents the order of the least squares polynomial.

*Initialization.* $ES_1$ and $ES_2$ respectively get random values $x_1$ and $x_2$, satisfying $x = x_1 + x_2$. In the process of initialization, $ES_1$ need to compute $f_0' \leftarrow C_0 + C_1 x_1$, $ES_2$ calculation $f_0'' \leftarrow C_1 x_2$. According to the polynomial exponent value, the iterative process shown below.

*Iteration.* In the process of iteration, we are mainly implemented by alternatively invoking $SecAdd(\cdot)$ and $SecMul(\cdot)$. First of all, $ES_1$ and $ES_2$ common computing $(g_0', g_0'') \leftarrow SecMul(C_2 x_1, C_2 x_2, x_1, x_2)$ and $f_1 \leftarrow SecAdd(f_0, g_0)$. Subsequently, $g_i$ can be calculated iteratively by similar calculation methods. And invoke the secure comparison function $SecCmp(i, n)$ to determine whether to achieve the required polynomial index. If the required polynomial order is reached, terminate the iteration and output $f_i'$ and $f_i''$. Otherwise, invoke the secure addition function to compute $SecAdd(f_{i-1}, g_{i-1})$.

### 4.3 Secure Tanh Function

*Tanh* is a hyperbolic tangent function, and the curves of the *Tanh* function and the *Sigmoid* function are relatively similar. The only difference is the output interval. The *Tanh* output interval is between $(-1, 1)$ and the full function center at 0. Therefore, we can also fit the *Tanh* function by least squares. Let $x$ be a function input, and the polynomial of the least squares fit of the $\tanh_{sec}(x)$ function is expressed as follows.

$$f(x) = \tanh(x) = \frac{e^x - e^{(-x)}}{e^x + e^{(-x)}} \approx \sum_{i=0}^{\infty} H_i x^i.$$

Also, since the *Initialization* and *Iteration* process of $\tanh_{sec}$ is similar to that of $\sigma_{sec}$, these processes are not repeated.

# 5    Lightweight Privacy-Preserving GRU for Encrypted HF Data

## 5.1    Secure Forward Propagation of GRU

Due to all the necessary security "components" have been constructed, the following work for the secure forward propagation of GRU is simply combining these security "components" appropriately to design a secure interactive subprotocol between the two edge servers $ES_1$ and $ES_2$. Note that in the following sections $[\![i]\!]$ stands for '" and '"'.

**Reset Gate.** The reset gate allows the hidden layer to delete any information that is not useful for future prediction. To achieve this, the input vector $x_i$ and information about the previous timestep $h_{t-1}$ are put into the sigmoid function after a series of linear operations. And, the final output will be between 0 and 1. Because matrixed weight $W_r$ and bias $b_r$ for the reset gate is not publicly known, at timestep $t$, $ES_1$ and $ES_2$ compute separately,

$$r_t^{[\![i]\!]} \leftarrow \sigma_{sec}(W_r^{[\![i]\!]} \cdot [h_{t-1}^{[\![i]\!]}, x_t^{[\![i]\!]}] + b_r^{[\![i]\!]}).$$
$$\tilde{h}_t^{[\![i]\!]} \leftarrow \tanh_{sec}^{[\![i]\!]}(W_{\tilde{h}}^{[\![i]\!]} \cdot [r_t^{[\![i]\!]} \odot h_{t-1}^{[\![i]\!]}, x_t^{[\![i]\!]}] + b_{\tilde{h}}^{[\![i]\!]}).$$

**Update Gate.** The update gate determines how much information from the previous time step and the current timestep needs to be transmitted. Given the input weight matrix $W_z$, input bias $b_z$ and timestep t, $ES_1$ and $ES_2$ consociation calculations,

$$z_t^{[\![i]\!]} \leftarrow \sigma_{sec}(W_z^{[\![i]\!]} \cdot [h_{t-1}^{[\![i]\!]}, x_t^{[\![i]\!]}] + b_z^{[\![i]\!]}).$$

The final output by invoking the secure multiplication function and the secure addition function. We let $ES_1$ and $ES_2$ respectively compute,

$$h_t^{[\![i]\!]} \leftarrow z_t^{[\![i]\!]} \odot h_{t-1}^{[\![i]\!]} + (1 - z_t^{[\![i]\!]}) \odot \tilde{h}_t^{[\![i]\!]}.$$

Both $h_t'$ and $h_t''$ are then sent to MU as feedback. And MU can decrypt the ciphertext by simply adding them together, $h_t = h_t' + h_t''$.

## 5.2    Back Propagation Based Training of GRU

It is assumed that the iterative forward propagation of the privacy protection GRU has been completed. Let $\delta_{t-1}$ represents the error term at time $t-1$. It can be calculated by the partial derivative function of the output $h_t$ at the timestep $t$. $ES_1$ and $ES_2$ combine calculates,

$$\delta_{t-1}^{[i]} \leftarrow \delta_{r,t}^{[i]} \cdot W_{rh}^{[i]} + \delta_{z,t}^{[i]} \cdot W_{zh}^{[i]} +$$
$$\delta_{\tilde{h},t}^{[i]} \cdot W_{hh}^{[i]} \odot r_t^{[i]} + \delta_{h,t}^{[i]} \odot (1 - z_t^{[i]}).$$

Respectively, $\delta_{r,t}$, $\delta_{z,t}$, $\delta_{\tilde{h},t}$ and $\delta_{h,t}$ denote the derivative with respect to $h_{t-1}$. Here, we present a calculation formula based on the addition secret sharing protocol. During this time, the values of $r_t$, $z_t$, and $\tilde{h}_t$ can be obtained by forwarding propagation.

$$\delta_{r,t}^{[i]} \leftarrow \delta_t^{[i]} \odot z_t^{[i]} \odot [1 - (\tilde{h}_t^{[i]})^2] \odot W_{hh}^{[i]} \odot h_{t-1}^{[i]} \odot r_t^{[i]} \odot (1 - r_t^{[i]}),$$
$$\delta_{z,t}^{[i]} \leftarrow \delta_t^{[i]} \odot (\tilde{h}_t^{[i]} - h_{t-1}^{[i]}) \odot z_t^{[i]} \odot (1 - z_t^{[i]}),$$
$$\delta_{\tilde{h},t}^{[i]} \leftarrow \delta_t^{[i]} \odot z_t^{[i]} \odot [1 - (\tilde{h}_t^{[i]})^2],$$
$$\delta_{h,t}^{[i]} \leftarrow \delta_t^{[i]}.$$

For the entire sample, its error is the sum of the errors at all times, and the gradient of the weights associated with the previous moment is equal to the amount of the gradients at all times, and the other weights do not have to be accumulated. Let $\gamma \in \{r, z, \tilde{h}\}$. We have,

$$\nabla W_{\gamma,h}^{[i]} \leftarrow \sum_{t=1}^{T} SecMul(\delta_{\gamma,t}^{[i]}, h_{t-1}^{[i]}),$$
$$\nabla W_{\gamma,x}^{[i]} \leftarrow SecMul(\delta_{\gamma,t}^{[i]}, x_t^{[i]}),$$
$$\nabla b_\gamma^{[i]} \leftarrow \sum_{t=1}^{T} \delta_{\gamma,t}^{[i]}.$$

Let $\alpha$ denote the learning rate of the gradient drop, and $\alpha$ is public. Then we can use the following formula to update the weight matrix and bias.

$$W_{new,\gamma}^{[i]} \leftarrow W_{old,\gamma}^{[i]} - \alpha \odot \nabla W_\gamma^{[i]},$$
$$b_{new,\gamma}^{[i]} \leftarrow b_{old,\gamma}^{[i]} - \alpha \odot \nabla b_\gamma^{[i]}.$$

Unlike forward propagation, after the backpropagation training complete, all updated encryption parameters are sent to the MSP instead of the MU. And MSP can decrypt the ciphertext by simply adding them together, $W_{new,\gamma} = W'_{new,\gamma} + W''_{new,\gamma}$, and $.b_{new,\gamma} = b'_{new,\gamma} + b''_{new,\gamma}$.

## 6   Theoretical Analysis

### 6.1   Correctness

Before the medical user uploads the heart failure specific data, the feature data $H$ is divided into $H = H_1 + H_2$. Then, under our security protocol built on the addition of secret sharing, a large number of linear and nonlinear operations are performed on $H$. Strictly speaking, the final output prediction result $F$ and the model parameter $NP$ may not be equal to the value of the original unencrypted algorithm. Here, we demonstrate through the theoretical derivation that the value of the output of our system is highly close to the original value.

First, some of the protocols [10] mentioned in Sect. 3.2 have been proven, and their output is still accurate no matter how many times they are invoked. Secondly, the security functions we construct are all approximated by polynomial. The operations used in these functions are only addition and multiplication. Therefore, in theory, as long as the edge server computing power is strong enough, we can achieve arbitrary calculation accuracy. As long as the accuracy reaches the precision required by GRU, it can be said that our proposed function is additive and correct as of the original function. In addition, since the activation function is composed of a combination of polynomials containing only addition and multiplication. This means that their output $\xi$ satisfies $\xi = \xi_1 + \xi_2$. Finally, we can draw some conclusions and give an arbitrary function $F$. We have $F = F_1 + F_2$ if and only if $F = f(\zeta_1, \zeta_2, \cdots)$, where $\zeta_i (i = 1, 2, \cdots)$ is a random linear mapping function and xi can be any of the security functions in this paper. Thus, based on the inference, we can ensure that $F = F_1 + F_2$ and $NP = NP_1 + NP_2$, because both forward and backward propagation can be considered as $F$.

## 6.2    Security

In this section, we analyse the safety of the proposed PHFP system. To prove the security of the system in this paper, we first need to define what is semi-honest security [9] formally.

**Definition 1.** *We say that a protocol s secure if there exists a probabilistic polynomial-time simulator $S$ that can generate a view for the adversary $\mathcal{A}$ in the real world and the view is computationally indistinguishable from its rear view.*

In addition to the Lemma 1 mentioned in Sect. 3, also need the following lemmas.

**Lemma 2** [9]. *If a random element $r$ is uniformly distributed on $\mathbb{Z}_n$ and independent from any variable $x \in \mathbb{Z}_n$, then $r \pm x$ is also uniformly random and independent from $x$.*

**Lemma 3** [10,12]. *The protocols SecAdd, SecMul, SecCmp and SecCon are secure in the semi-honest model.*

According to Lemma 3, we only need to verify the safety of other protocols. *The protocols $\sigma_{sec}$ and $\tanh_{sec}$ are secure in the semi-honest model.*

*Proof.* In $\sigma_{sec}$, given the order of the polynomial $n$, what $ES_1$ holds is receiver $Rec_1 = (u_1, G'_1, F'_1, \alpha')$, where $G'_1 = g'_0, g'_1, \cdots, g'_n$ and $F'_1 = f'_0, f'_1, \cdots, f'_{n-1}$. And $g'_i$ and $f'_i$ are respectively the outputs of *SecMul* and *SecAdd*. In the meantime, with $u_1$, they also compose the inputs of the next iteration. According to Lemma 3, it is guaranteed that $G'_1$ and $F'_1$ are sets of uniformly random values.

So they can all be correctly simulated by simulator $ES_1$, and are unable to distinguish by the adversary $\mathcal{A}$ in polynomial time. Similarly, $ES_2$ can also hold $Rec_2$ which is simulatable and distinguishable. In addition, $tanh_{sec}$ protocols are implemented by a similar polynomial composed of protocols and can be proved to be secure.

# 7    Performance Evaluation

To implement our framework, we utilise NumPy for parallel computation of matrixes in Python 3. All the data is encrypted on a personal computer with an Intel(R) Core (TM) i7-6700 CPU @3.40 GHz and 8.00 GB of RAM. Then, the ciphertexts respectively sent to two edge servers for privacy-preserving GRU training and pre-trained heart failure prediction. Each server is equipped with an Intel(R) Core (TM) i7-7700HQ CPU @2.80 GHz and 8.00 GB of RAM. Also, to obtain the correct pre-diagnosis results in the above evaluation environment, we considered a real data set from the UCI machine learning library called Arrhythmia to evaluate the accuracy and efficiency of our solution. The selected Arrhythmia dataset contains 452 instances, each of which includes 279 attributes (such as age, weight, gender, heart rate, QRS duration, P-R interval, Q-T interval, T interval, P interval, etc.)

## 7.1    Performance of Secure Sigmoid and Tanh Function

In the PHFP system, we tried two approaches to approximate the activation function in the GRU neural network. To avoid the local fitting problem of Taylor series, we finally use the least squares method to construct high-order polynomials to approximate the *Sigmoid* and *Tanh* functions. Since each hidden unit of the GRU contains two *Sigmoid* functions and one *Tanh* function, when our PHFP system has multiple hidden units, the secure *Sigmoid* and *Tanh* functions are invoked multiple times. Therefore, we evaluated the performance of the scheme II security function under different number of calls, and we also compared it with scheme I and OPSR scheme [12], as shown in Fig. 3(a)–(d). From the figure, we can see that scheme II is both accurate and efficient. It is better than the other two programs. The reasons summarise as follows: Firstly, since scheme II uses the least squares method to fit the activation function, the local fitting problem of the Taylor series is avoided, and the accuracy is improved to some extent. Secondly, scheme II adopts a scheme of directly constructing a security function, which prevents the time overhead caused by multiple invokes of security components. To sum up, scheme II is more suitable for our heart failure prediction system in terms of accuracy and efficiency.

## 7.2    Performance of PHFP

**Accuracy Evaluation.** To further evaluate the performance of the PHFP, we deployed the constructed safety components to our system to assess the accuracy of the system's forward propagation calculations. At the same time, we also

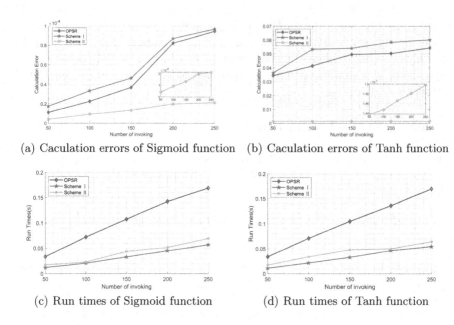

(a) Caculation errors of Sigmoid function    (b) Caculation errors of Tanh function

(c) Run times of Sigmoid function    (d) Run times of Tanh function

**Fig. 3.** Performance of secure Sigmoid and Tanh function

deployed the components built by [12] into our GRU neural network and used the same data set to evaluate the computational error of forwarding propagation. As shown in Fig. 4(a)–(b), since the numerical range of our dataset is not entirely concentrated on a certain point, the error of the scheme II we constructed in forwarding propagation is significantly better than the other two schemes. This benefit from the nature of the global fit of the least squares method. It is noteworthy that when medical users predict heart failure, only the process of forwarding propagation is needed, while the calculation error of forwarding propagation is controlled within $10^{-5}$, which can be neglected in actual heart failure prediction.

**Efficiency Evaluation.** In PHFP, the primary function of ESs is to calculate the user's data and train the model provided by the medical service provider. Both secure forward propagation and secure backpropagation are involved in training the model. However, the number of features of medical data, the number of medical instances and the number of GRU hidden layers have an essential impact on the computing cost of ESs. Accordingly, we first tested the computational overhead of ESs with a different number of features and a different number of medical cases. Here, we default the number of hidden layers of GRU to 20, and we compare scheme I and scheme II with OPSR. As shown in Fig. 4(c)–(d), since we adopted the idea of directly constructing sigmoid and tanh functions, and avoiding the overhead caused by repeated calls to multiple components, our two schemes are significantly better than the OPSR scheme in terms of computational cost. Besides, we noticed that in the scheme II adopted by the PHFP

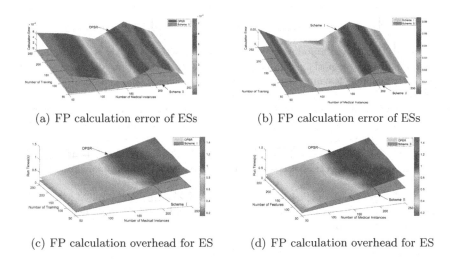

(a) FP calculation error of ESs     (b) FP calculation error of ESs

(c) FP calculation overhead for ES     (d) FP calculation overhead for ES

**Fig. 4.** ESs efficiency evaluation

system, although we let the ES perform the forward propagation calculation of 250 medical cases with 250 features, its calculation time is less than one second. At the same time, we also evaluated the computational overhead of backpropagation ESs.

## 8 Conclusion

In this paper, we proposed a privacy-preserving heart failure prediction system based on Secure Multiparty Computation and Gated Recurrent Unit, named PHFP. The PHFP system was adopted to protect the privacy of users' heart failure prediction data and the security of neural network parameters of medical service providers with high accuracy and low computing cost. Accurately, the program randomly split the heart failure prediction data and neural network parameters into secret sharing, and the edge server calculated the user data in the state of ciphertext. Therefore, the medical service provider cannot obtain the user's private data, and the user cannot receive any neural network parameter information of the medical service provider. Finally, we use a large number of experiments to prove the effectiveness of the system.

**Acknowledgment.** This research is supported by the key project of Anhui provincial department of education (Grant No. KJ2018A0031), the National Natural Science Foundation of China under Grant Nos. U1804263 and 61702105.

# References

1. Tripoliti, E.E., Papadopoulos, T.G., Karanasiou, G.S., Naka, K.K., Fotiadis, D.I.: Heart failure: diagnosis, severity estimation and prediction of adverse events through machine learning techniques. Comput. Struct. Biotechnol. J. **15**, 26–47 (2017)
2. Shoaib, A., et al.: Mode of presentation and mortality amongst patients hospitalized with heart failure? A report from the first euro heart failure survey. Clin. Res. Cardiol. **108**(5), 510–519 (2019)
3. Choi, E., Schuetz, A., Stewart, W.F., Sun, J.: Using recurrent neural network models for early detection of heart failure onset. J. Am. Med. Inform. Assoc. **24**(2), 361–370 (2016)
4. Machanavajjhala, A., Gehrke, J., Kifer, D., Venkitasubramaniam, M.: l-diversity: privacy beyond k-anonymity. In: 22nd International Conference on Data Engineering (ICDE 2006), pp. 24–24. IEEE (2006)
5. Narayanan, A., Shmatikov, V.: Myths and fallacies of "personally identifiable information". Commun. ACM **53**(6), 24–26 (2010)
6. Liu, X., Zhu, H., Lu, R., Li, H.: Efficient privacy-preserving online medical primary diagnosis scheme on Naive Bayesian classification. Peer-to-Peer Network. Appl. **11**(2), 334–347 (2018)
7. Ning, J., Xu, J., Liang, K., Zhang, F., Chang, E.-C.: Passive attacks against searchable encryption. IEEE Trans. Inf. Forensics Secur. **14**(3), 789–802 (2018)
8. Pullonen, P., Matulevičius, R., Bogdanov, D.: PE-BPMN: privacy-enhanced business process model and notation. In: Carmona, J., Engels, G., Kumar, A. (eds.) BPM 2017. LNCS, vol. 10445, pp. 40–56. Springer, Cham (2017). https://doi.org/10.1007/978-3-319-65000-5_3
9. Bogdanov, D., Laur, S., Willemson, J.: Sharemind: a framework for fast privacy-preserving computations. In: Jajodia, S., Lopez, J. (eds.) ESORICS 2008. LNCS, vol. 5283, pp. 192–206. Springer, Heidelberg (2008). https://doi.org/10.1007/978-3-540-88313-5_13
10. Huang, K., Liu, X., Fu, S., Guo, D., Xu, M.: A lightweight privacy-preserving CNN feature extraction framework for mobile sensing. IEEE Trans. Dependable Secure Comput. (2019)
11. Beaver, D.: Efficient multiparty protocols using circuit randomization. In: Feigenbaum, J. (ed.) CRYPTO 1991. LNCS, vol. 576, pp. 420–432. Springer, Heidelberg (1992). https://doi.org/10.1007/3-540-46766-1_34
12. Ma, Z., Liu, Y., Liu, X., Ma, J., Li, F.: Privacy-preserving outsourced speech recognition for smart IoT devices. IEEE Internet Things J. **6**, 8406–8420 (2019)
13. Greenspan, D.: Numerical Analysis. CRC Press, Boca Raton (2018)
14. Bos, J.W., Lauter, K., Naehrig, M.: Private predictive analysis on encrypted medical data. J. Biomed. Inform. **50**, 234–243 (2014)
15. Stoer, J., Bulirsch, R.: Introduction to Numerical Analysis, vol. 12. Springer, Heidelberg (2013)

# DAPS: A Decentralized Anonymous Payment Scheme with Supervision

Zhaoyang Wang[1(✉)], Qingqi Pei[1,2(✉)], Xuefeng Liui[1,2], Lichuan Ma[1,2], Huizhong Li[3], and Shui Yu[4]

[1] The State Key Laboratory of Integrated Services Networks, Xidian University, Xi'an, China
xd.zywang@gmail.com
[2] Shaanxi Key Laboratory of Blockchain and Secure Computing, Xi'an, China
{qqpei,liuxf}@mail.xidian.edu.cn, lcma@xidian.edu.cn
[3] Webank, Shenzhen, China
wheatli@webank.com
[4] School of Software, University of Technology Sydney, Sydney, Australia
Shui.Yu@uts.edu.au

**Abstract.** With the emergence of blockchain-based multi-party trading scenarios, such as finance, government work, and supply chain management. Information on the blockchain poses a serious threat to users' privacy, and anonymous transactions become the most urgent need. At present, solutions to the realization of anonymous transactions can only achieve a certain degree of trader identity privacy and transaction content privacy, so we introduce zero knowledge proof to achieve complete privacy. At the same time, unconditional privacy provides conditions for cybercrime. Due to the great application potential of the blockchain in many fields, supporting privacy protection and supervision simultaneously in the blockchain is a bottleneck, and existing works can not solve the problem of coexistence of privacy protection and supervision.

This paper takes the lead in studying the privacy and supervision in multi-party anonymous transactions, and proposes a distributed anonymous payment scheme with supervision (DAPS) based on zk-SNARK, signature, commitment and elliptic curve cryptography, which enables users to be anonymous under supervision in transactions. The advantages of DAPS are twofold: enhanced privacy and additional supervision. We formally discussed the security of the whole system framework provided by the zero-knowledge proof, and verified its feasibility and practicability in the open source blockchain framework BCOS.

**Keywords:** Blockchain · Zero-knowledge proof · Privacy protection · Supervision

## 1 Introduction

Blockchain is an integrated application for distributed digital storage, peer-to-peer transmission, consensus mechanisms and encryption algorithms, and is

© Springer Nature Switzerland AG 2020
S. Wen et al. (Eds.): ICA3PP 2019, LNCS 11945, pp. 537–550, 2020.
https://doi.org/10.1007/978-3-030-38961-1_46

considered to be the fifth revolutionary computing paradigm after large computers, personal computers, the Internet and mobile social networks. Now it has been widely used in finance, the Internet of Things, supply chain management, medicine, public welfare and other fields.

With the increasing number of one-to-one, many-to-one, many-to-many block-chain-based trading scenarios, the accompanying privacy leaks are becoming more and more prominent and must be fully valued [1–3]. Since different nodes need to perform the same verification on the same data, the data on the blockchain must be public. This increases data transparency and credibility, but it brings a new question: how to protect privacy of these data. If users' public transaction information is maliciously exploited and utilized, it will pose a serious threat to users' privacy. Therefore, privacy leaks must be addressed before the blockchain moves to be practical. Nakamoto et al. [4,5] can only achieve incomplete identity privacy; Monero Research Lab [6] can achieve full identity privacy but cannot guarantee content privacy; Sasson et al. [7] can achieve identity and content privacy simultaneously, but inefficiently. We propose to use zero-knowledge proof to achieve complete privacy.

After addressing the issues of privacy disclosure, unconditional privacy will also become a natural hotbed of cybercrime [8]. According to statistics, bitcoins money laundering funds amount to billions of dollars. Therefore, in addition to administrative means, it is necessary to study targeted regulatory techniques to check and control illegal activities. There are many companies and research institutes specializing in blockchain monitoring techniques, such as Chainalysis [9], Elliptic [10], and Blockchain Intelligence Group [11]. It can be seen that the great application potential of the blockchain in the financial field has made it urgent to support privacy protection and supervision. There is no solution to achieving privacy and supervision simultaneously. Therefore, it remains for researchers to design an effective digital trading system with privacy and supervision and solve the limitations of previous works.

In this paper, we solve the preceding challenging task by proposing a novel scheme, named DAPS, which combines zero-knowledge proof, asymmetric cryptography and other cryptographic primitives. Specifically, every user's identity is committed and associated cryptographically one-to-one with the currency he owns, making all identities and currency legal and valid; every party encrypts the transaction content, attaches the corresponding zero-knowledge proof and then publish it, so that other parties can verify the correctness of the transaction content without obtaining real content. Thereby, we achieve complete privacy of the transaction party's identity and transaction content. At the same time, the combination of zero-knowledge proof and asymmetric cryptography enables the transaction to be transparent to the supervisor, thus adding the supervision function. The final experiment results further confirm the feasibility of our scheme.

As a consequence, our scheme overcomes all privacy limitations in previous works, and outperforms these solutions in terms of supervision. A high-level comparison between our scheme and previous solutions is described in Table 1. The main contributions of our scheme are summarized as below:

- We propose a new multi-party digital trading system for realizing verifiable full privacy under supervision management by masterly combining zero-knowledge proof with public key cryptography. Our scheme can achieve perfect identify and transaction privacy for all users except the specific supervisor.
- To the best of our knowledge, our proposed DAPS scheme is the first digital trading system with supervision management, which supports regulatory authorities' supervision and prevents crimes resulting from unconditional privacy.
- We theoretically define and demonstrate the privacy guarantee that zero-knowledge proof brings to the entire system framework, and test the feasibility of our scheme in BlockChain OpenSource (BCOS) [12] to confirm its feasibility.

**Table 1.** The comparison among previous schemes and ours

| Scheme | Identity privacy | Content privacy | Supervision |
|--------|------------------|-----------------|-------------|
| [4] | Imperfect privacy | No privacy | No |
| [5] | Imperfect privacy | Imperfect privacy | No |
| [6] | Privacy | Privacy | No |
| [7] | Perfect privacy | Perfect privacy | No |
| Ours | Perfect privacy | Perfect privacy | Yes |

## 2   Related Work

Privacy in blockchains falls into two categories, namely identity privacy and content privacy.

**Identity Privacy:** Bitcoin [4] is the first application of the blockchain technology in the field of digital currency with "mix" mainly used. This defense method is widely used in digital currency fields, such as Bitlaunder [13], Bitcoin Fog [14] and Blockchain.info [15]. However, due to the need for third-party nodes to provide "hybrid", there are obvious shortcomings in identity privacy protection, and attackers can obtain identity privacy in a variety of ways. Many improved methods have emerged for these deficiencies. Bonneau et al. [16] proposed an improved centralized hybrid solution mixcoin. Based on mixcoin, Valenta and Rowan [17] further optimized the centralized hybrid scheme using the blind signature technology to ensure that third-party nodes can provide mixed services normally, but cannot map output addresses and input addresses simultaneously.

ShenTu and Yu [18] proposed a blind signature hybrid scheme based on elliptic curve, which can improve computational efficiency on the basis of guaranteeing anonymity. The anonymous digital currency Dash (DASH) [5] launched in 2015 is the first digital currency to protect privacy. The core technology is Darksend, an upgraded version of CoinJoin. It effectively avoids instability caused by the third-party and makes it impossible to track transactions. Monero (XMR) [6] uses ring signatures and stealth addresses to obfuscate the origins and destinations of all transactions, ensuring that all transactions remain fully irrelevant and untrackable. Zcash [7] designs the relationship of the user's public key to his private key, commitment of the trader's identity and corresponding zero-knowledge proof accompanied. So that other users can only verify that the trader's identity is legal but cannot obtain any information about his identity. In this way complete identity privacy is achieved.

**Content Privacy:** In the early blockchain digital currency applications, transaction contents were usually public and did not have any additional protections. Currently, there are two methods to achieve a certain degree of content privacy: data confusion and data encryption. There are two schemes that can truly achieve content privacy. Monero (XMR) [6] uses Ring CT [19,20] to obfuscate transaction contents; Zcash [7] uses a commitment function to encapsulate the contents of each transaction into a number of parameters and proves transactions using zero-knowledge proof technique—zk-SNARKs [21]. The proof process does not disclose any relevant information, so it can hide transaction contents.

Among all of the above work, only Zcash can simultaneously achieve identity and content privacy we expect, but the process of generating proof using zk-SNARKs is very long, making efficiency a bottleneck. And none has the supervision function.

## 3   Problem Statement

### 3.1   System Model

As we can see from Fig. 1, the system model of our scheme includes a currency issuer, a supervisor, and users making transactions. The issuer is used to convert users' money into tokens in the system and is fully-trusted. The user wants to conduct a transaction with somebody, but isn't willing to reveal both parties' identifies and transaction contents to others except the supervisor. Meanwhile, there should be a party playing a role as the supervisor, who has the right to view all the operations in this system. There are also two Merkle trees, one being users' public keys tree to store address public keys of all users who have registered into the system, and another being a commitment tree used to store all commitments of tokens that have been converted (Commitment will introduce detailly in Digital currency). Registration transactions, purchase transactions, and transfer transactions are published by the supervisor, currency issuers and users respectively, and publicly verified and recorded on the blockchain by all miners.

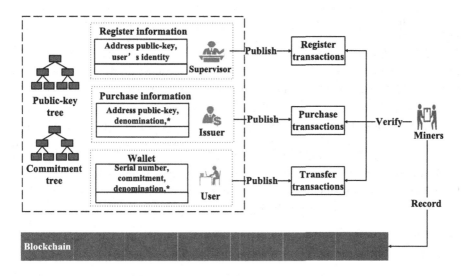

**Fig. 1.** System model.

## 3.2 Data Structure

### Digital Currency

In this section, we will introduce the structure of the digital currency we designed and how to verify.

- Structure

A digital currency in Fig. 2 owned by a user consists of four elements: denomination, serial number, commitment, and random number. These four elements are closely bound, inseparable, and can only be generated by the currency owner.

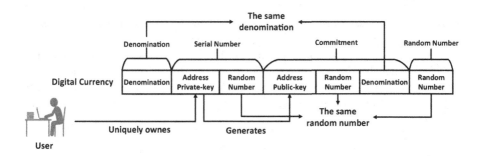

**Fig. 2.** System currency structure.

**Commitment:** It consists of the user's address public key, denomination and a random number, and is used to describe the binding relationship between the user's address public key and denomination. The structure of commitment is $CM = COMM(k, g^{(v+\rho)}, \rho)$, where COMM is a statistically-hiding commitment scheme, $k = H_{256}(U_{a_{pk}}||\rho||v)$, $U_{a_{pk}}$ is the user's address public key, $\rho$ is a random number with 256 bits, and v is the casting 64bits denomination of the digital currency. Once the currency is purchased, its corresponding commitment will be deposited into the system commitment tree.

**Serial Number:** It is the identifier of the user who effectively owns a digital currency. It is a hash value of the user's address public key and the random number. Its structure is $SN = H_{256}(U_{a_{sk}}||\rho)$, where $U_{a_{sk}}$ is user's address private key, and $\rho$ is a random number equal to the random number in commitment. Once the currency is spent, its corresponding serial number will be placed on the blockchain ledger L for everyone to query.

- Verification

A digital currency can be spent depends mainly on two aspects:

(a) Whether its commitment is in the system's commitment tree, and if not, the digital currency is considered nonexistent.
(b) Whether its serial number is already recorded on the blockchain ledger, if so, it means that the digital currency has been spent and cannot be spent any more. This is a way to solve *double spending*.

## 4   Our Construction

In this section, we describe the proposed decentralized anonymous payment scheme with supervision, named DAPS, which leverages a combination of five steps.

### 4.1   Overview

Our scheme proposes to introduce supervision on the basis of privacy protection, in view of various criminal activities in the current digital currency trading market resulting from the lack of supervision. In our design, a user must first **registers** with the supervisor, provides his own address public key and identity information. Then he **purchases** the digital currency from the issuer before **transferring** the transaction. Transaction contents are **verified** and recorded by miners.

### 4.2   Scheme Details

Here, we outline our construction in five incremental steps: register, purchase, transfer, receive and verify. We introduce our solution by taking one-to-one anonymous transactions as an example. Many-to-one, many-to-many anonymous transactions can be extended. Figure 3 gives the detail of each step.

- **Setup($1^\lambda$):**
  INPUT: security parameter $\lambda$
  OUTPUT: public parameters pp
  1)Construct: $C_{trans}$ for TRANS-FER at security $\lambda$;
  2)Compute:
  $(pk_{tr}, vk_{tr}) \leftarrow KeyGen(1^\lambda; C_{trans})$;
  3) Compute:
  $pp_{enc} \leftarrow G_{enc}(1^\lambda)$;
  4) Compute: $pp_{sig} \leftarrow G_{sig}(1^\lambda)$;
  5) Set $pp = (pk_{tr}; vk_{tr}; pp_{enc}; pp_{sig})$;
  6) Output pp.

- **CreatAddress:**
  INPUT: public parameters pp
  OUTPUT: public address pair: $(addr_{pk}, addr_{sk})$
  1) Compute
  $(e_{pk}; e_{sk}) \leftarrow Kenc(pp_{enc})$;
  2) Randomly sample a seed $a_{sk}$;
  3) Compute $a_{pk} \leftarrow H_{256}(a_{sk}, 0)$;
  4) Set $addr_{pk} = (a_{pk}; e_{pk})$;
  5) Set $addr_{sk} = (a_{sk}; e_{sk})$;
  6) Output public address pair: $(addr_{pk}, addr_{sk})$.

- **Register:**
  INPUT:
  · public parameters: $a_{pk}$;
  · user's real identify information.
  OUTPUT: register transaction $MSG_{reg}$.
  1) Compute the hash of the user's address public key
  $H_{pk} \leftarrow H_{256}(\rho, a_{pk})$;
  2) Sign $\sigma_G \leftarrow Sig_{G_{sk}}(\rho, H_{pk})$;
  3) Construct
  $MSG_{reg} = (ID_{reg}, Type_{reg}, \sigma_G, H_{pk}, rt_{pk})$;
  4) Output $MSG_{reg}$.

- **Purchase:**
  INPUT:
  · public parameters:pp;
  · coin value:v;
  · destination address public key

$addr_{pk}$.
OUTPUT: coin c and purchase transaction $MSG_{pur}$
1) Parse $addr_{pk}$ as $(a_{pk}; e_{pk})$;
2) Randomly sample a number $\rho$;
3) Compute $k \leftarrow H_{256}(a_{pk}, \rho, v)$;
4) Compute
$CM_{pur} \leftarrow COMM(k, g^{v+\rho}, \rho)$;
5) Compute $SN \leftarrow H_{256}(a_{sk}, \rho)$;
6) Set $c = (v; \rho, a_{pk}, CM_{pur}, SN)$;
7) Compute $\pi_p$;
8) Set
$MSG_{pur} = (ID_p, Type_p, CM_{pur}, \sigma_I, E_G, \pi_p, rt_{cm})$;
8) Output c and $MSG_{pur}$.

- **Transfer:**
  INPUT:
  · public parameters:pp;
  · coin value:v;
  · destination address public key $addr_{pk}$
  OUTPUT: coin c and transfer transaction $MSG_{trans}$
  1) For each $i \in \{S, R\}$,
  a. Query $SN^{old}$ and $CM_S^{old}$ from old coin;
  b. Query $rt_{cm}, rt_{pk}$;
  c. Randomly sample $\rho_S^{new}, \rho_R^{new}$;
  d. Compute
  $k_i^{new} \leftarrow H_{256}(U_{pk}^{new}, \rho_i^{new}, v)$;
  e. Compute
  $CM_i^{new} \leftarrow (k_i^{new}, g^{v_i^{new}+\rho_i^{new}}, \rho_i^{new})$;
  f. Set $c_i^{new}$ as $(v_i^{new}, \rho_i^{new}, U_{pk}^{new}, CM_i^{new}, SN_i^{new})$;
  2) Compute
  $G \leftarrow Enc_{G_{pk}}(v_R, S_{a_{pk}}, R_{a_{pk}})$;
  3) Set $E_R \leftarrow Enc_{R_{e_{pk}}}(v_R^{new}, \rho_R^{new})$;
  4) Compute $\pi_t$;
  5) Set
  $MSG_{trans} = (ID_{tr}, Type_{tr}, CM_S^{new}, CM_R^{new}, SN^{old}, G, E_R, \pi_t, rt_{cm})$;
  6) Output $c_1^{new}, c_2^{new}, MSG_{trans}$.

**Fig. 3.** Construction of our DAPS scheme (part I)

**Step1: Register**

When registering, the user first submits his address public key and identity information to the supervisor, and the supervisor records the information in the user information database maintained by the supervisor. Then the supervisor samples a random number r, calculates the hash of user's public key $H_{pk}$, signs $\sigma_G = Sig_{G_{pk}}(\rho, H_{pk})$, constructs and posts the register transaction information $MSG_{reg} = (ID_{reg}, Type_{reg}, \sigma_G, H_{pk}, rt_{pk})$, where $rt_{pk}$ is root of public key tree.

**Step2: Purchase system tokens**

A user U samples a random number $\rho$ and uses denomination v he wants to buy to generate a commitment $CM_{pur} = COMM(k, g^{(v+\rho)}, \rho)$, and a proof $\pi_p$, proving that $U_{a_{pk}} == H_{256}(U_{a_{sk}})$. The message $UMSG_{pur} = (U_{apk}, CM_{pur}, \pi_p, v, \rho)$ is then generated and sent to issuer. The issuer retrieves user's purchase denomination v' according to the his address public key $U_{a_{pk}}$, and verifies that:

- $v' \geq v$ (after the transaction is successful, the user's purchase denomination is $v' - v$);
- $CM_{pur}$ is correctly calculated;
- $\pi_p$ can pass verification to prove that the user who calculated commitment is the user who has the correct private key corresponding to the public key $U_{a_{pk}}$.

Then, issuer constructs and posts the purchase transaction information $MSG_{pur} = (ID_p, Type_p, CM_{pur}, \sigma_I, E_S)$.

**Step3: Transfer.** A sender S transfers money to receiver R. S uses a currency with denomination of $v_{old}$ (old coin) to pay R $v_R$, and the change's denomination is $v_S$ (two new coins). The corn component of transfer transaction is proof $\pi_t$ generated by S, as what follows:

- The formats of the old currency and the new currency commitment is complete. That is, $CM_y^x = COMM(k_y^x, g^{(v_x + \rho_x)}, \rho_x)$, where x = (new, old), y = (S, R).
- The real identity information of S corresponds to his address. That is, $S_{a_{pk}} == H_{256}(S_{a_{sk}})$;
- The old currency's serial number is calculated correctly. That is, $SN^{old} == H_{256}(S_{a_{sk}} || \rho_{old})$;
- The commitment of old currency belongs to commitment tree. That is, the node corresponding to $CM_R^{old}$ can be found on the tree by recalculating root value $rt'$ with $rt' == rt_{cm}$.
- The hash of address public keys of S and R belongs to user public key tree, similarly by calculating $rt'$ with $rt' == rt_{pk}$.
- The sum of old currency amounts equals to the sum of new currency amounts: $v_{old} == v_R + v_S$.
- The address public key and amount encrypted with the supervisor's public key is indeed address public key of S and R and payment amount: $Enc_{G_{pk}}(v_R, S_{a_{pk}}, R_{a_{pk}}) == G$.

- Receive:
  INPUT:
  · public parameters:pp;
  · recipient address key pair: $R_{a_{pk}}, R_{a_{sk}}$;
  · the current ledger L.
  OUTPUT: set of received coins.
  1) Decrypt $E_R$, get $v_R$ and $\rho_R$;
  2) Compute
  $CM' \leftarrow COMM(k_R, g^{(v_R+\rho_R)}, \rho_R)$;
  3) Verify whether
  $CM' == CM_R^{new}$;
  4) Verify whether $v_R$ is the amount that the payer S should pay to payee R;
  5) Compute
  $SN_R \leftarrow H_{256}(R_{a_{sk}}, \rho_R)$;
  6) Store $CM_R^{new}, v_R, \rho_R, SN_R$ into R's wallet.
- Verify:
  INPUT:
  · public parameters:pp;
  · a transaction message MSG;
  · the current ledger L.
  OUTPUT: bit b, equals 1 iff the transaction is valid.
  1.Register Transaction
  1) Parse $MSG_{reg}$ as
  $(ID_{reg}, Type_{reg}, \sigma_G, H_{pk}, rt_{pk})$;
  2) If $\sigma_G$ appears on L, output b=0;
  3) Compute $\sigma_G' = Sig_{G_{pk}}$;
  4) If $\sigma_G' \neq \sigma_G$, output b=0;
  5) Add $H_{pk}$ into the users'public

key tree, update every value of the nodes in the tree;
6) Record the root value of the public key tree $rt_{pk}$ and $MSG_{reg}$ into blockchain L.
2.Purchase Transaction
1) Parse $MSG_{pur}$ as
$(ID_p, Type_p, CM_{pur}, \sigma_I, E_G, \pi_p, rt_{cm})$;
2) Set $CM' \leftarrow H_{256}(v, k)$;
3) If $CM' \neq CM_{pur}$, output b=0;
4) If $\sigma_I$ appears on L, output b=0;
5) Compute
$\sigma_I' = Sig_{I_{pk}}(ID_p, CM_{pur})$;
6) If $\sigma_I' \neq \sigma_I$, output b=0;
7) Add the hash of $CM_{pur}, H_{CM_{pur}}$ into commitment tree, update every value of nodes in the tree;
8) Record the root value of the commitment tree $rt_{cm}$ and $MSG_{pur}$ into ledger L.
3.Transfer Transaction
1) Parse $MSG_{trans}$ as
$(ID_{tr}, Type_{tr}, CM_S^{new}, CM_R^{new}, SN^{old}, G, E_R, \pi_t, rt_{cm})$;
2) If $SN^{old}$ appears on L, output b=0;
3) If $rt_{cm}$ appears on L, output b=0;
4) Verify $\pi_t$.
5) Record the root value of the commitment tree $rt_{cm}$ and $MSG_{trans}$ into blockchain L

**Fig. 4.** Construction of our DAPS scheme (part II)

**Step4: Reiceive.** The payee R receives $MSG_{trans}$ and decrypts $E_R$ with its own private key $R_{e_{sk}}$ to obtain $v_R, \rho_R$. Then he calculates $CM' = H_{256}(k_R, g^{(v_R+\rho_R)}, \rho_R)$ and compares $CM'$ with CM. If it is equal, it further confirms that the received amount $v_R$ is indeed the amount that should be paid. Otherwise, R terminate the operation. Finally, R calculates serial number $SN_R$ of received cion with its own private key $R_{a_{sk}}$ and $\rho_R$ previously obtained. And R will put $CM_R^{new}, v_R, \rho_R, SN_R$ into his wallet.

**Step5: Verify**

Verification includes verification of three kinds of public transaction (register, purchase, and transfer) information by all miners. Specific details are shown in Fig. 4.

# 5  Security Definitions and Analyses

In this section, we analyze the security properties of our proposed scheme and show that it achieves the defined security goals.

## 5.1  Formal Security Definitions

We define the following security games between a challenger C and an adversary A. **Definition 1: (Indistinguishability).** A DAPS scheme in Fig. 3 satisfies ledger indistinguishability if for any no bounded adversary A, the probability of successfully distinguishing between two ledgers $L_0$ and $L_1$, constructed by A using queries to two DAPS scheme oracles is negligible. **Setup:** A challenger samples a random bit b and initializes two DAPS scheme oracles $O_0^{DAPS}$ and $O_1^{DAPS}$, maintaining ledgers $L_0$ and $L_1$.

Query: An adversary issues queries in pairs of the matching query type. There are four types that A can request: CreatAddress, Purchase, Transfer and Receive. If query type is CreateAddress, then the same address is generated at both oracles. If it is to Purchase, Transfer or Receive, then Q is forwarded to $L_0$ and to $L_1$. The adversary's queries are restricted in the sense that they must maintain the public consistency of the two ledgers. **Challenge:** The challenger provides the adversary with the view of both ledgers, but in randomized order: $L_{st} := L_b$ and $L_{nd} := L_{1-b}$.

**Guess:** A takes a guess $b'$ of b.

A wins when he can distinguish whether the view he sees corresponds to b = 0 or to b = 1. Ledger indistinguishability requires that A wins with probability at most negligibly greater than 1/2. **Definition 2: (Transaction non-malleability).** A DAPS scheme in Fig. 3 satisfies transaction non-malleability if for any no bounded adversary A, the probability of successfully modifying others' transactions before they are added to the ledger is negligible.

Adversary A wins the game if the result of interaction between A and scheme oracle $O^{DAPS}$ manages to modify some previous transfer transaction to spend the same coin in a different way. Transaction non-malleability requires that A wins with only negligible probability. More specifically, A adaptively interacts with a DAPS scheme oracle $O^{DAPS}$ and then outputs a transfer transaction $tx'$. Letting T denote the set of transfer transactions returned by $O^{DAPS}$, and L denote the final ledger, A wins the game if there exists tx ∈ T, such that (i) $tx' =$ tx; (ii) $tx'$ reveals a serial number contained in tx; and (iii) both tx and $tx'$ are valid with respect to the ledger L containing all transactions preceding tx on L.

## 5.2   Security Analyses

The use of zero-knowledge proof ensures that the user must follow all established principles when trading, including the legality of the parties of the transaction, the legality of the transaction currency, the total amount before and after the transaction, the transaction process in accordance with the system rules, etc. As long as there is a false behavior, the corresponding Zero-knowledge proof cannot be verified in the subsequent public validation phase.

Our scheme adds public key encryption to the NP statements based on zk-snark, and inherits the security of zk-snark and Zcash. For the proof of the first two properties, refer to the Zcash extended vision [22].

# 6   Performance Evaluation and Analyses

In this section, we use the experiment results to demonstrate the performance of our proposed scheme. We ran our scheme with the C language on a Linux server, selected BCOS [12] as the blockchain framework and evaluated its performance by running Ubuntu 16.04 with the Intel Core i7-7700 CPU. The PBC library and OpenSSL are leveraged for encryption and decryption involved in the scheme; *gadgetlib1* in *libsnark* [23] which is developed by the SCIPR-Lab project is used to implement zero-knowledge proof.

To measure the performance of our scheme, we ran several experiments. Table 2 reports performance characteristics of the resulting zk-SNARK for $\pi_t$ which had a single-thread performed on a desktop machine.

**Table 2.** Performance of our zk-SNARK for transfer transaction

| KeyGen | Time | 5 min32 s |
|--------|------|-----------|
|        | Proving key | 251 Mb |
|        | Verification key | 1023b |
| Prove  | Time | 3 min11 s |
|        | Proof | 1609b |
| Verify | Time | 59 ms |

Our KeyGen time depends on the number of tree layers and the number of hashes. The number of tree layers can be set by the user and we temporarily set it as 32 here. The final time of KeyGen is 5 min32 s, which is shorter than that of Zcash (7 min48 s). The reasons are that Zcash uses a full circuit, the number of hashes of the tree is more than 30 and there are other operations. Our Proof time is 3 min11 s, which is slightly longer than that of Zcash (2 min55 s), because we introduce zero- knowledge proof of elliptic curve encryption in $\pi_t$. But our scheme focuses mainly on adding supervision to distributed anonymous payment environments, so this time gap can be compromised in the face of new supervision

**Table 3.** Performance of our zk-SNARK for transfer transaction

| Intel Core i7-7700 CPU 3.60 GHZ with 4 GB of RAM | | |
|---|---|---|
| Setup | Time | 5 min3 s |
|  | PP | 251 Mb |
| Register | Time | 21 ms |
|  | U-apk | 256b |
|  | U-ask | 256b |
|  | MSGreg | 68b |
| Purchase | Time | 40 s |
|  | Coin c | 148b |
|  | MSGpur | 1906b |
| Transfer | Time | 3 min2 s |
|  | MSGtran | 1028b |
| Verify | Time | 59 ms |
| Receive | Time | 27 ms |

features. Our Verify time is 59 ms, which is determined to some degree by the equipment we used, and this time is still acceptable in practical payment.

In Table 3, we report performance characteristics for each of the six DAPS scheme algorithms in our scheme. Note that these values do not include the costs of maintaining the Merkle tree, because it's not the responsibility of these algorithms. Moreover, for Verify Transaction, we report the cost of verifying purchase and that of transfer transactions separately. Finally, for the case of Receive, we report the cost to process a given transfer transaction in L.

As we can see in Table 3, the Setup, Purchase and Transfer times are slightly longer, because there is a construction of proving and verifying the key pair in Setup and a construction of zk-SNARK in the last two steps, which involves the conversion of the calculation program to the arithmetic circuit, the construction of the constraint system and so on. In addition to the Register and Verify steps, other steps take a very short time to meet practical requirements.

## 7   Conclusion

To improve privacy protection in multi-party transfer scenarios based on the blockchain, we have developed a DAPS scheme by taking advantages of commitment, zero knowledge proof and asymmetric cryptography. By our scheme, we can achieve perfect identity and content privacy, and achieve supervision at the same time without losing privacy. Through theoretical analysis and experiments on BCOS, we demonstrate that our scheme can protect users' privacy and work efficiently. We also performed it and plan to improve its performance as our future work.

**Acknowledgment.** This work is supported by the Key Program of NSFC-Tongyong Union Foundation under Grant U1636209, the National Natural Science Foundation of China under Grant 61902292, the Key Research and Development Programs of Shaanxi under Grants 2019ZDLGY13-07 and 2019ZDLGY13-04.

# References

1. Zyskind, G., Nathan, O.: Decentralizing privacy: using blockchain to protect personal data. In: Security and Privacy Workshops (SPW), pp. 180–184. IEEE (2015)
2. Kosba, A., Miller, A., Shi, E., et al.: Hawk: the blockchain model of cryptography and privacy-preserving smart contracts. In: 2016 IEEE Symposium on Security and Privacy (SP), pp. 839–858. IEEE (2016)
3. Lazarovich, A.: Invisible Ink: blockchain for data privacy. Massachusetts Institute of Technology (2015)
4. Nakamoto, S.: Bitcoin: A Peer-to-Peer Electronic Cash System (2008). https:// bitcoin.org/bitcoin.pdf
5. Duffield, E., Diaz, D.: Dash: a privacy-centric crypto-currency. https://en. wikipedia.org/wiki/Dash_(cryptocurrency)
6. Monero: https://www.mendeley.com/catalogue/cryptonote--v--20/
7. Sasson, E.B., Chiesa, A., Garman, C., et al.: Zerocash: Decentralized Anonymous Payments from Bitcoin. In: Security and Privacy. IEEE (2014)
8. Wright, A., De Filippi, P.: Decentralized blockchain technology and the rise of lex cryptographia (2015)
9. Chainalysis: https://www.chainalysis.com/
10. Elliptic: https://www.elliptic.co/
11. Blockchain Intelligence Group: https://blockchaingroup.io/
12. BCOS: http://www.bcos.net.cn
13. Kaminsky, D.: Black Ops of TCP/IP 2011. https://dankaminsky.com/2011/08/ 05/bo2k11/
14. Biryukov, A., Khovratovich, D., Pustogarov, I.: Deanonymisation of clients in bitcoin P2P network. In: ACM SIGSAC Conference on Computer & Communications Security. ACM (2014)
15. Reid, F., Harrigan, M.: An analysis of anonymity in the bitcoin system. In: Altshuler, Y., Elovici, Y., Cremers, A., Aharony, N., Pentland, A. (eds.) Security and Privacy in Social Networks, pp. 197–223. Springer, New York (2011). https://doi. org/10.1007/978-1-4614-4139-7_10
16. Bonneau, J., Narayanan, A., Miller, A., Clark, J., Kroll, J.A., Felten, E.W.: Mixcoin: anonymity for bitcoin with accountable mixes. In: Christin, N., Safavi-Naini, R. (eds.) FC 2014. LNCS, vol. 8437, pp. 486–504. Springer, Heidelberg (2014). https://doi.org/10.1007/978-3-662-45472-5_31
17. Valenta, L., Rowan, B.: Blindcoin: blinded, accountable mixes for bitcoin. In: Brenner, M., Christin, N., Johnson, B., Rohloff, K. (eds.) FC 2015. LNCS, vol. 8976, pp. 112–126. Springer, Heidelberg (2015). https://doi.org/10.1007/978-3-662-48051-9_9
18. ShenTu, Q., Yu, J.: A blind-mixing scheme for Bitcoin based on an elliptic curve cryptography blind digital signature algorithm. arXiv preprint arXiv: 1510.05833, October 2015. https://arxiv.org/abs/1510.05833
19. Ring CT: https://eprint.iacr.org/2015/1098.pdf
20. Shen-Noether MRL. Ring CT for MONERO. https://pdfs.semanticscholar.org/ b9a3/8373a2fe3f224451b07ff3d7664e1b18b2b4.pdf

21. Ben-Sasson, E., Chiesa, A., Genkin, D., Tromer, E., Virza, M.: SNARKs for C: verifying program executions succinctly and in zero knowledge. In: Canetti, R., Garay, J.A. (eds.) CRYPTO 2013. LNCS, vol. 8043, pp. 90–108. Springer, Heidelberg (2013). https://doi.org/10.1007/978-3-642-40084-1_6
22. Ben-Sasson, E., et al.: Zerocash: decentralized anonymous payments from Bitcoin (extended version). Cryptology ePrint Archive (2014)
23. Libsnark: https://github.com/scipr-lab/libsnark

# An Approach of Secure Two-Way-Pegged Multi-sidechain

Jinnan Guo[1], Keke Gai[2(✉)], Liehuang Zhu[2], and Zijian Zhang[2,3]

[1] School of Information and Electronics, Beijing Institute of Technology,
Beijing, China
1120162383@bit.edu.cn
[2] School of Computer Science and Technology, Beijing Institute of Technology,
Beijing, China
{gaikeke,liehuangz}@bit.edu.cn
[3] School of Computer Science, University of Auckland, Auckland, New Zealand
zhang.alex@auckland.ac.nz

**Abstract.** As a temper-resistant ledger, blockchain ensures integrity of transaction information among trust-less participants in peer to peer network. Thus, blockchain has attracted enormous research interests in the past decade due to its prior application in cryptocurrency, financial auditing, supply chain management, etc. However, blockchain scalability limitations has impeded blockchain technology from large scale commercial applications. Since blockchain is low in throughput, blockchain in incapable of handling large scale asset transfer. To tackle this problem, in this paper, we propose a secure multi-sidechain system. The proposed approach can transfer assets simultaneously to increase throughput. In addition, security of assets during transfer was also ensured by implementing firewall property. Detailed adversarial analysis shows this proposed approach can (1) prevent double spending and transaction ordering dependence attacks during asset transfer, (2) protect mainchain from sidechain's catastrophic failure, (3) apply multi-sidechain model with different functions in each sidechain.

**Keywords:** Blockchain · Sidechain protocol · Scalability · High throughput · Global consensus

## 1 Introduction

Blockchain is a distributed, temper-resistant and jointly maintained ledger that is constructed in a decentralized manner. In the blockchain network, trust-less participants can transfer assets securely without trusted parties as long as the majority of participants are well-behaved. Thus, blockchain could eliminate the threat of single point failure, such as compromised central controller that disabled the whole system. In addition, incentive mechanisms are introduced to encourage miners in maintaining the distributed ledgers. Reward to miners, e.g. Bitcoin [1], leads to the emergence of the cryptocurrency market.

© Springer Nature Switzerland AG 2020
S. Wen et al. (Eds.): ICA3PP 2019, LNCS 11945, pp. 551–564, 2020.
https://doi.org/10.1007/978-3-030-38961-1_47

Bitcoin was invented by Nakamoto [1] in 2008 as the first widely used application of blockchain technology. Since then, blockchain-based cryptocurrency market has been explosively developed. Even one Bitcoin's value exceeded $1000 in the mid-2019. According to Morgan's technical report [2], the peak value of the whole cryptocurrency market has reached near $ 800 billion, with $ 2 trillion potential market. Attracted by the high-rewarding cryptocurrency, both Facebook and Morgan have participated in the cryptocurrency market [3].

The executable script i.e., smart contract, is a developer defined protocol that attaches to blockchain with computational capability. This breakthrough enables blockchain to possess the capability of handling complicated computational applications other than cryptocurrency. For example, in the field of data provenance [4–6], blockchain is applied to create temper-resistant operation log files without trusted auditors. In the payment management for outsourced services [7,8], blockchain ensures the fairness between service providers and users. Blockchain also has priori performance in edge computing [9], energy trading [10], voting [11], etc., due to its decentralized and trust-less property.

Despite numerous research interests in academia, blockchain rarely has large-scale applications as industrial products due to lack of scalability. Blockchain-related applications can hardly offer services to large scale of users. Blockchain's scalability issue has becoming a big drawback that impedes the blockchain from further development [12]. Two scalability issues are involved at current stage. The first challenge is the restriction of the low throughput considering high efficiency in demands of contemporary networking services [13]. Low throughput bottleneck limits blockchain's ability to simultaneously process asset transfers among large scale of users. Bitcoin network's throughput is 7 transactions per second (TPS), and Ethereum's throughput is around 50 TPS. However, the widely used visa system needs to process 50000 TPS. It implies that the blockchain's throughput is far behind the requirement to construct commercial payment system. The other challenge is the storage burden. The size of Bitcoin network has approached 210,557 MB in the beginning of the 2019 and its size has been continuously increasing as long as the Bitcoin is still working. Full nodes for verifying transactions need to store a complete copy of data in the Bitcoin network. It also implies that Bitcoin users who run the full nodes are encountering a severe storage challenge. One of the negative consequences is causing centralization when the storage burden discourages PC miners to participate in mining and results in the decrease of PC miners.

To address the scalability challenges of blockchain, in this paper, we introduce a secure multi-sidechain architecture. By this method, blockchain could dealing with different transactions in different sidechains in parallel to increase the throughput. Each node in this blockchain system only needs to store the block data generated by a few transactions on the specific blockchain. By this effort, the storage overhead is greatly reduced. In addition, different user groups manage transactions on their own sidechain. Since transactions are opened and transparent for all member in blockchain, using sidechain could mitigate the risk of privacy leakage since the sidechain is a permissioned chain. Meanwhile, some security issues emerge in the multiple sidechain system. In this work, we focus on securing the multi-sidechain model. Highlights of this work include:

1. This work proposes a secure asset transfer method to prevent double spending during the asset transfer between chains and protect parent chain from catastrophic failure of sidechain.
2. The proposed mechanism Implements a secure asset withdraw protocol to prevent transaction ordering dependence (TOD) attacks and denial of service (DOS) attack by selfish users in multi-sidechain system.
3. We have designed a novel multi-sidechain architecture. Transactions are processed in these sidechains in parallel to increase throughput. Different sidechains provide variable functions. Using permissioned sidechains can also ensure user's privacy.

**Paper Organization.** Section 2 briefly reviews related works to improve blockchain scalability. Section 3 provides the system design, while Sect. 4 explains the asset transfer algorithms. In Sect. 5 we do both simulation experiments and the security analysis to our proposed system. Finally, in Sect. 6, we draw conclusions of this work.

## 2   Related Work

In this section we reviewed previous approaches that aimed to tackle scalability challenge of blockchain.

Solutions to extend blockchain could be divided into two categories: extend on-chain and off-chain. In the on-chain extension, blockchain protocols, such as block structures or consensus mechanisms, were modified to increase the blockchain throughput. One simple way to improve scalable performance was to increase the size of the block. By doing so, more transactions can be processed in a given period of time. For example, Bitcoin SV applied dynamic 64MB size of block to increase throughput. BTC and BCH forked because BCH decided to improve scalability by increasing block size [14]. However, large block size leads to high costs in running full nodes. Thus, it sacrificed decentralization [15]. Sharding [16] was proposed in a divide-and-conquer manner to improve scalability. Luu et al. [16] introduced ELASTICO, a sharding protocol that could tolerate a quarter of Byzantine nodes in the network with 40 TPS. To further increase the throughput, Rapidchain et al. [17] used could tolerate 1/3 abnormal nodes with 7300 TPS.

Off-chain extension didn't modify blockchain protocols. By this method, majority of transactions were carried out off-chain. Sidechain technology was an important method to perform off-chain extension. Sidechain was explored by Back [15] in 2014. In this work, sidechain was realized by a two-way pegged method. In the pegged method, simplified verification proof was used to verify existence of certain transaction. To secure the coin transfer between sidechain and parent chain, conformation period and contest period were performed to lock the coin for a while in case of the potential threat such as DDoS attack and double spending.

BTC relay [18] applied one way peg technology to connect Bitcoin and Ethereum network. In this method, Ethereum was designed as the sidechain of Bitcoin. Relayers in this project relayed block headers in Bitcoin to smart contract in Ethereum. However, this project could not work due to the lack of relayers. Similarly, Teutsch et al. [19] built a dogethereum bridge to connect Ethereum and Dogecoin. Dogethereum bridge could make DOGE in Dogecoin and WOW token in Ethereum became interoperable. In the follow up study, Gazi et al. [20] proposed a proof-of-stake two way pegged sidechain based on Ouroboros consensus and Cardano blockchain. In this work, mainchain was used as the settlement layer. This layer was featureless to securely deposit money. Sidechain was the computation layer that used to run scripts to do complex computational work, such as executing smart contract. This work mathematically defined firewall property for the sidechain.

Another off-chain approach was implemented by state channel technology. State channel originated from the idea of payment channel in the Lighting Network [21]. In lighting network, both sides could finish their transaction off-chain via pre-defined payment channel. As a result, system TPS was increased by off-chain payment. State channel technology [22] could execute smart contract off-chain to further increase the blockchain scalability. However, our paper was based on two-way pegged sidechain technology. Discussions about state channel were out-of-scope in this paper. In this work, we focused on the two-way-pegged sidechain technology. We extended previous two-way-pegged single sidechain to multiple sidechain configuration.

## 3   System Design

In this section, we introduce our multi-sidechain model. We first define related terms in this multi-sidechain model. We then introduce the design purpose of this multi-sidechain system. In the third subsection we represent main phases in detail.

### 3.1   Definitions

**Definition 1** *(User set and user group).* $U = \{U_1, U_2, ..., U_i, ..., U_n\}, i \in [1, n]$ *is the total user set in this blockchain system. $n$ is the total amount of users. User group $UG_j \subseteq U, j \in [1, m]$ is the users on the jth blockchain.*

**Definition 2** *(Mainchain-sidechain relationship). In a two way pegged blockchain system, mainchain is the blockchain that can spontaneously and directly transfer assets to another blockchain that is called sidechain. In other word, mainchain could start the asset transfer between blockchains, while sidechain can only transfer asset back to mainchain after receive mainchain's asset. Set $C = \{C_1, C_2, ..., Cm\}$ is the blockchain set in this multi-sidechain system. We define $C_1$ is the mainchain, and $C_j, j \in (1, m]$ is the sidechain.*

**Definition 3** *(Conformation period). Conformation period $T_{comf}$ is the asset freeze period on the mainchain before the asset is sent to sidechain. Conformation time is the time duration of asset freezing on MC before transfer to SC. During conformation period, MC to SC transactions on the MC are verified by miners before payment. Thus, conformation period could ensure the asset is truly transferred to the locked output. We define the conformation period $T_{comf} = K_{conf} \times T_{block}$, where $K_{conf}$ is the conformation parameter set by user and $T_{block}$ is the average block generation time. For example, in Bitcoin, $K_{conf}$ can be set to 6.*

**Definition 4** *(Contest period). Contest period $T_{cont}$ is the asset locked time after assets arrived sidechain. During this period, newly arrived assets cannot be spent and delivered to other accounts in this sidechain. The reason we use the contest period is to avoid the risk of double spending that is caused by reorganization. $K_{cons}$ is the contest parameter. $T_{cons} = K_{conf} \times T_{block}$. Higher $K_{cont}$ and $K_{cons}$ can lessen the risk of double spending. However, system's efficiency is lower. Thus, trade-offs between system efficiency and security should be carefully considered.*

**Definition 5** *(Firewall). Firewall property means failure in the sidechain cannot be a threaten to the mainchain. It is because the pegged multi-sidechain system is a central hub configuration. Failure of mainchain can leads to the subsequent failure of other sidechain that connect to this mainchain. In this work, we applied the firewall property in [20]. To protect mainchain assets value, we define the total money from sidechain (SC) back to mainchain (MC) cannot exceed the money from MC to SC. Thus, for $\forall t$, asset transfers between MC and SC should satisfy Eq. (1).*

$$\int_{x=0}^{x=t} asset_{MC \to SC} dx \geq \int_{x=0}^{x=t} asset_{SC \to MC} dx. \tag{1}$$

## 3.2   System Description

In our multi-sidechain system, we apply multiple sidechains to process transactions in parallel to maximum the throughput. Mainchain and sidechains are configured in a star topology since one mainchain is connect to multiple sidechains. Figure 1 briefly shows the architecture of our multi-sidechain system. All blockchains in this system apply proof-of-work consensus [1].

This multi-sidechain system is composed of a single mainchain and multiple sidechains. Sidechains are all connected to the mainchain and form a star topology. Mainchain is responsible in asset deposit and transfer between all users in this system. Computational works such as running executable scripts cannot perform on mainchain. Such functional limitations can protect mainchain from adversarial attack against potential vulnerabilities in smart contract. Computational task can be finished on sidechain. Each sidechain is applied to perform

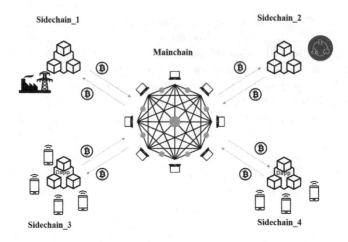

**Fig. 1.** System configuration.

specific function. For example, in Fig. 1, sidechain 1 is used to deal with power trading, sidechain 2 is applied for supply chain managing, and sidechain 3 and 4 responsible in running DApps.

It is worth noticing that each sidechain is also a permission chain. Only related stakeholders can access the transaction data on each sidechain. Thus, blockchain data privacy is protected by limiting data access from other honest but curious users that is not participate in this sidechain. Besides, multiple sidechain can process different tasks in parallel. Congestion in one sidechain cannot influence others. From the perspective of mainchain, computational works are offloaded to sidechains. This design can minimized the security risk on the mainchain. Firewall property ensures the failure from sidechain cannot affect mainchain. Thus, single sidechain failure cannot malfunction other parts in this system. In general, our multi-sidechain design makes the system robust, secure and high efficiency.

### 3.3   Threat Model

In our multi-sidechain system, we assume that majority of the miners on the mainchain are honest, i.e., dishonest miners cannot manage to launch 51% attack on the mainchain. We also assume that communications between blockchains are secure. Proof of proof-of-work informations cannot be tempered during the data packet transmission. Threats in this model can be categorized in three types: double spending attack, transaction ordering attack and sidechain 51% attack.

First, double spending attack can be launched on both mainchain and sidechain. In this model, attack is originated from the malicious asset sender who is dishonest about the transaction status after reorganization. By this effort, this sender sends money to the receiver while keep his money on the other chain. Thus, the transferred asset can be spent on both side of blockchain.

Second, transaction ordering dependence (TOD) [23] attack can also threat this multi-sidechain model. TOD attack is a race condition attack that happens during the asset withdraw process. One user start to withdraw assets from sidechain to mainchain by sending transaction. After discern this process, wealthy but malicious user withdraw high amount of asset that the sum of two withdraw asset amount is higher than the maximum amount of assets that can be unlocked on mainchain contract. Then malicious user call the withdraw function with a higher gas price to make sure its transaction can be executed before Alice's transaction. By doing so, Alice's withdraw action is failed. Thus, wealthy user can disable other users withdraw function by launching TOD attack.

Third, sidechain is under the threat of 51% attack. It is because as a subsystem, sidechain sometimes cannot obtain enough computation power to maintain its ledger. Thus, risk of 51% attacks on the sidechain is high. Under 51% attack, transactions can be counterfeit. So assets on 51% attacked sidechain is no longer valuable.

## 3.4   Main Phases

**System Initialization.** First, the system should be initialized. User accounts in this multi-sidechain system is generated and according addresses is obtained. Next, smart contracts are submitted by the transactions. During the smart contract establishment, we do the initial asset distribution to the contract account. All externally owned account as well as mainchain lock contract accounts do not have initial deposit. Sidechain deposit contracts are offered an initial deposit $D$. This initial deposits is the highest net asset transfer amount from mainchain to the $j^{th}$ sidechain. Then, mining process is activated and we can obtain the contract address after it is successfully written into blocks.

**Transfer from Mainchain to Sidechain.** After initialization, mainchain and sidechains are stepped into mining and block packaging process. Transfer assets from mainchain to sidechain contains three steps. Assume user $U_i$ with mainchain address $Add_{Si}$ is going to transfer $a_i$ amount of assets to $U_i$'s sidechain account with address $Add_{Ri}$. Asset lock addresses are mainchain deposit smart contract address for $j^{th}$ sidechain and $j^{th}$ sidechain's deposit smart contract address. Note that we only discuss interactions between mainchain and one sidechain since such interactions are occurred in parallel.

First, user $U_i$ transfer its assets $a_i$ to the smart contract address on the mainchain by invoking the contract as a send transaction manner. So a transaction $Tx$ is generated on the mainchain and broadcast to nodes in mainchain. The presence of this transaction in the blockchain is important to us to prevent double spending attacks. The contract deposits these assets after receiving them. Meanwhile, $U_i$ account on the sidechain interacting with the contract in the sidechain to inform it of the details of transaction.

Second, after the transaction is launched, the asset is locked for a conformation period. By doing so, we can make sure that assets are successfully transfered to the contract address by the aid of conformations in the following blocks.

Third, block headers on the mainchain are relayed from the mainchain to sidechain smart contract. Then the sidechain smart contract checks the validity of the transaction to make sure it is covered with enough proof-of-works. However, to make sure the relay information is still valid, any users on the mainchain are encouraged to send contest proofs during the $T_{cont}$ period. If the verification is valid, then the sidechain smart contract send the amount of $ai$ assets to the final destination address. At this point, assets successfully transfered from mainchain to sidechain.

**Transfer from Sidechain to Mainchain.** Transfer assets from sidechain back to mainchain is reverse procedure of steps mentioned in the above section. In this work, we especially focused in this process since insecure withdraw process can threaten the mainchain by TOD attack. To prevent TOD attack, before the asset transfer from sidechain to mainchain, user should first send a lock requirement to lock the two-way-pegged channel. Then the user can successfully implement the withdraw function without the threat of TOD attack. To implement this function, we deploy a control contract on the mainchain to realize the asset withdraw protocol.

However, abuse lock operation can impede withdraw process of other users. To tackle this issue, lock request should be sent with a collateral that cannot below the minimum requirement. A valid lock request can make sure only the request user can do the withdraw operation. If the user haven't start taking money from mainchain deposit contract after *challenge_time*, other users are motivated to call the challenge function to unlock the withdraw channel. It is because a successful challenge can make user earn all initial deposit that send along with the lock request.

User who successfully withdraw money from sidechain can also get its initial deposit back. Then the channel is unlocked. In addition, to prevent one user continuously occupy the withdraw channel, we define that one user cannot lock the channel again within *freeze_time* seconds.

## 4    Algorithms

### 4.1    Channel Lock and Challenge Algorithm

In this algorithm, we focus on the withdraw process i.e., transfer money from sidechain back to mainchain. In our proposed multi-sidechain model, different users might withdraw from the same deposit address. Hence the system is vulnerable to the TOD attacks. To prevent this, users can lock the withdraw channel. This algorithm is realized on the control contract in the mainchain in order to provide locking service to minimize the risk of transaction ordering attack.

The channel lock algorithm is activated during the interaction between external accounts and the control contract. To use this function, the external account that wants to lock the channel send a transaction to the control contract with collateral.

---

**Algorithm 4.1.** Channel Lock and Challenge Algorithm

---

**Require:** Sender Address $msg.sender$, collateral $msg.value$, current lock user $hold$, last lock user $lasthold$,minimum value of the collateral $coll\_require$, freeze time $freeze\_time$, lock start time $lock\_start$

**Ensure:** Current lock user $hold$, last lock user $lasthold$, lock start time $lock\_start$.

1: **function lockrequest()**
2: **if** $(hold! = address(0)) \vee (msg.value < coll\_require)$ **then**
3:    $revert()$;
4: **end if**
5: **if** $(lasthold == msg.sender) \wedge (now <= lock\_start + freeze\_time)$ **then**
6:    $revert()$;
7: **end if**
8: collateral $collateral \leftarrow msg.value$;
9: lock the channel $hold \leftarrow msg.sender$;
10: lock start time $lock\_start \leftarrow now$;
11: **end function**

**Require:** Sender Address $msg.sender$, collateral $collateral$, current lock user $hold$, last lock user $lasthold$, challenge time $challenge\_time$, lock start time $lock\_start$.

**Ensure:** Current lock user $hold$, last lock user $lasthold$, collateral $collateral$.

12: **function challenge()**
13: **if** $(hold == msg.sender) \vee (now < lock\_start + challenge\_time)$ **then**
14:    $revert()$;
15: **end if**
16: $lasthold \leftarrow hold$;
17: $hold \leftarrow address(0)$;
18: $add \leftarrow msg.sender$;
19: $add.transfer(collateral)$;
20: $collateral \leftarrow 0$;
21: **end function**

---

The algorithm firstly check the status of the current holder. If one user has already locked the channel, other users cannot lock the channel. Meanwhile, it checks the collateral carried along with the lock request. Lock requests with insufficient collateral is not valid.

Next, if the transaction sender account is the same as the last account that send lock request, the algorithm checks whether the time between two requests meets the freeze time requirement. A lock request transaction that meets all requirements above is a valid lock request. So contract record the new initial deposit value $collateral$, new user who lock the channel and the lock start time.

If the user lock channel, but not complete the withdraw process in time, other users call the challenge function to unlock the channel in case of the malicious channel occupying. First, the function checks if the time has entered the challenge

---

**Algorithm 4.2.** Withdraw Algorithm

---

**Require:** Sender Address *msg.sender*, current lock user *hold*, last lock user *lasthold*, collateral amount *collateral*, lock start time *lock_start*, deposit contract address *depo*, withdraw amount *amount*.

**Ensure:** Current lock user *hold*, last lock user *lasthold*, initial deposit value *collateral*.

1: **function takeback**(*depo,amount*)
2: **if** (*hold!* = *msg.sender*) **then**
3:    revert();
4: **end if**
5: *depo.withdraw*(*amount, hold*)
6: // function withdraw(unit amount, address hold)
7: // if (control.address == msg.sender) then
8: // hold.transfer(amount);
9: // else revert();
10: *hold.transfer*(*collateral*);
11: last lock user *lasthold* ← *hold*;
12: unlock system *hold* ← *address*(0);
13: collateral *collateral* ← 0;
14: **end function**

---

period. Then, all other users can send transaction to unlock the channel. These users are motivated to do so because they can not only get the collateral, but also release the channel.

## 4.2 Withdraw Algorithm

After the lock request is valid and funds are available to unlock, the user can then call the withdraw function to take his money back. As we can see from the algorithm, only current lock holder can use the withdraw function. After funds and the collateral are successfully send to the message sender, the system will be unlocked.

In this withdraw process, control contract called the function in the deposit contract to send asset back to the user. It is because the assets are stored in the deposit contract. Before asset transfer, deposit contract firstly check whether the function call is originated from the control contract.

## 5 Evaluations and Analysis

### 5.1 Experiment Evaluations

In this subsection we introduce our experiment environments and settings. We simulate the two algorithms and three functions proposed in the previous section. In this work, the experiment is carried out in a personal computer with Windows

10 operating system, a Intel Core i7-7500U CPU and a 8G RAM. We code and compile the smart contract on the Remix website. From the remix, we can get the gas cost of each function. Five times average gas cost when successfully executing three functions are shown the Sect. 4 is list as follows. In this simulation test, we set $challege\_time = 50\,$s and $freeze\_time = 100\,$s (Table 1).

**Table 1.** Functions gas cost

| Function | Transaction cost (gas) | Execution cost (gas) |
|----------|------------------------|----------------------|
| Lockrequest() | 47357 | 38085 |
| Challenge() | 30618 | 39346 |
| Takeback() | 60468 | 52532 |

Then, we deploy the complied contract on the geth console. By sending transactions to interact with our contracts, we can test functions of our algorithm. We check account status during the interactions, and the result shows our algorithm works well. From the perspective of average gas cost, we can see that the lockrequest() and takeback() function costs high amount of gas. Function challenge() costs smallest amount of gases. Such low gas cost of the challenge process is positive because users can be more motivated to challenge and unlock the system with lower cost. However, this takeback() function do not conclude SPV verification function, which will cause high amount of gas cost.

### 5.2  Security Analysis

**Double Spending Attack.** Double spending attacks could be launched by a malicious user. This kind of malicious user aims to find vulnerabilities in the transfer protocol in order to double spend money on both mainchain and sidechain. To prevent double spending attack, SPV proof as well as conformation and contest period is introduced. Assets are locked on the sender chain during conformation period, which can help the transaction get more conformations from other users in peer to peer network. This approach can increase the validity of the transactions on the mainchain. SPV proof checks the transaction validity after conformation period. However, reorganization on the sendchain can also in invalid the transaction. To avoid double spending cause by reorganization, asset is locked during contest period before circulating in receiver blockchain. Then, validity of transaction will be checked again under the governance of all miners in the mainchain during contest period.

**Transaction Ordering Dependence Attack.** Malicious users launch TOD attack by providing highest gasprice to make sure its withdraw transaction can be execute before others. By doing so, malicious user makes the amount of rest asset that can be unlocked insufficient to invalid other user's withdraw process. In our model, when user wants to withdraw money, they should firstly send a lock request to the control contract. During this process, malicious user cannot get the withdraw amount of others. Thus, they cannot get an accurate estimate of their withdraw value to successfully launch the attack. As a result, the possibility of TOD attack can be reduced significantly.

**51% Attack on Sidechain.** Since sidechain has fewer participants compared to mainchain, 51% attack on sidechain need to be considered. 51% attack can cause catastrophic failure on the sidechain. Attacker create arbitrary amount of assets on sidechain by 51% attack. As a result, sidechain asset depreciated significantly. However, under such failure, value of assets on mainchain is still under protected. It is because the firewall mechanism limit the amount of assets transferred from sidechain to mainchain. So mainchain assets will not be devalued by the monetary high caused by 51% attack on sidechain. In addition, since assets are not circulating between sidechains, catastrophic failure of one sidechain cannot affect any other blockchains in this system. Star topology and firewall property helps to segregated collapsed sidechain from others.

**DoS Attack.** As we discussed above, selfish users may continuously withdrawing small amount of asset to occupy the withdraw channel. Such actions severely infected the asset circulation between chains. From the view of most users, this two-way pegged system is degenerated to one-way peg since they cannot access to the withdraw service. In our model, after the successful lock request to the control contract, the deposit contract cannot access the control contract's service in the freezing period. This design eliminates the DoS attacks during withdraw process. In addition, malicious users might also lock the system by send lock request, but they do not unlock them. In addition, channel lock mechanism can also leads to DOS attack by malicious users. To prevent this DoS attack, user have to pay the collateral when they request to lock the channel. If user lock the channel but have no intention to unlock it (i.e., have not call the takeback() within *challenge_time*), other users can send challenge messages to earn the collateral and unlock the system. By doing so, malicious user is punished.

## 6    Conclusions

In this work, we proposed a secure multi-sidechain system. This multi-sidechain system realized different functions in separate sidechains to increase system throughout. To secure this system, we carefully designed asset transfer protocols. We then realize some of these protocols on the Ethereum platform. Exhaustive security analysis proved that our proposed multi-sidechain system can prevent four kinds of potential threats. In our future work, we will finish the construction of the whole multi-sidechain system on the Ethereum.

**Acknowledgement.** This work is supported by the National Natural Science Foundation of China (grant # 61972034).

# References

1. Nakamoto, S.: Bitcoin: a peer-to-peer electronic cash system (2008)
2. Morgan, J.P.: The next step for blockchain (2019). https://www.jpmorgan.com/global/research/blockchain-next-steps
3. Facebook. Libra write paper (2019). https://libra.org/en-US/white-paper/
4. Liang, X., Shetty, S., Tosh, D., Kamhoua, C., Kwiat, K., Njilla, L.: Provchain: a blockchain-based data provenance architecture in cloud environment with enhanced privacy and availability. In: Proceedings of the 17th IEEE/ACM International Symposium on Cluster, Cloud and Grid Computing, pp. 468–477. IEEE Press (2017)
5. Xia, Q., Sifah, E., Asamoah, K., Gao, J., Du, X., Guizani, M.: MeDShare: trust-less medical data sharing among cloud service providers via blockchain. IEEE Access **5**, 14757–14767 (2017)
6. Gai, K., Qiu, M.: Blend arithmetic operations on tensor-based fully homomorphic encryption over real numbers. IEEE Trans. Ind. Inf. **14**(8), 3590–3598 (2017)
7. Choudhuri, A., Green, M., Jain, A., Kaptchuk, G., Miers, I.: Fairness in an unfair world: fair multiparty computation from public bulletin boards. In: Proceedings of the 2017 ACM SIGSAC Conference on Computer and Communications Security, pp. 719–728. ACM (2017)
8. Hu, S., Cai, C., Wang, Q., Wang, C., Luo, X., Ren, K.: Searching an encrypted cloud meets blockchain: a decentralized, reliable and fair realization. In: IEEE Conference on Computer Communications (INFOCOM), pp. 792–800. IEEE (2018)
9. Gai, K., Wu, Y., Zhu, L., Xu, L., Zhang, Y.: Permissioned blockchain and edge computing empowered privacy-preserving smart grid networks. IEEE Internet Things J. **15**(6), 3548–3558 (2019)
10. Gai, K., Wu, Y., Zhu, L., Qiu, M., Shen, M.: Privacy-preserving energy trading using consortium blockchain in smart grid. IEEE Trans. Ind. Inf. (2019)
11. Zhu, L., Wu, Y., Gai, K., Choo, K.: Controllable and trustworthy blockchain-based cloud data management. Future Gener. Comput. Syst. **91**, 527–535 (2019)
12. Karame, G.: On the security and scalability of bitcoin's blockchain. In: Proceedings of the 2016 ACM SIGSAC Conference on Computer and Communications Security, pp. 1861–1862. ACM (2016)
13. Gai, K., Qiu, M., Zhao, H.: Energy-aware task assignment for mobile cyber-enabled applications in heterogeneous cloud computing. J. Parallel Distrib. Comput. **111**, 126–135 (2018)
14. Kwon, Y., Kim, H., Shin, J., Kim, Y.: Bitcoin vs. bitcoin cash: coexistence or downfall of bitcoin cash? In: Proceedings of the 2019 IEEE Symposium on Security and Privacy, pp. 1290–1306. IEEE Computer Society (2019)
15. Back, A., et al.: Enabling blockchain innovations with pegged sidechains. http://www.opensciencereview.com/papers/123/enablingblockchain-innovations-with-pegged-sidechains, p. 72 (2014)
16. Luu, L., Narayanan, V., Zheng, C., Baweja, K., Gilbert, S., Saxena, P.: A secure sharding protocol for open blockchains. In: Proceedings of the 2016 ACM SIGSAC Conference on Computer and Communications Security, pp. 17–30. ACM (2016)

17. Zamani, M., Movahedi, M., Raykova, M.: Rapidchain: scaling blockchain via full sharding. In: Proceedings of the 2018 ACM SIGSAC Conference on Computer and Communications Security, pp. 931–948. ACM (2018)

18. Chow, J.: BTC relay (2016). https://github.com/ethereum/btcrelay

19. Teutsch, J., Straka, M., Boneh, D.: Retrofitting a two-way peg between blockchains. Technical report (2018). https://people.cs.uchicago.edu

20. Gazi, P., Kiayias, A., Zindros, D.: Proof-of-stake sidechains. In: Proceedings of the 2019 IEEE Symposium on Security and Privacy, pp. 677–694. IEEE Computer Society (2019)

21. Poon, J., Dryja, T.: The bitcoin lightning network: scalable off-chain instant payments (2016)

22. Dziembowski, S., Faust, S., Hostáková, K.: General state channel networks. In: Proceedings of the 2018 ACM SIGSAC Conference on Computer and Communications Security, pp. 949–966. ACM (2018)

23. Luu, L., Chu, D., Olickel, H., Saxena, P., Hobor, A.: Making smart contracts smarter. In: Proceedings of the 2016 ACM SIGSAC Conference on Computer and Communications Security, pp. 254–269. ACM (2016)

# IoT and CPS Computing

# DCRRDT: A Method for Deployment and Control of RFID Sensors Under Digital Twin-Driven for Indoor Supervision

Siye Wang[1,2,4(✉)], Mengnan Cai[1,2(✉)], Qinxuan Wu[3], Yijia Jin[5],
Xinling Shen[1,2], and Yanfang Zhang[1,2]

[1] Institute of Information Engineering, Chineses Academy of Sciences, Beijing, China
{wangsiye,caimengnan,shenxinling,zhangyanfang}@iie.ac.cn
[2] School of Cyber Security, University of Chineses Academy of Sciences,
Beijing, China
[3] School of Computer Science and Technology, Zhejiang University, Hangzhou, China
wuqinxuan@zju.edu.cn
[4] School of Computer and Information Technology, Beijing Jiaotong University,
Beijing, China
[5] The Boeing Company, Seattle, USA

**Abstract.** In the field of indoor supervision based on RFID, the quality of monitoring is affected by how many and where the RFID sensors are deployed. Due to the limitation of time and workforce, It is a key problem to improve efficiency and to reduce the complexity of deployment. We propose a deployment & control scheme of RFID sensors based on digital-twin technology. The constructed digital-twin model can simulate the state, the performance, and the activity of physical entities. In this paper, we predict and analyze based on digital-twin technology to solve the problems of re-design & re-deployment. We further achieve the goal of saving deployment time & workforce through the intuition and virtual simulation of digital-twin. We take three problem scenarios to demonstrate the proposed RFID sensor deployment & control scheme is highly efficient and resource-saving.

**Keywords:** Indoor supervision · RFID sensors · Deployment & control · Digital-twin

## 1 Introduction

Sensor deployment & control schemes of indoor space have received extensive attention in the research field in the past two decades. The schemes achieve indoor monitoring by analyzing mass data collected through RFID, Bluetooth, WLAN, and ZigBee [1]. At present, researchers in the community mostly use

S. Wang and M. Cai—Both the authors were co-first author.

© Springer Nature Switzerland AG 2020
S. Wen et al. (Eds.): ICA3PP 2019, LNCS 11945, pp. 567–576, 2020.
https://doi.org/10.1007/978-3-030-38961-1_48

RFID which is suitable for the complex indoor environments to collect the location information in indoor space [2]. Researchers analyze the moving targets by the large-scale deployment of RFID sensors and accurate positioning technology.

The current researches mainly aim to achieve full regional coverage and high economic feasibility of deployment [3]. However, little researches considered the impact of indoor maps which are the basic element of sensor deployment. In the efficient sensor deployment & control scheme, the actual goals are to save time, save the workforce, and reduce the difficulty of deployment as much as possible. Due to the limitations of space requirements and other factors, most of the previous schemes are usually designed based on interior sketches provided by designers. The inconsistencies between the current actual situation and the design sketches affect the efficiency and complexity of deployment.

In our paper, we adopt digital-twin technology which can make the digital space and physical space co-evolve accurately. The accurate mapping function in digital-twin improve the inconsistencies between sketches and practical situations. The problems in current sensor deployment & control schemes, such as ambitious, laborious, unintuitive, and repeated construction are also solved. At the same time, every step in sensor deployment & control processes is monitored in real time. This strategy saves time and resources by making corrections at the end of each step rather than waiting until the whole deployment processes are finished.

Our contributions in this paper are summarized as follows:

(a) We formally define the problems in the actual RFID sensor deployment process within labile road network and sketches constraints.
(b) We design an appropriate scheme which solves the time-consuming and laborious problems using digital-twin technology.
(c) We propose a solution to control the life-cycle of physical entities by using digital-twin.
(d) We put forward a digital-twin-based sensor deployment algorithm which makes the comparison times and auxiliary monitoring nodes as small as possible.
(e) We conduct an empirical evaluation using fail-over scenarios, which proves the time-shorten effects and confirms the practicability, superiority, and efficiency of our proposed scheme.

The rest of the paper is organized as follows. In Sect. 2, we describe the related work of indoor RFID sensor deployment & control and digital-twin technology. In Sect. 3, we describe the proposed framework, including problem definitions, sensor deployment, and our framework. In Sect. 4, we do the demonstration and give out the evaluation results. Finally, we conclude in Sect. 5.

## 2    Background and Related Work

In the whole process of deployment & control strategy, digital-twin dynamically detects the logic problems and sensors states further to save deployment time and workforce. In this section, we introduce the relevant background of indoor RFID sensor deployment & control and digital-twin respectively.

## 2.1   Indoor RFID Sensor Deployment and Control

As for the deployment of indoor RFID sensors, various algorithms were proposed to maximize the coverage of monitoring areas by sensors [4]. Based on maximizing coverage, algorithms for the minimum sensor number were proposed. Various methods such as biological method [5] and novel normalization method [6] were used.

Most of the studies consider whether the coverage, the performance, and the number of deployed sensors are reasonable. However, few schemes take the complexity of the deployment, as well as the consumption of human resources and deploy-time into account. The site conditions and resource constraints in a real deployment environment are also needed to be considered. How to deploy the sensors reasonably and efficiently is a problem that needs to be solved.

## 2.2   Digital-Twin

The concept of digital-twin was first proposed by Professor Grieves in 2003 [7]. Researchers carried out many theoretical researches and practical explorations on how to achieve the interaction and integration of the physical world and information world [8,9]. Digital-twin played the role of predictive maintenance in the industry [10].

At present, studies based on digital-twin have made some progress in production and application. But few researches study on the control & supervision of interior space. We hold the point that the advantages of digital-twin technology in accurate mapping, real-time simulation, and digital debugging make it have great potential in the life-cycle management of physical entities.

## 3   The Proposed Framework

In this section, we first define the problems in most deployment sites. Next, we put forward our scheme to deploy sensors. Then, we propose our framework which is based on digital-twin technology. After that, we demonstrate the details of modules in the framework. The proposed deployment & control scheme of RFID sensors under digital twin-driven is called DCRRDT in our paper.

## 3.1   Problem Definitions

Our experience of deploying RFID sensors in large indoor venues suggests that the following unreasonable problems often exist in the actual sensor deployment process:

(a) The blind spots and logic problems in the thinking process when designing road network structure. That would result in irrational deployment schemes which are helpless for the monitoring areas and emergency response.
(b) The frequent changes in road network structure caused by the movements of objects. The locations of obstacles are often changed due to unpredictable factors. The changes lead to the variation of indoor monitoring areas and the need for the redeployment of sensors.

(c) The signal shielding areas or high-aesthetic areas are not allowed to deploy sensors. However, the areas are usually not known in advance through interior renderings.

(d) The deployed sensors are accidentally knocked off or damaged by moving targets or dropped the wire due to network ports and electrical ports are ill-contacted.

(e) The remote network ports and electric ports are ill-suited for sensor deployment in the actual site. However, since the construction of the indoor site has been completed, the locations of the power ports are disabled to be changed.

For these reasons, a flexible, intuitive, variable, and efficient scheme needs to be carried out. In our paper, the digital-twin is combined with indoor environmental characteristics to solve the unreasonable problems.

### 3.2    Schematic Diagram of Indoor RFID Sensor Deployment Based on Digital-Twin

We aim to create a deployment & control strategy of RFID sensors based on digital-twin. The goal of the strategy is to address the problems identified in Section *Problem Definitions*. The schematic diagram of indoor RFID sensor deployment & control based on digital-twin is called SDDDT for short in our paper, as shown in Fig. 1. In SDDDT, paths are firstly extracted from the actual deployment environment. The upper and lower boundaries of the path range are mainly extracted. Next, the path inflection points and the planned sensors are mapped. Then, the coverage of the planned sensors is indicated according to power values. Whether the coverage includes the upper and lower boundaries of all the paths is also ensured. Finally, the positions of power ports and the distance between ports and sensors are indicated.

When the road network changes due to movements of obstacles or some temporary situations, digital-twin will re-map the current road network in the

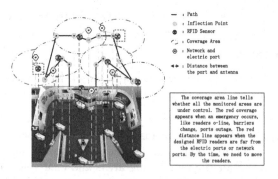

**Fig. 1.** Schematic diagram of indoor RFID sensor deployment & control based on digital-twin technology (Color figure online)

SDDDT. All the status will be re-checked. When RFID sensors are disconnected, the position of the sensors and sensor coverage will light red warning. When the distance between sensors and network ports or electrical ports exceeds a threshold value, the indicating distance will also light red warning, as shown in Fig. 1. The red light warns to modify deployment plans in time.

### 3.3 Sensor Deployment

After extracting the road network in SDDDT, the problem to be solved is how to deploy the sensors on the road network in SDDDT. In our paper, sensor nodes are divided into monitoring nodes and auxiliary monitoring nodes.

Monitoring nodes and auxiliary monitoring nodes are defined as follows:

---

**Algorithm 1.** Adjacent Monitoring Points

---

**Input:** The topological map G (V, E, M, NM).
**Output:** (a) All the adjacent monitor nodes of each monitor node. (b) Paths between each adjacent monitor nodes.
 1: V: The nodes set.
 2: E: The edges set.
 3: M: The monitoring points set.
 4: NM: The non-monitoring points set.
 5: **for** $var\ i = 0;\ i < M.length;\ i{+}{+}$ **do**
 6:    Stack push (M(i));
 7:    visited $[M[i]] = 1$;
 8:    **while** $Stack! = null$ **do**
 9:      a= the top of Stack;
10:      **for** $var j = 0;\ j < adj[a].length;\ j{+}{+}$ **do**
11:        **if** $Visited[adj[a][j]] == 0$ **then**
12:          Visited$[[a][j]]{=}1$;
13:          Stack push (adj[a][j]);
14:          **if** $adj[a][j]$ *is monitor node* **then**
15:            push adj[a][j] to the array of M[i] adjacent monitor nodes;
16:            push M[i] to the array of adj[a][j] adjacent monitor nodes;
17:            push all nodes in stack to Path$_{M[i]->adj[a][j]}$, Path$_{adj[a][j]->M[i]}$;
18:            Stack pop ();
19:          **end if**
20:        **end if**
21:        **if** $j >= adj[a].length$ **then**
22:          Stack pop ();
23:        **end if**
24:      **end for**
25:    **end while**
26:    **for** $var\ k = 0;\ k < V.length;\ k{+}{+}$ **do**
27:      Visited[k]=0;
28:    **end for**
29: **end for**

---

**Definition 1.** *For the road network graph G = (V, E) extracted according to SDDDT, V is the set of points and E is the set of edges. When monitoring nodes are added to SDDDT, paths can be distinguished. After the auxiliary monitoring points are added, it can still be ensured that the paths between the auxiliary monitoring points and the monitoring points in G can be distinguished.*

Monitoring nodes are mainly selected according to actual needs. The nodes usually locate at the vital indoor monitoring locations such as intersections and doorways [11]. Steps to select auxiliary monitoring nodes are as follows:

(a) Acquire path information between all adjacent monitoring points.
(b) Randomly select from a group of adjacent nodes as the first group of points.
(c) Calculate the occurrence frequency of other nodes in the paths except the monitoring points.
(d) Select the node whose occurrence frequency is close to half of the number of paths to dichotomize the path. The indistinguishable paths are divided into two groups of paths, one with the node and the other one without the node.
(e) Calculate the frequency of the nodes in two groups of paths respectively.
(f) Carry out the dichotomy on this group of indistinguishable paths, until the path between two adjacent monitoring points can be distinguished.

The nodes used in the above process to distinguish paths are selected as auxiliary monitoring points. In the deployment scheme, we obtain all the paths between any adjacent monitoring points firstly. Then we store the adjacent monitoring points which are next to monitoring points and the paths between adjacent monitoring points and monitoring points through *Algorithm 1*. Next, we execute from Step (b) to Step (f) in order.

In our paper, auxiliary monitoring points are selected for the indistinguishable path group between each pair of adjacent monitoring points. The selection of auxiliary monitoring points in the previous group of adjacent monitoring points affects the selection in a later group. In this paper, the idea of selecting the intermediate nodes makes the comparison less and is used to select less auxiliary monitoring points. This way of selection makes the number of deployed sensors is small.

### 3.4    Our Proposed Framework

Our proposed framework contains three layers which are responsible for different functions. The layers are: *Framework Layer*, *Engine Layer*, and *Actual Layer*. DCRRDT executes the designing, modeling, and simulation in *Framework Layer*. The results from *Framework Layer* are applied to deploy and adjust the equipment in *Engine Layer*. The intuitive monitoring interface can be observed in *Actual Layer*. Meanwhile, the three layers are constantly interacting with each other. The equipment in the *Engine Layer* collects the changing information from the *Actual Layer*, then passes the data to the *Framework Layer* to re-design, re-model, and re-simulate. The results are then applied to other layers in a cycle. The *Framework Layer* includes four components: *Modeling & Designing, Simulation, Maintenance & Control,* and *Visualization.*

The module *Modeling & Designing, Simulation* and *Maintenance & Control* in *Framework Layer* will be repeated to evaluate when the road network, obstacles, locations of equipment, etc. change. Our digital-twin model helps workers to implement parallel processing of abnormal problems in the whole processes. This essential advantage reduces the time of re-design, re-deploy, and re-manage greatly. In chapter 4, we verify the effectiveness and efficiency of our scheme.

# 4    Demonstration

The requirement in the demonstration is that only one worker is allowed to operate during one process. The worker is not allowed to perform other steps before completing the current step. Processes in different areas can be performed simultaneously. In each step, we first select the current task-free worker with the shortest completion time to execute.

In the demonstration, the basic scenario is as follows: Four workers need to execute the work and three divided deployment regions need to be controlled. According to the general sensor deployment & control strategy, each sensor needs to be carried out three processes: Design, Install, and Maintain (called D, I, M for short). When any problem occurs in the step of deployment & control, the corresponding repair step needs to be performed. In our paper, the corresponding repair steps are $D_r$, $I_r$, $M_r$. The time of each process is different to each worker, which is listed as in Table 1. The 1st column refers to the index of sensor region; The 2nd column refers to the index of the process; The 3rd to the 6th column is the processing time related to different workers. The angular $mn$ in the process label represents the n-th step in area m, such as $M_{23}$ represents process 3 (maintain process) in area 2.

**Table 1.** Case for DCRRDT.

| Sensor area | Process | Worker 1 | Worker 2 | Worker 3 | Worker 4 |
|---|---|---|---|---|---|
| 1 | $D_{11}$ | 2 | 3 | 1 | 3 |
|   | $I_{12}$ | 1 | 1 | 2 | 2 |
|   | $M_{13}$ | 2 | 2 | 1 | 1 |
| 2 | $D_{21}$ | 3 | 2 | 2 | 1 |
|   | $I_{22}$ | 1 | 2 | 1 | 2 |
|   | $M_{23}$ | 2 | 2 | 2 | 2 |
| 3 | $D_{31}$ | 2 | 1 | 2 | 2 |
|   | $I_{32}$ | 2 | 1 | 2 | 1 |
|   | $M_{33}$ | 3 | 1 | 2 | 2 |

Three scenes are provided to simulate the possible problems and the solution time. These three problematic scenes are shown in Table 2. The 1st column refers to the index of problem scene; The 2nd column refers to the index of the process in question which needs to re-work. The 3rd to the 6th column is the process of time related to different workers. The time to complete sensor deployment & control in normal circumstance is shown in Fig. 2. The respective time of sensor deployment & control in three problematic scenarios is shown in Fig. 3, 4 and 5.

**Table 2.** Case for problematic scene.

| Scene | Part | Worker 1 | Worker 2 | Worker 3 | Worker 4 |
|---|---|---|---|---|---|
| 1 | $I_{22r}$ | 2 | 1 | 3 | 3 |
|   | $D_{31r}$ | 1 | 2 | 3 | 2 |
| 2 | $D_{21r}$ | 3 | 2 | 4 | 1 |
|   | $I_{22r}$ | 4 | 3 | 3 | 2 |
|   | $M_{33r}$ | 3 | 4 | 3 | 3 |
| 3 | $I_{12r}$ | 4 | 2 | 3 | 4 |
|   | $M_{13r}$ | 2 | 2 | 2 | 3 |
|   | $M_{23r}$ | 2 | 4 | 3 | 4 |
|   | $D_{32r}$ | 2 | 2 | 3 | 3 |

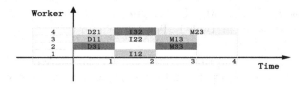

**Fig. 2.** Gantt chart for normal completion. The normal time without any emergency is four-time units under the condition in the demonstration.

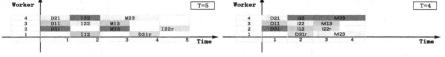

(a) Scheme without DCRRDT in Scene 1.    (b) Scheme with DCRRDT in Scene 1.

**Fig. 3.** Gantt chart in Scene 1. The scheme without DCRRDT takes five-time units. The scheme with DCRRDT takes four-time units.

We take scenario 3 as an example. In scene 3, problems happen in process $I_{12}$, $M_{13}$, $M_{23}$, and $D_{32}$, as shown in Table 2. Therefore, the repair process $I_{12r}$, $M_{13r}$, $M_{23r}$, and $D_{32r}$ need to be carried out after the above four steps. In the scheme without-DCRRDT, process D, I and M can only be carried out after $D_r$, $I_r$, and $M_r$. As shown in Fig. 5(a), process $D_{32r}$, $M_{13r}$, and $I_{12r}$ can be performed after

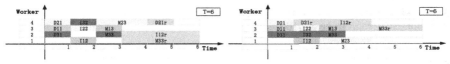

(a) Scheme without DCRRDT in Scene 2.    (b) Scheme with DCRRDT in Scene 2.

**Fig. 4.** Gantt chart in Scene 2. The scheme without DCRRDT takes six-time units. The scheme with DCRRDT takes six-time units.

process $M_{13}$ and $M_{33}$ are completed. After completing step $M_{23}$, process $M_{23r}$ can be proceeded. At this point, the scheme without-DCRRDT requires seven-time units. In the scheme DCRRDT, there is no need to wait for all the completion of process D, I, and M. Because different steps can be performed simultaneously. As shown in Fig. 5(b), process $I_{12}$, $I_{22}$, $I_{32}$, and $D_{21r}$ can be performed after process $D_{21}$, $D_{11}$, and $D_{31}$ are completed. After completing process $I_{12}$, $I_{22}$, $I_{32}$, and $D_{21r}$, process $M_{13}$, $M_{23}$, $M_{33}$, and $I_{12r}$ can be carried out. After process $M_{33}$ is finished, process $M_{33r}$ can be carried out. At this point, the DCRRDT scheme requires six units of time.

(a) Scheme without DCRRDT in Scene 3.    (b) Scheme with DCRRDT in Scene 3.

**Fig. 5.** Gantt chart in Scene 3. The scheme without DCRRDT takes seven-time units. The scheme with DCRRDT takes six-time units.

## 5    Conclusion

We integrated an emerging digital-twin technology into the scheme of deployment & control of RFID sensors which is called DCRRDT. The proposed DCR-RDT scheme improved the inconsistencies and saved the time & labor by its remarkable functions. Meanwhile, because of the intuition and synchronous simulation function of DCRRDT, we had no need to check & repair after all the steps in deployment & control were completed as in previous schemes. DCRRDT allowed us to find problems directly and to repair in every phase of processing timely. We also proposed an algorithm to reduce the number of deployed sensors based on full area coverage on our SDDDT. Our DCRRDT scheme effectively solved the unreasonable problems in sensor deployment schemes. DCR-RDT reduced the complexity of deployment and improved deployment efficiency. Excellent results showed that the DCRRDT-based scheme effectively saved the deployment & control time and had stable & reliable effectiveness. Digital-twin technology provides a new idea and method for the research of indoor space sensor deployment & control.

**Acknowledgment.** This work was supported by the National Key Research and Development Program (No. 61601459).

# References

1. Yilmaz, H., Nacar, O., Sezgin, Ö., Bostanci, E., Güzel, M.S., Sevínç, Ö.: Custom RFID location simulator. In: 2018 2nd International Symposium on Multidisciplinary Studies and Innovative Technologies (ISMSIT), pp. 1–7, Ankara (2018)
2. Sharma, V., Malhotra, S., Hashmi, M.: An emerging application centric RFID framework based on new web technology. In: 2018 IEEE International Conference on RFID Technology & Application (RFID-TA), pp. 1–6, Macau (2018)
3. Zahran, E.G., Arafa, A.A., Saleh, H.I., Dessouky, M.I.: Biogeography based optimization algorithm for efficient RFID reader deployment. In: 2018 13th International Conference on Computer Engineering and Systems (ICCES), pp. 454–459, Cairo, Egypt. (2018)
4. Dhillon, S.S., Chakrabarty, K.: Sensor placement for effective coverage and surveillance in distributed sensor networks. In: Proceedings Wireless Communication Networking Conference, New Orleans, LA, pp. 1609–1614 (2003)
5. Zhang, S., McCullagh, P., Zhou, H., Wen, Z., Xu, Z.: RFID network deployment approaches for indoor localisation. In: 2015 IEEE 12th International Conference on Wearable and Implantable Body Sensor Networks (BSN), pp. 1–6, Cambridge, MA (2015). https://doi.org/10.1109/BSN.2015.7299361
6. Yoon, Y., Kim, Y.H.: An efficient genetic algorithm for maximum coverage deployment in wireless sensor networks. IIEEE Trans. Cybern. **43**, 1473–1483 (2013)
7. Grieves, M.: Digital twin: manufacturing excellence through virtual factory replication. In: Melbourne: U.S. Florida Institute of Technology (2015)
8. Tao, F., Cheng, J., Qi, Q., Zhang, M., Zhang, H., Sui, F.: Digital twin-driven product design, manufacturing and service with big data. Int. J. Adv. Manuf. Technol. **94**, 3563–3576 (2018)
9. Qi, Q., Tao, F.: Digital twin and big data towards smart manufacturing and Industry 4.0: 360 degree comparison. IEEE Access **6**, 3585–3593 (2018)
10. Liu, Z., Meyendorf, N., Mrad, N.: The role of data fusion in predictive maintenance using Digital Twin. In: AIP Conference Proceedings, pp. 020023-1–020023-6 (2018)
11. Jensen, C.S., Lu, H., Yang, B.: Graph model based indoor tracking. In: Tenth International Conference on Mobile Data Management: Systems, Services and Middleware, pp. 122–131 (2009)

# A Binary Code Sequence Based Tracking Algorithm in Wireless Sensor Networks

Yang Zhang[1,2], Qianqian Ren[1,3(✉)], Yu Pan[3], and Jinbao Li[1,2,3(✉)]

[1] Key Laboratory of Database and Parallel Computing of Heilongjiang Province, Heilongjiang University, Harbin, China
{renqianqian,jbli}@hlju.edu.cn
[2] Key Laboratory of Electronic Engineering Colleges of Heilongjiang Province, Heilongjiang University, Harbin, China
[3] School of Computer Science and Technology, Heilongjiang University, Harbin, China

**Abstract.** This paper proposes a binary code sequence based tracking algorithm in wireless sensor network. The proposed algorithm can release the influence of sensed data on localization results via building the map between target's occurrence region and a binary code sequence. To solve the ambiguity problem existing in occurrence region determination, the paper further gives a Voronoi diagram based location refinement algorithm. The simulation results show the tracking results under difference trajectories.

**Keywords:** Target localization · RSSI · Voronoi

## 1 Introduction

Target tracking is widely applied in many applications such as intelligent transportation, battlefield surveillance and intrusion detection [1, 2]. In these applications, the target's location information need to report accurately and timely. However, the limits of wireless sensor networks including energy supply, computation capacity and storage capacity propose challenges for it. Considering theses limits, many research works have been done to solve the problem of target tracking and localization in wireless sensor networks. Received Signal Strength Indication (RSSI) is a popular technique and has been widely used in distance based localization as its simplicity and low cost [3–5]. However, the existence of obstacles, noise, signal fluctuation and environmental influence make it difficult to obtain accurate localization results. Xue *et al.* gave a selected RSSI mean value based localization algorithm, the given algorithm can solve the problem of signal instability existing in the traditional RSSI based localization technology [6]. Gao *et al.* proposed a RSSI quantization method based on Genetic Algorithm, which can reduce the amount of computation in the localization process effectively [7]. Fu *et al.* first constructed a feature scale model and then used continuous weight to assist localization [8]. Barsocchi *et al.* used the RSSI value obtained by fixed sensor nodes to estimate the distance to the target, and chose weighted RSSI values according to the intensity of RSSI measurements to locate the target [9]. Zafari *et al.* presented a particle filter and extended kalman filter based localization algorithm to

© Springer Nature Switzerland AG 2020
S. Wen et al. (Eds.): ICA3PP 2019, LNCS 11945, pp. 577–583, 2020.
https://doi.org/10.1007/978-3-030-38961-1_49

release the influence of environmental noise on RSSI [10]. Mizmizi *et al.* proposed a RSSI quantitation based method to reduce the computation cost during the localization procedure [11].

In this paper, we propose a binary code sequence based target tracking and localization algorithm. Instead of using RSSI value to estimate the distance between the target and sensors, we build the map between the target's occurrence region and a binary code sequence. According to RSSI value and the given threshold, we assign a binary code for each involved sensor, the sequence of binary codes from multiple sensors can determine the target's resident region. To validate the performance of the given algorithm, we construct a simulation environment and evaluate the tracking results under different trajectories.

The rest of this paper is organized as follows. Section 2 gives the sensing model used in this paper. Section 3 describes the algorithm in detail. Section 4 presents the simulation results. The conclusion is given in the last section.

## 2  Sensing Model

In most tracking applications, sensor nodes can estimate the distance from the target by measuring the received signal strength of the target. We assume that signal strength emitted by the target is $S$, which attenuates as distance increases. The signal strength a sensor samples can be defined as:

$$s_i = S \cdot f(d_i) \tag{1}$$

Where $d_i$ is the distance from a sensor to the target, $f(\cdot)$ is a signal decay function, which is a decreasing function satisfying $f(0) = 1$, $f(\infty) = 0$, and $f(x) = \Theta(x^{-b})$. Loss exponent $b$ is constant, it normally ranges from 2 to 5 [12].

Given a threshold $\eta_j$, sensor $i$ compares its sensed data with $\eta_j$, the result can be denoted as a bit as following:

$$q_i = \begin{cases} 0 & if \quad s_i < \eta_j; \\ 1 & otherwise \end{cases} \tag{2}$$

According to the quantization strategies in [7], we can set multiple thresholds. Let's take 3 thresholds as an example. The sensing disk of sensor $i$ is divided into 4 sub regions, the results can be denoted as a series of binary code, such as:

$$q_i = \begin{cases} 00, & s_i \in [0, \eta_1] \\ 01, & s_i \in [\eta_1, \eta_2] \\ 10, & s_i \in [0, \eta_3] \\ 11, & s_i \in [\eta_3, S] \end{cases} \tag{3}$$

Each binary code is corresponding to a sub-region, which is the region the target resides, and the target's estimated location can be represented as the centroid of the sub region. In order to shrink the residence area of the target to get more accurate localization result, the overlapping of sub regions from multiple sensors is utilized.

## 3 RSSI Threshold Based Localization Algorithm

This section gives the description of Binary Code Sequence based localization algorithm. In the rest of this section, sensor nodes are assumed to be deployed randomly and have unique IDs. Sensor nodes are further assumed to know their positions. The sensing area of each sensor node is a disk.

### 3.1 Approach Overview

Let $N = \{n_1, n_2, \ldots, n_m\}$ be a set of sensor nodes in a two dimensional plane. Figure 1 gives an overview of the approach. The yellow star denotes the target. After the deployment of sensor nodes and the choosing of thresholds, the map of the area under surveillance can be divided into a series of small regions. For a sensor nodes $n_i$, let $\eta = \{\eta_1, \eta_2, \eta_3\}$ be the given thresholds set, which divides the sensing disk of each sensor node into four sub regions, denoted as $A_{i1}$, $A_{i2}$, $A_{i3}$ and $A_{i4}$, respectively. $A_{i1}$ is a disk centered $n_i$ with the radius equals to $f^{-1}(\frac{\eta_3}{S})$. $A_{i2}$ is a ring centered at $n_i$ with inner radius equals to $f^{-1}(\frac{\eta_3}{S})$ and outer radius equals to $f^{-1}(\frac{\eta_2}{S})$, respectively. $A_{i3}$ is a ring centered at $n_i$ with inner radius equals to $f^{-1}(\frac{\eta_2}{S})$ and outer radius equals to $f^{-1}(\frac{\eta_1}{S})$, respectively. $A_{i4}$ is the area exclusive the circle centered $n_i$ with the radius $f^{-1}(\frac{\eta_1}{S})$.

**Table 1.** Binary Code Map.

| Area | Binary code of $n_i$ | Area | Binary code of $n_i$ |
|------|------|------|------|
| $A_{i1}$ | 11 | $A_{j1}$ | 11 |
| $A_{i2}$ | 10 | $A_{j2}$ | 10 |
| $A_{i3}$ | 01 | $A_{j3}$ | 01 |
| $A_{i4}$ | 00 | $A_{j4}$ | 00 |

When a mobile target enters into the monitored area, sensor nodes detect certain forms of physical signals emitted from the target. According to formula (2), the sensed data at each sensor node is converted into a binary code. The binary codes from multiple sensors gives us a sequence of binary code called *binary code sequence*, or for short *code sequence*. For simplicity, the code sequence is the combination of binary codes from sensors with the order of ascending sensor IDs. One fact about code sequence is that each sub region is corresponding to an unique code sequence. Therefore, with the pre-computed sensing disk division and a code sequence, the resident region of the target can be specified. For example, if code sequence is 0101(the binary code of $n_i$ and $n_j(i < j)$ is 01 and 01, respectively), we can conclude that the target is in the shadow area. Table 1 is the table of the map between a binary code and a sub region.

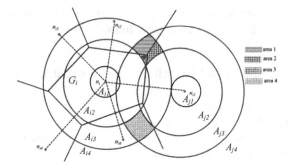

**Fig. 1.** An example of localization algorithm (Color figure online)

## 3.2   Voronoi Diagram Based Location Refinement

In the ideal case, a code sequence should be identical with one sub region. Let us consider the example shown in Fig. 1, where $n_i$ and $n_j$ are two neighbor nodes. If the code sequence is 0101, the target must be in the area $A_{i3} \cap A_{j3}$, that's the two shadow regions in the figure. In this situation, the localization result is considered ambiguous. To overcome this problem, we further give a Voronoi diagram based location refinement algorithm.

The feature of Voronoi diagram makes it a good choice for our localization. Initially, the Voronoi diagram of a collection of sensor nodes divide the map into a lot of polygons. The target in a given polygon is closer to the sensor nodes in this polygon than to any other sensor nodes outside the polygon. Figure 1 is an example of Voronoi polygons of sensor node $n_i$. We will present our algorithm from two steps: that's Vornio Diagram Construction and target Location Refinement.

1. **Vornio Diagram Construction**
   Initially, we construct the Voronoi polygons as the method in [13], that's each node first calculates the bisector of its neighbors and itself. All bisectors can form several polygons. Among these polygons, the smallest one enclosing the sensor node is its Voronoi polygon. In Fig. 1, $G_i$ is the Voronoi polygon of sensor node $n_i$, the set of Voronoi neighbor of $n_i$ can be denoted as $NN_i = \{n_{i0}, n_{i1}, n_{i2}, n_{i3}, n_{i4}\}$.

2. **Target Location Refinement**
   By geometry, the target in one polygon is closer to the sensor inside this polygon than the sensor nodes positioned outside the polygon. Therefore, we can specify the polygon the target resides via comparing the distance of the target and involved sensor nodes. Let's take Fig. 1 as an example to describe the procedure of refining the target's resident region as following:

   • At time $t$, the target enters a certain area of the network. The code sequence formed by involved sensor nodes $n_i$ and $n_j$ is 0101, then we can conclude that the target is in the shadow region, denoted as $A_{i3} \cap A_{j3}$. In order to shrink the resident region and increase the localization accuracy, we go to the next step.
   • All the sensors locate in Voronoi polygons that overlap with the shadow area report the distance to the target. As shown in Fig. 2, the polygons of sensor

nodes $n_i$, $n_{i1}$ and $n_{i2}$ overlap with the shadow area. Then we compute the distance between the target and $n_i$, $n_{i1}$, $n_{i2}$ respectively, the ascending order of distance value is denoted as $D = \{d_i, d_{i1}, d_{i2}\}$. Thus, the target's resident area is reduced to $A_{i3} \cap A_{j3} \cap G_i$, that's area 3 in the figure is the target's resident region.

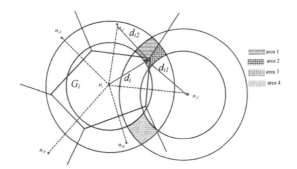

**Fig. 2.** Target's location refinement

## 4   Simulation and Evaluation

In order to validate the performance of the given algorithms, we construct a simulation platform via Matlab. In the simulations, 64 sensor nodes are deployed in the monitoring region. The sensing range and transmitting range are assumed to be the same among all nodes. We evaluate the tracking and locating results under different trajectories. The moving direction of the target are not known in advance. Figures 3 and 4 show the tracking results under curve trajectory and line trajectory, respectively. The horizontal and vertical coordinates represent the physical coordinates of the target position. We can conclude that the given algorithm can track the target efficiently, and the tracking results are stable.

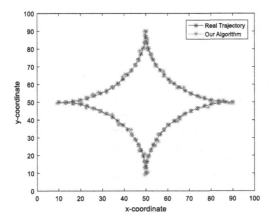

**Fig. 3.** Tracking results under curve trajectory

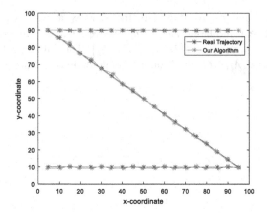

**Fig. 4.** Tracking results under line trajectory

# 5   Conclusion

In this paper, we present a range free localization algorithm. Instead of using RSSI value to estimate the distance between the target and sensors, we first compare it with given thresholds and encode the comparison results. Then we determine the target's resident region using the encoding results from multiple sensor nodes. To solve of problem of ambiguity existed in target's resident region determination, we further design a Vornio diagram based target location refinement algorithm. The tracking results under different trajectories are investigated under a simulation platform.

**Acknowledgment.** The work was supported in part by the Provincial Natural Science Foundation of Heilongjiang under Grant No. F2017022, the Fundamental Research Foundation of Universities in Heilongjiang Province for Youth Innovation Team under Grant No. RCYJTD201805, the Fundamental Research Foundation of Universities in Heilongjiang Province No. KJCX201815 and the Youth Reserve Talents Project of Harbin under Grant No. 2017RAQXJ131.

# References

1. Zheng, K., et al.: Energy-efficient localization and tracking of mobile devices in wireless sensor networks. IEEE Trans. Veh. Technol. **66**(3), 2714–2726 (2017)
2. Ahmadi, H., Viani, F., Bouallegue, R.: An accurate prediction method for moving target localization and tracking in wireless sensor networks. Ad Hoc Netw. **70**, 14–22 (2018)
3. Li, T., Chen, Y., Zhang, R., Zhang, Y., Hedgpeth, T.: Secure crowdsourced indoor positioning systems. In: IEEE INFOCOM 2018 - IEEE Conference on Computer Communications, pp. 1034–1042. IEEE (2018)
4. Chriki, A., Touati, H., Snoussi, H.: SVM-based indoor localization in wireless sensor networks. In: 2017 13th International Wireless Communications and Mobile Computing Conference (IWCMC), pp. 1144–1149. IEEE (2017)

5. Yiu, S., Dashti, M., Claussen, H., Perez-Cruz, F.: Wireless RSSI fingerprinting localization. Sig. Process. **131**, 235–244 (2017)
6. Xue, W., Qiu, W., Hua, X., Yu, K.: Improved Wi-Fi RSSI measurement for indoor localization. IEEE Sens. J. **17**, 2224–2230 (2017)
7. Gao, W., Nikolaidis, I., Harms, J.: RSSI quantization for indoor localization services. In: 2017 IEEE 28th Annual International Symposium on Personal, Indoor, and Mobile Radio Communications (PIMRC), pp. 1–7. IEEE (2017)
8. Fu, Y., Chen, P., Yang, S., Tang, J.: An indoor localization algorithm based on continuous feature scaling and outlier deleting. IEEE Internet Things J. **5**, 1108–1115 (2018)
9. Barsocchi, P., Lenzi, S., Chessa, S., Giunta, G.: A novel approach to indoor RSSI localization by automatic calibration of the wireless propagation model. In: VTC Spring 2009 - IEEE 69th Vehicular Technology Conference, pp. 1–5. IEEE (2009)
10. Zafari, F., Papapanagiotou, I., Hacker, T.: A novel Bayesian filtering based algorithm for RSSI-based indoor localization. In: 2018 IEEE International Conference on Communications (ICC), pp. 1–7. IEEE (2018)
11. Mizmizi, M., Reggiani, L.: Design of RSSI based fingerprinting with reduced quantization measures. In: 2016 International Conference on Indoor Positioning and Indoor Navigation (IPIN), pp. 1–6. IEEE (2016)
12. Xing, G., Tan, R., Liu, B., Wang, J., Jia, X., Yi, C.-W.: Data fusion improves the coverage of wireless sensor networks. In: Proceedings of the 15th Annual International Conference on Mobile Computing and Networking, pp. 157–168. ACM (2009)
13. Wang, G., Cao, G., La Porta, T.F.: Movement assisted sensor deployment. IEEE Trans. Mob. Comput. **5**(6), 640–652 (2006)

# Sampling Based Katz Centrality Estimation for Large-Scale Social Networks

Mingkai Lin[1], Wenzhong Li[1,2($\boxtimes$)] (iD), Cam-tu Nguyen[3], Xiaoliang Wang[1], and Sanglu Lu[1,2($\boxtimes$)]

[1] State Key Laboratory for Novel Software Technology, Nanjing University, Nanjing, China
mingkai@smail.nju.edu.cn, {lwz,waxili,sanglu}@nju.edu.cn
[2] Sino-German Institutes of Social Computing, Nanjing University, Nanjing, China
[3] Software Institute, Nanjing University, Nanjing, China
ncamtu@nju.edu.cn

**Abstract.** Katz centrality is a fundamental concept to measure the influence of a vertex in a social network. However, existing approaches to calculating Katz centrality in a large-scale network is unpractical and computationally expensive. In this paper, we propose a novel method to estimate Katz centrality based on graph sampling techniques. Specifically, we develop an unbiased estimator for Katz centrality using a multi-round sampling approach. We further propose **SAKE**, a **S**ampling based **A**lgorithm for fast **K**atz centrality **E**stimation. We prove that the estimator calculated by **SAKE** is probabilistically guaranteed to be within an additive error from the exact value. The computational complexity of **SAKE** is much lower than the state-of-the-arts. Extensive evaluation experiments based on four real world networks show that the proposed algorithm achieves low mean relative error with low sampling rate, and it works well in identifying high influence vertices in social networks.

**Keywords:** Social network · Katz centrality · Graph sampling

## 1 Introduction

With the rapid development of social network platforms such as Facebook and Twitter, there is a growing interest in network analysis and its applications in social networks [7]. One of important concepts in network analysis is the *Katz centrality* [12], which measures the "influence" of a vertex in a social network by recursively assessing the importance of its neighbors. Katz centrality has been used in a wide range of AI applications such as finding the most influential users in a social network [13], forming the word-of-mouth effect, promoting the adoption of innovation, widening the spread of public opinion, and viral marketing.

Katz centrality computes the relative influence of a vertex by measuring the number of routes from the vertex to the other vertices in the network multiplied

© Springer Nature Switzerland AG 2020
S. Wen et al. (Eds.): ICA3PP 2019, LNCS 11945, pp. 584–598, 2020.
https://doi.org/10.1007/978-3-030-38961-1_50

by an attenuation factor. Mathematically, Katz centrality of a vertex $i$ can be formulated as:

$$C_{katz}(i) = \sum_{k=1}^{\infty} \sum_{j=1}^{n} \alpha^k (A^k)_{ij}, \tag{1}$$

where $A$ is the adjacency matrix of the network and $\alpha \in (0,1)$ is the attenuation factor.

Computation of Katz centrality is non-trivial. According to the definition, the computation involves matrix multiplication. For a graph with $n$ nodes, the adjacency matrix $A$ is a $n \times n$ matrix, and the computational complexity of the exact Katz centrality is $O(n^3)$. A more efficient approach is to solve a linear system using Cholesky decomposition, which costs $O(n^2)$ [19]. Some approximation algorithms [6,9,18] were developed based on iterative methods that generally cost $O(m)$, where $m$ is the number of edges in the graph and $m \in O(n^2)$ in the worst case. Unfortunately, for large scale network where the number of vertex $n$ can be as large as millions or billions, even these approximation methods are still computationally expensive.

In this paper, we propose a novel sampling based method for Katz centrality estimation. Specifically, we sample a small subset of vertices and edges from the social network, and use the sampled data to estimate Katz centrality with limited walks $d$ (see Eq. 2), and tuning $d$ to achieve different level of approximation to the original Katz centrality. Since the estimation is based on a very small subset of the original network, the proposed method is computationally efficient and practical for large-scale networks.

We firstly develop a theoretical estimator for Katz centrality based on a multi-round graph sampling approach. Using the Horvitz-Thompson theory [11], we prove that the proposed estimator is an unbiased estimator of the Katz centrality (*Theorem* 1). We further propose **SAKE**, a **S**ampling based **A**lgorithm for **K**atz centrality **E**stimation. The main idea of **SAKE** is to use a random node sampling design, and compute the Katz centrality estimator by counting the number of routes in the sampled subgraph. We prove that the estimated Katz centrality computed by **SAKE** is probabilistically guaranteed to be upper-bounded by an additive error from the exact value (*Theorems* 3 and 4). Moreover, we show that the computational complexity of **SAKE** is $O(l(d-1)\frac{r^2}{n^2}m + lr)$, where $r$ is the size of the sampled subset and $l, d$ are small constants, which is more efficient than the state-of-the-arts. We conduct extensive evaluation experiments based on four datasets collected from real world social networks, which show that **SAKE** achieves low mean relative error, and it is efficient in identifying high influence vertices in social networks.

The contributions of the paper are summarized as follows.

– We are the first to propose the idea of sampling based Katz centrality estimation. To the best of our knowledge, using a sampled subgraph to infer the Katz centrality in the original graph has not been found in the literature, and our work address the fundamental problem of inferring vertex centrality via graph sampling theory, which has important implications in large-scale social networks.

- We propose an efficient algorithm called **SAKE** for Katz centrality estimation based on a multi-round sampling approach. The proposed algorithm has several advantages: (1) The proposed estimator is proved to be unbiased (*Theorem* 1); (2) The estimation error is guaranteed by a provable bound with high probability (*Theorems* 3 and 4); and (3) The computational complexity is $O(l(d-1)\frac{r^2}{n^2}m + lr)$, where $r$ is the size of the sampled subset and $l, d$ are small constants, which is more efficient than the state-of-the-arts.
- The efficiency and feasibility of the proposed algorithm is verified by extensive experiments based on four real world social networks.

## 2    Related Work

### 2.1    Computation of Katz Centrality

Katz centrality was introduced by Leo Katz in 1953 to measure the relative influence of a vertex in a network [12]. A few axioms for Eigenvector and Katz centralities had been proposed in [5,24]. Katz centrality had been shown to be useful in ranking users in social networks [19] and searching disease genes from gene expression and protein interaction networks [25].

For a graph with size $n$, Katz centrality can be calculated exactly by solving a linear system. In this case the expression $\overrightarrow{C}_{katz} = ((I - \alpha A^T)^{-1} - I)\overrightarrow{I}$ can be used to obtain Katz centrality where the complexity is $O(n^3)$. Later an approximate approach to calculating Katz centrality based on Cholesky decomposition costing $O(n^2)$ [19] was proposed. But in real world we usually observe that graphs of interest like Facebook are usually very large and sparse. Under such circumstances $n^2$ is much larger than $m$ where $m$ is the number of edges in the graph. Thus Foster et al. [9] presented a vertex-centric heuristic for Katz centrality by computing the recurrence $\overrightarrow{C}_{katz}^{new} = \alpha A(\overrightarrow{C}_{katz} + \overrightarrow{I})$ until reaching some specific condition. The algorithm performs well and is widely used in many toolkits for complex network. Furthermore, another algorithm [18] also used iterative method to approximate a personalized variant of Katz centrality with cost of $O(m)$.

However, the existing approximation algorithms operated on the adjacency matrix $A$ of the whole network, which are prohibitively computationally expensive for large networks with billions of vertices.

### 2.2    Social Network Sampling Techniques

Social network sampling techniques aim to obtain a smaller graph which can well represent the original network. There had been a rich literature in statistics, data mining, and physics on estimating graph properties using a small subsamples [2,16]. The works of [2,14,16] provided excellent surveys to graph sampling techniques. The study of [23] showed that different network sampling techniques are highly sensitive with regard to capturing the centrality measurement of nodes.

Several works adopted sampling techniques to infer different network characteristics. For instance, [4] presented a sampling algorithm to estimate the similarity matrix resulting from a bipartite graph stream projection. The fast approximations of Betweenness centrality using VC-dimension theory and Rademacher complexity were proposed in [20,21]. Approximation approach for degree distribution estimation using sublinear graph samples was further discussed in [8].

To the best of our knowledge, estimation of Katz centrality via sampling has not been addressed in the literature. Our work makes the first attempt to develop a sampling based method to achieve unbiased Katz centrality estimation.

## 3    Notations and Definitions

In this section we introduce the notations and definitions that will be used for analysis throughout the paper.

Let $G = (V, E)$ be a graph, where $V$ is the set of $n$ vertices and $E$ is the set of $m$ edges. Denote the $n \times n$ adjacency matrix $A$ of $G$ with entries $A_{ij} = 1$ if there exits an edge from vertex $i$ to $j$, and 0 otherwise.

We define *Katz centrality with limited walks d* as:

$$C_{katz}(i) = \sum_{k=1}^{d} \sum_{j=1}^{n} \alpha^k (A^k)_{ij} \tag{2}$$

where $d$ is the upper limit of the number of walks between a pair of vertices in the network. When $d \to \infty$, the above definition approaches the original Katz centrality.

Katz centrality quantifies the ability of a vertex to initiate walks around the network. The number of walks of length $k$ from vertex $i$ to $j$ is $(A^k)_{ij}$. Katz centrality of vertex $i$ counts the number of closed walks beginning at vertex $i$, while penalizing long walks by multiplying a fixed attenuation factor $\alpha \in (0, 1)$ for distant route. Unlike the conventional definition of Katz centrality that counts infinity walks, we consider Katz centrality with limited walks $d$, which is more practical and computable in real large-scale networks.

We denote $D_i^k(G)$ $(k = 1, 2, \cdots, d)$ as the set of all routes in graph $G$ beginning from $i$ with the length of walks $k$. Note that the number of walks of length $k$ from vertex $i$ to $j$ equals $(A^k)_{ij}$. Formally we can represent the Katz centrality of vertex $i$ with limited walks $d$ as:

$$Katz(i) = \sum_{k=1}^{d} \alpha^k \left| D_i^k(G) \right|, \tag{3}$$

where $\left| D_i^k(G) \right|$ is the cardinality of $D_i^k(G)$ that can be calculated by:

$$\left| D_i^k(G) \right| = \sum_{j \in V} (A^k)_{ij} = \sum_{v_1 \in V} \sum_{v_2 \in V} \cdots \sum_{v_k \in V} A_{iv_1} A_{v_1 v_2} \cdots A_{v_{k-1} v_k} \tag{4}$$

The term $A_{iv_1}A_{v_1v_2}\cdots A_{v_{k-1}v_k}$ in the right part of the above equation is a 0–1 indicator representing whether there exists a route from vertex $i$ to $v_k$ with length $k$. According to the definition of adjacency matrix, if the route $i \to v_1 \to v_2 \to \cdots \to v_k$ exists, $A_{iv_1}A_{v_1v_2}\cdots A_{v_{k-1}v_k}$ equals 1, and otherwise 0.

In this paper, we focus on computing a $(\epsilon, \delta)$-*approximation* of $Katz(i)$ with the following definition.

**Definition 1 (($\epsilon$, $\delta$)-approximation).** *Given* $\epsilon$, $\delta \in (0,1)$, *the estimator* $\widehat{Katz}(i)$ *is a* ($\epsilon$, $\delta$)*-approximation of* $Katz(i)$ *if it satisfies:*

$$Pr(|\widehat{Katz}(i) - Katz(i)| \le \epsilon) \ge 1 - \delta. \tag{5}$$

## 4   Katz Centrality Estimation

In this section, we propose the sampling approach and estimation algorithm for Katz centrality estimation.

### 4.1   Sampling and Estimation Method

According to the definitions in Eqs. (3) and (4), to calculate $\left|D_i^k(G)\right|$, it needs to compute the $d$-th power of the $n \times n$ adjacency matrix $A$, which complexity is $O(dn^3)$. In this paper, we explore the method of Katz centrality estimation based on sampling technique. The basic idea is to sample a small subset of vertices and edges from $G$, and based on which we can form an unbias estimation of Katz centrality for an objective vertex. The tool we use to estimate Katz centrality is Horvitz-Thompson estimator, which is introduced in the following.

The Horvitz-Thompson estimator [11] is a method for estimating the total and mean of a population from sampling. Suppose we have a population $U = \{1, 2, \cdots, N_u\}$ of $N_u$ units, and with each unit $j \in U$ there is an associated value $y_j$. Let $\tau = \sum_j y_j$ be the total value of $y$'s in the population. Let $U_s \subset U$ be an independent sample of $n_s$ distinct units from $U$. We observe $y_j$ for each $j \in U_s$, and suppose under the given sampling design each unit $j \in U$ has probability $\pi_j$ being included in $U_s$. The Horvitz-Thompson estimator gives an unbiased estimation of the total $\tau$ by $\hat{\tau} = \sum_{j \in U_s} \frac{y_j}{\pi_j}$.

Inspired by the Horvitz-Thompson estimator, we derive an unbiased estimator for $\left|D_i^k(G)\right|$ as follows. The key of deriving $\left|D_i^k(G)\right|$ is to estimate the number of routes from $i$ with length $k$. To achieve that, we adopt a $k$-round sampling design. We denote $\mathcal{R}(i)$ as the successors of vertex $i$ and in this $k$-round sampling design, we sample a set of vertices $S_1$ from $\mathcal{R}(i)$ in the first round and $S_j$ ($j = 2, \cdots, k$) from $V$ in the later rounds without replacement.

Let $\pi_v^{S_j}$ be the probability that a vertex $v$ is included in $S_j$ ($v \in \mathcal{R}(i)$ for $S_1$ and $v \in V$ for the others) during the sampling process. The value of $\left|D_i^k(G)\right|$ can be estimated by

$$\left|\widehat{D_i^k(G)}\right| = \sum_{v_1 \in S_1} \sum_{v_2 \in S_2} \cdots \sum_{v_k \in S_k} \frac{A_{iv_1}A_{v_1v_2}\cdots A_{v_{k-1}v_k}}{\pi_{v_1}^{S_1} \pi_{v_2}^{S_2} \cdots \pi_{v_k}^{S_k}}. \tag{6}$$

The numerator $A_{iv_1}A_{v_1v_2}\cdots A_{v_{k-1}v_k}$ in the right part of Eq. (6) indicates the existence of a route from vertex $i$ to $v_k$ with length $k$ in the sampling nodes, and the denominator $\pi_{v_1}^{S_1}\pi_{v_2}^{S_2}\cdots\pi_{v_k}^{S_k}$ is used for correcting the bias according to Horvitz-Thompson estimation. For the reason that the successors of objective vertex are easy to obtain and there is no need to estimate it, thus when $k = 1$, we have $|\widehat{D_i^1(G)}| = \sum_{v_1\in S_1}\frac{A_{iv_1}}{\pi_{v_1}^{S_1}} = \frac{|S_1|}{\pi_{v_1}^{S_1}} = |\mathcal{R}(i)| = |D_i^1(G)|$ which is a constant. Based on the $k$-round sampling design, for Katz centrality with limited walks $d$, the estimator of $Katz(i)$ can be derived by

$$\widehat{Katz(i)} = \alpha|\mathcal{R}(i)| + \sum_{v_1\in S_1}\sum_{v_2\in S_2}\frac{\alpha^2 A_{iv_1}A_{v_1v_2}}{\pi_{v_1}^{S_1}\pi_{v_2}^{S_2}} + \cdots +$$

$$\sum_{v_1\in S_1}\sum_{v_2\in S_2}\cdots\sum_{v_d\in S_d}\frac{\alpha^d A_{iv_1}A_{v_1v_2}\cdots A_{v_{d-1}v_d}}{\pi_{v_1}^{S_1}\pi_{v_2}^{S_2}\cdots\pi_{v_d}^{S_d}} \qquad (7)$$

$$= \sum_{k=1}^{d}\alpha^k|\widehat{D_i^k(G)}|.$$

The following theorem shows that $\widehat{Katz(i)}$ is an unbias estimation of $Katz(i)$.

**Theorem 1.** *To estimate Katz centrality with limited walks $d$, if we adopt the $d$-rounds independent sampling process $S_1, S_2, \cdots, S_d$ to construct the estimator $\widehat{Katz(i)}$ as Eq. (7), then we have $E(\widehat{Katz(i)}) = Katz(i)$.*

*Proof.* Firstly, we prove that $|\widehat{D_i^k(G)}|$ is an unbias estimator of $|D_i^k(G)|$. We introduce an 0–1 indicator to represent whether a vertex is included in the sampled set $S_j$:

$$I(v\in S_j) = \begin{cases} 1, & if\ vertex\ v\ is\ included\ in\ S_j \\ 0, & otherwise \end{cases} \qquad (8)$$

Since the $d$-round sampling processes are independent, we have

$$E(|\widehat{D_i^k(G)}|) = E(\sum_{v_1\in S_1}\sum_{v_2\in S_2}\cdots\sum_{v_k\in S_k}\frac{A_{iv_1}A_{v_1v_2}\cdots A_{v_{k-1}v_k}}{\pi_{v_1}^{S_1}\pi_{v_2}^{S_2}\cdots\pi_{v_k}^{S_k}})$$

$$= E(\sum_{v_1\in\mathcal{R}(i)}\sum_{v_2\in V}\cdots\sum_{v_k\in V}\frac{A_{iv_1}A_{v_1v_2}\cdots A_{v_{k-1}v_k}}{\pi_{v_1}^{S_1}\pi_{v_2}^{S_2}\cdots\pi_{v_k}^{S_k}}\cdot\prod_{j=1}^{k}I(v_j\in S_j))$$

$$= \sum_{v_1\in\mathcal{R}(i)}\sum_{v_2\in V}\cdots\sum_{v_k\in V}\frac{A_{iv_1}A_{v_1v_2}\cdots A_{v_{k-1}v_k}}{\pi_{v_1}^{S_1}\pi_{v_2}^{S_2}\cdots\pi_{v_k}^{S_k}}\cdot E(\prod_{j=1}^{k}I(v_j\in S_j))$$

$$= \sum_{v_1\in\mathcal{R}(i)}\sum_{v_2\in V}\cdots\sum_{v_k\in V}\frac{A_{iv_1}A_{v_1v_2}\cdots A_{v_{k-1}v_k}}{\pi_{v_1}^{S_1}\pi_{v_2}^{S_2}\cdots\pi_{v_k}^{S_k}}\cdot\prod_{j=1}^{k}E(I(v_j\in S_j))$$

$$= \sum_{v_1 \in \mathcal{R}(i)} \sum_{v_2 \in V} \cdots \sum_{v_k \in V} A_{iv_1} A_{v_1 v_2} \cdots A_{v_{k-1} v_k}$$

$$= \left| D_i^k(G) \right|.$$

Applying Eqs. (7) and (3), we have

$$E(\widehat{Katz(i)}) = E(\sum_{k=1}^{d} \alpha^k |\widehat{D_i^k(G)}|) = \sum_{k=1}^{d} \alpha^k E(|\widehat{D_i^k(G)}|)$$

$$= \sum_{k=1}^{d} \alpha^k \left| D_i^k(G) \right| = Katz(i),$$

## 4.2   Sampling Based Katz Computation

We in this paper propose an approximation algorithm named **SAKE**, a **S**ampling based **A**lgorithm for **K**atz centrality **E**stimation. It computes the Katz centralities for a set of vertices in batch based on the estimator derived in Theorem 1.

Assume $N_0 = \{v_1^{(0)}\}$ is the objective vertex that we want to estimate its Katz centrality. To explore the network connectivity condition up to $d$ hops, we conduct $d$ rounds of sampling and computation as follows. In the $j$-th sampling procedure, we generate a set of vertices $N_1 = \{v_1^{(1)}, v_2^{(1)}, \cdots, v_{|N_1|}^{(1)}\}$ sampling from $\mathcal{R}(i)(j = 1)$ and $N_j = \{v_1^{(j)}, v_2^{(j)}, \cdots, v_{|N_j|}^{(j)}\}$ $(j = 2, \cdots, d)$ by sampling from $V$ with probabilities $P_j = \{\pi_1^{(j)}, \pi_2^{(j)}, \cdots, \pi_{|N_j|}^{(j)}\}$ $(j = 1, \cdots, d)$, where $\pi^{(j)}$ is the probability that the corresponding vertex is included in $N_j$ according to the sampling method. In each round, we construct an adjacency matrix $B^{(j)}$ based on $N_{j-1}$ and $N_j$ $(j = 1, \cdots, d)$, where the element $B_{xy}^{(j)}$ $(x \in N_{j-1}, y \in N_j)$ is 1 if there is an edge from $x$ to $y$ in the graph $G$, and 0 otherwise. We further replace each number 1 in $B_{xy}^{(j)}$ $(x \in N_{j-1}, y \in N_j)$ with number $1/\pi_y^{(j)}$ for the reason of correcting bias. In this way, we can get $d$ matrices $B^{(j)}$ $(j = 1, 2, \cdots, d)$.

We define $B^{[k]} = \prod_{j=1}^{k} B^{(j)}$ as the product of the first $k$ adjacency matrices. As illustrated in Fig. 1, $B^{[k]}$ is a $|N_0| \times |N_k|$ matrix, where each row in the matrix represents the number of routes from the objective vertex in $N_0$ to the vertices in $N_k$. We define $b^{[k]}$ as the row sum of $B^{[k]}$, where the value $b^k = \sum_{x=1}^{|N_k|} B_x^{[k]}$ equals the estimator of Eq. (6) representing the total number of routes from $i$ to all vertices in $N_k$ with length $k$. According to Theorem 1, the Katz centrality estimators of the vertex $i$ can be calculated by $\alpha |\mathcal{R}(i)| + \sum_{k=2}^{d} \alpha^k b^{[k]}$.

In the above analysis, we do not make any restriction on the sampling methods. Theoretically, all the sampling methods can fit the proposed Katz centrality estimator. However, in practice we prefer the random node sampling method over the others due to the fact that the **SAKE** algorithm relies on $P_j$, the inclusion probabilities of the sampling nodes. In the other graph sampling methods, the

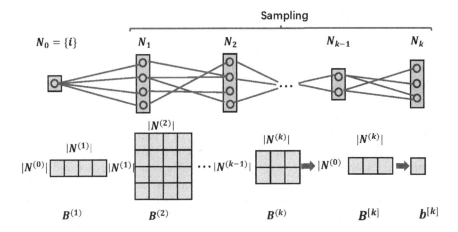

**Fig. 1.** Illustration of adjacency matrices in k-round sampling.

derivation of the inclusion probability for the sampled node is rather difficult. Therefore random node sampling is the chosen sampling method for the proposed algorithm.

To improve the computational efficiency, we make two simplifications in the sampling process: (1) We adopt the node sampling method and fix the sampling probability for each node. (2) We fix the number of sampled nodes in each round to $r$ (in the first sampling round the number is $\min\{r, |\mathcal{R}(i)|\}$). In this way the inclusion probability of each vertex can be represented by $\pi' = \frac{\min\{r, |\mathcal{R}(i)|\}}{|\mathcal{R}(i)|}$ for the first sampling round, and $\pi = \frac{r}{|V|}$ for the later sampling rounds. In this case, the computation of the Katz estimator $\widehat{Katz}(i)$ can be simplified greatly.

We further adopt a repeat estimation approach to improve the estimation accuracy. According to Theorem 1, the mean of the Katz centrality estimator approaches to its true value, therefore we can repeat the estimation process for several times and take the mean as the estimation value. Assuming the estimation process is repeated for $l$ times, we will show in theory that by carefully choosing the value of $l$, the estimation error can be bounded within an extent with high probability.

From the analysis we can find that the basic algorithm runs for $l$ loops, and in each loop, it computes the multiplication of two matrices (with sizes $|N_0| * r$ and $r * r$) for $d$ times. Therefore the total computation is $O(ld|N_0|r^2)$. Since $l$ and $d$ are small constants and $|N_0|$ is the number of objective vertices that is fixed, the computational complexity is $O(r^2)$. But sometimes the complexity $O(r^2)$ is also larger than $O(m)$ when the number of edges is small. Thus denoting $\overrightarrow{I_i}$ as $|N^{(i)}| \times 1$ vector of all 1s and according to the process in Fig. 1, we can reformulate $\widehat{Katz}(i)$ as:

$$\widehat{Katz(i)} = \alpha b^{[1]} + \alpha^2 b^{[2]} + \cdots + \alpha^k b^{[k]}$$
$$= \alpha B^{[1]} \overrightarrow{I_1} + \alpha^2 B^{[2]} \overrightarrow{I_2} + \cdots + \alpha^k B^{[k]} \overrightarrow{I_k}$$
$$= \alpha B^{(1)} \overrightarrow{I_1} + \alpha^2 B^{(1)} B^{(2)} \overrightarrow{I_2} + \cdots + \alpha^k B^{(1)} B^{(2)} \cdots B^{(k)} \overrightarrow{I_k} \qquad (9)$$
$$= \alpha B^{(1)} (\alpha B^{(2)} (\alpha B^{(3)} (\cdots (\alpha B^{(k)} \overrightarrow{I_k}) + \cdots + \overrightarrow{I_3}) + \overrightarrow{I_2}) + \overrightarrow{I_1})$$

From the equation, we can find that for the estimation of Katz centrality, we just need to each time calculate the product of a matrix and a vector which costs $O(r^2)$. What's more when using the CSR matrix data to represent $B^{(\cdot)}$, the complex for the first $d-1$ product processes can be further reduced to $O(l(d-1)\frac{r^2}{n^2}m)$ for the reason that we sample vertices randomly from the whole network. However the last product process costs $O(lr)$ for we get $N_1$ by sampling from successors of the objective vertex. As the consequence, we have the final complexity of $O(l(d-1)\frac{r^2}{n^2}m + lr)$ for the algorithm. The method to reduce complexity can be viewed as a variant of the method from [9], where Foster et al. estimate Katz centrality iteratively by computing partial sums with aggregating Katz values from the successors of vertices. The pseudo-code of **SAKE** is illustrated in Algorithm 1. Given the fact that $r \ll n$ is the size of sampled vertex set, the proposed **SAKE** algorithm reduces the computational complexity of Katz centrality dramatically. Even though the algorithm above only examines starting at a single vertex, our algorithm can be easily adapted to the case starting at

---

**Algorithm 1. SAKE**$(G, r, d, \alpha, N_0)$

---

**Input:**
$G$ : The original network
$r$ : Number of vertices to be sampled in each process
$d$ : The maximum number of walks
$\alpha$ : The attenuation factor
$i$ : The objective vertex to estimate Katz centrality
**Output:**
$\overline{Katz(i)}$: The estimated Katz centrality

  1: Let $l$ be the number of loops needed to obtain the desired error bound
  2: **for** $t = 1, 2 \cdots, l$ **do**
  3:     Let $N_0 = \{i\}$
  4:     Randomly sample $\min\{r, |\mathcal{R}(i)|\}$ nodes from $\mathcal{R}(i)$ with probability $\pi'$ to form $N_1$.
  5:     **for** $j = 2, 3 \cdots, d$ **do**
  6:         Randomly sample $r$ nodes from $V$ with probability $\pi$ to form $N_j$
  7:     $\widehat{Katz[t]} = \overrightarrow{0}_{|N_d|}$
  8:     **for** $j = d, d-1 \cdots, 1$ **do**
  9:         Construct matrices $B^{(j)}$ based on $N_{j-1}$ and $N_j$
10:         Compute $\widehat{Katz[t]} = \alpha B^{(j)} (\widehat{Katz[t]} + \overrightarrow{I_j})$
11: Let $\overline{Katz(i)} = \frac{\sum_{t=1}^{l} \widehat{Katz[t]}}{l}$
12: **return** $\overline{Katz(i)}$

---

multiple vertices. Instead of repeating the process of the algorithm **SAKE** for times independently to evaluate Katz Centralities of multiple objective vertices, we can construct the matrics $B^{(3)}, B^{(4)}, \cdots, B^{(k)}$ only once and reuse them in each process.

Next we show that the estimation error of **SAKE** can be bounded to some extent by carefully choosing the number of loops $l$.

**Theorem 2 (Hoeffding bound** [10]**).** *If $x_1, x_2, \cdots, x_k$ are independent random variables, where $a_t \leq x_t \leq b_t$ ($1 \leq t \leq k$), and $\mu = E[\sum_t x_t/k]$ is the expected mean, then for $\xi > 0$*

$$\Pr\left\{\left|\frac{\sum_{t=1}^{k} x_t}{k} - \mu\right| \geq \xi\right\} \leq 2e^{-2k^2\xi^2/\sum_{t=1}^{k}(b_t-a_t)^2}.$$

The Hoeffding bound provides a probabilistic bound for the mean of independent random variables, which can be applied to compute the error bound of the proposed **SAKE** algorithm as in Theorem 3.

**Theorem 3.** *Given a graph $G$ with $n$ vertices. Assume the **SAKE** algorithm runs $l$ loops to estimate Katz centrality by $\overline{Katz}(i) = \frac{\sum_{t=1}^{l} \widehat{Katz(i)}_t}{l}$, where $\widehat{Katz(i)}_t$ is the Katz estimator in the $t$-th loop. Let $\Delta$ be the maximum Katz centrality value of the vertices in the sampled graph, and $n_s$ be the size of a sampled graph. Let $\xi = \eta\Delta$ for $\forall \eta > 0$. If the number of loops $l \in \Omega(\frac{\log n_s}{\eta^2})$, then the estimation error $|\overline{Katz}(i) - Katz(i)| \leq \xi$ with high probability.*

*Proof.* Since $\Delta$ is the maximum Katz centrality value of the vertices in the sampled graph, the observed Katz centralities from the sampled graph satisfy $0 \leq \widehat{Katz(i)}_t \leq \Delta$ ($t = 1, 2, \cdots, l$) for all loops. Taking the $l$ estimators as random variables, we can apply the Hoeffding's bound with $x_t = \widehat{Katz(i)}_t$, $\mu = Katz(i)$, $a_t = 0$, and $b_t = \Delta$.

Theorem 1 has proved that $E(\widehat{Katz(i)}) = Katz(i)$. Thus the probability of the difference between the estimated Katz centrality $\overline{Katz}(i)$ and the actual Katz centrality $Katz(i)$ larger than $\xi$ is:

$$\Pr\{|\overline{Katz}(i) - Katz(i)| \geq \xi\} \leq 2e^{-2l^2\xi^2/\sum_{t=1}^{l}(b_t-a_t)^2} = 2e^{-2l\xi^2/\Delta^2}.$$

For $\xi = \eta\Delta$, if the number of loops $l = \frac{\log n_s}{2\eta^2}$, then

$$\Pr\{|\overline{Katz}(i) - Katz(i)| \geq \xi\} \leq \frac{2}{n_s}.$$

Therefore, if $l \in \Omega(\frac{\log n_s}{2\eta^2})$, the estimation error is less than $\xi$ with probability $1 - \frac{2}{n_s}$. For a graph with large number of vertices or with higher sampling ratio, the probability approaches 1 when $n_s$ increases. This proofs the theorem.

By letting $\epsilon = \xi$ and $\delta = \frac{2}{n_s}$, it is easy to verify the following theorem.

**Theorem 4.** *The estimator $\widehat{Katz}(i)$ computed by the **SAKE** algorithm is a $(\epsilon, \delta)$-approximation of $Katz(i)$.*

**Table 1.** Statistics of datasets

| Name | Type | Nodes | Edges |
|------|------|-------|-------|
| Livemocha [1] | Undirected | 104,103 | 2,193,083 |
| Pokec [22] | Directed | 1,632,803 | 30,622,564 |
| Livejournal [15] | Undirected | 5,204,176 | 49,174,464 |
| Wikipedia [3] | Directed | 18,268,992 | 172,183,984 |

## 5   Performance Evaluation

### 5.1   Datasets

The experiments are conducted on four real world networks: (1) *Livemocha* [1]: social network of an online language learning community where nodes represent users and edges represent friendships. (2) *Pokec* [22]: Pokec is the most popular online social network in Slovakia where nodes represent users and edges represent relationship. (3) *Livejournal* [15]: This is the social network of LiveJournal users and their connections where nodes represent users and edges represent connections. (4) *Wikipedia* [3]: The network is the hyperlink network of Wikipedia, as extracted in DBpedia. Nodes are pages and edges correspond to hyperlinks. Before applying these graphs we remove all self-loops in the networks.

What's more we drop all isolated nodes for the reason that their Katz centralities are always 0. The statistics of these networks after data cleaning are summarized in Table 1.

### 5.2   Experimental Setup

We present the setup of default system parameters. The number of walks is set to $d = 6$ by default. The attenuation factor $\alpha = \frac{0.85}{\|A\|_2}$ according to the literature [18] where $\|A\|_2$ denotes the the 2-norm of matrix $A$. The number of loops in Algorithm 1 is set to $l = 20$, which is an empirical value from the experiments. The default sampling ratio (the inclusion probability of a vertex) is set to 0.01, that is, only 1% of the total vertices are used for Katz centrality estimation. In each configuration, the experiments are repeated for 20 times to obtain the mean value and error bars. We apply Algorithm 1 to estimate the Katz centralities of a set of random vertices in the networks.

### 5.3   Numerical Results

**Estimation Performance.** We show the estimation performance in Fig. 2. Figure 2(a) compares the deviation between the estimated value and the ground truth in a normalized scale $(0, 1)$. It shows that all points are in or near the diagonal line, which means that the estimation is relatively accurate. Figure 2(b) shows the *MRE*s varying with the number of walks $d$. From the figure, when

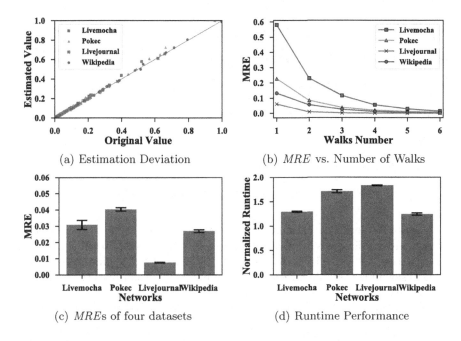

(a) Estimation Deviation

(b) *MRE* vs. Number of Walks

(c) *MRE*s of four datasets

(d) Runtime Performance

**Fig. 2.** Estimation performance for random vertices.

the walks increases, the *MRE* decreases dramatically, which means increasing $d$ can improve the estimation accuracy significantly. Figure 2(c) compares the *Mean Relative Error (MRE)* of estimation in different datasets. It shows that the *MRE* values are below 0.05 for all datasets. The datasets Livejournal has lower *MRE*s less than 0.01. According to the error bars in Fig. 2(c), the deviations of errors are small for all datasets. In Fig. 2(d) we report the computational efficiency of our algorithm in these four datasets. The algorithm Foster we compare with comes from [9] which is widely used in many complex network toolkits such as networkx. In the experiment, the running time of our algorithm is taken as baseline and the running time of the algorithm Foster is reported relative to this baseline. Figure 2(d) shows that our algorithm outperforms Foster and in some networks like Pokec and Livejournal, our algorithm saves nearly half the time. Note that sampling vertices and constructing matrices cost most of the time in the algorithm **SAKE**. Hence our algorithm will outperform more greatly without such prepared work.

**Influence of the Parameters.** We here evaluate the algorithm performance under various system parameters. Figure 3(a). shows the *MRE*s when varying the number of loops $l$ in Algorithm 1 from 1 to 20, With the increasing of loops, the *MRE*s of all datasets declines gradually. It is shown that when $l \geq 5$, the downtrend becomes slow, which implies the algorithm converges. The sampling ratio (the inclusion probability of a vertex in the sampled subset) is an important

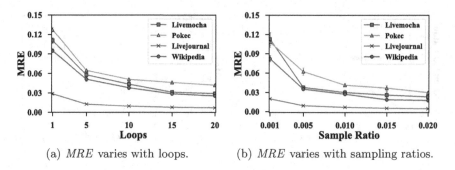

(a) *MRE* varies with loops.    (b) *MRE* varies with sampling ratios.

**Fig. 3.** Influence of parameters for random vertices.

parameter to influence the estimation accuracy. Figure 3(b) shows the estimation accuracy under different sampling ratios from 0.001 to 0.02. When the sampling ratio increases, the value of *MRE* declines gradually at the beginning, and then decreases slowly for sampling ratio larger than 0.005. This implies that sampling these vertices from the original network yields good approximation result. For larger sampling ratio, some datasets achieve very low *MRE* smaller than 0.03.

**Ability of Preserving Vertex Rankings.** We further explore the ability of the proposed algorithm to preserve the vertex rankings, i.e., whether the ranking orders of vertices' Katz centralities are preserved in sampling based estimation. Table 2 compares *Jaccard, Precision, MAP,* and *nDCG,* which are well-known performance metrics in recommendation system [17]. It is shown that the vertices' Katz centrality rankings for top 100 are well preserved with the proposed algorithm. The precision is higher than 97% for all dataset, which means more than 97% high Katz centrality vertices are correctly identified in the top 100 results.

**Table 2.** The ability of preserving vertex rankings (top 100).

| Dataset | Jaccard | Precision | MAP | nDCG |
|---|---|---|---|---|
| Livemocha | 0.970 | 0.985 | 0.997 | 0.997 |
| Pokec | 0.980 | 0.995 | 0.998 | 0.998 |
| Livejournal | 0.990 | 1.000 | 0.998 | 0.999 |
| Wikipedia | 0.951 | 0.975 | 0.992 | 0.992 |

# 6    Conclusion

Katz centrality is an important concept to measure the influence of a vertex in a social network. In this paper, we combined sampling technique with approximation analysis to develop an algorithm **SAKE**, which can estimate Katz centrality based on a small set of samples from the network. The estimation was proved to be unbiased, and the estimation error could be bounded with high probability. Extensive experiments based on four real social networks showed that **SAKE** achieved low estimation error and low complexity, and it performed well in identifying the most influential vertices in social networks.

**Acknowledgment.** This work was partially supported by the National Key R&D Program of China (Grant No. 2018YFB1004704), the National Natural Science Foundation of China (Grant Nos. 61972196, 61672278, 61832008, 61832005), the Key R&D Program of Jiangsu Province, China (Grant No. BE2018116), the science and technology project from State Grid Corporation of China (Contract No. SGSNXT00YJJS-1800031), the Collaborative Innovation Center of Novel Software Technology and Industrialization, and the Sino-German Institutes of Social Computing.

# References

1. Livemocha network dataset - KONECT, April 2017. http://konect.uni-koblenz.de/networks/livemocha
2. Ahmed, N.K., Neville, J., Kompella, R.: Network sampling: from static to streaming graphs. ACM Trans. Knowl. Discov. Data (TKDD 2014) **8**(2), 7 (2014)
3. Auer, S., Bizer, C., Kobilarov, G., Lehmann, J., Cyganiak, R., Ives, Z.: DBpedia: a nucleus for a web of open data. In: Aberer, K., et al. (eds.) ASWC/ISWC -2007. LNCS, vol. 4825, pp. 722–735. Springer, Heidelberg (2007). https://doi.org/10.1007/978-3-540-76298-0_52
4. Balkanski, E., Singer, Y.: Approximation guarantees for adaptive sampling. In: International Conference on Machine Learning (ICML 2018), pp. 393–402 (2018)
5. Boldi, P., Vigna, S.: Axioms for centrality. Internet Math. **10**(3–4), 222–262 (2014)
6. Bonchi, F., Esfandiar, P., Gleich, D.F., Greif, C., Lakshmanan, L.V.: Fast matrix computations for pairwise and columnwise commute times and Katz scores. Internet Math. **8**(1–2), 73–112 (2012)
7. David, E., Jon, K.: Networks, Crowds, and Markets: Reasoning About a Highly Connected World. Cambridge University Press, New York (2010)
8. Eden, T., Jain, S., Pinar, A., Ron, D., Seshadhri, C.: Provable and practical approximations for the degree distribution using sublinear graph samples. In: Proceedings of the 27th International Conference on World Wide Web (WWW 2018), pp. 449–458 (2018)
9. Foster, K.C., Muth, S.Q., Potterat, J.J., Rothenberg, R.B.: A faster Katz status score algorithm. Comput. Math. Organ. Theory **7**(4), 275–285 (2001)
10. Hoeffding, W.: Probability inequalities for sums of bounded random variables. J. Am. Stat. Assoc. **58**(301), 13–30 (1963)
11. Horvitz, D.G., Thompson, D.J.: A generalization of sampling without replacement from a finite universe. J. Am. Stat. Assoc. **47**(260), 663–685 (1952)
12. Katz, L.: A new status index derived from sociometric analysis. Psychometrika **18**(1), 39–43 (1953)
13. Kempe, D., Kleinberg, J., Tardos, É.: Maximizing the spread of influence through a social network. In: ACM SIGKDD International Conference on Knowledge Discovery and Data Mining (KDD 2003), pp. 137–146 (2003)
14. Leskovec, J., Faloutsos, C.: Sampling from large graphs. In: Proceedings of the 12th International Conference on Knowledge Discovery and Data Mining (KDD 2006), pp. 631–636. ACM (2006)

15. Leskovec, J., Lang, K.J., Dasgupta, A., Mahoney, M.W.: Statistical properties of community structure in large social and information networks. In: Proceedings of the 17th International Conference on World Wide Web, pp. 695–704. ACM (2008)

16. Maiya, A.S., Berger-Wolf, T.Y.: Benefits of bias: towards better characterization of network sampling. In: Proceedings of the 17th International Conference on Knowledge Discovery and Data Mining (KDD 2011), pp. 105–113. ACM (2011)

17. Manning, C., Raghavan, P., Schütze, H.: Introduction to information retrieval. Nat. Lang. Eng. 16(1), 100–103 (2010)

18. Nathan, E., Bader, D.A.: Approximating personalized Katz centrality in dynamic graphs. In: Wyrzykowski, R., Dongarra, J., Deelman, E., Karczewski, K. (eds.) PPAM 2017. LNCS, vol. 10777, pp. 290–302. Springer, Cham (2018). https://doi.org/10.1007/978-3-319-78024-5_26

19. Nathan, E., Sanders, G., Fairbanks, J., Bader, D.A., et al.: Graph ranking guarantees for numerical approximations to Katz centrality. Procedia Comput. Sci. 108, 68–78 (2017)

20. Riondato, M., Kornaropoulos, E.M.: Fast approximation of betweenness centrality through sampling. Data Min. Knowl. Discov. 30(2), 438–475 (2016)

21. Riondato, M., Upfal, E.: ABRA: approximating betweenness centrality in static and dynamic graphs with Rademacher averages. ACM Trans. Knowl. Discov. Data (TKDD 2018) 12(5), 61 (2018)

22. Takac, L., Zabovsky, M.: Data analysis in public social networks. In: International Scientific Conference and International Workshop Present Day Trends of Innovations, vol. 1 (2012)

23. Wagner, C., Singer, P., Karimi, F., Pfeffer, J., Strohmaier, M.: Sampling from social networks with attributes. In: Proceedings of the 26th International Conference on World Wide Web (WWW 2017), pp. 1181–1190 (2017)

24. Was, T., Skibski, O.: An axiomatization of the eigenvector and Katz centralities. In: Proceedings of the 32nd AAAI Conference on Artificial Intelligence (AAAI 2018) (2018)

25. Zhao, J., Yang, T.H., Huang, Y., Holme, P.: Ranking candidate disease genes from gene expression and protein interaction: a Katz-centrality based approach. PLoS ONE 6(9), e24306 (2011)

# Location Prediction for Social Media Users Based on Information Fusion

Gaolei Fei$^{(\boxtimes)}$ , Yang Liu , Yong Cheng , Fucai Yu, and Guangmin Hu

University of Electronic Science and Technology of China, Chengdu 611731, China
fgl@uestc.edu.cn , liuy@std.uestc.edu.cn

**Abstract.** The real locations of social media users have always been a hot spot for people. However, considering personal privacy and other factors, most locations provided by users are ambiguous, missing or wrong. In order to get users' real location, we collect various types of geographic related information from users in social networks and propose an information fusion network model to organize the information efficiently. After that, we take advantage of the iterative-based information fusion method to process the geographical related information in the information fusion network and the outputs are used as users' geographical location. Finally, the experimental results show that our research method can greatly improve the prediction accuracy and reduce the corresponding distance error.

**Keywords:** Information fusion · Social media · User location · Relationship strength

## 1 Introduction

The rapid development of online social media such as Twitter, LinkedIn, and Facebook has brought great convenience to people's interaction and information sharing. In social media, an account is the main part for its owner to carry out network activities, so we can analyse accounts in the social media to obtain information of users who own these accounts. Location prediction for social media users is the process of identifying the geographical location of a person by means of obtaining and analyzing registration, publishing and friendship information in its social media account. Geographical locations can be used in many areas such as regional event detection, virus propagation tracking and group language analysis. Hence, location prediction for social media users plays an important role in the control, management and optimization of social networks.

Location prediction for social media users is one of the important contents which belong to social network data mining research. The previous methods can be mainly divided into the following categories. The first type is modeled by the distance and the probability of becoming a friend. Backstrom et al. [1] proposed a method for calculating the probability of a user at a specific location by using this idea, which estimates the location with the greatest probability as the user's

© Springer Nature Switzerland AG 2020
S. Wen et al. (Eds.): ICA3PP 2019, LNCS 11945, pp. 599–612, 2020.
https://doi.org/10.1007/978-3-030-38961-1_51

location. The work of Kong et al. [2] is extremely relevant to Backstrom's. On the basis of their work, Kong et al. regard social relationships as a continuous feature and introduce the concept of social intimacy. Through the relationship between social intimacy and geographical distance, the estimation is significantly improved. In the second category, only the plain text content is used to obtain the user's city-level estimation model. For example, Chandra et al. [3] construct a probability framework from word distribution in plain text content to predict the city-level location of a Twitter user and give a way to estimate the top K possible cities of a user. Cheng et al. [4] also used a plain text to construct a city-level estimation model. The difference is that they used geography-related words in the tweet as features and used classification methods to obtain city-level geographic locations. In addition, some scholars [5–7] use the topic model in machine learning methods to mine potential geographic location information.

Most existing methods used to predict locations for social media users only model with a single or a few types of information. However, social media information usually has a lot of noise, and a single or a few types of information can't accurately describe the user's location, resulting in a low accuracy of user location prediction. In fact, there are a variety of information about a user's location in social media. For example, tweets published by users, personal descriptions, registered locations, etc. may contain the user's location information. If we conduct a fusion analysis of this information, we can effectively improve the accuracy of our prediction.

Aiming at the above problems, this paper proposes a method for predicting social media users' locations based on information fusion. Firstly, we mine and process a variety of information related to users' location as input data for information fusion. Because this information is interrelated, we need to quantitatively describe the relationship in social media. Therefore, we use the social relationship between users to build a network analysis model, mapping input data to the corresponding location of the model. After that, we propose an iterative-based fusion geolocation algorithm to fuse multiple input data in the network analysis model and finally output the main active area of the user. Experiments show that, through our method, the correct rate of 60% and 80% can be achieved under the two levels of city and state.

## 2    Problem Definition and Data Description

### 2.1    Problem Definition

In order to better understand our research, we firstly define the problem target. We use $U$ to represent a set of geo-related information published by users and their friends. For example, the location information filled in when the user registers and the geographic information contained in the user description can be used as elements in the set $U$. As we all know, the analysis and processing of any data is inseparable from the specific environmental background. For the set $U$, we use the graph $G <V, E>$ to indicate the social network environment in which it is located. Finally, we choose a suitable information fusion geolocation algorithm $F$, taking the set $U$ and the graph $G<V, E>$ as input parameters, then the problem target can be expressed as:

$$D = F(U, G<V, E>),$$

where $D$ represents the main active area of the user. According to different partition granularity, the city or state level can be used to describe the user's main activity area.

## 2.2   Data for Geolocation

Users' raw data collected in social media is the basis for geolocation. Twitter which is one of the top ten websites with the largest Internet traffic in the world contains rich social information. In addition, the reliable and stable API allows us to easily access it. This paper takes Twitter as an example to study the geolocation for social media users, but the method is not limited to Twitter and can be extended to other social media. There are lots of information used for geolocation in Twitter. In general, it can be divided into two aspects: background attribute information and social relationship information.

**Background Attribute Information.** The background attribute information refers to a geographically relevant information set published by the user or his friends. Background attribute information can be divided into two categories, one is direct location information which directly describes the user's location or the place where the tweet is published. For instance, location information filled in by the user when registering and GPS information when posting a tweet are direct location information. The direct location information in Twitter is shown in Table 1. Another is status description information describing the user's past or current status. Some of it includes geographic information, which cannot be ignored for geolocation. The description information and the text information below exactly belong to state description information.

**Table 1.** Direct location information in Twitter

| Information tag | Descriptions |
|---|---|
| *Location* | The geographical location information filled in by the user in the user profile |
| *Geo* | The tag has been deprecated and replaced by the Coordinates |
| *Coordinates* | The tag indicates the GPS coordinates of the tweet, which is automatically added by Twitter |
| *Place* | The tag indicates the details of the tweet location, Including the location name, administrative level and the location boundary |

We tried to predict the users geographic location only with the information in Table 1, but the data analysis shows the missing rate of it is very high. The statistical results are shown in Fig. 1.

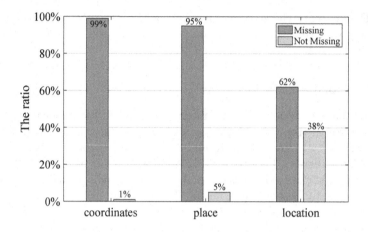

**Fig. 1.** Missing status of direct location information

Both *coordinates* and *place* are automatically filled in based on GPS information, so they have a high degree of credibility. However, the missing ratio of the two kinds of information is over 95%. Therefore it is impossible to predict the user's location by them. Although the proportion of *location* not missing is 38%, this part is freely filled by users. According to statistics, the error rate of *location* is 23%. So the correct rate is roughly 29% using *location* directly via calculation, which does not meet the accuracy we expect. However, this information is still valuable and the correct rate can be used as a reference for judging other methods.

Considering that there is less direct location information available, we introduce two types of status descriptions information as follows:

– **description:** The tag filled in by the user describes the user background and may involve the information about the surrounding environment.
– **text:** The tag represents tweet contents and some topics in it involve geographic location information.

Through statistical analysis of the *description* and *text*, we find as long as geographical location information appears in the *description*, this information is closely related to the user's place of birth, learning and working environment, and is highly correlated with our problem target. In addition, a user account can correspond to a large number of tweets. Although the proportion of texts with geographic information is low, because of the large base, the geographic information that the *text* can provide is still rich.

**Social Relationship Information.** In addition to the background attribute information, there is also a piece of information describing the social relationships between users in social media, and we call them social relationship information. While we would like to believe that our social options are endless, human relationships are constrained in many ways [1]. Considering time, energy and money, people are more inclined to choose long-term stable social friendships with people who are close to their geographical location. Therefore, social relationship information is actually highly correlated with geographic location. This is also the basis for us to build a network and organize different types of geographic location information by social relationship information. We divide the social relationship information into two categories: information flow and intrinsic relationship. The process of referring and replying between users can be used to indicate the information flow. And intrinsic relationship can be reflected in the users friends and followers.

## 3   Modeling

In order to use the social relationship information to associate background attribute information, we need to establish a suitable network analysis model. The main relationship network and the information fusion network are two types of network analysis model which we will introduce in the following.

### 3.1   Main Relationship Network

A good network relationship model should meet the following conditions. On the one hand, it can accurately describe the differences in relationships between users. On the other hand, it needs good anti-noise performance and convergence. The main relationship network is a network that meets the above conditions.

**Model Description.** The main relationship network (MRN) is an undirected weighted hierarchical graph $G'(V', E')$ where $V'$ represents the set of the user and his friends and the weight of $E'$ indicates the strength of the relationship between users. Besides, in the network, we remove the weak links and the edges that would cause the results to oscillate, so the main relationship network retains the research user's trunk friendship and has strong anti-noise performance.

The right subgraph in Fig. 2 is a main relationship network where the vertices are numbered hierarchically. The node $x$ in the $i$th layer and the node $y$ in the $(i + 1)$th layer constitute a node pair $P_i(x, y)$. For example, the $V_3$ in the 1th layer and the $V_2$ in the 3th layer constitute the node pair $P_1(3, 2)$. The direction of information flow between the node pair will be hidden. However, the information interaction and intrinsic relationship between the node pair $P_i(x, y)$ will be reflected in the way of the weight $w_i'(x, y)$. So it can be used to measure the strength of the relationship between user $x$ and user $y$.

**Network Construction.** From the above, we can see that in the main relationship network, the information interaction and intrinsic relationship are uniformly mapped on the edge weight. What's more, the weak connections and the edges that would cause the data to oscillate are deleted. Therefore, for the construction of the main relationship network, it can be divided into three parts: solving edge weight, extracting main relationship and optimizing network.

*(a) Solving Edge Weight.* We make $O_x^i$, $O_{xy}^i$ respectively represent the amount of all information flowing out by the node $x$ in the $i$th layer and the amount of information flowing out from the node $x$ in the $i$th layer to the node $y$ in $(i + 1)$th layer. In Twitter, the flow of information between nodes we consider is mainly the process of mentioning and replying between users. We use $|at_{xy}^i|$, $|reply_{xy}^i|$ respectively to indicate the times node $x$ refers to and replies to node $y$, then

$$\begin{cases} O_{xy}^i = |at_{xy}^i| + |reply_{xy}^i| \\ O_x^i = \sum_j |at_{xj}^i| + |reply_{xj}^i|, \end{cases}$$

where $j \in V'$ and $j \neq x$. The frequency of information interaction between $x$ and $y$ can be expressed as

$$freq_i(x, y) = \frac{min(O_{xy}^i, O_{yx}^i)}{\sqrt{O_x^i \times O_y^i}}. \tag{1}$$

In addition to the frequency of information, the intrinsic relationship is also an important consideration. Through intrinsic relationship, we can find out whether the relationship networks between two users are similar, and the relationship between users with high similarity networks will be closer. Next, we quantify this friendship similarity.

$$fs_i(x, y) = \frac{|F_x^i \cap F_y^i|}{\sqrt{|F_x^i| \times |F_y^i|}}, \tag{2}$$

where $F_j^i$ represents a set of friends who are fans with each other for the user $j$ in $i$th layer, $|F_j^i|$ indicates the number of elements in the $F_j^i$. After that, we need to unify the (1) and (2) to get the normalized formula of $w_i^*(x, y)$:

$$w_i^*(x, y) = \frac{freq_i(x, y) + fs_i(x, y)}{\sum_j (freq_i(x, y) + fs_i(x, y))}.$$

*(b) Extracting Main Relationship.* $w_i^*(x, y)$ reflects the degree of intimacy between $x$ and $y$. The higher the intimacy is, the closer the geographical location between them may be. Using the meaning of $w_i^*(x, y)$, we can remove some weakly connected edges to further simplify the network. Here we think that for node $x$, the edges that satisfy the following inequality will be deleted.

$$w_i^*(x, y) < \frac{1}{N} (\sum_j w_i^*(x, j) + \sum_k w_{i-1}^*(k, x)),$$

where N is the number of all edges $e_i(x, j)$ satisfying $w_i^*(x, j) \neq 0$ and edges $e_{i-1}(k, x)$ satisfying $w_{i-1}^*(k, x) \neq 0$.

*(c) Optimizing Network.* In addition to the extraction of the main relationship by deleting the weakly connected edges, we also need to remove the edges that cause the data to oscillate to optimize the network. We introduce the idea of the layered graph to solve it. The layered graph is based on the original graph, deleting the edges between the same layer nodes and the edges from the upper layer to the lower layers.

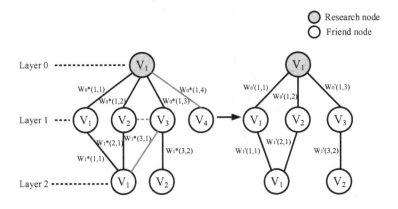

**Fig. 2.** Construction of the main relationship network (Color figure online)

The red solid edges in Fig. 2 belong to the weakly connected edges. And the red dashed edges represent the same layer edges. We choose to delete the two types of edges because the weakly connected edges will bring lots of noise users and the same layer edges will bring data convergence. Then we normalize the weight again and get the main relationship network.

## 3.2   Information Fusion Network

With the main relationship network, we have been able to describe the strength of the relationship between users. However, due to the simplification of the network, users will have fewer friends around them. At this time, it is very limited to predict the geographic location of users from surrounding friends. Therefore, we need to integrate additional information into the existing network and propose the information fusion network model.

**Model Description.** The information fusion network (IFN) is an undirected weighted layered graph based on the main relationship network which can adapt to the input of multiple types of information and we use $G^*(V^*, E^*)$ to indicate it. In the information fusion network, in addition to the immanent friends, the rest of the valuable information is regarded as friends into the network. As shown in Fig. 3, the triangle node is a friend derived from the node directly connected to it, and their essence is the set $R$ with valuable information around the directly connected node.

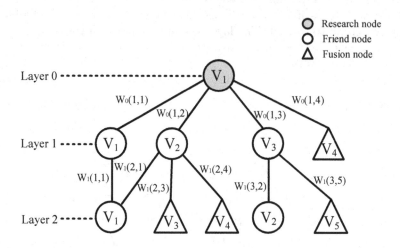

**Fig. 3.** Information fusion network

**Network Construction.** The information fusion network is built on the basis of the main relationship network by adding fusion nodes and corresponding edges. Therefore, adding fusion nodes and adjusting edge weight become two important parts of building network.

*(a) Adding Fusion Nodes.* Adding a fusion node needs to meet certain rules. Before introducing the rules, we need to introduce the concepts of non-edge nodes and edge nodes. In the main relationship network, for node $x$ in the $i$th layer. If there exists the node $y$ in the $(i + 1)$th layer which makes $w_i'(x, y) \neq 0$, then we call $x$ the non-edge node in the $i$th layer. The non-edge nodes of all layers in the graph form a set $V_c$, and the remaining nodes constitute the edge node set $V_e = C_{V'}V_c$. In other words, edge node set is the complement set of $V_c$ in full set $V'$. In fact, the fusion nodes derived from the main relationship network are also edge nodes in essence. Therefore, we do not mount any fusion nodes for the original edge nodes in the main relationship network.

*(b) Adjusting Edge Weight.* In addition to adding fusion nodes, the information fusion network adds corresponding edge to form a new undirected weighted graph $G^*(V^*, E^*)$. Therefore, the weight of the information fusion network needs to be adjusted accordingly. Considering that the fusion node is not a true friend, the weight should not be personalized. Also, the relative value of the weight between

the original nodes should not be changed. We follow the following principles to adjust network weights.

In the main relationship network, for node $x$ in the $i$th layer. Then we express its directly connected node set in the $(i+1)$th layer with $V'_{ix}$. If the number of elements in $V'_{ix}$ is $|V'_{ix}| = n$, from the weight relationship of the main relationship network, it is not difficult to get $\sum_{y=1}^{n} w'_i(x,y) = 1$. When evolving from the main relationship network to the information fusion network, the set of fusion nodes derived from the node $x$ is represented by $R_{ix}$ and $|R_{ix}| = m$. Then, the weight update formula of the network is as follows.

$$w_i(x,y) = \begin{cases} \frac{w'(x,y)}{1+\frac{m}{n}}, & y \in V'_{ix} \\ \frac{1}{n} \times \frac{1}{1+\frac{m}{n}}, & y \in R_{ix}. \end{cases}$$

So far, we have completed the evolution from the main relationship network to the information fusion network.

# 4  Information Fusion Geolocation Method

With the background attribute information as the input data set $U$ and the information fusion network $G^*(V^*, E^*)$ as our network analysis model, we need to choose a suitable fusion algorithm $F$ to finally predict the user location. Considering that our information fusion network is a hierarchical structure, there is an obvious iterative relationship between the upper user and the lower user. And the hierarchical iterative process is also a layer-by-layer denoising process, which can greatly improve the accuracy of the prediction. Therefore, we use an iterative-based information fusion geolocation algorithm to fuse the user's surrounding friends' information to the research node layer by layer. The whole algorithm contains the following three important components, and the final algorithmic process will be given in Algorithm 1.

## 4.1  Network Initialization

The initialization of the network is actually the initialization of the algorithm input. The initialization object is a node with the edge feature, and refers to the original edge node and the fusion node in the information fusion network.

- For the original edge node, we check in turn whether the values of GPS (*coordinates* or *place*), *location* and the highest frequency place in the *text* are missing. Then we take the first information that is not empty and convert it into latitude and longitude. If the above three candidate information are all empty, we mark the node as a missing status and assign it to NULL.
- For the fusion node, we extract the set of place names in *description* and *text*, and also convert it into latitude and longitude. Then we initialize the fusion node with the latitude and longitude.

## 4.2  DBSCAN Density Clustering

Clustering is the process of fusing the geographic information of the user's friends to the user and removing noise information. DBSCAN is chosen because the number of categories of user friends is unknown. DBSCAN does not need the number of categories and can find clusters with arbitrary shapes. Moreover, the number of friends of the user is small, and this paper has a fixed parameter selection. Our main goal is not to pay attention to the quality of the whole clustering effect, but to find the cluster that meets the parameter requirements perfectly. Even if the density difference is large, it will not affect the results seriously. The iterative use of DBSCAN can finally integrate the multi-layer network information into the research node to predict the user location.

## 4.3  Select the Target Cluster

After density clustering, we need to select the appropriate cluster as the target cluster $C_d$. The quality of the target cluster will directly affect our final result. The most intuitive idea is to use the cluster with the largest number of members as the target cluster. When the number of members in multiple clusters is the same, the most dense cluster is the optimal solution we think. In order to quantitatively describe the intensity of a cluster, we first introduce the concept of a regional center. We assume that the latitude and longitude set in the cluster $C_i$ is $\{(lat_1, lng_1), (lat_2, lng_2), \cdots, (lat_n, lng_n)\}$, then the regional center can be expressed as:

$$center = (lat_c = \frac{1}{n} \times \sum_{i+1}^{n} lat_i, lng_n = \frac{1}{n} \times \sum_{i+1}^{n} lng_i).$$

We use the average spherical distance $\bar{d}$ of all the points in the set to the center of the area $(lat_c, lng_c)$ as an indicator to measure the cluster density. The smaller the $\bar{d}$, the denser the cluster. When the geographic location information distribution is too scattered, the clustering will be a series of clusters with only a single member. At this time, it is impossible to judge with the intensity of the cluster. And we select the single cluster that is most closely related to the current user as the target cluster:

$$\begin{cases} y' = \underset{y}{max}\, w_i(x, y) \\ C_d = \{(lat_{y'}, lng_{y'})\}. \end{cases}$$

However, when there are other auxiliary information, we can use it to correct some of the wrong target clusters. TopK auxiliary judgment is a method which can correct some of the wrong target clusters using the Location information and the idea is as follows.

**Step 1:** We assume that there are $N$ clusters after clustering. Sort in descending order according to the number of members in the cluster to get the sequence $\{C_1, C_2, \cdots, C_n\}$.

**Step 2:** Only retain the front $\lceil N/K \rceil$ cluster and we take $K = 3$ by default. If the latitude and longitude of the place in the Location is within the range of the cluster $C_i$, we output $C_i$ as the target cluster, otherwise $C_1$ will be the target cluster.

As we can see, the high accuracy of the *location* information can weaken the decisive influence of the number of members, and the number of members can in turn suppress the noise of the *Location* information.

Algorithm 1 indicates that we firstly initialize the elements in the edge node set $V_e$ and the fusion node set $R$ in the network. After that, we use density clustering layer by layer from bottom to top, each node will correspond to a set $L'$ composed of a series of clusters. We obtain the target cluster of the node according to the TopK auxiliary judgment and two special clusters. Then, we use the regional center of the target cluster as the initialization information of the next density cluster, and iterate until obtain $C_{d1}^0$ which is the target cluster of the node in layer 0. Finally, we map the regional center $(lat_1^0, lng_1^0)$ of $C_{d1}^0$ into the user's main active area $D$ according to the accuracy requirement.

---

**Algorithm 1.** Infomation Fusion Geolocation Algorithm

**Input:** Network initialization data, Network depth $H$
**Output:** The user's main activity area $D$

1: **for** node in $\{V_e, R\}$ **do**
2:     Initialization(node);
3: $L = \{\}$
4: **for** $i = H - 1$ to $0$ **do**
5:     **for** $x = 1$ to $|V_i'|$ **do**
6:         **for** $y = 1$ to $|V_{i+1}'|$ **do**
7:             **if** $w_i(x, y) \neq 0$ **then**
8:                 $L$.append($(lat_y^{i+1}, lng_y^{i+1})$)
9:         $L' = \text{DBSCAN}(L)$
10:        $L = \{\}$
11:        **if** $Location_x^i \neq \text{NULL}$ **then**
12:            $C_{dx}^i = \text{TopK}(L', Location_x^i)$
13:        **else**
14:            $C_{dx}^i = \text{select}(L')$
15:        $(lat_x^i, lng_x^i) = \text{Center}(C_{dx}^i)$
16: $(lat_1^0, lng_1^0) \rightarrow D$
17: **return** $D$

---

## 5 Simulation and Analysis

### 5.1 Simulation Data and Environment

For data collection, we use Twitter's official API to get 10000 Twitter users' information. We filter out the spammers, celebrities, and officially authenticated users among them. Because our target users should be the ordinary users. In addition to filtering the above users, we also need to filter out users whose the most frequently occurring GPS information is less than 3. Because we take the GPS information as the true value of the user's geographic location, GPS information with high frequency of occurrence is highly reliable. In the end, we have a total of 1046 Twitter users. Considering the factors of computing and storage resource, we only got their second-level friendship.

For experimental tools and parameter settings, we use the named entity recognition extracting place names from text sequences. For the conversion between place names and latitude and longitude, the Google Maps API is used. For the parameter setting of DBSCAN, according to the literature [4], it is found that the location of two friends is more likely to be within 10 km. So we regard the two geographical locations not more than 10 km as the same geographical area, and divide them into the same classes. Besides, the least number of classes can contain only one geographic location.

### 5.2 Experimental Results and Analysis

For the experimental results, we analyze the two aspects of correct rate and distance error. According to the different granularity, we can have the correct rate under three kinds of granularity: city-level accuracy, state-level accuracy and Acc@161 [8]. Acc@161 is a good accuracy indicator which means if the predicted place is within the radius of 161 km of the real place, then the prediction is considered correct. The results of the three indicators in the four experimental categories are shown in Table 2.

**Table 2.** Three indicators in the four experimental categories

| Category | City-level accuracy | State-level accuracy | Acc@161 |
|----------|---------------------|----------------------|---------|
| MRN | 52.39% | 75.05% | 66.83% |
| MRN+TopK | 59.27% | 81.93% | 75.14% |
| IFN | 55.26% | 78.49% | 70.75% |
| IFN+TopK | 60.33% | 82.89% | 75.62% |

It is not difficult to find that the information fusion network has a 3% to 4% improvement in overall performance over the main relationship network, which is the result of IFN containing more abundant network information. The TopK auxiliary judgment increases the overall accuracy of the two networks by about 5%. The combination of IFN and TopK achieves the accuracy of 60% and 80% at the city and state levels respectively.

From the perspective of accuracy, we have an overall understanding of the experimental results of the four methods. Next, we need to understand them more specifically from the perspective of the distance error distribution. In Fig. 4,

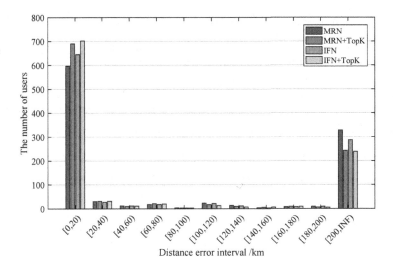

**Fig. 4.** Distance error distribution

we can see that the user error is mainly distributed within 20 km. Among them, the IFN+TopK has the best effect, and the number of people with low error segments ($\leq 100$ Km) is more than the number of the other three methods, especially within the 20 km error, the number exceeds 700.

## 6 Conclusion

In this paper, we extract and analyze the information of Twitter users and their friends, construct corresponding network analysis models that link the acquired data, and finally predict the main activity areas of users through means of information fusion. After our final experiments, it is proved that information fusion network and the auxiliary decision method can predict the user position with high accuracy, and can effectively reduce the distance error.

## References

1. Backstrom, L., Sun, E., Marlow, C.: Find me if you can: improving geographical prediction with social and spatial proximity. In: International Conference on World Wide Web (2010)
2. Kong, L., Liu, Z., Huang, Y.: Spot: locating social media users based on social network context. Proc. VLDB Endow. **7**(13), 1681–1684 (2014)
3. Chandra, S., Khan, L., Muhaya, F.B.: Estimating Twitter user location using social interactions-a content based approach. In: IEEE Third International Conference on Privacy (2012)
4. Cheng, Z., Caverlee, J., Lee, K.: You are where you Tweet: a content-based approach to geo-locating Twitter users. Cikm **19**(4), 759–768 (2010)

5. Hao, Q.: Equip tourists with knowledge mined from travelogues. In: International Conference on World Wide Web (2010)
6. Hong, L.: Discovering geographical topics in the Twitter stream. In: International Conference on World Wide Web (2012)
7. Mei, Q.: A probabilistic approach to spatiotemporal theme pattern mining on weblogs. In: International Conference on World Wide Web ACM (2006)
8. Rahimi, A., Vu, D., Cohn, T.: Exploiting text and network context for geolocation of social media users (2015)

# Performance Modelling and Evaluation

# Concurrent Software Fine-Coarse-Grained Automatic Modeling Method for Algorithm Error Detection

Tao Sun$^{(\boxtimes)}$, Jing Zhang, and Wenjie Zhong

College of Computer Science, Inner Mongolia University,
Hohhot, Inner Mongolia, China
cssunt@imu.edu.cn, Zhangj_vi@163.com,
zhongwenjie@mail.imu.edu.cn

**Abstract.** Concurrent software state space explodes, which makes algorithm error detection difficult. This paper proposes a fine-coarse-grained automatic modeling method. Based on the JAVA concurrent program, we generate the HCPN (Hierarchical Coloured Petri Net) fine-coarse-grained model that in accordance with the behavior of the source program automatically. The goal is to detect the algorithm errors in the program through the model checking technology. We complete the modeling of interactive, property-related and specific structure statements through fine-grained method and complete the modeling of other statements through coarse-grained method. Avoid the state space explosion effectively under the premise of retaining the interaction behavior and the property-related behavior execution path. This paper verifies the effect of fine-coarse-grained automatic modeling method by comparing and analyzing the experimental results.

**Keywords:** HCPN · Concurrent software · Coarse-grained modeling · Algorithm error detection

## 1 Introduction

Software defect detection is an indispensable part in software development. The concurrent behavior in concurrent software has led the execution of program become more complicate, and increased the difficulty of algorithm error detection. Software model checking [1] is a technology for detecting defects in a program. The main idea is to convert the source program into a model, then search the counter-example that not satisfied this property in the full state space of model. If a counter-example is found, there is a defect in source program. Due to the model checking technology needs to be performed in a finite state space, so the state space explosion problem has always been an important part of the research in the field of model detection.

Currently, most software model checkers are implemented through the method of Abstract-Verification-Abstract Refinement, such as SLAM [2], MAGIC [3], and BLAST [4], etc. These tools are used to verify the safety properties of C/C++ program. They convert the source program into a smaller abstract model by using the predicate abstraction technique, and then search a counter-example by verify the abstract model.

S. Wen et al. (Eds.): ICA3PP 2019, LNCS 11945, pp. 615–623, 2020.
https://doi.org/10.1007/978-3-030-38961-1_52

Abstract technique reduces the size of model and state space effectively, but the abstract model lacks partial information relative to the source program, which may lead to some invalid counterexamples. In this case, the abstract part of the model needs to be refined to eliminate invalid counterexamples, and then verify the model whether satisfies the properties again. The above three tools reduce the state space in different degrees through abstraction technology, but the invalid counter-example caused by abstraction still need to perform multiple abstraction and verification operations on the program, which requires a lot of time and resources.

Literature [5] proposed the fine-grained automatic modeling method based on HCPN (Hierarchical Coloured Petri Net). HCPN is a high-level Petri net. It realized hierarchical model on the basis of CPN (Coloured Petri Net), which can describe the complex systems with call relationships more efficiently. This method is different from the above tools, it does not use the abstraction, and that is, the execution of each statement in the program remains in the model. So the modeling and verification process only needs to be done once and does not lead to invalid counter-examples. However, this method lead to too many transitions in the model, and the state space is too large that need to be simplified for carry out model checking.

Only the property-related parts in the HCPN model are concerned when performs model checking. This paper proposes the fine-coarse-grained automatic modeling method. Use the fine-grained modeling method retains the execution path of the interactive statements and the property-related statements related to model checking; use the coarse-grained modeling method reduces the generation of the information about other unrelated statements. Thereby reducing the size of the model and state space and ensuring that model checking results is not affected. Finally, the HCPN model is verified to determine if there is a corresponding algorithm error in the program. We divide the fine-coarse-grained automatic modeling method into two parts: *Source program labeling and storage* and *Construct model*. They are detailed in Sects. 2 and 3 respectively.

## 2  Source Program Labeling and Storage

The fine-coarse-grained modeling methods are based on HCPN, so the definitions of CPN and HCPN are given here:

**Definition 1.** A non-hierarchical Coloured Petri Net is a nine-tuple CPN = (P, T, A, $\sum$, V, C, G, E, I) [6].

P: set of places; T: set of transitions; A: set of arcs; $\sum$: set of colour sets; V: set of variables; C: colour set function (assigns colour sets to places); G: guard function (assigns guards to transitions); E: arc expression function (assigns arc expressions to arcs); I: initialisation function (assigns initial markings to places).

**Definition 2.** A hierarchical Coloured Petri Net Module is a four-tuple HCPN = (CPN, Tsub, Pport, PT) [7].

CPN = (P, T, A, $\sum$, V, C, G, E, I) is a non-hierarchical Coloured Petri Net.

Tsub ⊆ T is a set of substitution transitions. Each substitution transition designates a subpage that gives more precise and detailed description of the activity represented by the substituted transition.

Ports are the interface at which the subpage is plugged into the upper level.

PT: Pport → {IN, OUT, I/O} is a port type function that assigns a port type to each port place. The surrounding places of a substitution transition are called socket places represented by Psock (t).

The algorithm error in this paper means the property descriptions that can be expressed by the ASK_CTL property formula. We assume that the program contains the following statements: order statements (assignments, inputs, etc.), branch statements, loop statements, object-oriented and multi-threading. Meanwhile, the JAVA program has been passed the compiler, because compile-time errors are not included in the algorithm errors defined in this article. We want to convert the source program to a model that is as small as possible and does not affect model checking. Therefore, we label the source program to find out the relevant statements that need to be retained, and the remaining statements are coarse-grained modeling to reduce the useless information. At the same time, it is necessary to ensure the logical relationship in the model as same as the source program. So we use the class table and the statement binary tree to store the logical relationship between the classes and the statements respectively. The source program needs to be processed in the following three steps:

**Label Program:** In this article, we use the ASK_CTL formula to describe property, ASK_CTL consist of the information of place or transition in HCPN model and the temporal operators. In order to perform model checking successfully, the places and transitions involved in ASK_CTL formula must exist in the model. These places and transitions correspond to some statements (these statements are called *property-related statement*, the rest of the statements are called *property-unrelated statement*) in the program. We need to find these property-related statements in order to construct a small model accurately without deleting the necessary places and transitions.

**Construct Class Table:** The information stored in the class table is divided into three categories: member variable *Var*; member function *Fun* (function declaration) and static variable *Svar*. Construct class tables have two advantages. Firstly, class table record the corresponding model fragment after the conversion of functions or variables. If the function or static variable is called multiple times, the model fragment generated by the first time is recorded in the class table, and it can be read directly from the class table without re-conversion when the next call. Secondly, class table can be used to distinguish homonymous functions in different classes. Homonymous functions are converted into different model fragments and recorded in different class tables. When calling homonymous functions in different classes, reading the corresponding class tables.

**Construct Statement Binary Tree:** In addition to using the class table to store the member information of the class, we also use the statement binary tree to store the statements in the function. Each function corresponds to a statement binary tree, and the left and right children of the statement binary tree respectively representing the nested relationship and the non-nested relationship between the statements, which can facilitate the conversion process. We need to set the **Type** and the conversion **granularity** of each node during construct the statement binary tree. Type is statement type, for example, *if, else, declarations,* etc. Conversion granularity is divided into two types: *fine* and *coarse*. The granularity setting principles are as follows:

(1) For property-related statement, it must be set to fine-grained.
(2) Nodes are set to fine-grained which corresponding the statements that include inter-process interaction or object-oriented, such as shared variables, creating objects, etc. Coarse-grained conversion converts multiple statements into a complex arc expression, while HCPN only supports partial type statements, and the rest of the statements need to be represented by model elements (fine-grained conversion).
(3) Node n1 with coarse-grained, then the granularity of the node n2 in its left subtree must be coarse-grained. If n2 is fine-grained, then the node n1 should be reset to fine-grained.
(4) Node n1 with fine-grained and n1 corresponds to a branch statement, then the node corresponding to a statement that matching this branch statement should be set to fine-grained. The setting of rule (3), (4) is because the coarse-grained modeling is converted in units of multiple lines of statement (code segments), that is, all the statement in the code segment must be coarse-grained.

## 3   Construct Model

We consider the execution of the statement in program to be the ignition of the transition in the model. That is to say, the value of the variable before/after the execution of a line of statement (or a statement segment) in the program corresponds to the value of the variable before/after the ignition of a certain transition in the model, and the statement (or a statement segment) can be converted into an model fragment (or an arc expression of the arc that connecting the transition and place). The difference is that, each variable is allocated a storage space in the program, but in HCPN model, there is no memory for storing variables. So the value of a variable is stored by a *token* in HCPN, and the circulation of token is treated as a transmission of a variable value. The color set of the place is determined by the declared variable.

The construction process is: preorder traversal the statement binary tree and judge the granularity of the node. If it is fine-grained, use the fine-grained modeling method, the conversion unit is a single statement; if it is coarse-grained, the coarse-grained modeling method is used, and the conversion unit is multiple statements (code segment). In general, there are four structures in the program: ① *branch* ② *loop* ③ *order (Ordinary)* ④ *order (Object-oriented)*. There are multiple types of statements in each structure. Type ① ② ③ can use both coarse-grained and fine-grained modeling methods, while type ④ can only use fine-grained modeling methods due to its particularity (Sect. 2).

**Fine-Grained Conversion:** Fine-grained modeling method have been described in literature [5], and briefly introduced here. The conversion unit of fine-grained is a single statement; each statement is converted into different model fragments, and finally connected according to the relationship between the statements to form a complete structure. As shown in Fig. 1, (a) is a branch structure, output transitions (T1, ..., Tn) of place (P1) are used to represent N branches. For the loop structure, we can use the ring structure in (b) to represent. The two output transitions T1, T3 after place P1 represent entering loop and exiting loop respectively. Other statements except *branch* and *loop* statements are *order* statements. Such as multi-threading and function calls statements, Fig. 1(c), (d) are the model fragments of them respectively.

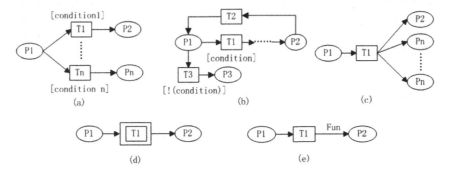

**Fig. 1.** Model fragments

**Coarse-Grained Conversion:** Coarse-grained modeling converts in units of a code segment. The code segment contains the combination of the above-mentioned structures (*order (ordinary)*, *branch* and *loop*). When modeling, each structure is converted into a function, the order in which the function is called is the order of execution in the code segment. The result of the function call is consistent with the result of the code segment execution. Finally, the code segment is converted into a function expression. As shown in Fig. 1(e), Fun is the function expression, and the expression as an arc expression for an arc connecting the transition T1 and the place P2 in the model. Since the color set of the place must be indicated at each place, the color set of the place is determined according to the arc expression of the arc that connected the place. If the arc expression is a function, the color set at that place is the composition of the variable returned by the function. If the arc expression only passes the variable, the color set of the place is constructed by the variable in the arc expression.

To facilitate understanding, we present the definition of "NEXT" here. During conversion, NEXT is used to construct the function. Whether the function will generate the next calling function or generate a variable tuple that returns the current value is determined by the NEXT of a statement:

**Definition 3.** NEXT: NEXT is used to describe the next step that needs to be performed after a certain statement in the code segment is executed. It can be a tuple consisting of multiple variables or a function. When it is a tuple, means that it needs to return the current value of variables. When it is a function, means that the next step is to executing the next structure.

In coarse-grained conversion, multiple statements are converted into a complex arc expression. The algorithm for building the complex arc expressions is given in Fig. 2. In a statement binary tree, the left child of the node indicates nested, so we judges whether the statement has a nested by judging whether the node has a left child or not. If there is a nest, create a new function and call it. If there is no nested, research NEXT as the next step of this statement. **cfun** represents the function currently being converted, **addExp** adds the type and expression of the current statement to cfun, indicates that the statement is recorded in the function being converted. Function **order_conversion** (Fig. 2 right) is used to process order structure, use different conversion methods depending on the type of statement. According to whether there are other structures after an order structures, we divided into two types of the end flag of order structures, indicating that the current function is processed. The algorithm handles the two flags differently.

```
public void Coarseg_conversion(Treenode cnode){
    cfun.type(cnode.type);
    addNEXT(NEXTstack,cnode);
    if(cnode.leftchild!=null){
    //Node has left child, nested structure conversion
        branchnum++;
        Fun fun1=createnewfun(varlist);
        if(cnode.type=="if"||cnode.type=="while"||
                            cnode.type=="for"){
            cfun.addExp("condition",cnode.exp);
            cfun.addExp("then",fun1.getName());
        }
        else if(cnode.type=="case"||cnode.type=="default"){
            cfun.addExp("value",cnode.getexp());
            cfun.addExp("exp",fun1.getName());
        }else{ //else and else if statements
            cfun.addExp("else",fun1.getName());
        }
        cfun=fun1;
    }
    else{ //node has no leftchild, looking for NEXT as the next step
        if(cnode.type=="else if"){
            Fun fun1=createnewfun(varlist);
            cfun.addExp("else",Fun1.getname());
            cfun=fun1;
        }
        else{
            NEXT next=getNEXT(NEXTstack);
            if(cnode.type=="if"||cnode.type=="while"||"for"){
                cfun.addExp("condition",cnode.exp);
                cfun.addExp("then",next.name+next.var);
            }else if(cnode.getType=="case"
                            ||cnode.getType=="default"){
                cfun.addExp("value",cnode.exp);
                cfun.addExp("exp",next.name+next.var);
            }else{
                cfun.addExp("else",next.name+next.var); }
        }
    }
}
```

```
public void order_conversion(Treenode cnode){
    if(cfun.type==null)
        cfun.type="order";
    if(cnode.type=="declare")
        VarList.addVar(cnode.exp,branchnum);
    if(cnode.type == continue or return ){
        cnode.righrchild()=null;
        cfun.addExp(cnode.type,createvar(varlist));
    }else {
        cfun.addExp(cnode.type,cnode.exp);
    }
    String type=Cnode.rightchild.type;
    if(cnode.rightchild==null||type=="if"||type=="elseif"
            ||type=="else"||type=="for"||type=="while"
            ||type=="switch"){
    //Two cases of termination of the sequence structure
        if(cnode.rightchild==null){
            if(cnode.type!="continue"
                        &&cnode.type!="return"){
                NEXT next=getNEXT(NEXTstack);
                cfun.addExp("in",next.name+next.var);    }
                if(NEXTstack.lastElement().type!="for")
                    funstack.pop();
                    cfun=funstack.lastElement();
        }else{
            cfun=funstack.pop();
            Fun fun1=createnewfun(varlist);
            cfun.addExp("in",fun1.getname());
            cfun=fun1;
        }//else
    }//if
}//ChangeOrderType
```

**Fig. 2.** Algorithm of arc expression construction

**CPN Tools Standard Model File:** The information such as place, transition and arc generated during the conversion are recorded in the corresponding table. When converted to an XML file, the information is read from the table and converted into corresponding tags in the XML file. The XML file can be opened with the CPN Tools. Model checker takes the HCPN model and the ASK_CTL formula as inputs. If the output is *true*, the model satisfies the property, and the program does not have the algorithm error; If the output is *false*, the model does not satisfy the property, there is an algorithm error in the program.

# 4 Application

In this section, we compare the fine-grained and the fine-coarse-grained modeling method with an example of Producer & Consumer. The JAVA program is shown in Fig. 3.

```
1   public class Test {
2       public static int count=0;
3       private static String LOCK = "lock";
4       public static void main(String[] args){
5           new Thread(new Producer()).start();
6           new Thread(new Consumer()).start();
7       }
8       static class Producer implements Runnable {
9           @Override
10          public void run() {
11              int sum=0;
12              for(int i=0;i<20;i++){
13                  try {
14                      Thread.sleep(500);
15                  } catch (Exception e) {
16                      e.printStackTrace();
17                  }
18                  synchronized (LOCK) {
19                      if(count>=10)
20                          continue;
21                      count++;
22                      sum++;
23                      if(sum>=15)
24                          break;
25                  }
26              }
27              System.out.println(sum);
28          }//run()
29      }//Producer
30      static class Consumer implements Runnable{
31          @Override
32          public void run(){
33              int sum=0;
34              for(int i=0;i<20;i++){
35                  try {
36                      Thread.sleep(500);
37                  } catch (Exception e) {
38                      e.printStackTrace();
39                  }
40                  synchronized (LOCK) {
41                      if(count<=0)
42                          continue;
43                      else
44                      {   count--;
45                          sum++;  }
46                  }
47              }
48              System.out.println(sum);
49          }//run()
50      }//Consumer
51  }
```

**Fig. 3.** Producer & Consumer

There are two threads (Producer, Consumer) in this program. The products produced are put into a buffer (variable *count*), and the Consumers take the products from the buffer. The property to be verification is the number of products produced by Producer is equal to the sum of the number of products consumed by Consumer and the number of products remaining in the buffer (*sum(Consumer) + count = sum(Producer)*). Firstly, we model the program using the fine-coarse-grained modeling method, and we get a fine-coarse-grained model that including 32 places and 22 transitions. It has 110444 states pace nodes. Then we using the fine-grained modeling method in literature [5] to model the program, and obtain the fine-grained model that includes 40 places and 28 transitions. It has 154630 states pace nodes. The state space reduction rate of fine-coarse-grained method is 28.58% compared with the fine-grained method. Finally, we use CPN Tools own tool to verify the two models. The results are shown in Fig. 4(a), (b) are the model checking result of fine-coarse-grained model and fine-grained model respectively, and the results of are the same (both true), indicate that the program satisfies the property.

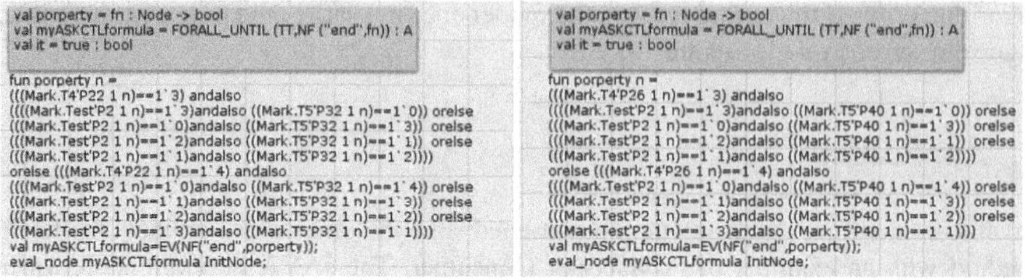

```
val porperty = fn : Node -> bool
val myASKCTLformula = FORALL_UNTIL (TT,NF ("end",fn)) : A
val it = true : bool

fun porperty n =
(((Mark.T4'P22 1 n)==1`3) andalso
((((Mark.Test'P2 1 n)==1`3)andalso ((Mark.T5'P32 1 n)==1`0)) orelse
((Mark.Test'P2 1 n)==1`0)andalso ((Mark.T5'P32 1 n)==1`3)) orelse
((Mark.Test'P2 1 n)==1`2)andalso ((Mark.T5'P32 1 n)==1`1)) orelse
((Mark.Test'P2 1 n)==1`1)andalso ((Mark.T5'P32 1 n)==1`2))))
orelse (((Mark.T4'P22 1 n)==1`4) andalso
((((Mark.Test'P2 1 n)==1`0)andalso ((Mark.T5'P32 1 n)==1`4)) orelse
((Mark.Test'P2 1 n)==1`1)andalso ((Mark.T5'P32 1 n)==1`3)) orelse
((Mark.Test'P2 1 n)==1`2)andalso ((Mark.T5'P32 1 n)==1`2)) orelse
((Mark.Test'P2 1 n)==1`3)andalso ((Mark.T5'P32 1 n)==1`1))))
val myASKCTLformula=EV(NF("end",porperty));
eval_node myASKCTLformula InitNode;
```
(a)

```
val porperty = fn : Node -> bool
val myASKCTLformula = FORALL_UNTIL (TT,NF ("end",fn)) : A
val it = true : bool

fun porperty n =
(((Mark.T4'P26 1 n)==1`3) andalso
((((Mark.Test'P2 1 n)==1`3)andalso ((Mark.T5'P40 1 n)==1`0)) orelse
((Mark.Test'P2 1 n)==1`0)andalso ((Mark.T5'P40 1 n)==1`3)) orelse
((Mark.Test'P2 1 n)==1`2)andalso ((Mark.T5'P40 1 n)==1`1)) orelse
((Mark.Test'P2 1 n)==1`1)andalso ((Mark.T5'P40 1 n)==1`2))))
orelse (((Mark.T4'P26 1 n)==1`4) andalso
((((Mark.Test'P2 1 n)==1`0)andalso ((Mark.T5'P40 1 n)==1`4)) orelse
((Mark.Test'P2 1 n)==1`1)andalso ((Mark.T5'P40 1 n)==1`3)) orelse
((Mark.Test'P2 1 n)==1`2)andalso ((Mark.T5'P40 1 n)==1`2)) orelse
((Mark.Test'P2 1 n)==1`3)andalso ((Mark.T5'P40 1 n)==1`1))))
val myASKCTLformula=EV(NF("end",porperty));
eval_node myASKCTLformula InitNode;
```
(b)

**Fig. 4.** Result of model checking.

## 5   Conclusion

Algorithm error detection of concurrent software is difficult. Convert the program to a model and perform model checking in model can detect the algorithm error in program. Base on the state space explosion problem in fine-grained modeling methods, this paper proposes a fine-coarse-grained automatic modeling method for algorithm error detection. Reduce the scale of state space of the model without affecting the results of the model test. The experiment shows that use fine-coarse-grained modeling method is significantly reduced the size of model and state space than use fine-grained modeling method, and improves the efficiency of model checking. We would like to research the model checking method in the future. For example, generate state space on-the-fly and search it by heuristic method, thereby improving the efficiency of model checking.

**Acknowledgement.** This work was supported by National Natural Science Foundation of China under Grant No. 61562064 and No. 61661041.

## References

1. Holzmann, G.J., Smith, M.H.: Software model checking. In: Wu, J., Chanson, S.T., Gao, Q. (eds.) Formal Methods for Protocol Engineering and Distributed Systems. IAICT, vol. 28, pp. 481–497. Springer, Boston, MA (1999). https://doi.org/10.1007/978-0-387-35578-8_28
2. Ball, T., Levin, V., Rajamani, S.K.: A decade of software model checking with SLAM. Commun. ACM **54**(7), 68–76 (2011)
3. Chaki, S., Clarke, E.M., Groce, A., Jha, S., Veith, H.: Modular verification of software components in C. IEEE Trans. Softw. Eng. **30**(6), 388–402 (2004)
4. Beyer, D., Henzinger, T.A., Jhala, R., Majumdar, R.: The software model checker BLAST: applications to software engineering. Int. J. Softw. Tools Technol. Transf. **9**(5), 505–525 (2007)
5. Sun, T., Liu, Y.Y.: A hierarchical CPN model automatically generating method aiming at multithreading program algorithm error detection. In: 17th IEEE International Symposium on Parallel and Distributed Processing with Applications. IEEE, Melbourne (2018)
6. Jensen, K., Kristensen, L.M.: Coloured Petri Nets: Modelling and Validation of Concurrent Systems. Springer, Heidelberg (2009). https://doi.org/10.1007/b95112

7. Jensen, K., Kristensen, L.M.: Formal definition of hierarchical coloured petri nets. In: Jensen, K., Kristensen, L.M. (eds.) Coloured Petri Nets, pp. 127–149. Springer, Heidelberg (2009). https://doi.org/10.1007/b95112_6
8. Huang, W., Hong, M., Yang, Q.H., et al.: C/C++ program memory leaked detection based on bounded model checking. Appl. Res. Comput. **33**(06), 1762–1766 (2016)
9. Wei, O., Shi, Y.F., Xu, B.F., Huang, Z.Q., Chen, Z.: Abstract modeling formalisms in software model checking. J. Comp. Res. Dev. **52**(7), 1580–1603 (2015)
10. Fehnker, A., Huuck, R.: Model checking driven static analysis for the real world: designing and tuning large scale bug detection. Innov. Syst. Softw. Eng. **9**(1), 45–56 (2013)
11. Wang, B., Wu, T.W., Hu, P.P.: Research on software defect classification and analysis. Comput. Sci. **40**(9), 16–20 (2013)

# EC-ARR: Using Active Reconstruction to Optimize SSD Read Performance

Shuo Li[1], Mingzhu Deng[2], Fang Liu[3(✉)], Zhiguang Chen[3], and Nong Xiao[1]

[1] College of Computer, National University of Defense Technology, Changsha, China
lishuo17@nudt.edu.cn
[2] College of International Studies, National University of Defense Technology, Nanjing, China
dk_nudt@126.com
[3] School of Data and Computer Science, SUN YAT-SEN University, Guangzhou, China
liufang25@mail.sysu.edu.cn

**Abstract.** Solid State Drive (SSD) has been becoming mainstream storage for its high performance, affordability proportional to its growing storage capacity. However, some inborn characteristics still limit its widespread application: (1) It wears out easily with increasing times of being written/erased. Therefore, SSDs are generally equipped with dedicated Erasure Coding (EC) modules for reliability concerns. However, the EC modules are only statically useful in the sheer scenarios of data loss. In other words, the EC module is never exploited in the regular access situations of dominating frequency, where data is unharmed and intact. (2) Huge latency differences exist among its three basic operations of reading, writing, and erasing, which could lead to performance degradation if there is no proper I/O scheduling. (3) SSD has excellent internal parallelism, which offers a strong possibility to further boost I/O performance if exploited properly.

Therefore, this paper proposes EC-ARR (Active-Reconstruction-Read), which exploits in a broader sense both its EC module and channel-level parallelism in combination to achieve better read performance. It is able to not only guard against data loss but also assist in normal data reads where data is intact, with active use of data reconstruction of the EC module. Additionally, to further this active reconstruction method in terms of channel-level parallelism, the static stripe with a length smaller than the number of channels and the data placement scheme with channel-wear-aware are adopted.

Simulation experiment based on SSDsim [1] shows that compared with conventional channel-RAID5 SSD, ARR-enabled SSD can increase the read performance by up to 18.5% without significant write performance degradation or storage overhead.

**Keywords:** Erasure code · NAND-flash SSD · Active-Reconstruction-Read · Load balance · Read latency

© Springer Nature Switzerland AG 2020
S. Wen et al. (Eds.): ICA3PP 2019, LNCS 11945, pp. 624–641, 2020.
https://doi.org/10.1007/978-3-030-38961-1_53

# 1  Introduction

In recent years, with the rapid development of semiconductor technology, the cost performance of SSD has been significantly improved, which makes SSD more widely used in the supercomputer, data center, and even PC [2–4]. More users can work directly on computers with SSDs equipped.

To ensure data reliability, SSDs are generally equipped with specialized EC modules. Unfortunately, these powerful EC modules are only used passively when there are data errors, which leads to a huge waste of resources. In this paper, we argue that this insufficient use of EC modules can be improved with an idea of additional active usage of EC modules in normal data reads, where there are no data errors and data is unharmed.

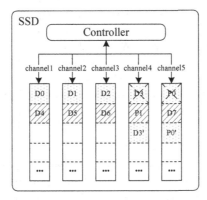

**Fig. 1.** Channel-level parallelism inside SSD, and the updating of the conventional RAID5. D represents data and P represents parity. The new data D3' and the new parity P0' are written back to the channels where the old data D3 and the old parity P0 located.

Abundant internal parallelism is another important feature of SSD [5]. There are four different parallelism levels in SSD: channel-level (see Fig. 1), chip-level, die-level, and plane-level [6]. Yang Hu has demonstrated experimentally that channel-level parallelism is the most efficient way of the four levels of parallelism. Although many pieces of research in this area have been carried out, there is still space and value for further investigation. For example, how to utilize the channel-level parallelism in terms of active reconstruction of an EC module has not been studied yet.

The performance of SSD is also limited by its media characteristics of NAND flash. The time taken of three basic operations in NAND flash varies widely when processing the same quantity of data. Taking Micron 256 Gb NAND Flash as an example, the delay of reading a page is about 50 μs, the delay of writing a page is 900 μs, and the block erasing time is up to 3 ms. Therefore, random small writes and garbage collections, which including lots of write/erase operations, can seriously affect the read performance of SSD.

To sum up, we argue in this paper the three main problems regarding SSD include: (1) EC module is under-exploited, specifically in normal data-reading scenarios. (2) Although some work has taken advantage of channel-level parallelism, there is still room for better utilization. (3) It is vital to do targeted scheduling for the three basic operations of flash to achieve better read performance.

Therefore, this paper proposes an ARR (Active-Reconstruction-Read)-enabled SSD, which exploits in a broader sense both its EC module and channel-level parallelism in combination to achieve better read performance. It is able to not only guard against data loss but also assist in data read when data is intact by serving data access with active use of data reconstruction of the EC module. Further, to make this new scheme of active adoption of EC module use of more channel-level parallelism, fixed stripe length that smaller than channel numbers and data placement scheme with channel-wear-aware are adopted.

This paper considers SSD with conventional channel-level RAID5 as a baseline approach. ARR-enabled SSD and channel-level-RAID5 SSD are both simulated by SSDsim which is a popular open-sourced SSD simulator in academia. Experimental results with real-world traces indicate ARR-enabled SSD achieves better read performance in general, as compared to the channel-level-RAID5 SSD. In the best case, ARR-enabled SSD can increase the read performance by up to 18.5% without significant write performance degradation or storage overhead. We also use synthesized workloads to explore the impact of different configuration parameters on ARR-enabled SSD.

Our contributions can be summarized as follows:

1. **An idea of the active use of data reconstruction of the EC module in SSD to serve regular read requests is proposed.** Different from the current practices, in which the EC module is used passively only as there are data errors, this paper proposes an idea of using the EC module actively to serve the regular read requests. It can avoid resources wasted while significantly improve read performance.
2. **A hybrid Active-Reconstruction-Read (ARR) scheme with dynamic judgment is put forward to accommodate read requests with different lengths and states.** Different from traditional direct read (DR) without active reconstruction (DR for all) or taking active reconstruction read for all requests (ARR for all), this scheme can dynamically judge and take the most time-saving solution for each read request according to its current state (DR or ARR).
3. **A setting of static stripe length that is smaller than the number of channels is adopted.** In most existing systems, the update is written back to the same channel (see Fig. 1), because the stripe length is set to the number of channels. This fixed-channel-update increases the times of garbage collection and erasing operation, which degrades SSD's performance. Dynamic stripe length can be used to update in a non-fixed-channel way. However, the shorter the stripe is, the lower the degree of parallelism in reading will be, which will fail to provide good reading parallelism for ARR. Different from both above, our setup can maintain non-fixed-channel-update while guarantee the maximum reading parallelism.
4. **A channel-wear-aware data placement strategy is proposed.** Different from the simple round-robin data layout algorithm that popular in modern storage systems,

under our channel-wear-aware data placement strategy, data is always preferred to be placed in the channel that has the least number of writes, which as a way to optimize channel-level wear leveling.

The remainder of this paper is structured as follows: Sect. 2 introduces the motivation of this work. Section 3 describes the detailed design of EC-ARR. Section 4 evaluates the EC-ARR and Sect. 5 concludes this paper.

## 2 Motivation

In this chapter, we will introduce the research motivations of this work. It includes the opportunities presented by the current technological developments in SSDs, as well as the inadequacies of previous research works.

### 2.1 Opportunities

**The Importance of Read Performance.** Users are much more sensitive to the read delay than write delay. The reason is that, for write operations, the users hand over instructions and data into computers, and disregard of when the data will be written back to the storage device; while for read operations, the response time determines when the users can proceed the next step. Therefore, this article is dedicated to improving read performance in SSD.

**Opportunity in Erasure Code.** In the field of storage, erasure codes are usually used to ensure the reliability of data [7] until Rashmi [8] put forward EC-Cache in OSDI 16. EC-Cache faces to the key-value store. During writes, the individual objects are split up and erasure-coded. Wherein obtaining any k out of (k + r) splits of an object are sufficient, during reads. EC-Cache demonstrates the effectiveness of erasure code for a new setting – improving load balancing and latency characteristics.

**Opportunity in Controller.** The controller is the main component in SSD. Its primary role is to command, calculate, and coordinate. The encoding/reconstruction of EC is generally done on a separate EC module in the controller. With the development of integrated circuit technology and firmware algorithms, the performance of current controllers can guarantee the efficiency of encoding/reconstruction [9], which means that in most normal cases, we do not have to worry about the impact of encoding/reconstruction overhead on SSD's productivity.

**Opportunity in Parallelism.** The rich parallel structure inside SSD is the key to improve IO performance. Although many pieces of research in this area have been carried out [10, 11], there is still a lot of space for further exploit.

### 2.2 Why not Other Schemes

CR5M [12] (Mirroring-Powered Channel-RAID5) is a RAID architecture proposed to improve the performance of channel-level RAID5. It adds an extra chip to each channel to store data of small writes as a mirroring chip. But obviously, it comes at the cost of increased storage overhead. Meanwhile, erasure codes are still used passively and inefficiently here.

Some scholars have come up with a self-balancing striping scheme [13]. The striping system encodes a small piece of data when SSD is idle and calculates redundancy in a 1:1 ratio. Then, during the reading, only the small part of data can be read by reconstruction. In this scheme, not all data's reliability can be guaranteed, the ratio of redundant data is not cost-effective, the scheme does not tell that which data should be encoded.

Many cache replacement strategies are proposed at the flash translation layer (FTL) level to improve the read performance of SSD, such as the classical LRU, NRU [14]. But no matter which schemes, the principle of temporal locality and spatial locality are utilized. In the long run, due to cache capacity limitations, there is still a lot of data that needs to be written back to or read from the storage devices. Therefore, the cache replacement strategies can only minimize the number of unnecessary write back operations, but cannot solve the problem of reading delay caused by write operations radically.

## 3    EC-ARR Design

For the convenience of description, related concepts in this paper are listed in Table 1:

**Table 1.** The illustration of several concepts.

| | |
|---|---|
| Stripe length | The number of channels involved in a stripe |
| Stripe depth | The quantity of data carried by a stripe on one channel |
| Stripe size | Stripe size = Stripe length * Stripe depth |
| Sub-request | The data that added to the channel wait queues |
| Valid channel | The channel where the valid data located |
| Original data | Data contained in a read request |
| Requested stripe | The stipe where original data located |
| Free page | The pages without valid/invalid data |

### 3.1    Overall Architecture

This section provides a high-level overview of EC-ARR's architecture. EC-ARR is an optimization I/O scheme to provide high read performance as well as data reliability in SSD. It uses the EC module actively when data is intact, so it can achieve better read performance based on advanced utilization of SSD internal parallelism. Figure 2 compares the (a) conventional SSD system without ARR (Active-Reconstruction-Read) and (b) our SSD system with ARR. In (a), the EC module is only used passively when there are data errors; in (b), the EC module is used actively during reading when data is intact. As shown in the figure, EC-ARR is able to actively reconstruct the original data required by read request through the unblocked data, thus avoiding waiting for the blocked data.

Implemented on SSD's controller, the EC module is in charge of 3 jobs: (1) encoding during writes, (2) reading from splits, and (3) active reconstructing during reads. FTL is responsible for address mapping. The controller also takes charge of the other judgment, calculation, scheduling work.

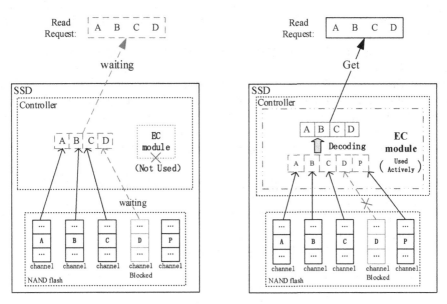

(a) Conventional system *without* ARR. EC-module will *not be used* here because there is no data error.

(b) EC-ARR system *with* ARR. EC module will *be used actively* during reading.

**Fig. 2.** The comparison between SSD without ARR and SSD with ARR. P is parity. The read request needs data A, B, C, D which are stored in different channels. The channel where D is located is blocked, the blocking may be due to the garbage collections or write operations. Because there is no data error, EC-module will not be used in (a), so, the read request needs to wait for data D. But in (b), EC-module is actively used in this reading process, the original data will be get from the active reconstruction. Thus, the read request is able to be served without waiting for the blocked data D.

## 3.2 Choice of Erasure Code

For simplicity, this article uses the classic Reed-Solomon code (RS) as an example. The reason for this choice is RS has the property of MDS, it can recover the original data from any k blocks in (k + r) blocks, which has minimal storage overhead, at the same time makes us more flexible in avoiding blocked channels. And the RS calculation module has been integrated into many existing controller chips so that we do not need to worry about the delay caused by the encoding or reconstruction process. Other erasure code with MDS properties can also be used, RS code is just used here as an example.

### 3.3   Data Layout

Data has to be written before it can be read. Therefore, in this section, we will introduce the data placement method. In order to make ARR (Active-Reconstruction-Read) accelerate the reading better, we hope that the data placement scheme can achieve the following two purposes: (1) increase read parallelism, and (2) balance the wear between channels, thereby reducing the garbage collection times of the channel where the hot data is located.

Here, we consider the I/O requests come in order. The stripes are organized based on logical address. The parity will be calculated when all original data in the stripe are arrived. The stripe width is set to 1 page. The stripe length is fixed and less than the number of channels. Considering the storage overhead, we set the stripe length to: the Number of Channel - 1.

The data/parity is placed into the channel according to the following three principles (The priority of the three principles is: Principle 1 > Principle 2 > Principle 3):

*Principle 1:* To guarantee the maximum degree of parallelism while reading stripe, data in the same stripe cannot appear on the same channel.
*Principle 2:* Considering channel-level wear leveling: Select the channel with the largest number of free pages from remaining channels.
*Principle 3:* After the above two steps, if there are still multiple channels to be selected, select the channel with a smallest index.

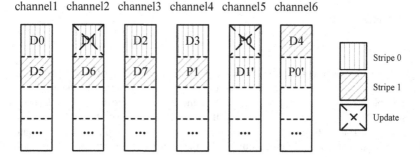

**Fig. 3.** Data layout

*Case Study*

As shown in Fig. 3, the 6-channel SSD, RS (5, 4) code is taken as an example. D0, D1, D2, D3, and P0 are form stripe 0, which are sequentially written to the channels.

(1)   When the data D4 arrives (index of the stripe where D4 located is 1), since no data has been written in the stripe1, consider principle 2. Select channel 6 which with the most quantity of free pages (other channels have already written a page).

(2)   When D5 arrives (index of the stripe where D5 located is 1), consider principle 1, channel 6 where D4 is located will be excluded first. After that, there are still multiple alternative channels (all channels have the same number of free pages).

Consider principle 3, select channel 1, which is the channel with the smallest index number.

(3) When it is necessary to update D1, P0, the new data D1'and P0' is sequentially placed according to the above steps, and details are not described again. (After D1 is updated, the old data D1 becomes invalid data, but it still exists on the device and has not been completely deleted, so the page where D1 is located is not a free page).

## 3.4   Active Reconstruction Read

In this section, we introduce hybrid-ARR, a hybrid active-reconstruction-read scheme with dynamic judgment. The scheme is mainly divided into the following two stages.

**Preprocessing Stage**

When a read request comes, all the data/parity in the requested stripe are added to the tail of the waiting queue of the corresponding channel. Each data/parity becomes a sub-request of the read request. If the sub-request already exists in the channel's waiting queue, it will not be added again. Also, if the requested data is in an incomplete stripe or there is an update operation for the same data in the waiting queue, then use the direct read for this read request to ensure data consistency. If the direct read is adopted, the extra sub-request will not be added.

**Processing Stage**

Then, the sub-requests will be processed by the channel and returned to the SSD controller. Each time the controller receives a sub-request, it will judge the current situation of the sub-requests that already received according to the three steps given below, until it can decide which way does the read request should take, direct read or active reconstruction read?

*Step 1: Have all the original data been received?*

If yes, the direct read will be adopted. All the unserved sub-requests from the read request are kicked out from the waiting queues, and the original data will be returned to the host. The read request ends now.

If not, continue to step 2.

*Step 2: Does the number of received data meet the reconstruction conditions?*

If yes, ARR (active-reconstruction-read) will be adopted. All the unserved sub-requests are kicked out from the waiting queue. At the same time, the controller reconstructs the original data by received data and returns them to the host. The read request ends now.

If not, continue to step 3.

*Step 3:* The controller will not make any decisions. When the next sub-request arrives, the judgment process will be restarted.

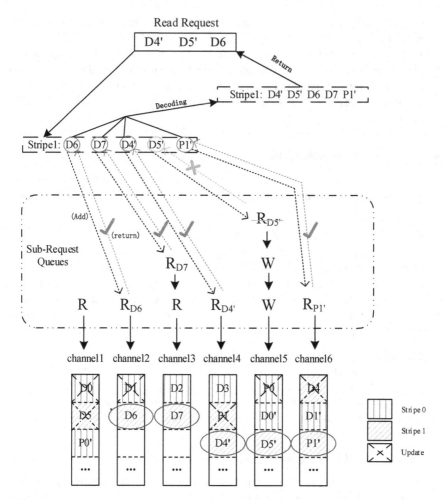

**Fig. 4.** Hybrid-ARR (Active-Reconstruction-Read)

*Case Study*

Still take 6-channel, RS (5, 4) code as an example. As shown in Fig. 4, this read request needs data D4', D5', D6. And they belong to stripe1. D4', D5', D6, D7, and P1' of stripe1 are added to the request queue. R/W in the figure indicates sub-requests that come from other requests earlier.

(1) The controller will first receive D6, D4', P1' as there are no other sub-requests in the queue before them. Then we start to judge according to the above three steps.

(2) Original data D5' has not been received, so go to step 2.
Here are only three chunks that have been received. Because (5,4) RS code needs four chunks at least to reconstruct, the reconstruction cannot be carried out.

(3) Go to step 3: wait for the next sub-request.

The next data that the controller receives is D7. The judgment process restart. At this time, the controller has received four data: D6, D4', P1', and D7, which satisfy the reconstruction condition. Therefore, ARR will be chosen. D5', the sub-request that hasn't been served by channel will be kicked out from the waiting queue. Then, the reconstruction will be performed. The original data D4', D5', D6, will be returned to the host. Read request completed.

In this way, we avoid reading blocked data D5', and complete the read request by actively reconstructing D5' through other data in the stripe. Therefore, the response time of the read request is greatly reduced.

For read requests with a small number of original data, readers can deduce the ARR process by yourself. For most requests, this method can always find the fastest way to complete the read request.

## 4 Simulation and Evaluation

In this section, we first introduce our simulator, configuration, and then perform experiments and analyze the results.

### 4.1 Simulator and Configuration

SSDsim is a simulation software for SSD which is developed by HUST [8]. It provides a set of powerful verification tools that are able to emulate hardware structures and software algorithms effectively. It has good plasticity and flexibility. A series of tests can be performed with specific input parameters or files. Compared with other SSD simulators, such as FlashSim [15] developed by PSU, the SSD extension for disksim simulation environment developed by Microsoft Research [16], SSDsim is well modularized and its simulation results are more accurate. In this paper, we choose SSDsim for our experiments.

SSDsim implements the major components of FTL and many advanced commands. As these advanced commands are irrelevant to our research object, they are not used in this experiment. We modified SSDsim's FTL module (including address allocation, mapping table, and other related modules), and read/write request processing to achieve the functions of EC-ARR. In addition, we also modified the pre-read module so that the data can be pre-read according to the EC-ARR scheme.

This paper considers SSD with conventional channel-level RAID5 as the baseline approach, in which EC module is only used passively when there is data error in SSD. ARR-enabled SSD and channel-level-RAID5 SSD are both simulated by SSDsim. Table 2 shows the configuration of the SSDsim in the experiments.

**Table 2.** Parameter Configuration of SSDsim

| Parameters | Values | Parameters | Values | Parameters | Values |
|---|---|---|---|---|---|
| Page read | 20 µs | Page size | 2 KB | Blocks per plane | 2048 |
| Page write | 200 µs | Chips per channel | 4 | Planes per die | 2 |
| Block erase | 1.5 ms | Dies per chip | 4 | Encoding/Decoding | 20 µs |

All experiments in this paper were performed on a Lenovo desktop computer equipped with a 3.60 GHz Intel Core i7-7700 CPU, a 256G SATA SSD, and 16 GB RAM. Its operating system is Ubuntu16.04.

Experiments are conducted with both real-world traces and synthesized workloads, to investigate the efficacy of EC-ARR. All response time in this paper has been standardized based on the results of the channel-RAID5 SSD.

### 4.2   Effectiveness

In this section, we evaluate the effectiveness of ARR-enabled SSD (for convenience, hereinafter referred to as ARR). The baseline is channel-RAID5 SSD (for convenience, hereinafter referred to as RAID5). The configuration parameters for both SSDs are shown in Table 3. Table 4 shows the characteristics of the five workloads used in the experiment. Among them, Financil1 and Financil2 are derived from the financial server. PC1 and PC2 are collected from the PC of the NTFS file system using the DiskMon tool. ATTO is generated through the test software ATTO.

**Table 3.** Two types of SSD

| SSD | Channel number | Stripe length | Parity number per stripe |
|---|---|---|---|
| Channel-level RAID5 SSD | 12 | 12 | 1 |
| ARR-enabled SSD | 12 | 11 | 1 |

**Table 4.** The characteristics of workload

| Trace name | Read ratio | Average size of read | Average size of write |
|---|---|---|---|
| Financial 1 | 23% | 2.25 KB | 3.73 KB |
| Financial 2 | 82% | 2.28 KB | 2.92 KB |
| PC 1 | 39% | 37.17 KB | 9 KB |
| PC 2 | 46% | 26.32 KB | 2.28 KB |
| ATTO | 51% | 175 KB | 220 KB |

Figures 5, 6 and 7 are the average response time of read requests, write requests and all requests. As shown in the Fig. 5, ARR has an excellent performance in the first four workloads, and it reduces the average read request response time by 9.8%, 8.2%, 13.3%, and 18.5%, respectively compared to RAID5. For read requests, the

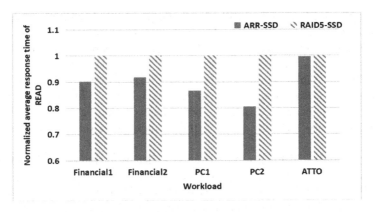

**Fig. 5.** Normalized average response time of READ requests

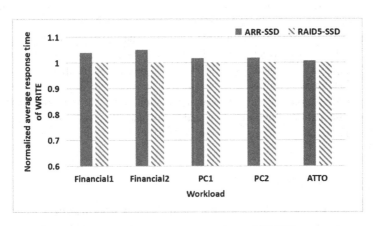

**Fig. 6.** Normalized average response time of WRITE requests

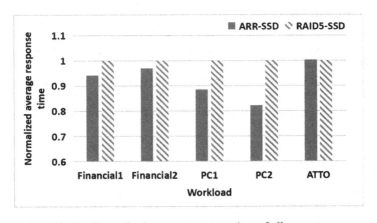

**Fig. 7.** Normalized average response time of all requests

performance on the workloads of PC is, in general slightly better than the Financial. And in two PC workloads, PC2 has a lower read latency than PC1. As shown in Fig. 6 of write requests, the response time of ARR is slightly increased. Compared to RAID5, the average response time of write requests of ARR increased by 3.97%, 5.15%, 1.73%, 0.018%, 0.72% under the five workloads, respectively. This is due to the extra sub-requests that are added to the system. As shown in Fig. 7, Although the response time for write requests is increased, the overall request response time for the system is still decreased. Under PC2, the overall request response time was 17.9% lower than in RAID5.

Read performance is improved because ARR can avoid congestion. Like the barrel principle, for a read request, its response time depends on the last sub-request served. When ARR is activated, originally otherwise straggling sub-requests are replaced with some "unblocked" sub-requests to complete the request. This is equivalent to make the response time of the last sub-request get early, thus boosting the response time of the read request as a whole.

However, ARR does not always prevail, for example under ATTO. Compared to RAID5, it shows an only 0.5% decrement in the average response time of read requests, with a 0.72% increment in write requests. The overall average response time was increased by 0.19%, which is clearly not worth the loss. This is because that most of the IO requests generated by ATTO are large and sequential. This character determines that there is almost no internal I/O performance degradation caused by individual channel congestion. Under this situation, ARR is not only cannot take advantage of the feature of avoiding congestion but also adds additional sub-requests to the system. Therefore, ARR that encodes with MDS is not suitable for the workload like ATTO.

### 4.3   Impact of Read Ratio

The synthetic workload is different from the real-world workload that collected from the real system. It is usually generated by the synthetic workload generation tools according to certain rules and parameters. Compared with the real-world workload, the synthetic workload can control parameters precisely. Therefore, it is possible to explore the relationship between system performance and a certain parameter by adjusting the target parameter but keeping other parameters unchanged. For example, in this paper, we explored the impact on EC-ARR of the ratio of read requests, the average size of read requests, and the parity number.

This section uses 7 synthetic workloads to test the impact of the ratio of read requests on ARR. The characteristics of the synthetic workloads are shown in Table 5. The requests in these workloads are random, with an average read request size of 20 KB and an average write request size of 2 KB. The ratio of read request is 20%, 30%, 40%, 50%, 60%, 70%, and 80% respectively.

From Fig. 8, we can see that as the read ratio increases, there are both lightly increasing tendencies in write latency and read latency. However, although ARR's latency is increased, it is still lower than RAID5's latency. For example, when the proportion of read request changes from 20% to 80%, the write delay increases from 1.0032 to 1.0298, the read delay increases from 0.8134 to 0.8401, and the overall response delay increases from 0.8326 to 0.8744, but this is still less than 1 which is the average request response time of RAID5 under the same conditions.

**Table 5.** Characteristics of synthetic workload

| Workload | Read ratio | Average size of reads | Average size of writes |
|---|---|---|---|
| Workload 1 | 20% | 20 KB | 2 KB |
| Workload 2 | 30% | | |
| Workload 3 | 40% | | |
| Workload 4 | 50% | | |
| Workload 5 | 60% | | |
| Workload 6 | 70% | | |
| Workload 7 | 80% | | |

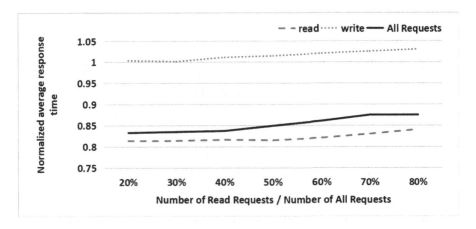

**Fig. 8.** Impact of read ratio

This increasing trend is attributed to extra sub-requests. In the active re-construction process, we add all the data/parity belongs to one stripe to the waiting queue, thus, the non-original data become the extra sub-requests. These extra sub-requests will degrade the IO performance. The higher the percentage of read requests, the more additional sub-requests in the system, and the longer the average response time we get.

So why is the overall read latency of ARR-SSD still lower than RAID5-SSD under the existence of extra sub-requests? There are two reasons. One is that the latency produced by extra sub-requests are less than the latency reduced by ARR, so the overall performance is still increased; the other is that some of the unserved sub-requests will be kicked out from the waiting queues (when the amount of unblocked data can meet the reconstruction condition, the sub-requests in the waiting queues which still not be served will be kicked out), so there are only a small part of the extra sub-requests actually take up system time.

## 4.4    Impact of Read Size

This section uses the synthetic workloads to explore the impact of the average size of read request on ARR. The stripe length is 11. The stripe width is 1. The average size of random writes in the synthetic workloads is controlled to 2 KB (1 page = 2 KB), and the average size of read requests is set to 4 KB–48 KB. The read request ratio is 60%. Figure 9 is the normalized average read/write response time.

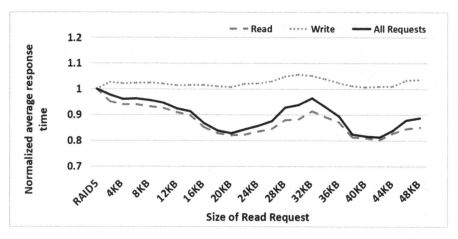

**Fig. 9.** Impact of read size

As we can see in Fig. 9, read performance gains are highest when the read request size is coupled to or is a multiple of the stripe size. For example, when the size of the read request is 20 KB (in this section, the stripe length is 11 and the stripe size is 22 KB), the response time of the read request is the local minimum which is 0.82, and when the size of the read request is 44 KB, the response time of the read request reaches the global minimum which is 0.81. This is because that the number of extra sub-requests is the fewest when the read request size is coupled to or is a multiple of the stripe size.

## 4.5    Impact of Parity Number

In this section, we evaluate the impact of numbers of parity on ARR. We set 14 channels, and show the I/O response time in the case of 1 parity, 2 parity, 3 parity, and 4 parity.

As shown in Fig. 10, when there is one parity in ARR, the read response time can be significantly reduced, but when the number of parity increases, it almost no longer works and even hurts I/O performance. The response time of ARR with one parity is 0.86, but when there is 4 parity, the overall response time increases to 1.29 which is even worse than RAID5.

This is because too much parity will increase the internal bandwidth of the SSD, which will increase the congestion and take up too much CPU resource to compute for encoding/decoding.

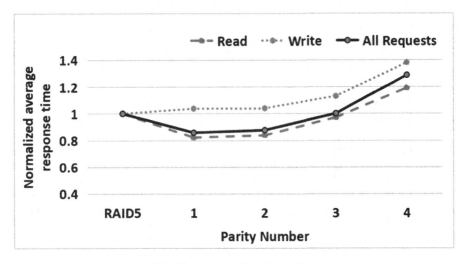

**Fig. 10.** Impact of parity number

### 4.6    Summary

In most cases, EC-ARR is effective. And it has the best read-performance under the workload with small random writes and large sequential reads without significant write performance degradation or storage overhead. This is because EC-ARR is able to reconstruct the original data required by read request actively through the unblocked data, thus avoiding waiting for the blocked data. And EC-ARR performs best when there is only one parity. As the parity number increases, the performance improvement will be smaller and smaller. This is because too much parity will increase the internal bandwidth of the SSD, which will increase the congestion and take up too much CPU resource to compute for encoding/decoding. Although there are extra sub-requests that are added to the system, the overall read latency of ARR-SSD is still better than RAID5-SSD. This is because that the latency produced by extra sub-requests are less than the latency reduced by ARR, and there is actually only a small part of extra sub-requests taking up system time (there is a part of unserved sub-requests that are kicked out from the waiting queues).

## 5    Conclusion and Future Work

This paper proposes EC-ARR, which utilizes ARR to improve the read performance of SSD. This scheme takes advantage of Erasure-Code reconstruction and the parallelism inside SSD, to ensure data reliability, balance read access load, reduce read latency, and optimize channel-level wear leveling. Experiments verified the effectiveness of EC-ARR.

In the future, we will consider more coding methods to optimize for different types of workloads. For example, a coding method with a smaller recovery bandwidth can be used to adapt a sizeable sequential workload like ATTO. The subsequent work will continue on the combination of cache replacement strategies and EC-ARR. The cache replacement strategies can only minimize the number of unnecessary write back operations, but not able to solve the problem of reading delay caused by writing operations radically. However, the above experimental results have been able to fully demonstrate that EC-ARR can reduces the read delay caused by writing. Therefore, we have every reason to believe that SSDs will show better performance under the combination of EC-ARR and the appropriate cache replacement strategies.

**Acknowledgment.** We appreciate all anonymous reviewers for valuable suggestions to improve this paper. This work is supported by The National Key Research and Development Program of China (2016YFB1000302), National Natural Science Foundation of China (61832020, 61702569), Natural Science Foundation of Guang Dong Province (2018B030312002), and Key-Area Research and Development Program of Guang Dong Province (2019B010107001). NSFC: 61872392, U1611261 Supported by the Program for Guangdong Introducing Innovative and Entrepreneurial Teams under Grant NO. 2016ZT06D211, and the Pearl River S & T Nova Program of Guangzhou under Grant NO. 201906010008.

# References

1. Hu, Y., Jiang, H., Feng, D., Tian, L., Luo, H., Ren, C.: Exploring and exploiting the multilevel parallelism inside SSDs for improved performance and endurance. IEEE Trans. Comput. **62**, 1141–1155 (2012)
2. Ni, Y., Jiang, J., Jiang, D., Ma, X., Xiong, J., Wang, Y.: S-RAC: SSD friendly caching for data center workloads. In: Proceedings of the 9th ACM International on Systems and Storage Conference, p. 8. ACM (2016)
3. Narayanan, I., et al.: SSD failures in datacenters: What? When? and Why? In: Proceedings of the 9th ACM International on Systems and Storage Conference, p. 7. ACM (2016)
4. Simon, W., Lauer, A., Wien, A.: FDTD simulations with 10 11 unknowns using AVX and SSD on a consumer PC. In: Antennas and Propagation Society International Symposium (2012)
5. Du, Y.-M., Xiao, N., Liu, F., Chen, Z.-G.: CSWL: cross-SSD wear-leveling method in SSD-based RAID systems for system endurance and performance. J. Comput. Sci. Technol. **28**, 28–41 (2013)
6. Hu, Y., Jiang, H., Feng, D., Tian, L., Luo, H., Zhang, S.: Performance impact and interplay of SSD parallelism through advanced commands, allocation strategy and data granularity. In: Proceedings of the International Conference on Supercomputing, pp. 96–107. ACM (2011)
7. Deng, M.-Z., Xiao, N., Yu, S.-P., Liu, F., Zhu, L., Chen, Z.-G.: RAID-6Plus: a comprised methodology for extending RAID-6 codes. Mob. Inform. Syst. (2017)
8. Rashmi, K., Chowdhury, M., Kosaian, J., Stoica, I., Ramchandran, K.: EC-Cache: load-balanced, low-latency cluster caching with online erasure coding. In: 12th USENIX Symposium on Operating Systems Design and Implementation (OSDI 2016), pp. 401–417 (2016)

9. Cheong, W., et al.: A flash memory controller for 15 μs ultra-low-latency SSD using high-speed 3D NAND flash with 3 μs read time. In: 2018 IEEE International Solid-State Circuits Conference-(ISSCC), pp. 338–340. IEEE (2018)
10. Lin, Z., Zuo, S., Zhao, X., Zhang, Y., Wu, W.: SSD accelerated parallel out-of-core higher-order method of moments and its large applications. Appl. Comput. Electromagn. Soc. J. **33** (2018)
11. Song, K., Kim, J., Lee, D., Park, S.: MultiPath MultiGet: an optimized multiget method leveraging SSD internal parallelism. In: Lee, W., Choi, W., Jung, S., Song, M. (eds.) Proceedings of the 7th International Conference on Emerging Databases. LNEE, vol. 461, pp. 138–150. Springer, Singapore (2018). https://doi.org/10.1007/978-981-10-6520-0_15
12. Wang, Y., Wang, W., Xie, T., Pan, W., Gao, Y., Ouyang, Y.: CR5M: a mirroring-powered channel-RAID5 architecture for an SSD. In: 2014 30th Symposium on Mass Storage Systems and Technologies (MSST), pp. 1–10. IEEE (2014)
13. Chang, Y.-B., Chang, L.-P.: A self-balancing striping scheme for NAND-flash storage systems. In: Proceedings of the 2008 ACM Symposium on Applied Computing, pp. 1715–1719. ACM (2008)
14. Zhou, J., Han, D., Wang, J., Zhou, X., Jiang, C.: A correlation-aware page-level FTL to exploit semantic links in workloads. IEEE Trans. Parallel Distrib. Syst. **30**, 723–737 (2019)
15. Kim, Y., Tauras, B., Gupta, A., Urgaonkar, B.: FlashSim: a simulator for nand flash-based solid-state drives. In: 2009 First International Conference on Advances in System Simulation, pp. 125–131. IEEE (2009)
16. Prabhakaran, V., Wobber, T.: SSD extension for DiskSim simulation environment. Microsoft Reseach (2009)

# Research of Benchmarking
# and Selection for TSDB

Feng Ye[1,4(✉)], Zihao Liu[2], Songjie Zhu[1], Peng Zhang[3],
and Yong Chen[4]

[1] Hohai University, Nanjing 211100, Jiangsu, People's Republic of China
yefeng1022@hhu.edu.cn
[2] Jiangsu University of Science and Technology, Zhenjiang 212003,
Jiangsu, People's Republic of China
[3] Jiangsu Province Water Resources Department, Nanjing 210029,
Jiangsu, People's Republic of China
[4] Nanjing Longyuan Micro-Electronic Company, Nanjing 211106,
Jiangsu, People's Republic of China

**Abstract.** With the increasing use of sensor and IoT technologies, sensor
stream data is generated and consumed at an unprecedented scale. Traditional
storage mechanisms represented by relational database systems become more
and more difficult to adapt to the store, query, update and other operations of
large-scale sensor stream data. This, in turn, has led to the emergence of a new
kind of complementary non-relational data store subsumed under the term time
series database (TSDB). However, the heterogeneity and diversity of numerous
TSDBs impede the well-informed comparison and selection for a given application context. A thorough survey shows that current benchmarks for TSDBs
are few and they still need improvement in workload implementation based on
real business requirements, data generator based on real-world data and fine-
grained performance metrics. How to implement a benchmarking tool for
TSDBs according to different tradeoffs in IoT scenarios becomes a key challenge, which will be addressed in this paper. Firstly, we propose a benchmarking
platform TS_Store_Test, which integrates five well-known TSDBs using the
micro-services mechanism. Meanwhile, we integrated and extend Prometheus to
capture the performance metrics in a refined manner. Based on TS_Store_Test,
the execution efficiency of some workloads from technical and business perspectives is tested using the real hydrological sensor data. Experimental results
demonstrate the usability and scalability of TS_Store_Test, and also show the
performance differences of different TSDBs for sensor stream data. Finally,
TS_Store_Test is compared with other NoSQL benchmarking suits.

**Keywords:** Sensor stream data · Micro-services · Time series databases ·
Benchmarking · Workload

## 1 Introduction

As the world gets more instrumented and connected, we are witnessing a flood of
digital data generated from sensors in the form of data streams. Real-time sensor stream
data are consumed at an unprecedented speed, serving environmental protection, flood

© Springer Nature Switzerland AG 2020
S. Wen et al. (Eds.): ICA3PP 2019, LNCS 11945, pp. 642–655, 2020.
https://doi.org/10.1007/978-3-030-38961-1_54

prevention and other business scenarios. However, the traditional storage represented by relational database is difficult to deal with large-scale sensor stream data effectively. The disadvantages of traditional databases mainly manifest in the following aspects: (1) There is more and more data in a single table schema, so it is inefficient in processing them; (2) The evolution of relational databases is costly due to complex data transformation/migration; (3) They do not parallelize well for heterogeneity of data and are difficult to improve database performance by scaling up the machine clusters [1]. To explore the value of big data, the first consideration is how to manage big data reasonably. Nowadays, the fast-evolving time series databases (TSDBs) [2] provide a referential solution, and often more characteristics apply such as schema-free, easy query support, simple API, eventually consistent. As is known to all, some well-known TSDBs have been adopted for big data applications in different fields, including IoTDB [3], Druid [4], Riak TS [5] and so on. Moreover, different TSDBs usually have the different characteristics [6]. However, it is still worth studying which time series solution is more appropriate for large-scale sensors stream data in various scenarios [7, 8]. Especially, existing research [9–11] shows that current benchmarks for TSDBs are few and they still need improvement in workload implementation based on real business requirements, data generator based on real-world data and fine-grained performance metrics.

In view of the problems above, at first, we proposed and implemented a benchmarking platform [12] named TS_Store_Test that integrates multiple well-known TSDBs using the micro-services mechanism [13] and Prometheus [14]. The purpose of integrating and extending Prometheus is to measure the workloads of the cluster where the TSDBs are located in a refined manner. Based on TS_Store_Test, the execution efficiency of some workloads for hydrological data processing was tested using the real hydrological sensor data acquired from the Chuhe River. Through comparative analysis, the result of the experiments shows that different TSDB varies greatly in different scenarios and Druid performs better overall. Compared with other benchmarking tools, TS_Store_Test has the advantage of supporting more types of TSDBs in an extensible and comparable way and providing the system level monitoring of resources in an integrated and a fine-grained manner.

The following contents are organized as follows: Sect. 2 discusses the research work related to this paper; Sect. 3 introduces the framework and key components of the proposed benchmarking platform TS_Store_Test. In Sect. 4, using the hydrological sensor data from Chuhe river, the execution efficiency and workloads of operations of different TSDBs are verified and compared experimentally on TS_Store_Test. In Sect. 5, we further qualitatively compare TS_Store_Test with other benchmarking suits. Finally, the summary and prospect are given.

## 2  Related Works

TSDBs belong to NoSQL databases and offer to store multiple time series such that queries to retrieve data from one or a few time series for a particular time range are particularly efficient. At present, according to db-engines.com, there are nearly thirty different TSDBs.

Different TSDBs have different characteristics and applicability, and it is hard to grasp where they excel, where they fail or even where they differ, as implementation details change quickly and feature sets evolve over time. Some literatures, such as [6–8], discussed NoSQL databases from different perspectives, and surveyed a concise and up-to-date comparison of NoSQL engines, identifying their most beneficial use case scenarios. However, the above work is not specific to the introduction and comparison of TSDBs. In [15], the authors presented a thorough analysis and classification of TSDBs developed through academic or industrial research and documented through publications. In [16], a short overview of time series storage and processing in a cloud environment was provided. In [17], the authored aimed for a complete list of all available TSDBs and a feature list of popular open source TSDBs, and then compared twelve most prominent open source TSDBs. According to [15–17], we can see that due to the lack of benchmarks for TSDBs, it is still a challenge for architects or designers who want to select the most appropriate one for the applications.

Some researches focused on NoSQL stores from a quantitative comparison perspective. Some well-known benchmarks for NoSQL stores were summarized in [9], such as YCSB [18, 19], YCSB++ [20], BG [21], CloudSuite [22] and BigDataBench [23]. In [18, 19], YCSB was adopted as the benchmarking framework for HBase [24] and MongoDB [25]. YCSB++ extended YCSB to evaluate the advanced features (e.g. ingest speed-up techniques) of NoSQL stores. BG provided workloads to emulate social network actions and can be used to compare NoSQL. CloudSuite provided popular scale-out workloads to evaluate different NoSQL stores deployed in cloud architectures. BigDataBench was an open-source big data and AI benchmark suite, providing MongoDB and HBase implementations for NoSQL benchmarking. Unfortunately, current benchmarks for TSDBs are few, and the benchmarks above still need improvement in workload implementation based on real-world sensor data and fine-grained performance metrics. For example, YCSB, YCSB++ and BG are not used for realistic IoT applications, hence they are different to benchmark TSDBs. Moreover, because they lack the ability to integrate performance metrics, system level monitoring of resources has to be performed to identify bottleneck separately. Up till now, although CloudSuite and BigDataBench have rich functions for benchmarking, none of them provide real sensor datasets and workloads for benchmarking TSDBs. In general, a large number of TSDBs have not yet been fully and fairly measured and compared. Compared to the achievements of traditional relational database benchmark, there is still a lot of room for improvement.

Other representative work is [26, 27]. In [26], the authors tested both collection speed and aggregation speed for reasonable data streams of sensor data, and then used relational databases, key-value stores, column stores, self-tuning databases, as well as TSDB systems for performing the test. Their experiments confirmed that column stores and key-value stores perform better than relational databases, while time-series databases outperform all the others. However, the performance metrics of various NoSQL stores are neglected in the benchmarking process, and the difference between different TSDBs is not quantified. Shah [27] presented a framework and methodology for benchmarking NoSQL data stores in the context of large-scale modelling applications. However, we think it is necessary to utilize real sensor dataset for benchmarking various TSDBs.

To sum up, the main shortcomings of the work above lie in the following two aspects: (1) There are few comprehensive and quantitative studies on which TSDB is more appropriate for large-scale sensor stream data in real scenarios. (2) Current benchmarks for TSDBs still need improvement in workload implementation based on real business requirements, data generator based on real-world sensor data and fine-grained performance metrics.

## 3   The Proposed Benchmarking Platform

### 3.1   The Framework of TS_Store_Test

The framework of the benchmarking platform TS_Store_Test adopts a hierarchical model, which can be divided into five layers, namely infrastructure layer, data storage layer, message transport layer, workloads implementation layer and user interface layer. The framework of TS_Store_Test is shown in Fig. 1.

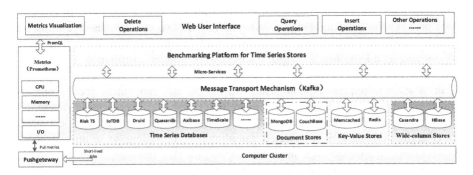

**Fig. 1.** The framework of the benchmarking platform for TSDBs

In infrastructure layer, the essence is a computer cluster or cloud computing environment, and it provides the hardware foundation or virtual machine runtime environment for TSDBs. Various TSDBs are installed or deployed on the cluster or virtual machine cluster. For example, IoTDB only supports single machine mode, but Druid can be deployed to the entire cluster. The configuration of virtual machines or machines within a cluster is consistent, which ensures fairness for benchmarking TSDBs at the infrastructure level. In addition, resource consumption of infrastructure layer is also an indispensable part of benchmark, but it is still lacking in the present research. We introduce Prometheus to realize resource monitoring of system hardware which regularly pulls short-lived job from infrastructure layer and provided it on Pushgateaway. The mechanism described above allows us to integrate our TS_Store_Test with metric in a uniform normalized format. Specific metric details are provided in Sect. 3.2. Considering that Prometheus is a short-term job, the integration of Prometheus has slight impact on hardware resources, so it was negligible. In this way, fairness is guaranteed.

The data storage layer is composed of integrated TSDBs, and five well-known TSDB instances have been integrated in TS_Store_Test including IoTDB, Riak TS, Druid, QuasarDB [28], and TimeScale [29]. More TSDBs will be integrated later, such as InfluxData [30], AXIBASE [31], OpenTSDB [32], kdb+ [33], KairosDB [34] and SiteWhere [35]. Certainly, we have also successfully integrated other types of databases, such as MongoDB and HBase.

The core is message transport layer. In order to implement the scalability for different TSDBs, the key is adopting the micro-services mechanism. Using micro-services approach is loosely coupled and can provide greater flexibility. Currently, using message transport mechanism Kafka [36, 37] is a typical design pattern for implementing micro-services, and they can also provide a rich feature for supporting stream data processing, caching and transmission. If we need to extend the new connection to new TSDBs or even other NoSQL databases, the method is to write a corresponding service implementation according to standard interface method, thus each TSDB has an interface with Kafka. In addition, more importantly, in order to fairly perform the benchmarking of different TSDBs, we use the message transport mechanism to simulate a variety of real sensor data stream scenarios. For example, we can use and send hydrological sensor datasets with different distribution rules to various TSDBs in synchronous data transmission modes.

In workloads implementation layer, we design and implement multi-scenarios setting according to the requirement of hydrological sensor data processing. These test scenarios include: (1) Sensor data loading or data insertion with different data sizes; (2) Querying the data of a selected sampling site according to a time interval; (3) According to a certain time interval, querying the data of multiple sampling sites; (4) Increasing the number of queries as sensor stream data is continuously injected into the NoSQL stores; (5) Increasing the amount of data sampled by the sensors while keeping the number of queries; (6) Querying the data of a sampling site according to multiple time intervals; (7) Aggregation query using GroupBy; (8) Indexing query.

The user interface layer is mainly responsible for three aspects: (1) Based on the navigation tree, users can choose TSDBs for benchmarking; (2) According to the TSDBs selected by users, TS_Store_Test provides UI for listing the benchmarking operations; (3) It shows the performance results and workloads of benchmarking TSDBs. The presentation of user interface is shown in Fig. 2.

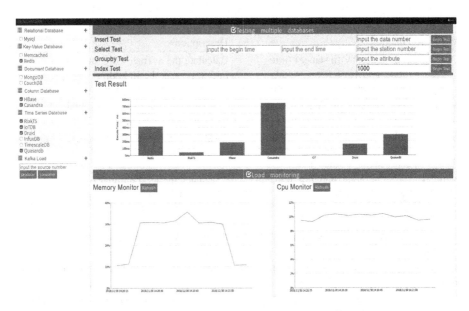

**Fig. 2.** The web user interface of TS_Store_Test

## 3.2   Performance Metrics Acquisition Based on Prometheus

Existing tools for benchmarking like YCSB or YCSB++ require that you separately do system level monitoring of resources to identify bottleneck. To acquiring the performance metrics in a fine-grained way, we choose and extend Prometheus as the performance metrics mechanism.

Specifically speaking, Prometheus is an open-source, scalable systems monitoring and alerting toolkit. It fits both machine-centric monitoring as well as monitoring of highly dynamic service-oriented architectures, because it uses carefully engineered data structures and algorithms to achieve very low per-node overheads and high concurrency. The implementation is robust, and has been ported to an extensive set of operating systems and processor architectures. In view of the aforementioned advantages, TS_Store_Test integrates Prometheus, and collects the performance metrics of various TSDBs in different workloads in a refined manner. We also extend its API for more fine-grained monitoring including different CPU indicators, memory indicators and so on. For example, MySQL and Influxdb provide the corresponding plug-in, but RiakTS lacks the corresponding tools to monitor. So, fairness is not guaranteed. For this, we adopt a unified computing method to monitor resources. A series of functions provided by Prometheus are used for calculation to obtain the metric needed indirectly. Table 1 lists the corresponding calculation methods for CPU, memory, and network I/O.

**Table 1.** Monitoring metrics.

| Metric | Function | Description |
|--------|----------|-------------|
| CPU | 100 – (avg by (instance) (irate (node_cpu {instance = "xxx", mode = "idle"} [5 s])) * 100) | Calculate the average CPU utilization in 5 s. (xxx is the IP address) |
| Memory | node_memory_MemTotal_bytes-node_memory_Buffers_bytes-node_memory_Cached_bytes-node_memory_MemFree_bytes-node_memory_Slab_bytes | The memory consumption is obtained by subtracting the cache and free memory usage from the total memory |
| Network | sum(irate(node_network_transmit_bytes_total[5 s])) | Calculates the total number of bytes transferred over the network in 5 s |

After configuring Prometheus, it collects metrics from monitored target machines by scraping metrics HTTP endpoints using Node Exporter. The Node Exporter exposes an extensive set of machine-level metrics on Linux and other Unix systems. Occasionally, we need to monitor components which cannot be scrapped. The Prometheus Pushgateway helps us to push time series from short-lived service-level batch jobs to an intermediary job. Therefore, based on Prometheus, three important types of metrics are collected including CPU, memory, and network bandwidth in TS_Store_Test.

### 3.3   The Execution Mode of TS_Store_Test

When the user starts performing the benchmark, she/he only needs three steps using user interface. At first, she/he needs to select the TSDBs and sensor data generation mode. There are two data generation modes: stream data generator and batch data generator.

In order to generate highly simulated sensor data, we use the Max-Min classification and Markov chain to obtain simulation model of sensor data by inputting seed file. We select the first-row data in the sensor data as the starting data. After passing through the simulation model, the data will not only be output, but also be used as the next inputting data. This allows you to continuously generate an infinite flow of data through a looping statement and write to the target database. User can control the beginning and end of the stream generator. For batch data generation, we assume it as an extension of stream data generation. In other words, the storage target is no longer the database but the HDFS file system. When running the test, the database under test will read the previously generated batch data from HDFS.

Before benchmarking, the selected TSDBs will be checked whether they can execute the selected workload. If not, TS_Store_Test will give some prompts. Then, she/he selects the specific workload and fills in the parameters required for the workload. We designed eight workloads to simulate common operations in a database based on the criteria of having or not having additional workloads.

Additional workload means that while the user performs an action, the database is operated at the same time. For this, we designed two workloads. One is to increase the number of queries while data is inserted. More specific, while Kafka writes data streams to the database, the user continues to increase the number of queries. The other is that the number of queries remains the same while the number of topics in Kafka continues to increase.

For not having additional workload, they are pure database operations. The included workloads are: (1) Data import operation; (2) Selecting station data according to a time interval; (3) Selecting data from multiple stations according to a time interval; (4) Selecting data from the same station according to multiple time intervals; (5) Down-frequency aggregated query; (6) Index query.

At last, TS_Store_Test executes the logic for the workload, and then the user interface layer shows the results through visual mechanisms. When the user clicks on the different databases, the resource monitoring module below will also make corresponding changes. Through the refresh button, the user can see the current resource consumption of the selected database in real time.

## 4 Experiments and Result Analysis

### 4.1 Sensor Dataset and Experimental Setting

The data structure of the hydrological sensor data is shown in Table 2 below, which is derived from the data of more than 70+ hydrological sensor sampling points in the Chuhe river basin in the year of 2015–2017, with a total number of 30 million pieces of data.

**Table 2.** The structure of hydrological sensor data.

| Field name | Type and length | isNull | Field description |
| --- | --- | --- | --- |
| id | int(11) | No | Primary key |
| stcd | varchar(20) | No | Hydrological site |
| tm | varchar(20) | No | Sampling time |
| rz | varchar(20) | No | Water level |
| rfrom | varchar(20) | No | Hydrological sensor |

According to the size of the sensor dataset, the whole benchmarking system TS_Store_Test is deployed in a cluster using four PCs and the hardware environment is: Intel(R) Xeon(R) E5645@2.40 GHz dual-core 24 CPU; Kingston DDR3 1333 MHz 8G, 500 GB SSD Flash memory. Operating system tools are Ubuntu 16.04 64-bit, Kafka 2.20 and Linux 3.11.0 kernel. Certainly, to further verify different dataset and the result, we can increase the cluster scale and input the data scale at any time.

In order to show the difference between TSDBs and functions of TS_Store_Test, our experiments tested different five types of TSDBs respectively. The TSDBs used for the result presentation are IoTDB 0.7.0, Riak TS 1.5.2, Druid 2.7.8, Timescale 1.2.2 and QuasarDB 2.7.0.

## 4.2   Experiments and Analysis

**Experimental Scenario 1.** Based on Kafka, we firstly create 70 topics to consume the data from 70 hydrological sensor sampling points. Every topic produces sampling data at the rate of 10 pieces per second, and the data is concurrently and continuously injected into different TSDBs. In this process, we set the different number of query requests, and then monitor the CPU or memory status in real-time manner.

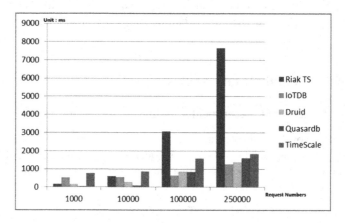

**Fig. 3.** The performance of TSDBs in scenario 1

Figure 3 shows the performance of time series NoSQL stores. We can obviously see that in this scenario setting, the performance of TSDBs is different. When number of query requests exceeds 100,000, the performance of Riak TS has deteriorated significantly. IoTDB and Druid are suitable for such scenario from performance standpoint, but Riak TS's performance is greatly affected by the increasing the number of queries.

Figure 4 shows the CPU monitoring results of different TSDBs. As the number of requests increases, Druid and QuasarDB outperformed other TSDBs. Timescale and Riak TS are even more affected. By comparison, we can have a deeper understanding of the importance of choosing the right TSDBs. However, for memory utilization ratio, the above TSDBs make little difference. Thus, based on TS_Store_Test, we can further investigate the optimal use of memory of some TSDBs.

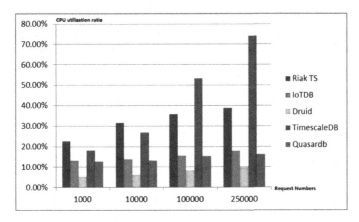

**Fig. 4.** CPU utilization ratio monitoring results of different TSDBs in scenario 1

**Experimental Scenarios 2.** At first, we maintain 50,000 query requests for the latest water level. Then, based on Kafka, we respectively create 500, 1000 and 5000 topics to simulate increasing number of sensor sampling points. Every topic produces sampling data at the rate of 10 pieces per second, and the data is concurrently and continuously injected into different TSDBs. In this process, we observe and compare the performance of different TSDBs and meanwhile monitor the CPU or memory situations in real-time manner.

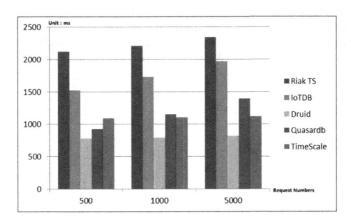

**Fig. 5.** The performance of TSDBs in scenario 2

In Fig. 5, it shows the performance of TSDBs in scenarios 2. By comparison, in this scenarios setting, the performance advantages of time series NoSQL stores are still contrasting. Druid demonstrates good performance against increasing sensor stream data. It's worth noting that MongoDB also performs very well, and its performance is similar to that of Druid.

Figure 6 shows the memory utilization ratio monitoring results of different TSDBs using TS_Store_Test in this scenario. As the number of sensor sampling points increases, TSDBs are significantly better than other NoSQL databases at memory utilization ratio.

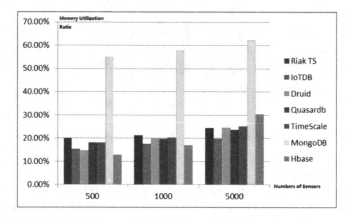

**Fig. 6.** Memory utilization ratio monitoring results of different TSDBs in scenario 2

**Experimental Scenarios 3.** Further, we select a certain time interval, then query the data of multiple sampling sites from different TSDBs. The selected time intervals are day, week, month and year. Figure 7 shows the result of querying the data of multiple sampling sites. We can see that as the query size increases, the performance of QuasarDB and TimeScale declines dramatically, and Druid demonstrates best performance. When the time interval is year, Riak TS crashed because the query object is too large.

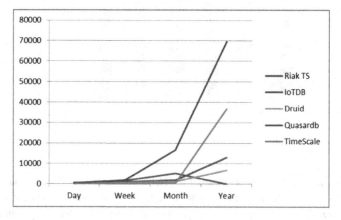

**Fig. 7.** The result of querying the data from multiple sampling sites

To sum up, compared with document database and key value database, TSDBs have more performance advantages for sensor stream data in various application scenarios. In addition, different TSDB performance varies greatly in different scenarios and Druid's performance is the best overall. Thus, based on TS_Store_Test, we can further investigate the availability enhancement mechanism of some TSDBs.

# 5   Comparison with YCSB and BigDataBench

As an open-source specification and program suite, YCSB is often used to compare relative performance of NoSQL database management systems. It has been used in scholarly or tutorial discussions, particularly for Apache HBase. Also, it has been used for multiple-product comparisons by industry observers such as Cassandra, MongoDB, Couchbase, Aerospike, OrientDB, Redis and Riak. BigDataBench is an open-source big data and AI benchmark suite. Now it covers seven workload types including AI, online services, offline analytics, graph analytics, data warehouse, NoSQL, and streaming from important application domains.

Compared with the two representative benchmarking platforms above, we compare and summarize Table 3 from seven aspects. From a software perspective, all three benchmarking platforms have good extensibility and usability, but we think YCSB and BigDataBench are more mature. From the perspective of NoSQL support, YCSB now supports the largest number and types of NoSQL databases, but TS_Store_Test can support more TSDBs, which became the highlight of our work. In terms of existing benchmarking capabilities, YCSB and BigDataBench are very strong, and TS_Store_Test is still relatively lacking. Certainly, we have submitted the tool and some of the data to GitHub. In particular, we choose and utilize Prometheus as the performance metrics mechanism, thus we can capture the performance metrics in a refined manner.

**Table 3.** The comparison for benchmarking tools.

| Names | YCSB | BigdataBench | TS_Store_Test |
|---|---|---|---|
| Extendibility | High | High | High |
| Open source<br>Software<br>maturity<br>Usability | Yes<br>High<br>High | Yes<br>High<br>High | Yes<br>Common<br>High |
| Supported NoSQL databases | **Key Value Stores** (Memcached/Redis/DynaoDB/Voldemort/Aerospike/Riak/ Tarantool/Voldemort/Aerospike/Tarantool/Riak)<br>**Wide Column Stores** (HBase/Cassandra/Hypertable)<br>**Document Stores** (MongoDB/Couchbase)<br>**Multi-model Databases** (OrientDB)<br>**Cloud Database** (Infinispan/GemFire) | **Document Stores** (MongoDB)<br>**Wide Column Stores** (HBase) | **Key Value Stores** (Redis/Memcached)<br>**Wide Column Stores** (HBase/Cassandra)<br>**Document Stores** (MogoDB/CouchBase)<br>**TSDBs** (QusarDB/IoTDB/Riak TS/Druid/TimeScale) |
| Bencmarks<br>Performance metrics | Rich<br>Weak | Rich<br>Good | Abundant<br>Good |

## 6  Summary and Prospect

Each NoSQL technology is suited for specific use cases and data models. The importance of selecting the correct TSDBs solution for the environment and the data is often overlooked. The decision will have a huge impact on performance and supported functionality for user's environment. A standardized benchmark that can be used to evaluate the performance of different TSDBs can greatly help organizations choose the right solution. Therefore, we designed and implemented a benchmarking platform integrating multiple well-known TSDBs. Based on the hydrological sensor data obtained from Chuhe river, we tested and compared the execution efficiency of common operations of various time series mechanisms. Based on the results, the feasible storage solutions in the field of water resources information are summarized.

In the future, we will focus on integrating more NoSQL storage mechanisms and test sensor stream data storage and processing to accommodate more business scenarios. In addition, we will explore the elasticity and dependability of NoSQL stores.

## References

1. Qin, X., Wang, H., Du, X., Wang, S.: Big data analysis-competition and symbiosis of RDBMS and MapReduce. J. Softw. **23**(1), 32–45 (2012)
2. Dunning, T., Friedman, E.: Time Series Databases-New Ways to Store and Access Data. O'Reilly Media, Sebastopol (2015)
3. IoTDB Homepage. http://tsfile.cn/index. Accessed 21 Apr 2019
4. Druid Homepage. http://druid.io/. Accessed 21 Apr 2019
5. Riak TS Homepage. http://basho.com/products/riak-ts/. Accessed 21 Apr 2019
6. Davoudian, A., Chen, L., Liu, M.: Survey on NoSQL stores. ACM Comput. Surv. **51**(2), 1–43 (2018)
7. Gessert, F., Wingerath, W., Friedrich, S., Ritter, N.: NoSQL database systems: a survey and decision guidance. Comput. Sci. Res. Dev. **32**, 353–365 (2016)
8. Lourenço, J., Cabral, B., Carreiro, P., Vieira, M., Bernardino, J.: Choosing the right NoSQL database for the job: a quality attribute evaluation. J. Big Data **2**, 1–26 (2015)
9. Han, R., John, L.K., Zhan, J.: Benchmarking big data systems: a review. IEEE Trans. Serv. Comput. **11**(3), 580–595 (2018)
10. Zhou, X., Qin, X., Wang, Q.: Big data benchmarks: state-of-art and trends. J. Comput. Appl. **35**(4), 1137–1142 (2015)
11. Qian, W., Xia, F., Zhou, M., Jin, C., Zhou, A.: Challenges and progress of big data management system benchmarks. Big Data Res. **1**, 1–15 (2015)
12. Gregg, B.: Systems Performance: Enterprise and the Cloud. Prentice Hall, Ann Arbor (2013)
13. Kai, J.: Research on reliable-oriented adapation on microservice system. Shanghai University, Shanghai (2016)
14. Prometheus Homepage. https://prometheus.io/. Accessed 21 Apr 2019
15. Jensen, S.K., Pedersen, T.B., Thomsen, C.: Time series management systems: a survey. IEEE Trans. Knowl. Data Eng. **29**(11), 2581–2600 (2017)
16. Wlodarczyk, T.W.: Overview of time series storage and processing in a cloud environment. In: 4th IEEE International Conference on Cloud Computing Technology and Science, pp. 625–628. IEEE Computer Society, Taipei (2012)

17. Bader, A., Kopp, O., Michael, F.: Survey and comparison of open source time series databases. In: Mitschang, B., et al. (eds.) BTW 2017. LNI, pp. 249–268. Gesellschaft für Informatik, Bonn (2017)
18. Gandini, A., Gribaudo, M., Knottenbelt, W.J., Osman, R., Piazzolla, P.: Performance evaluation of NoSQL databases. In: Horváth, A., Wolter, K. (eds.) EPEW 2014. LNCS, vol. 8721, pp. 16–29. Springer, Cham (2014). https://doi.org/10.1007/978-3-319-10885-8_2
19. Matallah, H., Belalem, G., Bouamrane, K.: Experimental comparative study of NoSQL databases: HBase versus MongoDB by YCSB. Comput. Syst. Sci. Eng. **32**(4), 307–317 (2017)
20. Patil, S., et al.: YCSB++: benchmarking and performance de-bugging advanced features in scalable table stores. In: SOCC 2011, Article No. 9. ACM, Cascais (2011)
21. Alabdulkarim, Y., Barahmand, S., Ghandeharizadeh, S.: BG: a scalable benchmark for interactive social networking actions. Future Gener. Comput. Syst. **85**, 29–38 (2018)
22. Ferdman, M., et al.: Clearing the clouds: a study of emerging scale-out workloads on modern hardware. In: International Conference Architectural Support for Programming Languages and Operating Systems, ASPLOS 2012, pp. 37–48. ACM, London (2012)
23. Zhan, J.F., et al.: BigDataBench: an open-source big data benchmark suite. Chin. J. Comput. **39**(1), 196–210 (2016)
24. MongoDB Homepage. https://www.mongodb.com/. Accessed 21 Apr 2019
25. HBase Homepage. https://hbase.apache.org/. Accessed 21 Apr 2019
26. Pungilă, C., Fortiş, T., Aritoni, O.: Benchmarking database systems for the requirements of sensor readings. IETE Tech. Rev. **26**(5), 342–349 (2009)
27. Shah, S.M., Wei, R., Kolovos, D.S., Rose, L.M., Paige, R.F., Barmpis, K.: A framework to benchmark NoSQL data stores for large-scale model persistence. In: Dingel, J., Schulte, W., Ramos, I., Abrahão, S., Insfran, E. (eds.) MODELS 2014. LNCS, vol. 8767, pp. 586–601. Springer, Cham (2014). https://doi.org/10.1007/978-3-319-11653-2_36
28. QuasarDB Homepage. https://www.quasardb.net/. Accessed 21 Apr 2019
29. Timescale Homepage. https://www.timescale.com/. Accessed 21 Apr 2019
30. InfluxData Homepage. https://www.influxdata.com/. Accessed 21 Apr 2019
31. AXIBASE Homepage. https://axibase.com/products/axibase-time-series-database/. Accessed 21 Apr 2019
32. OpenTSDB Homepage. http://opentsdb.net/. Accessed 21 Apr 2019
33. kdb+ Homepage. https://kx.com/. Accessed 21 Apr 2019
34. KairosDB Homepage. http://kairosdb.github.io/. Accessed 21 Apr 2019
35. SiteWhere Homepage. https://github.com/sitewhere/sitewhere. Accessed 21 Apr 2019
36. Dunning, T., Friedman, E.: Streaming Architecture: New Designs Using Apache Kafka and MapR Streams. O'Reilly Media, Sebastopol (2016)
37. Lu, R., Wu, G., Xie, B., Hu, J.: Stream bench: towards benchmarking modern distributed stream computing frameworks. In: IEEE/ACM 7th International Conference of Utility and Cloud Computing, pp. 69–78. IEEE, London (2014)

# HDF5-Based I/O Optimization for Extragalactic HI Data Pipeline of FAST

Yiming Ji[1], Ce Yu[1], Jian Xiao[1(✉)], Shanjiang Tang[1], Hao Wang[1], and Bo Zhang[2]

[1] College of Intelligence and Computing, Tianjin University, Tianjin, China
{jiym,yuce,xiaojian,tashj,imwh}@tju.edu.cn
[2] CAS Key Laboratory of FAST, NAOC, Chinese Academy of Sciences, Beijing, China
zhangbo@nao.cas.cn

**Abstract.** The Five-hundred-meter Aperture Spherical Radio Telescope (FAST), which is the largest single-dish radio telescope in the world, has been producing a very large data volume with high speed. So it requires a high performance data pipeline to covert the huge raw observed data to science data product. However, the existing solutions of pipelines widely used in radio data processing cannot tackle this situation efficiently. The paper proposes a pipeline architecture for FAST based on HDF5 format and several I/O optimization strategies. First, we design the workflow engine driving the various tasks efficiently in the pipeline; second, we design a common radio data storage specification on the top of HDF5 format, and also developed a fast converter to map the original FITS format to the new HDF5 format; third, we apply several concrete strategies to optimize the I/O operations, including chunks storage, parallel reading/writing, on-demand dump, and stream process etc. In the experiment of processing 700 GB of FAST data, the results show that HDF5 based data structure without other optimizations was 1.7 times faster than original FITS format. If chunk storage and parallel I/O optimization are applied, the overall performance can reach 4.5 times as the original one. Moreover, due to the good expansibility and flexibility, our solution of FAST pipeline can be adapted to other radio telescopes.

**Keywords:** FAST · FITS · HDF5 · Pipeline in parallel · High performance I/O

## 1 Introduction

Data pipeline, a key procedure for modern observational astronomy, is to convert the raw observed data to science product, which astronomers can use directly to make new discoveries and theoretical testing. As the continuous improvement of capacity and resolution of telescopes, the observed data volume explosively increases. So the performance of astronomical data pipeline becomes one of the

© Springer Nature Switzerland AG 2020
S. Wen et al. (Eds.): ICA3PP 2019, LNCS 11945, pp. 656–672, 2020.
https://doi.org/10.1007/978-3-030-38961-1_55

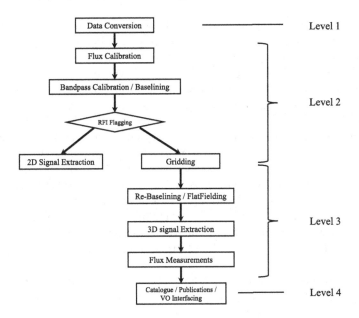

**Fig. 1.** The main steps of pipeline

biggest challenges for modern large telescopes. For example, the Five-hundred-meter Aperture Spherical Radio Telescope (FAST) [15] in Guizhou, China, is the largest single-dish radio telescope in the world. Currently, typical data rate of FAST is as high as 3 GB/s. FAST will output about 10 TB of raw data per hour. Its tasks include neutral hydrogen survey and pulsar detection, which will record a huge amount of data. Meanwhile, the 19-beam receiver, the most frequently used receiving system of FAST [23], will produce 19 times data size as the single beam receiver. These observed data provides more opportunities to new scientific discoveries, meanwhile it also brings unprecedented pressure to the traditional solution of data pipeline.

Figure 1 shows a classical work flow of radio data pipeline. It is usually divided into four levels. Level 1 is the data format conversion, converting the original format to an inner format throughout the whole pipeline. Level 2 is the necessary calibrations to eliminate the noise from the universe, the earth and the device itself to a tolerable level, including flux calibration, bandpass correction, baseline subtraction, and radio frequency interference (RFI) mitigation [8] etc. The main purpose of level 3 is to generate the data cube for extracting signals [18]. Level 4 refers to publish the data product to astronomers for further research. For FAST, the process flow is almost the same as Fig. 1. In terms of the overall architecture of pipeline, data exchange between different steps is the efficiency bottleneck to process data.

On the other hand, unlike optical telescopes, the radio telescope have a very wide frequency coverage, so each radio telescope has its unique physical features, though they can share a relative common process flow, but the details of each steps may be quite different. Therefore, there is none existing universal solution or mature software for various radio telescopes. For example, CASA [14] aims to adapt all radio telescopes, but is used mainly by VLA and ALMA until now, which both are radio array or similar structure instead of single dish. CLASS [21] is designed for (sub-)millimeter single dishes. CLASS does not support parallel processing because it only serves small and medium telescopes. Therefore, it is not suitable for processing tasks with large amounts of data.

There are several research works aiming to improve the data pipeline efficiency. Data Activated Liu Graph Engine (DALiuGE) is an execution framework for processing large astronomical datasets at a scale required by the Square Kilometre Array Phase 1 (SKA1) [27]. It can be considered as one representation of next generation pipeline architectures, but is more like a task deployment and scheduling tool, lack of consideration of data layout and exchange. In addition, AST3 daemon [10,28] is a light-weight pipeline engine for Antarctic Schmidt Telescopes, and it has been running in Antarctic dome A for 5 years normally. While AST3 also lacks of support for cluster environment, which is necessary for FAST.

At present, most of existing solutions still use FITS file as the default data format during the data processing [16,26]. FITS is a traditional astronomical data format with fixed format specifications and complex content packaging. According to our experiment, FITS related I/O operations take roughly 25%–30% of the whole process. Obviously the I/O efficiency has become one of bottlenecks of the modern radio data pipelines. Faced with rapidly increasing volume of data, the FITS performance seems to be inadequate. Motivated by data volumes, Hierarchical Data Format version 5 (HDF5) [9] has been implemented for the LOFAR radio telescope [1], the CCAT telescope [22], and the CHIME pathfinder [13]. Compared with FITS format, HDF5 format has more concise structure. So it's possible to improve I/O efficiency by changing the layout of data in HDF5 format. In addition, the pipeline has separate processing steps and the intermediate results are saved frequently. The large volume of FAST's data makes I/O cost become a vital factor impacting the overall performance. Therefore, reducing I/O time can improve efficiency significantly.

So in order to process FAST's continuous huge output data with high speed and accuracy, an end-to-end solution including parallel pipeline engine, fast inner data exchange format, high speed I/O interface, and deeply optimized algorithm of each steps is necessary. In this paper, we focus on the common aspects of building the qualified pipeline for FAST, that is providing an optimized pipeline framework with high performance I/O support. Various calibration algorithms, RFI mitigation and data cube generation modules can be easily integrated into the pipeline through the interface, and all of them can exchange data with high efficiency within the pipeline. The work content of this paper is described as follows. We propose the data layout specification for pipeline on the top of HDF5 format, and also develop a fast converter to map the original FITS format to the new HDF5 format. In addition, we apply several concrete strategies to optimize the I/O operations, including chunks storage, parallel reading/writing, on-demand dump, and stream process etc. Furthermore, we implement the FAST pipeline engine, which supports concurrency, real-time monitoring and error reporting, in-memory execution, and asynchronous storage of resulting data. In the experiment of processing 700 GB of FAST data, the results show that merely HDF5-based data structure is 1.7 times faster than original FITS format. With chunk storage and parallel I/O optimization, the overall performance can be 4.5 times as the original one. The key contributions are:

- We design a parallel data pipeline engine to tackle FAST huge data volume, and it can efficiently drive the pipeline to process large intermediate result in stream style.
- We define a general data layout for single dish radio data based on HDF5, which is used for fast data exchange between various tasks of pipeline.
- We explore several optimization strategies such as chunks storage, parallel I/O, etc. We also integrate main processing steps into the pipeline, and make a comprehensive evaluation based on real observed data.

This article is organized as follows. In Sect. 2, we provide some related work, including the characteristics of FITS and HDF5 file formats, the performance of some radio telescope pipelines, and I/O method of pipelines. In Sect. 3, we present a description of our overall work; we also present a mapping of FAST's raw data (FITS) into intermediate data (HDF5) and the details of the pipeline. Section 4 gives the experimental results to verify the efficiency of our proposed format and the concrete implementation of pipeline. Some general summaries, practical applications for FAST and possible future extensions are given in Sect. 5.

## 2 Related Work

In this section, we discuss related work from three aspects: the format of astronomical data, the pipeline of radio telescopes and file I/O.

### 2.1 Data Format

**FITS Model.** The Flexible Image Transport System (FITS) has enjoyed several decades of usage among the field of Astronomy [16], and it was stipulated

**Table 1.** Example of FITS HDU

| Key = | Value/comment |
|---|---|
| SIMPLE = | T/File does confirm to FITS standard |
| BITPIX = | 16/ Number of bits per data pixel |
| NAXIS = | 2/Number of data axes |
| NAXIS1 = | 320/Number of pixels along the fastest changing axis |
| NAXIS2 = | 512/The number of pixels along the sub-fast changing axis |
| END | |

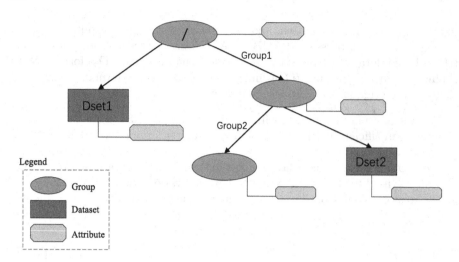

**Fig. 2.** Example of HDF5 object [17]

as the unified standard format for data transmission and exchange between different observatories established by the International Astronomical Union (IAU) in 1982.

FITS file consists of a set of header data units (HDUs), which are ASCII headers followed by consecutive blocks of data (binary or ASCII encoding). HDUs contain some descriptive variables, whose format is "Key = Value/Comment". There are some indispensable keywords listed in Table 1, such as SIMPLE, BITPIX, NAXIS, NAXISn and END. In general, there are other keywords indicating related information of observation data, such as date, telescope, observer and so on. In astronomy, FITS is a classical format for pictorial data and spectral data.

**HDF5 Model.** The hierarchical data format 5 (HDF5) is the latest among the series. HDF5 consists of data format specification and library implementation. Compared with the old version of HDF, Its related support has been extended. In addition, its hierarchical structure and supported libraries can reflect the advantages of storage, reading and writing [7].

HDF5 format file is organized in a hierarchical structure, which contains three main elements: Dataset, Group and Attribute.

- Dataset: Multidimensional arrays of data elements and support for metadata.
- Group: The grouping structure contains instances of zero or more objects, groups or datasets; it's supported for metadata.
- Attribute: User-defined metadata information, which can be attached to Datasets and Groups.

The Group is the root of HDF5 object. A separate HDF5 object can perform as its substructure. The dataset mainly stores data array, whose dimensions can be set from 1 to N (any positive integer). At the same time, the attributes can be linked to different nodes of HDF5 object. Attributes annotate temperature, time and other information defined by users. Figure 2 shows hierarchical structure of HDF5 object and relation of Groups, Datasets and Attributes.

## 2.2  Pipelines Introduction

Pipeline is one kind of dataflow computation model, which was initially proposed [19] to express programs as Directed Acyclic Graphs (DAG), where the vertices are the stateless computational tasks that compose the program, and edges connect the output of one task with the input of another. In astronomy, the pipeline implements the data processing algorithm in the order of dataflow. Each step of the pipeline is independent. The output of the previous step will flow into the next step.

At present, many telescopes all over the world have their own data processing pipelines. Davis [5,6] proposed the ALMA prototype science pipeline in 2004. As of 2014, it had already supported distributed processing. The shooting speed of ALMA determines the pipeline speed to be 6M/s, which is not suitable for massive data processing of FAST.

The Transients Key Science Project (TKP) [24] is developed for LOFAR and can study all variable sources detected by LOFAR. Its functions include the study of transient and variable low-frequency radio sources with an extremely broad science case ranging from relativistic jet sources to pulsars, exoplanets, flare stars, radio bursts at cosmological distances, the identification of gravitational wave sources and even SETI. As Lofar and FAST have different research goals, TKP's approach cannot be transplanted into FAST's supporting environment.

The Very Large Telescope (VLT) [3] is a collection of eleven instruments. For each of them, European Southern Observatory (ESO) provides automatic data reduction facilities in the form of instrument pipelines developed in collaboration with the instrument consortia. The Multi Unit Spectroscopic Explorer (MUSE, Bacon et al. [2]) is one of four second generation instruments being built for the ESO VLT. MUSE can process 150 GB of raw data per night and support two modes, online and offline. Online mode focuses on timeliness, while offline mode tries to optimize results and minimize user interaction.

## 2.3  File I/O

At present, the data file performs as the I/O unit of FAST pipeline. To optimize the I/O of FAST pipeline, this paper provides several studies that concentrate on optimizing the parallel file-I/O performance of HPC applications. Y.Chen and R.Thakur proposed libraries and parallel file systems respectively, and exploit advancements in storage technologies [4,25]. This system has some constraints on files. It require data to be stored in its own file system. The method is not suitable for FAST pipeline because specific file systems or libraries are not suitable for FAST neutral hydrogen data. Other optimization techniques include exploiting access patterns to assist file system prefetching, data sieving, and caching [12], overlapping computation with I/O [11], and employing asynchronous prefetching [20]. FAST pipeline is a fixed process that does the same execution to each file and the same file is not read repeatedly. Thus data sieving, caching, overlapping computation and so on cannot provide FAST pipeline with improvement.

From preceding part of this section, both of the formats are used in the astronomy field. By contrast, the simplicity and flexibility of the HDF5 format show an advantage. With HDF5 being used as intermediate data format to replace FITS in the pipeline, I/O overhead will be reduced. As a result, overall efficiency is improved. In addition, concurrent processing of pipeline provides favor to process massive FAST data. By the analysis of the existing work, no method can fully satisfy the performance demand of FAST data processing. Based on the technical environment, the paper proposes the implementation of workflow engine and optimization strategy for FAST pipeline.

## 3    HDF5-Based I/O Optimization in Pipeline

The Fig. 3 shows the architecture of the FAST data pipeline with HDF5-based I/O optimization. The whole solution includes a work flow engine to manage the execution of pipeline, a highly optimized I/O interface to read/write intermediate data during the process, the HDF5 data model of FAST, and a dedicated mapper for converting raw data from FITS to HDF5 format.

The data produced by the FAST is recorded in FITS files. Since the FITS format is a multi-layer encapsulated table structure, the process of parsing FITS files takes much time. HDF5 format is chosen as the alternative to reduce the overhead of parsing FITS. In order to maximize compatibility with existing astronomical software and share data, we keep the archived raw files in FITS format and use HDF5 format as an intermediate format to improve I/O performance. HDF5 has two distinct characteristics: hierarchical structure and attributes. Hierarchical structure highlights the hierarchy and ownership among different parts. In our proposed HDF5 specification, the data is grouped according to the scanning sequence or polarization number, which makes it easier for programmers and astronomers to clarify the data content. It's straightforward to determine the location of the target data according to the hierarchy. In addition, the attributes is added to the specified parts of the HDF5 object. This combination determines its self-explanatory advantages for each grouping pair. By adding attributes, it's visual to specify the content of the object. There are some descriptive variables like time, shooting equipment, shooting status and so on, which need to be recorded as float or string variables. It's unnecessary to create new datasets. Attributes work in this situation, which indicate these variables in the corresponding location. In this way, the efficiency has been saved and the relationship has been described clearly. Compared with the traditional format FITS, our proposed data layout based on HDF5 performs better structurally.

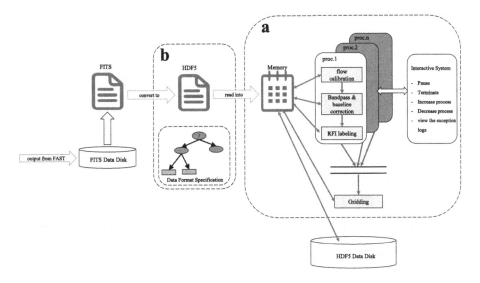

**Fig. 3.** An overview of the design, divided into two parts – **a** and **b**

## 3.1  Transformation and Mapping from FITS to HDF5

For all the content contained in the FITS file, we design the transformation and mapping from FITS to HDF5, showed in Fig. 4. In the raw FITS file, PrimaryHDU stores the descriptive information of FAST data, and the observed content are independently stored in binTableHDUs. The size of binTableHDU is 2048(rows) × 21(columns). Each row is the scanning serial number according to time. The first 20 columns record the time, coordinates, external conditions, etc. The 21st column named "DATA" is a two-dimensional array (65536 × 4). 65536 is the number of channels, and 4 is the polarization serial number. As shown in the Fig. 4, the entire HDF5 object contains only one group as its root directory. The original FITS file's PrimaryHDU is transformed into a set of attributes directly connected to the root directory, which includes BIT-PIX, NAXIS, EXTEND, ORIGIN, DATE. Then 2048 datasets are corresponding to the "DATA" columns of FITS binTableHDUs. Each binTableHDU's first 20 columns are turned into a set of attributes for each dataset. In the intermediate data, attributes contain the following information: frequency of observation center, bandwidth, number of spectral channels, start/end time of data recording in the file, Angle of the telescope (azimuth Angle and zenith Angle) after the correction of heliocentric system, and information about the definition standard of coordinate reference system.

## 3.2  HDF5 Optimization Strategies

Outside of the transformation from FITS to HDF5, there are some improvement on the underlying field. HDF5 data is stored linearly in memory by default, so

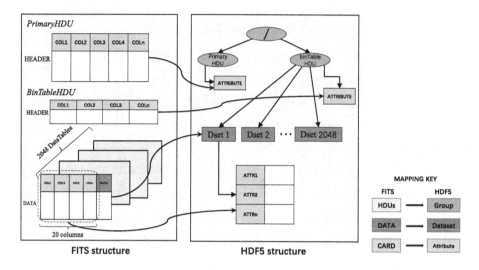

**Fig. 4.** The details of mapping FITS to HDF5

the reading process has to go through all the content to find the target. If a smaller sub-block is the target, accessing parts outside the block is invalid. In order to avoid these invalid operations, the high frequency accessed areas are supposed to be obtained directly.

In the process of converting data from FITS to HDF5, we specify the dataset as a two-dimensional float array. Each dataset in HDF5 data uses a type system similar to the Numpy module in Python. At the time of reading and storing, the concrete array is processed as the Numpy array, and the datatype has been mapped to the dtype of Numpy.

**Chunks Storage.** A two-dimensional array has two dimensions in the mathematical sense, but virtually all the dimensional data in the computer's memory is stored in a linear continuum. Sequential storage is suitable for reading all data at one time. However, when we only need to read one or more blocks in many cases, such as reading the sub-dataset [2048 : 8192, 0 : 2] from an array of 65536 × 4, the program will read from beginning to end under sequential storage. There will be a lot of overhead outside the target region. It indicates that sequential storage does not match most sub-block reading directly. By default, the HDF5 datasets are sequentially stored, which causes unnecessary I/O cost for pipeline.

For the datasets, the N-dimensional shape is specified in the chunks storge, which fits the access mode best. When data needs to be written into disk, it will be split into blocks of the specified shape and written into memory in blocks. These blocks are stored in the file whose coordinates are indexed by a B-tree. Because pipeline involves a large number of array readings, and the size of sub-block is determined by the algorithm in the pipeline. The pipeline proposed in this paper supports both the default size and manually specifying the size.

**In-memory Cache.** Pipeline involves many times data reading and writing. If each reading or writing is directly sent to disk, the overall running time will be spent primarily on reading and writing rather than computing. To overcome this disadvantage, the data file is kept in the memory until it is processed completely. By this strategy, the pipeline can fetch target data from memory and every step can exchange the data directly. As an HDF5 object is generated, new space in memory will be created to maintain it. As the in-memory file is closed, its contents are saved to disk. As long as the entire file is put into memory, the pipeline process only needs to read and write to the disk once per process. Subsequent data reading and writing, attributes creation, and other operations needn't occupy disk I/O at all.

**Data Operation in Parallel.** According to the above data specification, each file contains 2048 datasets, and each dataset is an array of 65536 × 4. Parallel I/O for data of this size is a great way to increase efficiency. Common parallel operations are multi-threaded and multi-process. In our early exploration experiments, thread-level concurrency for HDF5 object takes a lot of technical development. This will shift the focus of our research and it's uncertain whether we can meet our expectations by this way. And using HDF5 in multi-threaded programs does not improve efficiency. If multi-process is used directly to manipulate a single HDF5 file, it is easy to conflict with the process of the pipeline hierarchy and the structure appears redundant. Considering the above points, Message Passing Interface (MPI) is a superior choice. In the MPI program environment, one HDF5 object can be accessed by multiple processes. This method supports frequent communication between processes and collection of final results. The process created for each HDF5 file is coordinated by the MPI library and does not conflict with the upper process. In the process, we specify the MPI driver and an MPI communicator. The MPI communicator is responsible for communication between different processes. For example, the process keeping a single data file creates 4 sub-processes. Each sub-process is responsible for calculating one chunk of data, so that four sub-blocks are computed simultaneously. In theory, the performance of this scheme is four times better than that of the serial scheme with the same computing power.

## 3.3  The Implementation of FAST Pipeline

Our work is based on the real neutral hydrogen data of FAST. FAST pipeline in parallel is implemented based on the proposed optimization details above. The main function of the pipeline is to process FAST raw data and archive processed data. The sub-systems of the pipeline can be described as follows:

- Data processing system: the pipeline starts a specified number of processes, each responsible for one dataflow.
- Data I/O system: the pipeline's data read-write system is responsible for reading data files into memory and writing them to disk. When the raw data

files are read for the first time, they will be converted to the designed HDF5 specification.

– Fault-tolerant system: the pipeline's fault-tolerant system logs all exceptions occurring in the middle of the process and re-executes from this step.
– Interactive system: the pipeline's interactive system displays the current total amount of unprocessed tasks in real time, and the current task progress percentage of each process.

As showed in part **a** of Fig. 3, the result of FAST surveying is stored in FITS files disk in real time. When the pipeline starts, it creates a certain number of processes to monitor FITS files disk. If the processes are more than the files, the existing files enter the corresponding number of processes, and other processes are idle. Then the newly generated files enter the waiting processes. If the processes are less than the files, all the processes start working and some files enter the task queue. Processes in working mode are locked. The only access to the processes is the exception handling and enforcement commands issued by the interacting system. When the task is finished, the process will be unlocked and load the next task.

In each independent process, flow calibration, bandpass correction, baseline correction and RFI labeling are conducted in turn. When processed data files reach to a certain amount, Gridding is executed. The input to the first step is an in-memory HDF5 data file, and the input to the next step is the output from the previous step. The pipeline's data stream is executed entirely in memory. Typically, each data file requires only one time disk reading into memory, and I/O of subsequent steps are memory-based. Since the size of a single HDF5 file is 2 GB, the total memory consumption of the pipeline is 2n GB when the number of processes is n. This trade-off for time efficiency at the expense of physical memory consumption is reasonable in large scientific projects. According to the requirements of FAST staff, the results of the intermediate steps should be backed up and saved. The size and dimensions of the stored data files are the same as the proposed HDF5 specification. After each step in the process ends, saving the intermediate results and processing the next step data are executed in parallel.

If an exception occurs in the pipeline, the exception information will be written into the log file. And the pipeline will be re-executed from the appropriate step. Supposing the same exception occurs three times in a row, the task of the data file will be cleared from the process. Then it loads a new file from the task queue.

During the execution of the pipeline, the interactive system can display the total number of tasks in the queue and the schedule of all processes in real time. At the same time, interactive systems support command operations, such as: pause, terminate, increase or decrease the process and view the exception logs.

## 3.4    Functional Module of the Pipeline

To be used for research, the raw files need to be processed. The pipeline implements four main FAST data processing algorithms, which are introduced in the form of modules.

**Spectral Data Flow Calibration.** The raw data recorded by FAST is presented in the form of mechanical records, which does not have direct physical significance. The purpose of flow calibration is to correspond mechanical records to physical units so that one can measure the flow of celestial sources. There are two steps among the transformation: first, the mechanical records are converted into the source's bright temperature $T_{source}$ (K), and then the bright temperature is converted into the source's physical flow (Jy) through the gain coefficient.

**Bandpass and Baseline Correction.** The ultimate goal of spectral observation is to extract spectral signals from celestial bodies and to measure the physical information of spectral lines. After the calibration mentioned previously, the observed data have been corresponded with the actual flow density of the celestial body. But it have still not met the demand of spectral data observation, which can be explained by two reasons. Firstly, even for ON/OFF observations in tracking mode, the calibrated data will inevitably contain the information of continuous spectrum radiation in the sky; secondly, the frequency response and time evolution properties of the device itself need to be considered. We can introduce two vital concepts, bandpass and baseline. Bandpass indicates the frequency response of the observation instruments. Baseline is the influence of background continuous spectrum. Correcting baseline and bandpass is to minimize the impact of these two factors.

**RFI Labeling.** Radio astronomical observation faces a huge challenge from the ubiquitous radio frequency transmission in modern society, such as radio, mobile communications, satellite signal raking, navigation, military/civilian, various daily electronic equipment and so on. These radio frequency radiation levels are often much stronger than the celestial signals measured. Therefore, in order to map the large-scale sky survey data accurately in the later stage, screen and mark the radio frequency interference is of great importance during the pre-processing period.

**Gridding.** As a data product obtained from sky survey, the three-dimensional data cube (three dimensions are the right ascension, declination, and the frequency or spectral line velocity) for final analysis needs to have uniform spacing between the right longitude and the right latitude. Radio data processing process usually requires Gridding to convert raw data from irregular sampling space to regular grid space with uniform spacing, so as to carry out scientific research using scanning data. In the process of FAST data pre-processing, it often needs to accumulate multiple scans of data before Gridding. When the data can cover a large area of sky, the Gridding will be carried out.

# 4   I/O Performance Evaluation

The experimental data was generated during the trial operation of FAST sky survey on September 18, 2018. The total volume of data is 700 GB, and a single data file is 2 GB. The experimental environment is a 4-core 8 GB Tencent Cloud Server.

## 4.1   Efficient Data Format Conversion

The first step of our approach is transforming FAST raw data from FITS format to HDF5 format. The size of a single Fast raw data file is 2 GB, and the converted HDF5 file is 2 GB. The conversion time is 2.91 s, which accounts for a small proportion in the whole pipeline. Thus the conversion step is assumed to have no effect on overall efficiency.

## 4.2   FITS and HDF5 Reading-Writing Comparisons

This paper provides the results for the read-write performance with native FITS and HDF5. FITS file contains 2048 arrays (65536, 4). The corresponding HDF5 file contains the 2048 datasets to store these arrays. Figure 5a shows the performance of reading sub-blocks from it.

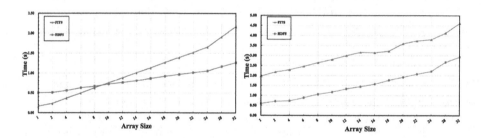

(a) Reading arrays of different sizes from the FITS    (b) Writing arrays of different sizes into the FITS and
and HDF5 files directly                                                   HDF5 files directly

**Fig. 5.** FITS and HDF5 read-write comparisons without other optimizations

The horizontal axis denotes the size factor of the selected block and the vertical axis denotes the time. The size factor is n, which denotes the actual size of the array is (2048 $\times$ n, 4). As shown in Fig. 5a, when the size of the read block is small(n < 9), the speed of FITS-based method is faster than that of HDF5; when n = 9, the reading performance of both them is equal; when the block is larger (n > 9), the HDF5-based method is faster. When the whole arrays (65536 $\times$ 4) are read, the speed of HDF5-based method is 1.7 times as that of FITS. HDF5 performs better than FITS in large arrays reading. FITS format encapsulates data in a more complex way. The tree structure of HDF5 format is beneficial to the retrieval of intersection components.

Figure 5b shows the performance of writing 2048 different size arrays into FITS and HDF5. The sizes of new arrays are (2048 × n, 4). The horizontal axis denotes n and the vertical axis denotes time. As shown in Fig. 5a, HDF5-based method consistently performs better than FITS-based when writing new arrays, whose size is from (2048 × 4) to (65536 × 4). When the size is (65536 × 4), the speed of HDF5-based method is 1.7 times as that of FITS.

### 4.3   HDF5 Performance with Different Chunks

Based on the data layout we specified, this paper studies the performance of different chunks storage schemes with the file driver of HDF5. Figure 6 shows the test results of reading different size of blocks under four chunks schemes. The four schemes are respectively chunks-free, automatic chunks (chunks = TRUE), chunks = (4096, 2), and chunks = (32768, 2). These four schemes set the size of chunks of dataset in memory. It has been proved that in this experiment, the scheme of automatic chunks splits the dataset into chunks of size (4096, 1).

When chunks of the dataset are not specified (the line marked with triangles in the figure), the time of reading an array of (x, 4) is less than that of (x, 1), (x, 2) or (x, 3). This is because the default sequential storage is on the basis of row order. When chunks = True, the system automatically sets the chunks of dataset to (4096, 1). Reading sub-blocks of (4096,1), (8192, 1), (32768, 1) and (65536, 1) performs better than others. When chunks = (4096, 2), reading sub-blocks of (4096, 2), (8192, 2) performs better than others. When chunks = (32768,2), reading sub-blocks of (32768, 2) and (65536, 2) performs better than others. Where chunks = (4096, 2) goes the same way as chunks = (32768, 2), because (32768, 2) is an integer multiple of (4096, 2). When chunks are set up largely, the performance of reading small sub-blocks will be sacrificed because it needs to shred the whole chunk of storage. The scheme of chunks = (32768, 2) performs worst in reading sub-blocks smaller than (32768, 1). The four columns of FAST data respectively represent four polarizations. The relation between the polarizations in calculation is lower than that of the rows, so the connection between columns is not the main concern. Combining the curves of each scheme, we choose the schemes of chunks = True as the best strategy.

### 4.4   HDF5 Performance with MPI

Selecting the most suitable chunks scheme, we apply it to three main algorithms in pipeline for evaluation. The three algorithms are used for flow calibration, bandpass correction and RFI marking. The RFI marking uses the classic algorithm SumThreshold. Figure 7 shows the comparison results in FITS format, HDF5 format, and MPI-based HDF5. It illustrates performance gaps among the FITS-based method, HDF5-based method and parallel hdf5-based method.

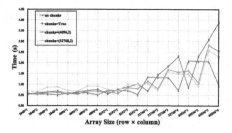

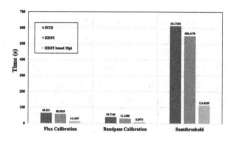

**Fig. 6.** HDF5 reading performance for different chunks storage strategies

**Fig. 7.** Performance comparisons of different strategies applied to the three algorithms

In three modes, three algorithms are implemented respectively to record the time. The time of HDF5-based is slightly less than that of FITS in three algorithms, but the extent of the improvement is not obvious. There was a significant improvement with a data file is operated in parallel based on MPI. In the experiment, there are 4 processes working simultaneously (all four cores of the CPU in the experimental environment are fully operational). In flux calibration, the speed of HDF5-based with MPI driver is 4.60 times faster than that of FITS-based; in bandpass calibration, the speed of HDF5-based with MPI driver is 4.79 times faster than that of FITS-based; in Sumthreshold, the speed of HDF5-based with MPI driver is 5.34 times faster than that of FITS-based. As experimental environment configuration is limited, this paper only implements four processes to work simultaneously. While the field environment will achieve a larger amount of calculation. This paper shows that the performance has been improved in existing experimental environment.

## 5   Conclusion and Future Work

In view of the problem that FITS format cannot meet the performance requirements of FAST neutral hydrogen data pipeline, this paper proposes the method of using HDF5 as an intermediate format to optimize data I/O. We propose the efficient file format conversion scheme, in which we use the dataset of HDF5 to store the data in FITS's binTables, and use the attributes of HDF5 to store FITS variables in form of $\langle key, value \rangle$. In addition, we implement the workflow engine for extragalactic HI data pipeline of FAST. HDF5 format performs better than FITS in the pipeline. Without any drivers, HDF5's performance is 2.1 and 2.5 times as that of FITS in reading and writing respectively. Furthermore, we improve the performance of HDF5 through chunks storage and MPI driver, and improve the performance of flux calibration, bandpass correction and RFI marking by 4 times, 5 times and 6 times respectively.

In addition, the strategies of converting FITS to HDF5 may also work well with other telescope pipelines to improve I/O performance. At the same time, the MPI driver of our approach can support the distributed system conveniently.

**Acknowledgement.** This work is supported by the Joint Research Fund in Astronomy (U1731125, U1731243, 11903056) under cooperative agreement between the National Natural Science Foundation of China (NSFC) and Chinese Academy of Sciences (CAS), the National Natural Science Foundation of China (11573019). BZ is supported by Open Project Program of the Key Laboratory of FAST, NAOC.

# References

1. Anderson, K., Alexov, A., Baehren, L., Griessmeier, J.M., Renting, A.: LOFAR and HDF5: toward a new radio data standard. Astron. Data Anal. Softw. Syst. XX **442**, 53–56 (2010)
2. Bacon, R., et al.: The second-generation VLT instrument muse: science drivers and instrument design. In: Proceedings of SPIE - The International Society for Optical Engineering, pp. 1145–1149 (2004)
3. Ballester, P., et al.: Data reduction pipelines for the very large telescope. Proc. SPIE - Int. Soc. Opt. Eng. **22**(2), 85–98 (2006)
4. Chen, Y., Winslett, M., Yong, C., Kuo, S.W.: Automatic parallel I/O performance optimization in Panda. In: Proceedings of Annual ACM Symposium on Parallel Algorithms and Architectures, pp. 108–118 (1998)
5. Davis, L.E.: An overview of the ALMA pipeline system. In: Astronomical Data Analysis Software and Systems XVIII ASP Conference Series, vol. 411, p. 306 (2009)
6. Davis, L.E., Glendenning, B.E., Tody, D.: The ALMA prototype science pipeline. Astron. Data Anal. Softw. Syst. XIII **314**, 89 (2004)
7. Folk, M., Heber, G., Koziol, Q., Pourmal, E., Robinson, D.: An overview of the HDF5 technology suite and its applications. In: EDBT/ICDT Workshop on Array Databases, pp. 36–47 (2011)
8. Fridman, P.A., Baan, W.A.: RFI mitigation methods in radio astronomy. Astron. Astrophys. **378**, 327–344 (2001)
9. Group, H.: The board of trustees of the University of Illinois: "introduction to HDF5" (2006). http://web.mit.edu/fwtools_v3.1.0/www/H5.intro.html
10. Yan, J., et al.: Optimized data layout for spatio-temporal data in time domain astronomy. In: Ibrahim, S., Choo, K.-K.R., Yan, Z., Pedrycz, W. (eds.) ICA3PP 2017. LNCS, vol. 10393, pp. 431–440. Springer, Cham (2017). https://doi.org/10.1007/978-3-319-65482-9_30
11. Ma, X., Jiao, X., Campbell, M.T., Winslett, M.: Flexible and efficient parallel I/O for large-scale multi-component simulations. In: International Parallel and Distributed Processing Symposium (2003)
12. Madhyastha, T.M., Reed, D.A.: Exploiting Global Input/Output Access Pattern Classification. In: Supercomputing, ACM/IEEE Conference (1997)
13. Masui, K., et al.: A compression scheme for radio data in high performance computing. Astron. Comput. **12**, 181–190 (2015)
14. McMullin, J.P., et al.: CASA architecture and applications. In: Astronomical Data Analysis Software and Systems XVI, Vol. 376 (2007)
15. Nan, R.: Five hundred meter aperture spherical radio telescope (FAST). Sci. China **49**(2), 129–148 (2006)
16. Pence, W.D., Chiappetti, L., Page, C.G., Shaw, R.A., Stobie, E.: Definition of the flexible image transport system (FITS), version 3.0. Astron. Astrophys. **524**, 10 (2010)

17. Price, D.C., Barsdell, B.R., Greenhill, L.J.: HDFITS: porting the FITS data model to HDF5. Astron. Comput. **12**, 212–220 (2015)
18. Luo, G., et al.: HyGrid: a CPU-GPU hybrid convolution-based gridding algorithm in radio astronomy. In: Vaidya, J., Li, J. (eds.) ICA3PP 2018. LNCS, vol. 11334, pp. 621–635. Springer, Cham (2018). https://doi.org/10.1007/978-3-030-05051-1_43
19. Rodrigues, J.E., Rodriguez Bezos, J.E.: A graph model for parallel computation. Massachusetts Institute of Technology (1969)
20. Sanders, P.: Asynchronous scheduling of redundant disk array. IEEE Trans. Comput. **52**(9), 1170–1184 (2000)
21. Bardeau, S., Pety, J.: CLASS: continuum and line analysis single-dish software, a GILDAS software. https://www.iram.fr/IRAMFR/GILDAS/doc/html/class-html/. Accessed 21 Nov 2006
22. Schaaf, R., Brazier, A., Jenness, T., Nikola, T., Shepherd, M.: A new HDF5 based raw data model for CCAT. Eprint Arxiv (2014)
23. Smith, S., Dunning, A., Bowen, M., Hellicar, A.D.: Analysis of the five-hundred-metre aperture spherical radio telescope with a 19-element multibeam feed. In: IEEE International Symposium on Antennas and Propagation, pp. 383–384 (2016)
24. Swinbank, J.D., et al.: The lofar transients pipeline. Astron. Comput. **11**, 25–48 (2015)
25. Thakur, R., Gropp, W., Lusk, E.: Data sieving and collective I/O in ROMIO. In: Symposium on the Frontiers of Massively Parallel Computation (1999)
26. Wells, W.D., Greisen, E.W., Harten, R.H.: FITS-a flexible image transport system. Astron. Astrophys. Suppl. Ser. **44**, 363 (1981)
27. Wu, C., et al.: DALiuGE: a graph execution framework for harnessing the astronomical data deluge. Astron. Comput. **20**, 1–15 (2017)
28. Zichao, Y., et al.: An energy efficient storage system for astronomical observation data on dome A. In: International Conference on Algorithms and Architectures for Parallel Processing, pp. 33–46 (2015)

# Understanding the Resource Demand Differences of Deep Neural Network Training

Jiangsu Du$^{(\boxtimes)}$, Xin Zhu, Nan Hu, and Yunfei Du

School of Data and Computer Science, Sun Yat-Sen University, Guangzhou, China
dujs@mail2.sysu.edu.cn

**Abstract.** More deep neural networks (DNN) are deployed in the real world, while the heavy computing demand becomes an obstacle. In this paper, we analyze the resource demand differences of DNN training and help understand its performance characteristic. In detail, we study both shared-memory and message-passing behavior in distributed DNN training from layer-level and model-level perspectives. From layer-level perspective, we evaluate and compare basic layers' resource demand. From model-level perspective, we measure parallel training of representative models then explain the causes of performance differences based on their structures. Experimental results reveal that different models vary in resource demand and even a model can have very different resource demand with different input sizes. Further, we give out some observations and recommendations on performance improvement of on-chip training and parallel training.

**Keywords:** Deep neural network training · Performance · Resource demand differences

## 1 Introduction

Over the last few years, deep learning (DL) achieves great success in many domains. New deep learning (DL) applications are constantly developed and deployed to real-world utility [4]. New requirement that provides high performance under limited budgets is emerged.

In this paper, we uncover resource demand differences of DNN models from which people can understand the resource demand features of all kinds of models. In order to have a comprehensive understanding, we analyze models in a divide-conquer style, from layer-level and model-level perspectives. From layer-level perspective, we first abstract training process of DNN and measure the floating point operands (FLOPs), memory consumption, and communication amount of basic layers. Moreover, two metrics are designed to compare their resource demand differences. From model-level perspective, we evaluate overall throughput with different batch sizes and interconnection networks. Then an

© Springer Nature Switzerland AG 2020
S. Wen et al. (Eds.): ICA3PP 2019, LNCS 11945, pp. 673–681, 2020.
https://doi.org/10.1007/978-3-030-38961-1_56

analysis is given based on their structure. We provide readers a comprehensive insight on DNN training, and make some important observations and recommendations on performance improvement of DNN on-chip computing and parallel computing.

## 2  Methodology

### 2.1  Training Simplification

Based on a careful technical survey and the node usage of a commercial V100 GPU cluster locating at national supercomputing center in Guangzhou, we choose data parallelism, synchronous stochastic gradient descent (SGD), and all-reduce methods as our experimental object since their effectiveness and popularity. Notably, our focus is the feature of models, and the simplification is to make our analysis intuitive. As shown in Fig. 1(a), under the configuration above, DNN training can be divided into on-chip computation and off-chip communication. In each iteration, each device has a complete model copy and runs both feed-forward and back-propagation locally. After all devices complete computing, updates will be aggregated for next iteration.

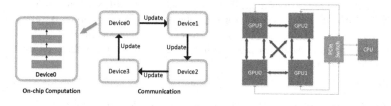

(a) Training Simplification          (b) Topology of DGX Station

**Fig. 1.** Parallel Training and DGX Topology (Color figure online)

### 2.2  Layer-Level Perspective

The training of DNN includes two processes, feed-forward and back-propagation. Generally, a DNN is made up by several different layers and the computing process between upper and lower layers is independent. Thus, the next layer cannot start operating until finishing the previous layer. Therefore, the overall training process can be divided by layers and studied independently.

Basic layers mainly used today are Fully Connected Layer (FCL), Convolutional Layer (CONVL) and Recurrent Layer (RCL). Their static structure can be easily learned from online resources today. For RCL, the basic RCL and two variants, LSTM and GRU are considered.

Here we analyze runtime resource demand which is floating point unit, memory and interconnection. So we measure **FLOPs, memory consumption, and communication amount** to reflect the demand. At first, we identify what resource demand should be included for a layer. For FLOPs demand, it is easy to distinguish. For memory demand, the boundary is not that clear. The memory of a layer consists of input placeholder, newly requested memory by operations, and weights. Input placeholder is also the output of the upper or lower layer, so it is not counted in. In this way, we only include newly requested memory and weights as a layer's memory demand. Also it is common to reuse memory requested in feed-forward for back-propagation, so here we don't count twice. Notably, the intermediate result produced by feed-forward is called feature map and that of back-propagation is called gradient map. As for communication demand, because almost all communication overhead comes from weight synchronization, so we represent communication demand by weight amount.

We implement each basic layer in Tensorflow and evaluate using TFprofiler. Because minibatch SGD can largely increase the concurrency in today's multi-core or many-core architecture, we evaluate the demand with different batch sizes. Additionally, different configurations of layers are taken into account.

Moreover, two metrics are defined to compare resource demand. The first metric is based on two facts. The first fact is that the memory access intensity of dominant operations in these layers are similar. Second, because of the compute dependency, the FLOPs can not directly determine the running time. However, for RCL, the influence of dependency goes weaker as batch size becomes large. The first metric is floating point operands per weight (FOPP). It can reflect how sensitive is a model to interconnect performance. The mathematical expression is as follows:

$$FOPP = \frac{floating\,point\,operands}{weights \times batch\,size} \tag{1}$$

The second metric is instant floating point operands per memory (IFOPM), it reflects the demand ratio of floating point unit and memory size. As for RCL, it should be additionally divided by time step since the FLOPs of different time step cannot be computed simultaneously. The mathematical expression is as follows:

$$IFOPM = \frac{floating\,point\,operands}{memory\,usage\,(\times time\,step)} \tag{2}$$

### 2.3   Model-Level Perspective

We analyze the resource demand differences of models selected through observing their performance change with different memory usage and interconnects. Memory usage is achieved by setting different batch sizes. As for different interconnect, we switch between NvLink and PCIe. Figure 1(b) displays the topology of DGX Station. It has 1 CPU and 4 GPUs. Each GPU can access other GPUs by NvLink (green lines) or PCIe Gen3 ×16 (orange lines). According to our measurement, the bandwidth of Nvlink is about 5× of PCIe and latency is only

1 eighth. Obviously, there is a great difference between these two interconnection networks.

# 3   Evaluation and Analysis

## 3.1   Environmental Setup

The software we use: Ubuntu 16.04.4 LTS, CUDA 10, NCCL 2.4.2, cuDNN 7.4.2, Tensorflow v1.11, Pytorch 1.0. DGX Station is equipped with a Intel Xeon E5-2698 V4 CPU and 4 Tesla V100 (32 GB) with NVLink.

## 3.2   Basic Layer Result

**Fully Connected Layer.** We configure FCL with different batch size and neuron number, and evaluate corresponding weight amount, FLOPs and memory usage. From the result, we can observe that FLOPs increase proportionally with both layer size and batch size. The dominant operation in FCL is matrix multiplication which accounts for more than 99%. For memory, it increases proportionally with layer size and a little with batch size. Memory demand for FCL is from memory newly requested by matrix multiplication and weights. For matrix multiplication, it needs one copy of weight and only request new memory for feature map in feed-forward. When using larger batch size, only feature map will increase. However, the variable number of feature map is only equal to layer size, so the memory consumed by weights is thousands of times larger than that of feature map.

**Insights:** Weights occupy most memory demand in FCL training and it only brings a little memory increase with larger batch size.

**Convolutional Layer.** We configure CONVL with different batch size and kernel size, and evaluate corresponding weights, FLOPs and memory usage. Our result presents that FLOPs and memory demand are almost proportional to batch size. For FLOPs, the dominant operation is Conv2D which accounts for more than 99.5%. For memory demand, feature map occupied most of newly requested memory. Not like FCL, memory usage of feature map is much larger than that of weights in CONVL.

**Insights:** Intermediate result occupies most memory demand in CONVL training. If memory size becomes a limitation for DNN training in GPU. CONVL can be the primary structure to be considered when reducing memory demand by re-calculating feature maps.

**Recurrent Layer.** We configure RCL with different batch size, neuron number and time step, and evaluate corresponding weights, FLOPs and memory usage. It can be observed that different RCLs demonstrate very similar trends on FLOPS and memory. For weight number, they are only influenced by hidden layer size. For FLOPs, matrix-related operations dominate the overall computational complexity and they occupy more than 99%. Notably, the weight amount and FLOPs of these three RCLs are about 1:3:4. For memory, it is much more complicated than FCL and CONVL. Memory is not mainly requested by a single operation. In RCLs, newly requested memory comes from element-wise, matrix-vector multiplication, and data movement operations. The increase of memory is only proportional to time step and slower than a linear relation with hidden layer size and batch size. Based on the profiling result, these implementations will take three copies of weights. Even so, weights only contribute to a small percentage of memory usage and the intermediate result is dominant.

**Comparison.** The comparison uses metrics, FOPP and IFOPM, raised above. To make the result easy-observable, values are normalized.

For FOPP, FCL and CONVL fluctuate at a stable value. FCL is about 0.006 and CONVL is about 6. In terms of RCL, the metric of basic RCL, GRU, and LSTM only change with time step. If we divide FOPP of RCL by time step, they are similar with FCL at 0.006. As we investigate in complete applications, time step is the length of human sentence in general, so FOPP of these layers: $FCL >> BasicRNN \approx LSTM \approx GRU >> CONVL$.

**Insights:** FCL or RCL, especially FCL, usually contribute to more weights and less computation comparing to CONVL.

For IFOPM, all these layers change in a wide range. We explore their range based on evaluation and theoretical analysis. Firstly, variables of a FCL are input size, output size, and batch size. As claimed above, both memory demand and FLOPs are proportional to input size. FLOPs are proportional to output size and memory demand is almost not related to output size. For batch size, it ranges widely from 16 to 1024 or even larger. So, IFOPM of FCL is approximately from 12 to 756 (even larger and mainly around 100). Secondly, variables of a CONVL are kernel size, kernel number, batch size, input size. We can know that input size, batch size, and kernel number only slightly influence this metric. For kernel size, it is quadratic to FLOPs and only influence memory demand a little. The biggest kernel size yet we know is 11 and it cannot be smaller than 2. Also, a kernel is sometimes 3 dimensions and IFOPM should be multiplied with the channel number. So, IFOPM of CONVL is approximately from 2 to 183 (even larger). Thirdly, for RCL, IFOPM of three variants are similar and always $LSTM > GRU > BasicRCL$ when using same configuration. We consider hidden layer size is from 16 to 1024, batch size is from 16 to 512, and time step is from 5 to 40. Then, basic RCL is from 0.17 to 23, GRU is from 0.17 to 24.5, and LSTM is from 0.26 to 25.7. Comparing these three layer types, the rank of IFOPM is $FCL \geq CONVL > LSTM > GRU > BasicRCL$ in most circumstances.

**Insights:** A layer with different input size or configuration can vary in resource demand. FCL has much more weights than CONVL, but the IFOPM of FCL can be similar or even larger than CONVL, which is different to our initial impression. Additionally, for a device, RCL occupies larger memory then it can use up floating point unit.

## 3.3 Model Result

This part we evaluate representative models that achieve competitive accuracy in their domains. Models are listed in Table 1.

**Table 1.** Domains, models, datasets, and frameworks

| Domains | Models | Dominant layer | Framework | Dataset |
|---|---|---|---|---|
| Image classification | AlexNet | CONVL, FCL | Tensorflow | ImageNet-1k |
| | Vgg16 | CONVL, FCL | Tensorflow | ImageNet-1k |
| | ResNet50 | CONVL | Tensorflow | ImageNet-1k |
| | InceptionV3 | CONVL | Tensorflow | ImageNet-1k |
| Object detection | SSD [2] | CONVL | Pytorch | COCO |
| Recommendation | NCF [1] | FCL | Pytorch | MovieLens |
| Adversarial learning | DCGAN [3] | CONVL | Pytorch | LSUN |
| Machine translation | Seq2Seq [5] (GRU) | GRU | Pytorch | WMT16 |
| | Seq2Seq (LSTM) | LSTM | Pytorch | WMT16 |

We display our results in Fig. 2 and further extract features in Fig. 3(a) and Fig. 3(b). Figure 3(a) is the performance improvement rate when expanding batch size. In other words, it is the ratio of performance with two neighborhood batch size when using a single GPU. Figure 3(b) shows the performance ratio of 4 GPUs with different interconnection networks.

**AlexNet, Vgg16, InceptionV3, ResNet50.** We first compare the results of image classification models. For AlexNet, it has 8 layers (5 convolutional layers and 3 fully connected layers). After simple calculation, almost all the weights come from fully connected layer. For Vgg16, it has 16 layers (3 FCLs) and most of weights still come from fully connected layer. For ResNet50, only one fully connected layer is used in ResNet. In this way, weights are not mainly contributed by fully connected layer. For InceptionV3, it completely remove fully connected layer.

Moving on to the evaluation, Fig. 2(a), (b), (c), and (d) show the result of these four models. Initially, we focus on the on-chip performance with different batch size. When batch size is small, the expansion of batch size can bring considerable performance improvement, and it becomes weak when batch size goes large. As for AlexNet, because of simplicity, it obtains good performance improvement at beginning then declines quickly. The growing rate of other three models is very limited, especially for Vgg16. From the evaluation, we can predict there exist a saturation point of floating point unit and the performance will not keep increasing with batch size. In other words, it uses up floating point unit.

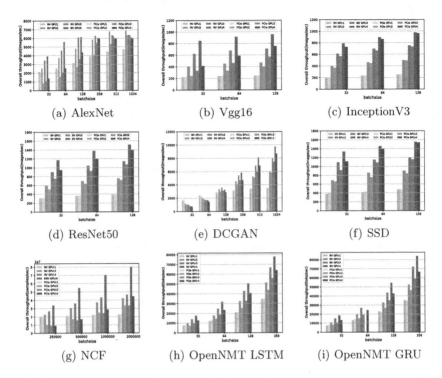

(a) AlexNet        (b) Vgg16        (c) InceptionV3

(d) ResNet50        (e) DCGAN        (f) SSD

(g) NCF        (h) OpenNMT LSTM        (i) OpenNMT GRU

**Fig. 2.** Model performance display

Moving on to multi-GPU training, we can observe that all these models gain performance improvement when using large batch size, because the communication frequency is relatively reduced. However, the batch size cannot be increased infinitely since it will damage convergence speed. From Fig. 3(b), models show different sensitivity to interconnection. For AlexNet, it is influenced largely switching to weak interconnection. For Vgg16, the influence of weak interconnection is also huge (about 21%) but much better than AlexNet. For InceptionV3 and ResNet50, the bad interconnection performance damages only a little performance (about 1.4% and 7.6%). These ratio is calculated when using 4 GPUs and the largest batch size.

**DCGAN.** DCGAN uses two CNNs as the core of model. Although the training process is more complicated than pure CNN models, its training can be simply considered as the addition of two CNNs. The implementation uses four convolutional layers as generator network and five convolutional layers as discriminator network. It removes all fully connected layer and pooling layer. Figure 2(e) displays the evaluation result. For single GPU training, its performance improvement rate is quite high at the beginning comparing with other CNN based models since it is a very small model which can hardly consume much resource. Then the improvement rate gradually reduces to a normal level. For multi-GPU training,

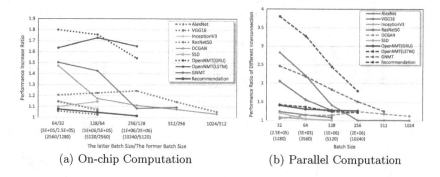

(a) On-chip Computation                  (b) Parallel Computation

**Fig. 3.** Performance change summary

it is not very sensitive to interconnect performance. It experiences about 11% performance decline when using 4 GPUs at batch size 1024.

**Single Shot MultiBox Detector.** SSD can use ResNet, Vgg, and other classical CNN as its backbone. In our implementation, ResNet34 is used. Besides ResNet34, there exists some other structure which contributes extra time.

Figure 2(f) displays the evaluation of SSD. Obviously, in Fig. 3(a) and (b), it shows very similar trend with InceptionV3 and ResNet50.

**Neural Collaborative Filtering.** NCF can be divided into 4 layer types: input layer, embedding layer, neural CF layer, and output layer. In our implementation, all these layer are substantially FCLs. For single GPU training, as is shown by Fig. 3(a), it only improves a little, which can validate that FCL has very high IFOPM. As for multi-GPU training, even using large batch size, it experiences a 44.4% decline when switching interconnection. It is extremely sensitive to interconnect since it is almost fully made up by FCLs.

**Insights:** All previous models use CONVL and FCL as their main structure. FCL usually contributes most weights and CONVL contributes most FLOPs in a CNN. For on-chip performance, they quickly occupy all floating point unit when increasing batch size. For parallel performance, FCL hugely influences the scalability of training and CONVL-dominant models show slight performance decline when switching to weak interconnection. For FCL-dominant models, data parallelism can gain even no improvement if only PCIe provided.

**Seq2Seq (GRU), Seq2Seq (LSTM).** These two models are dominated by RCLs. Here we use the Seq2Seq demo provided officially by OpenNMT. The demo uses 2 RCLs as encoder and another 2 RCLs as decoder (500 hidden size). Also, users can choose RCL type, LSTM or GRU.

Figure 2(i), and (h) demonstrate the results. From Fig. 3(a), they occupy the top 2 places. Although they experience a decline when batch size increases, the improvement rate is still very high. In other words, they gain more improvement when increasing batch size. In other words, it is difficult for RCLs to occupy all floating point units with small batch size. For multi-GPU training, even with large batch size, weak interconnection still damage overall performance a lot (20.8% for OpenNMT (GRU), 17.8% for OpenNMT (LSTM)). 3-GPU training with Nvlink sometimes is even better than 4-GPU training with PCIe. Additionally, comparing Seq2Seq(GRU) and Seq2Seq(LSTM), it validates that LSTM consumes more resources than GRU.

**Insights:** For on-chip performance, RCL based model needs a large batch size to occupy all floating point unit which leads to higher memory requirements. For parallel performance, RNN heavily depends on interconnection performance. Additionally, for all models, a frequently mentioned but very important insight is that increasing batch size can largely improve the on-chip running time and decrease communication frequency.

## 4  Conclusion

DNN training has stepped into a new stage, which raised new challenges on improving performance and reducing cost. We try to uncover resource demand differences of DNN training. The work focuses on both on-chip computation and off-chip communication. To have an insight on the demand, we analyze from layer-level and model-level perspectives. The results reveal that there exist huge resource demand differences among models. In detail, FCL and RCL should contribute to much more communication overhead. FCL has much more weights than CONVL but it has similar or even larger FOPP. For RCL, because of computing dependency, it will create more intermediate results and RCL needs larger memory size to use up device's floating point unit. Based on these results, we make several important observations, which can provide guidance for designing software and hardware or simply purchasing new hardware.

**Acknowledgement.** This research was supported by the Natural Science Foundation of China under Grant NO. U1811464 and the Program for Guangdong Introducing Innovative and Enterpreneurial Teams under Grant NO. 2016ZT06D211.

## References

1. He, X., Liao, L., Zhang, H., Nie, L., Hu, X., Chua, T.S.: Neural collaborative filtering. In: Proceedings of the 26th International Conference on World Wide Web, pp. 173–182. International World Wide Web Conferences Steering Committee (2017)
2. Liu, W., et al.: SSD: single shot multibox detector. In: Leibe, B., Matas, J., Sebe, N., Welling, M. (eds.) ECCV 2016. LNCS, vol. 9905, pp. 21–37. Springer, Cham (2016). https://doi.org/10.1007/978-3-319-46448-0_2
3. Radford, A., Metz, L., Chintala, S.: Unsupervised representation learning with deep convolutional generative adversarial networks. arXiv preprint arXiv:1511.06434 (2015)
4. Ratner, A., et al.: SysML: the new frontier of machine learning systems. arXiv preprint arXiv:1904.03257 (2019)
5. Sutskever, I., Vinyals, O., Le, Q.V.: Sequence to sequence learning with neural networks. In: Advances in Neural Information Processing Systems, pp. 3104–3112 (2014)

# Twitter Event Detection Under Spatio-Temporal Constraints

Gaolei Fei$^{(\boxtimes)}$ , Yong Cheng , Yang Liu , Zhuo Liu, and Guangmin Hu

University of Electronic Science and Technology of China, Chengdu 611731, China
fgl@uestc.edu.cn, chengyong@std.uestc.edu.cn

**Abstract.** Billions of data spread on Twitter every day, which carries a lot of information. It is meaningful to mine the useful information and make it valuable. The purpose of Twitter event detection is to detect what happened in our real life from these unstructured data. We introduce the spatio-temporal information of tweets into event detection. The event detection can be divided into three steps in this paper. First, we use the space difference between event words and noise words and introduce the relationship between words, then we can build a model to separate event words and noise words. Then we define the similarity between event tweets from three different aspects, which make up for the shortcomings of existing methods. Finally, we construct a graph based on the similarity between tweets, and the graph can be divided into different event clusters to complete the event detection. Our method has achieved good results and can be applied to event detection in actual life.

**Keywords:** Twitter event detection · Noisy words identification · Spatio-temporal constraints · Condtional probability

## 1 Introduction

The number of active users in Twitter reaches 400 million, hundreds of million tweets are sent every day. These tweets record the details of events at the first moment. In these unstructured data, the goal of event detection is to find out tweets which describing events and extract the information we need from these event tweets, such as location, time and the key word of the event.

In the Twitter event detection, many researchers have proposed various methods. The general idea of these methods is to cluster the keywords or the text of tweets so that each tweet cluster corresponds to an event. For example, Doulamis et al. [1] uses tweet's sending time and the influence of Twitter users to define the similarity between words in tweets, and implements event detection by dividing words into different events. Dong et al. [2] combines the time and space information of tweets to get the similarity between tweets to complete event detection. This paper follows their idea that noisy words should obey the homogeneous Poisson process, and besides this, we establish a word network to extract noisy

© Springer Nature Switzerland AG 2020
S. Wen et al. (Eds.): ICA3PP 2019, LNCS 11945, pp. 682–694, 2020.
https://doi.org/10.1007/978-3-030-38961-1_57

words more accurately. Ifrim et al. [3] uses the length and structural characteristics of tweets to cluster tweets. Caverlee et al. [4] regards the spatio-temporal
information about tweets as signal, analyzes the characteristics of these noise signals, and applies many noise filters to remove noisy tweets, which can improve
the quality of event detection.

However, the above methods base on text similarity and clustering can not
applied to actual social media event detection. There are some problems in these
methods.

First, in actual, more than 90% of tweets do not contain event information.
many tweets are used to record the user's own life, express their own emotions
and so on. These "noisy tweets" will affect the event detection results if we do
not filter out them.

Second, there are a lot of tweets that do not contain event information, but
their text may be similar to the tweets that describe the event, and these tweets
may be clustered together with event tweets, which lead to a large number of
"noisy tweets" in the clustering result.

Third, there also have some tweets describe the same event, but they do not
contain common words and are considered completely different in their text.
For example, $Tweet_1$ = "please do what you can to help the victims of the
campfire in Paradise", $Tweet_2$ = "It breaks my heart to hear about people and
animals losing their lives due to the California wildfires", although they have
no common words, they all describe California fire events. However, existing
methods usually can not cluster these tweets together, which may cause the
mission of event information. How to solve this problem? We know that there
may exist some tweets that describe the same event and have common words
with $Tweet_1$ and $Tweet_2$, such as $Tweet_3$ = "Paradise, CA #wildfire #campfire
@Paradise, California". We can build a model to measure the similarity between
$Tweet_1$ and $Tweet_2$ through the intermediate tweet ($Tweet_3$), the problem can
be solved in this way.

Aiming to solve these problems, we have proposed some methods. First, we
study the difference between noisy words and event words, and find that noisy
words are independent to each other and appear randomly in space. Using these
features, we can separate noise words and event words to achieve the purpose
of identifying noise tweets. Second, we use the spatiotemporal information of
tweets as an important constraint on the measurement of tweets similarity. In
this way, we can measure the similarity between tweets more comprehensively
and accurately, and solve the second problem. Finally, to measure the similarity
of tweets that are different in their text but belong to the same event, we introduce the co-occurrence similarity, using the co-occurrence of different words to
measure the similarity of different tweets.

## 2    Problem Formulation

The input of Twitter event detection is tweet stream $T = \{T_1, T_2, ..., T_n\}$, where
$T_i$ denotes one tweet. The purpose of event detection is to divide $T$ into different

clusters so that each tweet cluster can correspond to an event in actual life. The idea of our method can be divided into three steps. First, filtering out "noisy tweets" in $T$ and remain "event tweets". Secondly, We define the similarity between "event tweets" from three different perspectives. Finally, we construct a tweet similarity graph $G = (V, E)$, where graph vertex $V$ denotes tweet, graph edge $E$ denotes the similarity between tweets. By dividing this graph, the "event tweet" can be divided into different clusters, and each cluster can correspond to an event.

## 3   Noisy Words Identification

The first step in event detection is to remove noisy tweets. If all the words in $T_i$ are noise words, then $T_i$ can be regarded as a noise tweet. Hence, we need to identify noisy words first.

Noise words have two completely different characteristics from event words. Firstly, the appearance of noise words in different tweets are independent of each other. In contrast, the occurrence of words describing the same event is interrelated. Secondly, noise words appear in different regions with the same probability, that is, noisy words appear randomly in space. However, event words concentrates on the place where the event occurs. Therefore, we deem that noise words follow the homogeneous Poisson process in space, while event words do not follow. In short, if a word follows the homogeneous Poisson process, then we think this word is a noise word, and vice versa.

To measure whether a word $w_i$ follows the homogeneous Poisson process, we can use Ripley's K function [5] to quantify. The Ripley's K function is as follows:

$$\widehat{K}(s) = V(A) \sum_{i \neq j} N(d_{ij} < s)/n^2 \tag{1}$$

Where $s$ denotes the distance threshold between tweets, which is a experience value and the setting of $s$ will be elaborated in the experimental part. $V(A)$ denotes the size of area A where the tweet stream is located, $d_{ij}$ represents the Eucidean distance between two tweets that both contain $w_i$, and $n$ is the number of tweets containing $w_i$. If $w_i$ follows to the homogeneous Poisson process, then the result calculated by Eq. (1) will be $\pi s^2$. Because the result is related to s and not easy to measure, so we use the standardized $K$-funcation: $\widehat{L}(s) = \sqrt{\widehat{K}(s)/\pi} - s$ to standardize, and $\widehat{L}(s, w_i)$ is approximately equal to 0 if $w_i$ follows the homogeneous Poisson process approximately. Thus, the proximity of $\widehat{L}(s, w_i)$ to 0 can be employed for evaluating how similar $w_i$ follows the homogeneous Poisson process. We can define a threshold $l$ and a tolerance limit $\beta$. If $\widehat{L}(s, w_i) < l - \beta$, then $w_i$ follows the homogeneous Poisson process approximately and can be regarded as a noisy word. If $\widehat{L}(s, w_i) > l + \beta$, then $w_i$ can be regarded as an event word. Finally, we think we cannot judge wether wi is a noisy word or an event word if $l - \beta \leq \widehat{L}(s, w_i) \leq l + \beta$.

It is not enough to judge whether $w_i$ is a noise word by simply calculating it's standardized Replay's K function value. In our experiment, we also find that the selection of s value has some influence on the event word, but has little effect on the noise word. In this case, some event words may be mistakenly judged as noise words.

We know that words describing the same event are related to each other. If an event word is misjudged as a noise word, this misjudgment can be saved by other words that related to it. For a word $w_i$, we can use the conditional probability $P(i,j) = P(w_i|w_j)$ as the correlation strength of $w_j$ to $w_i$, where $P(i,j)$ means the occurrence probability of $w_i$ when $w_j$ appears. In this way, we can create a graph $G_w = (V,E)$ to show the relationship between words. In graph $G_w = (V,E)$, $V$ is vertices collection and each vertex represents a word, $E$ is the edges between words and each edge denotes the correlation strength $P(i,j)$ between two words. In this way, $V_i$ represents word $w_i$, and we can use $V_i$ to judge wether $w_i$ is a noisy word. We set the initial value of each vertex $V_i$ to $\widehat{L}(s,w_i)$, then we update $V_i = V_i + \sum_{j=1}^{k} p(i,j) * V_j$, where $V_j$ represents $w_j$ related to $w_i$. In this way, whether a word is a noise word is not only affected by the $\widehat{L}(s,w_i)$ value, but also by the word associated with it. The specific algorithm is as follows. If $w_i$ is an event word, then words related to it are most

---

**Algorithm 1.** Word attribute division

---

1: **input:**
   $T = \{T_1, T_2, ...\}$:tweet stream
   $T_i = \{geo, words\}$:each tweet information
   $n$:number of words (usually set to 3w-5w)
   $l, \beta$:$l$ is $\widehat{L}(s)$ value threshold, $\beta$ is the fuzzy bound.
2: Take out the most frequently occurring $n$ words and calculate the conditional probability $P(i,j) = P(w_i|w_j)$.
3: Take each word as a vertex $n_i$, the weight between the vertices is $P(i,j)$.
   vertex initial value $V_i = \widehat{L}(n_i, s)$
4: If $V_i \geq l + \beta$, set $V_i = 1$, indicating $n_i$ is an event word.
   If $V_i \leq l - \beta$, set $V_i = -1$, indicating $n_i$ is a noisy word.
   Otherwise set $V_i = 0$, indicating $n_i$ is unable to judge.
5: Starting from one vertex $n_i$, find all the vertexes $\{N_k\}$ that connected to $n_i$, update $V_i = V_i + \sum p(i,k) * V_k$ , repeat this process until all $V_i$ value have been updated.
6: Find $V_i \in [-1,1]$ and continue with process 5 until all $V_i \notin [-1,1]$.
7: **output:**
   $V_i(0 \leq i \leq n)$. $V_i < -1$ means $n_i$ is a noise word, otherwise $n_i$ is an event word.

---

likely event words that belong to the same event. Even if the $\widehat{L}(s)$ value of $w_i$ "drop in" the scope of the noise word, according to the 5th step of Algorithm 1, this misjudgment can be saved.

## 4    Tweet Similarity

In Sect. 3, we can identify noisy words and event words. If $T_i$ is all consisted of noise words, we can confirm that $T_i$ does not contain event information and delete it, so that the remaining tweets are all event tweets. In this section, we define the similarity between tweets by merging three similarities in different aspects, that is the text similarity, the word time signal similarity under the spatio-temporal constraints, and the co-occurrence similarity.

### 4.1    Text Similarity

When calculating the similarity of the tweet text, we use TF-IDF (Term Frequency-Inverse Document Frequency) to assign weights to each word first, then convert each tweet into a vector, and finally use the cosine similarity to calculate the text similarity between tweets.

Supposing that two tweets $T_i, T_j$ have a TF-IDF weight vector $\mathbf{X}, \mathbf{Y}$, then the text similarity between them is

$$S_{text}(i,j) = \frac{\mathbf{X} \cdot \mathbf{Y}}{|\mathbf{X}| \cdot |\mathbf{Y}|} = \frac{\sum_i x_i \cdot y_i}{\sqrt{\sum x_i^2} \cdot \sqrt{\sum y_i^2}} \tag{2}$$

### 4.2    Spatio-Temporal Similarity

Another situation is that although two tweets are similar in their text, they may not belong to the same event. Just considering text similarity as tweet similarity is not enough in this case. Tweets belonging to different events tend to have large differences in time or space. Therefore, it is necessary to add spatio-temporal constraints to the measurement of similarity between tweets.

First, we need to construct tweet words' signal. For two tweets $T_1$ and $T_2$, supposing that they have a common word $w_i$. Respectively taking the two tweets as the regional center and the event scope $d$ as the radius, then counting the frequency that $w_i$ appears in all the tweets in the two regions in each time period. The length of the time period is the time resolution $\Delta t$. In this way, $w_i$ gets two time signals series from two tweets. If the two tweets describe the same event, then the two signals of $w_i$ should have some correlation. The coefficient $r^2$ is used to measure the similarity of the two signals, and the larger value of $r^2$ indicates the higher similarity between this two signals.

$$r^2 = \left( \frac{\sum_{i=1}^{n}(x_i - \overline{x})(y_i - \overline{y})}{\sqrt{\sum_{i=1}^{n}(x_i - \overline{x})^2(y_i - \overline{y})^2}} \right)^2 \tag{3}$$

Secondly, we add the spatio-temporal constraints to the similarity measurement. Assuming the distance between the two tweets is $\Delta d$. We connect $\Delta d$ and the time resolution $\Delta t$ together by adjusting $\Delta t$ with $\Delta d$. Specifically, if two tweets are far from each other, then they should not belong to the same event, and the similarity between them should be as low as possible. We can reduce the similarity between tweets by making $\Delta t$ smaller. That is, the bigger $\Delta d$ is, the smaller $\Delta t$ will be. In the same condition, the bigger the time resolution $\Delta t$ is, the higher of the similarity between tweets will be. As shown in Fig. 1, the statistics "fire" and "lose" appear in 64 h. If resolution $\Delta t = 1\,\text{h}$, then $r^2 = 0.345$, and if $\Delta t = 4\,\text{h}$, $r^2 = 0.693$. It can be seen that the time resolution $\Delta t$ can directly affect the similarity between word signals. In the same condition, the similarity between word signals can be smaller by reducing the time resolution. Therefore, the time resolution $\Delta t$ can be adjusted by calculating the distance between tweets, then we can achieve the purpose of adjusting the similarity.

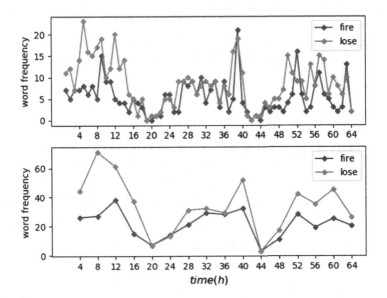

**Fig. 1.** Frequency of words at different time resolutions $\Delta t$

We define $D_{max}$ as the maximum distance between two tweets in tweet stream, define $T_{max}$ as the maximum time interval. According to the distance between two tweet $\Delta d$, then the time resolution $\Delta t$ can be defined as

$$\Delta t = \frac{T_{max}}{6}/(log_{10}^{\frac{D_{max}}{100}} + 1 - log_{10}^{\frac{D_{max}}{\Delta d}}) \qquad (4)$$

In Eq. (4), we set $\Delta d = 100\,\text{m}$ if $\Delta d < 100\,\text{m}$. If $\Delta d$ is the minimal value, $\Delta t = \frac{T_{max}}{6}$, which means we can divide $T_{max}$ into 6 segments at least. The word time signal contains at least 6 values, which guarantees the amount of basic information. If $\Delta d$ becomes larger, then $T_{max}$ is divided into more segments. For example, assuming $D_{max} = 100000$, $\Delta d = 10000$, $\Delta t = \frac{T_{max}}{24}$ will divide $T_{max}$ into 24 segments. In this way, the time resolution $\Delta t$ is adjusted by the space distance $\Delta d$.

After adding spatio-temporal constraints to the similarity measure, we also need to determine the value of event scope $d$. We are unable to get $d$ without knowing the current specific event. We set $d = \{100, 1000, 10000, \cdots, D_{max}\}$, and traverse these values in turn to calculate similarity between word signals. We take the biggest value as the similarity of two word signals. If the two tweets describe the same event, then under this value, the similarity of the word signal is the highest. The reason is that if the size of the statistics area is larger than the event region, noise is added. If smaller, useful information is lost. So when the similarity is the highest, $d$ is the value closest to the real event scope.

The last problem is that when two tweets have many common words, we take the highest similarity value as the similarity between the two tweets. The entire algorithm and implementation details are shown in Algorithm 2.

---

**Algorithm 2.** Calculating tweet similarity under spatio-temporal constraints

1: **input:**
   $T = \{T_1, T_2, ...\}$:tweet stream
   $T_i = \{timestamp, geo\}$:time and location information for
   each tweet
1: Calculate $D_{max}$ and $T_{max}$.
2: Find out $commonWords = \{w_i, i = 1, 2, \cdots, w_n\}$ for every two tweets in $T$, calculate their distance $\Delta d$, then calculate $\Delta t$ by equation (4).
3: For each common word $w_i$, respectively taking the two tweets as the regional center and the different event scope $d = \{100, 1000, \cdots, D_{max}\}$ as the radius, counting the frequency that $w_i$ appears on the time interval $\Delta t$, then we can get word signals in different scopes. Take the value with the highest similarity as the time signal similarity of $w_i$.
4: Perform the operation shown in step 3 for all common words in turn, taking the maximum value of the similarity as the similarity between the two tweets $S_{wordSignal}$.
5: **output:**
   Similarity between two pairs of tweets $S_{wordSignal}$

---

### 4.3   Co-occurrence Similarity

The main disadvantage of above method is that the similarity is always 0 if two tweets do not have a common word. That is to say, the calculation of similarity by text similarity or spatio-temporal similarity will become invalid in this situation. In fact, two tweets that are different in text may belong to the same event. Aiming at solving this problem, we propose another similarity measurement method called cooccurrence similarity—using the co-occurrence of different words to measure the similarity of different tweets.

For all tweets, we can use conditional probability to represent the strength of the association between two words. Let $T_i$ words list be $\{w_i, i = 1, 2, \cdots, m\}$, $T_j$ words list be $\{w_j, j = 1, 2, \cdots, n\}$. For all tweets, we calculate the $P(w_j|w_i)$ and $P(w_i|w_j)$ respectively, where $P(w_j|w_i)$ represents the occurrence probability of

$w_j$ when $w_i$ appears, $P(w_i|w_j)$ represents the occurrence probability of $w_i$ when $w_j$ appears. The strength of association between $w_i$ and $w_j$ is the maximum value of $P(w_j|w_i)$ and $P(w_i|w_j)$. With the strength of association between words, we can define the similarity between $T_i$ and $T_j$ which have no common words as $S_{prob}$

$$S_{prob} = \frac{1}{mn} \sum_{i=1}^{m} \sum_{j=1}^{n} max(P(w_j|w_i), P(w_i|w_j)) \qquad (5)$$

### 4.4   Comprehensive Measure of Tweet Similarity

In the above subsection, we have defined the tweet similarity measurements model which is applicable to every condition from three parts—text similarity, spatio-temporal similarity and co-occurrence similarity. Word signal similarity is a supplement to text similarity, and mainly used to this situation that two tweets are similar in their text but not belong to the same event. Therefore, these two similarity measurements need to be combined, we use the word signal similarity $S_{wordSignal}$. as the weight coefficient of the text similarity $S_{text}$.

Co-occurrence similarity is another supplement. It applies in this condition which the tweet text have no common words but they belong to the same event. We take the maximum value of $S_{text} * S_{wordSignal}$ and co-occurrence similarity $S_{prob}$ as the similarity between two tweets.

$$S = max(S_{text} * S_{wordSignal}, S_{prob}) \qquad (6)$$

## 5   Tweet Cluster Partition

In the above section, we have been able to calculate the similarity between two tweets and complete the second step of event detection. After defining the similarity between tweets, we can create a tweet similarity graph $G = (V, E)$, where $V$ denotes tweets, $E$ denotes the similarity between tweets. Using the Louvain algorithm to divide $G$, we can cluster the tweets that describe the same event.

Louvain is a community detection algorithm and it is very efficient. The time complexity of Louvain is $O(kN + E)$, where $N$ is the number of vertices, $E$ is the number of edges We use Louvain algorithm to divide G into multiple clusters. and each cluster is a description of an event. However, not all tweet clusters represent an event we need, we also need to filter out the tweet cluster that do not contain event information. First, if a cluster contain too few tweets (less than 3) should be deleted, because this is more likely to describe some small things or noise tweets, not the object of interest. Second, most of the tweets in the tweet cluster are sent from the same person should be deleted. In this case, it is likely to be an advertisement.

## 6  Simulation Results

### 6.1  Data Collection and Preprocessing

In order to evaluate the performance of our method, we use Twitter stream API to collect a total of $284k$ tweets in three days from 2018-11-17 to 2018-11-20 in California, USA. Each tweet is formed as $T_i = \{user\_id, text, user\_mentioned, hashtag, timestamp, geo, words\}$, where *user_mentioned* and *hashtag* are extracted from the tweet text, *geo* is the latitude and longitude information of the tweet sender, *timestamp* is the timestamp when the tweet is sent. *words* is obtained by tweet text segmentation, lemmatization, and filtering out stop words. We also need to remove some tweets that do not contain valid information. The rules are as follows

- The number of words in *words* is less than 2, which means the available information is too small.
- The words in *words* are all *user_mentioned* words, then the tweet does not describe the content of the event.

### 6.2  Filtering Out Noisy Words

we take the $n(n = 30000)$ words with the highest frequency to analysis. For each word $w_i$, we need to take out all the geographic coordinates that $w_i$ appears and calculate their distance in pairs. For two points $A(LatA, LonA)$ and $B(LatB, LonB)$ on the earth, the distance between them is

$$d = 2R \arcsin(\sqrt{\sin^2(\alpha) + \cos(LatA)\cos(LatB)\sin^2(\beta)}) \qquad (7)$$

where $\alpha = (LatA - LatB)/2$, $\beta = (LonA - LonB)/2$, $R = 6371\,\mathrm{km}$. For all tweets containing the word $w_i$, Eq. 7 can be used to calculate the distance $d_{ij}$ between two tweets.

The $\widehat{L}(s)$ value of the noise word is hardly affected by changing $s$, but the event word will be affected. The reason is that the distribution of event words is concentrated in the event occurrence area. The scope of event ranges from a few hundred meters to several tens of kilometers. Therefore, we select a list of s values, calculate $\widehat{L}(s)$ from 1 km to 36 km, and take the average value as the $\widehat{L}(s)$ value of the word $w_i$. Take California as an example, in Eq. (1), $V(A)$ denotes the area of California, $V(A) = 411000\,\mathrm{km}^2 s = \{1\,\mathrm{km}, 2\,\mathrm{km}, \cdots, 36\,\mathrm{km}\}$. For the noise words "love", "night" and the event words "wildfire", "death", the relationship between $\widehat{L}(s)$ and $s$ is shown in the Fig. 2. There was a big wildfire in California on November 17, so "wildfire", "fire" are event words, and their $\widehat{L}(s)$ values are all above 0.8. The value of $\widehat{L}(s)$ fluctuated with the change of s value. Conversely, "night", "love" are "noisy words", their $\widehat{L}(s)$ value is close to zero, and their $\widehat{L}(s)$ values are almost unaffected by s value. In the algorithm I, we set $l = 0.6, \beta = 0.15$, we can initially judge that "night" and "love" are noise words, "wildfire" and "fire" are event words. In the actual situation, there will be some words' $\widehat{L}(s)$ value in the interval $[l - \beta, l + \beta]$, and it is impossible

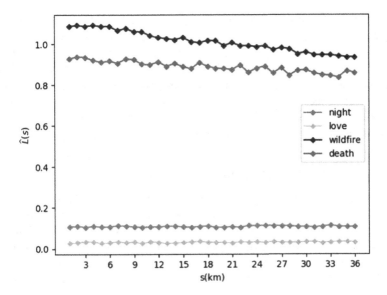

**Fig. 2.** $\widehat{L}(s)$ value for different s values

to judge whether the word is a noise word or cause misjudgment. We take the average of $\widehat{L}_s$ as the initial value of the word $w_i$, and bring it in the algorithm I to calculate the final value $V_i$ of the word $w_i$. After testing, algorithm I can effectively save this misjudgment under normal circumstances.

The experimental results show that among the 30,000 words with the highest frequency, only 1587 words are event words, and words over 94% are noise words. After removing the tweet without any event words, the number of tweets is reduced from $284k$ to $36k$.

## 6.3   Tweet Similarity

First, we calculate the tweet text similarity $S_{text}$, we use the TF-IDF method in *sklearn* to calculate the word weight, then bring the result into the Eq. (2) and calculate the text similarity between two tweets.

Secondly, we measure the similarity $S_{wordSignal}$ between tweets by constructing word signal sequence. The distance of the longest distance in the tweet stream is $D_{max} = 989.9$ km, and the longest time gap is $T_{max} = 72$ h. For $T_i, T_j$, suppose they have two common words $w_i, w_j$, the distance between them is $\Delta d = 1000$ m, then the time resolution $\Delta t = 6$ h according to the equation (4). For one word $w_i$, dividing $T_{max} = 72$ h into 12 segments, the event scope $d = \{0.1$ km, 1 km, 10 km, 100 km, $D_{max}\}$. Respectively taking the two tweets as the regional center and the event scope $d_i$ as the radius, then counting the frequency that $w_i$ appears in each time period $\Delta t$, finally we get time signal series of $w_i$. The similarity of these time series signals are calculated by the Eq. (3), and the maximum value of similarity under different $d_i$ is taken as the similarity value. Then we calculate the similarity of the word $w_j$, and take the maximum value as the similarity of $T_i$ and $T_j$.

Thirdly, we use conditional probability to calculate tweet similarity. The key point is to find out the probability of $w_j$ when the word $w_i$ appears. There are two ways to achieve this, one is to complete the statistic by traversing each tweet, and the speed is slow. The second is to use the idea of FP-growth to build a tree structure, which is fast. Finally, the similarity between the two tweets is determined by Eq. (6).

## 6.4   Tweet Partition and Event Extraction

The previous section achieve a measurement of similarity between two pairs of tweets. After denoising, tweets stream remain only $36k$ tweets. With each tweet as a vertex, the similarity between the tweets as the edge, and a tweet similarity graph is constructed. We delete edges with a similarity less than 0.05, which not only prevents the Louvain algorithm from combining the low-similar tweets, but also effectively reduces the amount of computation. In this way we construct

**Table 1.** California Top 3 event (with denoising + three similarity measures)

| event1 | Time | 2018-11-17 03:59 |
|---|---|---|
| | Location | $[-120.10, 35.14]$ |
| | Key words | fire, lose, paradise, california, wildfire, smoke, heart, forest, death, burn, campfire, angeles |
| | Key tweets | 1. Pray for the citizens of California, Fires to the east and south<br>2. #SanFrancisco #california #airquality #campfire @ San Francisco, California<br>3. It breaks my heart to hear about people losing their lives due to the wildfires |
| event2 | Time | 2018-11-18 00:05 |
| | Location | $[-118.12, 34.00]$ |
| | Key words | celebrate, birthday, november, day, mickey, mouse, 90th, happen, california, love, anniversary, great |
| | Key tweets | 1. Mickey Mouse turns 90 today! Happy Birthday Mickey!<br>2. Happy birthday Mickey! #mickey90 #happybirthdaymickey #mickeymouse #disney<br>3. Happy 90th Birthday Mickey Mouse! And Happy Birthday Minnie Mouse! |
| event3 | Time | 2018-11-18 06:20 |
| | Location | $[121.81, 39.73]$ |
| | Key words | trump, california, californiafires, impact, woolseyfire, presidential, visit, areas, diss, forest, management, again |
| | Key tweets | 1. PRESIDENTIAL VISIT: @realdonaldtrump toured areas impacted by the #CampFire<br>2. Trump in California and he dissed forest management again lmao<br>3. #makeamericarakeagain #californiafires #rake #trump @ Paradise, California |

**Table 2.** California Top 3 event (without denoising+use text similarity only)

| event1 | Time | 2018-11-17 04:41 |
|---|---|---|
| | Location | [−119.56, 35.31] |
| | Key words | fire, lose, paradise, california, day, forget, return, ag, smoke, heart, camp, air |
| | Key tweets | 1. Our hearts go out to those working to recover from the Woolsey Fire<br>2. This is what California calls...A Beautiful Disaster<br>3. A few shots from the former town of Paradise, wiped from existence by fire last week |
| event2 | Time | 2018-11-18 07:38 |
| | Location | [−118.26,34.51] |
| | Key words | celebrate, birthday, november, day, love, time, friend, mickey, night, 90th, happen, anniversary |
| | Key tweets | 1. Happy 90th Birthday Mickey! I'm so happy we could celebrate with you today in disneyland<br>2. Had such a wonderful day with friends celebrating Mickey<br>3. Celebrated Mickeys 90th Birthday at Walts Barn!! |
| event3 | Time | 2018/11/17 3:59 |
| | Location | [−118.28, 34.15] |
| | Key words | car, fire, right, lane, traffic, fairway, stop, delay, lose, ave, destroy, center |
| | Key tweets | 1. Car fire on the right shoulder in #Lynwood on 105 EB at Long Beach Blvd<br>2. !! sigalert !! the two right lanes are closed because of a car fire<br>3. Vehicle on fire in #Salida on Hwy 99 NB before Hammett Rd |

a sparse graph of $36k$ nodes and $356k$ edges. Using the Louvain algorithm to divide tweets into several tweet clusters, using the tweet cluster filtering and event information extraction methods in Sect. 5, we can get the event detection results as shown in Table 1. For comparison, Table 2 shows the results obtained by not using tweets denoising and using only text similarity as a measure. Take the largest top K tweet clusters as important events. Here we take the first three clusters, extract the event information, and compare it with the real event. The real events are

- On 2018.11.17, two vast wildfires ravaged parts of California, killing at least 66 people.
- 2018.11.18, the 90th birthday of Disney Mickey Mouse.
- US President Trump arrived in California on the afternoon of Saturday (17th) to learn about the serious damage caused by wildfires.

From the results in the table, we can see that we can extract keywords more accurately if we filter out noisy tweets and use three methods to measure tweets similarity, and all three things are successfully detected. In Table 2, the event detection result key words is mixed with a large number of words unrelated to the current event because of lacking denoising. At the same time, because the lack of comprehensive measurement of the similarity between tweets, event3 puts tweets that describe the traffic accidents together. In fact, these traffic accidents are not the same event.

# 7   Conclusion

This article focus on Twitter event detection, and put forward our own ideas in filtering out noise tweets and defining the similarity between tweets. In order to remove noise words more accurately, we not only quote the $Replay's K$ function to measure the spatial distribution of noise words, but also establish a word graph to comprehensively judge word attributes by their related words, which greatly reduces the probability of word misjudgment. In order to define the similarity of tweets, we quantify the influence of the spatiotemporal information of tweets, and propose how to measure the similarity between tweets when they have no common words. These ideas have also achieved good results with less noises and higher accuracy in practice.

# References

1. Doulamis, N.D., Doulamis, A.D., Kokkinos, P., et al.: Event detection in Twitter microblogging. IEEE Trans. Cybern. **46**, 1–15 (2015)
2. Dong, X., Mavroeidis, D., Calabrese, F., et al.: Multiscale event detection in social media. Data Min. Knowl. Discov. **29**(5), 1374–1405 (2015)
3. Ifrim, G., Shi, B., Brigadir, I.: Event detection in Twitter using aggressive filtering and hierarchical tweet clustering (2014)
4. Liang, Y., Caverlee, J., Cao, C.: A noise-filtering approach for spatio-temporal event detection in social media. In: Hanbury, A., Kazai, G., Rauber, A., Fuhr, N. (eds.) ECIR 2015. LNCS, vol. 9022, pp. 233–244. Springer, Cham (2015). https://doi.org/10.1007/978-3-319-16354-3_25
5. Dixon, D.P.M.: Ripley's K Function. Encyclopedia of Environmetrics (2006)
6. Patil, M., Chavan, H.K.: Event based sentiment analysis of Twitter data. In: IEEE Conference Record #42656; IEEE Xplore ISBN 978-1-5386-3452-3
7. Sato, K., Wang, J., Cheng, Z.: Credibility evaluation of Twitter-based event detection by a mixing analysis of heterogeneous data. IEEE Access **7**, 1095–1106 (2018). https://doi.org/10.1109/ACCESS.2018.2886312
8. Shi, L.-L., Liu, L., Wu, Y., Jiang, L., Hardy, J.: Event detection and user interest discovering in social media data streams. IEEE Access **5**, 20953–20964 (2017)
9. Shi, L., Wu, Y., Liu, L., Sun, X., Jiang, L.: Event detection and identification of influential spreaders in social media data streams. Big Data Min. Anal. **1**(1), 34–46 (2018). ISSN 2096-0654 03/06
10. Kala, T.: Event detection from text data. Department of Cybernetics Faculty of Electrical Engineering, Czech Technical University in Prague, May 2017

# Correction to: Reliability Enhancement of Neural Networks via Neuron-Level Vulnerability Quantization

Keyao Li, Jing Wang, Xin Fu, Xiufeng Sui, and Weigong Zhang

**Correction to:**
**Chapter "Reliability Enhancement of Neural Networks**
**via Neuron-Level Vulnerability Quantization" in: S. Wen et al.**
**(Eds.): *Algorithms and Architectures for Parallel Processing*,**
**LNCS 11945, https://doi.org/10.1007/978-3-030-38961-1_24**

In the version of this paper that was originally published the affiliation of the third author "Xin Fu" was incorrect. This has now been corrected.

---

The updated version of this chapter can be found at
https://doi.org/10.1007/978-3-030-38961-1_24

© Springer Nature Switzerland AG 2022
S. Wen et al. (Eds.): ICA3PP 2019, LNCS 11945, p. C1, 2022.
https://doi.org/10.1007/978-3-030-38961-1_58

# Author Index